B Vitamins and Folate
Chemistry, Analysis, Function and Effects

Food and Nutritional Components in Focus

Series Editor:
Professor Victor R. Preedy, *School of Medicine, King's College London, UK*

Titles in the Series:
 1: Vitamin A and Carotenoids: Chemistry, Analysis, Function and Effects
 2: Caffeine: Chemistry, Analysis, Function and Effects
 3: Dietary Sugars: Chemistry, Analysis, Function and Effects
 4: B Vitamins and Folate: Chemistry, Analysis, Function and Effects

How to obtain future titles on publication:
A standing order plan is available for this series. A standing order will bring delivery of each new volume immediately on publication.

For further information please contact:
Book Sales Department, Royal Society of Chemistry, Thomas Graham House, Science Park, Milton Road, Cambridge, CB4 0WF, UK
Telephone: +44 (0)1223 420066, Fax: +44 (0)1223 420247
Email: booksales@rsc.org
Visit our website at http://www.rsc.org/Shop/Books/

B Vitamins and Folate
Chemistry, Analysis, Function and Effects

Edited by

Victor R. Preedy
School of Medicine, King's College London, UK

RSCPublishing

Food and Nutritional Components in Focus No. 4

ISBN: 978-1-84973-369-4
ISSN: 2045-1695

A catalogue record for this book is available from the British Library

© The Royal Society of Chemistry 2013

All rights reserved

Apart from fair dealing for the purposes of research for non-commercial purposes or for private study, criticism or review, as permitted under the Copyright, Designs and Patents Act 1988 and the Copyright and Related Rights Regulations 2003, this publication may not be reproduced, stored or transmitted, in any form or by any means, without the prior permission in writing of The Royal Society of Chemistry or the copyright owner, or in the case of reproduction in accordance with the terms of licences issued by the Copyright Licensing Agency in the UK, or in accordance with the terms of the licences issued by the appropriate Reproduction Rights Organization outside the UK. Enquiries concerning reproduction outside the terms stated here should be sent to The Royal Society of Chemistry at the address printed on this page.

The RSC is not responsible for individual opinions expressed in this work.

Published by The Royal Society of Chemistry,
Thomas Graham House, Science Park, Milton Road,
Cambridge CB4 0WF, UK

Registered Charity Number 207890

For further information see our web site at www.rsc.org

Printed and bound in Great Britain by CPI Group (UK) Ltd, Croydon, CR0 4YY, UK

Preface

In the past three decades there have been major advances in our understanding of the chemistry and function of nutritional components. This has been enhanced by rapid developments in analytical techniques and instrumentation. Chemists, food scientists and nutritionists are, however, separated by divergent skills and professional disciplines. Hitherto this transdisciplinary divide has been difficult to bridge.

The series **Food and Nutritional Components in Focus** aims to cover in a single volume the chemistry, analysis, function and effects of single components in the diet or its food matrix. Its aim is to embrace scientific disciplines so that information becomes more meaningful and applicable to health in general.

The series **Food and Nutritional Components in Focus** covers the latest knowledge base and has a structured format.

B Vitamins and Folate has four major sections namely:

- B Vitamins and Folate in Context
- Chemistry and Biochemistry
- Analysis
- Function and Effects.

Coverage includes B vitamins and folate in the context of a historical background, disease, cardiovascular effects and the importance of vitamins in biochemistry as illustrated by a single vitamin. Thereafter there are chapters on the chemistry and biochemistry of thiamine, riboflavin, niacin, pantothenic acid, pyridoxine, biotin, folate and cobalamin. Methodical aspects include characterization and assays of B vitamins and folate in foods of all kinds, dietary supplements, biological fluids and tissues. The techniques cover solid-phase extraction, spectrofluorimetry, mass spectrometry, HPLC, enzymatic assay, biosensor and chemiluminescence. In terms of function and effects or health and disease, there are both overviews and focused chapters covering such topics such as cardiovascular disease, stroke, cognitive decline, dementia, epilepsy, micronutrient interventions, Wernicke's encephalopathy, neuronal calcium, obesity, diabetes, gene expression, beta-oxidation, lipoprotein disorders, pellagra, energy metabolism, immunity, kidney disease and many other areas.

Each chapter transcends the intellectual divide with a novel cohort of features namely by containing:

- an *Abstract* (appears in the eBook only)
- *Key Facts* (areas of focus explained for the lay person)
- *Definitions of Words and Terms*
- *Summary Points.*

B Vitamins and Folate is designed for chemists, food scientist and nutritionists, as well as health care workers and research scientists. Contributions are from leading national and international experts including contributions from world renowned institutions.

Professor Victor R. Preedy,
King's College London

Contents

B Vitamins and Folate in Context

Chapter 1 Historical Context of Vitamin B   3
Hideyuki Hayashi

1.1	Evidence for the Presence of Unidentified Factors Essential for Life	3
1.2	Establishment of the Concept of Vitamins	4
1.3	Resolution of Vitamin B	6
1.4	Discovery that Vitamins act as Coenzymes	6
1.5	Influence on the Research into other Vitamins	8
1.6	Microbial Nutritional Factors and Vitamins	9
1.7	Vitamin B_6	10
1.8	Anaemia and Vitamins	12
1.9	Concluding Remarks	14
	Summary Points	15
	Key Facts about Beriberi in the Russo-Japanese War (1905)	15
	List of Abbreviations	16
	References	16

Chapter 2 B Vitamins and Disease   21
Jutta Dierkes and Ottar Nygård

2.1	Introduction	21
2.2	The Role of Different Study Types	22
2.3	Treatment of Deficiency or Supplementation without Reference to Vitamin Status	23
2.4	Mechanisms	24
2.5	Cardiovascular Disease	25
	2.5.1 Folic acid, Vitamin B_{12} and Vitamin B_6	25
	2.5.2 Niacin in Cholesterol Reduction	26

Food and Nutritional Components in Focus No. 4
B Vitamins and Folate: Chemistry, Analysis, Function and Effects
Edited by Victor R. Preedy
© The Royal Society of Chemistry 2013
Published by the Royal Society of Chemistry, www.rsc.org

2.6 Cancer 27
2.7 Prevention of Age-related Cognitive
     Decline or Dementia 28
2.8 Renal Disease 28
2.9 Concluding Remarks 30
Summary Points 30
List of Abbreviations 31
References 31

**Chapter 3 Vitamins and Folate Fortification in the Context
of Cardiovascular Disease Prevention** 35
*Alexios S. Antonopoulos, Cheerag Shirodaria and
Charalambos Antoniades*

3.1 Introduction 35
3.2 Homocysteinaemia as a Risk Factor for
     Atherosclerosis 36
     3.2.1 Mechanisms of Homocysteine-mediated
            Vascular Disease 36
     3.2.2 Epidemiological Evidence on Homocysteine as
            a Cardiovascular Risk Factor 37
3.3 Cardiovascular Effects of Folic Acid, B_6 and B_{12}:
     Any Need for Folate Fortification? 37
     3.3.1 B Vitamins as Homocysteine Lowering Agents 37
     3.3.2 Effects on Proinflammatory Mechanisms 39
     3.3.3 Effects on Endothelial Function 39
     3.3.4 Further Mechanistic Insights in Human
            Vessels 40
     3.3.5 Effects on Atheroma Progression 40
3.4 Folate and B Vitamins in Cardiovascular Disease:
     Insights from Clinical Trials and Folate Fortification
     Programme 42
     3.4.1 Homocysteine Lowering and Cardiovascular
            Disease Prevention 42
     3.4.2 Criticism of Randomized Clinical Trials 43
     3.4.3 Is Folate Fortification and B Vitamins
            Administration Safe? 46
3.5 Concluding Remarks 47
Summary Points 47
Key Facts about B Vitamins as Therapeutic Agents in
Cardiovascular Disease 48
Definitions of Words and Terms 48
List of Abbreviations 49
References 50

Contents

| Chapter 4 | The Importance of Vitamins in Biochemistry and Disease as Illustrated by Thiamine Diphosphate (ThDP) Dependent Enzymes | 55 |

Shinya Fushinobu and Ryuichiro Suzuki

4.1	Reactions of ThDP-dependent Enzymes	55
	4.1.1 General Mechanism	55
	4.1.2 Pyruvate-processing Enzymes	57
	4.1.3 Transketolase and Phosphoketolase	59
4.2	Structures and Classification	61
	4.2.1 Structures of TK and PK	61
	4.2.2 Structures of Other ThDP-Dependent Enzymes	62
	4.2.3 Classification of ThDP-Dependent Enzyme Families	62
Summary Points		63
Key Facts		64
Key Facts about the History of Structural and Functional Studies on ThDP-dependent Enzymes		64
Key Facts about Metabolic Pathways Involving ThDP-dependent Enzymes		64
Definitions of Words and Terms		65
List of Abbreviations		65
References		66

Chemistry and Biochemistry

| Chapter 5 | The Chemistry, Biochemistry and Metabolism of Thiamin (Vitamin B_1) | 71 |

Lucien Bettendorff

5.1	Introduction	71
5.2	Chemical Properties of Thiamin	72
5.3	Thiamin Biosynthesis and Degradation	77
5.4	Riboswitches	77
5.5	Thiamin Transport	78
	5.5.1 Thiamin Transporters in Mammalian Cells	78
	5.5.2 Mitochondrial Transport of Thiamin Diphosphate	79
5.6	Distribution of Thiamin Derivatives in Living Organisms	79
	5.6.1 Occurrence of ThDP	80
	5.6.2 Occurrence of ThTP	80
	5.6.3 Occurrence of Adenylated Thiamin Derivatives	81

5.7	Metabolism of Thiamin Phosphates	81
	5.7.1 Free and Bound ThDP Pools Coexist in the Same Cell	81
	5.7.2 Synthesis of Thiamin Diphosphate	83
	5.7.3 Synthesis of ThTP	84
	5.7.4 Synthesis of AThTP	84
	5.7.5 Hydrolysis of ThMP	84
	5.7.6 Hydrolysis of ThDP	85
	5.7.7 Hydrolysis of ThTP	85
	5.7.8 Thiamin-binding Proteins	86
5.8	Concluding Remarks	87
Summary Points		87
Key Facts about Thiamin-related Diseases in Humans		88
Definitions of Words and Terms		88
List of Abbreviations		89
References		89

Chapter 6 Chemistry and Biochemistry of Riboflavin and Related Compounds — 93

Mariana C. Monteiro and Daniel Perrone

6.1	Chemistry of Riboflavin	93
	6.1.1 Structure and General Properties	93
	6.1.2 Modes of Degradation and Stability in Foods	95
6.2	Biochemistry of Riboflavin	97
	6.2.1 Digestion, Bioavailability, Absorption and Transport	97
	6.2.2 Metabolism, Storage and Excretion	100
	6.2.3 Biological Functions	101
	6.2.4 Requirements and Intakes	101
Summary Points		103
Key Facts about Digestion and Absorption		103
Definitions of Words and Terms		104
List of Abbreviations		105
References		105

Chapter 7 The Chemistry and Biochemistry of Niacin (B_3) — 108

Asdrubal Aguilera-Méndez, Cynthia Fernández-Lainez, Isabel Ibarra-González and Cristina Fernandez-Mejia

7.1	Introduction	108
7.2	Niacin Chemistry	109
7.3	Niacin Daily Requirement, Food Sources and Niacin Deficiency	110
7.4	Factors and Diseases Affecting Niacin Requirement	110

7.5	Metabolism	111
	7.5.1 Niacin *de novo* Synthesis from Tryptophan	111
	7.5.2 NAD Biosynthesis	113
	7.5.3 NADP Biosynthesis	114
	7.5.4 NAD Recycling	114
7.6	NAD/NAPD Chemical and Structural Proprieties	115
7.7	NAD/NADP Metabolic Actions	115
	7.7.1 Energy Metabolism and Oxidation Processes	115
	7.7.2 Protective Action of NADP	118
	7.7.3 Non-redox Adenosine Diphosphate (ADP)-Ribose Transfer Reactions	119
7.8	Concluding Remarks	121
Summary Points		122
Key Facts about Niacin History		122
Definitions of Words and Terms		123
List of Abbreviations		123
Acknowledgments		124
References		124

Chapter 8 The Chemistry of Pantothenic Acid (Vitamin B$_5$)   **127**
Katsumi Shibata and Tsutomu Fukuwatari

8.1	Introduction	127
8.2	Chemical Structure of Pantothenic Acid and its Related Compounds	127
8.3	Stereoisomers of Pantothenic Acid	128
8.4	Characteristic Properties of D-(+)-Pantothenic Acid	129
8.5	Stability of D-(+)-Pantothenic Acid	130
8.6	Analogues of Pantothenic Acid	130
	8.6.1 Panthenol	130
	8.6.2 Pantetheine	131
	8.6.3 Pantethine	131
	8.6.4 Coenzyme A (CoA)	132
	8.6.5 Iso-coenzyme A (Iso-CoA)	132
Summary Points		132
Key Facts		133
Definitions of Words and Terms		133
List of Abbreviations		133
References		134

Chapter 9 The Chemistry and Biochemistry of Vitamin B$_6$: Synthesis of Novel Analogues of Vitamin B$_6$ **135**
Dajana Gašo Sokač, Spomenka Kovač, Valentina Bušić, Colin R. Martin and Jasna Vorkapić Furač

9.1	Introduction	135

9.2	Discovery of Vitamin B_6	136
9.3	Function of Vitamin B_6	138
9.4	Diversity of Vitamin B_6 Derivatives	140
Summary Points		142
Key facts about Antidotes and Oxidative Stress		142
Definitions of Words and Terms		143
List of Abbreviations		143
References		143

Chapter 10 Biochemistry of Biotin 146
Janos Zempleni, Wei Kay Eng, Mahendra P. Singh and Scott Baier

10.1	Introduction	146
	10.1.1 History	146
	10.1.2 Biosynthesis	147
10.2	Catabolism of Biotin	147
10.3	Biochemistry of Biotin	148
	10.3.1 Biotin-dependent Carboxylases	148
	10.3.2 Biotinylation of Histones	150
	10.3.3 HLCS	151
	10.3.4 Biotinidase	152
	10.3.5 Cell Signalling	152
10.4	Biotin–Drug Interactions	153
Summary Points		153
Key Facts about Histones		154
Definitions of Words and Terms		154
List of Abbreviations		154
Acknowledgments		154
References		155

Chapter 11 The Chemistry of Folate 158
Abalo Chango

11.1	Introduction	158
11.2	Chemistry	158
	11.2.1 Extraction and Isolation	159
	11.2.2 Structure	160
	11.2.3 Physicochemical Properties	160
11.3	Analysis	160
11.4	Concluding Remarks	162
Summary Points		162
Key Facts		162
List of Abbreviations		162
Acknowledgements		162
References		163

Chapter 12 The Chemistry of Cobalamins — 164
Alexios S. Antonopoulos and Charalambos Antoniades

12.1	Introduction	164
12.2	Chemical Structure of Cobalamins	164
12.3	Natural Forms of Cobalamins	166
12.4	Biochemistry of Cobalamins	166
	12.4.1 Methionine Synthase	167
	12.4.2 Methylmalonyl-CoA Mutase	167
12.5	Concluding Remarks	168

Summary Points — 168
Key Facts about the Chemistry of Cobalamins — 168
Definitions of Words and Terms — 169
List of Abbreviations — 169
References — 170

Analysis

Chapter 13 Assay of B Vitamins and other Water-soluble Vitamins in Honey — 173
Marco Ciulu, Nadia Spano, Severyn Salis, Maria I. Pilo, Ignazio Floris, Luca Pireddu and Gavino Sanna

13.1	Honey	173
13.2	Vitamins in Honey	174
	13.2.1 B Group Vitamins in Honey	177
	13.2.2 Vitamin C: Ascorbic Acid	180
13.3	Assay of Water-soluble Vitamins in Honey	180
	13.3.1 Assay of B Group Vitamins in Honey	180
	13.3.2 Assay of Vitamin C in Honey	182
13.4	Validation	183
	13.4.1 LoD and LoQ	184
	13.4.2 Linearity	184
	13.4.3 Precision	186
	13.4.4 Trueness	186
13.5	Concluding Remarks	187

Summary Points — 187
Key Facts — 187
    Key Facts about Honey — 187
    Key Facts about the Vitamins in Honey — 188
    Key Facts about Assaying Water-soluble Vitamins in Honey — 188
    Key Facts about Assaying B Group Vitamins in Honey — 188
    Key Facts about Assaying Vitamin C in Honey — 188
    Key Facts about Validation Protocols — 188

Definitions of Words and Terms		188
List of Abbreviations		189
References		191

Chapter 14 Analytical Trends in the Simultaneous Determination of Vitamins B_1, B_6 and B_{12} in Food Products and Dietary Supplements   195
Anna Lebiedzińska and Marcin L. Marszałł

14.1	Introduction	195
14.2	Analysis of B Vitamins in Food Products and Dietary Supplements	197
14.3	Simultaneous Analysis of Vitamins B in Food Supplements and Food Products by HPLC	199
14.4	Simultaneous Analysis of Vitamins B_6 and B_{12} in Seafood Products	200
Summary Points		205
Key Facts about Electrochemical Detection		205
Definitions of Words and Terms		206
List of Abbreviations		206
References		207

Chapter 15 Spectrofluorimetric Analysis of Vitamin B_1 in Pharmaceutical Preparations, Bio-fluid and Food Samples   210
Sang Hak Lee, Mohammad Kamruzzaman and Al-Mahmnur Alam

15.1	Introduction	210
15.2	Fluorescence Determination of Vitamin B_1 using Various Catalysts	211
15.3	Extraction-based Determination	215
15.4	Nanomaterial-based Determination of Vitamin B_1	217
15.5	Flow Injection and Other Techniques	217
15.6	Concluding Remarks	219
Summary Points		219
Key Facts		220
Key Facts about Spectrofluorimetry		220
Key Facts about Nanomaterials		220
Key Facts about Oxidation		220
Definitions of Words and Terms		220
List of Abbreviations		222
References		222

Chapter 16 Measurement of Thiamine Levels in Human Tissue 227
Natalie M. Zahr, Mary E. Lough, Young-Chul Jung and Edith V. Sullivan

16.1	Introduction	227
16.2	Thiamine Measurements in Human Urine	228
16.3	Thiamine Measurements in Human Blood	230
	16.3.1 Erythrocyte Transketolase Activation and the TDP Effect	230
	16.3.2 High Performance Liquid Chromatography Methods	233
16.4	Thiamine Measurements in Other Human Tissue	234
	16.4.1 Thiamine Measurements in Human Heart, Liver, and Other Organs	234
	16.4.2 Thiamine Measurements in Human Cerebrospinal Fluid	239
	16.4.3 Thiamine Measurements in Human Brain	240
16.5	Concluding Remarks	241
	Summary Points	241
	Key Facts	241
	Key Facts about High Performance Liquid Chromatography (HPLC)	241
	Key Facts about Thiamine Diphosphate (TDP)	242
	Key Facts about Wernicke–Korsakoff Syndrome (WKS)	242
	Definition of Words and Terms	243
	List of Abbreviations	244
	References	244

Chapter 17 The Assay of Thiamine in Food 252
Henryk Zieliński and Juana Frias

17.1	Food as a Source of Thiamine	252
17.2	The Recommended Dietary Allowance for Thiamine	253
17.3	Overview of Analytical Methods	254
17.4	Analysis of Thiamine in Food	255
	17.4.1 Extraction Procedures	255
	17.4.2 Microbiological Methods for Thiamine Analysis	256
	17.4.3 Chemical Methods of Thiamine Analysis	257
	17.4.4 Electrochemical Methods	264
	17.4.5 Animal Assays (Biological Methods)	264
	Summary Points	265
	Key Facts	265

Definitions of Words and Terms		266
List of Abbreviations		266
References		267

Chapter 18 Assays of Riboflavin in Food using Solid-phase Extraction   271
Marcela A. Segundo, Marcelo V. Osório, Hugo M. Oliveira, Luísa Barreiros and Luís M. Magalhães

18.1	Introduction	271
18.2	Batch Solid-phase Extraction Methods	273
18.3	Automatic Flow Based Solid-phase Extraction Methods	274
	18.3.1 Off-line Method	275
	18.3.2 On-line Methods	276
18.4	Concluding Remarks	281
Summary Points		281
Key Facts		281
	Key Facts about Solid-phase Extraction	281
	Key Facts about Flow Injection Analysis	282
Definitions of Words and Terms		282
List of Abbreviations		283
References		283

Chapter 19 Isotope Dilution Mass Spectrometry for Niacin in Food   285
Robert J. Goldschmidt and Wayne R. Wolf

19.1	Introduction	285
19.2	Methodological Considerations	287
	19.2.1 Equilibration Requirement in IDMS	287
	19.2.2 Sample Preparation and Analysis	287
19.3	IDMS Calculations	289
19.4	Isotopic Effect in Fragmentation of Nicotinic Acid	290
19.5	Illustrative Results for Food Samples	290
19.6	Concluding Remarks	295
Summary Points		296
Key Facts about Isotope Dilution Mass Spectrometry		296
Definitions of Words and Terms		297
List of Abbreviations		299
References		300

Chapter 20 Analysis of Pantothenic Acid (Vitamin B_5)   302
Tsutomu Fukuwatari and Katsumi Shibata

20.1	Introduction	302
20.2	Extraction for Measurement	303

20.3	Analytical Methods		304
	20.3.1	Microbiological Assay	304
	20.3.2	Radioimmunoassay (RIA) and Enzyme-linked Immunosorbent Assay (ELISA)	307
	20.3.3	High-performance Liquid Chromatography (HPLC)	307
	20.3.4	Mass Spectrometry (MS)	310
	20.3.5	Optical Biosensor-based Immunoassay	311
20.4	Concluding Remarks		312
Summary Points			312
Key Facts about AOAC International			312
Definitions of Words and Terms			313
List of Abbreviations			314
References			314

Chapter 21 High-performance Liquid Chromatography Mass Spectrometry Analysis of Pantothenic Acid (Vitamin B_5) in Multivitamin Dietary Supplements 317
Pei Chen

21.1	Introduction		317
	21.1.1	Pantothenic Acid Supplementation: Aims and Problems	318
	21.1.2	Chemical Properties of Pantothenic Acid	318
	21.1.3	History of Analysis of Pantothenic Acid	318
21.2	Review of Analysis of Pantothenic Acid Using High-performance Liquid Chromatography		319
	21.2.1	Introduction to High-performance Liquid Chromatography	319
	21.2.2	Earlier HPLC Method for Pantothenic Acid Analysis	319
	21.2.3	Common Detection Techniques for HPLC	320
21.3	Analysis of Pantothenic Acid Using Liquid Chromatography Mass Spectrometry		320
	21.3.1	LC-MS Interfaces	320
	21.3.2	Ion Suppression	320
	21.3.3	Mass Analysers	321
	21.3.4	Stable Isotope Dilution Mass Spectrometry	322
	21.3.5	Mobile Phases for LC-MS	322
21.4	Analysis of Pantothenic Acid Using LC/MS in Multivitamin Dietary Supplements		322
	21.4.1	Reference Standard	322
	21.4.2	Sample Preparation	322

| | | 21.4.3 Chromatography | 323 |
| | | 21.4.4 LC-MS Analysis | 325 |

Summary Points 326
Key Facts 327
    Key Facts about the National Institute of Standards and Technology 327
    Key Facts about Standard Reference Material (SRM) 3280 327
Definitions of Words and Terms 327
List of Abbreviations 331
References 332

Chapter 22 Enzymatic HPLC Assay for all Six Vitamin B_6 Forms 335
Toshiharu Yagi

22.1 Current Status and Challenges of Vitamin B_6 Analysis 335
22.2 Enzymatic Reactions for Conversion of Vitamin B_6 Forms and PNG into 4-PLA 337
22.3 Materials and Methods 338
    22.3.1 Materials 338
    22.3.2 Enzyme Activity Determination 338
    22.3.3 HPLC System 339
    22.3.4 Enzymatic Conversion into 4-PLA and Calculation 339
    22.3.5 Sample Preparation 342
22.4 Chromatograms from Analyses of Standard 4-PLA and Pistachio Nuts, and Contents of Vitamin B_6 Forms and PNG in the Sample and Human Urine 342
    22.4.1 Chromatographic Analyses of Standard 4-PLA, Vitamin B_6 Forms and PNG 342
    22.4.2 Analysis of all Vitamin B_6 Forms and PNG in Pistachio Nuts 342
    22.4.3 Analysis of Vitamin B_6 Forms and PNG in Human Urine 348
22.5 Concluding Remarks 349
Summary Points 349
Key Facts 349
    Key Facts about Vitamin B_6 349
    Key Facts about Enzymes 350
Definitions of Words and Terms 350
List of Abbreviations 351
References 351

Chapter 23	**Analysis of Biotin (Vitamin B$_7$) and Folic Acid (Vitamin B$_9$): A Focus on Immunosensor Development with Liposomal Amplification**	**353**
	Ja-an Annie Ho, Yu-Hsuan Lai, Li-Chen Wu, Shen-Huan Liang, Song-Ling Wong and Jr-Jiun Liou	

    23.1   Introduction   353
    23.2   Current Available Analytical Methods for Biotin and Folate   355
          23.2.1 Analytical Methods for Biotin   355
          23.2.2 Analytical Methods for Folic Acid   357
    23.3   Application of Liposome in the Development of New Immunosensing Systems for Biotin and Folic Acid   358
          23.3.1 Immunosensors for Biotin   358
          23.3.2 Immunosensors for Folic Acid   364
    23.4   Concluding Remarks   367
    Summary Points   369
    Key Facts about Immunosensors   369
    Definitions of Words and Terms   370
    List of Abbreviations   371
    References   372

Chapter 24	**Biotin Analysis in Dairy Products**	**377**
	David C. Woollard and Harvey E. Indyk	

    24.1   Introduction   377
    24.2   Microbiological Methods   379
    24.3   Biological Methods   380
    24.4   Chromatographic and Physicochemical Methods   381
    24.5   Ligand-binding Methods   385
          24.5.1 Labelled Techniques   385
          24.5.2 Non-Labelled Techniques   387
    24.6   Biotin Forms and Concentration in Milk and Dairy Products   389
    Summary Points   391
    Key Facts   391
    Definitions of Words and Terms   392
    List of Abbreviations   392
    References   393

Chapter 25	**Quantitation of Folates by Stable Isotope Dilution Assays**	**396**
	Michael Rychlik	

    25.1   Folates   396
    25.2   Current Methods for the Analysis of Folates   398

	25.2.1	Microbiological Assays	398
	25.2.2	Binding Assays	399
	25.2.3	Chromatography	399
25.3	Stable Isotope Dilution Assays		399
25.4	Benefits and Limitations of Folate SIDAs		401
	25.4.1	Benefits	401
	25.4.2	Limitations	401
25.5	Application of SIDAs to Folate Analysis		403
	25.5.1	Stable Isotopologues of Folates	403
	25.5.2	SIDAs in Food Folate Analysis	405
	25.5.3	Tracer Folates and SIDAs in Folate Analysis of Clinical Samples	405
25.6	Method Comparisons		407
	25.6.1	Folates in Blood Plasma	408
	25.6.2	Comparisons of Folates Analyses in Foods	410
25.7	Perspectives		412
Summary Points			412
Key Facts			413
Definitions of Words and Terms			413
List of Abbreviations			414
Acknowledgment			414
References			414

Chapter 26 Analysis of Cobalamins (Vitamin B_{12}) in Human Samples: An Overview of Methodology — 419

Dorte L. Lildballe and Ebba Nexo

26.1	Introduction		419
26.2	Cobalamins in Humans		420
	26.2.1	Cobalamins in Human Serum	420
26.3	Analysis of Cobalamins in Human Serum		421
	26.3.1	Sample Preparation	422
	26.3.2	Calibrators	424
	26.3.3	Analytical Principles	425
	26.3.4	Choice of Method	430
26.4	Concluding Remarks		432
Summary Points			432
Key Facts			433
Key Facts about Assay Validation			433
Key Facts about Cobalamin Analysis in Human Samples			434
Definitions of Words and Terms			434
List of Abbreviations			436
References			436

Chapter 27 Assay by Biosensor and Chemiluminescence for Vitamin B_{12} 439
M.S. Thakur and L. Sagaya Selva Kumar

27.1	Introduction to Biosensors	439
27.2	Biosensor-based Assay for Vitamin B_{12}	440
27.3	Chemiluminescence	442
	27.3.1 Enzyme-based Enhanced CL	443
	27.3.2 Chemiluminescence of Luminol	443
	27.3.3 Chemiluminescence of the Dioxetane, CDP-*Star*	443
	27.3.4 Chemiluminescence of Acridans	444
	27.3.5 Electrochemiluminescence	445
27.4	Chemiluminescence Analysis of Vitamin B_{12}	445
	27.4.1 Acridium Ester Based Chemiluminescence	445
	27.4.2 Luminol Based Chemiluminescence	445
	27.4.3 Chemiluminescence Work at CFTRI	447
27.5	Concluding Remarks	452
Summary Points		452
Key Facts		453
Definition of Words and Terms		454
List of Abbreviations		454
References		455

Chapter 28 The Diagnostic Value of Measuring Holotranscobalamin (Active Vitamin B_{12}) in Human Serum: A Clinical Biochemistry Viewpoint   458
Fabrizia Bamonti and Cristina Novembrino

28.1	Introduction	458
28.2	HoloTC Analytical Performance	461
	28.2.1 Pre-analytical Phase	461
	28.2.2 Analytic Phase	462
	28.2.3 Post-analytical Phase: Clinical Studies	464
28.3	Clinical Significance and Utility	469
	28.3.1 Vitamin B_{12} Status of Elderly People	469
	28.3.2 Vitamin B_{12} Status of Vegetarians and Vegans	470
	28.3.3 Vitamin B_{12} Status in Different Countries and Ethnic Groups	471
	28.3.4 HoloTC in Subjects with Renal Failure	471
	28.3.5 Vitamin B_{12} Status in Pregnancy	471
28.4	Discussion	471
	28.4.1 Asymptomatic Subjects' HoloTC Concentrations	472
	28.4.2 Asymptomatic Subjects' 'Hcy Panel'	474
28.5	Concluding Remarks	474

	Summary Points	474
	Key Facts about Cobalamin Deficiency	475
	Definitions of Key Terms	475
	List of Abbreviations	476
	References	476

Function and Effects

Chapter 29 B Vitamins (Folate, B_6 and B_{12}) in Relation to Stroke and its Cognitive Decline — 481
Concepción Sánchez-Moreno and Antonio Jiménez-Escrig

29.1	Introduction	481
29.2	B Vitamins: Folate, B_{12}, B_6	484
29.3	Homocysteine	494
29.4	Concluding Remarks	495
	Summary Points	495
	Key Facts	496
	Key Features of B Vitamins: Folate, Vitamin B_6 and Vitamin B_{12}	496
	Key Features of Stroke	496
	Definitions of Words and Terms	497
	List of Abbreviations	498
	References	498

Chapter 30 Epilepsy and B Vitamins — 504
Terje Apeland, Roald E. Strandjord and Mohammad Azam Mansoor

30.1	Introduction		504
30.2	Infrequent B Vitamin Disorders may Induce Epilepsy		504
	30.2.1	Vitamin B_6	505
	30.2.2	Folate	506
	30.2.3	Thiamine	507
	30.2.4	Vitamin B_{12}	508
30.3	Antiepileptic Drugs and B Vitamins		508
	30.3.1	AEDs and Folate	509
	30.3.2	AEDs and Vitamin B_6	513
	30.3.3	AEDs and Vitamin B_{12}	514
	30.3.4	AEDs and Vitamin B_2	515
	30.3.5	AEDs and Vitamin B_7 (Biotin)	515
	30.3.6	AEDs and Vitamin B_1	516
30.4	AEDs and Vitamin Supplementation		516
	30.4.1	Folic Acid Supplements	517
	30.4.2	Supplements with Vitamins B_1, B_2, B_6, B_7 and B_{12}	517

	Summary Points	518
	Key Facts	518
	Key Features about Epilepsy	518
	Key Facts about Epilepsy Treatment	519
	Key Facts about the C677T Polymorphism of MTHFR	519
	Definitions of Words and Terms	520
	List of Abbreviations	520
	References	521
Chapter 31	**B Vitamins and Folate in Multiple Micronutrient Intervention: Function and Effects**	**524**
	Faruk Ahmed	
	31.1 Introduction	524
	31.2 Functions of B Vitamins	525
	31.3 Effect of Supplementation	526
	31.3.1 Haemoglobin Concentration	526
	31.3.2 Micronutrient Status	527
	31.3.3 Growth in Children	528
	31.3.4 Morbidity in Children	528
	31.3.5 Birth Outcomes	529
	31.3.6 Cognitive Function	531
	31.3.7 Cardiovascular Disease	532
	31.4 Concluding Remarks	532
	Summary Points	533
	Key Facts	533
	Key Facts about Multiple Micronutrient Intervention	533
	Key Facts about Anaemia (Low Haemoglobin in Blood)	534
	Key Facts about Homocysteine	534
	Definition of Words and Terms	534
	List of Abbreviations	535
	References	535
Chapter 32	**Wernicke's Encephalopathy caused by Thiamine (Vitamin B_1) Deficiency**	**538**
	Alan S. Hazell	
	32.1 Introduction	538
	32.2 Neuroanatomical Damage in Wernicke's Encephalopathy and Thiamine Deficiency	539
	32.3 Thiamine-dependent Enzymes	539
	32.4 Pathophysiology of Thiamine Deficiency	540
	32.5 Glucose Utilization in Thiamine Deficiency	541
	32.6 Excitotoxicity and Thiamine Deficiency	542

32.7	Oxidative Stress and Thiamine Deficiency	544
32.8	Inflammation in Thiamine Deficiency and Wernicke's Encephalopathy	544
32.9	Blood–Brain Barrier in Thiamine Deficiency	545
32.10	Neurodegenerative Disease and Thiamine Deficiency	545
32.11	Relative Contributions of Thiamine Deficiency and Alcohol to Wernicke's Encephalopathy	546
32.12	Concluding Remarks	547
	Summary Points	547
	Key Facts about Wernicke's Encephalopathy	548
	Definitions of Words and Terms	548
	List of Abbreviations	549
	Acknowledgements	549
	References	550

Chapter 33 Disturbances in Acetyl-CoA Metabolism: A Key Factor in Preclinical and Overt Thiamine Deficiency Encephalopathy   553
Andrzej Szutowicz, Agnieszka Jankowska-Kulawy and Hanna Bielarczyk

33.1	Introduction	553
33.2	Sources of Neuronal Susceptibility to Thiamine Diphosphate Deficiency	555
33.3	Brain Thiamine Diphosphate Levels and Encephalopathy	557
33.4	Transmitter Systems in Thiamine Deficiency Encephalopathy	557
	33.4.1 Glutamate	558
	33.4.2 Acetylcholine	559
33.5	Acetyl-CoA Metabolism a Primary Target for TDP Deficiency	560
	33.5.1 Sources of Acetyl-CoA in the Brain	560
	33.5.2 Acetyl-CoA Compartmentalization in Brain Cells	562
	33.5.3 Redistribution of Acetyl-CoA in Brain Subcellular Compartments in Pathological Conditions	563
33.6	Alcoholism and Subclinical Thiamine Deficiency	564
33.7	Screening for Minimal or Asymptomatic Thiamine Deficiency	565
	Summary Points	566
	Key Facts	566
	Definitions of Words and Terms	567
	List of Abbreviations	568

| | Acknowledgements | 569 |
| | References | 569 |

Chapter 34 Thiamine Deficiency and Neuronal Calcium Homeostasis — 572
Zunji Ke and Jia Luo

34.1	Introduction	572
34.2	Thiamine Deficiency Disrupts Homeostasis of Intracellular Ca^{2+}	574
34.3	Consequences of TD-induced Disruption of Ca^{2+} Homeostasis	575
	Summary Points	576
	Key Facts about Glutamate Receptors	576
	Definitions of Words and Terms	577
	List of Abbreviations	577
	References	578

Chapter 35 Role of Thiamine in Obesity-related Diabetes: Modification of the Gene Expression — 580
Yuka Kohda, Takao Tanaka and Hitoshi Matsumura

35.1	Introduction	580
35.2	Does Thiamine Intervention Decrease the Extent of Weight Gain?	581
35.3	Thiamine Intervention Decreases not only Body Weight but also Visceral Fat Mass and Adipocyte Size	581
35.4	Thiamine Decreases Hepatic Triglyceride Accumulation and Modulates PDH Activity	584
35.5	Thiamine Differentially Modifies Transcript Expression Levels of Genes Involved in Carbohydrate Metabolism, Lipid Metabolism, Vascular Physiology and Carcinogenesis in the Liver	585
35.6	Concluding Remarks	587
	Summary Points	587
	Key Facts	588
	Key Features of Otsuka Long–Evans Tokushima Fatty Rats	588
	Key Features of Thiamine Intervention in OLETF Rats	588
	Key Features of Obesity Worldwide	588
	Key Features of Microarray Analysis	589
	Definitions of Words and Terms	589
	List of Abbreviations	590
	References	590

Chapter 36	**Riboflavin Uptake**	**592**
	Magdalena Zielińska-Dawidziak	

	36.1	Introduction	592
	36.2	Recognized Mechanisms of Riboflavin Transport	594
		36.2.1 Passive Diffusion	594
		36.2.2 Carrier-mediated Transport	594
		36.2.3 Receptor-mediated Endocytosis	594
	36.3	Absorption and Transport of Riboflavin	595
		36.3.1 Small Intestine	595
		36.3.2 Large Intestine	598
		36.3.3 Blood	602
		36.3.4 Liver cells	602
		36.3.5 Brain	603
		36.3.6 Placenta	604
		36.3.7 Kidneys	604
		36.3.8 Ocular System	605
	36.4	Concluding Remarks	605
	Summary Points		606
	Key Facts about Enterocytes		606
	Definition of Words and Terms		606
	List of Abbreviations		607
	References		607

Chapter 37	**Riboflavin and β-oxidation Flavoenzymes**	**611**
	Bárbara J. Henriques, João V. Rodrigues and Cláudio M. Gomes	

	37.1	Riboflavin Metabolism and Chemistry	611
		37.1.1 Riboflavin Metabolism	612
		37.1.2 Flavin Chemistry and Flavoproteins	613
	37.2	Mitochondrial β-oxidation Flavoenzymes	613
		37.2.1 Overview of Mitochondrial β-oxidation	614
		37.2.2 The Flavoprotein Enzymatic Machinery	615
	37.3	Riboflavin Effects in Defective β-oxidation Flavoenzymes	621
		37.3.1 Proteomics Responses to Riboflavin Supplementation	622
		37.3.2 Molecular Basis for Effects of ETF Flavinylation	623
	37.4	Concluding Remarks	626
	Summary Points		626
	Key Facts		627
		Key Facts about Flavoproteins	627
		Key Facts about Inborn Errors of Fatty Acid Oxidation	628

Contents

Definitions of Words and Terms	628
List of Abbreviations	629
Acknowledgements	630
References	630

Chapter 38 Function and Effects of Niacin (Niacinamide, Vitamin B₃)   633
Ahmed A. Megan, Said O. Muhidin, Mahir A. Hamad and Mohamed H Ahmed

38.1	Food Sources of Niacin	633
38.2	Absorption, Excretion and Clinical Features of Niacin Deficiency	634
38.3	Niacin Overdose	634
38.4	Cellular Function and Effects of Niacin	634
	38.4.1 Niacin and Cellular Lipid Metabolism	636
	38.4.2 Non-oxidative and Reduction Reactions involving Niacin (NAD Substrate)	636
	38.4.3 DNA Repair	637
	38.4.4 Apoptosis and Necrosis	637
38.5	Metabolic Effects of Niacin	638
	38.5.1 Niacin and Lipid Metabolism and Atherosclerosis	638
	38.5.2 Effect of Niacin on Atherosclerosis, Inflammation and Vascular Reactivity	641
	38.5.3 Niacin and Dyslipidaemia and Hyperphosphatemia with Chronic Kidney Disease	643
	38.5.4 Effect of Niacin on Insulin Sensitivity and Glucose Metabolism	645
	38.5.5 Niacin and Cardiovascular Disease	646
Summary Points		648
Key Facts		649
List of Abbreviations		649
References		651

Chapter 39 Pharmacological Use of Niacin for Lipoprotein Disorders   660
John R. Guyton, Wanda C. Lakey, Kristen B. Campbell and Nicole G. Greyshock

39.1	Introduction	660
39.2	Mechanisms of Action	661
	39.2.1 Role of the Niacin Receptor (GPR109A) in Inhibiting Non-esterified Fatty Acid Release from Adipocytes	661
	39.2.2 Effects on Apolipoprotein B-containing Lipoproteins	662

	39.2.3	Niacin-Related Mechanisms to Increase HDL and Reverse Cholesterol Transport	662
	39.2.4	Niacin-Induced Cutaneous Flushing	663
	39.2.5	Niacin-Induced Reduction in Inflammation and Oxidative Stress	663
39.3	Lipoprotein Effects of Niacin		663
39.4	Adverse Effects and Drug Administration		664
	39.4.1	Flushing	664
	39.4.2	Hepatotoxicity	664
	39.4.3	Myopathy	665
	39.4.4	Insulin Resistance and Hyperglycemia	665
39.5	Randomized Trials with Cardiovascular and Clinical Endpoints		665
	39.5.1	Coronary Drug Project, the Only Large Monotherapy Trial	665
	39.5.2	Smaller Randomized Trials with Anatomic Endpoints	667
	39.5.3	The AIM-HIGH Study	667
Summary Points			668
Key Facts			669
Key Facts about Mechanisms in Niacin Pharmacology			669
Key Facts about Niacin's Effects on Lipoproteins			669
Key Facts about Niacin and Atherosclerosis Prevention			669
Definitions of Words and Terms			670
List of Abbreviations			671
References			671

Chapter 40 Pellagra: Psychiatric Manifestations 675
Ravi Prakash, Priyanka Rastogi and Suprakash Choudhary

40.1	Introduction	675
40.2	Common Early Psychiatric Features	676
40.3	Cognitive Deficits in Pellagra	676
40.4	Psychotic Spectrum Features in Pellagra	677
40.5	Case Vignette: A Unique case of Pellagra Delusional Parasitosis	678
40.6	Neurophysiological Mechanisms of Pellagra–Psychiatric Features	680
Summary Points		681
Key Facts		681
Key Facts about Pellagra		681
Key Facts about Neuropathophysiological Understanding of Pellagra		682

	Definitions of Words and Terms	682
	List of Abbreviations	683
	References	683

Chapter 41 Pantetheine and Pantetheinase: From Energy Metabolism to Immunity 685
Takeaki Nitto

	41.1	The Synthesis and Metabolism of Coenzyme A	685
	41.2	Enzymatic Features of Pantetheinase	686
	41.3	Pantetheinase Gene Family	687
		41.3.1 Pantetheinase/Vanin-1/VNN1	687
		41.3.2 GPI-80/VNN2	688
		41.3.3 Vanin-3/VNN3	689
	41.4	Role of Pantetheinase *in vivo*: Regulation of Inflammation rather than CoA Metabolism?	690
		41.4.1 Vanin-1 Deficient Mice	690
		41.4.2 Cysteamine: A Key Player in Inflammation and Host Defence	691
	41.5	Panetetheinase Family Genes and Proteins as Indicators of Human Diseases	692
	41.6	Studies on Pantetheinase in Future	692
	Summary Points		693
	Key Facts		693
		Key Facts about Neutrophils	693
		Key Facts about Cysteamine	694
	Definitions of Words and Terms		694
	List of Abbreviations		695
	References		695

Chapter 42 Function and Effects of Pyridoxine (Vitamin B_6): An Epidemiological Review of Evidence 699
Junko Ishihara and Hiroyasu Iso

	42.1	Overall Characteristics and Function	699
	42.2	Current Recommended Intake and Chronic Diseases	701
	42.3	Epidemiologic Evidence of Vitamin B_6 and Vascular Disease	702
		42.3.1 Findings on Dietary Intake in Prospective Observational Studies	702
		42.3.2 Findings on Blood Level in Prospective Observational Studies	703
		42.3.3 Findings from Clinical Trials	703
		42.3.4 Current Knowledge of the Effect on Vascular Disease	704

42.4	Epidemiologic Evidence of Vitamin B_6 and Cancer	704
	42.4.1 Findings on Dietary Intake in Observational Studies	704
	42.4.2 Findings for Blood Level in Observational Studies	706
	42.4.3 Findings from Clinical Trials	706
	42.4.4 Current Knowledge of Effect on Cancer	707
Summary Points		707
Key Facts: Levels of Evidence Reliability		707
Definitions of Words and Terms		708
List of Abbreviations		709
References		709

Chapter 43 Function and Effects of Biotin — 716
Jean-Jacques Houri, Philippe Mougenot, François Guyon and Bernard Do

43.1	Biotin and Biochemical Pathways	716
43.2	Biotin and Regulation of Gene Expression	718
43.3	Sources of Biotin and Human Biotin Requirements	720
43.4	Bioavailability	721
43.5	Pharmacological Effects of Biotin	721
43.6	Physiopathological Aspects of Biotin Deficiency	722
	43.6.1 Causes of Biotin Deficiency	722
	43.6.2 Biotin Deficiency Assessment	724
43.7	Consequences of Biotin Deficiency	725
Summary Points		727
Key Facts about Prosthetic Groups		727
Definitions of Words and Terms		727
List of Abbreviations		728
References		728

Chapter 44 The Importance of Folate in Health — 734
Abalo Chango, David Watkins and Latifa Abdennebi-Najar

44.1	Introduction	734
44.2	Folate Absorption, Transport and Metabolism	736
	44.2.1 Absorption	736
	44.2.2 Folate Transport	737
	44.2.3 Distribution, Storage and Excretion	738
	44.2.4 Metabolism	738
44.3	Biochemical Function, Consequences of Folate Deficiency and Health Alteration	740
	44.3.1 Biochemical Function	740
	44.3.2 Folate Deficiency and Health Alteration	742
44.4	Common Genetic Variation in Folate Metabolism	745

Contents xxxi

    44.5 Synthetic Folic Acid Use for Health:
           Supplementation, Fortification and Adverse
           Effect   746
    44.6 Folate Nutrigenetics, Nutrigenomics and Epigenetics
           For Future Investigations   747
    44.7 Concluding Remarks   747
    Summary Points   748
    Key Facts: Folate discovery and neural tube defects   748
    Definitions of Words and Terms   748
    List of Abbreviations   749
    Acknowledgments   750
    References   750

Chapter 45 Homocysteine and Vascular Disease: A Review of the Published Results of 11 Trials involving 52 260 Individuals   **754**
Robert Clarke and Jane Armitage

    45.1 Introduction   754
    45.2 Observational Studies of Homocysteine and
           Cardiovascular Disease   755
    45.3 Homocysteine, Folate and Cancer   756
    45.4 Trials of B Vitamins for Prevention of
           Cardiovascular Disease   757
           45.4.1 Homocysteine-lowering Trials for
                    Prevention of Cardiovascular Disease   757
           45.4.2 Methodological Considerations   758
    45.5 Effects of B Vitamins on Cardiovascular Disease,
           Cancer and Mortality   758
           45.5.1 Characteristics of the Participating Trials   758
           45.5.2 Effects on CHD and Stroke Outcomes   759
           45.5.3 Effects on Cancer and All-cause Mortality
                    Outcomes   760
    45.6 Analysis of the Role of B Vitamins for Prevention of
           Cardiovascular Disease   762
    Summary Points   764
    Key Facts   765
    List of Abbreviations   765
    References   765

Chapter 46 Vitamin B_{12} and Folate in Dementia   **769**
Rachna Agarwal

    46.1 Introduction   769
    46.2 Vitamin B_{12}, Folate and Homocysteine   771
           46.2.1 Biochemistry   771

	46.2.2	One-carbon Metabolism and Brain Functions	771
	46.2.3	Interrelationship between vitamin B_{12}, Folate and Homocysteine	772
46.3		Status of B Vitamins in Dementia	772
46.4		Old Age and Decline in Vitamin B Status	773
46.5		Vitamin B and Cognitive Decline Mechanism	775
46.6		Role of Laboratory Indicators in Detecting Vitamin B Status	777
46.7		Treatment-related Issues	777
	46.7.1	Vitamin B_{12} Supplementation	778
	46.7.2	Folate Supplementation	779
46.8		Concluding Remarks	779
Summary Points			780
Key Facts			780
Definition of Words and Terms			781
List of Abbreviations			781
References			782

Chapter 47 Cobalamin and Nutritional Implications in Kidney Disease 786
Katsushi Koyama

47.1		Intake and Absorption of Vitamin B_{12}	786
47.2		Intracellular Metabolism of Vitamin B_{12}	787
	47.2.1	Cyanocobalamin Trafficking Chaperone and CKD	787
	47.2.2	Abnormal Cyanide Metabolism in CKD and Oxidative Stress	787
47.3		Does CKD Induce Vitamin B_{12} Deficiency?	790
47.4		Vitamin B_{12}-related Biomarkers and CKD	793
	47.4.1	Homocysteine—the Best Known Biomarker Associated with Vitamin B_{12}	793
	47.4.2	Manifestation of Hyperhomocysteinemia	793
47.5		Association with Vitamin B_6	796
47.6		Concluding Remarks	798
Summary Points			798
Key Facts			799
		Key Facts about Chronic Kidney Disease	799
		Key Facts about Homocysteine	799
		Key Facts about Asymmetric Dimethylarginine	800
Definitions of Words and Terms			800
List of Abbreviations			801
References			802

Subject Index   **805**

B Vitamins and Folate in Context

CHAPTER 1
Historical Context of Vitamin B

HIDEYUKI HAYASHI

Department of Chemistry, Osaka Medical College, 2-7 Daigakumachi, Takatsuki, Osaka 569-8686, Japan
Email: hayashi@art.osaka-med.ac.jp

1.1 Evidence for the Presence of Unidentified Factors Essential for Life

From ancient times, people were aware of the presence of a specific type of disease, beriberi, which affected people mainly in the East and South Asia. As early as 2600 BC, a Chinese account described that beriberi was caused by long-term rice eating but could be prevented by taking rice bran simultaneously. This finding was not recognized in modern medicine. However, in a similar context, the antiscorbutic effect of citrus fruits was already empirically known in the 17th century when James Lind of the Royal Navy systematically carried out experiments to demonstrate the beneficial effect of orange and lemon. Although he never thought that citrus juice is the only solution to scurvy, the surgeons of the Royal Navy were by experience convinced of the efficacy of citrus juice even if the reason was unknown. In the early 1880s, the Surgeon General of the Imperial Japanese Navy, Kanehiro Takagi, noticed that beriberi was common among crews and lower rank officers but not among officers who ate a Western-style diet. He considered that the low-protein diet was the cause of beriberi and performed an experiment in 1884 in which crews of a battleship were given bread and meat during a nine-month mission. Only 16 out of 333 developed beriberi and no one died, compared with a similar mission in the previous year, in which 169 out of 376 developed the disease and 25 died. The Japanese Navy adopted the Western-style diet (bread was later replaced by rice cooked with barley) and eliminated beriberi after 1885. This important lesson, however, was dismissed by the surgeons of the Imperial Japanese Army, who considered beriberi as an infectious disease and criticized Takagi for insufficiency of data and lack of theory that could explain the results. As a result, the army lost 28 000 soldiers due to beriberi in the Russo-Japanese War.

Independently of this, in 1887 a Dutch physician Christiaan Eijkman found an animal model for beriberi when he was in Dutch East Indies (Indonesia). As a former student of Robert Koch, he had been trying to isolate and infect chickens with 'beriberi bacteria'. As he had expected, all the chickens developed polyneuritis gallinarum, the bird counterpart of human beriberi, but soon recovered spontaneously. He noticed that chickens were sick while they were fed polished rice, but recovered after the diet was inadvertently changed to unpolished rice. He conducted detailed experiments to exclude other possibilities, such as that polished rice promotes the growth of bacteria during storage, and concluded around 1895 that what made the difference was the presence or absence of the silver layer of rice. He hypothesized that rice contains some toxin and a substance in the silver layer—he called this the *antiberiberi factor*—neutralizes its virulence (Eijkman 1897). In 1901, Gerrit Grijns, who was an assistant to Eijkman, observed that chickens fed on raw meat did not develop polyneuritis whereas those exclusively fed on meat that had been heated long enough at 120 °C developed the disease. This clearly showed that neither polyneuritis nor beriberi was caused by some substances in rice. From this result, and by taking Takagi's observations into account, he interpreted Eijkman's results in another way: polyneuritis is a deficiency syndrome of a still unknown substance that is essential for life and destroyed by moist heat (Grijns 1901). His theory was later adopted by Eijkman in 1906.

1.2 Establishment of the Concept of Vitamins

In the 19th century, chemists knew that food contains carbohydrates, proteins and lipids. They imagined that it could be possible to make artificial food by properly mixing these nutrients. Ironically, it became a practical issue to test this idea during the Siege of Paris in 1870. A French chemist Jean Dumas made an artificial milk, but infants fed the milk did not survive. This observation attracted the attention of Gustav von Bunge who, believing that minerals are critical for preparing efficient artificial food, ordered his student Nicholas Lunin to study the effect of salt content on mice maintained on artificial food. The mice, however, could not survive for very long, irrespective of the salt content of the food. Lunin concluded that 'a natural food such as milk must therefore contain besides these known principal ingredients small quantities of unknown substances essential to life' (Lunin 1881). Unfortunately, his view was not supported by Bunge and the true significance of this report was not recognized in the scientific community, partly due to the title ('On the importance of inorganic salts in the diet of the animal') reflecting the thought of Bunge's school. In 1905, Cornelius Pekelharing, who had been a predecessor of Eijkman in Java, carried out similar experiments and reached the same conclusion. His report was published in Dutch and was not circulated widely. It was only in 1912 that the idea of 'unknown substances essential to life' was made widely known by the famous paper of Sir Frederick Hopkins (Hopkins 1912). Hopkins was studying the nutritional effect of tryptophan, which he had discovered in 1901, and noticed that animals fed with the tryptophan-deficient 'synthetic' diet could not live long, even if tryptophan was added to the diet. Perhaps without knowing the works of Lunin and Pekelharing (Hopkins 1929), he proposed the notion of 'deficiency diseases'—beriberi, scurvy and rickets as distinct entities. He further forecast that there were many other nutritional errors dependent on unknown dietary factors.

The next step was, undoubtedly, to isolate these substances. Umetaro Suzuki was probably the first to extract the anti-beriberi factor in a concentrated form from rice bran. He presented his results, in December 1910, at the meeting of the Tokyo Chemical Society, and published them in the January issue of the Society's journal in 1911. Because it was written in Japanese, his work was not known outside Japan. In December 1911, Casimir Funk published a paper describing the extraction of the factor using a similar method to Suzuki (Funk 1911). In the following year, Suzuki published in German a paper combining his collective works (Suzuki *et al.* 1912). At this point, he was aware that the material he had previously extracted was largely nicotinic acid and that this had misled him to name the compound 'aberic *acid*'. He could precipitate the effective component with picric acid and succeeded in concentrating the active substance. He then corrected the name to 'oryzanin' (from the Latin *oryza* meaning rice). Funk published a paper in the same year describing the concept of 'deficiency disease'. Although it was not different from that proposed by Hopkins, the name *vitamine* ('vital amine') he used in his paper was soon adopted by researchers in this field. However, as Funk himself admitted these essential substances (a fat-soluble substance necessary for preventing xerophthalmia had been found by Elmer McCollum) did not need to be organic bases. Therefore, in 1920 Jack Drummond recommended dropping the final 'e' of 'vitamine'. He also proposed discontinuing the use of the term 'fat-soluble A' and 'water-soluble B' suggested by McCollum as the unidentified substances necessary for growth, and instead calling these substances 'vitamin A', 'vitamin B', *etc.*, until their true structures were determined.

The pure crystalline vitamin was obtained in Java in 1927 by Barend Jansen and William Donath, who were both students of Eijkman (Jansen and Donath 1927). Robert Williams developed an effective method to isolate vitamin B from rice bran, proposed its structure, and confirmed it by synthesizing the compound (Williams and Cline 1936). The compound, then already given the name vitamin B_1, had a thiazole ring and therefore was named 'thiamine' (Figure 1.1). (As with 'vitamin', 'thiamine' later lost its 'e' to 'thiamin', although the spelling 'thiamine' is still frequently used.)

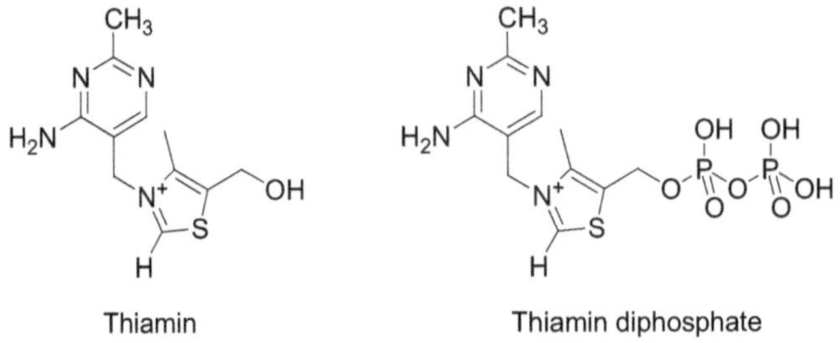

Figure 1.1 Thiamin (Vitamin B_1) and thiamin diphosphate. The diphosphate ester of thiamin is the coenzyme form of thiamin.

1.3 Resolution of Vitamin B

Although there had been a belief that the 'water-soluble B' necessary for growth was identical with the antineuritic vitamin, people became aware of the possibility that vitamin B is not a single entity. In 1927, following the crystallization of the antineuritic vitamin, the British Committee on Accessory Food Factors distinguished the heat-labile and heat-stable components of vitamin B, and named the former B_1 and the latter B_2.

In 1933 Richard Kuhn, Paul György and Theodor Wagner-Jauregg isolated from egg white a yellow pigment having vitamin B_2 activity, which they crystallized and named 'ovo-flavin' (Kuhn *et al.* 1933). They noticed that the absorption spectra of ovo-flavin resembled those of the 'yellow enzyme' isolated by Warburg and Christian (see below). In the same year they also crystallized lacto-flavin, which had been reported by P. Ellinger and W. Koschara, and showed it was identical with ovo-flavin. Alkaline hydrolysis of lumiflavin ($C_{13}H_{12}N_4O_2$), the photolysis product of lacto-flavin ($C_{17}H_{20}N_4O_6$) in alkaline medium, afforded urea and a compound having the chemical formula $C_{12}H_{12}N_2O_3$, which yielded on thermal decomposition CO_2 and $C_{11}H_{12}N_2O$. These reactions were similar to those of alloxazine, which gave urea, CO_2 and 2-hydroxyquinoxaline-3-carboxylic acid ($C_8H_6N_2O$), suggesting that lumiflavin is a trimethyl derivative of alloxazine. Combining the observation by E. Holiday and K. Stern of the spectral similarities between lacto-flavin and alloxazine and its derivatives (Holiday and Stern 1934), Kuhn proposed the structure of lumiflavin to be 7,8,10-trimethylisoalloxazine (Kuhn and Rudy 1934). In 1935, Kuhn's group (Kuhn *et al.* 1935) and Karrer's group (Karrer *et al.* 1935) synthesized independently 7,8-dimethyl-10-D-1'-ribitylisoalloxazine, and confirmed its identity with lacto-flavin. Once the structure (Figure 1.2) was finally determined, the vitamin was called riboflavin thereafter.

1.4 Discovery that Vitamins act as Coenzymes

The 1930s saw the dawn of coenzyme research. In 1932, Otto Warburg obtained a yellow enzyme from yeast and showed that the yellow dye reversibly

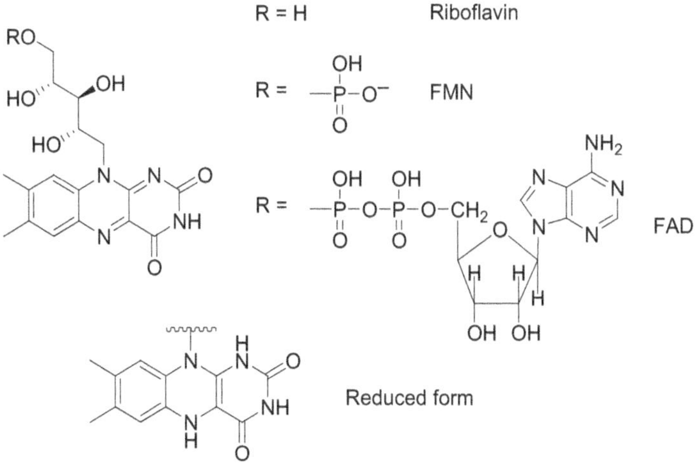

Figure 1.2 Riboflavin (Vitamin B_2) and its two coenzyme forms. The terminal OH group of riboflavin is either phosphorylated (FMN: flavin mononucleotide) or conjugated with ADP (FAD: flavin adenine dinucleotide) when it acts as coenzymes.

underwent oxidation and reduction while conducting the oxidation of glucose 6-phosphate (Warburg and Christian 1932). Hugo Theorell, working in Warburg's laboratory, crystallized the enzyme and showed that the pigment could be reversibly removed from the enzyme protein and the enzyme which lost the pigment was inactive (Theorell 1935). This was the first discovery of a 'coenzyme'. Theorell also determined the correct structure of the coenzyme, the phosphate ester of riboflavin, flavin mononucleotide (FMN) (Threorell 1937). The more abundant and complex form of the coenzyme was found by Warburg and Christian in D-amino acid oxidase, and was determined to be flavin adenine dinucleotide (FAD) (Warburg and Christian 1938).

Dating back to 1911, Neuberg had discovered that bacteria and plants have an enzyme that catalyses the decarboxylation of pyruvate to acetaldehyde and CO_2. He named the enzyme 'carboxylase', which in today's nomenclature refers to an enzyme undergoing carboxylation but at that time meant decarboxylase. In 1932, Ernst Anhagen observed that yeast 'carboxylase' lost activity when treated with alkali (Auhagen 1932). The activity was restored by the addition of a heated solution of yeast extract. He then speculated that the 'carboxylase' contains a non-proteinous low-molecular weight compound, named 'cocarboxylase'. At the same time, R. Peters and colleagues observed that the brain extract of pigeons deficient of thiamin showed a decreased rate of lactate degradation, but that the rate was restored by the addition of 'concentrated vitamin B_1' (crystalline thiamin was expensive) to the extract (Meiklejohn *et al.* 1932). Based on these observations, Lohman and Schuster isolated cocarboxylase from yeast and proposed its structure to be thiamin diphosphate (Lohman and Schuster 1937).

1.5 Influence on the Research into other Vitamins

The history of the pursuit for vitamin B_1 and B_2 paved the way for the research on other vitamins. Pellagra is a disease which was epidemic in southern Europe in the 19th century and in the southern USA in the early 20th century. In 1915, Joseph Goldberger reported that pellagra was not an infectious disease and could be ascribed to maize eating (Goldberger and Wheeler 1915). This view was consistent with the findings of Willcock and Hopkins in 1906 that mice fed with maize protein, which is low in tryptophan, as the sole protein source could not grow (Willcock and Hopkins 1906). Thus Goldberger suspected that tryptophan deficiency was the cause of pellagra. Although this is known to be correct today, the history of the pellagra-preventing factor took a curious path. In 1920, Carl Voegtlin and associates showed that pellagra could be cured by administering yeast extract, suggesting that vitamin B was the pellagra-preventing factor (Voegtlin *et al.* 1920). However, with the isolation of vitamin B_1 and B_2, it was soon acknowledged that neither of these vitamins had pellagra-preventing activity.

Independently of the pursuit of the pellagra-preventing factor, several studies on yeast fermentation provided for related fields of research. In 1906, Arthur Harden and William John Young found that fermentation by yeast juice required both a 'heat-labile and nondialysable fraction' and a 'heat-stable and dialysable fraction', the latter of which was termed 'co-ferment'. In the early 1930s, co-ferment was resolved into several components. One was shown by Otto Meyerhof and colleagues as ATP (Meyerhof *et al.* 1931). Warburg and colleagues showed that nicotinic acid was present in co-ferment in the form of niconitic acid amide conjugated with two pentoses, three phosphoric acids and an adenine (Warburg *et al.* 1935). This is known today as NADP (Figure 1.3). Shortly thereafter, another nicotinic acid derivative with two phosphoric acids, now known as NAD (Figure 1.3), was discovered by Warburg and Christian and independently by a rival group led by Hans von Euler-Chelpin, a notable student of Warburg's father, Emil Warburg (Schlenk and von Euler 1936; Warburg and Christian 1936).

About the same time, in 1937, Bert C.J.G. Knight found that nicotinic acid is a *Staphylococcus* growth factor. Inspired probably by these findings, Elvehjem and associates were able to cure black-tongue disease, a dog counterpart of pellagra, by administering nicotinic acid (Elvehjem *et al.* 1937). There remained, however, a mystery: diet that induced black tongue in dogs was richer in nicotinic acid or its amide than normal milk. In 1945, Elvehjem's group showed that tryptophan can fully substitute the vitamin action of nicotinic acid, indicating that nicotinic acid is synthesized from tryptophan (Krehl *et al.* 1945). Henderson and Ramasarma revealed that quinolinic acid is formed from 3-hydroxyanthranilic acid, a metabolite of tryptophan (Henderson and Ramasarma 1949). However, they were unable to explain how quinolic acid is converted to nicotinic acid. The solution to this problem was given in 1963 by Nishizuka and Hayaishi, who showed that quinolinic acid first reacts with 5-phosphoribosyl-1-pyrophosphate (PRPP) to form a nucleotide before

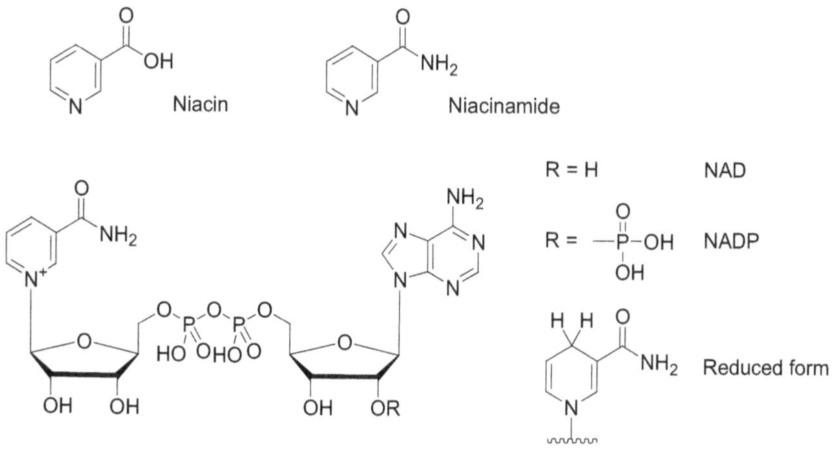

Figure 1.3 Niacin, Niacinamide, NAD, and NADP. Nicotinic acid and nicotinamide are now called niacin and niacinamide, in order to avoid confusion with nicotine. NAD: niacinamide adenine dinucleotide; NADP: niacinamide adenine dinucleotide phosphate.

being decarboxylated to form nicotinic acid mononucleotide (Andreoli *et al.* 1963). Thus contrary to the earlier belief, the direct biosynthetic product was not nicotinic acid but its phosphoribosylated form.

1.6 Microbial Nutritional Factors and Vitamins

The vitamins discovered in the 1930s were soon found to have profound physiological effects on microorganisms. In an opposite manner, pantothenic acid was initially discovered in microorganisms and later in mammals. Since the beginning of the 20th century, it was known that yeasts require "bios", which is present in malt, yeast extract, *etc.*, for growth. Bios was later resolved into I, IIa and IIb. Bios I was identified as *meso*-inositol. Roger Williams found in 1933 the ubiquitous occurrence of bios IIa in the tissues of many organisms; he named it pantothenic acid and isolated it in 1938 (Williams *et al.* 1938). Owing to the feature of 'universal biological occurrence', pantothenic acid was recognized as a vitamin (vitamin B_5) found earlier than pyridoxine (B_6). In the following year, T. Jukes and Elvehjem's group independently announced that deficiency in pantothenic acid caused skin lesion in chickens (Jukes 1939; Woolley *et al.* 1939).

Analogously to the 'artificial diet' for humans, microbiologists attempted to prepare a 'purified medium' for microorganisms comprising all the necessary nutrients in chemically pure form. In 1939, Esmond Snell observed that, when riboflavin was added to the medium for *Lactobacillus casei*, in which the peptone was treated with alkali in advance to destroy the inherent riboflavin, no growth was observed. This indicated that some nutritional factors were

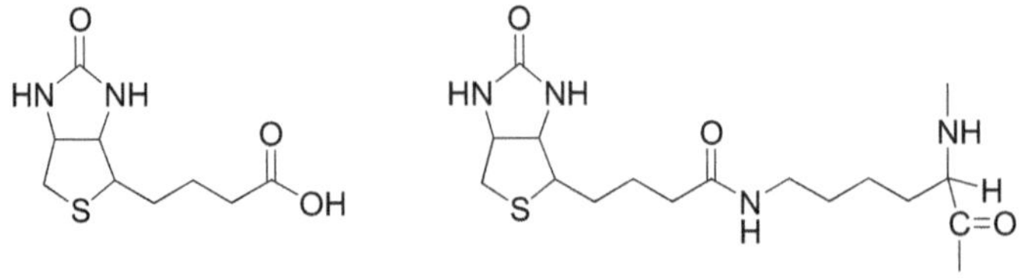

Figure 1.4 Biotin and its covalent attachment to proteins. In enzymes utilizing biotin as a coenzyme, biotin is bound to a specific Lys residue of the enzyme protein.

removed by the alkali treatment. In 1939 he purified the factor 1000-fold from liver extract and showed it to be identical with pantothenic acid. Based on this, he developed a microbiological assay of riboflavin (Snell and Strong 1939).

In 1936, F. Kögl and B. Tönnis succeeded in crystallizing 4 mg of bios IIb from 1000 duck egg yolks and named it biotin (Kögl and Tönnis 1936). The chemical structure (Figure 1.4) was determined by V. du Vigneaud and associates in 1942, and synthesized by Folkers' group in 1945. Paul György had been studying egg white injury and in 1939 partially purified 'vitamin H', which prevents the disease in rats, from bovine liver. The identity of vitamin H and biotin was shown in 1940 by a collaboration of György and du Vigneaud. In 1940, Snell developed a yeast assay system for biotin and using this Snell and Williams isolated the protein in egg white that tightly binds biotin and causes 'egg white injury' (Eakin *et al.* 1940). This was named 'avidin' because of its peculiar biotin-binding capacity.

Through these investigations the idea that microorganisms and animals share many vitamins in common became gradually accepted and the microbial bioassay system of vitamins became an essential tool in vitamin research. We will appreciate the significance of the method in the following sections.

1.7 Vitamin B_6

In 1934, György found that there is a specific type of skin lesions in rat which was protected by an unidentified 'vitamin B_6'. In 1938, Samuel Lepkovsky reported its crystallization. Slightly later, four other groups, Keresztesy and Stevens, György, Kuhn and Wendt, and Ichiba and Michi, independently announced the crystallization of vitamin B_6 from various sources (summarized in György 1964; György admits Lepkovsky, Keresztesy and he were in contact with each other just before publication). The structure of vitamin B_6 was solved independently by S.A. Harris and Folkers and Kuhn and associates in 1939 (György 1964). The compound, first named 'adermin' for its dermatitis-preventing activity, was soon renamed 'pyridoxine' derived from its chemical structure (Figure 1.5).

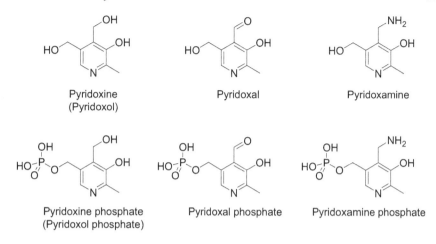

Figure 1.5 B_6 vitamers. All six forms have vitamin B_6 activity; hence they are called B_6 vitamers. Two of the phosphorylated forms, pyridoxal phosphate and pyridoxamine phosphate, act as coenzymes. Pyridoxine phosphate is considered to be the intermediate on the way from pyridoxine to pyridoxal phosphate.

While developing a microbiological assay system for pyridoxine, Snell noticed that the content of 'pyridoxine' in rat tissues as determined by growth of *Streptococcus faecalis* was several thousand times higher than those obtained by yeast growth, rat growth or colorimetric methods. Another important finding was that when media supplemented with pyridoxine was heat-sterilized, progressively smaller amounts of pyridoxine were required by *S. faecalis* as the autoclave period was increased. Snell proposed that a substance, which he called 'pseudopyridoxine', was formed from pyridoxine by autoclaving and that *S. faecalis* was much more sensitive to pseudopyridoxine than to pyridoxine. Pseudopyridoxine was also found to be formed from pyridoxine *in vivo* by human and rat bodies. In collaboration with Folkers, Snell showed that pseudopyridoxine was a mixture of the 4-aldehyde and 4'-amine derivatives of pyridoxine, which were named pyridoxal and pyridoxamine, respectively (Figure 1.5; Harris *et al.* 1944; Snell 1944). Using the differential effect of the three forms of vitamin B_6 on *S. faecalis* (pyridoxal and pyridoxamine are active), *Lactobacillus casei* (pyridoxal is active) and *Saccharomyces carlsbergensis* (all three forms are active), Snell developed a microbial quantification system to quantify the three forms separately.

The conversion of pyridoxal to pyridoxamine in heat-sterilized media was confirmed to be due to the reaction of pyridoxal with amino acids (Snell 1945). This was the first discovery of nonenzymatic transamination. The transamination reaction in animal tissues was first discovered in 1937 by Alexander Braunstein and associates as an amino group transfer between glutamate and alanine in pigeon muscle extract (Braunstein 1939). Irwin Gunsalus and associates showed that the tyrosine decarboxylase activity of *S. faecalis* was slightly stimulated by addition of pyridoxal but markedly stimulated by that of pyridoxal and ATP (Gunsalus *et al.* 1944). This led to the finding that pyridoxal phosphate (Figure 1.5) is the coenzyme. A detailed mechanistic study of pyridoxal was carried out in metal-ion catalysed model systems and a general mechanism of pyridoxal-catalysed diverse reactions, transamination, decarboxylation, aldol cleavage, β-replacement/elimination, γ-replacement/elimination, *etc.* was proposed (Metzler *et al.* 1954). These mechanisms were later found to operate in pyridoxal 5′-phosphate-dependent enzymes (Hayashi 1995; Eliot and Kirsch 2004).

1.8 Anaemia and Vitamins

In Bombay (now Mumbai) in 1930, Lucy Wills observed patients complicated with an unusual macrocytic anaemia (Wills 1931). The symptoms of the anaemia resembled those of pernicious anaemia, but unlike pernicious anaemia, it lacked neurological complications and quickly responded to yeast extract, which was then already known to be rich in 'vitamin B complex'. Liver was also efficacious when administered orally, but not when injected as liver extract, which was a promising remedy for pernicious anaemia. This indicated that the curative factor was removed by preparation of liver extract. Wills could reproduce the anaemia in monkeys fed with the local food deficient in vitamin B complex. At the same time, similar observations were reported by Paul L. Day, who named the protective factor in brewer's yeast 'vitamin M', but could not continue the research owing to the restrictions that come with an assay system using monkeys. Albert Hogan developed an assay system using chickens, and with the help of a large research team, could obtain crystals of the vitamin, which was alternatively named 'vitamin B_c' (Pfiffner *et al.* 1943).

Snell was at that time working on nutrition of *Lactobaccillus casei*. He found a factor necessary for the growth of *L. casei* in spinach, which was easily available in Austin, Texas. He and colleagues processed 4 tons of spinach and isolated the factor (Mitchell *et al.* 1941) and named the compound 'folic acid' (from the Latin *folium* meaning leaf). The ability of folic acid to prevent chicken anaemia was confirmed in 1942 by Elvehjem's group. On the other hand, E. L. Robert Stokstad crystallized the growth factor for *L. casei* from liver and showed its identity with vitamin B_c but not with Snell's 'folic acid' (Stokstad 1943). Later, he and colleagues discovered a bacterium that produced large amounts of a compound with vitamin B_c activity. Hydrolysis of the compound yielded a pteridine derivative, *p*-aminobenzoic acid, and glutamic acid. Reconstitution studies showed that the nature of vitamin B_c was pteroylglutamic acid ((Figure 1.6) Waller *et al.* 1948). Elucidation of the structure enabled explanation for the action of sulfa drugs; they act as analogues for *p*-aminobenzoic acid. 'Folic acid' present in plants was later found to be pteroylpolyglutamic acid, explaining the difference in the effect of Stokstad's and Snell's preparations on growth of *L. casei*. However, the name 'folic acid' became used as a term that encompasses both pteroylglutamic acid and pteroylpolyglutamic acid.

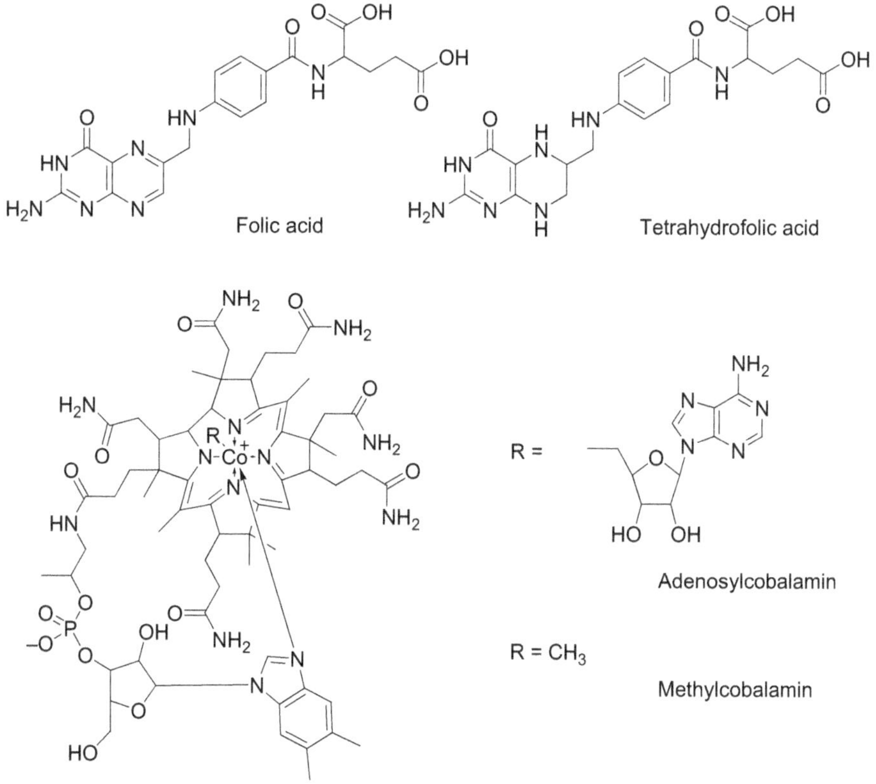

Figure 1.6 Folic acid and cobalamins. Tetrahydrofolic acid is the active form of folic acid and carries C_1 compounds such as methanol, formaldehyde, formic acid, *etc*. In mammals, methionine synthase and methylmalonyl-CoA mutase are the only known B_{12} enzymes, using methylcobalamin and adenosylcobalamin, respectively, as coenzymes.

Pernicious anaemia was first described by James Combe in 1824. Thomas Addison described in detail the disease in 1849, although he posited some connection with the adrenal gland. No significant progress had been made until in 1897, when F. Martius and O. von Lubarsch reported an association of pernicious anaemia with achlorhydria. Thereafter, more attention was paid to the role of gastric function in the pathogenesis of the anaemia.

In 1925, Whipple found by chance that severe anaemia in dogs induced by exsanguination could be cured by administering liver. This observation was immediately applied by Minot and Murphy to the treatment of pernicious anaemia and a remarkable response was observed for patients eating large amounts of liver (Minot *et al*. 1928). Subsequently, William Castle found that ground beef partially digested in the gastric juice of healthy men was effective as a remedy for the anaemia. However, when beef or gastric juice was given alone, or when given successively, no remission was observed (Castle *et al*. 1930). He then postulated that haematopoiesis requires simultaneous administration of an 'extrinsic factor' present in meat and liver, and an 'intrinsic factor' secreted by the stomach. He also showed that the intrinsic factor was neither HCl nor pepsin, and was heat-labile.

A tough quest for the two factors began. Owing to the lack of appropriate animal models, isolation of the vitamin was difficult. However, after Mary Shorb showed that the vitamin, now called vitamin B_{12}, was also a growth factor for *Lactobacillus lactis* Dorner, a microbial bioassay system was developed. In 1948, Folkers' group succeeded in crystallizing vitamin B_{12} (Rickes *et al.* 1948), and showed it to be a cyano complex of Co^{3+}. In today's view, it is considered that cyanocobalamin formed with internal cyanide source was preferentially obtained since it is the most stable vitamin B_{12}. The complete structure of vitamin B_{12} (Figure 1.6) was finally solved in 1955 by Dorothy Hodgkin using X-ray crystallography (summarized in Hodgkin *et al.* 1956). Microbial bioassay system was, however, inappropriate for detecting the intrinsic factor activity. It was only in 1972 when the intrinsic factor was purified using the newly developed purification method, affinity chromatography, with vitamin B_{12}-Sepharose (Allen and Majerus 1972).

The function of vitamin B_{12} was first recognized in 1958 when Horace Barker found that it acts as a coenzyme for interconversion between L-glutamate and 3-methylaspartate (Barker *et al.* 1958). The structure of the coenzyme, adenosylcobalamin, was solved again by Hodgkin in 1961.

The finding in 1946 by T. Spies and colleagues that thymine can substitute the functions of folic acid and vitamin B_{12} led to the understanding that both folic acid and vitamin B_{12} are involved in methyl transfer reactions (Vilter *et al.* 1950). Donaldson and Keresztesy showed that folic acid can exist in various forms with one-carbon group attached. The coenzyme methylcobalamin was discovered in 1964 (Lindstrand 1964).

1.9 Concluding Remarks

The history of B vitamin research, like that of other vitamins, began with the investigations of the aetiology of various diseases that were related to dietetic problems. However, in the 19th century, the glorious works of Robert Koch and others might have induced people to think that most causes of diseases could be ascribed to pathogens such as bacteria and toxins. It was considered that nutritional problems were largely solved by the discovery of three major nutrients and minerals, and the remaining problem was to find an ideal ratio of these nutrients. Even after the discovery of an essential factor to prevent beriberi, time was required to make Eijkman start to believe that this was not a detoxificant but a factor essential for life. This gives us a lesson: as we all know, there can be no theory without experimental evidence—but we often forget this especially when our minds are caught by some great ideas or concepts.

Why are B vitamins necessary? If we consider the action of B vitamins as coenzymes, the answer is clear. There is no electrophilic functional group in the 20 amino acids of proteins. Coenzymes are therefore required for enzymes to carry out electrophilic and radical catalysis. Thiamin diphosphate is by itself a nucleophile, but it generates an electrophilic centre once it reacts with a carbonyl compound. The exceptions are CoA and tetrahydrofolate, which are more properly regarded as substrates rather than catalysts. However, since they are recycled, they are not usual substrates and share 'reusability' as a common property with other catalytic coenzymes.

Today, the catalytic mechanisms of coenzymes are understood in detail. However, only few, namely those of folic acid and vitamin B_{12}, can explain the symptoms of their deficiency. With regard to beriberi, our knowledge has not much improved since the time of Takagi and Eijkman. Certainly, B vitamins have functions other than those as coenzymes. The discovery of NAD-dependent ADP ribosylation was, in this sense, a breakthrough. Interestingly, the electrophilic nature of NAD is again exploited here. However, it may be appropriate to think that we should not be dazzled by the fantastic chemical properties of vitamins, for we learnt a lesson that too much adherence to theory caused the tragedy of unthinkable loss of life to beriberi in the early 20th century.

Summary Points

- This chapter focuses on the historical context of B vitamins.
- Beriberi was known to be related to diet in ancient China but was not recognized in early modern medicine.
- There was a debate over the cause of beriberi in Japan in the 19th century, causing large numbers of victims of the disease comparable to those of wars.
- The prevailing idea in the 19th century was that diseases were caused by extraneous pathogens such as bacteria and toxins.
- It was not until the early 20th century that the notion of essential micronutrients for life, vitamins, was established.
- Almost all the B vitamins were discovered in the 1920s to 1940s.
- A microbiological assay system contributed much to the isolation of vitamins.
- B vitamins were found to function as coenzymes in enzymatic reactions and metabolism.
- Only a part of the symptoms of B vitamin deficiency can be explained by coenzyme functions of vitamins.
- Exploring the functions of B vitamins other than as coenzymes that account for their deficiency symptoms is a goal for future work.

Key Facts about Beriberi in the Russo-Japanese War (1905)

1. In the Russo-Japanese War, the death toll of the Imperial Japanese Army was 48 400 from battle and 37 200 from disease. Among the latter, 27 800 died of beriberi. The total number of beriberi patients in the Army reached 250 000. On the other hand, the Imperial Russian Army suffered from scurvy, which led to the fall of Port Arthur.

2. The efficacy of barley-blended rice against beriberi was known to the chief medical officers of many army divisions. However, their request to adopt barley-blended rice as army provision was neglected by Rintaro (Ogai) Mori, a famous novelist and the head of the Second Army Medical Corps, Tadanori Ishiguro, a former director already retired but still influential on the Medical Department of the Japanese Army, and Tanemichi Aoyama, Professor of Tokyo Imperial University.
3. The reason for the Japanese army not having adopted barley-blended rice is a complicated combination of several factors.
4. Some critics ascribe the reason to the aggressive and stubborn personality of Rintaro Mori, who did not admit that 'beriberi bacillus theory' was wrong until his death, although he seemed to be aware of it. They blame Mori as 'the officer who killed more Japanese soldiers than any Russian officer'.
5. Others point out that it was the revenge of Ishiguro who was forced to resign after the Sino-Japanese War (1895) taking the responsibility for the beriberi pandemic among the soldiers, and the pride of Aoyama who considered himself as the authority of bacteriology in Japan.
6. There was also a logistic problem: production of barley was not enough to support soldiers whose number was several tens larger than that of sailors. Furthermore, barley is not as stable as polished rice on storage.
7. Most of the soldiers of the Japanese army were sons of poor peasants and eating polished rice was their desire even at the cost of their lives. It was probably the 'mercy' of army officers to let their soldiers eat polished rice as much as they wanted before leaving for the battlefield. Even in the Japanese navy, there was recurrence of beriberi in the period around World War II, because barley was discarded by cooks who disliked its flavour.

List of Abbreviations

FAD   flavin adenine dinucleotide
FMN   flavin mononucleotide
NAD   nicotinamide adenine dinucleotide
NADP   nicotinamide adenine dinucleotide phosphate

References

Allen, R.H., and Majerus, P.W., 1972. Isolation of vitamin B12-binding proteins using affinity chromatography. I. Preparation and properties of vitamin B12-Sepharose. *Journal of Biological Chemistry*. 247: 7695–7701.
Andreoli, A.J., Ikeda, M., Nishizuka, Y., and Hayaishi, O., 1963. Quinolinic acid: a precursor to nicotinamide adenine dinucleotide in *Escherichia coli*. *Biochemical and Biophysical Research Communications*. 18: 92–97.

Auhagen, E., 1932. Co-carboxylase, ein neues Co-enzyme der alkoholischen Gärung. *Hoppe-Seyler's Zeitschrift für physiologische Chemie.* 204: 149–167.

Barker, H.A., Weissbach, H., and Smyth, R.D., 1958. A coenzyme containing pseudovitamin B12. *Proceedings of the National Academy of Sciences of the United States of America.* 44: 1093–1097.

Braunstein, A.E., 1939. Die enzymatische Umaminierung der Aminosäuren und ihre physiologische Bedeutung. *Enzymologia.* 7: 25–52.

Castle, W.B., Townsend, W.C., and Heath, C.W., 1930. Observations on the etiologic relationship of achylia gastrica to pernicious anemia. III. *American Journal of Medical Science.* 180: 305–335.

Eakin, R.E., Snell, E.E., and Williams, R.J., 1940. A constituent of raw egg white capable of inactivating biotin *in vitro*. *Journal of Biological Chemistry.* 136: 801–802.

Eijkman, C., 1897. Eine beri-beri-ähnliche Krankheit der Hühner. *Virchows Archiv für pathologische Anatomie und Physiologie und für klinische Medizin.* 148: 523–532.

Eliot, A.C., and Kirsch, J.F., 2004. Pyridoxal phosphate enzymes: mechanistic, structural, and evolutionary considerations. *Annual Review of Biochemistry.* 73: 383–415.

Elvehjem, C.A., Madden, R.J., Strong, F.M., and Woolley, D.W., 1937. Relation of nicotinic acid and nicotinic acid amide to canine black tongue. *Journal of American Chemical Society.* 59: 1767–1768.

Funk, C., 1911. On the chemical nature of the substance which cures polyneuritis in birds induced by a diet of polished rice. *Journal of Physiology.* 43: 395–400.

Goldberger, J., and Wheeler, G.A., 1915. Experimental pellagra in the human subject brought about by a restricted diet. *Public Health Reports.* 30: 3336.

Grijns, G., 1901. Over polyneuritis gallinarum. *Geneeskundig Tijdschrift voor Nederlands Indie.* 49: 216–231.

Gunsalus, I.C., Bellamy, W.D., and Umbreit, W.W., 1944. A phosphorylated derivative of pyridoxal as the coenzyme of tyrosine decarboxylase. *Journal of Biological Chemistry.* 154: 685–686.

György, P., 1964. *The history of vitamin B6. Introductory remarks*. In: Harris, R.S. (ed.) *Vitamins and Hormones.* Vol. 22. Academic Press, New York and London, pp. 361–366.

Hayashi, H., 1995. Pyridoxal enzymes: mechanistic diversity and uniformity. *Journal of Biochemistry.* 118: 463–473.

Harris, S.A., Heyl, D., and Folkers, K., 1944. The structure and synthesis of pyridoxamine and pyridoxal. *Journal of Biological Chemistry.* 154: 315–316.

Henderson, L.M., and Ramasarma, G.B., 1949. Quinolinic acid metabolism. III. Formation from 3-hydroxyanthranilic acid by rat liver preparations. *Journal of Biological Chemistry.* 181: 687–692.

Hodgkin, D.C., Kamper, J., Mackay, M., Pickworth, J., Trueblood, K.N., and White, J.G., 1956. Structure of vitamin B12. *Nature.* 178: 64–66.

Holiday, E.R., and Stern, K.G., 1934. Über das spektrale Verhalten des Photoflavins, des Alloxazins und verwandter Verbindungen: Einfluß der Wasserstoff-Ionen-Konzentration und der zweistufigen Reduktion. *Berichte der Deutschen chemischen Gesellschaft.* 67: 1352–1358.

Hopkins, F.G., 1912. Feeding experiments illustrating the importance of accessory factors in normal dietaries. *Journal of Physiology.* 44: 425–460.

Hopkins, F.G., 1929. Nobel lecture: the earlier history of vitamin research. Available at: http://nobelprize.org/nobel_prizes/medicine/laureates/1929/hopkins-lecture.html. Accessed 17 June 2011.

Jansen, B.C.P., and Donath, W.F., 1927. Over de isoleering van het anti-beriberi-vitamine. *Geneeskundig Tijdschrift voor Nederlands Indie.* 66: 810–827.

Jukes, T.H., 1939. Pantothenic acid and the filtrate (chick anti-dermatitis) factor. *Journal of the American Chemical Society.* 61: 975–976.

Karrer, P., Schöpp, K., and Benz, F., 1935. Synthesen von Flavinen IV. *Helvetica Chimica Acta.* 18: 426–429.

Kögl, F., and Tönnis, B., 1936. Über das Bios-Problem. Darstellung von krystallisiertem Biotin aus Eigelb. *Hoppe-Seyler's Zeitschrift für physiologische Chemie.* 242: 43–73.

Krehl, W.A., Teply, L.J., Sarma, P.S., and Elvehjem, C.A., 1945. Growth-retarding effect of corn in nicotinic acid-low rations and its counteraction by tryptophane. *Science.* 101: 489–490.

Kuhn, R., György, P., and Wagner-Jauregg, T., 1933. Über neue Klasse on Naturfarbstoffen. *Berichte der Deutschen chemischen Gesellschaft.* 66: 317–320.

Kuhn, R., and Rudy, H., 1934. Über den alkali-labilen Ring des Lacto-flavins. *Berichte der Deutschen chemishcen Gesellschaft.* 67: 892–898.

Kuhn, R., Reinemund, K., Weygand, F., and Ströbele, R., 1935. Über die Synthese des Lactoflavins (Vitamin B2). *Berichte der Deutschen chemischen Gesellschaft.* 68: 1765–1774.

Lindstrand, K., 1964. Isolation of methylcobalamin from natural source material. *Nature.* 204: 188–189.

Lohmann, K., and Schuster, P., 1937. Untersuchungen über die Cocarboxylase. *Biochemische Zeitschrift.* 294: 188–214.

Lunin, N., 1881. Ueber die Bedeutung der anorganischen Salze für die Ernährung des Thieres. Hoppe-Seyler's *Zeitschrift für physiologische Chemie.* 5: 31–39.

Meiklejohn, A.P., Passmore, R., and Peters, R.A., 1932. The independence of vitamin B1 deficiency and inanition. *Proceedings of the Royal Society B: Biological Sciences.* 111: 391–395.

Metzler, D.E., Ikawa, M., and Snell, E.E., 1954. A general mechanism for vitamin B6-catalyzed reactions. *Journal of the American Chemical Society.* 76: 648–652.

Meyerhof, O., Lohmann, K., and Meyer, K., 1931. Über das Koferment der Milchsäure-bildung im Muskel. *Biochemische Zeitschrift.* 237: 437.

Minot, G.R., Murphy, W.P., and Stetson, R.P., 1928. The response of the reticulocytes to liver Therapy: particularly in pernicious anemia. *American Journal of Medical Science.* 175: 581–598.

Mitchell, H.K., Snell, E.E., and Williams, R.J., 1941. The concentration of 'folic acid'. *Journal of the American Chemical Society.* 63: 2284.

Pfiffner, J.J., Binkley, S.B., Bloom, E.S., Brown, R.A., Bird, O.D., Emmett, A.D., Hogan, A.G., and O'Dell, B.L., 1943. Isolation of the antianemia factor (vitamin Bc) in crystalline form from liver. *Science.* 97: 404–405.

Rickes, E.L., Brink, N.G., Koniuszy, F.R., Wood, T.R., and Folkers, K., 1948. Crystalline vitamin B_{12}. *Science.* 107: 396–397.

Schlenk, F., and von Euler, H., 1936. Cozymase. *Naturwissenschaften.* 24: 794–795.

Snell, E.E., and Strong, F.M., 1939. A microbiological assay for riboflavin. *Industrial & Engineering Chemistry*, Analytical Edition. 11: 346–350.

Snell, E.E., 1944. The vitamin activities of 'pyridoxal' and 'pyridoxamine'. *Journal of Biological Chemistry.* 154: 313–314.

Snell, E.E., 1945. The vitamin B6 group. V. The reversible interconversion of pyridoxal and pyridoxamine by transamination reactions. *Journal of the American Chemical Society.* 57: 194–197.

Stokstad, E.L.R., 1943. Some properties of a growth factor for *Lactobaccillus casei*. *Journal of Biological Chemistry.* 149: 573–574.

Suzuki, U., Shimamura, T., and Odake, S., 1912. Über Oryzanin, ein Bestandteil der Reiskleie und seine physiologische Bedeutung. *Biochemische Zeitschrift.* 43: 89–153.

Theorell, H., 1935. Das gelbe Oxydationsferment. *Biochemische Zeitschrift.* 278: 263–290.

Theorell, H., 1937. Die freie Eiweisskomponente des gelben Ferments und ihre Kupplung mit Lacto-flavinphosphorsaure. *Biochemische Zeitschrift.* 290: 293–303.

Vilter, R.W., Horrigan, D., Mueller, J.F., Jarrold, T., Vilter, C.F., Hawkins, V., and Seaman, A., 1950. Studies on the relationships of vitamin B12, folic acid, thymine, uracil and methyl group donors in persons with pernicious anemia and related megaloblastic anemias. *Blood.* 5: 695–717.

Voegtlin, C., Neill, M.H., and Hunter, A., 1920. The influence of the vitamines on the course of pellagra. *Public Health Reports.* 35: 1435.

Waller, C.W., Hutchings, B.L., Mowat, J.H., Stokstad, E.L.R., Boothe, J.H., Angier, R.B., Semb, J., SubbaRow, Y., Cosulich, D.B., Fahrenbach, M.J., Hultquist, M.E., Kuh, E., Northey, E.H., Seeger, D.R., Sickels, J.P., and Smith, J.M., 1948. Synthesis of pteroylglutamic acid (liver L. casei factor) and pteroic acid. I. *Journal of the American Chemical Society.* 70: 19–22.

Warburg, O., and Christian, W., 1932. Über ein neues Oxydationsferment und sein Absorptionsspektrum. *Biochemische Zeitschrift.* 254: 438–458.

Warburg, O., Christian, W., and Griese, A., 1935. Wasserstoffübertragendes Co-Ferment, seine Zusammensetzung und Wirkungsweise. *Biochemische Zeitschrift.* 282: 157–165.

Warburg, O., and Christian, W., 1936. Pyridin, der wasserstoffübertragende Bestandteil von Gärungsfermenten (Pyridin-Nucleotide). *Biochemische Zeitschrift.* 287: 291–328.

Warburg, O., and Christian, W., 1938. Isolierung der prosthetischen Gruppe der d-Aminosäureoxydase. *Biochemische Zeitschrift*. 298: 150–155.

Willcock, E.G., and Hopkins, F.G., 1906. The importance of individual amino acids in metabolism; observations on the effect of adding tryptophan to a dietary in which zein is the sole nitrogenous constituent. *Journal of Physiology*. 35: 88–102.

Williams, R.R., and Cline, J.K., 1936. Synthesis of Vitamin B1. *Journal of the American Chemical Society*. 58: 1504–1505.

Williams, R.J., Truesdail, J.H., Weinstock, H.H. Jr., Rohrmann, E., Lyman, C.M., and McBurney, C.H., 1938. Pantothenic acid. II. Its concentration and purification from liver. *Journal of the American Chemical Society*. 60: 2719–2723.

Wills, L., 1931. Treatment of pernicious anaemia of pregnancy and tropical anaemia. *British Medical Journal*. 1: 1059–1064.

Woolley, D.W., Waisman, G.A., and Elvehjem, C.A., 1939. Studies on the structure of the chick antidermatitis factor. *Journal of Biological Chemistry*. 129: 673–679.

CHAPTER 2
B Vitamins and Disease

JUTTA DIERKES* AND OTTAR NYGÅRD

Institute of Medicine, University of Bergen, Postbox 7804, N-5020 Bergen, Norway
*Email: jutta.dierkes@med.uib.no

2.1 Introduction

The group of B vitamins includes eight chemically distinct compounds, which have all proven to be essential for humans: vitamin B_1 (thiamine), vitamin B_2 (riboflavin), pantothenic acid, biotin, niacin, vitamin B_6 (pyridoxine), folate and vitamin B_{12} (cobalamin). While some of these vitamins are hardly investigated in clinical research, others have attracted huge attention during the last decades. These includes especially folate or folic acid, cobalamin and pyridoxine, which have been used in large-scale randomized clinical trials for the (secondary) prevention of major chronic diseases. In addition, niacin is used for the treatment of dyslipidemia and therefore, also in prevention and treatment of cardiovascular disease (CVD). As this chapter focuses on major chronic diseases, other B vitamins (*e.g.* thiamine which has been used for the treatment of Wernicke encephalopathy) will not be discussed here but elsewhere. These chronic diseases include coronary heart disease and stroke, some types of cancers and neurodegenerative diseases like dementia, and age-related decline of cognitive function. Folic acid has also been used for the prevention of neural tube defects and other congenital malformations when used periconceptionally.

2.2 The Role of Different Study Types

When discussing the health effects of nutrients, different levels of evidence have to be recognized:

1. A biological plausible mechanism should explain why a (mild) deficiency of a certain nutrient can be associated with a chronic disease that usually develops over years or decades. Cell studies and animal studies, together with modern molecular biology methods are invaluable to elucidate these mechanisms.
2. There should be sound and consistent evidence for an association of the nutrient status with a disease from epidemiological studies. Nowadays, the major chronic diseases are difficult to study in classic case-control studies and this study design is no longer regarded as an appropriate one for most diseases. This is because diet, behaviour and the nutrient of interest are studied after the diagnosis of disease in case-control studies, and both the disease may affect the metabolism of a nutrient and the diagnosis may affect the diet, behaviour and supplement use of the patients. In epidemiological research, large population-based cohort studies are preferred to gain knowledge on the association of a nutrient with a disease. Examples for these study types include the Nurses' Health Study, the Physicians' Health follow-up study and the European Investigation into Cancer and Nutrition. These studies included large numbers of participants who were healthy at the start of the study and who have been followed for many years. These studies are observational; that means that they observe, among other variables, dietary intake at baseline and correlate this to the occurrence of disease later in life. These studies usually focus on intake of B vitamins from food and from supplements, and many investigations distinguish between these forms.
3. While epidemiological cohort studies show associations, they cannot prove causal relations. To proof the hypothesis that a nutrient is associated with a disease, randomized controlled trials (RCTs) are needed that can show a clear dose–response effect. The general hypothesis underlying the concept of RCTs is that large amounts of a certain vitamin or a certain vitamin combination will result in lower risk of a certain disease. Examples for these studies are the numerous trials that have been conducted using folic acid, vitamin B_6 and B_{12} for the prevention of cardiovascular disease, folic acid for the prevention of cancer, and folic acid, vitamin B_6 and B_{12} for the prevention of cognitive decline and dementia. RCTs are generally regarded as the gold standard in clinical medicine to provide a proof of causality between a disease and a treatment.

Advantages of RCTs are that they study single nutrients or fixed combinations of nutrients, at a given dose, and for a defined period of follow up. Disadvantages are that many of the trials are secondary prevention trials, which means that they include patients who already have the disease and that the recurrence of an event or worsening of the disease should be prevented. Thus, it may be that the trials are too late during the course of a disease to reverse the disease process. Other disadvantages are that they may be of too short a duration, that they use doses of vitamins that are either too low or too high, or that the effect of the nutrient intervention may be modified by disease therapy. As a speciality of RCTs in nutrition research, it should be mentioned that in case of essential nutrients, there is no true placebo group, as the patients who receive the study placebo also ingest nutrients from food, fortified foods or over-the-counter supplements. In the case of folic acid, there can be a substantial contamination in the placebo groups in trials conducted in those countries with mandatory folic acid supplementation of staple foods (*e.g.* the US, Canada and Chile).

Thus, epidemiological cohort studies can provide associations between vitamin intake either from food or from fortified food or supplements and a specific disease, and the RCTs can provide a proof whether this association is causal or not. The major differences are that the cohort studies usually include healthy people at baseline, while the RCTs usually include patients who suffer from the disease (secondary prevention). Observational studies usually have longer follow-up periods and they assess food intake, from which vitamin intake is calculated from. It may also be mentioned that an observational study may either find an increased disease risk at low intake or low plasma levels of a nutrient (usually in the lowest quartile or quintile of the cohort), which is opposite to the finding of a reduced disease risk at high intake or high plasma levels. RCTs, however, aim to find a reduced disease risk at high intake levels that is achieved through the nutrient supplement used. This difference is discussed in more detail below.

2.3 Treatment of Deficiency or Supplementation without Reference to Vitamin Status

The criteria for the diagnosis of vitamin deficiency are specific for every single vitamin. However, some common rules can be applied to most of the B vitamins. First, it has to be stated that, in clinical practice, some vitamins (biotin, pantothenic acid, niacin) are hardly ever measured and therefore data on vitamin status are hardly available. Other vitamins like folate and cobalamin are measured regularly using commercial assays.

Since concentration measurements in blood may not reflect true vitamin status, the measurement of metabolites or activity of enzymes that use a coenzyme form of a B vitamin has been suggested. For example, the measurement of either homocysteine or methylmalonic acid has been suggested as a marker for cobalamin status, and the erythrocyte transaminase coefficient has been suggested as a marker for vitamin B_6 status. However, there are no accepted definitions for the diagnosis of vitamin deficiency in the case of many vitamins.

As already mentioned, some of the B vitamins, especially folic acid, vitamin B_{12} and vitamin B_6 have been used during the last two decades for the secondary prevention of cardiovascular disease, colorectal adenomas or age-related cognitive decline. In these studies, usually large amounts of these vitamins have been given to patients suffering from the disease. However, in most studies, vitamin status was not a criterion for inclusion in the studies. Only a few studies used elevated concentrations of the metabolite homocysteine as an inclusion criterion. Thus, these studies investigated the effect of defined doses of single vitamins or combinations of vitamins (following the 2×2 factorial study design) in patients with a particular disease, but without vitamin deficiency.

2.4 Mechanisms

There are several biological mechanisms that may explain a role of B vitamins in the etiology of chronic diseases. B vitamins act as coenzymes and can therefore affect a huge number of biochemical reactions. Among others, these reactions are central in protein and amino acid metabolism, fatty acid and lipoprotein metabolism, DNA biosynthesis and repair, and carbohydrate and energy metabolism. In many reactions, more than one B vitamin derived coenzyme is involved, which may also explain joint actions by different vitamins.

One of the most investigated B vitamins is folate, which occurs in many different forms in metabolism and which is involved in the synthesis of the purine bases adenosine and guanosine, the pyrimidine nucleoside thymidine and, of utmost interest, in many methylation reactions. Within the methionine cycle, 5-methyltetrahydrofolate re-methylates homocysteine to methionine. A high folate intake contributes to a low circulating homocysteine level, which itself has been associated with increased risk of coronary heart disease, cerebrovascular disease and cognitive decline (Calvaresi and Bryan 2001). Methionine is further metabolized to S-adenosylmethionine (Figure 2.1), the principal methyl donor in most cellular reactions. In addition, the cellular S-adenosylmethionine concentration may control gene transcription by cytosine methylation in the DNA and also enzyme activity. Methylation of DNA and histones has been shown to be important for gene expression and subsequently protein synthesis (Handy et al. 2011). The formation of thymidine from uracil is dependent on 5,10-methylenetetrahydrofolate. In folate deficiency, less thymidine is formed and uracil is misincorporated into the DNA molecule. This can result in DNA strand breakage and malignant transformation (Duthie 2010).

In the methionine cycle, cobalamin also plays a central role, as the methyl group of 5-methyltetrahydrofolate is first transferred to the cobalamin molecule and then further transferred to homocysteine to form methionine. This reaction explains the close connection of folate and cobalamin metabolism, as 5-methyltetrahydrofolate cannot be used for other reactions over than the methylation of homocysteine. Mild deficiency of folate and cobalamin is also associated with cognitive decline due to the neurotoxic effects of increased

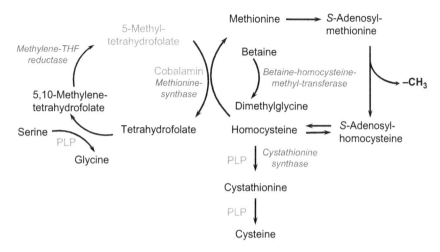

Figure 2.1 Vitamins involved in homocysteine metabolism. Vitamins involved in metabolism are shown in grey. Enzymes are shown in italics. PLP = pyridoxal phosphate.

levels of homocysteine, increased oxidative stress and decreased synthesis of neurotransmitters, phospholipids and myelin (Calvaresi and Bryan 2001).

2.5 Cardiovascular Disease

2.5.1 Folic acid, Vitamin B_{12} and Vitamin B_6

In the early 1990s, elevated blood concentrations of the amino acid homocysteine were associated with increased risk of cardiovascular disease. Supplementation with folic acid, vitamin B_{12} and B_6 can lower homocysteine blood concentrations and therefore randomized clinical trials using high amounts of folic acid, either single or in combination with vitamin B_{12} and vitamin B_6, were initiated. These studies followed the hypothesis that large amounts of folic acid would be effective in reducing elevated homocysteine concentrations (used as an intermediate endpoint) and therefore also reduce the risk of coronary heart disease or stroke. Doses in these trials ranged from 0.8 to 5 mg folic acid daily, 400–1000 μg vitamin B_{12} and 25–50 mg vitamin B_6; median follow-up periods were between 2 and 7.3 years. One trial in renal patients even used higher vitamin doses (40 mg folic acid, 2 mg vitamin B_{12} and 100 mg vitamin B_6). Recently, a meta-analysis of eight trials involving about 37 000 patients has been reported in which it was obvious that supplementation with these vitamins neither reduced cardiovascular morbidity or mortality nor total mortality (Clarke et al. 2010).

The results of the RCTs are in contrast to a number of observational cohort studies that report either a reduced risk for those participants ingesting high amounts of food folate or supplemental folic acid, or those having the highest

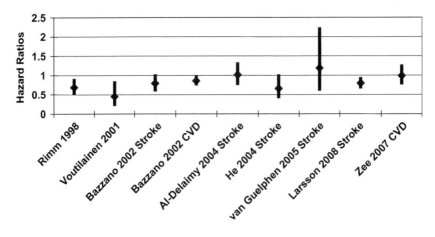

Figure 2.2 Observational studies* on folate intake and future risk of cardiovascular events. The point estimate is given, together with the 95% confidence interval. Usually the risk estimate is figured for the highest quartile or quintile of folate intake (also including intake of folic acid by dietary supplements) compared to the lowest quartile or quintile. Figure based on Al-Delaimy *et al.* (2004), Bazzano *et al.* (2002), He *et al.* (2004), Larsson *et al.* (2008), Rimm *et al.* (1998), Van Guelpen *et al.* (2005), Voutilainen *et al.* (2001) and Zee *et al.* (2007).

blood concentrations of folate, or an increased risk for those in the lowest quartile or quintile of folate intake or folate blood concentrations (Drogan *et al.* 2006, Figure 2.2). Similar results were reported for vitamin B_{12} (Weikert *et al.* 2007) and, at least in some studies, for vitamin B_6 (Dierkes *et al.* 2007).

Reasons for the divergent results achieved by observational studies and randomized clinical trials are largely unknown. Among possible reasons, there may be substantial residual bias in the observational studies (Lawlor *et al.* 2004); it is important to consider lifelong vitamin intake instead of short-term vitamin supplementation, different chemical structures of vitamins in food *vs.* in supplements (most pronounced for folic acid in supplements *vs.* the various forms of folate in food), and the differences in treating patients already suffering from a disease (as in the randomized clinical trials) from observing healthy subjects (as in the observational studies). Additionally, in the majority of studies, most patients randomized were without overt vitamin deficiency and the interactive effects of other nutrients have not usually been considered.

2.5.2 Niacin in Cholesterol Reduction

The cholesterol reducing effect of megadoses of niacin has been known for more than 50 years (Altschul *et al.* 1955). While the daily allowance for niacin is in the order of 15 mg per day, the cholesterol reducing effect requires doses of about 1.5–3 g of nicotinic acid per day (Knopp 1999). In these doses, niacin reduces the atherogenic very low-density lipoprotein (VLDL) and low-density lipoprotein (LDL) cholesterol, but raises the level of high-density lipoprotein

(HDL) cholesterol, which is believed to be protective. The clinical effect of niacin has been tested in several randomized clinical trials and a recent meta-analysis revealed that niacin treatment reduces the risk of cardiovascular events in secondary prevention (Bruckert *et al.* 2010). However, the treatment has also been associated with increased risk of diabetes. Because the trials usually have been small, the results of ongoing large-scale RCTs must be awaited before firm conclusions can be made.

2.6 Cancer

B vitamin intake or vitamin blood concentrations have also been related to various cancers. The cancer types that have been best investigated with respect to folic acid are colon cancer and colorectal cancer. Convincing evidence from observational studies led to the initiation of randomized controlled trials with folic acid in colorectal adenomas. However, similar to cardiovascular disease, there seems to be a discrepancy between the observational epidemiological studies that reported in the majority an inverse association of folate and cancer risk, and the effect of folic acid supplementation in the randomized controlled trials, which reported no effect of folic acid on recurrence of colorectal adenoma risk.

The association between folate intake and colorectal cancer in 27 observational epidemiological studies was analysed in a recent meta-analysis (Kennedy *et al.* 2011). This meta-analysis, which included 18 case-control studies and nine cohort studies, showed that high folate intake was associated with reduced risk of colorectal adenomas. The association was stronger in case-control studies [relative risk (RR) 0.85; 95% confidence interval (CI) 0.74–0.99] than in cohort studies (RR 0.92; 95% CI 0.81–1.05).

Three large and some smaller RCTs have investigated the effect of folic acid on the prevention of the recurrence of colorectal adenomas. Recent meta-analyses showed that the larger trials (involving more than 1300 patients treated with placebo and more than 1300 patients treated with folic acid, with folic acid doses of 0.5–1.0 mg/d and follow-up periods up to 42 months) showed no effect of folic acid on the recurrence of any adenoma (RR 0.98, 95% CI 0.82–1.17) or advanced adenoma (RR 1.06, 95% CI 0.81–1.39). Analyses of other outcomes (number of adenomas, adenoma size or location) also showed no difference between placebo and folic acid (Figueiredo *et al.* 2011). Another meta-analysis, which also included some smaller trials, derived the similar conclusion that folic acid has no effect on the recurrence of colorectal adenomas (RR 1.08, 95% CI 0.87–1.33) (Ibrahim *et al.* 2010).

In addition, cancer incidence and mortality was reported in seven of the eight randomized controlled trials in CVD prevention (Clarke *et al.* 2010). It appeared that supplementation with folic acid (0.8–40 mg/d) had no effect on overall cancer incidence (RR 1.05; 95% CI 0.98–1.13) and mortality (RR 1.00; 95% CI 0.85–1.18). It has to be noted that none of these trials was specifically designed to investigate cancer incidence or mortality, and that there was no analysis for specific cancers. A combined analysis of two of these trials conducted in Norway, however, reported an increased cancer incidence and increased mortality due to cancer during an extended follow-up period of 78 months (Ebbing *et al.* 2009).

2.7 Prevention of Age-related Cognitive Decline or Dementia

Worldwide, the prevalence of dementia is increasing and is a serious public health concern. Age-related cognitive changes include mild memory loss, cognitive impairment and dementia. There is a substantial degree of overlap among these conditions, and although there are a number of widely accepted tests of cognitive performance, it has to be acknowledged that the different tests measure different domains of cognitive function and complicate direct comparisons of studies. In observational studies, low folate status and elevated homocysteine levels were associated with poor cognitive test performance, as summarized by Raman *et al.* 2007. These authors also stated that there is large inconsistency in both the applied cognitive tests and definitions of low folate status, which precludes a formal meta-analysis of the observational trials.

Subsequently, several randomized controlled trials have been conducted to test the hypothesis that folic acid may prevent cognitive decline. So far, nine RCTs have been conducted that have been analysed in a meta-analysis, including nearly 3000 participants (Wald *et al.* 2010). Folic acid doses varied between 0.2 and 15 mg per day, and duration of treatment was between one month and three years. This meta-analysis concluded that folic acid alone or in combination with other vitamins has no effect on the prevention of age-related cognitive decline within three years. This lack of effect was consistent across different elements of cognitive function (memory, language, processing speed, decision-making). Neither the age of participants, treatment with other B vitamins, dose of folic acid, nor duration of treatment had a significant effect on the outcome. However, one important conclusion of this meta-analysis was that more studies of longer duration are needed to draw final conclusions on the long term effect of B vitamin supplementation on cognitive function. Other meta-analyses on this topic came to similar conclusions (Dangour *et al.* 2010; Ligthart *et al.* 2010). The addition of vitamin B_{12} to folic acid or the supplementation with vitamin B_{12} alone does not seem to alter either the results or the conclusion of the randomized trials (Eussen *et al.* 2006; Malouf and Grimley Evans 2008).

2.8 Renal Disease

As the kidney has a major role in B vitamin metabolism, it is plausible that chronic kidney disease may affect vitamin status to a clinically significant extent. This holds especially true in end-stage renal disease, when the dialysis process may cause additionally vitamin losses (Heinz *et al.* 2008). Although vitamin supplementation among patients with end-stage renal disease is widely practised, the scientific evidence for doing so was, until recently, very vague. And contrary to common beliefs, supplementation with B vitamins in patients

Table 2.1 Overview of evidence for risk estimate of several clinical endpoints from randomized clinical trials with B vitamins.

Disease/outcome	Vitamin	Proposed mechanisms	Level of evidence	References
Mortality	Folic acid, vitamin B_{12} vitamin B_6	Homocysteine reduction	Meta-analysis (MA) of randomized controlled trials (RCTs): no evidence of reduced risk	Clarke et al. (2010)
Coronary heart disease	Folic acid Cobalamin (vitamin B_{12}) Pyridoxine (vitamin B_6)	Homocysteine reduction	MA of RCTs: supplementation in patients does not affect risk of a recurrent event	Clarke et al. (2010)
	Nicotinic acid	Reduction of VLDL and LDL cholesterol, increase in HDL cholesterol	MA shows risk reduction in secondary prevention at high doses	Bruckert et al. (2011)
Stroke	Folate/folic acid Cobalamin (vitamin B_{12}) Pyridoxine (vitamin B_6)	Homocysteine reduction	MA of RCTs: supplementation in patients does not affect risk of a recurrent event	Clarke et al. (2010)
Colorectal cancer	Folate/folic acid	Methylation	RCTs with folic acid did not show reduced risk of recurrence	Figueiredo et al. (2011)
Venous thrombosis	Vitamin B_6	Homocysteine reduction	RCT: no reduction of risk in disease recurrence	den Heijer et al. (2007)
Neural tube defects	Folic acid		RCTs showed reduction in recurrence and occurrence when folic acid was taken preconceptionally	MRC study (1991), Czeizel and Dudás (1992)
Cognitive decline	Folic acid Vitamin B_{12}	Methylation, homocysteine reduction	RCTs with folic acid did not improve cognitive function RCTs with vitamin B_{12} did not improve cognitive function	Wald et al. (2010) Eussen et al. (2006)

with either chronic kidney disease or end-stage renal disease affected neither cardiovascular morbidity and mortality, nor total mortality (Clarke *et al.* 2010; Heinz *et al.* 2010; Bostom *et al.* 2011).

2.9 Concluding Remarks

B vitamin supplementation is widely used by many healthy people with the aim of reducing the burden of chronic disease. Although mechanistic studies have provided plausible mechanisms for a contribution of vitamin deficiency to biochemical dysfunctions, and observational studies have suggested associations between B vitamin status and disease risk, B vitamins were not effective in reducing disease risk in randomized controlled trials (Table 2.1). There is only one exception to this conclusion, and that is the pre- and periconceptional use of folic acid with the aim to reduce the numbers of foetuses affected by neural tube defects and other congenital malformations (MRC Vitamin Study Research Group 1991). This effect of folic acid was the reason to introduce mandatory folic acid fortification of staple foods in the USA and Canada in 1998. However, the uncontrolled use of B vitamin supplementation by healthy people beyond the reproductive age must be seriously questioned in the light of current scientific data. Obviously, vitamin supplementation has different effects than consuming a healthy diet containing vitamin-rich foods. This remains a challenge to nutrition research to elucidate the discrepancies between vitamin supplementation and dietary vitamin intake.

Summary Points

- B vitamins include thiamine, riboflavin, pyridoxine, niacin, pantothenic acid, biotin, folate and cobalamin.
- Status assessment in humans includes assessment of vitamin intake, measurement of vitamin concentration in blood or urine, and/or measurement of metabolites/enzyme activity.
- Folate, cobalamin and pyridoxine have been studied with respect to homocysteine metabolism, a suspected cardiovascular risk factor.
- Observational epidemiological studies suggested a protective effect of increased intake of folate for various cancers and cardiovascular disease.
- Randomized controlled trials tested the effect of folic acid on the recurrence of colorectal cancer, but did not show a protective effect of folic acid.
- Randomized controlled trials tested the effect of folic acid on recurrence of coronary heart disease and stroke, but did not show a protective effect of folic acid.
- Therefore, RCTs did not confirm a protective effect of folic acid, as suggested by observational studies.
- However, the protective effect of supplemental preconceptional folic acid for the prevention of neural tube defects in women of childbearing age is proven. This seems to be the only proven protective effect of supplemental folic acid intake.

List of Abbreviations

CI	confidence interval
CVD	cardiovascular disease
DNA	deoxyribonucleic acid
HLD	high-density lipoprotein
LDL	low-density lipoprotein
MA	meta-analysis
PLP	pyridoxal phosphate
RCT	randomized controlled trial
RR	relative risk
VLDL	very low-density lipoprotein

References

Al-Delaimy, W.K., Rexrode, K.M., Hu, F.B., Albert, C.M., Stampfer, M.J., Willett, W.C., and Manson, J.E., 2004. Folate intake and risk of stroke among women. *Stroke.* 35: 1259–1263.

Altschul, R., Hoffer, A., and Stephen, J.D., 1955. Influence of nicotinic acid on serum cholesterol in man. *Archives of Biochemistry and Biophysics.* 54: 558–589.

Bazzano, L.A., He, J., Ogden, L.G., Loria, C., Vupputuri, S., Myers, L., and Whelton, P.K., 2002. Dietary intake of folate and risk of stroke in US men and women: NHANES I Epidemiologic Follow-up Study. National Health and Nutrition Examination Survey. *Stroke.* 33: 1183–1188.

Bostom, A.G., Carpenter, M.A., Kusek, J.W., Levey, A.S., Hunsicker, L., Pfeffer, M.A., Selhub, J., Jacques, P.F., Cole, E., Gravens-Mueller, L., House, A.A., Kew, C., McKenney, J.L., Pacheco-Silva, A., Pesavento, T., Pirsch, J., Smith, S., Solomon, S., and Weir, M., 2011. Homocysteine-lowering and cardiovascular disease outcomes in kidney transplant recipients: primary results from the Folic Acid for Vascular Outcome Reduction in Transplantation trial. *Circulation.* 123: 1763–1770.

Bruckert, E., Labreuche, J., and Amarenco P., 2010. Meta-analysis of the effect of nicotinic acid alone or in combination on cardiovascular events and atherosclerosis. *Atherosclerosis.* 210: 353–361.

Calvaresi, E., and Bryan, J., 2001. B vitamins, cognition, and aging: a review. *Journals of Gerontology, Series B: Psychological Sciences and Social Science.* 56: P327–P339.

Clarke, R., Halsey, J., Lewington, S., Lonn, E., Armitage, J., Manson, J.E., Bønaa, K.H., Spence, J.D., Nygård, O., Jamison, R., Gaziano, J.M., Guarino, P., Bennett, D., Mir, F., Peto, R., and Collins, R., 2010. B-Vitamin Treatment Trialists' Collaboration. Effects of lowering homocysteine levels with B vitamins on cardiovascular disease, cancer, and cause-specific mortality: meta-analysis of 8 randomized trials involving 37 485 individuals. *Archives of Internal Medicine.* 170: 1622–1631.

Czeizel, A.E., and Dudás, I., 1992. Prevention of the first occurrence of neural-tube defects by periconceptional vitamin supplementation. *New England Journal of Medicine.* 327: 1832–1835.

Dangour, A.D., Whitehouse, P.J., Rafferty, K., Mitchell, S.A., Smith, L., Hawkesworth, S., and Vellas, B., 2010. B vitamins and fatty acids in the prevention and treatment of Alzheimer's disease and dementia: a systematic review. *Journal of Alzheimer's Disease.* 22: 205–224.

den Heijer, M., Willems, H.P., Blom, H.J., Gerrits, W.B., Cattaneo, M., Eichinger, S., Rosendaal, F.R., and Bos, G.M., 2007. Homocysteine lowering by B vitamins and the secondary prevention of deep vein thrombosis and pulmonary embolism: a randomized, placebo-controlled, double-blind trial. *Blood.* 109: 139–144.

Dierkes, J., Weikert, C., Klipstein-Grobusch, K., Westphal, S., Luley, C., Möhlig, M., Spranger, J., and Boeing, H., 2007. Plasma pyridoxal-5-phosphate and future risk of myocardial infarction in the European Prospective Investigation into Cancer and Nutrition Potsdam cohort. *American Journal of Clinical Nutrition.* 86: 214–220.

Drogan, D., Klipstein-Grobusch, K., Dierkes, J., Weikert, C., and Boeing, H., 2006. Dietary intake of folate equivalents and risk of myocardial infarction in the European Prospective Investigation into Cancer and Nutrition (EPIC)–Potsdam study. *Public Health Nutrition.* 9: 465–471.

Duthie, S.J., 2011. Folate and cancer: how DNA damage, repair and methylation impact on colon carcinogenesis. *Journal of Inherited Metabolic Disease.* 34: 101–109.

Ebbing, M., Bønaa, K.H., Nygård, O., Arnesen, E., Ueland, P.M., Nordrehaug, J.E., Rasmussen, K., Njølstad, I., Refsum, H., Nilsen, D.W., Tverdal, A., Meyer, K., and Vollset, S.E., 2009. Cancer incidence and mortality after treatment with folic acid and vitamin B12. *JAMA, the Journal of the American Medical Association.* 302: 2119–2126.

Eussen, S.J., de Groot, L.C., Joosten, L.W., Bloo, R.J., Clarke, R., Ueland, P.M., Schneede, J., Blom, H.J., Hoefnagels, W.H., and van Staveren, W.A., 2006. Effect of oral vitamin B-12 with or without folic acid on cognitive function in older people with mild vitamin B-12 deficiency: a randomized, placebo-controlled trial. *American Journal of Clinical Nutrition.* 84: 361–370.

Figueiredo, J.C., Mott, L.A., Giovannucci, E., Wu, K., Cole, B., Grainge, M.J., Logan, R.F., and Baron, J.A., 2011. Folic acid and prevention of colorectal adenomas: a combined analysis of randomized clinical trials. *International Journal of Cancer.* 129: 192–203.

Handy, D.E., Castro, R., and Loscalzo, J., 2011. Epigenetic modifications: basic mechanisms and role in cardiovascular disease. *Circulation.* 123: 2145–2156.

He, K., Merchant, A., Rimm, E.B., Rosner, B.A., Stampfer, M.J., Willett, W.C., and Ascherio, A., 2004. Folate, vitamin B6, and B12 intakes in relation to risk of stroke among men. *Stroke.* 35: 169–174.

Heinz, J., Domröse, U., Westphal, S., Luley, C., Neumann, K.H., and Dierkes, J., 2008. Washout of water-soluble vitamins and of homocysteine during

haemodialysis: effect of high-flux and low-flux dialyser membranes. *Nephrology.* 13: 384–389.
Heinz, J., Kropf, S., Domröse, U., Westphal, S., Borucki, K., Luley, C., Neumann, K.H., and Dierkes, J., 2010. B vitamins and the risk of total mortality and cardiovascular disease in end-stage renal disease: results of a randomized controlled trial. *Circulation.* 121: 1432–1438.
Ibrahim, E.M., and Zekri, J.M., 2010. Folic acid supplementation for the prevention of recurrence of colorectal adenomas: metaanalysis of interventional trials. *Medical Oncology.* 27: 915–918.
Kennedy, D.A., Stern, S.J., Moretti, M., Matok, I., Sarkar, M., Nickel, C., and Koren, G., 2011. Folate intake and the risk of colorectal cancer: a systematic review and meta-analysis. *Cancer Epidemiology.* 35: 2–10.
Knopp, R.H., 1999. Drug treatment of lipid disorders. *New England Journal of Medicine.* 341: 498–511.
Larsson, S.C., Männistö, S., Virtanen, M.J., Kontto, J., Albanes, D., and Virtamo, J., 2008. Folate, vitamin B6, vitamin B12, and methionine intakes and risk of stroke subtypes in male smokers. *American Journal of Epidemiology.* 167: 954–961.
Lawlor, D.A., Davey Smith, G., Kundu, D., Bruckdorfer, K.R., and Ebrahim, S., 2004. Those confounded vitamins: what can we learn from the differences between observational *versus* randomised trial evidence? *Lancet.* 363(9422): 1724–1727.
Ligthart, S.A., Moll van Charante, E.P., Van Gool, W.A., and Richard, E., 2010. Treatment of cardiovascular risk factors to prevent cognitive decline and dementia: a systematic review. *Vascular Health and Risk Management.* 6: 775–785.
Malouf, R., and Grimley Evans, J., 2008. Folic acid with or without vitamin B12 for the prevention and treatment of healthy elderly and demented people. *Cochrane Database System Review.* 8: CD004514.
MRC Vitamin Study Research Group, 1991. Prevention of neural tube defects: results of the Medical Research Council Vitamin Study. *Lancet.* 338(8760): 131–137.
Raman, G., Tatsioni, A., Chung, M., Rosenberg, I.H., Lau, J., Lichtenstein, A.H., and Balk, E.M., 2007. Heterogeneity and lack of good quality studies limit association between folate, vitamins B-6 and B-12, and cognitive function. *Journal of Nutrition.* 137: 1789–1794.
Rimm, E.B., Willett, W.C., Hu, F.B., Sampson, L., Colditz, G.A., Manson, J.E., Hennekens, C., and Stampfer, M.J., 1998. Folate and vitamin B6 from diet and supplements in relation to risk of coronary heart disease among women. *JAMA, the Journal of the American Medical Association.* 279: 359–364.
Van Guelpen, B., Hultdin, J., Johansson, I., Stegmayr, B., Hallmans, G., Nilsson, T.K., Weinehall, L., Witthöft, C., Palmqvist, R., and Winkvist, A., 2005. Folate, vitamin B12, and risk of ischemic and hemorrhagic stroke: a prospective, nested case-referent study of plasma concentrations and dietary intake. *Stroke.* 36: 1426–1431.

Voutilainen, S., Rissanen, T.H., Virtanen, J., Lakka, T.A., and Salonen, J.T., 2001. Low dietary folate intake is associated with an excess incidence of acute coronary events: The Kuopio Ischemic Heart Disease Risk Factor Study. *Circulation.* 103: 2674–2680.

Wald, D.S., Kasturiratne, A. and Simmonds, M., 2010. Effect of folic acid, with or without other B vitamins, on cognitive decline: meta-analysis of randomized trials. *American Journal of Medicine.* 123: 522–527.e2.

Weikert, C., Dierkes, J., Hoffmann, K., Berger, K., Drogan, D., Klipstein-Grobusch, K., Spranger, J., Möhlig, M., Luley, C., and Boeing, H., 2007. B vitamin plasma levels and the risk of ischemic stroke and transient ischemic attack in a German cohort. *Stroke.* 38: 2912–2918.

Zee, R.Y., Mora, S., Cheng, S., Erlich, H.A., Lindpaintner, K., Rifai, N., Buring, J.E., and Ridker, P.M., 2007. Homocysteine, 5,10-methylenetetrahydrofolate reductase 677C>T polymorphism, nutrient intake, and incident cardiovascular disease in 24 968 initially healthy women. *Clinical Chemistry.* 53: 845–851.

CHAPTER 3
Vitamins and Folate Fortification in the Context of Cardiovascular Disease Prevention

ALEXIOS S. ANTONOPOULOS, CHEERAG SHIRODARIA AND CHARALAMBOS ANTONIADES*

Department of Cardiovascular Medicine, University of Oxford, Oxford, UK
*Email: antoniad@well.ox.ac.uk

3.1 Introduction

Over the past two decades attention has focused on the potential use of B vitamins in cardiovascular disease (CVD) prevention. Although classic antioxidant vitamins have failed to prevent vascular events in clinical trials, interest in B vitamins arose due to their homocysteine (Hcy) lowering properties. Strong epidemiological and mechanistic evidence suggests that Hcy is an independent cardiovascular risk factor (Antoniades *et al.* 2009a). In addition, this hypothesis confirms the original observation that patients with homocystinuria develop premature atherosclerosis and thromboembolic events early in life.

Nevertheless, despite the decline of cardiovascular risk in North America after the introduction of folate food fortification (Yang *et al.* 2006), large clinical trials in patients with coronary heart disease (CHD) failed to demonstrate any benefit from B vitamins administration (Clarke *et al.* 2010). Despite the rather disappointing results of Hcy lowering clinical trials, B vitamins may still have an Hcy-independent role in CVD prevention.

In this chapter we discuss the existing basic and clinical data on B vitamins and CVD prevention, focusing on the Hcy-lowering B vitamins (folic acid, B_6 and B_{12}). First we review the mechanisms of Hcy-mediated CVD and highlight the significance of B vitamins as Hcy lowering agents. Furthermore we discuss the beneficial cardiovascular effects of B vitamins *per se* at a molecular level as well as the lessons learned from randomized clinical trials (RCTs). The possible risks and adverse effects by administration of B vitamins at pharmacological doses and the issue of folate fortification are also discussed.

3.2 Homocysteinaemia as a Risk Factor for Atherosclerosis

3.2.1 Mechanisms of Homocysteine-mediated Vascular Disease

Animal and cell studies have confirmed the deleterious effects of Hcy on the vascular wall. Hcy contains a free sulfydryl group which, when oxidized to a disulfide, produces superoxide anions and hydrogen peroxide. Furthermore, homocysteinaemia upregulates the expression of the $p22^{phox}$ subunit of NADPH oxidase in rats' aorta leading to increased superoxide generation, while transfection of vascular wall cells with siRNA for $p22^{phox}$ or Rac1 effectively abolishes Hcy-mediated ROS generation (Antoniades *et al.* 2009a). In addition, Hcy induces NADPH oxidase activity in monocytes by upregulating $p47^{phox}$ and $p67^{phox}$ subunits.

Homocysteinaemia also induces endothelial nitric oxidase synthase (eNOS) uncoupling. First, in acute homocysteinaemia, post-methionine loading asymmetric dimethylarginine (ADMA) levels are increased, while folate administration for Hcy lowering also augments ADMA synthesis by enhancing protein methyltransferase mediated conversion of L-arginine to ADMA. Increased circulating ADMA levels lead to eNOS uncoupling and diminish nitric oxide (NO) bioavailability in human vessels (Antoniades *et al.* 2009b). Secondly, the increased peroxynitrite formation leads to oxidative degradation of the critical eNOS cofactor tetrahydrobiopterin (BH_4) (Antoniades *et al.* 2006). Thirdly, studies on endothelial cells have also demonstrated that Hcy affects eNOS phosphorylation status and its translocation to endothelial caveolae. In contrast, Hcy in high concentrations upregulates inducible nitric oxidase synthase (iNOS) expression in both vascular smooth muscle cells and macrophages, which further aggravates vascular nitrosative stress in homocysteinaemia.

The overall effects of Hcy on vascular reactive oxygen species (ROS) generation adversely modulate vascular redox state and activate critical redox-sensitive transcriptional factors such as nuclear-factor kappa B or activator protein-1, which leads to a vicious cycle of inflammation – oxidative stress – endothelial dysfunction favouring vascular disease development.

Other mechanisms of Hcy-associated vascular disease include a prothrombotic and dysfibrinolytic state. Although the exact biochemical background is not fully explored, the increased risk of vascular thrombosis may be derived from the vascular oxidative injury and the modification of physiological antithrombotic mechanisms (Antoniades *et al.* 2009a).

It remains though unclear whether variations of plasma Hcy levels in the general population are sufficient to directly cause endothelial dysfunction; experimental models have used millimolar concentrations of Hcy, which are 100-fold higher than those observed in subjects with moderate homocysteinaemia (15–30 μmol/L). In intermediate (30–100 μmol/L) or severe (>100 μmol/L) homocysteinaemia, endothelial dysfunction has been well documented. Nevertheless, evidence is conflicting in moderate homocysteinaemia, and not all observational studies have reported a significant correlation between plasma Hcy and flow mediated dilation (FMD) of the brachial artery.

3.2.2 Epidemiological Evidence on Homocysteine as a Cardiovascular Risk Factor

Severe homocysteinaemia accompanying the genetic defects of homocystinuria is closely linked with recurrent vascular thrombosis. Half of the patients with homocystinuria will suffer a vascular event before the age of 30. This increased vascular risk involves mainly the veins and the increased risk of thromboembolic events in homocysteinaemia has been confirmed in observational studies.

In prospective clinical studies, plasma Hcy is an independent predictor of overall mortality in CHD patients after adjustment for confounding factors. A meta-analysis of prospective and observational studies suggested that a 25% reduction in Hcy levels (about 3 μmol/L) would confer an 11% CHD risk and 19% stroke risk reduction (Anonymous 2002). Another meta-analysis of genetic studies of methyl-tetrahydrofolate reductase (MTHFR) and prospective studies similarly suggested that lowering Hcy levels by 3 μmol/L in the general population would reduce the risk of CHD by 16% (11–20%), deep vein thrombosis by 25% (8–38%), and stroke by 24% (15–33%) (Wald *et al.* 2002). Observational data have also consolidated Hcy as an independent risk factor for stroke (Bostom *et al.* 1999; Kelly *et al.* 2002). These strong epidemiological data provided firm evidence in support of the potentially causal role of Hcy in atherosclerosis. This led to the conduction of large Hcy lowering clinical trials and to the issue of folate fortification.

3.3 Cardiovascular Effects of Folic Acid, B_6 and B_{12}: Any Need for Folate Fortification?

3.3.1 B Vitamins as Homocysteine Lowering Agents

Folic acid and vitamins B_6 and B_{12} are essential cofactors in Hcy metabolism (Figure 3.1). Folate administration consistently reduces plasma Hcy levels even in healthy individuals without homocysteinaemia. It is estimated that oral

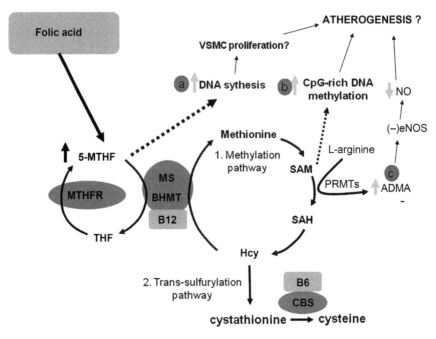

Figure 3.1 Homocysteine lowering with B vitamins administration and possible proatherogenic effects. Hcy production is dependable on S-adenosyl-methionine (SAM), which is responsible for multiple intracellular methylation reactions. The by-product of these reactions is S-adenosyl-homocysteine (SAH), which is hydrolysed to Hcy. Metabolization of Hcy can be achieved *via* two distinct pathways. Under conditions of methionine deficiency, Hcy is metabolized *via* the methylation pathway (1). Hcy is remethylated in the liver *via* betaine-homocysteine-methyltransferase (BHMT); however, in most tissues Hcy is remethylated into methionine by methionine synthase (MS), which uses vitamin B12 as cofactor and 5-methyltetrahydrofolate (5-MTHF) as substrate. 5-MTHF synthesis is catalysed by methyltetrahydrofolate-reductase (MTHFR), which uses tetrahydrofolate (THF) as substrate. On the contrary, when there is an excess of methionine, Hcy is metabolized *via* the pathway of trans-sulfurylation (2), producing cystathionine and cysteine in turn. The responsible enzyme for the transformation of Hcy to cystathionine is cystathionine beta synthase (CBS) which requires vitamin B_6 as an essential cofactor. By administrating folate the methylation pathway is favoured with possible adverse effects on atherogenesis: (a) DNA synthesis is increased as folic acid is used as substrate for thymidine synthesis; this may have possible adverse effects on vascular smooth muscle cells' (VSMC) proliferation. (b) methylation of Gp-rich DNA domains in the promoter region of proatherogenic genes may upregulate their expression. (c) SAM conversion to SAH is dependent on the action of protein methyl-transferases (PRMTs) which increase asymmetric dimethyl-arginine (ADMA) levels with adverse effects on eNOS coupling and NO bio-availability (unpublished).

administration of folic acid (0.4–5.0 mg/day) reduces fasting Hcy levels by 25–30%, while supplementation with vitamin B_{12} (0.02–1 mg/d) yields an additional 7% reduction in Hcy levels. Vitamin B_6 has no effect on fasting Hcy levels but significantly lowers post-methionine loading homocysteinaemia. Treatment modalities according to homocysteinaemia severity and cause are reviewed elsewhere (Antoniades et al. 2009a). Addition of 150 μg of folic acid per 100 g of flour in the USA resulted in an average 7.5% decrease in Hcy levels of the general population during the first three months of the mandatory fortification program.

3.3.2 Effects on Proinflammatory Mechanisms

Both acute and chronic, experimentally-induced homocysteinaemia are associated with a proinflammatory, pro-oxidant state. In animal models, treatment with B vitamins suppresses vascular wall inflammation and reverses Hcy-mediated expression of adhesion molecules and matrix metalloproteinases in the vasculature (Hofmann et al. 2001). Clinical studies, however, have yielded conflicting results. Folic acid administration has been shown to reduce the release of chemokines from mononuclear cells in subjects with homocysteinaemia, but other studies have failed to document any effect of Hcy lowering treatment on circulating levels of proinflammatory cytokines.

3.3.3 Effects on Endothelial Function

Folic acid administration in patients with CHD improves brachial artery FMD, a well-established surrogate marker of endothelial function. Folic acid administration (5 mg/day) for three or four months (Thambyrajah et al. 2001; Title et al. 2000) consistently improves brachial artery endothelium dependent dilatation. Higher doses of folic acid (10 mg/day) for six weeks increase FMD in patients post acute myocardial infarction (Moens et al. 2007). However, it is debatable whether this improvement of endothelial function is mediated *via* changes in Hcy metabolism, since it takes place too early after folic acid intake before any changes occur in plasma Hcy levels (Doshi et al. 2003). Observational studies have indicated that 5-methyl-tetrahydrofolate (5-MTHF) levels are a stronger predictor of FMD than plasma Hcy. Intravenous administration of 5-MTHF improves brachial artery FMD, an effect that is almost completely blocked by monomethyl L-arginine (LNMMA), a non-selective inhibitor of nitric oxide synthase (NOS) (Doshi et al. 2003). A meta-analysis of conducted RCTs suggested that folic acid dose-dependently increases FMD by 1.08% [95% confidence interval (CI): 0.57–1.59) (de Bree et al. 2007). A *post hoc* analysis of the VITATOPS trial suggested that B vitamins significantly increased FMD (Potter et al. 2008); however, in the meta-analysis of randomized studies, the effect of B vitamins on FMD (1.4%, 95% CI 0.7–2.1%) was significant only in those studies with a shorter intervention period. Longer-term administration of B vitamins in studies with an extended follow-up period had no effect on endothelial function (Potter *et al.* 2008). Similarly three years' supplementation with folic acid in middle-aged individuals with Hcy $\geq 13\,\mu\text{mol/L}$ did not slow arterial stiffening despite Hcy lowering (Durga *et al.* 2011).

3.3.4 Further Mechanistic Insights in Human Vessels

Evidence suggests an Hcy-independent effect of B vitamins on endothelium. Our studies indicate that folate administration in patients undergoing a coronary artery by-pass grafting operation leads to a global improvement of the function of arterial and vein grafts which is independent of any changes in vascular Hcy levels. These beneficial vascular effects are mediated by an increase in plasma and vascular 5-MTHF levels. 5-MTHF, which has structural similarities to BH_4, scavenges peroxynitrite radicals and protects BH_4 from oxidative degradation. The anti-oxidant effects of 5-MTHF increase BH_4 bioavailability, improve eNOS dimerization, activity and coupling, and lead to a significant improvement of endothelial function in patients with advanced atherosclerosis (Antoniades *et al.* 2006). An overview of folic acid's beneficial effects is presented in Figure 3.2.

Importantly, our studies indicate that these vascular effects of folic acid (mediated *via* vascular 5-MTHF) take place even when folic acid is administered at low doses, equal to the Recommended Dietary Allowance (RDA). In a randomized clinical trial (Shirodaria *et al.* 2007), we compared the vascular effects of high (5 mg/day) *vs.* low (0.4 mg/day) oral folic acid dosage for seven weeks in coronary patients before a coronary artery by-pass grafting operation. Both dosages were associated with improved vasomotor responses of the grafts, reduced vascular superoxide production and improved eNOS coupling. Along with these effects on endothelium, folic acid administration improved arterial distensibility in human aorta and carotid artery suggesting a strong effect on global vascular function.

However, the maximum beneficial vascular effect of folic acid was observed even with 0.4 mg/day, while the higher dose failed to induce any further benefit. Despite the fact that plasma 5-MTHF increases in accordance with the dose of folic acid, vascular intracellular 5-MTHF levels do not exhibit any further increase with high doses compared with low doses of folic acid (Shirodaria *et al.* 2007). This is possibly explained by the saturation of human vascular wall with 5-MTHF after low-dose treatment (equivalent to the amount received by folate fortification of grain in North America) and the inability of any further increase of circulating folate levels to result into respective elevation of vascular 5-MTHF.

3.3.5 Effects on Atheroma Progression

Existing data on the effects of B vitamins on atheroma progression are conflicting. Folic acid, B_6 and B_{12} administration in middle-aged individuals

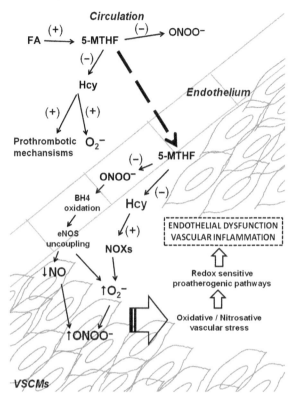

Figure 3.2 Beneficial effects of folic acid on vascular wall. Folic acid circulates in human body as 5-methyltetrahydrofolate (5-MTHF). 5-MTHF lowers circulating homocysteine (Hcy) levels, thus reducing systemic oxidative stress and Hcy-induced activation of prothrombotic mechanisms. In addition, vascular 5-MTHF has a favourable effect on intracellular Hcy metabolism, attenuating Hcy-induced activation of NADPH oxidase isoforms (NOXs) in the vascular wall. Furthermore vascular 5-MTHF scavenges *per se* peroxynitrite ($ONOO^-$) radicals in the vascular wall preventing the oxidation of vascular tetrahydrobiopterin (BH_4) associated with endothelial nitric oxide synthase (eNOS) uncoupling and diminished vascular nitric oxide (NO) bioavailability. In total through these effects 5-MTHF lowers vascular oxidative and nitrosative stress. Thus by modulating vascular redox, 5-MTHF inhibits activation of proinflammatory pathways which orchestrate vascular wall inflammation and perpetuate endothelial dysfunction and atherogenesis development (unpublished).

without CVD decreased carotid artery intima-media thickness (cIMT) (Hodis *et al.* 2009). Moreover, the combination of B vitamins for six months in patients with coronary atherosclerosis post angioplasty significantly reduced plasma Hcy and was associated with clinical benefits (Schnyder *et al.* 2001). Patients under Hcy lowering treatment had reduced need of re-vascularization of the target lesion (10.8% *vs.* 22.3%, $p=0.047$), and reduced rate of restenosis (19.6% *vs.* 37.6%, $p=0.01$). Notably, this study recruited patients that were treated with both stenting and balloon angioplasty, and there was a trend towards increased stents' use in the control group which could have affected the trial's findings (Schnyder *et al.* 2001). The mechanism of restenosis differs between stenting and balloon angioplasty. The main mechanism leading to in-stent restenosis is considered to be the proliferation of smooth muscle cells and matrix formation; restenosis after balloon angioplasty is thought to be due to thrombus formation and vascular remodelling, which are more susceptible to the effects of Hcy-lowering (Lange *et al.* 2004). A later study supported the finding that the administration of B vitamins had adverse effects on the risk of restenosis post coronary stenting in all subgroups except for women, patients with diabetes, and patients with markedly elevated Hcy levels at baseline ($>15\,\mu mol/L$) (Lange *et al.* 2004). Unfortunately, both these studies did not use intravascular ultrasound in order to assess the pathophysiological mechanism of B vitamins effects, *i.e.* possible effects on intimal hyperplasia or vascular remodelling (Schnyder *et al.* 2001; Lange *et al.* 2004). A more recent RCT suggested that folic acid (1 mg/day) supplementation post coronary stenting did not decrease the rate of in-stent restenosis within six months (Namazi *et al.* 2006). Therefore, despite the initial enthusiasm about the possible anti-atherogenic effect of homocysteine lowering treatment with folate and B vitamins, more recent data appear rather disappointing.

3.4 Folate and B Vitamins in Cardiovascular Disease: Insights from Clinical Trials and Folate Fortification Programme

3.4.1 Homocysteine Lowering and Cardiovascular Disease Prevention

In 1998, the USA and Canada mandated a large-scale flour fortification programme with folate (150 µg folate per 100 g flour). Folate fortification increased mean folate concentrations and reduced mean plasma Hcy in the general population and this was associated with an accelerated reduction in stroke mortality in the USA and Canada (Yang *et al.* 2006). No such accelerated reduction in stroke mortality was observed in England and Wales which did not introduce fortification of flour with folate during the same time period. These epidemiological observations strongly supported the hypothesis that folate fortification contributes to the reduction of stroke mortality, at least at the level of primary prevention (Yang *et al.* 2006).

However large RCTs, such as VISP (Toole *et al.* 2004) or VITATOPS (Anonymous 2010) have failed to document any significant effects of B vitamins on stroke prevention. *Post hoc* analyses, such as that of HOPE-2 (Saposnik *et al.* 2009) have suggested that folate, B_6 and B_{12} could reduce stroke risk in patients with known CVD, mainly if patients were younger than 70 years, recruited from regions without folic acid food fortification, with higher baseline cholesterol and Hcy levels, and not under antiplatelet or lipid-lowering treatment. A recent meta-analysis suggested that folic acid supplementation did not have any significant effect on stroke prevention, although combined administration of folate with B vitamins could confer potential mild benefits in the primary prevention of stroke in male patients (Lee *et al.* 2010). However, these findings derived from *post hoc* analyses cannot justify the use of B vitamins for stroke prevention.

Besides, landmark RCTs such as NORVIT (Bonaa *et al.* 2006) and SEARCH (Armitage *et al.* 2010) trials in myocardial infarction (MI) survivors or the HOPE-2 trial in patients with known vascular disease or diabetes (Lonn *et al.* 2006) have also failed to document any benefits with folic acid and B vitamins administration on CHD prevention. Similarly a recent large meta-analysis (Clarke *et al.* 2010) also failed to document any effect of B vitamins on overall CVD risk reduction. Surprisingly though, some studies have indeed reported a small but significant increase in CVD risk with B vitamins; for example, in HOPE-2, B vitamins were associated with increased risk for unstable angina [relative risk (RR($=1.24$; 95% CI: 1.04–1.49] (Lonn *et al.* 2006) and in NORVIT with increased overall CVD risk (RR $= 1.22$; 95% CI: 1.00–1.50)) (Bonaa *et al.* 2006). These findings warrant further investigation.

Homocysteine lowering treatment in the high risk patients enrolled in HOPE-2 (with known CVD or diabetes) had no effect on reducing the risk of venous thromboembolism (VTE). Nevertheless, data on VTE history at enrollment were not available (Ray *et al.* 2007). The VITRO trial is currently the largest trial examining the effects of Hcy lowering on thrombosis risk. VITRO enrolled patients with high baseline plasma Hcy and history of deep vein thrombosis, but demonstrated that high doses of B vitamins had no effects on secondary prevention of VTE (den Heijer *et al.* 2007).

Several recent RCTs have focused on the effects of Hcy lowering treatment in high risk populations, such as end-stage renal disease (ESRD) patients. However treatment of ESRD patients with high doses of B vitamins has failed to confer any benefits on cardiovascular prognosis (Heinz *et al.* 2010). Additionally in a large RCT in kidney transplant recipients, B vitamins did not reduce the incidence of cardiovascular events (Bostom *et al.* 2011). Interestingly, however, a recent trial concluded that pretreatment with B vitamins along with intravenous 5-MTHF administration significantly improved survival of hemodialysis patients (Cianciolo *et al.* 2008). A summary of all important RCTs on the effects of B vitamins on vascular outcomes is provided in Table 3.1.

3.4.2 Criticism of Randomized Clinical Trials

Various limitations have been highlighted in the design of existing RCTs that could account for their failure to report any cardiovascular benefit with B vitamins administration. The VISP trial has received criticism for not having included a placebo group. In the NORVIT study, the events were mainly during the first post-infarction year, a period for which even statins cannot provide adequate protection and the effects may be masked due to other drugs with strong pleiotropic effects (Antoniades *et al.* 2009a). The statistical power

Table 3.1 Folic acid, B_6, B_{12} and cardiovascular outcomes. This table summarizes all the important double-blind randomized clinical trials with the use of folic acid, B_6 and B_{12} vitamins for cardiovascular disease prevention. 5-MTHF: 5-methyl tetrahydrofolate; CKD: chronic kidney disease; CVD: cardiovascular disease; DVT: deep vein thrombosis; ESRD: end stage renal disease; FA: folic acid; f/u: follow-up; MI: myocardial infarction; RR: relative risk; UA: unstable angina.

Trial	Study population	N	Folate status	f/u (y)	Treatment (mg/day)	Results
VISP (Toole et al. 2004)	Stroke patients	3680	fortified + unfortified	2.0	2.5 mg FA + 25 mg B_6 + 0.4 mg B_{12}; low dose regiment	No effect on stroke, coronary events or death
VITATOPS (Anonymous 2010)	Stroke patients	8164	unfortified	3.4	2.0 mg FA + 25 mg B_6 + 0.5 mg B_{12}; placebo	RR for the composite end-point of MI, stroke or vascular death: 0.91 (0.82–1.00).
HOPE-2 (Lonn et al. 2006)	Vascular disease or diabetes	5522	fortified + unfortified	5.0	2.5 mg FA + 50 mg B_6 + 1 mg B_{12}; placebo	No effect on CVD mortality, MI. RR for stroke: 0.75 (0.54–0.97). RR for UA: 1.24 (1.04–1.49).
NORVIT (Bonaa et al. 2006)	MI survivors	3749	unfortified	3.3	0.8 mg FA + 40 mg B_6 + 0.4 mg B_{12}; placebo	No effect on CVD. Combined FA/B_6/B_{12} increased CVD RR: 1.22 (1.00–1.50)
WAFACS (Albert et al. 2008)	Women with CVD or high-risk women	5442	fortified	7.3	2.5 mg FA + 50 mg B_6 + 1 mg B_{12}; placebo	No effect on MI, stroke, coronary revascularization, or CVD mortality

Trial	Population	N	Fortification	Duration (y)	Intervention	Outcome
WENBIT (Ebbing et al. 2008)	Individuals undergoing coronary angiography	3096	unfortified	3.2	0.8 mg FA + 40 mg B_6 + 0.4 mg B_{12}; placebo	No effect on cardiovascular events
SEARCH (Armitage et al. 2010)	MI survivors	12 064	unfortified	6.7	Simvastatin 80 mg vs. 20 mg 2.0 mg FA + 1 mg B_{12}; placebo (2×2 factorial)	No effects of B vitamins on cardiovascular events
VITRO (den Heijer et al. 2007)	Patients with DVT history and increased Hcy levels	701	unfortified	2.5	5.0 mg FA + 50 mg B_6 + 0.4 mg B_{12}; placebo	No effect on venous thrombosis
FAVORIT (Bostom et al. 2011)	CKD patients	4110	fortified	4.0	5.0 mg FA + 50 mg B_6 + 1 mg B_{12}; low dose regiment	No effect on CVD, all-cause mortality, or dialysis-dependent kidney failure
(Heinz et al. 2010)	ESRD patients	650	unfortified	2.0	5.0 mg FA + 50 mg B_6 + 0.05 mg B_{12}; low dose regiment (three times a week)	No effect on mortality or vascular events
(Cianciolo et al. 2008)	ESRD patients	341	fortified	4.6	5.0 mg FA + 300 mg B_6 + 1 mg B_{12}; 50 mg 5-MTHF IV + 300 mg B_6 + 1 mg B_{12}	5-MTHF administration associated with improved survival
HOST (Jamison et al. 2007)	Advanced CKD or ESRD patients with increased Hcy	2056		3.2	40.0 mg FA + 100 mg B_6 + 2 mg B_{12}; placebo	No effects on survival or vascular events

of HOPE-2 has also been questioned, since it did not secure low baseline folate levels in the study population. As already discussed, the vascular wall is saturated with 5-MTHF even with low folic acid dose (400 µg/day). Therefore higher doses of folic acid may proportionally increase plasma 5-MTHF levels, but vascular 5-MTHF remains constant and no additional benefit in vascular function is conferred. This may explain the failure of HOPE-2 to document any effect of B vitamins on clinical outcome in populations already receiving the RDA of folic acid (Antoniades *et al.* 2009a). Other strategies to achieve optimal Hcy metabolism such as increasing vascular 5-MTHF levels, increasing Hcy renal clearance or reducing ADMA levels might be potentially useful.

3.4.3 Is Folate Fortification and B Vitamins Administration Safe?

Folic acid is a stable, inexpensive and safe molecule. Folic acid has no toxic effect even at extreme dosages and the main safety concern lies in the fact that it can mask the diagnosis of pernicious anemia. This is because high folic acid levels correct the anemia but allow the neuropathy to progress undiagnosed, leading eventually to an irreversible degeneration of the spinal cord. It is therefore important to measure B_{12} levels before the start of folate treatment.

Recently, however, some major concerns have been raised on the issue of folate fortification, after the publication of data implying that B vitamins administration at therapeutic doses may adversely affect cancer risk. These concerns originated from clinical studies, where subjects receiving high doses of B vitamins had a borderline increase in the incidence of cancers (Bayston *et al.* 2007). Folate is an essential cofactor for the *de novo* biosynthesis of purines and thymidines, and in this capacity, folate plays an important role in DNA synthesis and replication. Therefore, even though it is unlikely that B vitamins are causally related with cancer development, high doses of B vitamins may possibly accelerate tumor development and growth in cases of undiagnosed underlying cancer. However, a recent meta-analysis (Clarke *et al.* 2010) concluded that B vitamins administration for a median of five years had no effect on cancer incidence, cancer mortality and all-cause mortality. Although overall data suggest that folate fortification is safe (Bayston *et al.* 2007), the debate on whether B vitamins prevent or promote cancer seems to hold on and deserves further investigation.

Another important observation derived from RCTs was the unexpected association of B vitamins with increased cardiovascular risk. Evidence suggests an interaction between folic acid and baseline Hcy, with potential harmful effects in patients with plasma Hcy > 12 µmol/L. Although there is no clear explanation of these findings, it can be hypothesized that B vitamins may in some way induce proatherogenic pathways by Hcy-independent mechanisms (Figure 3.1). Despite these experimental data, it remains questionable whether B vitamins administration can adversely affect plaque progression since clinical studies have reported beneficial or at least neutral effects of B vitamins on carotoid intima-media thickness (cIMT). However, other data (Lange *et al.* 2004), supporting an adverse effect of B vitamins on in-stent restenosis after coronary stenting, suggest that B vitamins administration may be safe only in certain subgroups such as diabetics, women and patients with marked homocysteinaemia.

Overall, folate, B_6 and B_{12} should be considered as safe compounds, without toxic effects. However the recent epidemiological data linking B vitamins with increased cardiovascular and cancer risk question the rationale of using high doses of B vitamins in clinical trials. Folic acid and B vitamins enhance Hcy metabolism and their administration consistently lowers plasma Hcy levels, while adherence to a Mediterranean diet, which includes high B vitamins and folate consumption, is associated with lower plasma Hcy. It has been clearly demonstrated that folic acid intake at the RDA provides the maximum beneficial vascular effects (Shirodaria et al. 2007). Until more convincing evidence is available, folic acid intake (400 µg/day), either by folate fortification or dietary supplements, should be considered safe and unlikely to pose any threat (Bayston et al. 2007). Besides, folic acid supplementation prevents neural tube defects and may protect against adverse pregnancy outcomes (including placental abruption) and cognitive decline. Finally, even though evidence from large RCTs is missing (since mainly patients with mild Hcy elevations were recruited in clinical studies), folate $\pm B_{12}$ and B_6 vitamins supplementation should be considered in patients with severe and/or intermediate homocysteinaemia to reduce the well-described thomboembolic and atherosclerotic risk (Wilcken and Wilcken 1997; Yap et al. 2001). Key facts of B vitamins as therapeutic agents are summarized below.

3.5 Concluding Remarks

Despite the strong evidence for the role of plasma Hcy in cardiovascular risk prediction, clinical trials have consistently shown that Hcy lowering strategies in either primary of secondary prevention are ineffective. However, Hcy lowering treatment with folate and B vitamins is still considered to be an effective means to reduce vascular and thrombotic complications of homocystinuria, while folate fortification may be beneficial in selected groups of patients, such as those with reduced folate levels due to dietary or specific genetic characteristics.

Summary Points

- This chapter focuses on the role of B vitamins in preventing cardiovascular disease.
- Folic acid with or without B_{12} and B_6 have been used as a 'homocysteine lowering treatment'.
- Experimental data suggests that B vitamins exert beneficial vascular effects.
- Folic acid/B vitamins improve clinical outcome in severe homocysteinaemia.

- Epidemiological data indicate a benefit with folate fortification on stroke prevention.
- Clinical trials have failed to document any significant clinical benefit on cardiovascular risk with B vitamins administration.
- Concerns have been raised on possible risks that food fortification with folate carries.
- Further studies are needed to fully assess the risks/benefits with B vitamins treatment in various groups of patients with cardiovascular disease.

Key Facts about B Vitamins as Therapeutic Agents in Cardiovascular Disease

- Homocysteine lowering with B vitamins has been investigated as a cheap and safe way to prevent atherosclerosis.
- Administration of folic acid (usually 0.4–5.0 mg) alone or in combination with B_{12} (usually 0.05–1.00 mg) $\pm$ B_6 (usually 25–100 mg) efficiently lowers plasma homocysteine levels.
- Folic acid and B_{12} enhance the conversion of homocysteine to methionine and thus lower fasting plasma homocysteine, while B_6 enhances homocysteine lowering post-methionine loading.
- Numerous randomized clinical trials have been already conducted with B vitamins for primary or secondary prevention of cardiovascular events in fortified or unfortified population (*i.e.* populations already receiving the RDA of folic acid).
- Only small benefits in vascular outcomes have been documented by the use of B vitamins, despite efficient lowering of plasma homocysteine, which has led to questions about the homocysteine hypothesis of atherosclerosis.
- Concerns have been raised regarding possible adverse effects of B vitamins on cardiovascular or cancer risk in certain patients.

Definitions of Words and Terms

C677T mutation of MTHFR gene. A common polymorphism that has been used in many clinical studies as a genetic model of homocysteinaemia. The 677TT genotype is associated with a thermolabile variant of the enzyme with an almost 50% reduced enzymatic activity.

Endothelial nitric oxide synthase (eNOS). The enzyme responsible for nitric oxide production in human endothelium. Nitric oxide (NO) is a gas that relaxes vascular smooth muscle cells, induces vasodilatation and is critical for global vascular homeostasis. eNOS is a homodimer enzyme that 'couples' oxygen consumption to L-arginine use to produce NO. In its uncoupled form, eNOS is rendered into a source of superoxides instead of NO.

Homocysteinaemia. Increased circulating plasma homocysteine levels (>15 μmol/L). According to fasting levels, homocysteinaemia is classified as moderate (15–30 μmol/L), intermediate (30–100 μmol/L) and severe (>100 μmol/L).

Homocystinuria. An inherited autosomal recessive disorder of homocysteine metabolism (1/200 000 births) caused by mutations in the cystathione beta synthase (CBS) gene that disrupts the normal activity of CBS. Homocystinuria patients have highly elevated plasma homocysteine levels and are at increased early risk of thrombotic/vascular events. Homocystinuria can be more rarely caused by mutations in other genes involved in homocysteine metabolic pathways.

5-Methyltetrahydrofolate (5-MTHF). The circulating form of folic acid in humans. 5-MTHF is produced by 5,10-methylenetetrahydrofolate *via* the action of MTFH reductase (MTHFR). 5-MTHF scavenges peroxynitrites, the main BH_4 oxidant, and helps to BH_4 regeneration inside the human vascular wall. It is considered the key mediator of folic acid's vascular effects (as in the presence of the C677T mutation in MTHFR gene that reduces enzyme's activity almost by half).

Recommended dietary allowance. The daily intake of an essential nutrient considered to be adequate to meet the known needs of practically all healthy people.

Redox. A term used to describe all reactions involving electrons transfer (reduction–oxidation reactions).

Redox sensitive transcriptional factors. Important intracellular pathways that control the expression of multiple genes. Their activation status is modulated by changes in intracellular redox state. Such transcriptional factors are the nuclear factor kappa-b or activating protein-1 that, at the level of the vascular wall, regulate the expression of proinflammatory genes and orchestrate vascular inflammation.

Tetrahyrobiopterin (BH_4). A critical cofactor required for eNOS 'coupling'. BH_4 is prone to oxidation, mainly by peroxynitrite. Advanced atherosclerosis is characterized by low vascular BH_4 bioavailability that results in eNOS uncoupling and endothelial dysfunction.

Vascular redox. Is determined by the activity of oxidative sources (NADPH oxidase, uncoupled NOS, xanthine oxidase, *etc*.) and antioxidant enzymes (superoxide dismutase, glutathione peroxidase) at the vascular wall level.

List of Abbreviations

ADMA	asymmetrical dimethylarginine
BH_4	tetrahydrobiopterin
BHMT	betaine-homocysteine-methyltransferase
CBS	cystathione beta synthase
CHD	coronary heart disease
CI	confidence interval
cIMT	carotid intima-media thickness
CVD	cardiovascular disease
eNOS	endothelial nitric oxidase synthase
ESRD	end-stage renal disease
FMD	flow mediated dilatation

Hcy	homocysteine
iNOS	inducible nitric oxidase synthase
LNMMA	monomethyl L-arginine
MI	myocardial infarction
5-MTHF	5-methyl-tetrahydrofolate
MTHFR	methyl-tetrahydrofolate reductase
NOS	nitric oxide synthase
NOX	NADPH oxidase isoform
ONOO$^-$	peroxynitrite
PRMT	protein methyl-transferase
RCT	randomized clinical trials
RDA	Recommended Dietary Allowance
RR	relative risk
ROS	reactive oxygen species
SAH	S-adenosyl-homocysteine
SAM	S-adenosyl-methionine
THF	tetrahydrofolate
VSMC	vascular smooth muscle cells
VTE	venous thromboembolism

References

Albert, C.M., Cook, N.R., Gaziano, J.M., Zaharris, E., MacFadyen, J., Danielson, E., Buring, J.E., and Manson, J.E., 2008. Effect of folic acid and B vitamins on risk of cardiovascular events and total mortality among women at high risk for cardiovascular disease: a randomized trial. *JAMA, the Journal of the American Medical Association.* 299: 2027–2036.

Anonymous, 2002. Homocysteine and risk of ischemic heart disease and stroke: a meta-analysis. *JAMA, the Journal of the American Medical Association.* 288: 2015–2022.

Anonymous, 2010. B vitamins in patients with recent transient ischaemic attack or stroke in the VITAmins TO Prevent Stroke (VITATOPS) trial: a randomised, double-blind, parallel, placebo-controlled trial. *Lancet Neurology.* 9: 855–865.

Antoniades, C., Shirodaria, C., Warrick, N., Cai, S., de Bono, J., Lee, J., Leeson, P., Neubauer, S., Ratnatunga, C., Pillai, R., Refsum, H., and Channon, K.M., 2006. 5-Methyltetrahydrofolate rapidly improves endothelial function and decreases superoxide production in human vessels: effects on vascular tetrahydrobiopterin availability and endothelial nitric oxide synthase coupling. *Circulation.* 114: 1193–1201.

Antoniades, C., Antonopoulos, A.S., Tousoulis, D., Marinou, K., and Stefanadis, C., 2009a. Homocysteine and coronary atherosclerosis: from folate fortification to the recent clinical trials. *European Heart Journal.* 30: 6–15.

Antoniades, C., Shirodaria, C., Leeson, P., Antonopoulos, A., Warrick, N., Van-Assche, T., Cunnington, C., Tousoulis, D., Pillai, R., Ratnatunga, C.,

Stefanadis, C., and Channon, K.M., 2009b. Association of plasma asymmetrical dimethylarginine (ADMA) with elevated vascular superoxide production and endothelial nitric oxide synthase uncoupling: implications for endothelial function in human atherosclerosis. *European Heart Journal*. 30: 1142–1150.

Armitage, J.M., Bowman, L., Clarke, R.J., Wallendszus, K., Bulbulia, R., Rahimi, K., Haynes, R., Parish, S., Sleight, P., Peto, R., and Collins, R., 2010. Effects of homocysteine-lowering with folic acid plus vitamin B_{12} vs placebo on mortality and major morbidity in myocardial infarction survivors: a randomized trial. *JAMA, the Journal of the American Medical Association*. 303: 2486–2494.

Bayston, R., Russell, A., Wald, N.J., and Hoffbrand, A.V., 2007. Folic acid fortification and cancer risk. *Lancet*. 370: 2004.

Bonaa, K.H., Njolstad, I., Ueland, P.M., Schirmer, H., Tverdal, A., Steigen, T., Wang, H., Nordrehaug, J.E., Arnesen, E., and Rasmussen, K., 2006. Homocysteine lowering and cardiovascular events after acute myocardial infarction. *New England Journal of Medicine*. 354: 1578–1588.

Bostom, A.G., Rosenberg, I.H., Silbershatz, H., Jacques, P.F., Selhub, J., D'Agostino, R.B., Wilson, P.W., and Wolf, P.A., 1999. Nonfasting plasma total homocysteine levels and stroke incidence in elderly persons: the Framingham Study. *Annals of Internal Medicine*. 131: 352–355.

Bostom, A.G., Carpenter, M.A., Kusek, J.W., Levey, A.S., Hunsicker, L., Pfeffer, M.A., Selhub, J., Jacques, P.F., Cole, E., Gravens-Mueller, L., House, A.A., Kew, C., McKenney, J.L., Pacheco-Silva, A., Pesavento, T., Pirsch, J., Smith, S., Solomon, S., and Weir, M., 2011. Homocysteine-lowering and cardiovascular disease outcomes in kidney transplant recipients: primary results from the folic acid for vascular outcome reduction in transplantation trial. *Circulation*. 123: 1763–1770.

Cianciolo, G., La Manna, G., Coli, L., Donati, G., D'Addio, F., Persici, E., Comai, G., Wratten, M., Dormi, A., Mantovani, V., Grossi, G., and Stefoni, S., 2008. 5-Methyltetrahydrofolate administration is associated with prolonged survival and reduced inflammation in ESRD patients. *American Journal of Nephrology*. 28: 941–948.

Clarke, R., Halsey, J., Lewington, S., Lonn, E., Armitage, J., Manson, J.E., Bonaa, K.H., Spence, J.D., Nygard, O., Jamison, R., Gaziano, J.M., Guarino, P., Bennett, D., Mir, F., Peto, R., and Collins, R., 2010. Effects of lowering homocysteine levels with B vitamins on cardiovascular disease, cancer, and cause-specific mortality: meta-analysis of 8 randomized trials involving 37 485 individuals. *Archives of Internal Medicine*. 170: 1622–1631.

de Bree, A., van Mierlo, L.A., and Draijer, R., 2007. Folic acid improves vascular reactivity in humans: a meta-analysis of randomized controlled trials. *American Journal of Clinical Nutrition*. 86: 610–617.

den Heijer, M., Willems, H.P., Blom, H.J., Gerrits, W.B., Cattaneo, M., Eichinger, S., Rosendaal, F.R., and Bos, G.M., 2007. Homocysteine lowering by B vitamins and the secondary prevention of deep vein thrombosis and pulmonary embolism: a randomized, placebo-controlled, double-blind trial. *Blood.* 109: 139–144.

Doshi, S., McDowell, I., Moat, S., Lewis, M., and Goodfellow, J., 2003. Folate improves endothelial function in patients with coronary heart disease. *Clinical Chemistry and Laboratory Medicine.* 41: 1505–1512.

Durga, J., Bots, M.L., Schouten, E.G., Grobbee, D.E., Kok, F.J., and Verhoef, P., 2011. 'Effect of 3 y of folic acid supplementation on the progression of carotid intima-media thickness and carotid arterial stiffness in older adults1,3. *American Journal of Clinical Nutrition.* 93: 941–949.

Ebbing, M., Bleie, O., Ueland, P.M., Nordrehaug, J.E., Nilsen, D.W., Vollset, S.E., Refsum, H., Pedersen, E.K., and Nygard, O., 2008. Mortality and cardiovascular events in patients treated with homocysteine-lowering B vitamins after coronary angiography: a randomized controlled trial. *JAMA, the Journal of the American Medical Association.* 300: 795–804.

Heinz, J., Kropf, S., Domrose, U., Westphal, S., Borucki, K., Luley, C., Neumann, K.H., and Dierkes, J., 2010. B vitamins and the risk of total mortality and cardiovascular disease in end-stage renal disease: results of a randomized controlled trial. *Circulation.* 121: 1432–1438.

Hodis, H.N., Mack, W.J., Dustin, L., Mahrer, P.R., Azen, S.P., Detrano, R., Selhub, J., Alaupovic, P., Liu, C.R., Liu, C.H., Hwang, J., Wilcox, A.G., and Selzer, R.H., 2009. High-dose B vitamin supplementation and progression of subclinical atherosclerosis: a randomized controlled trial. *Stroke.* 40: 730–736.

Hofmann, M.A., Lalla, E., Lu, Y., Gleason, M.R., Wolf, B.M., Tanji, N., Ferran, L.J. Jr., Kohl, B., Rao, V., Kisiel, W., Stern, D.M., and Schmidt, A.M., 2001. Hyperhomocysteinemia enhances vascular inflammation and accelerates atherosclerosis in a murine model. *The Journal of Clinical Investigation.* 107: 675–683.

Jamison, R.L., Hartigan, P., Kaufman, J.S., Goldfarb, D.S., Warren, S.R., Guarino, P.D., and Gaziano, J.M., 2007. Effect of homocysteine lowering on mortality and vascular disease in advanced chronic kidney disease and end-stage renal disease: a randomized controlled trial. *JAMA, the Journal of the American Medical Association.* 298: 1163–1170.

Kelly, P.J., and Furie, K.L., 2002. Management and prevention of stroke associated with elevated homocysteine. *Current Treatment Options in Cardiovascular Medicine.* 4: 363–371.

Lange, H., Suryapranata, H., De Luca, G., Borner, C., Dille, J., Kallmayer, K., Pasalary, M.N., Scherer, E., and Dambrink, J.H., 2004. Folate therapy and in-stent restenosis after coronary stenting. *New England Journal of Medicine.* 350: 2673–2681.

Lee, M., Hong, K.S., Chang, S.C., and Saver, J.L., 2010. Efficacy of homocysteine-lowering therapy with folic Acid in stroke prevention: a meta-analysis. *Stroke.* 41: 1205–1212.

Lonn, E., Yusuf, S., Arnold, M.J., Sheridan, P., Pogue, J., Micks, M., McQueen, M.J., Probstfield, J., Fodor, G., Held, C., and Genest, J. Jr., 2006.

Homocysteine lowering with folic acid and B vitamins in vascular disease. *New England Journal of Medicine.* 354: 1567–1577.

Moens, A.L., Claeys, M.J., Wuyts, F.L., Goovaerts, I., Van Hertbruggen, E., Wendelen, L.C., Van Hoof, V.O., and Vrints, C.J., 2007. Effect of folic acid on endothelial function following acute myocardial infarction. *American Journal of Cardiology.* 99: 476–481.

Namazi, M.H., Motamedi, M.R., Safi, M., Vakili, H., Saadat, H., and Nazari, N., 2006. Efficacy of folic acid therapy for prevention of in-stent restenosis: a randomized clinical trial. *Archive of Iranian Medicine.* 9: 108–110.

Potter, K., Hankey, G.J., Green, D.J., Eikelboom, J., Jamrozik, K., and Arnolda, L.F., 2008. The effect of long-term homocysteine-lowering on carotid intima-media thickness and flow-mediated vasodilation in stroke patients: a randomized controlled trial and meta-analysis. *BMC Cardiovascular Disorders.* 8: 24.

Ray, J.G., Kearon, C., Yi, Q., Sheridan, P., and Lonn, E., 2007. Homocysteine-lowering therapy and risk for venous thromboembolism: a randomized trial. *Annals of Internal Medicine.* 146: 761–767.

Saposnik, G., Ray, J.G., Sheridan, P., McQueen, M., and Lonn, E., 2009. Homocysteine-lowering therapy and stroke risk, severity, and disability: additional findings from the HOPE 2 trial. *Stroke.* 40: 1365–1372.

Schnyder, G., Roffi, M., Pin, R., Flammer, Y., Lange, H., Eberli, F.R., Meier, B., Turi, Z.G., and Hess, O.M., 2001. Decreased rate of coronary restenosis after lowering of plasma homocysteine levels. *New England Journal of Medicine.* 345: 1593–1600.

Shirodaria, C., Antoniades, C., Lee, J., Jackson, C.E., Robson, M.D., Francis, J.M., Moat, S.J., Ratnatunga, C., Pillai, R., Refsum, H., Neubauer, S., and Channon, K.M., 2007. Global improvement of vascular function and redox state with low-dose folic acid: implications for folate therapy in patients with coronary artery disease. *Circulation.* 115: 2262–2270.

Thambyrajah, J., Landray, M.J., Jones, H.J., McGlynn, F.J., Wheeler, D.C., and Townend, J.N., 2001. A randomized double-blind placebo-controlled trial of the effect of homocysteine-lowering therapy with folic acid on endothelial function in patients with coronary artery disease. *Journal of the American College of Cardiology.* 37: 1858–1863.

Title, L.M., Cummings, P.M., Giddens, K., Genest, J.J. Jr., and Nassar, B.A., 2000. Effect of folic acid and antioxidant vitamins on endothelial dysfunction in patients with coronary artery disease. *Journal of the American College of Cardiology.* 36: 758–765.

Toole, J.F., Malinow, M.R., Chambless, L.E., Spence, J.D., Pettigrew, L.C., Howard, V.J., Sides, E.G., Wang, C.H., and Stampfer, M., 2004. Lowering homocysteine in patients with ischemic stroke to prevent recurrent stroke, myocardial infarction, and death: the Vitamin Intervention for Stroke Prevention (VISP) randomized controlled trial.' *JAMA, the Journal of the American Medical Association.* 291: 565–575.

Wald, D.S., Law, M., and Morris, J.K., 2002. Homocysteine and cardiovascular disease: evidence on causality from a meta-analysis. *British Medical Journal*. 325: 1202.

Wilcken, D.E., and Wilcken, B., 1997. The natural history of vascular disease in homocystinuria and the effects of treatment. *Journal of Inherited Metabolic Disease*. 20: 295–300.

Yang, Q., Botto, L.D., Erickson, J.D., Berry, R.J., Sambell, C., Johansen, H., and Friedman, J.M., 2006. Improvement in stroke mortality in Canada and the United States, 1990 to 2002. *Circulation*. 113: 1335–1343.

Yap, S., Boers, G.H., Wilcken, B., Wilcken, D.E., Brenton, D.P., Lee, P.J., Walter, J.H., Howard, P.M., and Naughten, E.R., 2001. Vascular outcome in patients with homocystinuria due to cystathionine beta-synthase deficiency treated chronically: a multicenter observational study. *Arteriosclerosis, Thrombosis, and Vascular Biology*. 21: 2080–2085.

CHAPTER 4

The Importance of Vitamins in Biochemistry and Disease as Illustrated by Thiamine Diphosphate (ThDP) Dependent Enzymes

SHINYA FUSHINOBU*[a] AND RYUICHIRO SUZUKI[b]

[a] Department of Biotechnology, The University of Tokyo, 1-1-1 Yayoi, Bunkyo-ku, Tokyo 113-8657, Japan; [b] Department of Biological Production, Faculty of Bioresource Sciences, Akita Prefectural University, Akita 010-0195, Japan
*Email: asfushi@mail.ecc.u-tokyo.ac.jp

4.1 Reactions of ThDP-dependent Enzymes

4.1.1 General Mechanism

The biologically active form of thiamine (vitamin B_1) is thiamine diphosphate (ThDP), in which a diphosphate group is attached to the thiazole ring of thiamine. ThDP was formally known as 'cocarboxylase', which is required for the decarboxylation reaction of pyruvate decarboxylase (PDC) from yeast. In 1937, Lohmann and Schuster established that the coenzyme 'cocarboxylase' is ThDP (Lohmann and Schuster 1937). Enzymes requiring ThDP as a cofactor

---

Food and Nutritional Components in Focus No. 4
B Vitamins and Folate: Chemistry, Analysis, Function and Effects
Edited by Victor R. Preedy
© The Royal Society of Chemistry 2013
Published by the Royal Society of Chemistry, www.rsc.org

are called 'ThDP-dependent enzymes'. ThDP-dependent enzymes are ubiquitously present in all organisms and catalyse various essential reactions in metabolic pathways. They generally catalyse oxidative conversion or non-oxidative transfer of 2-keto acids (or acyloins) and also require divalent cations (usually Mg^{2+} or Ca^{2+}).

In 1958, Breslow proposed that formation of ThDP ylide is the first activation step of ThDP (Breslow 1958) and it is now established as the essential step common in all ThDP-dependent enzymes (Kluger and Tittmann 2008). The ThDP ylide is a C2 ionized state (C2 carbanion) of the thiazolium ring (Figure 4.1a). Deprotonation of the C2 atom is accomplished by the exocyclic N4' imino group of the pyrimidine ring. When ThDP is bound to enzymes, it adopts a twisted conformation called 'V-conformation' (Figure 4.1b). The V-conformation puts the N4' imino group close to the reactive C2 carbon, enabling proton transfer within the same cofactor molecule. In most cases of ThDP-dependent enzymes, a glutamate residue is positioned near to the endocyclic N1' atom of the pyrimidine. The glutamate residue facilitates the ylide formation by stabilizing the N4' imino tautomer. A divalent cation is required to neutralize the negative charge of the diphosphate group to help its binding to the protein. After formation of the ThDP ylide, the substrate carbonyl is attacked by the thiazolium C2 carbanion to form a tetrahedral substrate-ThDP adduct.

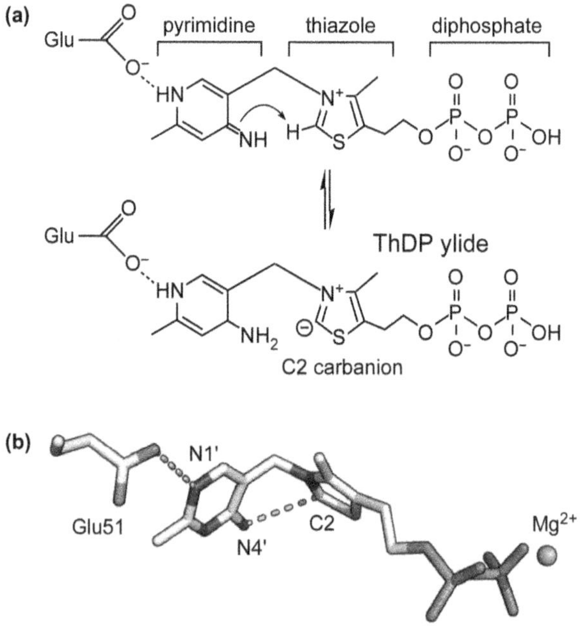

Figure 4.1 Formation of ThDP ylide. (a) Proton transfer from the C2 atom to the N4' imino group forms the ThDP ylide, which is the first activation step of all ThDP-dependent enzymes. (b) Structure of ThDP bound to pyruvate decarboxylase. Glu51 residue facilitates ylide formation.

4.1.2 Pyruvate-processing Enzymes

Most of ThDP-dependent enzymes catalyse the processing of 2-keto acids such as pyruvate, branched-chain keto acids and ketoglutarate. Among them, pyruvate is processed by various ThDP-dependent enzymes (Figure 4.2). Its carbonyl group is attacked by the ThDP ylide, yielding a tetrahedral 2-(2-lactyl)-ThDP adduct. The reactions followed by decarboxylation can be roughly classified into oxidative or nonoxidative ones.

4.1.2.1 PDC

PDC is a representative of ThDP-enzymes that catalyses a simple nonoxidative conversion of pyruvate—the decarboxylation of pyruvate into acetaldehyde and carbon dioxide. PDC is widely distributed in plants and fungi, but it is rarely found in bacteria and not found in animals. PDC catalyses a key reaction in alcohol fermentation by yeast. The structure and function of PDCs from

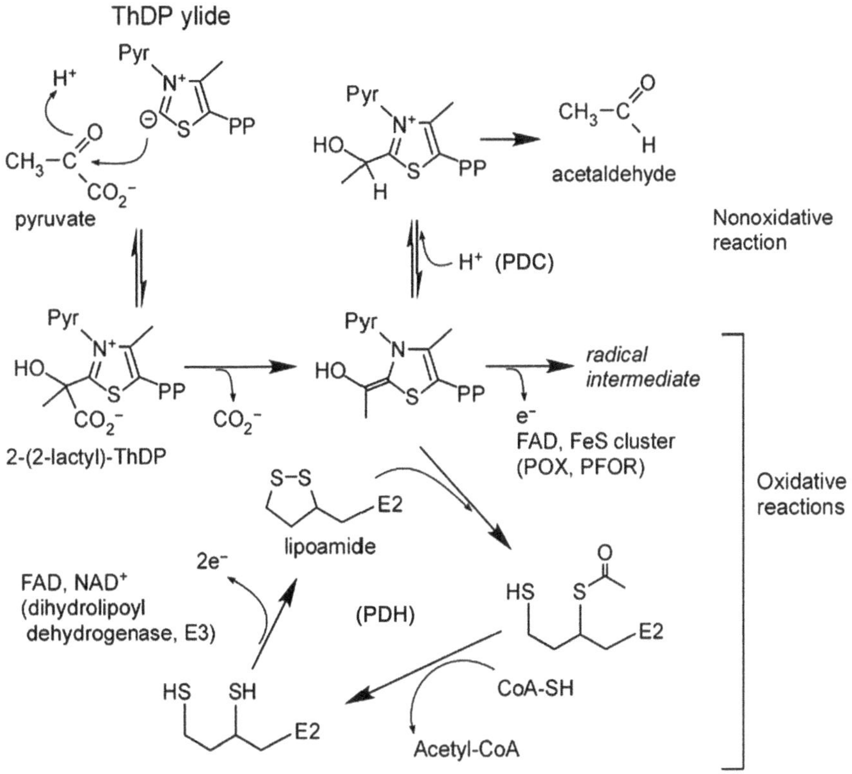

Figure 4.2 Reactions of pyruvate-processing enzymes: pyruvate decarboxylase (PDC), pyruvate oxidase (POX), pyruvate:ferredoxin oxidoreductase (PFOR) and pyruvate dehydrogenase (PDH). Their reactions diverge after the decarboxylation step. Pyr and PP represent 2,5-dimethyl-4-aminopyrimidine and the ethyl diphosphate tail, respectively.

yeast and an anaerobic bacterium, *Zymomonas mobilis*, have been much studied (Candy and Duggleby 1998).

4.1.2.2 Pyruvate Dehydrogenase Multienzyme Complex

Pyruvate dehydrogenase (PDH) is located at a key junction point in sugar metabolism between the glycolysis and the tricarboxylic acid (TCA) cycle (Patel and Roche 1990). It catalyses the irreversible conversion of pyruvate, CoA and NAD^+ into CO_2, NADH and acetyl-CoA. Acetyl-CoA serves as a precursor for the TCA cycle and the biosynthesis of fatty acids and steroids, and NADH provides the reducing equivalents into the respiratory chain for oxidative phosphorylation.

PDH forms a huge multienzyme complex with molecular masses of $5–10 \times 10^6$ and diameters of up to 50 nm, significantly larger than a ribosome. The multienzyme complex consists of three components: ThDP-dependent pyruvate dehydrogenase (E1), dihydrolipoyl acetyltransferase (E2), and dihydrolipoyl dehydrogenase (E3). The E2 component carries a covalently attached lipoic acid. It is linked to a 'swinging arm' region of the E2 protein through an amide bond between its carboxyl group and the terminal N^6 amino group of lysine residues—and so is called lipoamide. E3 has a strongly bound flavin adenine dinucleotide (FAD) cofactor. At first, E1 catalyses the decarboxylation of pyruvate and subsequent transfer of the acyl group to the lipoyl group of E2. E2 is responsible for the transfer of the acyl group to CoA. Finally, E3 regenerates the disulfide bridge in the lipoyl group and transfers the two reducing equivalents to NAD^+ through the FAD cofactor. Not only the swinging arm but also a domain of E2 that carries the swinging arm ('swinging domain') is structurally flexible in the multienzyme complex, so the lipoyl group can move across the distant active sites to couple the multiple reactions at different components (Perham 2000).

Because of its importance in metabolic pathways, PDH complexes in all organisms are subject to strict allosteric regulation of the rate-limiting E1 component, for which pyruvate and acetyl-CoA are positive and negative effectors, respectively. Moreover, mammalian PDH complexes are regulated by phosphorylation (inhibition) and dephosphorylation (reactivation) of the E1 component. The PDH multienzyme complex is one of the most interesting and the most important ThDP-dependent enzymes, but its unusually large size hampers complete understanding of the dynamic nature of the sequential reactions within it.

4.1.2.3 Radical-forming Enzymes

Several ThDP-dependent enzymes, *e.g.* pyruvate oxidase (POX) and pyruvate:ferredoxin oxidoreductase (PFOR), form a transient radical intermediate after one-electron oxidation by neighbouring redox-active cofactors such as flavins (in the case of POX) and Fe_4S_4 clusters (in the case of PFOR) (Tittmann 2009). The radical intermediates are sometimes stable even at room temperature and can be detected by electron paramagnetic resonance (EPR) spectroscopy. 'Snapshots' of the reaction intermediates of a radical-forming enzyme, POX from *Lactobacillus plantarum*, have been captured by X-ray crystallography and nuclear magnetic resonance (NMR) spectroscopy, and its detailed chemical mechanism has been elucidated (Wille *et al.* 2006).

There are two types of POXs: acetyl phosphate-producing and phosphate-independent ones. Acetyl phosphate-producing POX from *Lactobacillae* transfers the electrons to dioxygen to yield hydrogen peroxide:

$$\text{Pyruvate} + \text{phosphate} + O_2 + H^+ \rightarrow \text{acetyl phosphate} + CO_2 + H_2O_2$$

However, phosphate-independent POX from *Escherichia coli* transfers the electrons to a membrane-bound electron carrier ubiquinone 8 (Q_8):

$$\text{Pyruvate} + Q_8 + OH^- \rightarrow \text{acetate} + CO_2 + Q_8H_2$$

PFOR transfers the electrons to a ferredoxin to yield acetyl-CoA (Ragsdale 2003):

$$\text{Pyruvate} + \text{CoA} + 2\text{Fd}_{ox} \Leftrightarrow \text{acetyl-CoA} + CO_2 + 2\text{Fd}_{red}$$

The reaction catalysed by PFOR is similar to that of the PDH multienzyme complex. However, PFOR has a far simple domain structure with molecular masses of $1-2.5 \times 10^5$ and its reaction is reversible.

4.1.3 Transketolase and Phosphoketolase

4.1.3.1 *Transketolase*

Transketolase (TK) is involved in anaerobic carbohydrate metabolisms such as the nonoxidative phase of the pentose phosphate pathway. In plants and photosynthetic bacteria, TK is involved in the Calvin–Benson cycle. TK catalyses the transfer of a 2-carbon dihydroxyethyl group from a ketose phosphate (donor substrate such as D-xylulose 5-phosphate) to the C1 position of an aldose phosphate (acceptor substrate such as D-ribose 5-phosphate) (Figure 4.3) (Schneider and Lindqvist 1998). The first product is an aldose phosphate released from the donor (such as glyceraldehyde 3-phosphate) and the second is a ketose phosphate (such as sedoheptulose 7-phosphate), in which the 2-carbon fragment is attached to the acceptor. Examples of the substrates and the products mentioned above are for the first reaction of the pentose phosphate pathway. In the second reaction of the same pathway, the acceptor is D-erythrose 4-phosphate and the second product is D-fructose 6-phosphate. A 'snapshot' X-ray crystallographic study revealed that an α-carbanion/enamine α,β-dihydroxyethyl ThDP is formed as a key intermediate (Fiedler *et al.* 2002). Then, a nucleophilic attack of the α-carbanion intermediate on the acceptor substrate occurs.

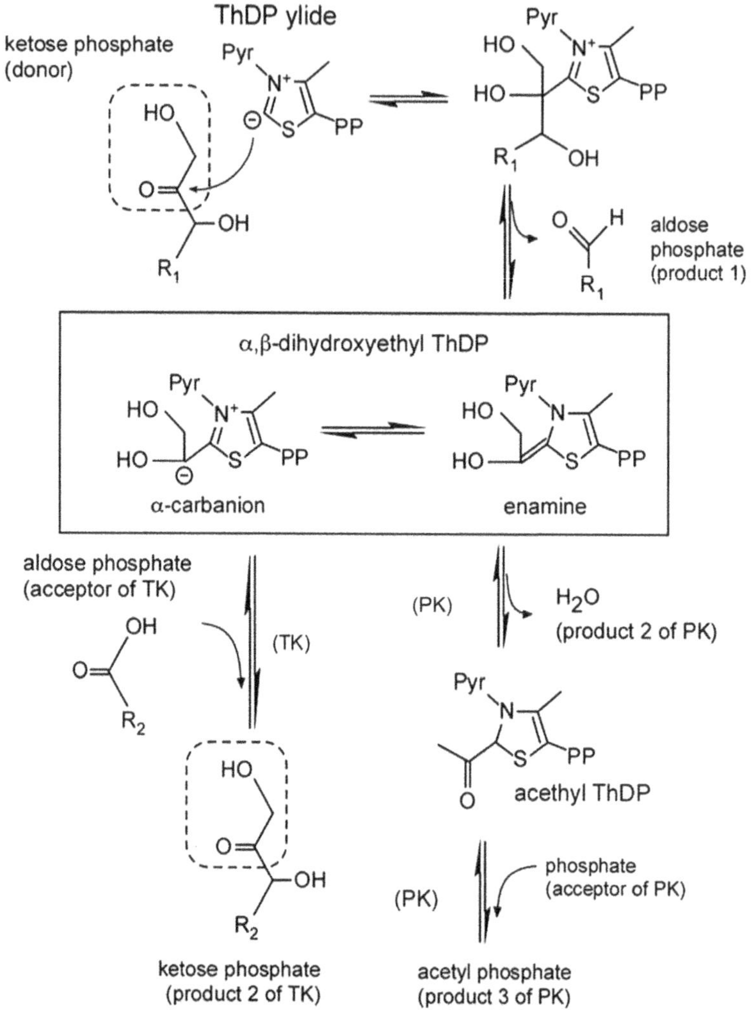

Figure 4.3 Reactions of transketolase (TK) and phosphoketolase (PK). The first half of the reactions of TK and PK are identical. TK catalyses the transfer of the 2-carbon fragment to an acceptor, whereas PK catalyses dehydration and subsequent nucleophilic attack of phosphate to produce acetyl phosphate. For the first reaction of TK in the pentose pathway, where donor, acceptor, products 1 and 2 are xylulose 5-phosphate, ribose 5-phosphate, glyceraldehyde 3-phosphate and sedoheptulose 7-phosphate, $R_1 = R_2 = $ -CHOH-CH$_2$-OPO$_3^-$. Pyr and PP represent 2,5-dimethyl-4-amino-pyrimidine and the ethyl diphosphate tail, respectively.

4.1.3.2 Phosphoketolase

Various microbes including filamentous fungi and yeasts have a pentose catabolism involving phosphoketolase (PK), which is called the 'PK pathway'. In addition, heterofermentative lactic acid bacteria including the genera

Lactobacillus, *Leuconostoc* and *Bifidobacterium* employ the PK pathway as the central fermentative pathway. PK catalyses the cleavage of a ketose phosphate (donor substrate such as D-xylulose 5-phosphate), utilizing an inorganic phosphate (acceptor substrate) to produce an aldose phosphate released from the donor (first product such as glyceraldehyde 3-phosphate), water (second product) and acetyl phosphate (third product) (Figure 4.3). The first half of the reaction of PK is identical to that of TK. However, the subsequent reaction catalysed by PK is distinct from TK (Yevenes and Frey 2008). PK releases a water molecule (dehydration) from the α,β-dihydroxyethyl ThDP intermediate to form the acetyl ThDP intermediate, and then a nucleophilic attack of the acceptor substrate (phosphate) on the acetyl ThDP intermediate yields the third product (acetyl phosphate). The crystal structures of the intermediates before and after the dehydration step have been reported (Suzuki *et al.* 2010).

4.2 Structures and Classification

4.2.1 Structures of TK and PK

TK was one of the first ThDP-dependent enzymes whose three-dimensional structure to be determined by X-ray crystallography (Lindqvist *et al.* 1992). TK consists of three α/β-fold domains; an *N*-terminal domain that binds the pyrophosphate group of ThDP (PP domain), a middle domain that binds the pyrimidine group (Pyr domain) and a *C*-terminal domain (Figure 4.4a). The PP and

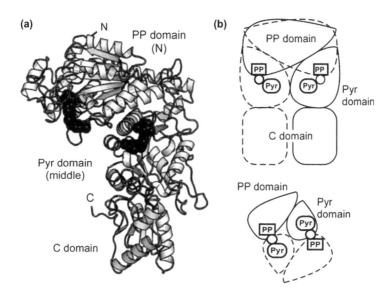

Figure 4.4 Structure of transketolase. (a) Monomer structure. The two ThDP molecules, which interact with the PP and Pyr domain of the subunit, are shown as black spheres. (b) Schematic drawing of interactions of homodimeric transketolase and bound ThDP. Front view (top) and top view (bottom). One subunit is shown with dotted lines.

Pyr domains share a similar topology of secondary structures, suggesting a common evolutionary origin. TK is a homodimer and each subunit packs through extensive interactions between the PP and Pyr domains (Figure 4.4b). ThDP is bound at the interface, the pyrophosphate and the pyrimidine groups being held by different subunits. Recently, the crystal structure of PK was determined (Suzuki *et al.* 2010), and its overall structure is basically similar to that of TK.

4.2.2 Structures of Other ThDP-Dependent Enzymes

The three-dimensional structures of PDC from *Lactobacillus plantarum* (Muller and Schulz 1993) and POX from yeast (Dyda *et al.* 1993) revealed that they share a basically similar domain architecture. These proteins consist of three α/β-fold domains of similar size; an *N*-terminal Pyr domain, a middle domain (FAD-binding in the case of POX) and a *C*-terminal PP domain (Figure 4.5). The pyrimidine-binding Pyr and pyrophosphate-binding PP domains of TK/PK and PDC/POX are structurally similar. However, the *C*-terminal domain of TK/PK and the middle domain of PDC/POX are completely different, suggesting that they have distinct evolutionary origins.

The E1 component of PDH from *E. coli* also consists of three α/β-fold domains of similar size, and the *N*-terminal and middle domains are PP and Pyr domains, respectively (Arjunan *et al.* 2002), like in the case of TK. PFOR from *Desulfovibrio africanus* consists of seven domains (Chabrière *et al.* 1999), and the first and the sixth domains of the protein are Pyr and PP domains, respectively.

4.2.3 Classification of ThDP-Dependent Enzyme Families

Structures of various ThDP-dependent enzymes have been reported. In 2006, Duggleby proposed classifying ThDP-dependent enzymes into five broad families, based on their structural (domain) relationships and possible evolutionary routes (Table 4.1) (Duggleby 2006).

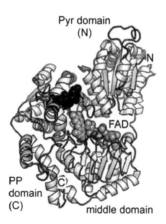

Figure 4.5 Structure of pyruvate oxidase. Monomer structure. ThDP and FAD molecules are shown as black and grey spheres, respectively.

Table 4.1 ThDP-dependent enzyme families. Prepared according to Duggleby (2006).

Family	Enzymes
DC	POX, PDC, indolepyruvate decarboxylase, benzoylformate decarboxylase, oxalyl-CoA decarboxylase, acetohydroxyacid synthase, acetolactate synthase, benzaldehyde aldolase, N^2-(2-carboxyethyl)arginine synthase, glyoxylate carboligase, CDP-4-aceto-3,6-dideoxygalactose synthase, sulfoacetaldehyde acetyltransferase, phenylpyruvate decarboxylase, 2-ketoisovalerate decarboxylase, 2-hydroxyphytanoyl-CoA lyase, and (1R,6R)-2-succinyl-6-hydroxy-2,4-cyclohexadienecarboxylate synthase
TK	TK, PK, dihydroxyacetone synthase and 1-deoxy-D-xylulose-5-phosphate synthase
OR	PFOR, 2-ketoglutarate:ferredoxin oxidoreductase, and 2-ketoisovalerate:ferredoxin oxidoreductase
KD (K1)	PDH E1 from Gram-negative bacteria (including *E. coli*) and actinomycetes
KD (K2)	PDH E1 from eukaryotes and Gram-positive bacteria, branched-chain 2-ketoacid dehydrogenase E1 from human, 2-ketoglutarate dehydrogenase E1, and acetoin dehydrogenase

The DC family contains enzymes catalysing the decarboxylation of a 2-keto acid and it is the largest family including at least 16 enzymes. Of these, POX is atypical member of the DC family, because it catalyses a redox reaction, whereas others do not. The TK family consists of nonoxidizing enzymes such as TK and PK.

The OR family consists of PFOR and its variant enzymes that have substrate specificity for other 2-keto acids than pyruvate, *e.g.* 2-ketoglutarate or 2-ketoisovalerate. The OR family members are found in all *Archae*a and some anaerobic bacteria, and they have variations in numbers of subunits and domains (Zhang *et al.* 1996). The 2-keto acid dehydrogenases, such as the E1 component of PDH, form a KD superfamily in this classification. The KD superfamily is divided into K1 and K2 families. The K1 family members, such as *E. coli* PDH E1, consist of a single chain, whereas the K2 family members consist of two chains (α plus β). Acetoin dehydrogenase in the K2 family is atypical because it does not act on a 2-keto acid.

Summary Points

- This chapter focuses on ThDP-dependent enzymes.
- ThDP-dependent enzymes are ubiquitously present in all organisms.
- ThDP-dependent enzymes catalyse various essential reactions in metabolic pathways.
- ThDP-dependent enzymes generally catalyse oxidative conversion or nonoxidative transfer of 2-keto acids.
- The first essential activation step common in all ThDP-dependent enzymes is formation of ThDP ylide.

- The ThDP ylide is a C2 ionized state of the thiazolium ring that attacks the substrate carbonyl group.
- Many enzymes catalyse decarboxylation of pyruvate but subsequent reactions are different.
- Pyruvate decarboxylase catalyses a simple nonoxidative conversion of pyruvate to acetaldehyde and carbon dioxide.
- The E1 component of pyruvate dehydrogenase multienzyme complex catalyses the decarboxylation of pyruvate and transfer of the acyl group to the lipoyl group of the E2 component.
- Pyruvate oxidase and pyruvate:ferredoxin oxidoreductase catalyse the oxidative decarboxylation of pyruvate and forms a radical intermediate.
- Transketolase catalyses transfer of a 2-carbon fragment between sugar phosphates.
- Phosphoketolase catalyses cleavage of sugar phosphates utilizing an inorganic phosphate to produce water, acetyl phosphate, and a shortened sugar phosphate.
- ThDP-dependent enzymes are classified into five families.

Key Facts

Key Facts about the History of Structural and Functional Studies on ThDP-dependent Enzymes

- In 1937, Lohmann and Schuster established that the coenzyme in pyruvate decarboxylase is ThDP.
- In 1958, Breslow proposed the formation of ThDP ylide, which is the common activation step of ThDP-dependent enzymes.
- In 1992–1993, the crystal structures of transketolase, pyruvate oxidase and pyruvate decarboxylase were reported as the earliest structures of ThDP-dependent enzymes.
- In 1999, Chabrière and colleagues reported the crystal structure of pyruvate:ferredoxin oxidoreductase, which is the sole structure of the OR family.
- In 2002, Arjunan and colleagues reported the crystal structure of a functional E1 component of PDH complex from *E. coli*.
- In 2002, Fiedler and colleagues trapped the key intermediate structure of transketolase.
- In 2006, Wille and colleagues took 'structural snapshot' pictures of the key intermediates in the radical-forming reaction of pyruvate oxidase.
- In 2006, Duggleby proposed a classification system of ThDP-dependent enzymes.

Key Facts about Metabolic Pathways Involving ThDP-dependent Enzymes

- Pyruvate dehydrogenase catalyses the key junction point between the glycolysis and the TCA cycle.

- The E1 component of pyruvate dehydrogenase complexes is subject to allosteric regulation by pyruvate and acetyl-CoA, and it is additionally regulated by phosphorylation in mammals.
- Pyruvate decarboxylase catalyses a key reaction in alcohol fermentation by yeast, which greatly contributes to the brewing industry and bioethanol production.
- Transketolase is involved in the non-oxidative phase of pentose phosphate pathway.
- Transketolase is also involved in the Calvin–Benson cycle of photosynthesis by plants and some bacteria.
- Phosphoketolase is involved in special types of pentose catabolism of fungi and yeasts, and the central fermentative pathway of commensal heterofermentative lactic acid bacteria.
- Phosphoketolase is involved in a unique and effective glycolysis of Bifidobacteria called 'bifid shunt'.
- Phosphoketolase and a certain type of pyruvate oxidase produce an acetyl phosphate that is a high-energy metabolite and produce ATP by acetate kinase during anaerobic growth of bacteria.

Definitions of Words and Terms

Acyloin. Organic compounds containing adjacent carbonyl and hydroxyl groups. Acyloins are also referred as ketols. Many of the substrates of ThDP-dependent enzymes, such as ketose and aldose, have an acyloin structure.

Aldose. Sugars containing an aldehyde group, such as glucose.

Cofactor. A non-protein compound that assists functions of proteins and/or enzymes. A loosely bound cofactor is called 'coenzyme'.

Ferredoxin. Redox-active small proteins containing one or two iron-sulfur cluster(s), such as [4Fe-2S], [3Fe-4S] or [4Fe-4S].

α/β-Fold. A class of protein fold (polypeptide chain folding pattern) consisting of combinations of β-α-β motifs. The β-α-β motif forms a predominantly parallel β sheet surrounded by α helices.

2-Keto acids. Organic compounds containing adjacent carboxylic acid and ketone groups. They include pyruvate, 2-ketoglutarate and 2-ketoisovalerate.

Ketose. Sugars containing a ketone group, such as fructose.

Phosphorolysis. An enzymatic reaction that cleaves a substrate compound by adding an inorganic phosphate into two products, one of which has a phosphate group.

ThDP ylide. A form of ThDP, whose C2 atom on the thiazolium ring is deprotonated to form a carbanion.

X-ray crystallography. A method of determining the three-dimensional structures of proteins and macromolecules.

List of Abbreviations

EPR   electron paramagnetic resonance
FAD   flavin adenine dinucleotide

NMR   nuclear magnetic resonance
PDC   pyruvate decarboxylase
PDH   pyruvate dehydrogenase
PFOR   pyruvate:ferredoxin oxidoreductase
PK   phosphoketolase
POX   pyruvate oxidase
ThDP   thiamine diphosphate
TK   transketolase

References

Arjunan, P., Nemeria, N., Brunskill, A., Chandrasekhar, K., Sax, M., Yan, Y., Jordan, F., Guest, J.R., and Furey, W., 2002. Structure of the pyruvate dehydrogenase multienzyme complex E1 component from *Escherichia coli* at 1.85 Å resolution. *Biochemistry*. 41: 5213–5221.

Breslow, R., 1958. On the mechanism of thiamine action: IV. Evidence from studies on model systems. *Journal of the American Chemical Society*. 80: 3719–3726.

Candy, J.M., and Duggleby, R.G., 1998. Structure and properties of pyruvate decarboxylase and site-directed mutagenesis of the *Zymomonas mobilis* enzyme. *Biochimica et Biophysica Acta*. 1385: 323–338.

Chabrière, E., Charon, M.H., Volbeda, A., Pieulle, L., Hatchikian, E.C., and Fontecilla-Camps, J.C., 1999. Crystal structures of the key anaerobic enzyme pyruvate:ferredoxin oxidoreductase, free and in complex with pyruvate. *Nature Structural Biology*. 6: 182–190.

Duggleby, R.G., 2006. Domain relationships in thiamine diphosphate-dependent enzymes. *Accounts of Chemical Research*. 39: 550–557.

Dyda, F., Furey, W., Swaminathan, S., Sax, M., Farrenkopf, B., and Jordan, F., 1993. Catalytic centers in the thiamin diphosphate dependent enzyme pyruvate decarboxylase at 2.4 Å resolution. *Biochemistry*. 32: 6165–6170.

Fiedler, E., Thorell, S., Sandalova, T., Golbik, R., Konig, S., and Schneider, G., 2002. Snapshot of a key intermediate in enzymatic thiamin catalysis: crystal structure of the alpha-carbanion of (alpha,beta-dihydroxyethyl)-thiamin diphosphate in the active site of transketolase from *Saccharomyces cerevisiae*. *Proceedings of the National Academy of Sciences of the United States of America*. 99: 591–595.

Kluger, R., and Tittmann, K., 2008. Thiamin diphosphate catalysis: enzymic and nonenzymic covalent intermediates. *Chemical Reviews*. 108: 1797–1833.

Lindqvist, Y., Schneider, G., Ermler, U., and Sundstrom, M., 1992. Three-dimensional structure of transketolase, a thiamine diphosphate dependent enzyme, at 2.5 Å resolution. *The EMBO Journal*. 11: 2373–2379.

Lohmann, K., and Schuster, P., 1937. Untersuchungen über die Cocarboxylase. *Biochemische Zeitschrift*. 294: 188–214.

Muller, Y.A., and Schulz, G.E., 1993. Structure of the thiamine- and flavin-dependent enzyme pyruvate oxidase. *Science*. 259: 965–967.

Patel, M.S., and Roche, T.E., 1990. Molecular biology and biochemistry of pyruvate dehydrogenase complexes. *FASEB Journal*. 4: 3224–3233.

Perham, R.N., 2000. Swinging arms and swinging domains in multifunctional enzymes: catalytic machines for multistep reactions. *Annual Review of Biochemistry*. 69: 961–1004.

Ragsdale, S.W., 2003. Pyruvate ferredoxin oxidoreductase and its radical intermediate. *Chemical Reviews*. 103: 2333–2346.

Schneider, G., and Lindqvist, Y., 1998. Crystallography and mutagenesis of transketolase: mechanistic implications for enzymatic thiamin catalysis. *Biochimica et Biophysica Acta*. 1385: 387–398.

Suzuki, R., Katayama, T., Kim, B.J., Wakagi, T., Shoun, H., Ashida, H., Yamamoto, K., and Fushinobu, S., 2010. Crystal structures of phosphoketolase: thiamine diphosphate-dependent dehydration mechanism. *The Journal of Biological Chemistry*. 285: 34279–34287.

Tittmann, K., 2009. Reaction mechanisms of thiamin diphosphate enzymes: redox reactions. *FEBS Journal*. 276: 2454–2468.

Wille, G., Meyer, D., Steinmetz, A., Hinze, E., Golbik, R., and Tittmann, K., 2006. The catalytic cycle of a thiamin diphosphate enzyme examined by cryocrystallography. *Nature Chemical Biology*. 2: 324–328.

Yevenes, A., and Frey, P.A., 2008. Cloning, expression, purification, cofactor requirements, and steady state kinetics of phosphoketolase-2 from *Lactobacillus plantarum*. *Bioorganic Chemistry*. 36: 121–127.

Zhang, Q., Iwasaki, T., Wakagi, T., and Oshima, T., 1996. 2-Oxoacid:ferredoxin oxidoreductase from the thermoacidophilic archaeon, *Sulfolobus* sp. strain 7. *The Journal of Biochemistry*. 120: 587–599.

Chemistry and Biochemistry

CHAPTER 5
The Chemistry, Biochemistry and Metabolism of Thiamin (Vitamin B_1)

LUCIEN BETTENDORFF

GIGA-Neurosciences, B36 University of Liege, Tour de Pathologie 2, Avenue de l'Hôpital, B-4000 Liege 1 (Sart Tilman), Belgium
Email: L.Bettendorff@ulg.ac.be

5.1 Introduction

Thiamin (thiamine, vitamin B_1), also formerly called aneurin (antineuritic factor), is a water-soluble vitamin of the B group. Like other B vitamins, thiamin is the precursor of an important coenzyme, thiamin diphosphate (ThDP), required for the oxidative decarboxylation of 2-oxo acids. However, in contrast to other B vitamins, non-cofactor roles have been proposed for thiamin derivatives. These could be mediated by two triphosphate derivatives, thiamin triphosphate (ThTP) and the recently discovered adenosine thiamin triphosphate (AThTP) (Bettendorff *et al.* 2007).

Microorganisms, fungi and plants synthesize thiamin but animals have to rely on exogenous dietary sources. In the absence of adequate thiamin intake, thiamin deficiency results in specific diseases. The typical thiamin deficiency syndrome in humans is beriberi, characterized by polyneuritis and paralysis of the lower limbs. In addition, to this so-called 'dry' beriberi, the rarer 'wet' (Shoshin) beriberi results in oedema and cardiac insufficiency. Until the end of the 19th century, beriberi was a major health problem in East Asian countries relying on polished rice as staple food. Indeed, the vitamin is mainly present in the bran but not in the endosperm. Nowadays, beriberi remains a problem in some isolated or displaced populations. In Western populations, thiamin deficiency is mainly observed in chronic alcoholics, who may develop Wernicke–Korsakoff syndrome. Wernicke's encephalopathy, the acute manifestation of thiamin deficiency in alcoholic patients, presents with ataxia, ophthalmoplegia and a global confusional state. If not treated promptly by pharmacological doses of thiamin, the syndrome will result in coma and death. Korsakoff psychosis, sometimes considered the 'chronic phase' of Wernicke's encephalopathy is characterized by retrograde and anterograde amnesia. It is linked to irreversible lesions in the thalamus and the mammillary bodies. Subacute thiamin deficiency is also observed in elderly people, pregnant or lactating women, AIDS or diabetic patients.

ThDP plays a crucial role as coenzyme for several enzymes and enzyme complexes such as transketolase (EC 2.2.1.1) and the enzyme complexes pyruvate (EC 1.2.4.1) and 2-oxoglutarate (EC 1.2.4.2) dehydrogenases, present in nearly all organisms. They play important catabolic roles and are key actors in cell energy metabolism (Figure 5.1). Reduced activity of these enzymes as a consequence of thiamin deficiency results in decreased glucose oxidation. As the brain heavily relies on oxidative metabolism, it is more severely affected by thiamin deficiency than other organs.

Many other ThDP-dependent enzymes are found in microorganisms. Among these, pyruvate decarboxylase (EC 4.1.1.1) is of particular importance, as it is the first step in alcohol fermentation in yeast. This enzyme is therefore of fundamental importance in the production of alcoholic beverages. Another ThDP-requiring enzyme in bacteria is the flavoprotein pyruvate oxidase (EC 1.2.3.3) catalysing the reaction pyruvate $+ P_i + O_2 + H_2O \Leftrightarrow$ acetyl phosphate $+ CO_2 + H_2O_2$. A few studies have revealed the existence of hydroxyethylthiamin in microorganisms. This compound probably originates from hydrolysis of hydroxyethyl thiamin diphosphate, released from pyruvate decarboxylase or pyruvate oxidase.

5.2 Chemical Properties of Thiamin

Thiamin consists of a pyrimidine moiety (2-methyl-4-amino-pyrimidine) and a thiazolium moiety [5-(2-hydroxyethyl)-4-methylthiazolium] linked through a methylene bridge (Figure 5.2). It was first chemically synthesized in 1936 by R.R. Williams and J.K. Cline at Columbia University in New York. Thiamin is a relatively hydrophilic molecule: one gram of thiamin chloride hydrochloride ($C_{12}H_{17}ClN_4OS \cdot HCl$, molecular mass $337.27 \, g \, mol^{-1}$) dissolves in ~ 1 ml of water. Thiamin is poorly soluble in organic solvents, including ethanol. Aqueous

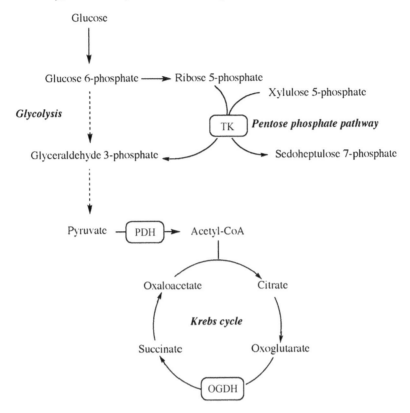

Figure 5.1 Role of thiamin diphosphate-dependent enzymes in cell energy metabolism. In most cells, glucose is degraded *via* glycolysis to pyruvate, which is transformed to acetyl-CoA by the pyruvate dehydrogenase complex. Acetyl-CoA is then oxidized in the Krebs cycle. In eukaryotes, the pyruvate dehydrogenase complex and the Krebs cycle enzymes are localized inside the mitochondrial matrix. The pentose monophosphate shunt is an alternative pathway of glucose degradation, leading to the generation of NADPH for fatty acid synthesis and ribose 5-phosphate required for nucleic acid synthesis. Thiamin diphosphate-dependent enzymes are encircled. PDH, pyruvate dehydrogenase complex; OGDH, 2-oxoglutarate dehydrogenase complex; TK, transketolase.

solutions are most stable between pH 2 and 4. We can distinguish three different functional parts of the thiamin molecule:

- the catalytic aromatic thioazole heterocycle;
- the pyrimidine aromatic heterocycle helping in catalysis and important for recognition by enzymes;
- the hydroxyethyl group for the attachment of phosphate groups.

The catalytic properties of thiamin have been reviewed by Kluger and Tittmann (2008). The particular properties of thiamin result from the planar

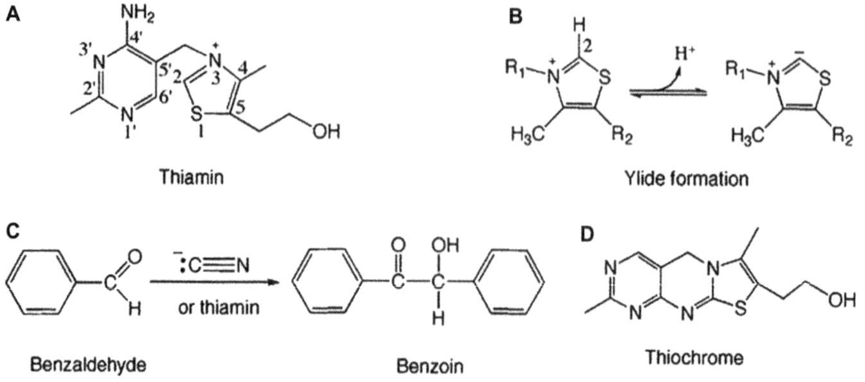

Figure 5.2 Structural formula of thiamin and its chemical properties. (A) Structural formula of thiamin. (B) The C-2 carbon of the thiazolium moiety can lose its proton yielding a highly nucleophilic ylide analogous to cyanide. (C) Cyanide and thiamin catalysed condensation of benzaldehyde to benzoin. (D) Oxydation of thiamin in strongly alkaline medium yields the highly fluorescent thiochrome, a property used for the experimental determination of thiamin and its phosphate esters.

thiazole, a heterocycle rarely found in biological molecules except as a fusion product with benzene in benzothiazoles such a luciferin. Thiazoles are related to oxazoles and imidazoles but the sulfur atom, replacing either an oxygen or a nitrogen atom, is responsible for a larger π-electron delocalization than in oxazoles or imidazoles. The important site is C-2, which can be deprotonated to form a carbanion (Figure 5.2B). The alkylation of the N-3 of the thiazole heterocycle to form a thiazolium increases the acidic character of C-2 by stabilizing the carbanion as a ylide (Breslow 1958). This carbanion is a strong nucleophile and, as noted by Breslow, an analogue of cyanide, able to catalyse benzoin condensation. The rate of catalysis of free thiamin is, however, low and in ThDP-dependent enzyme catalysis, the enzymatic environment accelerates the reactions by at least a factor 10^{12} (Kluger and Tittmann 2008).

Even in relatively strong alkaline medium, the ylide carbanion is never observed in aqueous solutions of thiamin because the estimated pK_a of the C-2 proton is too high (in the range from 14 to 20). Nonetheless, a relative lability of the C-2 proton is demonstrated by the observation that it can be rapidly exchanged with deuterium.

The pyrimidine heterocycle is chemically more inert than the thiazole heterocycle. At pH <4, the N-1' of the pyrimidine ring is protonated, while the N-3' is protonated only under extreme conditions such as in concentrated sulfuric acid. The 4'-amino group has no basic character and is not protonated even in strong acids. However, it plays an important role in positioning the cofactor ThDP in the active site of the apoenzymes.

Thiamin is unstable in alkaline solutions (Figure 5.3A): in the pH range 8–10, hydroxyl attack on C-2 of the thiazolium ring leads to the relatively slow formation of a 'pseudobase' intermediate (compound 2 in Figure 5.3) and the

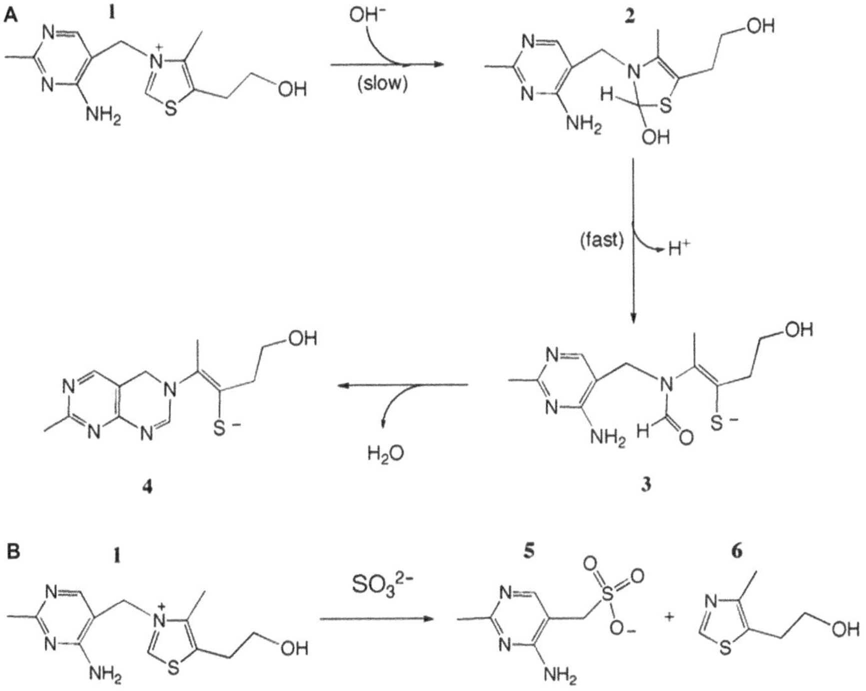

Figure 5.3 Alkali-induced transformations of thiamin and cleavage by sulfite. (A) In alkaline medium, thiamin (1) can be slowly transformed to the so-called yellow form (4). This transformation involves the formation of a pseudobase (2) and opening of the thiazolium ring to form a thiolate (3).

subsequent fast opening of the thiazolium ring. This yields a thiolate form (compound 3 in Figure 5.3) that can lose a water molecule to form the so-called 'yellow' form (compound 4 in Figure 5.3).

Sulfites cleave thiamin at the level of the methylene bridge (Figure 5.3B) between the two heterocycles yielding (6-amino-2-methylpyrimid-5-yl)methanesulphonic acid (compound 5 in Figure 5.3) and 5-β-hydroxyethyl-4-methylthiazole (compound 5 in Figure 5.3) (Leichter and Joslyn 1969). This reaction is of particular interest as sulfites are widely used as food preservatives and can thus be responsible for thiamin cleavage during storage, in particular in aqueous solutions even at low temperature. It is also destroyed by ultraviolet (UV) irradiation; it is heat-labile and is partially lost on cooking. At high temperatures, thiamin participates strongly in Maillard-type reactions.

The hydroxyethyl group of thiamin can be esterified to form phosphorylated derivatives: thiamin monophosphate (ThMP), thiamin diphosphate (ThDP) and thiamin triphosphate (ThTP) (Figure 5.4). While ThMP is a phosphoester, ThDP and ThTP are phosphoanhydrides and are 'energy-rich' compounds in the same sense as nucleoside di- and triphosphates: they have a high potential for phosphate group transfer. This property was demonstrated in particular for

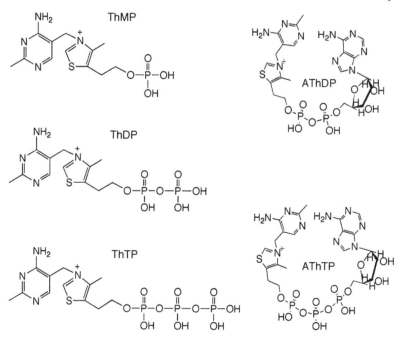

Figure 5.4 Structural formulae of thiamin phosphate esters. At present, five natural thiamin phosphate derivatives have been described: thiamin monophosphate (ThMP); thiamin diphopshate (ThDP); thiamin triphosphate (ThTP); adenosine thiamin diphosphate (AThDP); and adenosine thiamin triphosphate (AThTP). Catalytic intermediates, such as for instance α-hydroxyethyl thiamine diphosphate formed by the action of yeast pyruvate decarboxylase (EC 4.1.1.1), are not considered here.

ThTP through its ability to phosphorylate proteins (Nghiêm *et al.* 2000). Thiamin is the only known natural organic molecule that is not a nucleotide but forms mono-, di- and triphospho-derivatives. Recently, even thiamin adenylated derivatives have been described: adenosine thiamin diphosphate (AThDP) and adenosine thiamin triphosphate (AThTP) (Bettendorff *et al.* 2007; Frédérich *et al.* 2009).

In the coenzyme ThDP, the C-2 proton of the thiazolium ring is particularly labile and is completely exchanged with deuterium within a few minutes. This is also the case in thiamin and in the other phosphate derivatives ThMP and ThTP. In all three compounds, the chemical shift is ∼9.60 ppm (pH 7.4), values higher than expected for aromatic protons (in general 7.5–8.5 ppm) (Frédérich *et al.* 2009). The presence of phosphate groups has no direct effect on the catalytic properties of the thiazole heterocycle. Therefore, thiamin, ThMP, ThDP and ThTP should be interchangeable as catalysts but apoenzymes recognize only the diphosphorylated derivative.

In AThDP and AThTP, the chemical shift of the C-2 proton (9.14/9.18 ppm) is lower than in ThMP, ThDP or ThTP, indicating a relative stabilization of this proton through interaction with the adenine moiety and suggesting a

U-fold structure of adenylated thiamin derivatives in solution. These derivatives are therefore probably not active as catalysts, at least in solution. Molecular modelling suggested that a U-folded conformation is energetically possible for adenylated thiamin derivatives (Frédérich *et al.* 2009).

5.3 Thiamin Biosynthesis and Degradation

Thiamin biosynthesis occurs in prokaryotes, fungi and plants through complex pathways (Jurgenson *et al.* 2009). Synthesis of the thiazole moiety requires the precursors glycine, tyrosine, cysteine and glyceraldehyde 3-phosphate. The pyrimidine moiety is synthesized from hydroxymethyl pyrimidine phosphate, derived from the purine biosynthetic pathway. The thiazole and pyrimidine moieties are synthesized separately before assembling to form ThMP, which is either hydrolysed to thiamin or phosphorylated to ThDP. As thiamin is required for the survival of any known organism, the enzymes involved in thiamin biosynthesis could be interesting antimicrobial drug targets.

Thiamin-degrading enzymes (thiaminases) exist in many organisms. Thiaminase I (EC 2.5.1.2), a pyrimidine transferase, is present in microorganisms, but also in some higher multicellular eukaryotes such as fern, shellfish and fish. Consumption of raw parts of these organisms by animals or man can lead to serious thiamin deficiency syndromes, such as cortical necrosis in ruminants and beriberi in humans. Thiaminase II (EC 3.5.99.2), a hydrolase, might not be involved in the breakdown of thiamin as previously thought, but rather in the hydrolysis of aminopyrimidine to hydroxypyrimidine, a building block for the biosynthesis of thiamin by some bacteria (Jenkins *et al.* 2007).

There are only a few studies concerning the degradation and elimination of thiamin in animals. Most excess thiamin is eliminated as such by the kidneys. In a couple of older studies (Neal and Pearson 1964; Pearson *et al.* 1966), radioactive thiamin was administered in rats and the urines were analysed. Several radioactive degradation products of thiamin (2-methyl-4-amino-5-pyrimidine carboxylic acid and 4-methyl-thiazole-5-acetic acid), resulting from the cleavage between the thiazole and the pyrimidine moieties, were excreted in the urine. Other products were also detected but not identified. No enzymes specifically involved in thiamin degradation in mammals have been identified.

5.4 Riboswitches

Riboswitches are non-coding RNA elements involved in metabolite-sensing (Baird *et al.* 2010). They can bind small metabolites, thereby regulating the expression of metabolite-synthesizing enzymes. Riboswitches have been found in bacteria, fungi and plants for several metabolites, such as for instance thiamin, *S*-adenosylmethionine or lysine. The binding of the metabolite leads to a conformational change of the mRNA, affecting protein expression. The ThDP riboswitch (Thi-box) is the most widespread riboswitch known today and can be found in many bacteria, fungi and plants. In *Escherichia coli*, ThDP binding to the Thi-box leads to the formation of a terminator hairpin, halting the transcription of several thiamin biosynthetic enzymes. The mechanism is different in fungi and plants where the Thi-box is involved in the control of RNA splicing rather than transcription or translation. The ThDP riboswitch is the only riboswitch so far described in eukaryotes.

5.5 Thiamin Transport

Thiamin transport has been nearly exclusively studied in mammalian cells. However, a few studies carried out in bacteria show that thiamin can be transported *via* ABC transporters. These transporters comprise a periplasmic thiamin binding protein in gram-negative bacteria such as *E. coli* and *Salmonella typhimurium*. These proteins are generally not specific for thiamin as they also bind ThMP and ThDP. Examination of the three-dimensional structure of an *E. coli* thiamin binding protein revealed structural similarities with thiaminase I, suggesting a common ancestor (Soriano *et al.* 2008).

5.5.1 Thiamin Transporters in Mammalian Cells

Thiamin transport across cell membranes requires specific transporters, mainly studied in mammalian cells (Ganapathy *et al.* 2004). At low concentrations ($<1\,\mu M$), cells absorb the vitamin *via* an active process, while at high concentrations a passive process seems to prevail. Passive thiamin transport has not been investigated in detail, though it might be pharmacologically relevant when high therapeutic doses of the vitamin are administered. It is possible that the passively diffusing form of thiamin is the open uncharged thiol form (see Figure 5.3). As a result of the two different mechanisms, a larger proportion of small doses of thiamin is absorbed than of high doses. A saturable low-affinity thiamin transport has also been described. In cultured neuroblastoma cells, this saturable low-affinity passive transport ($K_m = 0.8\,mM$), was inhibited by thiamin antagonists, as the high affinity mechanism, suggesting the involvement of a transport protein (Bettendorff and Wins 1994).

High-affinity thiamin transporters belong to the *SLC19A* solute carrier family (the folate/thiamin transporter family), containing 12 transmembrane spanning domains, with a molecular mass of approximately 55 kDa. The first member of this gene family, *SLC19A1*, encodes a reduced folate carrier (RFC-1) that also transports ThMP, present in plasma and cerebrospinal fluid. RFC-1 may thus present an alternative route for cells to capture thiamin. The second member of this family, *SLC19A2* coding for thiamin transporter 1 (ThTR1), has a K_m in the low micromolar concentration range. Mutations in *SLC19A2* leading to a non-functional transporter cause thiamin-responsive megaloblastic anaemia (TMRA), a condition characterized by megaloblastic anaemia, sensorineural hearing loss and diabetes mellitus. Relief of the symptoms is generally obtained by high-dose thiamin treatment.

Another thiamin transporter, ThTR2, is the product of the *SLC19A3* gene. It transports thiamin with a very high apparent affinity ($K_m = 20$–$100\,nM$).

In humans, mutations in this gene are associated with often very severe encephalopathies that can be overcome by high-dose thiamin or, for yet unknown reason, by biotin therapy (Zeng et al. 2005; Debs et al. 2010).

Thiamin is transported at a relatively slow rate and, even when blood-circulating vitamin levels are high, a net increase of thiamin derivatives in brain is difficult to achieve. For that reason, thiamin derivatives with higher biological bioavailability were developed. The prototype of these derivatives is thiamin allyl disulfide (allithiamin), formed in crushed garlic from thiamin and allicin (diallyl thiosulphinate). Other synthetic thiamin disulfide derivatives, such as sulbutiamine (o-isobutyrylthiamin disulfide) and fursultiamine (thiamin tetrahydrofurfuryl disulfide), were developed later. These molecules have a lipophilic character and cross membranes far more easily than thiamin. Inside the cells, they are easily transformed into the vitamin after reduction of the disulfide bond and closing of the thiazole ring. Another synthetic thiamin precursor, benfotiamine (S-benzoylthiamin O-monophosphate) has given promising results in animal models by preventing diabetic complications such as retinopathy (Hammes et al. 2003). Recently, benfotiamine was also shown to improve cognitive functions and to dramatically decrease amyloid plaques and neurofibrillary tangles in a mouse model of Alzheimer's disease (Pan et al. 2010). Benfotiamine is a thioester and the presence of a phosphate group makes it lipid insoluble. After absorption, it is hydrolysed to the more lipophilic S-benzoylthiamin by intestinal alkaline phosphatase. S-benzoylthiamin easily crosses membranes and generates thiamin after hydrolysis by liver thioesterases.

5.5.2 Mitochondrial Transport of Thiamin Diphosphate

In animal cells, thiamin entering the cells is pyrophosphorylated to ThDP, which can then be transported into mitochondria. The mitochondrial ThDP transporter (SLC25A19) seems to act as an antiporter, probably exchanging extramitochondrial ThDP against intramitochondrial nucleotides (Lindhurst et al. 2006). Point mutations in *SLC25A19* lead to its inability to transport the cofactor into the mitochondrial matrix. This causes the inactivation of pyruvate dehydrogenase (PDH) and 2-oxyglutarate dehydrogenase (OGDH) complexes and impairment of oxidative metabolism. This condition is extremely severe and the children affected are born with a very small head and an underdeveloped brain (Amish lethal microcephaly); life expectancy is less than six months.

5.6 Distribution of Thiamin Derivatives in Living Organisms

High-performance liquid chromatography (HPLC) is the method of choice for the separation and determination of thiamin derivatives in complex mixtures and tissue extracts. Various methods have been described using different columns and separation conditions, but all have in common the oxidation of thiamin compounds to the corresponding highly fluorescent thiochrome derivatives (Figure 5.2D). Thiochromes are planar molecules that share an excitation optimum at 360 nm and an emission maximum at 430–435 nm. The thiochrome derivatives of thiamin, ThMP, ThDP and ThTP, have approximately the same fluorescence intensity, while the those of AThTP and AThDP have a twice higher fluorescence intensity (Frédérich *et al.* 2009).

In most organisms, the coenzyme ThDP is the most abundant thiamin compound. Free thiamin and ThMP generally amount to 5–15% of total thiamin. Thiamin is generally present in small amounts: in *E. coli* it is hardly detectable (ThMP is the first thiamin derivative synthesized in this organism), while in mammalian tissues it may represent several percent of total thiamin. The triphosphorylated derivatives ThTP and AThTP are generally minor compounds in animal tissues, but they can be major thiamin compounds in *E. coli* under particular conditions of starvation (Bettendorff *et al.* 2007). No data on the distribution of thiamin derivatives are available for bacteria other than *E. coli*.

5.6.1 Occurrence of ThDP

In animal cells, such as neurons, ThDP is found in highest amounts in mitochondria where it is mostly bound to PDH and OGDH complexes. Smaller amounts are found in the cytosol (partly free and partly bound to transketolase). ThDP is also present in peroxisomes, where it is bound to 2-hydroxyacyl-CoA ligase (Fraccascia *et al.* 2007). In liver and skeletal muscle, however, ThDP is mostly free and cytosolic. It is thought that liver and, to some extent, red blood cells can store excess thiamin under the form of ThDP. When free circulating thiamin levels are decreased during prolonged fasting, liver ThDP is dephosphorylated to thiamin, which can then be released and enter the circulating free thiamin pool. Hepatocytes can therefore act as a thiamin buffer and contribute to the homeostasis of thiamin levels (Volvert *et al.* 2008; Gangolf *et al.* 2010a). Such a mechanism would contribute to the protection of the brain cells, which are highly sensitive to thiamin deficiency. Indeed, brain total thiamin levels remain relatively constant and decrease only after severe thiamin deficiency.

Strangely, cellular ThDP levels are 2–3 times lower in humans than in rodents, and this is also true for the brain (Gangolf *et al.* 2010a). This is probably linked to very low circulating thiamin levels in humans, possibly because intestinal thiamin absorption is less efficient. This would explain the higher sensitivity of humans to thiamin deficiency, in particular in association with alcohol consumption: partial inhibition of intestinal thiamin uptake by alcohol would be sufficient to critically affect thiamin absorption.

5.6.2 Occurrence of ThTP

ThTP is always found in nervous tissue, skeletal muscle and the ontogenetically muscle-derived electric organs; this led to the hypothesis that it might play a role in neurotransmitter release. However, it was later shown to be present in non-excitable animal tissues, and also in plants and bacteria (Makarchikov et al. 2003). This suggests that it may have a more basic cell physiological role. In most eukaryotic cell types, ThTP concentrations do not exceed 10^{-7} to 10^{-5} M. However, in some tissues ThTP levels can be very high, reaching over 50% of total thiamin. This is the case for chicken and pig skeletal muscle as well as for *Electrophorus electricus* electric organ. It appears that ThTP levels are always inversely correlated with the presence of a functional 25 kDa thiamin triphosphatase (ThTPase) (see below). Indeed, this enzyme is largely inactive in the domestic pig and seems to be absent in birds and electric organs.

In *E. coli*, ThTP is not found under optimal growth conditions, but it can accumulate significantly (>20% of total thiamin) during amino acid starvation and is thought to play a role in the adaptation of the bacteria to amino acid starvation.

The subcellular distribution of ThTP in animal cells varies with the tissue studied. In rodent brain, it is mainly localized in mitochondria, while it is mostly cytoplasmic in skeletal muscle. Those differences are probably related to different mechanisms of ThTP synthesis (see below).

5.6.3 Occurrence of Adenylated Thiamin Derivatives

Adenylated thiamin derivatives have only been recently discovered (Bettendorff et al. 2007; Frédérich et al. 2009). AThTP is found in several tissues (mammalian liver and heart, roots of higher plants, *etc*.) but it is not as widespread as ThTP. It is very often found in cultured cells. In *E. coli*, it is produced in response to carbon starvation and may reach up to 15% of total thiamin (Gigliobianco et al. 2010). AThDP was only found in low amounts in rodent liver and in *E. coli*.

5.7 Metabolism of Thiamin Phosphates

5.7.1 Free and Bound ThDP Pools Coexist in the Same Cell

ThDP synthesized in the cytosol has several possible metabolic fates. It may bind with high affinity to transketolase, a cytosolic enzyme. Bound cytosolic ThDP may therefore be relatively abundant in cell types that are rich in transketolase such as hepatocytes and erythrocytes. Part of the cytosolic ThDP may remain free in the cytosol and another part is transported into the mitochondrial matrix, where it may bind to apoenzymes (mostly PDH and OGDH complexes), but part of it remains free in the matrix.

ThDP can be hydrolysed to ThMP by more or less specific ThDPases and other phosphohydrolases, and a variable fraction can be converted to ThTP.

After injection of [^{14}C]-labelled thiamin to rats, it was found that, in the brain, the incorporation of radioactivity was slower into ThDP than into ThMP and ThTP (Gaitonde and Evans 1983). As ThDP is the most likely precursor of the two other derivatives (thiamin kinase activity has never been found in brain), these results can only be understood if there exists a smaller, faster pool of precursor ThDP, which in rat brain is essentially cytosolic. In cultured neuroblastoma cells, the slow and fast ThDP pools coexist within the same cell type and are not the result of a different thiamin metabolism in glia and neurons for instance (Bettendorff 1994). This fast turnover pool is free ThDP, precursor of ThMP and ThTP, while the slow turnover ThDP pool corresponds to apoenzyme-bound cofactor (Figure 5.5).

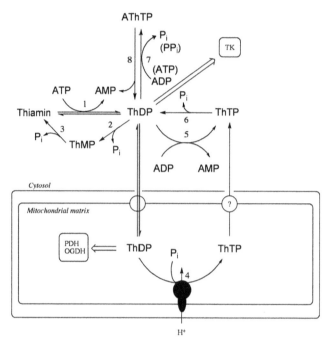

Figure 5.5 Thiamin metabolism in an idealized eukaryotic cell. Thiamin diphosphate (ThDP) is formed by pyrophosphorylation of thiamin. ThDP can bind to cytosolic transketolase or be transported (in eukaryotic cells) into mitochondria, where it can bind to pyruvate dehydrogenase complex (PDH) or oxoglutarate dehydrogenase complex (OGDH). Free ThDP is the precursor for the synthesis of thiamin monophosphate (ThMP), thiamin triphosphate (ThTP) and adenosine thiamin triphosphate (AThTP). Hydrolysis of ThDP yields ThMP, which can in turn be hydrolysed to thiamin. ThTP can be formed in the cytosol by adenylate kinase and in mitochondria by a chemiosmotic mechanism, possibly involving F_oF_1-ATP synthase. AThTP formation is catalysed by a cytosolic adenylyl thiamin diphosphate transferase, either from ATP or ADP. Similar reactions are involved in thiamin metabolism in most prokaryotes, except that the mitochondrial part takes place in the cytosol. 1, thiamin pyrophosphokinase; 2, thiamin diphosphatase; 3, thiamin monophosphatase; 4, ThTP synthase (could be identical to ATP synthase); 5; adenylate kinase; 6, 25 kDa thiamin triphosphatase; 7, ThDP adenylyl transferase (can use both ATP and ADP as substrate); 8, AThTP hydrolase.

In cultured neuroblastoma cells, the slow ThDP pool represents 90–95% of the total cellular ThDP and has a turnover of 6–20 h. For the smaller fast pool, the turnover is of the order of 1–3 h (Bettendorff 1994). The bulk of the ThDP bound to apoenzymes (transketolase, PDH and OGDH) is considered the 'slow turnover ThDP pool', though the stabilities of the different coenzyme-apoenzyme complexes vary: ThDP dissociates faster from PDH than from OGDH and transketolase.

Thiamin phosphate derivatives can be readily interconverted (Figure 5.5), but the enzymes involved in these reactions remain poorly characterized. Only two of them, the essential thiamin pyrophosphokinase and the 25 kDa thiamin triphosphatase, have been characterized at the molecular level.

5.7.2 Synthesis of Thiamin Diphosphate

In Arabidopsis, two TPK genes were identified *AtTPK1* and *AtTPK2* (Ajjawi *et al.* 2007). The two corresponding proteins share 93% sequence identity. While single gene knockout plants were viable, the double mutant had a seedling lethal phenotype, underscoring the importance of ThDP, even in plants. No TPK mutant phenotype is known in humans, probably because ThDP production is indispensable for embryonic development.

In those organisms able to synthesize thiamin, the endpoint of synthesis is ThMP. In enterobacteria, such as *E. coli*, ThMP can then be phosphorylated to ThDP by thiamin phosphate kinase (EC 2.7.4.16). Thiamin taken up from the extracellular medium is first phosphorylated to ThMP by thiamin kinase (EC 2.7.1.89). In thiamin-synthesizing eukaryotes such as fungi and plants, ThMP is dephosphorylated to thiamin by phosphatases. Thiamin is then pyrophosphorylated to ThDP by TPK.

In animals, thiamin transported into the cells is rapidly pyrophosphorylated into ThDP according to the reaction: thiamin + ATP ThDP + AMP, catalysed by thiamin diphosphokinase or thiamin pyrophosphokinase (TPK, EC 2.7.6.2). TPK was first partially purified from baker's yeast. The enzyme has a high apparent affinity for thiamin (K_m in the submicromolar range) but a low affinity for ATP ($K_m = 1$–40 mM). TPK requires a metal ion activator such as Mg^{2+}. It may not be very specific for ATP as diphosphate donor, and according to the origin of the enzyme, GTP or UTP are also good substrates. The equilibrium of this reaction lies far towards the reactants. However, inside the cells the equilibrium is shifted to the right because of the binding of ThDP to cytosolic transketolase and transport into mitochondria. It may also be shifted to the right as AMP is converted to ADP by the adenylate kinase reaction, particularly in muscle cytoplasm. TPK is one of the best-characterized enzymes of thiamin metabolism. It has no amino acid sequence similarity with other known proteins. It is a homodimer of subunits of 23–36 kDa and its three-dimensional structure is known.

5.7.3 Synthesis of ThTP

The mechanism of ThTP synthesis and its physiological role are very controversial. The first mechanism proposed, a reaction catalysed by a soluble ThDP-kinase (ATP + ThDP $\leftrightarrows$ ADP + ThTP), was never substantiated.

Cytosolic adenylate kinases may synthesize ThTP at a slow rate, both in prokaryotes and in animals. This mechanism is probably only relevant in tissues with high AK activity, such as skeletal muscle and electric organs. Therefore, in some species where, in addition to a high AK activity, soluble 25 kDa ThTPase is either absent or inactive (*E. electric* organ, pig and chicken skeletal muscle, see above), ThTP may accumulate in the cytosol, even becoming the major thiamin compound.

A third mechanism has recently been described in the rat brain, where ThTP is synthesized inside mitochondria according to the reaction ThDP + P_i $\leftrightarrows$ ThTP by a chemiosmotic mechanism resembling ATP synthesis by oxidative phosphorylation (Gangolf *et al.* 2010b). ThTP can be released from mitochondria in exchange for P_i. Therefore, cytosolic ThTP may have two different origins, either by exit from mitochondria or synthesis by cytosolic adenylate kinase. The chemiosmotic mechanism occurs in brain mitochondria, but not in mitochondria isolated from rat liver for instance, suggesting a tissue-specific regulation of ThTP synthesis.

5.7.4 Synthesis of AThTP

In *E. coli*, AThTP can be synthesized by a soluble ThDP adenylyl transferase, according to the reaction ADP (ATP) + ThDP $\leftrightarrows$ $P_i(PP_i)$ + AThTP (Makarchikov *et al.* 2007). Though both ATP and ADP may be the phosphate donor for this reaction, ADP is probably the more physiological donor, as AThTP is synthesized under conditions of energy stress and low energy charge in *E. coli*.

5.7.5 Hydrolysis of ThMP

It has been known for a long time that ThMP can be hydrolysed to thiamin by phosphohydrolases present in brain and other organs. However, no phosphatase specific for ThMP has so far been characterized. ThMPase activity, a classical marker of small-diameter dorsal root ganglia neurons, is probably identical to the transmembrane isoform of prostatic acid phosphatase (Csillik *et al.* 2008). This enzyme, in fact an ecto-5'-nucleotidase, hydrolyses adenosine monophosphate to adenosine, which can then activate A_1-adenosine receptors, suppressing pain. In those cytochemical studies of spinal cord, it is of advantage to use ThMP as substrate since, in contrast to β-glycerophosphate or *p*-nitrophenyl phosphate, it is not a good substrate for other acid phosphatases.

Recently, a ThDP/ThMP phosphatase was characterized from *ZEA mays* seedlings (Rapala-Kozik *et al.* 2009). This enzyme has a somewhat lower K_m for

ThDP and ThMP compared with other phosphate esters, but the catalytic efficiency was more or less the same for all substrates tested.

5.7.6 Hydrolysis of ThDP

Enzymatic hydrolysis appears to be the main metabolic fate of ThDP in the brain. The alternative route (phosphorylation to ThTP) is quantitatively important only in skeletal muscles and electric organs, provided the soluble 25 kDa ThTPase is absent. Several phosphatases may hydrolyse ThDP, but until now, no specific thiamin diphosphatase (ThDPase) has been characterized at the molecular level. ThDPases studied so far also hydrolyse nucleoside diphosphates. A phosphohydrolase that is relatively specific for ThDP has been purified and characterized from brain (Sano et al. 1988). GDP, IDP and UDP are also good substrates, while ATP and ADP are not. The enzyme requires Mg^{2+} as cofactor. It must be solubilized by non-ionic detergents before purification. Its molecular mass was estimated to be 75 kDa and the K_m value for ThDP was around 0.66 mM, well above the concentration of free ThDP in the cytoplasm of neurons. This brain (B)-type ThDPase is a membrane-bound enzyme, presumably associated with the endoplasmic reticulum and Golgi apparatus of neurons. ThDPase has been considered a marker of the Golgi apparatus, where it might play a role in protein glycosylation (Roth and Berger 1982). More recent studies suggest that B-ThDPase is a marker of brain phagocytes (Walski and Gajkowska 1998).

In the liver, ThDP can be hydrolysed by a different enzyme called L-type ThDPase which also hydrolyses nucleoside diphosphates and hence is also known as nucleoside diphosphatase (NDPase, EC 3.6.1.6). It catalyses the hydrolysis of GDP, IDP and UDP, but not ADP. With ThDP, the activity was low at neutral pH but higher at pH 9.

The presence of a ThDPase activity in isolated rat liver mitochondria has also been reported (Barile et al. 1998).

5.7.7 Hydrolysis of ThTP

Cytosolic and membrane-associated thiamin triphosphatases (ThTPase, EC 3.6.1.28) were described in the early 1970s. Membrane-associated ThTPases were partially characterized from rat brain, *Electrophorus* electric organ and mammalian skeletal muscle, but none has so far been purified. The electric organ and muscle enzymes are strongly activated by anions. Despite numerous studies, the role of membrane-associated ThTPases remains unclear. As they could not be obtained in pure form, their specificity for ThTP remains unproven and their catalytic efficiency is probably low: the estimated K_m values ranged from 0.25 mM in *Electrophorus* electric organ to 1–2 mM in rat brain. Those values are well above the physiological concentrations of ThTP in most tissues. It is thus doubtful that membrane-associated ThTPases could play a significant role in the regulation of cellular ThTP levels.

Finally, it should be pointed out that a number of other phosphohydrolases such as myosin ATPase, nucleoside triphosphatases, and acid or alkaline phosphatases have been shown to catalyse the hydrolysis of ThTP with variable but low degree of efficiency. Two different enzymes hydrolysing ThTP have been reported in *E. coli*, but they have not been characterized.

However, the soluble 25 kDa mammalian ThTPase is very specific for ThTP: neither nucleoside triphosphates nor ThDP were hydrolysed by this enzyme. It is a 25 kDa monomer with high catalytic efficiency ($k_{cat}/K_m = 6 \times 10^6 \, \text{s}^{-1} \cdot \text{M}^{-1}$). The mRNA for human ThTPase (hThTPase) was found to be widely expressed in human tissues, though at relatively low levels (Lakaye *et al.* 2002). Furthermore, tissues with high 25 kDa hThTPase activity have low cytosolic ThTP levels, while in tissues devoid of this activity, ThTP levels can be relatively high, suggesting that this enzyme is involved in the control of cytosolic ThTP levels in animal cells.

The catalytic domains of human 25 kDa hThTPase and CyaB-like adenylyl cyclase from *Aeromonas hydrophilia* define a novel superfamily of domains that should bind organic phosphates (Iyer and Aravind 2002). This superfamily of proteins was therefore called 'CYTH' (CYaB-THiamin triphosphatase), and the presence of orthologs was demonstrated in all three superkingdoms of life. This suggested that CYTH is an ancient enzymatic domain and that a representative must have been present in the last universal common ancestor (LUCA) of all extant life forms. It appears that the 'CYTH' superfamily includes enzymes with various catalytic properties (adenylyl cyclase or inorganic triphosphatase in some bacteria, RNA triphosphatase in yeast, ThTPase in animals) but with important common features:

1) The activity always requires divalent metal cations (Mg^{2+} or Mn^{2+}).
2) There is a specificity for substrates containing a triphosphate group and the active site is located in a β-barrel-like structure.
3) Ca^{2+} is inhibitory.
4) The pH optimum is always in the alkaline range.

ThTPase may represent a relatively divergent acquisition of a new catalytic activity, which, so far, appears to be restricted to mammalian tissues.

The solution structure of recombinant mouse 25 kDa ThTPase has been determined (Song *et al.* 2008) and the residues responsible for binding Mg^{2+} and ThTP were determined from nuclear magnetic resonance (NMR) titration experiments. While the free enzyme has an open cleft form, the enzyme-ThTP complex tends to adopt a closed tunnel-fold, resembling the β-barrel structure of yeast RNA triphosphatase.

5.7.8 Thiamin-binding Proteins

Thiamin-binding proteins have been described in many animal and plant tissues, but none has been characterized at the molecular level. They exist in eggs or plant seeds, probably as thiamin storage proteins. Thiamin phosphate binding proteins were reported in rat brain synaptosomes, but their role remains enigmatic.

5.8 Concluding Remarks

It was long considered that the only reason for the existence of thiamin was to form the cofactor ThDP, an argument also based on the observation that thiazole, a heterocycle absolutely required for catalysis, is rarely observed in biological molecules.

Thiamin derivatives, common to all organisms, are highly diverse, and one can only wonder what the significance of this diversity may be. Strictly speaking, the thiazole ring, whose presence could be considered as a way to keep the nucleophilic properties of cyanide without its toxicity, would be enough for catalysis. The pyrimidine moiety confers additional specificity in the interaction with apoenzymes and thiamin-metabolizing enzymes. One could argue that phosphorylation is important for the cellular sequestration of the cofactor. In that case, the monophosphate derivative (or any other type of acidic group forming an ester bond with the hydroxyl) would be sufficient to ensure the specific binding of the coenzyme.

Therefore, though it remains true that the main functions of this vitamin are linked to catalysis by ThDP, the existence of triphosphorylated derivatives, such as ThTP and AThTP whose levels are highly specifically regulated, strongly suggests that these compounds play an ancient biological role, at least in bacteria. It would be worth investigating their possible role(s) in archaea, protists and plants, and various animal phyla. Thiamin compounds form a family of molecules which, with respect to their degree of phosphorylation, is reminiscent of the world of nucleotides (though thiamin, lacking a sugar moiety, is not a nucleotide). This suggests that in addition to the classical cofactor role of ThDP, some thiamin derivatives have non-cofactor roles.

Summary Points

- Thiamin (vitamin B_1) is present in all living organisms and is required for all life forms.
- Thiamin is synthesized through complex pathways in bacteria, fungi and plants.
- In animals, insufficient thiamin uptake affects the nervous system and leads to severe syndromes.
- Thiamin is composed of a pyrimidine moiety linked to a thiazole moiety by a methylene bridge.
- Thiamin is unstable in alkaline solutions and is cleaved by sulfite.
- A hydroxyethyl function on the thiazole ring can be modified by phosphorylation.

- Thiamin is the precursor of several phosphorylated derivatives, the most important of which is thiamin diphosphate (ThDP).
- The chemical properties of ThDP as a cofactor are linked to the possibility of the thiazole ring to form a highly nucleophilic carbanion ylide analogous to cyanide.
- ThDP is a coenzyme of several key enzymes in cell energy metabolism as well as several other enzymes in catabolism.
- In animals, ThDP is formed from thiamin by a thiamin pyrophosphokinase and, in enterobacteria, from thiamin monophosphate by a thiamin phosphate kinase.
- Other thiamin phosphate derivatives are thiamin triphosphate (ThTP) and the recently discovered adenosine thiamin triphosphate (AThTP).
- ThTP is synthesized by a chemiosmotic mechanism from ThDP and inorganic phosphate and is hydrolysed by a specific soluble thiamin triphosphatase.
- Transporters play a major role in thiamin homeostasis and mutations in these transporters leads to severe phenotypes in humans and animals.

Key Facts about Thiamin-related Diseases in Humans

1. Thiamin-related diseases are the result of either insufficient thiamin intake (thiamin deficiency), poisoning by antithiamins, or of mutations in thiamin transporters or thiamin diphosphate-dependent enzymes.
2. The archetype of primary thiamin deficiency disease is beriberi, a polyneuritis, rapidly reversible upon thiamin administration.
3. Wernicke–Korsakoff syndrome results from alcohol-induced thiamin deficiency. It affects the central nervous system. The patients suffer from irreversible memory loss as a result of severe mesencephalic lesions.
4. Mutations in thiamin transporters lead to various disorders ranging from megaloblastic thiamin-responsive anaemia to severe, sometimes lethal, encephalopathies.
5. Mutations in thiamin diphosphate dependent enzymes, in particular pyruvate dehydrogenase, may lead to Leigh's disease or subacute necrotizing encephalomyelopathy. This severe, inherited syndrome is characterized by movement disorders and, at later stages, by lactic acidosis. The children affected have a life expectancy of less than three years.

Definitions of Words and Terms

ABC transporter. ABC transporters for 'ATP-binding cassette' transporters are an ancient family of proteins found in most organisms from bacteria to mammals. They contain a transmembrane segment and an intracellular ATP binding motive, and are involved in the transmembrane transport of various metabolites. The energy for translocation comes from ATP hydrolysis.

Coenzyme. A coenzyme is a small molecule, generally a derivative of a B-type vitamin, required for the catalytic activity of some enzymes. The coenzyme is generally tightly bound to the enzyme. The complex enzyme–coenzyme forms the holoenzyme, while the free enzyme is the apoenzyme. Sometimes, the term 'coenzyme' is used interchangeably with the term 'cofactor'. The latter includes in addition to coenzymes metal factors such as Mg^{2+} or Ca^{2+} that are often required for enzyme activity.

Encephalopathy. Encephalopathy refers to a disease resulting in brain dysfunction. It can have various etiologies. It can be relatively reversible (Wernicke's encephalopathy) or irreversible (spongiform encephalopathies caused by prions).

Mitochondria. Mitochondria are cellular organelles present in nearly all eukaryotic cells. Most of the cellular ATP is produced in mitochondria by a chemiosmotic mechanism and therefore they are sometimes referred to as the power plants of the cell. They are surrounded by a double membrane, the inner membrane; this is strongly invaginated and contains the machinery for ATP synthesis.

2-Oxo acid. The most important oxo acids in cell energy metabolism are pyruvate and 2-oxoglutarate. They are sometimes called α-keto acids, and contain a ketone function adjacent to a terminal carboxyl group.

Synaptosomes. Synaptosomes are formed after homogenization of brain tissue and can be isolated by centrifugation. They are composed of the terminal parts of axons containing the synapses and form closed compartments containing either the presynaptic machinery, the postsynaptic machinery or both.

List of Abbreviations

AThDP	adenosine thiamin diphosphate
AThTP	adenosine thiamin triphosphate
hThTPase	human thiamin triphosphatase
OGDH	oxoglutarate dehydrogenase complex
PDH	pyruvate dehydrogenase complex
ThDP	thiamin diphosphate
ThMP	thiamin monophosphate
ThTP	thiamin triphosphate
ThTPase	thiamin triphosphatase
THTR	thiamin transporter
TMRA	thiamin-responsive megaloblastic anaemia
TPK	thiamin pyrophosphokinase

References

Ajjawi, I., Rodriguez Milla, M.A., Cushman, J., and Shintani, D.K., 2007. Thiamin pyrophosphokinase is required for thiamin cofactor activation in Arabidopsis. *Plant Molecular Biology*. 65: 151–162.

Baird, N.J., Kulshina, N., and Ferre-D'Amare, A.R., 2010. Riboswitch function: flipping the switch or tuning the dimmer? *RNA Biology*. 7: 328–332.

Barile, M., Valenti, D., Brizio, C., Quagliariello, E., and Passarella, S., 1998. Rat liver mitochondria can hydrolyse thiamine pyrophosphate to thiamine monophosphate which can cross the mitochondrial membrane in a carrier-mediated process. *FEBS Letters*. 435: 6–10.

Bettendorff, L., 1994. The compartmentation of phosphorylated thiamine derivatives in cultured neuroblastoma cells. *Biochimica Biophysica Acta*. 1222: 7–14.

Bettendorff, L., and Wins, P., 1994. Mechanism of thiamine transport in neuroblastoma cells. Inhibition of a high affinity carrier by sodium channel activators and dependence of thiamine uptake on membrane potential and intracellular ATP. *Journal of Biological Chemistry*. 269: 14379–14385.

Bettendorff, L., Wirtzfeld, B., Makarchikov, A.F., Mazzucchelli, G., Frédérich, M., Gigliobianco, T., Gangolf, M., De Pauw, E., Angenot, L., and Wins, P., 2007. Discovery of a natural thiamine adenine nucleotide. *Nature Chemical Biology*. 3: 211–212.

Breslow, R., 1958. On the mechanism of thiamine action. IV.1 Evidence from studies on model systems. *Journal of the American Chemical Society*. 80: 3719–3726.

Csillik, B., Mihaly, A., Krisztin-Peva, B., Farkas, I., and Knyihar-Csillik, E. 2008. Mitigation of nociception *via* transganglionic degenerative atrophy: possible mechanism of vinpocetine-induced blockade of retrograde axoplasmic transport. *Annals of Anatomy*. 190: 140–145.

Debs, R., Depienne, C., Rastetter, A., Bellanger, A., Degos, B., Galanaud, D., Keren, B., Lyon-Caen, O., Brice, A., and Sedel, F., 2010. Biotin-responsive basal ganglia disease in ethnic Europeans with novel SLC19A3 mutations. *Archives of Neurology*. 67: 126–130.

Fraccascia, P., Sniekers, M., Casteels, M., and Van Veldhoven, P.P., 2007. Presence of thiamine pyrophosphate in mammalian peroxisomes. *BMC Biochemistry*. 8: 10.

Frédérich, M., Delvaux, D., Gigliobianco, T., Gangolf, M., Dive, G., Mazzucchelli, G., Elias, B., De Pauw, E., Angenot, L., Wins, P., and Bettendorff, L., 2009. Thiaminylated adenine nucleotides. Chemical synthesis, structural characterization and natural occurrence. *FEBS Journal*. 276: 3256–3268.

Gaitonde, M.K., and Evans, G.M., 1983. Metabolism of thiamin in rat brain. *Biochemical Society Transactions*. 11: 695–696.

Ganapathy, V., Smith, S.B., and Prasad, P.D., 2004. SLC19: the folate/thiamine transporter family. *Pflügers Archives*. 447: 641–646.

Gangolf, M., Czerniecki, J., Radermecker, M., Detry, O., Nisolle, M., Jouan, C., Martin, D., Chantraine, F., Lakaye, B., Wins, P., Grisar, T., and Bettendorff, L., 2010a. Thiamine status in humans and content of phosphorylated thiamine derivatives in biopsies and cultured cells. *PLoS One*. 5: e13616.

Gangolf, M., Wins, P., Thiry, M., El Moualij, B., and Bettendorff, L., 2010b. Thiamine triphosphate synthesis in rat brain occurs in mitochondria and

is coupled to the respiratory chain. *Journal of Biological Chemistry.* 285: 583–594.
Gigliobianco, T., Lakaye, B., Wins, P., El Moualij, B., Zorzi, W., and Bettendorff, L., 2010. Adenosine thiamine triphosphate accumulates in Escherichia coli cells in response to specific conditions of metabolic stress. *BMC Microbiology.* 10: 148.
Hammes, H.P., Du, X., Edelstein, D., Taguchi, T., Matsumura, T., Ju, Q., Lin, J., Bierhaus, A., Nawroth, P., Hannak, D., Neumaier, M., Bergfeld, R., Giardino, I., and Brownlee, M., 2003. Benfotiamine blocks three major pathways of hyperglycemic damage and prevents experimental diabetic retinopathy. *Nature Medicine.* 9: 294–299.
Iyer, L.M., and Aravind, L., 2002. The catalytic domains of thiamine triphosphatase and CyaB-like adenylyl cyclase define a novel superfamily of domains that bind organic phosphates. *BMC Genomics.* 3: 33.
Jenkins, A.H., Schyns, G., Potot, S., Sun, G., and Begley, T.P., 2007. A new thiamin salvage pathway. *Nature Chemical Biology.* 3: 492–497.
Jurgenson, C.T., Begley, T.P., and Ealick, S.E., 2009. The Structural and Biochemical Foundations of Thiamin Biosynthesis. *Annual Review of Biochemistry.* 78: 569–603.
Kluger, R., and Tittmann, K., 2008. Thiamin diphosphate catalysis: enzymic and nonenzymic covalent intermediates. *Chemical Reviews.* 108: 1797–1833.
Lakaye, B., Makarchikov, A.F., Antunes, A.F., Zorzi, W., Coumans, B., De Pauw, E., Wins, P., Grisar, T., and Bettendorff, L., 2002. Molecular characterization of a specific thiamine triphosphatase widely expressed in mammalian tissues. *Journal of Biological Chemistry.* 277: 13771–13777.
Leichter, J., and Joslyn, M.A., 1969. Kinetics of thiamin cleavage by sulphite. *Biochemical Journal.* 113: 611–615.
Lindhurst, M.J., Fiermonte, G., Song, S., Struys, E., De Leonardis, F., Schwartzberg, P.L., Chen, A., Castegna, A., Verhoeven, N., Mathews, C.K., Palmieri, F., and Biesecker, L.G., 2006. Knockout of Slc25a19 causes mitochondrial thiamine pyrophosphate depletion, embryonic lethality, CNS malformations, and anemia. *Proceedings of the National Academy of Sciences of the United States of America.* 103: 15927–15932.
Makarchikov, A.F., Lakaye, B., Gulyai, I.E., Czerniecki, J., Coumans, B., Wins, P., Grisar, T., and Bettendorff, L., 2003. Thiamine triphosphate and thiamine triphosphatase activities: from bacteria to mammals. *Cellular and Molecular Life Sciences.* 60: 1477–1488.
Makarchikov, A.F., Brans, A., and Bettendorff, L., 2007. Thiamine diphosphate adenylyl transferase from *E. coli*: functional characterization of the enzyme synthesizing adenosine thiamine triphosphate. *BMC Biochemistry.* 8: 17.
Neal, R.A., and Pearson, W.N., 1964. Studies of thiamine metabolism in the Rat. I. Metabolic products found in urine. *Journal of Nutrition.* 83: 343–350.
Nghiêm, H.O., Bettendorff, L., and Changeux, J.P., 2000. Specific phosphorylation of *Torpedo* 43K rapsyn by endogenous kinase(s) with thiamine triphosphate as the phosphate donor. *FASEB Journal.* 14: 543–554.

Pan, X., Gong, N., Zhao, J., Yu, Z., Gu, F., Chen, J., Sun, X., Zhao, L., Yu, M., Xu, Z., Dong, W., Qin, Y., Fei, G., Zhong, C., and Xu, T.L., 2010. Powerful beneficial effects of benfotiamine on cognitive impairment and beta-amyloid deposition in amyloid precursor protein/presenilin-1 transgenic mice. *Brain.* 133: 1342–1351.

Pearson, W.N., Hung, E., Darby, W.J.J., Balaghi, M., Neal, and R.A., 1966. Excretion of metabolites of 14C-pyrimidine-labeled thimine by the rat at different levels of thiamine intake. *Journal of Nutrition.* 89: 133–142.

Rapala-Kozik, M., Golda, A., and Kujda, M., 2009. Enzymes that control the thiamine diphosphate pool in plant tissues. Properties of thiamine pyrophosphokinase and thiamine-(di)phosphate phosphatase purified from *Zea mays* seedlings. *Plant Physiology and Biochemistry.* 47: 237–242.

Roth, J., and Berger, E.G., 1982. Immunocytochemical localization of galactosyltransferase in HeLa cells: codistribution with thiamine pyrophosphatase in trans-Golgi cisternae. *Journal of Cell Biology.* 93: 223–229.

Sano, S., Matsuda, Y., and Nakagawa, H., 1988. Type B nucleoside-diphosphatase of rat brain. Purification and properties of an enzyme with high thiamin pyrophosphatase activity. *European Journal of Biochemistry.* 171: 231–236.

Song, J., Bettendorff, L., Tonelli, M., and Markley, J.L., 2008. Structural basis for the catalytic mechanism of mammalian 25-kDa thiamine triphosphatase. *Journal of Biological Chemistry.* 283: 10939–10948.

Soriano, E.V., Rajashankar, K.R., Hanes, J.W., Bale, S., Begley, T.P., and Ealick, S.E., 2008. Structural similarities between thiamin-binding protein and thiaminase-I suggest a common ancestor. *Biochemistry.* 47: 1346–1357.

Volvert, M.L., Seyen, S., Piette, M., Evrard, B., Gangolf, M., Plumier, J.C., and Bettendorff, L., 2008. Benfotiamine, a synthetic S-acyl thiamine derivative, has different mechanisms of action and a different pharmacological profile than lipid-soluble thiamine disulfide derivatives. *BMC Pharmacology.* 8: 10.

Walski, M., and Gajkowska, B., 1998. Ultrastructural localization of thiamine pyrophosphatase in plasma membranes of brain phagocytes long time after cardiac arrest. *Journal für Hirnforschung.* 39: 183–191.

Zeng, W.Q., Al-Yamani, E., Acierno, J.S., Jr., Slaugenhaupt, S., Gillis, T., MacDonald, M.E., Ozand, P.T., and Gusella, J.F., 2005. Biotin-responsive basal ganglia disease maps to 2q36.3 and is due to mutations in SLC19A3. *American Journal of Human Genetics.* 77: 16–26.

CHAPTER 6

Chemistry and Biochemistry of Riboflavin and Related Compounds

MARIANA C. MONTEIRO[a] AND DANIEL PERRONE*[b]

[a] Laboratório de Química e Bioatividade de Alimentos, Instituto de Nutrição, Universidade Federal do Rio de Janeiro, Av. Carlos Chagas Filho, 373, CCS, J2-16, Cidade Universitária, Ilha do Fundão, Rio de Janeiro, RJ, 21941-590, Brazil; [b] Laboratório de Bioquímica Nutricional e de Alimentos, Departamento de Bioquímica Instituto de Química, Universidade Federal do Rio de Janeiro, Av. Athos da Silveira Ramos, 149, CT, Bl. A, 528A, Cidade Universitária, Ilha do Fundão, Rio de Janeiro, RJ, 21941-909, Brazil
*Email: danielperrone@iq.ufrj.br

6.1 Chemistry of Riboflavin

6.1.1 Structure and General Properties

Formerly known as vitamin B_2, riboflavin (Figure 6.1) is an organic compound present in most living systems. The term 'riboflavin' derives from the sugar alcohol present in its structure (ribitol) and from its yellow colour (*flavus* is the Latin word for yellow). Riboflavin was first isolated in 1879 and its chemical structure was determined in 1933. In the past, this vitamin was also designated as lactoflavin, ovoflavin and uroflavin, depending on its isolation source. Riboflavin was also referred to as vitamin G in the early part of the 20th century because it was recognized as a dietary factor needed for growth.

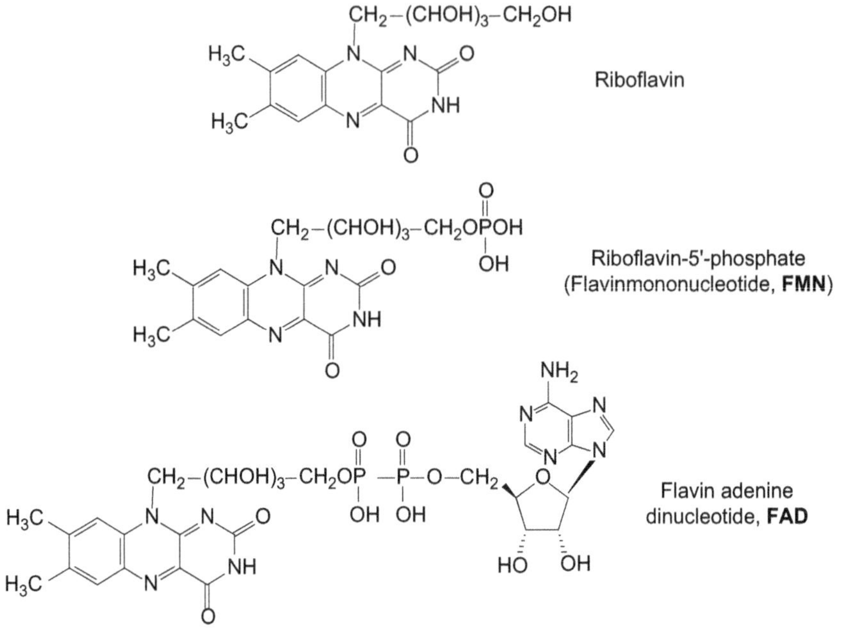

Figure 6.1 Structures of riboflavin, flavin mononucleotide and flavin dinucleotide.

Chemically, riboflavin may be described as a tricyclic, nitrogen-containing isoalloxazine ring-system with a ribitol side chain. Its systematic name is 7,8-dimethyl-10-(1-D-ribityl)-benzo[g]pteridin-2,4-dion, with an empirical formula of $C_{17}H_{20}N_4O_6$ and molecular weight of 376.36 g/mol.

Riboflavin is the biologically active component of the prosthetic group of many enzymes and proteins collectively known as flavoproteins. In most biological materials, including food, flavins are broadly distributed in tissues, but little is present as free riboflavin. They are predominantly found in the forms of flavocoenzymes, mainly as flavin adenine dinucleotide (FAD) and in a lesser extend as flavin mononucleotide (FMN) (Figure 6.1), the common name of riboflavin-5′-phosphate, although riboflavin is not truly a nucleoside, since the linkage between the ribityl chain and the N_{10} of the isoalloxazine ring is not glycosidic.

Flavins are widely recognized by their ability of participate in both one- and two-electron transfer processes, since these compounds can exist in three different redox states: oxidized (quinone), one-electron reduced (semiquinone) and two-electron reduced (hydroquinone). The redox potential for the complete reduction of oxidized flavins is about −200 mV, but this value may largely vary in flavoproteins, as a consequence of the protein activity site environment, ranging from −400 mV to +60 mV (Fraaije and Mattei 2000). Flavins may transfer single electrons, hydrogen atoms and hydride ions. In addition, N_5 and $C_{4\alpha}$ of the oxidized flavin molecule are susceptible sites for

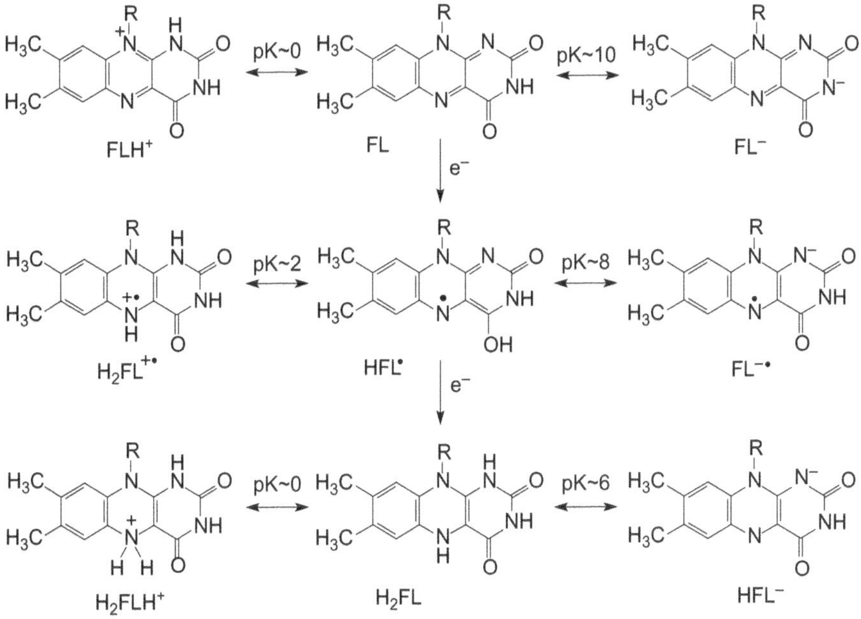

Figure 6.2 Redox and acid–base equilibria of flavins (as proposed by Heelis 1982).

nucleophilic attack. Therefore, flavoenzymes are very versatile catalysts in terms of substrate and type of reactions, which partially explains their ubiquity in living systems.

In an aqueous solution, free flavin species exist in a pH-dependent equilibrium, shown in Figure 6.2, as proposed by Heelis (1982). Among the nine forms presented (derived from different redox and protonation states), at least six are physiologically relevant according to their pK_a values. At neutral pH, only approximately 5% of flavins are found in an equimolar mixture of completely oxidized and reduced flavins, with most being found as the semiquinone in neutral or anionic forms ($pK_a \sim 8.5$). Upon binding to a protein, such equilibrium changes drastically. In some enzymes, flavins may be found almost completely in the semiquinone forms, while in other enzymes these forms are not observed. Moreover, the pK_a of flavoenzymes may be shifted up or down if the protein can stabilize the neutral radical species or the semiquinone anion, respectively (Miura 2001).

6.1.2 Modes of Degradation and Stability in Foods

Riboflavin is one of the most stable vitamins, showing stability towards acidic pH and relatively small losses at neutral pH. Although riboflavin is rapidly degraded in alkaline pH, this condition is uncommon in food products and therefore of less concern. The typical mechanism for riboflavin degradation is

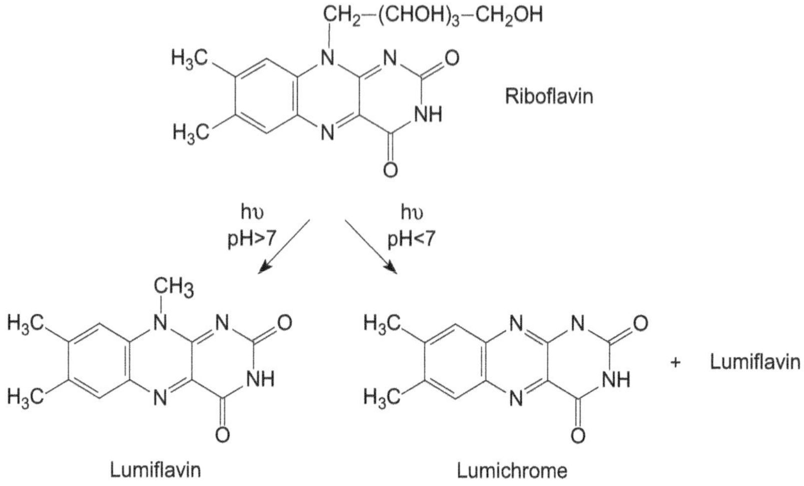

Figure 6.3 Photodegradation of riboflavin to lumiflavin and lumichrome.

photochemical, yielding two biologically inactive compounds lumichrome (acidic pH) and lumiflavin (neutral and alkaline pH) (Figure 6.3), in addition to a variety of free radicals. This degradation route is of less importance in opaque food such as meat, but may be crucial for milk. Studies showed that 30–50% of the riboflavin content of milk may be lost as a consequence of direct exposure to bright sunlight, while boiling for 30 min led only to a 12% decrease (Wishner 1964). In pasta, light-induced degradation of riboflavin is also observed and more than 50% of the riboflavin content may be lost within one day (Woodcock et al. 1984).

Riboflavin retention during various food processing and storage conditions has been extensively studied. Riboflavin in dehydrated foods, such as milk powder and dried fruits and vegetables, was stable for a storage period of one year at 54 °C, when metal containers with air, N_2 or CO_2, and paper cartons where employed (Heberlein and Clifcorn 1944). In contrast, carotene, ascorbic acid, folic acid, thiamin, vitamin B_6 and pantothenic acid were unstable under the same storage conditions. Irradiation of foods with γ-rays did not significantly affect riboflavin contents (Lee and Hau 1996). Apparently, proteins to which riboflavin is commonly bound in food protect the prosthetic group from the direct or indirect attack by the high energy γ-irradiation. Pulsed electric fields, a non-thermal preservation technology increasingly employed to maintain the quality of certain fresh foods, was shown not to affect riboflavin retention (Rivas et al. 2007).

On the other hand, riboflavin retention is affected by oxygen concentration, other components such as metal sulfates and amino acid chelates, water activity, and most of all, light exposure. Riboflavin photodegradation in liquid systems such as milk follows first order degradation mechanisms and is dependent on light intensity, exposure time and wavelength (Choe et al. 2005).

Sunlight is more detrimental to riboflavin retention than light from fluorescent lamps and at 450 nm, the absorption maximum of riboflavin, destruction is maximized. Agitation and sample location also have a significant impact on riboflavin loss in milk and cheese exposed to light during storage.

Riboflavin is involved in the photosensitized degradation of many food components, as triplet riboflavin has a higher reduction potential than amino acids, proteins, lipids and many vitamins (Silva *et al.* 1998; Lu *et al.* 2000; Choe *et al.* 2005). Aliphatic amino acids are degraded to aldehydes by riboflavin-photosensitized oxidation, in a reaction similar to the Strecker degradation. Tyrosine and tryptophan are also subjected to riboflavin-photosensitized oxidation, producing mainly bityrosine and a mixture of indole-, flavin- and indole-flavin-type aggregates as photoproducts. Aspartame, the methyl ester of the aspartic acid/phenylalanine dipeptide used as an artificial sweetener, is also destroyed by the combination of riboflavin and light. In relation to proteins, riboflavin photosensitization promotes cross-linking reactions and consequent aggregation of collagen. Moreover, photosensitized modification of proteins by riboflavin inactivates some enzymes, such as catalase and lysozyme.

Unsaturated bonds in fatty acids can undergo riboflavin-photosensitized cis-trans isomerization. Lipid oxidation is also an important reaction which is catalysed by riboflavin and light and, together with the deamination of methionine to form methanal and dimethyl sulfide, it plays a major role in the development of light-induced off-flavour in milk. The demise of the home delivery of milk in glass bottles and the employment of alternative package materials, such as paper carton, reduced the incidence of sunlight-flavoured milk.

Destruction of vitamins A, C, D and E induced by riboflavin-photosensitized oxidation has been reported. Vitamin A and its esters, along with carotenoids with pro-vitamin A activity, undergo ring opening upon sunlight exposure in the presence of riboflavin. Although an excellent antioxidant, ascorbic acid is rapidly photooxidized in the presence of riboflavin. Even a small decrease in riboflavin content in milk due to photosensitization can lead to virtual complete destruction of ascorbic acid. Fortunately, milk is not an important source of vitamin C in most diets.

6.2 Biochemistry of Riboflavin

Riboflavin is a widely distributed vitamin, being found at least in small amounts in almost all animal and plant derived foods. Its occurrence is very similar to that of thiamin and, as a rule, protein-rich foods of animal origin are considerable sources with good bioavailability. The content of riboflavin in various foods is given in Table 6.1.

6.2.1 Digestion, Bioavailability, Absorption and Transport

As mentioned before, only a small amount of riboflavin is present in foods as the free form or as riboflavin phosphate. In the diet, riboflavin occurs

Table 6.1 Riboflavin content of foods[a] and their contribution to RDA.

Food	Content (mg per 100g)	Serving size (g)	% of RDA (per serving)[b]
Herbs and spices			
Paprika, powder	1.74	4.0	5.4
Coriander, fresh	0.80	8.0	4.9
Parsley, fresh	0.80	8.0	4.9
Basil, fresh	0.21	8.0	1.3
Meat and meat products			
Beef liver, grilled	2.69	100.0	206.9
Chicken liver, grilled	0.56	100.0	43.1
Beef (lean), cooked	0.15–0.32	100.0	11.5–24.6
Chicken eggs, fried	0.32	50.0	12.3
Chicken eggs, boiled	0.30	45.0	10.4
Salami	0.23	30.0	5.3
Milk and dairy products			
Low fat milk, fluid	0.26	200.0	40.0
Whole milk, fluid	0.24	200.0	36.9
Yogurt, plain	0.22	170.0	28.8
Whole milk, powder	1.03	30.0	23.8
Camembert	0.70	40.0	21.5
Low fat milk, powder	1.20	20.0	18.5
Roquefort	0.57	40.0	17.5
Brie	0.52	40.0	16.0
Parmesan	0.44	30.0	10.2
Condensed milk	0.33	20.0	5.1
Cottage cheese	0.15	30.0	3.5
Fish			
Salmon, cooked	0.75	120.0	69.2
Trout, cooked	0.42	120.0	38.8
Cod, salted	0.21	135.0	21.8
Crab, cooked	0.31	50.0	11.9
Cereals			
Breakfast cereal, corn	1.02	50.0	39.2
French bread	0.67	50.0	25.8
Wheat bran	0.58	20.0	8.9
Vegetables and legumes			
Tomato, sun-dried	0.49	40.0	15.1
Cabbage, boiled	0.31	40.0	9.5
Spinach, boiled	0.21	50.0	8.1
Broccoli, boiled	0.18	30.0	4.2
Fruits and nuts			
Pequi (*Caryocar brasiliense*)	0.48	50.0	18.4
Avocado	0.15	65.0	7.5
Plum	0.22	40.0	6.8
Almond	0.16	10.0	1.2
Other			
Yeast extract spread (Marmite)	7.00–14.3	8.0	43.1–88.0
Black tea, infusion	0.48	200.0	73.8
Milk chocolate	0.22	30.0	5.1
Baker's yeast	0.36	7.0	1.9

[a]According to Souci *et al.* (2000) and TACO (2011).
[b]Considering the RDA (DRI 2010) for adult males (1.3 mg/day).

Figure 6.4 Intestinal transformation of flavin adenine dinucleotide (FAD) and flavin mononucleotide (FMN) to riboflavin.

predominantly in the form of coenzymes, mainly FAD and, at lesser extent, FMN which is bound non-covalently to proteins. In contact with gastric acids these forms are released from proteins, being this the first step of riboflavin absorption. The second step in the absorption of dietary riboflavin is the hydrolysis of FAD and FMN to riboflavin. This hydrolysis step occurs inside the intestinal lumen by the concerted action of FAD pyrophosphatase, which converts FAD to FMN, and FMN phosphatase, which converts FMN to riboflavin (Figure 6.4). Additionally, it is thought that intestinal phosphatases (nucleotide diphosphatase and alkaline phosphatase) also hydrolyse the riboflavin phosphate group. Riboflavin is also formed by the normal intestinal microflora, being absorbed in this region. This formation route depends on the type of the diet, being favoured by the ingestion of vegetable in comparison to meat consumption. It is worth noting that not all bound riboflavin is hydrolysed and available for absorption. Since a small amount (less than 10%) of FAD is covalently bound to histidine or cysteine, it is not released during digestion and therefore cannot function in the body. These riboflavin-amino acid complexes are excreted unchanged in the urine.

Some divalent metals (*e.g.* copper, zinc, iron and manganese), caffeine, theophylline, acid ascorbic, niacin and tryptophan are able to chelate riboflavin and FMN thereby inhibiting its absorption. Alcohol consumption also impairs riboflavin digestion and absorption since ethanol inhibits FAD pyrophosphatase and FMN phosphatase (Figure 6.4). The presence of food in the gut increases the absorption of riboflavin by decreasing gastric emptying and consequently increasing the contact of the vitamin with the absorptive surface. Bile salts also increase riboflavin absorption.

Recently, studies have showed that the mechanism of riboflavin absorption in the small and large intestines involves an efficient and specific Na^+-independent carrier (Said 2004; Said and Seetharam 2006). The absorption rate is proportional to dose (saturation level is achieved with about 25–30 mg) and riboflavin half-life is 1.1 h. Three human riboflavin transporters (hRFT), hRFT-1, hRFT-2 and hRFT-3, have been described, all of them being expressed in the human intestine. Very recently, Subramanian *et al.* (2011) suggested a predominant role for the hRFT-2 in riboflavin intestinal absorption.

In humans, riboflavin, FAD and FMN are transported in blood plasma mainly through weak associations with albumin and some immunoglobulins. Other riboflavin-binding proteins are specific to pregnancy, including the classic case of the estrogen-induced egg-white protein. These proteins have at least some portion of the binding domain in common and are essential to normal foetal development. In humans, placental transfer of riboflavin involves binding proteins that help transport the vitamin and therefore enhances the supply to the foetus. Only a modest amount of the vitamin circulates through the enterohepatic circulation.

6.2.2 Metabolism, Storage and Excretion

Within the cellular cytoplasm of intestinal cells, riboflavin is phosphorylated to FMN in a reaction catalysed by flavokinase, which utilizes Zn^{2+} and require ATP (Figure 6.5).

At the serosal surface, most of the FMN is dephosphorylated to riboflavin, probably by a nonspecific alkaline phosphatase, then entering into portal blood to be transported to the liver. Riboflavin is carried to the liver where it is again converted by flavokinase to FMN which will be converted to FAD by FAD synthetase utilizing Mg^{2+} (Figure 6.5). FAD is the predominant flavocoenzyme present in tissues, where it is mainly found complexed with numerous flavoprotein dehydrogenases and oxidases. The greatest concentrations of riboflavin are found in the liver, kidney and heart. Unbound flavins are relatively labile, being rapidly hydrolysed to free riboflavin, which then diffuses from cells and is excreted.

The biosynthesis of flavocoenzymes is regulated by the supply of riboflavin, competition for ATP and hormonal balances. Increasing riboflavin concentrations in the intestinal lumen decreases the rate of vitamin absorption, while riboflavin deficiency leads to increased absorption efficiency. The hormones thyroxine and triiodothyronine stimulate FMN and FAD synthesis in mammals by increasing the activity of flavokinase and for this reason hypothyroidism leads to reduced tissue levels of flavins. In the kidney, aldosterone promotes an increase in the activity of flavokinase and therefore an increase of flavins.

Being a water-soluble vitamin, when riboflavin intake is higher than tissue requirements, the excess is excreted in the urine, mainly as free riboflavin (60–90%) or other metabolites, such as 7-hydroxymethylriboflavin (3–7%), 8α-sulfonylriboflavin (2–15%), 10-hydroxyethylflavin (1–7%), 8-hydroxymethylriboflavin (1–8%), riboflavinyl peptide ester (up to 5%), with traces of lumiflavin and, sometimes, the 10-formylmethyl- and carboxylmethylflavins (Figure 6.6).

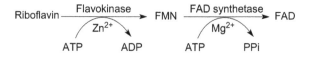

Figure 6.5 Interconversion of riboflavin to flavin mononucleotide (FMN) and flavin adenine dinucleotide (FAD).

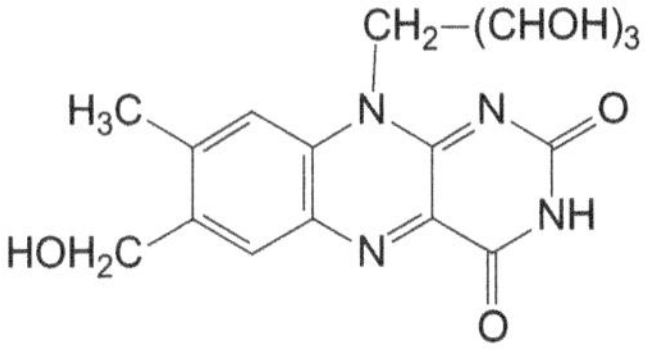

7-Hydroxymethylriboflavin

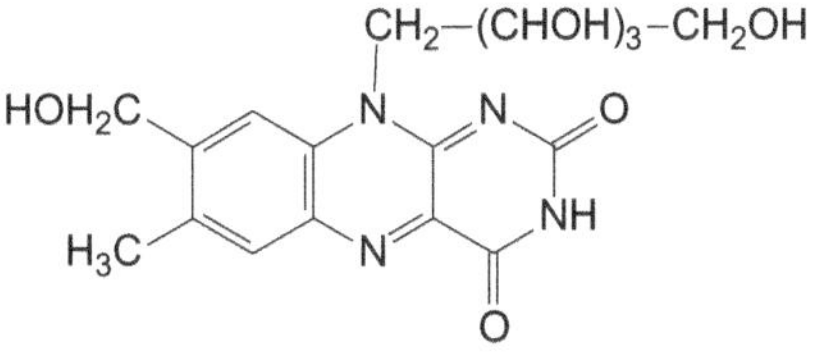

8-Hydroxymethylriboflavin

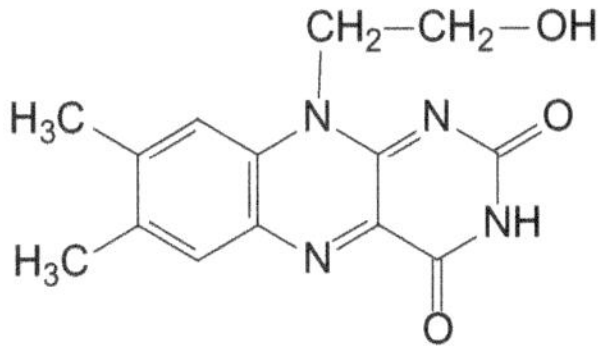

10-Hydroxyethylflavin

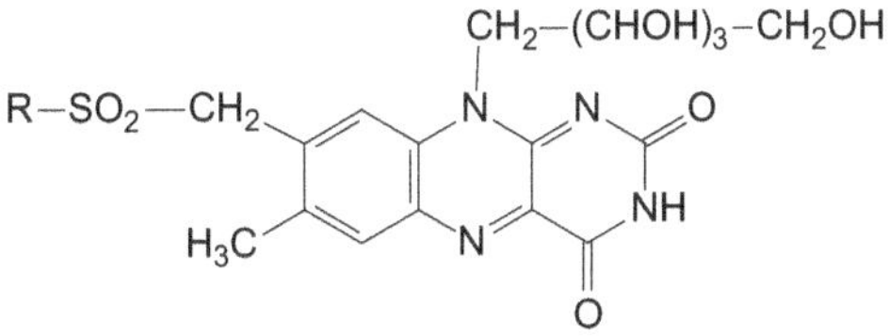

8α-Sulfonylriboflavin

Figure 6.6 Urinary metabolites of riboflavin.

6.2.3 Biological Functions

Riboflavin in its coenzyme forms (FMN and FAD) plays key metabolic roles in biological oxidation–reduction reactions involving carbohydrates, amino acids and lipids, and in energy production *via* the respiratory chain. These coenzymes also act in cellular metabolism of other water-soluble vitamins through the production and activation of folate and pyridoxine (vitamin B_6) to their respective coenzyme forms and in the synthesis of niacin (vitamin B_3) from tryptophan. In addition, some neurotransmitters and other amines require FAD for their metabolism. Recently, Chocano-Bedoya *et al.* (2011) suggested a possible benefit of high intakes of riboflavin (about 2.5 mg/day) from food sources on the reduction of incidence of premenstrual syndrome.

6.2.4 Requirements and Intakes

The Recommended Dietary Allowance (RDA) of riboflavin ranges from 0.5 mg/day for children to 1.1 mg/day and 1.3 mg/day for adult females and males, respectively. During pregnancy and lactation, an additional of 0.3 mg/day and 0.5 mg/day, respectively, is recommended. These values are similar to previous recommendations made by the World Health Organization (WHO) in 1974 and the European Population Reference Intake in 1993. For infants of 0–12 months the Adequate Intake (AI) is adopted. The detailed values of RDA and AI according to the life stage group are given in Table 6.2.

The inadequate intake of riboflavin is usually associated with populations whose diet is lacking dairy products and meat. In fact, Hughes and Sanders (1979) reported that vegetarians have a lower intake of riboflavin compared

Table 6.2 Recommended Dietary Allowances and Adequate Intakes of riboflavin.[a]

Life stage group	RDA/AI (mg/day)
Infants	
0–6 months	0.3
7–12 months	0.4
Children	
1–3 years	0.5
4–8 years	0.6
Males	
9–13 years	0.9
14 years and older	1.3
Females	
9–13 years	0.9
14–18 years	1.0
19 years and older	1.1
Pregnancy (all ages)	1.4
Lactation (all ages)	1.6

[a]Adapted from DRI (2010). The Recommended Dietary Allowances (RDAs) apply to children and adults, and the Adequate Intakes (AIs) to infants.

with non-vegetarians. Despite this, using the data from the US National Health and Nutrition Examination Survey (NHANES) in the period 1999–2004, Farmer *et al.* (2011) recently reported that the mean intake of riboflavin in US adults was higher for vegetarians when compared with non-vegetarians.

The Health Survey for São Paulo, accomplished with adolescents, showed that 8% and 12% of Brazilian male and female adolescents, respectively, have inadequate usual intake of riboflavin. Those adolescents who came from lower-income families and were overweight had a higher prevalence of inadequate intake (Junior *et al.* 2011). In contrast, data obtained from the Korean National Health and Nutrition Examination Survey (KNHANES) in 2008 showed that the intake of riboflavin is inadequate for all Korean adolescents (Cho *et al.* 2011).

When evaluating elderly (aged 65 years and older) Lebanese men and women, Sibai *et al.* (2003) observed that riboflavin intake is above that recommended by the RDA. The same results were reported in Irish elderly (Madigan *et al.* 1998). However, Bailey *et al.* (1997) observed that 25% of men and 11% of women in the elderly of Norwich (UK) had inadequate intake of riboflavin.

Using data from NHANES (1999–2004), Buchols *et al.* (2011) found that overall, 3.7% of children aged 3–5 years have an intake of riboflavin below the RDA. By dividing the group, it was observed that 8.2% of children from low-income families had inadequate intakes of riboflavin compared with only 2.7% of children from those of non-low-income. For children of 7–9 years of age, the estimated prevalence of inadequate intake of riboflavin widely varied according to the country, with 1.6% in Kenya, 16.3% in Egypt and 83.4% in Mexico (Murphy *et al.* 1995).

The inadequate intake of riboflavin seems to be the main cause for the deficiency of this vitamin, being common in populations whose diet lack dairy products and meat, and in anorexic individuals. Digestion and intestinal absorption disorders are other causes of disability, as observed in individuals with lactose intolerance, tropical sprue, coeliac disease and intestinal resection, as well as gastrointestinal and biliary obstruction. Other disorders such as diarrhoea, infectious enteritis and irritable bowel syndrome can cause poor absorption by increasing intestinal motility. Riboflavin deficiency also occurs in conditions such as chronic alcoholism, diabetes mellitus and inflammatory bowel diseases.

Deficiency of riboflavin leads to a variety of clinical abnormalities that include neurological disorders, anaemia, growth retardation and skin abnormalities. Moreover, inadequate intakes of riboflavin lead to disturbances in intermediary metabolism. Severe riboflavin deficiency can also affect the conversion of vitamin B_6 to its coenzyme and even decrease conversion of tryptophan to niacin. Recently, Nakano *et al.* (2011) demonstrated that riboflavin deficiency in humans results in altered cell turnover in the duodenal crypt.

Due to its limited absorption in the gastrointestinal tract, riboflavin shows no toxicity by oral ingestion. Parenteral administration of extremely high doses (400 mg/kg body weight) may lead to crystallization of riboflavin in the kidney. The tolerable upper intake levels for riboflavin were not defined due to lack of data on their adverse effects.

Summary Points

- This chapter focuses on riboflavin chemistry and biochemistry.
- Riboflavin is a fairly stable vitamin, mainly degraded by photochemical mechanisms.
- Riboflavin is widely distributed in foods, with dairy and meat as the main contributors for its dietary intake.
- Fluorimetry is the standard technique for riboflavin analysis in food.
- As a coenzyme, riboflavin plays key metabolic roles in many biological oxidation–reduction reactions.
- Absorption of riboflavin occurs in the intestine by a specific Na^+-independent carrier.
- Riboflavin is transported in plasma associated with albumin and immunoglobulins.
- Intestinal enzymes interconvert riboflavin to their biological active forms.
- Only populations whose diet is lacking dairy products and meat are at risk of riboflavin deficiency.

Key Facts about Digestion and Absorption

1. Digestion and absorption of nutrients occurs in the digestive tract, which consists of the mouth, the oesophagus, the stomach, the small and large intestines, and the rectum.

2. In the stomach, hydrochloric acid and enzymes, such as amylases and proteases, help the digestion of nutrients.
3. Digestion of nutrients in the small intestine is promoted by bile salts and the pancreatic juice, which contains enzymes.
4. Villi are the sites in which the absorption of nutrients occurs in the intestines.
5. The enterocytes are the cells responsible for the absorption of nutrients in the intestines. These cells are composed of an apical and a basolateral surface, which face the intestinal lumen and blood, respectively.
6. Absorption of nutrients in the small intestine may happen by passive transport (or diffusion) or active transport (involving a protein carrier).
7. Riboflavin is digested in the stomach and in the small intestine by the action of hydrochloric acid and specific enzymes, respectively. Absorption occurs in the small intestine by a Na^+-independent carrier.

Definitions of Words and Terms

Adequate Intake (AI). Estimated recommendation of a nutrient, based on survey of scientific data on healthy subjects, when the RDA cannot be determined.

Coeliac disease. Autoimmune disease that affects the small intestine in genetically predisposed individuals, caused by eating foods that contain gluten. The disease causes villous atrophy of the intestinal mucosa, causing reduced absorption of nutrients.

Enterohepatic circulation. Refers to the circulation of substances from the liver to the small intestine *via* bile salts and their return to the liver *via* the portal system.

Fluorescence. A physical phenomenon characterized by the emission of light by a substance stimulated by the absorption of incident radiation of shorter wavelength.

Food irradiation. Technology for controlling food spoilage and eliminating pathogens through exposure to ionizing radiation, usually γ-rays. The result is similar to conventional pasteurization and is often called 'cold pasteurization' or 'irradiation pasteurization'.

Hypothyroidism. Disorder defined by the deficiency in production of hormones thyroxine and tri-iodothyronine produced by thyroid gland.

Infectious enteritis. Inflammation of the small intestine caused by bacteria, viruses or other microbiological agents.

Irritable bowel syndrome. Functional disorder characterized most commonly by cramping, abdominal pain, bloating, constipation and diarrhoea. Its symptoms can be controlled with diet, stress management and medications.

Lactose intolerance. Disorder characterized by the incapacity of the organism in digesting and metabolizing lactose, a sugar present in milk. It is caused by a lack of lactase, required to break down lactose in the digestive tract.

Photosensitization. Process in which a reaction is initiated by transferring the energy of light absorbed by a substance to the reactants.

Prosthetic group. The very tightly or even covalently bound non-amino acid part of a conjugated protein, which usually plays an important role in the protein's biological function.

Recommended Dietary Allowance (RDA). Daily dietary intake level of a nutrient considered sufficient by the Food and Nutrition Board of the Institute of Medicine of the US National Academies to meet the requirements of nearly all (97–98%) healthy individuals in each life-stage and gender group.

Tropical sprue. An intestinal malabsorption syndrome of infectious origin commonly found in the tropical regions. This syndrome is characterized by acute or chronic diarrhoea and consequently weight loss and malabsorption of nutrients.

List of Abbreviations

AI	Adequate Intake
ATP	adenosine triphosphate
FAD	flavin adenine dinucleotide
FMN	flavin mononucleotide
hRFT	human riboflavin transporter
KNHANES	Korean National Health and Nutrition Examination Survey
NHANES	National Health and Nutrition Examination Survey
RDA	Recommended Dietary Allowances

References

Bailey, A.L., Maisey, S., Southon, S., Wright, A.J.A., Finglas, P.M., and Fulcher, R.A., 1997. Relationships between micronutrient intake and biochemical indicators of nutrient adequacy in a 'free-living' elderly UK population. *The British Journal of Nutrition.* 77: 224–242.

Buchols, E.M., Desai, M.M., and Rosenthal, M.S., 2011. Dietary intake in head start vs. non-head start preschool-aged children: results from the 1999–2004 national health and nutrition examination survey. *Journal of the American Diet Association.* 111: 1021–1030.

Cho, K.O., Nam, S.N., and Kim, Y.S., 2011. Assessments of nutrient intake and metabolic profiles in Korean adolescents according to exercise regularity using data from the 2008 Korean National Health and Nutrition Examination Survey. *Nutrition Research and Practice.* 5: 66–72.

Chocano-Bedoya, P.O., Manson, J.E., Hankinson, S.E., Willet, W.C., Johnson, S.R., Chasan-Taber, L., Ronnenberg, A.G., Bigelow, C., and Bertone-Johnson, E.R., 2011. Dietary B vitamin intake and incident premenstrual syndrome. *The American Journal of Clinical Nutrition.* 93: 1080–1086.

Choe, E., Huang, R., and Min, D., 2005. Chemical reactions and stability of riboflavin in foods. *Journal of Food Science.* 70: R28–R36.

Farmer, B., Larson, B.T., Fulgoni, V.L., Rainville, A.J., and Liepa, G.U., 2011. A vegetarian dietary pattern as a nutrient-dense approach to weight management: an analysis of the National Health and Nutrition Examination Survey 1999–2004. *Journal of the American Diet Association.* 111: 819–827.

Fraaije, M.W., and Mattevi, A., 2000. Flavoenzymes: diverse catalysts with recurrent features. *Trends in Biochemical Sciences.* 25: 126–132.

Heberlein, D.G., and Clifcorn, L.E., 1944. Vitamin content of dehydrated foods: effect of packaging and storage. *Journal of Industrial and Engineering Chemistry.* 36: 912–917.

Heelis, P.F., 1982. The photophysical and photochemical properties of flavins (isoalloxazines). *Chemical Society Reviews.* 11: 15–39.

Hughes, J., and Sanders, T.A., 1979. Riboflavin levels in the diet and breast milk of vegans and omnivores. *The Proceedings of the Nutrition Society.* 38(2): 95A.

Institute of Medicine, Food and Nutrition Board, 2010. *Dietary Reference Intakes: RDA and AI for Vitamins and Elements.* National Academic Press, Washington DC, USA.

Junior, V.E., Cesar, C.L.G., Fisberg, R.M., and Marchioni, D.M.L., 2011. Socio-economic variables influence the prevalence of inadequate nutrient intake in Brazilian adolescents: results from a population-based survey. *Public Health Nutrition.* 14: 1533–1538.

Lee, K.F., and Hau, L.B., 1996. Effect of γ-irradiation and post-irradiation cooking on thiamine, riboflavin and niacin contents of grass prawns (*Penaeus monodon*). *Food Chemistry.* 55: 379–382.

Lu, C.Y., Lin, W.Z., Wang, W.F., Han, Z.H., Zheng, Z.D., and Yao, S.D., 2000. Kinetic observation of rapid electron transfer between pyrimidine electron adducts and sensitizers of riboflavin, flavin adenine dinucleotide (FAD) and chloranil: a pulse radiolysis study. *Radiation Physics and Chemistry.* 59: 61–66.

Madigan, S.M., Tracey, F., McNulty, H., Eaton-Evans, J., Coulter, J., McCartney, H., and Strain, J.J., 1998. Riboflavin and vitamin B-6 intakes and status and biochemical response to riboflavin supplementation in free-living elderly people. *The American Journal of Clinical Nutrition.* 68: 389–395.

Miura, R., 2001. Versatility and specificity in flavoenzymes: control mechanisms of flavin reactivity. *The Chemical Record.* 1: 183–194.

Murphy, S.P., Calloway, D.H., and Beaton, G.H., 1995. Schoolchildren have similar predicted prevalences of inadequate intakes as toddlers in village populations in Egypt, Kenya and Mexico. *European Journal of Clinical Nutrition.* 49: 647–657.

Nakano, E., Mushtaq, S., Heath, P.R., Lee, E-S., Bury, J.P., Riley, S.A., Powers, H.J., Corfe, B.M., 2011. Riboflavin Depletion Impairs Cell Proliferation in Adult Human Duodenum: Identification of Potential Effectors. *Digestive Diseases and Science.* 56: 1007–1019.

Rivas, A., Rodrigo, D., Company, B., Sampedro, F., and Rodrigo, M., 2007. Effects of pulsed electric fields on water-soluble vitamins and ACE inhibitory peptides added to a mixed orange juice and milk beverage. *Food Chemistry*. 104: 1550–1559.

Said, H.M., 2004. Recent advances in carrier-mediated absorption of water-soluble vitamins. *Annual Review of Physiology*. 66: 419–446.

Said, H.M., and Seetharam, B., 2006. Intestinal absorption of vitamins. In: Johnson, L.R., Barrett, K.E., Ghishan, F.K., Merchand, J.M., Said, H.M., and Wood, J.D. (ed.) *Physiology of the Gastrointestinal Tract*, 4th edn. Academic Press, New York, USA, Vol. 2, pp. 1811–1812.

Sibai, A., Zard, C., Adra, N., Baydoun, M., and Hwalla, N., 2003. Variations in nutritional status of elderly men and women according to place of residence. *Gerontology*. 49: 215–224.

Silva E., Gonzalez T., Edwards, A.M., and Zuloaga, F., 1998. Visible light induced lipoperoxidation of a parenteral nutrition fat emulsion sensitized by flavins. *Journal of Nutritional Biochemistry*. 9: 149–154.

Souci, S.W., Fachmann, W., and Kraut, H., 2000. *Food Composition and Nutrition Tables*. Medpharm Scientific, Stuttgart, Germany. pp. 1182.

Subramanian, V.S., Subramanya, S.B., Rapp, L., Marchant, J.S., Ma, T.Y., and Said, H.M., 2011. Differential expression of human riboflavin transporters -1, -2, and -3 in polarized epithelia: a key role for hRFT-2 in intestinal riboflavin uptake. *Biochimica et Biophysica Acta*. 1808: 3016–3021.

TACO, 2011. Tabela brasileira de composição de alimentos, 4th edn. NEPA-UNICAMP, Campinas, Brazil. pp. 161.

Wishner, L.A., 1964. Light induced oxidation in milk. *Journal of Dairy Science*. 47: 216–221.

Woodcock, E.A., Warthesen, J.J., and Labuza, T.P., 1982. Riboflavin photochemical degradation in pasta measured by high performance liquid chromatography. *Journal of Food Science*. 47: 545–555.

CHAPTER 7

The Chemistry and Biochemistry of Niacin (B_3)

ASDRUBAL AGUILERA-MÉNDEZ,[a] CYNTHIA FERNÁNDEZ-LAINEZ,[b] ISABEL IBARRA-GONZÁLEZ[c] AND CRISTINA FERNANDEZ-MEJIA*[c]

[a] Universidad Michoacana de San Nicolás de Hidalgo, Ciudad Universitaria, CP 58030 Morelia, Michoacan, Mexico; [b] Laboratorio de Errores Innatos del Metabolismo y Tamiz, Instituto Nacional de Pediatría, Av. del Iman #1, 9th floor, Mexico City, CP 04530, Mexico; [c] Unidad de Genética de la Nutrición, Instituto de Investigaciones Biomédicas, Universidad Nacional Autónoma de México/Instituto Nacional de Pediatría, Av. del Iman #1, 4th floor, Mexico City, CP 04530, Mexico
*Email: crisfern@biomedicas.unam.mx

7.1 Introduction

Niacin, also known as vitamin B3, nicotinic acid or vitamin PP, is a water-soluble B-complex vitamin (Table 7.1). This vitamin is the generic descriptor for two vitamers: niacin and niacinamide. In the research literature the terms nicotinic acid/nicotinamide are most commonly used, while in medical practice niacin/niacinamide are preferred. The vitamin is obtained from the diet in the form of nicotinic acid, nicotinamide and tryptophan, which are transformed to nicotinamide adenine dinucleotides, NAD and NADP. These compounds participate in cellular oxidation–reduction reactions that are critical for energy production. NAD and NADP also participate in a wide variety of

---

Food and Nutritional Components in Focus No. 4
B Vitamins and Folate: Chemistry, Analysis, Function and Effects
Edited by Victor R. Preedy
© The Royal Society of Chemistry 2013
Published by the Royal Society of Chemistry, www.rsc.org

Table 7.1 The chemistry of niacin and niacinamide. Vitamin B_3 vitamers: niacin and niacinamide are composed of a pyrimidine ring bound to a carboxylic group or to a carboxamide group.

NIACIN	IUPAC nomenclature[a] CAS number[b] Molecular weight[a]	Pyridine 3-carboxylic acid 59-67-6 122.101 g/mol
NIACINAMIDE	IUPAC nomenclature[a] CAS number[b] Molecular weight[a]	Pyridine 3-carboxamide 98-92-0 122.124 g/mol

[a]Source: PubChem.
[b]Source: ChemIndustry.

ADP-ribosylation reactions such as DNA repair, calcium mobilization and deacetylation (Kirkland 2009). In addition, at pharmacological concentrations, niacin is an effective agent for the treatment of dislipidemias and atherosclerosis (Prousky *et al.* 2011). Furthermore, evidence exists that niacin ameliorates acute migraine, chronic tension-type headaches, depression and schizophrenia (Prousky *et al.* 2011). This chapter focuses on chemical and biochemical aspects of the vitamin.

7.2 Niacin Chemistry

Nicotinic acid and nicotinamide (Table 7.1) are colourless crystalline substances; each is insoluble or only sparingly soluble in organic solvents. Nicotinic acid is slightly soluble in water and ethanol; nicotinamide is very soluble in water and moderately soluble in ethanol. The two compounds have similar absorption spectra in water, with absorption maxima at 262 nm.

Nicotinic acid is zwitterionic in nature; at high pH it is negatively charged at the carboxylic function, while at low pH it is positively charged at the pyridinyl nitrogen. Thus it is considered an amphoteric molecule because it forms salts with acids as well as bases (Mullangi and Srinivas 2011) (Figure 7.1).

Both nicotinic acid and nicotinamide are very stable in dry form (Gonçalves *et al.* 2011); in solution nicotinamide is hydrolysed by acids and bases to yield nicotinic acid. The standard molar enthalpy of formation of nicotinic acid involved in protonation/deprotonation equilibrium of its three species at infinite dilution is $\Delta_f H°_m$ (HN^+ C_5H_4COOH ∞H_2O, aq) = (328.2 ± 1.2) kJ/mol, $\Delta_f H°_m$ (HN^+ $C_5H_4COO^-$ ∞H_2O, aq) = (325.0 ± 1.2) kJ/mol, and $\Delta_f H°_m$ (N^+ $C_5H_4COO^-$ ∞H_2O, aq) = (313.7 ± 1.2) kJ/mol (Gonçalves *et al.* 2011).

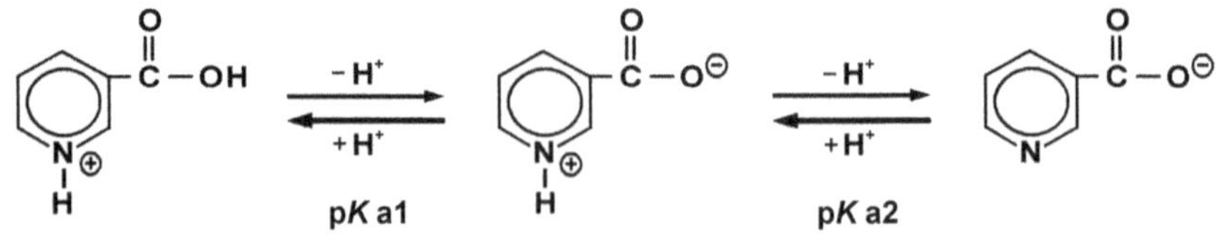

Figure 7.1 Nicotinic acid is an amphiprotic system composed of three species in equilibrium. Nicotinic acid at high pH is negatively charged at the carboxylic function, while at low pH it is positively charged at the pyridinyl nitrogen.

7.3 Niacin Daily Requirement, Food Sources and Niacin Deficiency

The vitamin is obtained from the diet in the form of nicotinic acid, nicotinamide, NAD/NADP and tryptophan. The Recommended Dietary Allowance (RDA) for adults is 16 mg/day of niacin equivalents for men and 14 mg/day for women (Food and Nutrition Board 1998).

Niacin in mature cereal grains, particularly in corn, is largely bound and is poorly available; alkali treatment of the grain increases the percentage absorbed. Meat and fish have the scarce free form of niacin and niacinamide but contain high levels of NAD/ NADP, which are available as niacinamide after digestion (Prousky *et al.* 2011). Fortification of flour and cereal products adds up to 20 mg of the free form of niacin per serving to items such as breakfast cereals (Food and Nutrition Board 1998).

Subclinical niacin deficiency is still present in developing countries and frank deficiency occurs in cancer patients, alcoholics and anorexia nervosa (Kirkland 2007; Prousky *et al.* 2011).

Niacin deficiency, named Pellagra, is characterized by diarrhoea, dermatitis, dementia and death, which usually appear in this order. The clinical expressions of pellagra are diverse (Prousky *et al.* 2011). Diagnosis was, and still is, difficult due to the unpredictable appearance of the different signs and symptoms in individual patients (Prousky *et al.* 2011). Pellagra can be divided into primary and secondary forms. Primary pellagra results from inadequate niacin and/or tryptophan in the diet. Secondary pellagra occurs when other diseases or factors affect niacin requirements.

7.4 Factors and Diseases Affecting Niacin Requirement

The requirement for preformed niacin tends to be lower with higher tryptophan intakes, while the requirement for preformed niacin is increased by factors that reduce the conversion of tryptophan to niacin. These factors include low tryptophan intake and inadequate iron, riboflavin or vitamin B_6 status, which participate in the conversion of tryptophan to niacin (Food and Nutrition Board 1998). Other cases of reduced conversion of tryptophan to niacin are

Hartnup disease, an autosomal recessive trait that interferes with the absorption of tryptophan, and carcinoid syndrome in which the amino acid is preferentially oxidized to 5-hydroxytryptophan and serotonin. Prolonged treatment with the drug isoniazid, which competes with pyridoxal 5′-phosphate (a vitamin B_6-derived coenzyme required in the tryptophan-to-niacin pathway), also reduces the conversion of tryptophan to niacin. Oral contraceptives that contain high doses of estrogen increase tryptophan conversion efficiency (Braidman and Rose 1971).

Niacin requirements are also affected by diseases or conditions interfering with niacin absorption and/or processing, such as prolonged diarrhoea, chronic dialysis treatment, chronic colitis (particularly ulcerative colitis), cirrhosis of the liver, tuberculosis of the gastrointestinal tract, malignant carcinoid tumour and chronic alcoholism (Food and Nutrition Board 1998). Substantial individual differences (about 30%) in the conversion efficiency of tryptophan to niacin have been reported (Horwitt *et al.* 1981).

7.5 Metabolism

To date there is no consensus with respect to the NAD/NADP preferred synthesis substrate. Feeding of rats with tryptophan, nicotinic acid or nicotinamide showed that the first resulted in the highest NAD concentration in liver, suggesting that this is the favoured substrate at least in this organ. In tissues that lack the complete *de novo* NAD biosynthesis pathway, nicotinamide is thought to be chosen over nicotinic acid as the main precursor for NAD biosynthesis (Houtkooper *et al.* 2010). Excess niacin is methylated in the liver to N^1-methyl-nicotinamide, which is excreted in the urine along with the 2- and 4-pyridone oxidation products of N^1-methyl-nicotinamide (Mrochek *et al.* 1976).

7.5.1 Niacin *de novo* Synthesis from Tryptophan

The kynurenine–anthranilate pathway, which is part of tryptophan catabolism, is involved in the liver in the conversion of tryptophan to nicotinamide (Figure 7.2). The first step is catalysed by either tryptophan 2,3 dioxygenase or indolamine-pyrrole 2-3 dioxygenase—iron porphyrin metalloproteins which oxidize the pyrrole moiety of L-tryptophan and represent the rate-limiting enzyme of the kynurenine pathway (Thomas and Stocker 1999). In mammals, tryptophan 2, 3 dioxygenase is the major enzyme contributing to NAD biosynthesis in the liver, while in extrahepatic tissues, indolamine-pyrrole 2,3 dioxygenase plays an important role. Dioxygenases catalyse cleavage of the indole ring with incorporation of two atoms of molecular oxygen, forming *N*-formylkynurenine. Tryptophan 2,3 dioxygenase is positively induced in liver by cortisol and by tryptophan (Oxenkrug 2010). Tryptophan dioxygenase is feedback-inhibited by nicotinic acid derivatives, including NADPH (Michal 1999). In a second step, the hydrolytic removal of the formyl group of *N*-formylkynurenine, catalysed by arylformamidase, produces L-kynurenine, which is then hydroxylated by kynurenine-3 monooxygenase resulting in

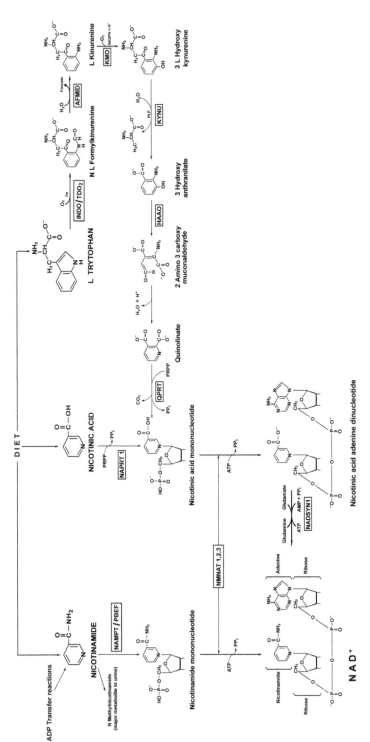

Figure 7.2

3-L-hydroxykynurenine. Hydroxylation requires molecular oxygen in a NADPH-dependent reaction. Further transformation of 3-L-hydroxykynurenine in 3-hydroxyanthranylate and alanine is catalysed by kynureninase, a pyridoxal phosphate enzyme. A deficiency of vitamin B_6 results in reduced niacin synthesis due to the failure to catabolyse these kynurenine derivatives. Kynureninase is positively regulated by tryptophan. In a subsequent step, the action of 3-hydroxyanthranilate 3,4-dioxygenase causes ring opening. The resulting semialdehyde undergoes a spontaneous condensation and rearrangement to form quinolinate, which is the initial compound for NAD and NADP biosynthesis (Figure 7.2).

7.5.2 NAD Biosynthesis

7.5.2.1 From Tryptophan

Quinolinate decarboxylation and conversion to nicotinic acid mononucleotide is catalysed by quinolinate phosphoribosyltransferase, a rate-limiting enzyme in the conversion of tryptophan to NAD; the reaction requires Mg^{2+} and is negatively regulated by nicotinamide. Next the transfer of adenylate from ATP by an intermediate of nicotinamide/nicotinate-mononucleotide-adenyltransferases isoenzymes (NMNAT, see below) yields nicotinic acid adenine

Figure 7.2 Nicotinamide, nicotinic acid and tryptophan are utilized through distinct metabolic pathways to form NAD and NADP. Tryptophan is converted to NAD in the eight-step *de novo* pathway. First, tryptophan is converted to *N*-formylkynurenine by the gene products of either indoleamine dioxygenase (INDO) or tryptophan dioxygenase (TDO2). From *N*-L-formylkynurenine, arylformamidase (AFMID) forms kynurenine. Kynurenine is then used as a substrate of kynurenine monooxygenase (KMO) and forms 3-hydroxykynurenine. Kynureninase (KYNU) then forms 3-hydroxyanthranilate, which is converted to 2-amino-3-carboxymucoaldehyde by 3-hydroxyanthranilate dioxygenase (HAAO). The semialdehyde undergoes a spontaneous condensation and rearrangement to form quinolinate, which is converted to nicotinic acid mononucleotide by quinolinate phosphoribosyltransferase (QPRT). Nicotinic acid mononucleotide is then adenylylated by NMNAT1-3 genes to form nicotinic acid adenine dinucleotide, which is converted to NAD by glutamine-dependent NAD synthetase (NADSYN1). Nicotinic acid is utilized in the three-step Preiss–Handler pathway. Nicotinic acid phosphoribosyltransferase (NAPRT1) forms nicotinic acid mononucleotide by addition of the 5-phosphoribose. In two steps shared with the *de novo* pathway, nicotinic acid mononucleotide is then converted to nicotinic acid adenine dinucleotide and finally to NAD *via* activity of nicotinamide/nicotinate-mononucleotide-adenyltransferases isoenzymes: NMNAT1-3 and NAD synthase-1 NADSYN1. Nicotinamide, from the diet and from ADP transfer reactions, is converted to NAD *via* nicotinamide phosphoribosyltransferase (NAMPT/PBEF1), which catalyses the addition of a phosphoribose moiety onto nicotinamide to form nicotinamide mononucleotide, this latter is subsequently converted to NAD by NMNAT1-3.

dinucleotide. NAD negatively regulates this step. Finally, nicotinic acid adenine dinucleotide is transformed by the catalytic action of NAD synthase-1 in the amide, NAD (Figure 7.2).

7.5.2.2 Salvage Pathways from Niacin and Niacinamide

Humans use both nicotinic acid and nicotinamide to synthesize NAD coenzymes, but utilize different salvage pathways to achieve this. An enzyme in common between the pathways is the adenylation enzyme, NMNAT. This enzyme has three isoforms in humans (NMNAT-1, NMNAT-2 and NMNAT-3). NMNAT-1 is the most proficient catalyst as determined by catalytic velocity (V_{max}) and efficiency (V_{max}/K_m); the enzyme is localized in the nuclei. Isoenzymes NMNAT-2 and -3 are localized in the Golgi and in mitochondria, respectively (Sauve 2008). All isoforms exhibit dual specificity for both nicotinic acid mononucleotide and nicotinamide mononucleotide (Sauve 2008) (Figure 7.2).

7.5.2.2.1 Salvage Pathway from Niacin. In humans, nicotinic acid is converted to nicotinic acid mononucleotide by the catalytic action of nicotin phosphorybosyltransferase-1. By the action of NMNAT(s) and NAD synthase-1, this compound is then transformed to NAD (Figure 7.2).

7.5.2.2.2 Salvage Pathway from Nicotinamide. Although an activity that converted nicotinamide to NAD independently of nicotinic acid had been indicated for some time, the enzyme was not identified until recently (Rongvaux *et al.* 2002). Nicotinamide is coupled directly to phosphorybosyl pyrophosphate to form nicotinamide mononucleotide by nicotinamide phosphoribosyltransferase, an enzyme that was previously identified as a cytokine pre-B-cell colony-enhancing factor (PBEF) and controversially claimed as an insulin-mimetic hormone visfatin (Imai 2009) (Figure 7.2).

7.5.3 NADP Biosynthesis

The generation of NADP is catalysed by NAD kinase, which transfers a phosphate group from ATP onto the 2′-hydroxyl group of the ribose moiety of NAD. A single mammalian NAD kinase has so far been identified; NAD kinase activity is essential for cell survival (Agledal *et al.* 2010).

7.5.4 NAD Recycling

Humans have metabolic pathways that are able to recycle NAD. The main pathway is catalysed by ADP-ribosyltransferases, which participate in nonredox adenosine diphosphate (ADP)–ribose transfer reactions. These NAD-consuming enzymes break down NAD to nicotinamide and ADP-ribosyl product. Nicotinamide is then retransformed to NAD by the enzymatic action of nicotinamide phosphoribosyltransferase (NAMP/PBEF) and NMATs (Figure 7.3).

Recently, a recycling pathway independent of nicotinamide was found to be broadly conserved in bacteria, yeast and mammals (Bieganowski and Brenner 2004) The pathway leads from the metabolite nicotinamide riboside, the dephosphorylated form of nicotinamide mononucleotide. A highly biologically conserved nicotinamide riboside kinase is able to use nicotinamide riboside as a substrate and can convert nicotinamide riboside to nicotinamide mononucleotide (Bieganowski and Brenner 2004). Subsequently, nicotinamide mononucleotide is converted to NAD by the catalytic action of NMNATs. In humans, two isoforms of the kinase (Nrk1 and Nrk2) have been cloned, although little is known about the biochemical properties and regulation of these enzymes (Sauve 2008) (Figure 7.3).

7.6 NAD/NAPD Chemical and Structural Proprieties

NAD and NADP are composed of two nucleotides joined through their phosphate groups by a phosphoanhydride bond (Table 7.2). NADP differs from NAD in the presence of an additional phosphate group on the 2' position of the ribose ring (Table 7.2). These coenzymes are white amorphous powders that are hygroscopic and highly water-soluble.

NAD and NADP act as electron acceptors during the enzymatic removal of hydrogen atoms from specific substrate molecules. One hydrogen atom from the substrate is transferred as hydride ion to the nicotinamide portion of the oxidized forms of these coenzymes to yield the reduced coenzymes NADH or NADPH, respectively; the other hydrogen atom from the substrate becomes a hydrogen ion (Figure 7.4). The reduced nucleotides absorb light at 340 nm but the oxidized forms do not. This difference in absorption is used to assay reactions involving these coenzymes.

Most dehydrogenases that use NAD or NADP bind the cofactor in a conserved protein domain called the Rossmann fold. This structure results in a relative loose binding that permits the coenzyme diffusion from one enzyme to another.

7.7 NAD/NADP Metabolic Actions

7.7.1 Energy Metabolism and Oxidation Processes

NAD and NADP play a central role in energy metabolism, both being involved in many cellular oxidation–reduction reactions. NAD functions in energy-producing reactions involving the catabolism of innumerable metabolites. NAD formed from oxidation reacts at the point of Complex I of the mitochondrial electron transport chain. Each mole of NADH consumed by the mitochondria furnishes energy for the formation of three moles of ATP. NADH is highly depleted in the cytosol. The phosphorylated dinucleotide,

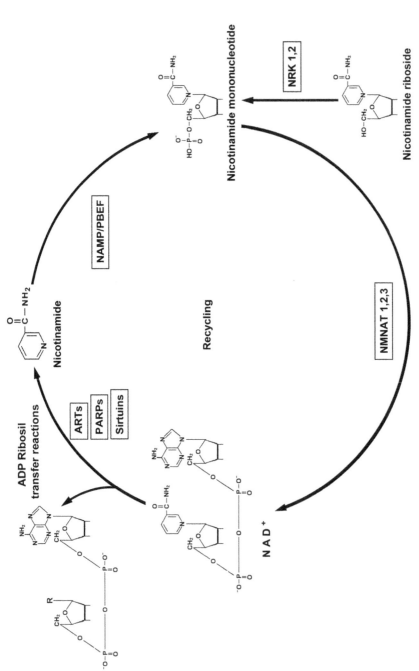

Figure 7.3 NAD recycling. Humans have two metabolic pathways that are able to recycle nicotinamide. NAD-consuming enzymes (ARTs, PARPs, sirtuins) break down NAD to nicotinamide and ADP-ribosyl product. Nicotinamide by the enzymatic action of nicotinamide phosphoribosyltransferase (NAMP/PBEF) and nicotinamide/nicotinate-mononucleotide-adenylyltransferases isoenzymes (NMAT1-3) is then retransformed to NAD. In a second pathway, nicotinamide riboside is phosphorylated by nicotinamide riboside kinase (NRK 1,2) to nicotinamide mononucleotide. Subsequently, nicotinamide mononucleotide is converted to NAD by the catalytic action of NMNATs.

Table 7.2 The chemistry of NAD and NADP. Nicotin adenine dinucleotides: NAD and NADP are composed of two nucleotides joined through their phosphate groups by a phosphoanhydride bond.

NAD+	IUPAC name	[[(2R,3S,4R,5R)-5-(6-aminopurin-9-yl)-3,4-dihydroxyoxolan-2-yl]methoxy-hydroxy-phosphoryl][(2R,3S,4R,5R)-5-(3-carbamoylpyridin-1-ium-1-yl)-3,4-dihydroxy-oxolan-2-yl]methyl hydrogen phosphate
	CAS number (chem. Industry)	53-84-9
	Chemical formula (pubchem)	C21H28N7O14P2+
	Molecular weight (pubchem)	664.433042 [g/mol]
NADP+	IUPAC name	[[(2R,3R,4R,5R)-5-(6-aminopurin-9-yl)-3-hydroxy-4-phosphonooxyoxolan-2-yl]methoxy-hydroxyphosphoryl][(2R,3S,4R,5R)-5-(3-carbamoylpyridin-1-ium-1-yl)-3,4-dihydroxy-oxolan-2-yl]methyl hydrogen phosphate
	CAS number (chem. Industry)	53-59-8
	Chemical formula (pubchem)	C21H29N7O17P3+
	Molecular weight (pubchem)	744.412943 [g/mol]

NADP, functions in biosynthetic (anabolic) reactions such as the synthesis of fatty acids cholesterol, bile acids, steroid hormones and the amino acids glutamate and proline. NADPH + H is also essential for the reduction of ribonucleotides to deoxyribonucleotides and is thereby indirectly involved in DNA synthesis.

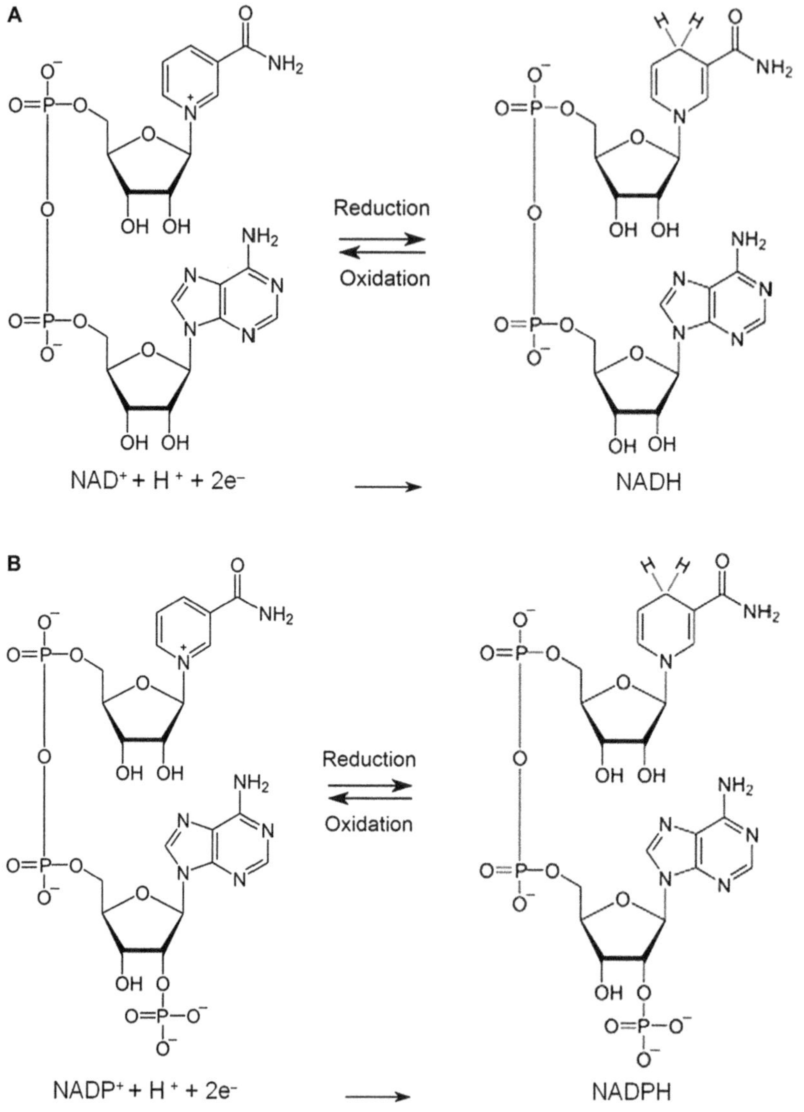

Figure 7.4 NAD/NADP redox reactions. NAD and NADP act as electron acceptors during the enzymatic removal of hydrogen atoms from specific substrate molecules. One hydrogen atom from the substrate is transferred as a hydride ion to the nicotinamide portion of the oxidized forms of these coenzymes to yield the reduced coenzymes NADH or NADPH, respectively; the other hydrogen atom from the substrate becomes a hydrogen ion.

7.7.2 Protective Action of NADP

In addition to its biosynthetic role, NADPH is the unique provider of reducing equivalents to maintain or regenerate the cellular detoxifying and antioxidant defence systems. NADPH is an important cofactor for P450 enzymes that detoxify xenobiotics in reactions converting relatively insoluble organic compounds to hydrophilic ones, thereby facilitating their breakdown and secretion. In oxidative defence, NADPH acts as a terminal reductant for glutathione reductase, which maintains reduced glutathione. In some cell types, NADPH is required for the reactivation of catalase, when the enzyme is inactivated by H_2O_2. In addition, NADPH serves as a substrate for NADPH oxidase, which generates peroxides for release in oxidative burst processes of the immune system (Agledal et al. 2010).

7.7.3 Non-redox Adenosine Diphosphate (ADP)-Ribose Transfer Reactions

NAD/NADP have also shown to be required for important non-redox adenosine diphosphate (ADP)-ribose transfer reactions involved in DNA repair, calcium mobilization and deacetylation reactions (Lautier et al. 1993; Pollak et al. 2007) (Figure 7.5).

Three classes of enzymes cleave the β-*N*-glycosylic bond of NAD to free nicotinamide and catalyse the transfer of ADP-ribose in non-redox reactions (Lautier et al. 1993). The first class consists of ADP ribose transferases: mono-ADP-ribosyltransferases and poly-ADP-ribose polymerase, which catalyse ADP-ribose transfer to proteins. The second class correspond to ADP-ribosyl cyclases—enzymes that promote the formation of cyclic ADP-ribose, a compound that mobilizes calcium from intracellular stores in many types of cells (Pollak et al. 2007). The third class of NAD consuming enzymes consists of sirtuins—proteins that possess either histone deacetylase or mono-ribosyltransferase activity.

7.7.3.1 Mono-ADP-ribosyltransferases

These enzymes catalyse the ADP-ribose moiety of NAD transfer to an acceptor amino acid. Five mammalian ADP-ribosyltransferases have been cloned and expression is restricted to tissues such as cardiac and skeletal muscle, leukocytes, brain and testis. ADP-ribosyltransferases-1 and -2 are glycosylphosphatidylinositol (GPI)-anchored ectoenzymes. ADP-ribosyltransferase-5 appears not to be GPI-linked and may be secreted. In skeletal muscle and lymphocytes, ADP-ribosyltransferases-1 modifies specific members of the integrin family of adhesion molecules, suggesting that ADP-ribosylation affects cell-matrix or cell-cell interactions (Okazaki and Moss 1999).

7.7.3.2 Poly-ADP-ribose Polymerases

These enzymes catalyse the attachment of ADP-ribose polymers to target proteins and represent an elaborate protein modification. The most studied human enzyme among 17 members of this family, poly-ADP-ribose polymerase-1 (PARP-1), is involved in DNA repair and apoptotic pathways. PARP-1, a nuclear enzyme which detects DNA damage, binds to DNA single

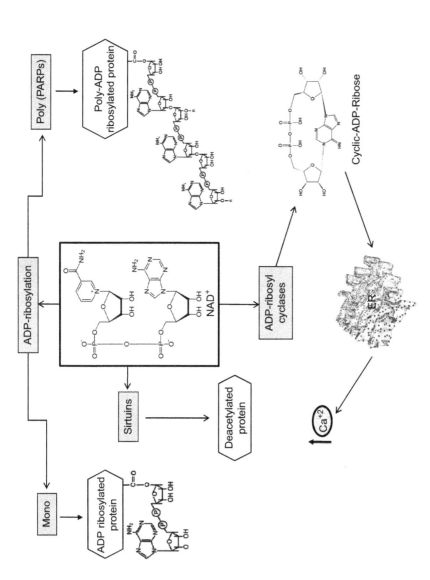

Figure 7.5 Adenosine diphosphate (ADP)-ribosylation biochemical reactions. Mono-ADP-ribosyltransferases (ARTs) and poly-ADP-ribose polymerases Poly (PARPs) catalyse the ADP-ribose moiety of NAD transfer to amino acid residues. ADP-ribosyl cyclases generate cyclic ADP-ribose and 2-phospho-cyclic ADP-ribose from NAD and NADP, respectively. Both molecules trigger cyclic ADP-ribose cytosolic Ca^{2+} elevation, presumably by activating the ryanodine receptor in the endoplasmic/sarcoplasmic reticulum (RER). SIRT1 catalyses a reaction that couples protein deacetylation to NAD hydrolysis.

or double strand breaks, and then catalyses the transfer of many ADP-ribose units from NAD to an acceptor protein and also to the enzyme itself. The intensity of DNA damage determines cellular pathways: survival, apoptosis, or necrosis. In the case of mild DNA damage, poly(ADP-ribosylation) enhances DNA repair and thus cell survival. When the damage is beyond repair, PARP-1 facilitates apoptosis, preventing ATP depletion. Severe DNA damage leads to PARP-1 over-activation, cellular energy depletion and necrotic cell death (Surjana et al. 2010). Poly-ADP-ribosylation has also important functions in telomere dynamics, transcriptional regulation, cell division and trafficking of endosomal vesicles (Pollak et al. 2007).

7.7.3.3 ADP-ribosyl Cyclases

These enzymes generate cyclic ADP-ribose and 2-phospho-cyclic ADP-ribose from NAD and NADP, respectively. Both molecules trigger cyclic ADP-ribose cytosolic Ca^{2+} elevation, presumably by activating the ryanodine receptor in the endoplasmic/sarcoplasmic reticulum (Pollak et al. 2007). In addition to cyclic ADP-ribose, niacin adenine dinucleotide phosphate, a metabolite of NADP, can also mobilize Ca^{2+} stores. The release mechanism and the stores on which niacin adenine dinucleotide phosphate acts are from lysosomal Ca^{2+} stores, which are independent of the stores of activated by cyclic ADP-ribose or inositol 1,4,5-trisphosphate (Yamasaki et al. 2004).

7.7.3.4 Sirtuins

Sirtuins are a highly conserved family of protein deacetylases and ADP-ribosyltransferases, whose distinguishing characteristic is a requirement for the oxidized form of NAD (Lomb et al. 2010). The dependence of sirtuins upon NAD plays a major role in connecting their enzymatic activity to the grade of energy of the cell by means of the cellular NAD/NADH ratio. Seven mammalian orthologs (SIRT1–7) have been described, with SIRT 1 the best studied. SIRT1 catalyses a reaction that couples lysine deacetylation to NAD hydrolysis. The deacetylation reaction of sirtuins consists of two steps. In the first, sirtuins cleave NAD and produce nicotinamide, and in the second step, the acetyl group is transferred from the substrate to the ADP-ribose moiety of NAD to generate O-acetyl-ADP ribose and the deacetylated substrate. Nicotinamide is a strong inhibitor of SIRT1 deacetylase activity. Fasting has been found to increase SIRT1 protein levels and activity. SIRT1 has also found to be activated in response to oxidative stress, low glucose availability and endurance exercise (Lomb et al. 2010). Importantly, SIRT-1 regulates the activity of several transcription factors and cofactors by modulating their acetylation status, and by this means links the metabolic state of the cell to transcriptional regulation.

7.8 Concluding Remarks

In this chapter we have summarized several interesting findings that extend the actions of NAD/NADP from oxidoreductase cofactors to signalling molecules.

Furthermore, novel data have unveiled the link between the cellular redox status and the control of metabolism, cellular viability and transcriptional events. Continued progress will be needed to understand NAD/NADP metabolism and signalling in health and disease.

Summary Points

- Niacin is also known as vitamin B_3, nicotinic acid or vitamin PP. It is the generic descriptor for two vitamers: niacin and niacinamide.
- The vitamin is obtained from the diet in the form of nicotinic acid, nicotinamide and tryptophan.
- Niacin is a precursor to NAD and NADP, coenzymes whose classical role is to participate in redox reactions.
- In addition to their classical role as coenzymes, NAD and NADP participate in a wide variety of ADP-ribosylation reactions, which are involved in cell signalling and the control of many cell processes.
- ADP ribose transferases: mono- and poly-ADP-ribose polymerases catalyse ADP-ribose transfer to proteins.
- ADP-ribosyl cyclases promote the formation of cyclic ADP-ribose, a compound that mobilizes calcium from intracellular stores in many types of cells.
- Sirtuins possess either histone deacetylase or mono-ribosyltransferase activity. The dependence of sirtuins activity upon NAD plays a major role in connecting their enzymatic activity to the grade of energy of the cell by means of the cellular NAD/NADH ratio.

Key Facts about Niacin History

- American natives knew that corn needed to be processed to be nutritious in a process known nixtamalization (alkaline treatment commonly used in Mexico and Central America in the preparation of tortillas, tamales and hominy) or ingested as immature ears. This knowledge was lost when Christopher Columbus returned from the New World with samples of the new seed.
- The use of corn without processing spread over time and gradually the clinical signs of niacin deficiency became recognized as a disease.
- In 1771, Francesco Frapolli made the first reference to the disease as pellagra, meaning 'rough skin' in Italian. He linked it with poverty and subsistence on nutritionally marginal corn-based diets.
- In 1937, Conrad A. Elvehjem isolated the P-P (pellagra preventive) factor, identified it as nicotinic acid (niacin).
- The term 'niacin' was coined to avoid the associations of nicotinic acid with acids and nicotine, which shocked the public.

Definitions of Words and Terms

ADP-ribosylation reactions: Reaction whereby an ADP-ribose moiety is linked covalently to another compound.

Apoptosis: Programmed cell death.

De novo synthesis: Synthesis of complex molecules from simple molecules such as sugars or amino acids, as opposed to their being recycled after partial degradation.

DNA repair: A collection of processes by which a cell identifies and corrects damage to the DNA molecules that encode its genome.

Ectoenzymes: Catalytic membrane proteins with their active sites outside the cell.

Necrosis: Uncontrolled cell death that leads to lysis of cells and inflammatory responses.

Nixtamalization: Refers to a process for the preparation of maize (corn) or other grain in which the grain is soaked and cooked in an alkaline solution, usually limewater, and hulled.

Orthologs: Orthologs, or orthologous genes, are genes in different species that are similar to each other because they originated by vertical descent from a single gene of the last common ancestor.

Recommended Dietary Allowances: These refer to the recommended daily levels of nutrients to meet the needs of nearly all healthy individuals in a particular age and gender group.

Redox: Oxidation–reduction.

Salvage pathway: A pathway that utilizes compounds formed in catabolism for biosynthetic purposes, even though these compounds are not true intermediates of the corresponding normal biosynthetic pathway.

Telomere: Repetitive DNA sequences at the end of a chromosome, which protects the end of the chromosome from deterioration or from fusion with neighbouring chromosomes.

Toxicokinetics: The description of what rate a chemical will enter the body and what happens to it once it is in the body.

Vitamers: One or multiple related chemical compounds having a given vitamin activity.

Xenobiotics: A compound that is foreign to an organism. It can also cover substances that are present in much higher concentrations than are usual.

Zwitterionic: A molecule carrying both a positive and a negative charge.

List of Abbreviations

AFMID	arylformamidase
ARTs	mono ADP-ribosyltransferases
$\Delta_f H^\circ_m$	standard molar enthalpy of formation
GPI	glycosylphosphatidylinositol
HAAO	3-hydroxyanthranilate dioxygenase

INDO	indoleamine dioxygenase
IUPAC	International Union of Pure and Applied Chemistry
kJ/mol	kilo Joule per mol
KMO	kynurenine monooxygenase
KYNU	kynureninase
NAD	nicotinamide adenine dinucleotide
NADP	nicotinamide adenine diphosphate
NADSYN1	NAD + synthetase-1
NAMP	nicotinamide phosphoribosyltransferase
NMNAT	nicotinamide/nicotinate mononucleotide adenyltransferases
NAPRT1	nicotinic acid phosphoribosyl transferase-1
NRK 1,2	nicotinamide ribiside kinase 1 and 2
PARP-1	poly-ADP-ribose polymerase-1
PBEF	pre-B-cell colony-enhancing factor
QPRT	quinolinate phosphoribosyltransferase
RDA	Recommended Dietary Allowance
RER	endoplasmic/sarcoplasmic reticulum
SIRT	silent information regulator proteins
TDO2	tryptophan dioxygenase

Acknowledgments

The authors are grateful to Luis Rodriguez-Villela for illustrations. Their work was supported by research grants from the Consejo Nacional de Ciencia y Tecnología 99294-M and the Dirección General de Asuntos del Personal Académico, Universidad Nacional Autónoma de México IN214811. Asdrubal Aguilera is recipient of the CONACyT scholarship number 91634 and PROMEP, folio UMSNH-208.

References

Agledal, L., Niere, M., and Ziegler, M., 2010. The phosphate makes a difference: cellular functions of NADP. *Redox Report.* 15: 2–10.

Bieganowski, P., and Brenner, C., 2004. Discoveries of nicotinamide riboside as a nutrient and conserved NRK genes establish a Preiss-Handler independent route to NAD+ in fungi and humans. *Cell.* 117: 495–502.

Braidman, I.P., and Rose, D.P., 1971. The effect of sex hormones on the activity of tryptophan oxygenase and other corticosteroid-inducible enzymes in rat liver. *Biochemical Journal.* 122: 28P.

Food and Nutrition Board, 1998. Dietary reference intakes for thiamine, riboflavin, niacin, vitamin B_6, folate, vitamin B_{12}, panthotenic acid, biotin and choline. National Academy Press, Washington DC, USA, pp. 374–389.

Gonçalves, E., Rego, T., and Minas da Piedade, M., 2011. Thermochemistry of aqueous pyridine-3-carboxylic acid (nicotinic acid). *Journal of Chemical Thermodynamics.* 43: 974–979.

Horwitt, M.K., Harper, A.E., and Henderson, L.M., 1981. Niacin-tryptophan relationships for evaluating niacin equivalents. *The American Journal of Clinical Nutrition.* 34: 423–427.

Houtkooper, R.H., Canto, C., Wanders, R.J., and Auwerx, J., 2010. The secret life of NAD+: an old metabolite controlling new metabolic signaling pathways. *Endocrine Reviews.* 31: 194–223.

Imai, S., 2009. Nicotinamide phosphoribosyltransferase (NAMPT): a link between NAD biology, metabolism, and diseases. *Current Pharmaceutical Design.* 15: 20–28.

Kirkland, J.B., 2007. Niacin. In: Rucker, R., Zempleni, J., Suttie, J.W., McCormick, D.B. (eds.), *Handbook of vitamins*, 4th edn. Taylor and Francis, Boca Raton, USA, pp. 191–232.

Kirkland, J.B., 2009. Niacin status, NAD distribution and ADP-ribose metabolism. *Current Pharmaceutical Design.* 15: 3–11.

Lautier, D., Lagueux, J., Thibodeau, J., Menard, L., and Poirier, G.G., 1993. Molecular and biochemical features of poly (ADP-ribose) metabolism. *Molecular and Cellular Biochemistry.* 122: 171–193.

Lomb, D.J., Laurent, G., and Haigis, M.C., 2010. Sirtuins regulate key aspects of lipid metabolism. *Biochimica Biophysica Acta.* 1804: 1652–1657.

Michal, G., 1999. Cofactors and Vitamins. *Biochemical Pathways: An Atlas of Biochemistry and Molecular Biology.* Wiley, New York, USA, pp. 108–121.

Mrochek, J.E., Jolley, R.L., Young, D.S., and Turner, W.J., 1976. Metabolic response of humans to ingestion of nicotinic acid and nicotinamide. *Clinical Chemistry.* 22: 1821–1827.

Mullangi, R., and Srinivas, N.R., 2011. Niacin and its metabolites: role of LC-MS/MS bioanalytical methods and update on clinical pharmacology. An overview. *Biomedical Chromatography.* 25: 218–237.

Okazaki, I.J., and Moss, J., 1999. Characterization of glycosylphosphatidylinositiol-anchored, secreted, and intracellular vertebrate mono-ADP-ribosyltransferases. *Annual Review of Nutrition.* 19: 485–509.

Oxenkrug, G.F., 2010. Tryptophan kynurenine metabolism as a common mediator of genetic and environmental impacts in major depressive disorder: the serotonin hypothesis revisited 40 years later. *Israeli Journal of Psychiatry and Related Sciences.* 47: 56–63.

Pollak, N., Dolle, C., and Ziegler, M., 2007. The power to reduce: pyridine nucleotides--small molecules with a multitude of functions. *Biochemistry Journal.* 402: 205–218.

Prousky, J., Millman, C., and Kirkland, J., 2011. Pharmacologic use of niacin. *Journal of Evidence-Based Complementary & Alternative Medicine.* 16: 91–101.

Rongvaux, A., Shea, R.J., Mulks, M.H., Gigot, D., Urbain, J., Leo, O., and Andris, F., 2002. Pre-B-cell colony-enhancing factor, whose expression is up-regulated in activated lymphocytes, is a nicotinamide phosphoribosyltransferase, a cytosolic enzyme involved in NAD biosynthesis. *European Journal of Immunology.* 32: 3225–3234.

Sauve, A.A., 2008. NAD+ and vitamin B3: from metabolism to therapies. *Journal of Pharmacology and Experimental Therapeutics*. 324: 883–893.

Surjana, D., Halliday, G.M., and Damian, D.L., 2010. Role of nicotinamide in DNA damage, mutagenesis, and DNA repair. *Journal of Nucleic Acids*. 2010: Article ID 157591, 13 pp. doi:10.4061/2010/157591.

Thomas, S.R., and Stocker, R., 1999. Redox reactions related to indoleamine 2,3–dioxygenase and tryptophan metabolism along the kynurenine pathway. *Redox Report*. 4: 199–220.

Yamasaki, M., Masgrau, R., Morgan, A.J., Churchill, G.C., Patel, S., Ashcroft, S.J., and Galione, A., 2004. Organelle selection determines agonist-specific Ca2+ signals in pancreatic acinar and beta cells. *Journal of Biological Chemistry*. 279: 7234–7240.

CHAPTER 8
The Chemistry of Pantothenic Acid (Vitamin B_5)

KATSUMI SHIBATA* AND TSUTOMU FUKUWATARI

Department of Food Science and Nutrition, School of Human Cultures, The University of Shiga Prefecture, Hikone, Shiga 522-8533, Japan
*Email: kshibata@shc.usp.ac.jp

8.1 Introduction

Pantothenic acid was discovered in 1933 (Williams *et al.* 1933). Its name originates from the Greek 'pantos' meaning 'everywhere', and small quantities of pantothenic acid are found in nearly every food. It was first isolated from liver cells in 1938 and first synthesized in 1940 (Stiller *et al.* 1940). The principal biologically active forms of pantothenic acid are coenzyme A (CoA) (Lipmann *et al.* 1950) and 4′-phosphopantetheine in acyl carrier protein (Majerus *et al.* 1965).

Pantothenic acid, also called vitamin B_5, is a water-soluble vitamin. The average adequate intake of pantothenic acid for adults is ~ 5 mg/day (Food and Nutrition Board 1998).

8.2 Chemical Structure of Pantothenic Acid and its Related Compounds

The International Union of Pure and Applied Chemistry (IUPAC) name for pantothenic acid is 3-[[(2R)-2,4-dihydroxy-3,3-dimethyl-butanoyl]amino]propanoic

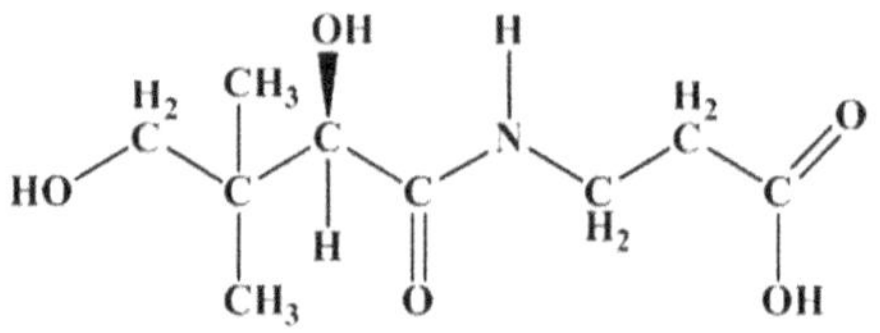

Figure 8.1 Structure of pantothenic acid.

Pantoic acid   *β-Alanine*

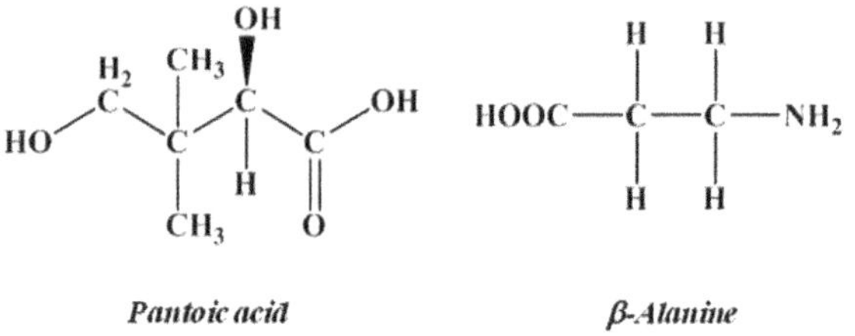

Figure 8.2 Structures of pantoic acid and β-alanine.

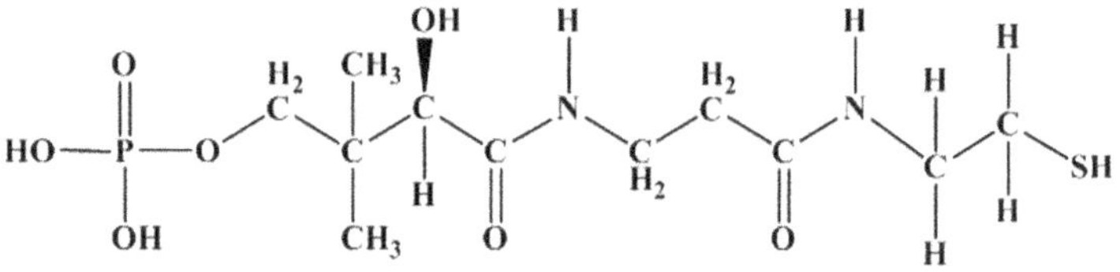

Figure 8.3 Structure of 4′-phosphopantetheine.

acid and its CAS number is 137-08-6. The chemical structure of pantothenic acid is shown in Figure 8.1. Pantothenic acid is composed of β-alanine and 2,4-dihydroxy-3,3-dimethylbutyric acid (common name, pantoic acid), which is acid amide-linked (Figure 8.2). Pantothenic acid is biosynthesized to 4′-phosphopantetheine (Figure 8.3) and CoA (Figure 8.4), which have integral roles in the biosynthesis and oxidation of fatty acids. CoA consists of 4′-phosphopantetheine, adenosine monophosphate (AMP) and phosphate. Therefore, it should be renamed to phosphopantetheinyladenylate phosphate, abbreviated PPAP.

8.3 Stereoisomers of Pantothenic Acid

Pantothenic acid occurs in nature only as the D-(+)- or the (R)-enantiomer; the L(−)- or (S)-form has no vitamin activity (Sarett and Barboriak 1963). The CAS number of L(−)-acid is 37138-77-5 and the IUPAC name is 3-[[(2S)-2,4-dihydroxy-3,3-dimethyl-butanoyl]amino]propanoic acid.

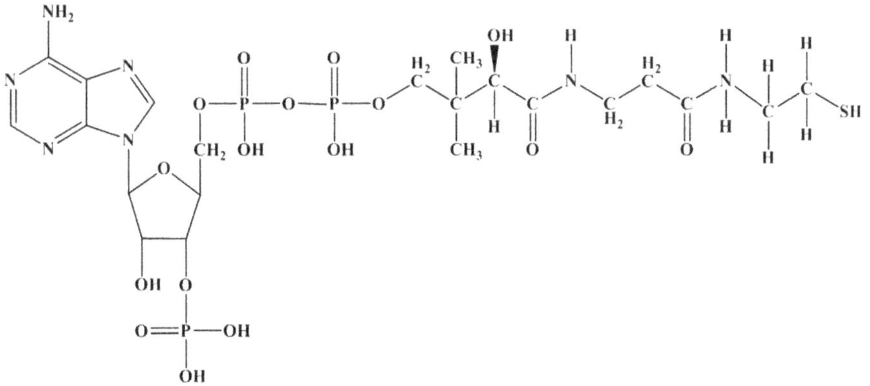

The Chemistry of Pantothenic Acid (Vitamin B_5) 129

Figure 8.4 Structure of coenzyme A.

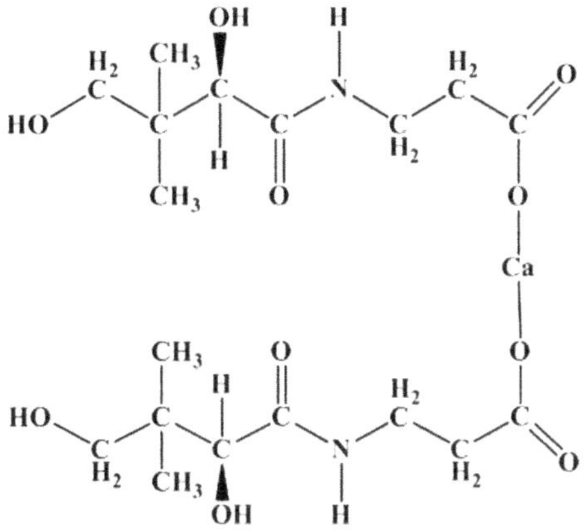

Figure 8.5 Structure of calcium pantothenate.

8.4 Characteristic Properties of D-(+)-Pantothenic Acid

Pantothenic acid is a yellow, viscous, oily liquid which is readily soluble in water, alcohols and dioxane. It is slightly soluble in diethyl ether and acetone, and virtually insoluble in benzene and chloroform.

Pantothenic acid *per se* is hygroscopic and unstable. Its more stable sodium salt and, in particular, calcium salt, are synthesized chemically and used pharmaceutically, mainly in solid multivitamin preparations and as an additive compound for some foods and domestic animal feeds. Calcium pantothenate (Figure 8.5) is most stable in almost neutral media (pH 6–7). Salts of pantothenic acid are colourless crystals and less hygroscopic than pantothenic acid (particularly the calcium salt). The solubility of calcium pantothenate in water at 25 °C is 0.356 g/mL, whereas the sodium salt is also very soluble in water.

8.5 Stability of D-(+)-Pantothenic Acid

Pantothenic acid is stable under neutral conditions, but is readily destroyed by heat in alkaline or acid solutions. It is hydrolytically cleaved to yield pantoic acid (Figure 8.2) and its salts, respectively, and β-alanine (Figure 8.2). In acidic solutions, pantoic acid spontaneously eliminates one molecule of water, forming (R)-2-hydroxy-3,3-dimethyl-4-butanolide (α-hydroxy-β,β-dimethyl-γ-butyrolactone), which is referred to as pantoyl lactone or pantolactone (Figure 8.6). Up to 50% may be lost during cooking (probably due to leaching) and ≤80% as a result of food processing and refining (*e.g.* canning, freezing, milling). Pasteurization of milk causes only minor losses.

8.6 Analogues of Pantothenic Acid

8.6.1 Panthenol

An alcohol related to pantothenic acid (and referred to as D-panthenol or D-pantothenyl alcohol) also possesses vitamin activity (Figure 8.7) because it is readily oxidized to form pantothenic acid in organisms. Only D-panthenol is biologically active. The IUPAC name is (2R)-2,4-dihydroxy-N-(3-hydroxypropyl)-3,3-dimethylbutanamide and the CAS number is 81-13-0. It is

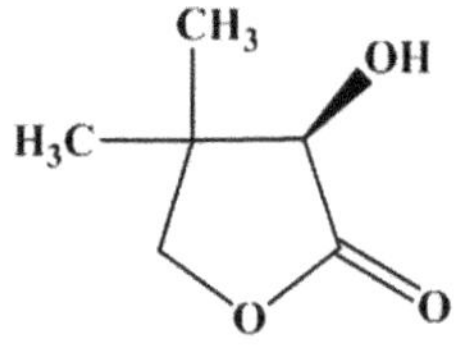

Figure 8.6 Structure of D(+)-pantolactone.

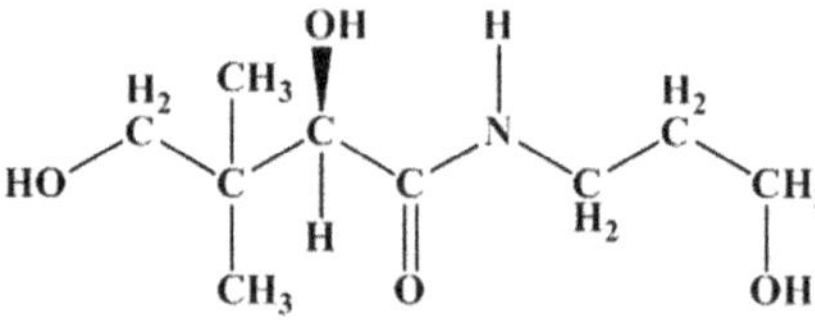

Figure 8.7 Structure of D-panthenol.

commonly used in liquid pharmaceutical and cosmetic preparations. Panthenol is a hygroscopic viscous oil that can be crystallized. Panthenol is very slightly soluble in water but very soluble in alcohol. The products of the hydrolysis of panthenol are pantoic acid and 3-amino-1-propanol (β-alanol).

8.6.2 Pantetheine

The IUPAC name is (2R)-2,4-dihyrdoxy-3,3-dimethyl-N-[2-(2-sulfanylethyl-carbamoyl)ethyl]butanamide and the CAS number is 496-65-1 (Figure 8.8). Pantetheine is a syrup, glassy substance.

8.6.3 Pantethine

The IUPAC name is N-[2-[2-[3-(2,4-dihydroxy-3,3-dimethyl-butanoyl)amino-propanoylamino]ethyldisulfanyl]ethylcarbamoyl]ethyl]-2,4-dihydroxy-3-3-dimethyl-butanamide and the CAS number is 16816-67-4 (Figure 8.9). Pantethine is a dimeric form of pantetheine and is formed by the oxidation of pantetheine. Pantethine is glassy, colourless to light-yellow substance which is freely soluble in water. It is less soluble in ethanol, and practically insoluble in ether, acetone, ethylacetate and chloroform.

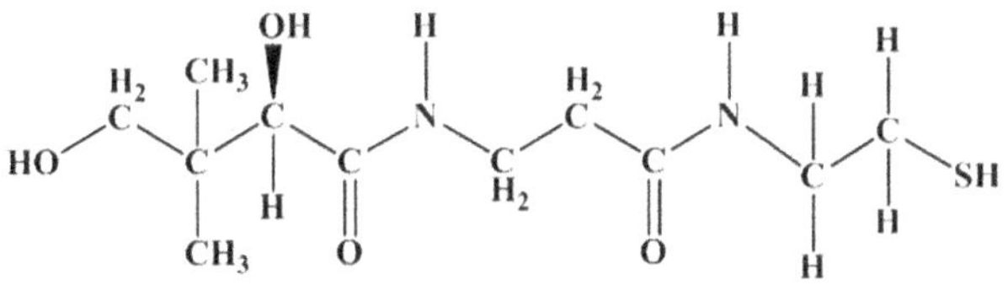

Figure 8.8 Structure of pantetheine.

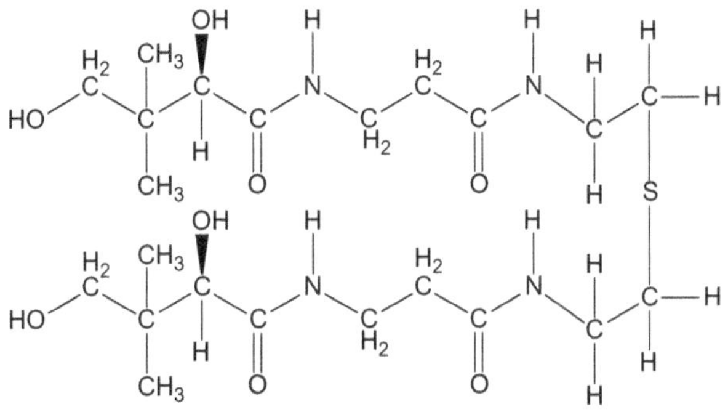

Figure 8.9 Structure of pantethine.

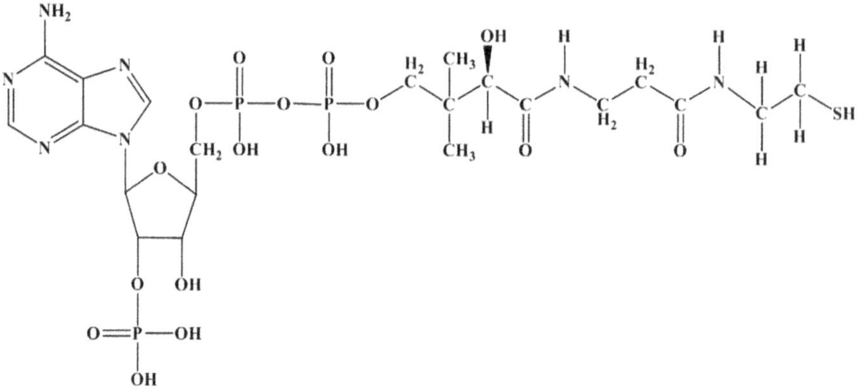

Figure 8.10 Structure of iso-CoA.

8.6.4 Coenzyme A (CoA)

CoA (Figure 8.4) is white powder and is soluble in water. It is a fairly strong acid. The pK_a of the thiol is 9.6, that of the secondary phosphate is 6.4, and that of adenine (NH_3^+) is 4.0. It is practically insoluble in ethanol, ether and acetone, and has a characteristic thiol odour. Pure dry CoA is best stored in evacuated ampules at room temperature. CoA is readily oxidized by air to the catalytically inactive disulfide. Solutions of CoA are relatively stable at pH 2–6.

8.6.5 Iso-coenzyme A (Iso-CoA)

Iso-CoA is an isomer of CoA in which the monophosphate is attached to the 2′-carbon of the ribose ring (Figure 8.10) (Burns *et al.* 2005). Iso-CoA can act as a substrate in enzymatic acyl transferase reactions (Burns *et al.* 2005).

Summary Points

- This chapter focuses on the structure of pantothenic acid and its related compounds.
- Pantothenic acid (also called vitamin B_5) is a water-soluble vitamin.
- Pantothenic acid occurs in nature only as the D-(+)-enantiomer. The L-(−)-form does not have vitamin activity.
- Calcium D-(+)-pantothenate is synthesized chemically and used pharmaceutically.
- Calcium D-(+)-pantothenate is most stable in almost neutral media (pH 6–7).
- Calcium D-(+)-pantothenate is readily destroyed by heat in alkaline or acidic solutions.
- Panthenol, pantetheine, and pantetheine are analogs of pantothenic acid; all have vitamin B_5 activity.

The Chemistry of Pantothenic Acid (Vitamin B_5)

Key Facts

- Pantothenic acid (also called vitamin B_5) is a water-soluble vitamin. The average adequate intake of pantothenic acid for adults is 5 mg/day.
- Deficiencies in pantothenic acid are manifested as dermal, hepatic, thymic and neurological changes.
- Synthetic pantothenic acid is a racemic compound comprising D(+) and L(−) forms. Pantothenic acid occurs naturally as the D(+) isomer and is biologically active. L(−)-Pantothenic acid is biologically inactive and the racemic compound, DL-pantothenic acid, has half the biological activity of D(+)-pantothenic acid. However, large excess administration of L(−)-pantothenic acid to young rats causes growth retardation.
- The D(+)-pantothenic acid analog D(+)-panthenol is a provitamin of B_5 which is converted into pantothenic acid in the body. D(+)-Panthenol is used for cosmetic purposes.

Definitions of Words and Terms

IUPAC. The main purpose of the nomenclature is to unambiguously identify a chemical species. For further information, see http://www.acdlabs.com/iupac/nomenclature/.

CAS number. CAS numbers are numerical identifiers assigned to each chemical by the Chemical Abstracts Service.

Stereoisomer. Stereoisomers have the same molecular formula and sequence of bonded atoms and differ only in the three-dimensional orientations of their atoms in space.

λ_{max} (nm). λ (Lambda) max is the wavelength at which the maximum fraction of light is absorbed by a solution. λ is a Greek letter that scientists use as the symbol for wavelength.

pKa. An ability to release hydrogen ions.

Specific optical rotation for the sodium D line at 25 °C, as denoted by $[\alpha]^{25}_D$. In stereochemistry, the specific rotation of a chemical compound [α] is defined as the observed angle of optical rotation α when plane-polarized light is passed through a sample with a path length of 1 decimetre and a sample concentration of 1 gram per 1 millilitre.

List of Abbreviations

AMP	adenosine monophosphate
CAS	Chemical Abstracts Service
CoA	coenzyme A
IUPAC	International Union of Pure and Applied Chemistry
PPAP	phosphopantetheinyladenylate phosphate

References

Burns, K.L., Gelbaum, L.T., Sullards, M.C., and Bostwick, D.E., 2005. Isocoenzyme A. *The Journal of Biological Chemistry*. 280: 16550–16558.

Food and Nutrition Board, 1998. Dietary reference intakes for thiamin, riboflavin, niacin, vitamin B_6, folate, vitamin B_{12}, pantothenic acid, biotin, and choline. National Academy Press, Washington DC, USA. pp. 357–373.

Lipmann, F., Kaplan, N.O., Novelli, D., and Tuttle, L.C., 1950. Isolation of coenzyme A. *The Journal of Biological Chemistry*. 186: 235–243.

Majerus, P.W., Alberts, A.W., and Vagelos, P.R., 1965. Acyl carrier protein, IV. The identification of 4′-phosphopantetheine as the prosthetic group of the acyl carrier protein. *Proceedings of the National Academy of Science of the United States of America*. 53: 410–417.

Sarett, H.P., and Barboriak, J.J., 1963. Inhibition of D-pantothenate by L-pantothenate in the rats. *The American Journal of Clinical Nutrition*. 13: 378–384.

Stiller, E.T., Harris, S.A., Finkelstein, J., Keresztesy, J.C., and Folkers, K., 1940. Pantothenic acid. VIII. The total synthesis of pure pantothenic acid. *Journal of the American Chemical Society*. 62: 1785–1790.

Williams, R.J., Lyman, C.M., Goodyear, G.H., Truesdail, H.T., and Holaday, D., 1933. 'Pantothenic acid,' a growth determinant of universal biological occurrence. *Journal of the American Chemical Society*. 55: 291–292.

CHAPTER 9
The Chemistry and Biochemistry of Vitamin B_6: Synthesis of Novel Analogues of Vitamin B_6

DAJANA GAŠO SOKAČ,*[a,b] SPOMENKA KOVAČ,[a,b] VALENTINA BUŠIĆ,[b] COLIN R. MARTIN[c] AND JASNA VORKAPIĆ FURAČ[d]

[a] Faculty of Food Technology, University J.J. Strossmayer in Osijek, Kuhačeva 18, 31000 Osijek, Croatia; [b] Department of Chemistry, University J.J. Strossmayer in Osijek, Kuhačeva 20, 31000 Osijek, Croatia; [c] Faculty of Education, Health and Social Sciences, University of the West of Scotland, University Campus Ayr, KA8 0SR, UK; [d] Faculty of Food Technology and Biotechnology, University of Zagreb, Pierottieva 6, 10000 Zagreb, Croatia
*Email: dajana.gaso@ptfos.hr

9.1 Introduction

Vitamin B_6 is water-soluble and multimodal with respect to its numerous and diverse physiological functioning, metabolism and chemistry. Vitamin B_6 is the recommended generic descriptor for the group of 3-hydroxy-2-methyl-5-hydroxymethylpyridine derivatives exhibiting the biological activity of pyridoxine (Fennema 1996). The various forms of vitamin B_6 differ according to the nature of substituent at 4-position of the pyridine ring.

The name pyridoxine (PN) is a trivial designation of one vitamin B_6 component in which the substituent is an hydroxymethyl [3-hydroxy-4,5-bis(hydroxymethyl)-2-methylpyridine]. The biologically active analogues are

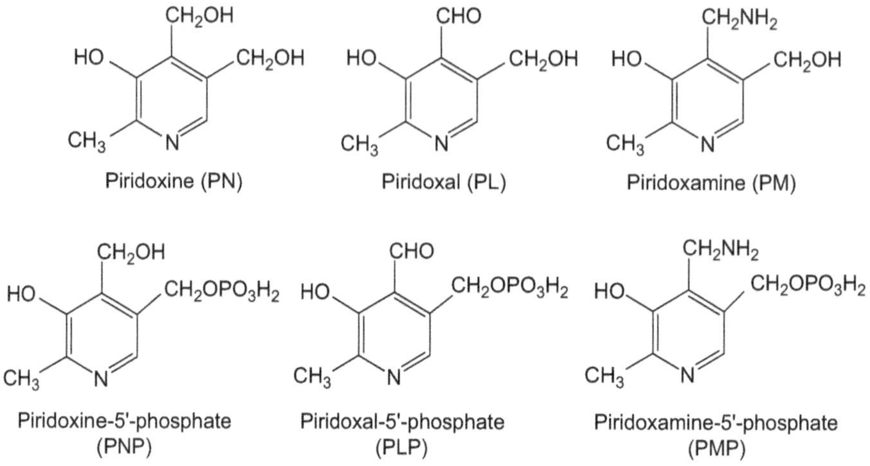

Figure 9.1 Chemical structures of vitamin B_6 that comprise mixture of vitamers.

pyridoxal (aldehyde, PL) and pyridoxamine (amine, PM) (Figure 9.1). The three basic forms can also be phosphorylated at the 5′-hydroxymethyl group to give pyridoxine-5′-phosphate (PNP), pyridoxal-5′-phosphate (PLP), pyridoxamine-5′-phosphate (PMP).

Vitamin B_6 occurs widely in meat, whole grains, legumes, leafy vegetables, potatoes and bananas. All forms of vitamin B_6 are colourless crystals at room temperature. They are readily soluble in water, but they are minimally soluble in organic solvents. Directly relevant to food producers and consumers is the relative stability of vitamin B_6, all components being generally labile, though there are differences in the degree to which each is degraded. B_6 component sensitivity is influenced by pH, but although they remain relatively heat stable in an acidic medium, they are heat labile in alkaline medium. Storage and processing lead to vitamin B_6 losses, particularly the application of dry heat (baking). Vitamin B_6 loss is 45% in the cooking of meat and 20–30% in the cooking of vegetables. During milk sterilization, a reaction with cysteine transforms the vitamin into an inactive thiazolidine derivative and this reaction may account for vitamin losses also in other heat-treated foods. The most stable form of the vitamin is pyridoxal which is used for vitamin fortification of food.

9.2 Discovery of Vitamin B_6

The formula of vitamin B_6 was first published by Ohdake (1932) at the same time as several other scientific groups were working on the elucidation of the vitamin B family members. György in 1934 first described vitamin B_6 as the active 'rat pellagra prevention factor' and subsequently named the substance pyridoxine due to its structural homology to pyridine (György 1938). It was

The Chemistry and Biochemistry of Vitamin B_6

also demonstrated during this era that vitamin B_6 could exist in other chemical forms; pyridoxine was first isolated in crystalline form from yeast (Kuhn and Wendt 1938) and from rice polishing (Ichiba and Michi 1938; Keresztesy and Stevens 1938).

The first total synthesis was accomplished in 1939 by three groups of scientists (Harris and Folkers 1939; Ichiba and Michi 1939; Kuhn *et al.* 1939). Harris and Folkers also described a synthesis of pyridoxine that was later augmented and technically tailored for industrial purposes. Synthesis was obtained by the condensation of 1-ethoxypentane-2,4-dione and cyanacetamide, Scheme 9.1 (Harris *et al.* 1939). According to Harris and Folkers synthetic vitamin B_6 hydrochloride is chemically identical with the natural vitamin B_6 hydrochloride, also defined by it being biologically active.

Procedures for pyridoxine synthesis are divided into two classes: synthesis from the substrate with formed ring and those who need ring-forming procedures. The ring skeleton can be built up from one, two or three segments. The Diels–Alder reaction is useful in organic synthesis and is widely used for the preparation of six-membered compounds. Oxazoles that have a conjugate azadiene system in the reaction of addition to olefins produce substituted pyridines. This methodology was used in the synthesis of natural products, among them vitamin B_6 (pyridoxal and pyridoxyl alkaloids). A novel series of pyridine derivatives and analogues of vitamin B_6 have been synthesized *via* one-pot Diels–Alder reactions of 2,4-dimethyl-5-methoxyoxazole with different types of dienophiles (Bondock 2005).

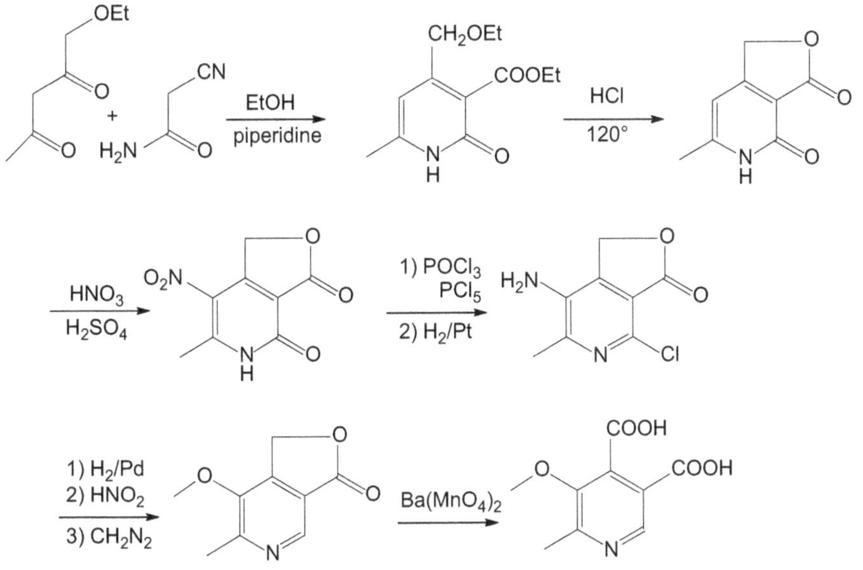

Scheme 9.1 First total synthesis of dibasic acid obtained by Harris, Stiller and Folkers.

9.3 Function of Vitamin B_6

The phosphorylated and non-phosphorylated forms of vitamin B_6 have various physical and chemical properties. Vitamin B_6 in the form of pyridoxal-5′-phosphate (PLP) and to a lesser extent, pyridoxamine-5′-phosphate (PMP), functions as a coenzyme in over 100 enzymatic reactions. All the forms of vitamin B_6 possess vitamin activity because they can be converted *in vivo* to pyridoxal. PN, PM and PL are converted to 5′-phosphate by a single kinase enzyme which in the brain and liver is most active with zinc. PNP and PMP are then converted to PLP by flavin dependent oxidase; this is the reason why vitamin B_2 deficiency causes a fall in available PLP (Holman 1995). Human cells can synthesize PLP from three vitamers *via* the B_6 salvage pathway but cannot synthesize PLP *de novo* and must obtain it from dietary sources.

Two pathways for PLP biosynthesis *de novo* are known in plant and microorganisms. The first was extensively studied in *Escherichia coli*. This pathway is articulated in two branches which join in a ring closure reaction catalysed by PNP synthase. One branch started from pyruvate and glyceraldehyde 3-phosphate and the other from 4-phosphohydroxy-L-threonine (derived from erythrose 4-phosphate). The second route is when PLP is formed from glutamine, either ribose 5-phosphate or ribulose 5-phosphate and either dihydroxyacetone phosphate or glyceraldehyde 3-phosphate by action of the PLP synthase complex (Roje 2007). This pathway was discovered in fungi and it has become clear that it is much more widely distributed than the first pathway. It exists in *Archaea*, most eubacteria and plants.

The remarkably versatile chemistry of PLP is due to its ability to form stable Schiff's base adducts with amino groups and to act as an effective electron sink to stabilize reaction intermediates. PLP is covalently bound to enzymes *via* a Schiff's base with the ε-amino group of lysine in the active site. PLP exist under physiological conditions in two isomeric forms (Figure 9.2).

PLP dependent enzymes belong to five enzyme classes: transferases, hydrolases, lyases, isomerases and oxidoreductases. Nearly half of PLP-dependent enzymes are involved in important steps of amino acid metabolism (decarboxylation, α,β-elimination reactions, racemization and transamination). During the reaction of amino acids (and other amino compounds), the

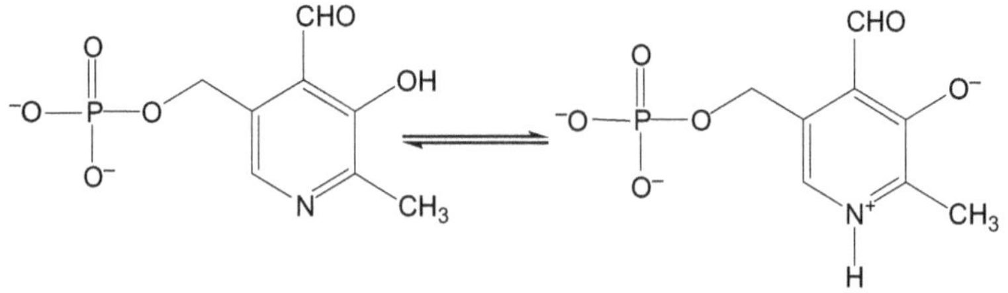

Figure 9.2   Isomeric forms of PLP.

carbonyl group of the coenzyme forms a Schiff's base with the amino group of the substrate, leading to labilization of bonds around the α-carbon, followed by decarboxylation to the amine, amino transfer or a variety of reactions of the side-chain of the amino acid.

The PLP-dependent enzymes are also involved in hemoglobin formation (δ-aminolevulinic acids synthase) and chlorophyll biosynthesis (δ-glutamate-1-semialdehyde aminomutase). The PLP represents an important cofactor for the degradation of storage carbohydrates (PLP-dependent glycogen phosphorylase). The PLP in the reaction of glycogen phosphorylase remains tightly bound through its carbonyl group to a lysine residue in the enzyme. The reactive group is the phosphate and acts as a proton buffer to facilitate the phosphorolysis of glycogen by inorganic phosphate (Palm *et al.* 1990). PLP is involved in fatty acid metabolism and, together with zinc and magnesium, is required for the action of δ-6-desaturase.

Apart from its function as a cofactor in enzymes, over the past decade researchers found that vitamin B_6 could have the role in preventing oxidative stress of cells and that it is an important quencher of singlet oxygen, with quenching rates comparable to those of vitamins C and E (Ehrenshaft *et al.* 1999; Bilski *et al.* 2000).

Previous research has demonstrated that pyridoxine plays a role in the resistance of the fungus *Cercospora nicotianae* to its own abundantly produced strong photosensitizer of singlet molecular oxygen ($^1O_2$), cercosporin. Pyridoxine appears to contribute strongly to the unusual resilience of fungi to photooxidative stress. Pyridoxine can function in photosensitizer resistance by quenching $^1O_2$, but research also indicates that $^1O_2$ degrades pyridoxine. Oxidative degradation of vitamin B_6 can occur during processes in which $^1O_2$ is produced, such as food processing and storage, and during photosensitization in the skin. Investigations have revealed that vitamin B_6 is a strong antioxidant with potential importance during the plant pathogen defence response (Bilski *et al.* 2000; Danslow *et al.* 2005).

Low levels of vitamin B_6 may cause hyperhomocysteinaemia since vitamin B_6 acts as a cofactor for *trans*-sulfation of cysteine (Siri *et al.* 1998). There is evidence that an elevated homocysteine level is a risk factor for heart disease and stroke. Endo *et al.* (2006) reported that vitamin B_6 deficiency induced the oxidant stress which accelerates atherosclerosis. They also highlighted that the antioxidant activity of vitamin B_6 may suppress the homocysteine-induced atherosclerosis. Vitamin B_6 levels may be important in the prevention of coronary heart disease; intake of vitamin B_6 above the current recommended dietary allowance has been shown to be instrumental in the primary prevention of coronary heart disease among women (Rimm *et al.* 1998).

Vitamin B_6 has a crucial role in 1-carbon metabolism, which involves DNA synthesis and DNA methylation, and can modulate gene expression. Electrochemical and spectroscopic studies (Liu *et al.* 2008) have shown that vitamin B_6 can interact with DNA. These findings are helpful in understanding the biological functions of vitamin B_6.

9.4 Diversity of Vitamin B_6 Derivatives

The chemistry of pyridoxal and some of its phosphorylated derivatives has been the subject of continuous research directed toward the elucidation of the relationship between structure and biological activity. A variety of different derivatives of vitamin B_6 have been described. For many of them the precise function is not understood, yet these derivatives have novel functions and could be crucial to fully understanding the biological relevance of vitamin B_6.

Ginkgotoxin (4′-O-methylpyridoxine) (**1**) from the tree *Ginko biloba* and African tree *Albizia julibrisson* is one vitamin B_6 derivative (Figure 9.3). *Albizia julibrissin* also synthesizes other, more complex, derivatives but the biological function mechanism is not fully understood. A likely possibility is that they serve as protecting compounds against pathogens due to their toxicity.

Several vitamin B_6 derivatives, especially pyridoxine, exist in plants as ß-D-glucosides. The major form of glycosylated vitamin B_6 is pyridoxine-5′-ß-D-glucoside (**2**), which comprises between 5 and 80% of the total vitamin B_6 in various plants. Glycosylated forms are abundant and detected in soybean, rice and the *Ginko biloba* tree. There is no explanation yet for the high amount of glycosylated vitamin B_6 found, but a plausible account is that it might serve as storage compounds for the vitamin, or even carbohydrates, which can be used upon demand. Isotopic studies have indicated that dietary PN-glycoside exhibits ∼25% bioavailability in rats and ∼58% in humans, relative to simultaneously ingested pyridoxine. Investigations have also revealed that

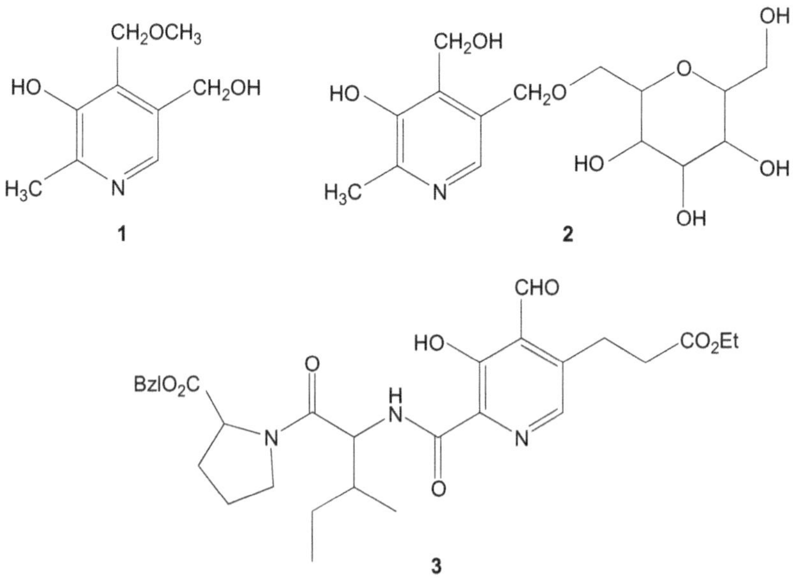

Figure 9.3 Structure of some novel derivaties of vitamin B_6 which possess biological activity.

PN-glucoside serves as a source of nutritionally available vitamin B_6, with 50% bioavailability relative to PN (Maeno *et al.* 1997; Nakano *et al.* 1997).

During the past decade researchers discovered new functions of vitamin B_6. Researchers synthesized various artificial vitamin B_6 derivatives and tested the specific inhibitory effects for various cathepsins. The pyridoxal derivate (**3**, Figure 9.3) showed selective inhibition for cathepsin K. Therefore, investigators suggested that intra-lysosomal protein degradation caused by cathepsins may be suppressed by administration of these and similar vitamin B_6 derivatives. The findings suggest that vitamin B_6 derivatives might also induce the suppression of other physiological functions mediated by the cathepsins (Matsui *et al.* 2000).

Since pyridoxal (PL) has an aldehyde group that could easily be converted to an oxime group and oximes derived from pyridinium aldehyde are used as antidotes in the therapy of poisoning with organophosphorus compounds (nerve agents, insecticides), researchers investigated oxime derivatives of vitamin B_6. The dioximes derived from vitamin B_6 were tested and their reactivating potency on acetylcholinesterase (AChE) from human erythrocytes inhibited by sarin, soman, tabun and VX poison were determined. Findings show that oximes have promising reactivating potency, and it has been pointed out that vitamin B_6 and its derivatives have diverse function with human functioning (Milatović and Vorkapić-Furač 1989).

New derivatives of pyridoxal (**4**, Figure 9.4) have been synthesized and tested as reactivators of AChE inhibited by the chemical warfare agent tabun and the insecticide paraoxon (Gašo-Sokač *et al.* 2010). To assess the potential antimicrobial properties of new vitamin B_6 derivatives, they have been tested against a variety of pathogens including Gram-positive (*Bacillus subtilis, Enterococcus faecilis, Listeria monocytogenes, Staphylococcus aureus*) and Gram-negative bacteria (*Escherichia coli, Salmonella enteritidis, Serratia marceancens, Yersinia entrocolitica*) (Gašo-Sokač, unpublished results). Results show that new derivatives of vitamin B_6 have an antimicrobial effect especially on *Bacillus subtilis* and *Yersinia entrocolitica*.

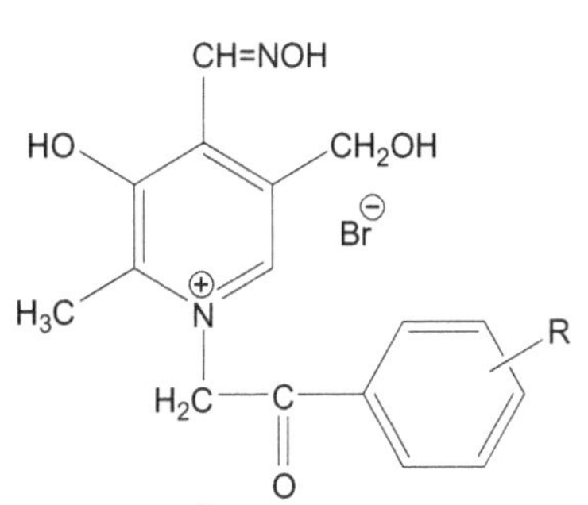

Figure 9.4 Derivatives of pyridoxal that can reactivate AChE inhibited with nerve agents and insecticides.

Summary Points

- This chapter focuses on vitamin B_6.
- Vitamin B_6 exists in three basic chemical forms: pyridoxine (an alcohol), pyridoxal (an aldehyde) and pyridoxamine (a primary amine). The three major forms can be phosphorylated at the 5′-hydroxymethyl group.
- Vitamin B_6 is an essential participant in numerous metabolic processes and has a role as a cofactor in enzyme-catalysed reactions.
- The first total synthesis of vitamin B_6 was obtained by the condensation of acetylacetone and cyanacetamide.
- Alternative methods for the synthesis of pyridoxine analogues have been described, which include Diels–Alder reactions of substituted oxazoles.
- All living beings need vitamin B_6 for their existence. However, only microorganisms and plants are able to synthesize it *de novo*.
- Two independent *de novo* biosynthetic routes are known.
- Vitamin B_6 is essential for more than 100 enzymes involved in protein metabolism. Pyridoxal-5′-phosphate is covalently bound to enzymes *via* a Schiff's base with an ε-amino group of lysine in the enzyme.
- PLP-dependent enzymes are involved in important steps of amino acid metabolism, hemoglobin formation and chlorophyll biosynthesis. Vitamin B_6 could have a role in preventing oxidative stress of the cells and is an impotent quencher of singlet oxygen.
- A deficiency of vitamin B_6 in the human diet encompasses the full spectrum of clinical disorders such as hyperhomocysteinaemia and increases the risk for atherosclerosis.
- It has been shown that vitamin B_6 can interact with DNA.

Key facts about Antidotes and Oxidative Stress

- Organophosphorus compounds are widely used in agriculture as insecticides, in industry and in military technology as chemical warfare agents. They are extremely potent inhibitors of the enzyme AChE, which is responsible for the termination of the action of acetylcholine (ACh) at cholinergic synapses.
- The inhibition of AChE leads to the accumulation of the neurotransmitter ACh in synapses of the central and peripheral nervous systems and overstimulation of post-synaptic cholinergic receptors. Exposure to even small amounts of an organophosphorus compound can be fatal as the poison causes seizures, convulsions and lesions of the central nervous system. The current standard treatment for poisoning usually consists of combined administration of anticholinergic drugs (preferably atropine) and AChE reactivators (called oximes).
- Commonly used reactivators are characterised by the presence of several structural features: functional oxime group, quaternary nitrogen group and different length of linking chain between two pyridinium rings in the case of bispyridinium reactivators.

- As well as being antidotes, B_6 derivatives can act against oxidative stress.
- Oxidative stress is mainly caused by the presence of the reactive oxygen species. Reactive oxygen species is a term collectively describing radical (OH•, NO•, RO_2•) and other non-radical oxygen derivatives (peroxynitrite $ONOO^-$, hydrogen peroxide H_2O_2, singlet oxygen, ozone).
- Free radicals react with lipids, proteins and DNA; oxidation can damage them, disturb normal function and contribute to a variety of disease states.
- Oxidative stress occurs when the generation of reactive oxygen species in an organism exceeds its ability to neutralize and eliminate them.

Definitions of Words and Terms

Alkaloid. A natural occurring compound isolated from a plant source that contains a basic nitrogen.
Antioxidant. A compound that stops an oxidation.
De novo. Referring to the synthesis of complex molecules from simple starting materials.
Diels–Alder reaction. An addition reaction between a 1,3-diene and a dienophile to form a cyclic compound.
Diene. A hydrocarbon that contains two carbon–carbon double bonds.
Glycoside. A monosaccharide in which the hemiacetal group has been converted to an acetal with an alkoxy group bonded to an anomeric carbon.
In vivo. Performing an experiment with a living organism.
Olefin. An organic compound possessing a carbon–carbon double bond.
Oximes. An organic compound having the general structure $R_2C=NOH$.
Schiff's base. Another name for an imine, a compound of the type $R_2C=NR'$.
Singlet molecular oxygen. The lowest excited state of the dioxygen molecule.

List of Abbreviations

ACh	acetylcholine
AChE	acetylcholinesterase
DNA	deoxyribonucleic acid
PL	pyridoxal
PLP	pyridoxal-5'-phosphate
PM	pyridoxamine
PMP	pyridoxamine-5'-phosphate
PN	pyridoxine
PNP	pyridoxine-5'-phosphate

References

Bilski, P., Li, M.Y., Ehrenshaft, M., Daub, M.E., and Chignell, C.F., 2000. Vitamin B_6 (pyridoxine) and its derivatives are efficient singlet oxygen quenchers and potential fungal antioxidants. *Photochemistry and Photobiology.* 71: 129–134.

Bondock, S., 2005. One-pot synthesis of pyridine derivatives *via* Diels-Alder reactions of 2,4-dimethyl-5-methoxyoxazole. *Heteroatom Chemistry.* 16: 49–55.

Danslow, S.A., Walls, A.M., and Daub, E.M., 2005. Regulation of biosynthetic genes and antioxidant properties of vitamin B_6 vitamers during plant defense responses. *Physiological and Molecular Plant Pathology.* 66: 244–255.

Ehrenshaft, M., Bilski, P., Li, M.Y., Chignell, C.F., and Daub, M.E., 1999. A highly conserved sequence is a novel gene involved in de novo vitamin B_6 biosynthesis. *Proceedings of the National Academy of Sciences of the United States of America.* 96: 9374–9378.

Endo, N., Nishiyama, K., Otsuka, A., Kanouchi, H., Taga, M., and Oka, T., 2006. Antioxidant activity of vitamin B_6 delays homocysteine-induced atherosclerosis in rats. *British Journal of Nutrition.* 95: 1088–1093.

Fennema, O.R., 1996. *Food Chemistry*, 3rd edition. Marcel Dekker Publisher, New York, USA, pp. 580–588.

Gašo-Sokač, D., Katalinić, M., Kovarik, Z., Bušić, V., and Kovač, S., 2010. Synthesis and evaluation of novel analogues of vitamin B6 as reactivators of tabun and paraoxon inhibited acetylcholinesterase. *Chemico-Biological Interactions.* 187: 234–237.

György, P., 1938. Crystalline vitamin B_6. *Journal of the American Chemical Society.* 60: 983–984.

Harris, S.A., and Folkers, K., 1939. Synthetic vitamin B_6. *Science.* 89: 347.

Harris, S.A., Stiller, T.E., and Folkers, K., 1939. Structure of vitamin B_6. *Journal of the American Chemical Society.* 61: 1242.

Holman, P., 1995. Pyridoxine—vitamin B_6. *Journal of Australian College of Nutritional & Environmental Medicine.* 14: 5–16.

Ichiba, A., and Michi, K., 1938. Crystalline vitamin B_6. *Scientific Papers of the Institute of Physical and Chemical Research.* 34: 623–626.

Ichiba, A., and Michi, K., 1939. Isolation of vitamin B. *Scientific Papers of the Institute of Physical and Chemical Research.* 36: 173.

Keresztesy, J.C., and Stevens, J.R., 1938. Vitamin B-6. *Proceedings of the Society for Experimental Biology and Medicine.* 38: 64–65.

Kuhn, R., and Wendt, G., 1938. Über das antidermatitische Vitamin der Hefe. *Berichte der Deutschen Chemischen Gesellschaft.* 71: 780.

Kuhn, R., Westphal, K., Wendt, G., and Westphal, O., 1939. Synthesis of adermin. *Naturwissenschaften.* 27: 469.

Liu, S.-Q., Cao, M.-L., and Dong S.-L., 2008. Electrochemical and ultraviolet-visible spectroscopic studies on the interaction of deoxyribonucleic acid with vitamin B_6. *Bioelectrochemistry.* 74: 164–169.

Maeno, M., Morimoto, Y., Hayakawa, T., Suzuki, Y., and Tsuge, H., 1997. Feeding experiments of pyridoxine derivatives as vitamin B_6. *International Journal for Vitamin and Nutrition Research.* 67: 444–449.

Matsui, A., Tsuzuki, H., Murata, E., Tada, Y., Asao, T., and Katunuma, N., 2000. Inhibition of cathepsin activities by vitamin B_6 derivatives. *BioFactors.* 11: 117–120.

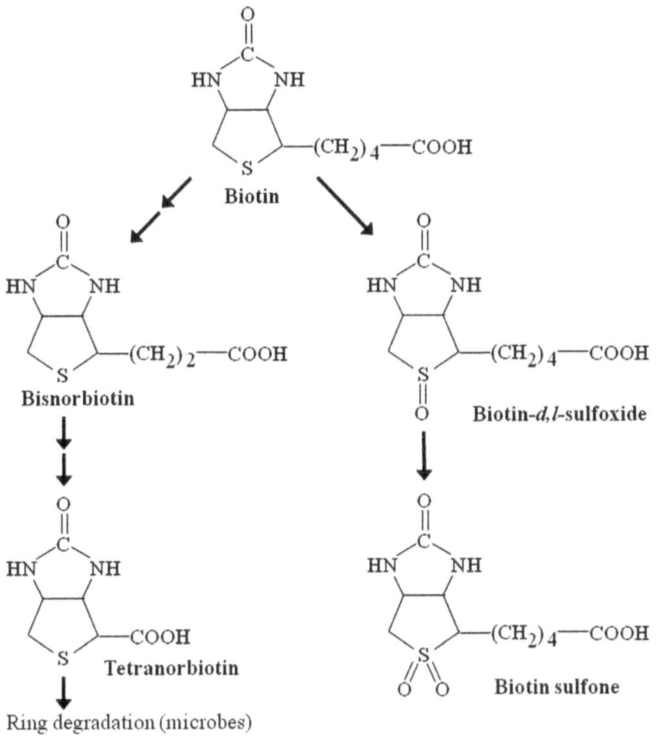

Biochemistry of Biotin

Figure 10.1 Biotin is catabolized by both sulfur oxidation and β-oxidation. (Data from McCormick and Wright 1971; Lee et al. 1972; Zempleni et al. 2009).

10.1.2 Biosynthesis

Humans cannot synthesize biotin and depend on dietary biotin originating in microbial and plant biosynthetic pathways. The route of biosynthesis of biotin was largely elaborated by Rolfe and Eisenberg (1968) working with *Escherichia coli*. In this pathway, dethiobiotin is formed from pimelyl-CoA (which can be synthesized from oleic acid) and carbamyl phosphate. Sulfur is incorporated into dethiobiotin in a synthase-dependent step, generating biotin (reviewed by Camporeale and Zempleni 2006). Figure 10.1 shows the chemical structure of biotin.

10.2 Catabolism of Biotin

McCormick and coworkers identified two pathways of biotin catabolism using microbial and rat models (McCormick and Wright 1971; Lee et al. 1972) (Figure 10.1). These pathways are conserved in humans (Zempleni et al. 2009), with degradation of the heterocyclic ring being the notable exception (Lee et al. 1972) (Table 10.1). The valeric acid side chain of biotin is catabolized by

Table 10.1 Serum concentrations and urinary excretions of biotin and catabolites.[a]

Compound	Serum (pmol/L)	Urine (nmol/24 h)
Biotin	244 ± 61	35 ± 14
Bisnorbiotin	189 ± 135	68 ± 48
Biotin-d,l-sulfoxide	15 ± 33	5 ± 6
Bisnorbiotin methyl ketone	N.D.[b]	9 ± 9
Biotin sulfone	N.D.[b]	5 ± 5
Total biotin	464 ± 178[c]	122 ± 66

[a]Means ± SD are reported ($n = 15$ for serum; $n = 6$ for urine). Data from Zempleni *et al.* (2009).
[b]N.D., not determined.
[c]Including three unidentified biotin catabolites.

β-oxidation, where the repeated cleavage of two-carbon units leads to the formation of bisnorbiotin, tetranorbiotin and intermediates known from the β-oxidation of fatty acids (*i.e.* α,β-dehydro-, β-hydroxy- and β-keto-intermediates). β-Ketobiotin and β-ketobisnorbiotin are unstable and may decarboxylate spontaneously to form bisnorbiotin methyl ketone and tetranorbiotin methyl ketone (McCormick and Wright 1971). Microorganisms, but not mammals, cleave and degrade the heterocyclic ring of tetranorbiotin (McCormick and Wright 1971). Mammals are capable of oxidizing the sulfur in the heterocyclic ring in biotin to produce biotin-l-sulfoxide, biotin-d-sulfoxide and biotin sulfone (McCormick and Wright 1971). Likely, sulfur oxidation in the biotin molecule occurs in the smooth endoplasmic reticulum in a reaction that depends on nicotinamide adenine dinucleotide phosphate (NADP). Biotin is also catabolized by a combination of both β-oxidation and sulfur oxidation, producing compounds such as bisnorbiotin sulfone.

10.3 Biochemistry of Biotin

10.3.1 Biotin-dependent Carboxylases

Biotin serves as a covalently bound coenzyme for acetyl-CoA carboxylases (ACC) 1 and 2, pyruvate carboxylase (PC), propionyl-CoA carboxylase (PCC) and 3-methylcrotonyl-CoA carboxylase (MCC) in mammals and other metazoans (Zempleni *et al.* 2009). Additional carboxylases exist in microbes (Knowles 1989). The attachment of biotin to the ε-amino group of a specific lysine residue in holocarboxylases is catalysed by holocarboxylase synthetase (HLCS) or microbial orthologs such as BirA. Biotinylation of carboxylases requires ATP to produce the energy-rich intermediate biotinyl-5'-AMP (Zempleni *et al.* 2009).

Holocarboxylases mediate the covalent binding of bicarbonate (not carbon dioxide) to organic acids, using 1'-N-carboxybiotinyl as an HLCS-bound carboxyl donor (Knowles 1989). Both the cytoplasmic ACC1 and the mitochondrial ACC2 catalyse the binding of bicarbonate to acetyl-CoA to generate malonyl-CoA (Figure 10.2), but the two isoforms play distinct roles in intermediary metabolism (Kim *et al.* 1997). ACC1 produces malonyl-CoA for the synthesis of fatty acid synthesis in the cytoplasm and ACC2 is an important regulator of fatty acid oxidation in mitochondria. The malonyl-CoA produced by ACC2 inhibits mitochondrial uptake of fatty acids for β-oxidation.

PC, PCC and MCC localize in mitochondria. PC is a key enzyme in gluconeogenesis. PCC catalyses an essential step in the metabolism of propionyl-CoA, which is produced in the metabolism of some amino acids, the cholesterol side chain and odd-chain fatty acids. MCC catalyses an essential step in leucine metabolism (Figure 10.2). Both PCC and MCC are composed of non-identical subunits, *i.e.* biotinylated α subunits and nonbiotinylated β subunits, which are encoded by distinct genes.

Proteolytic degradation of holocarboxylases leads to the formation of biotinyl peptides and biocytin (biotin-ε-lysine). These compounds are further degraded by biotinidase to release biotin, which is then recycled in holocarboxylase synthesis (Wolf *et al.* 1985).

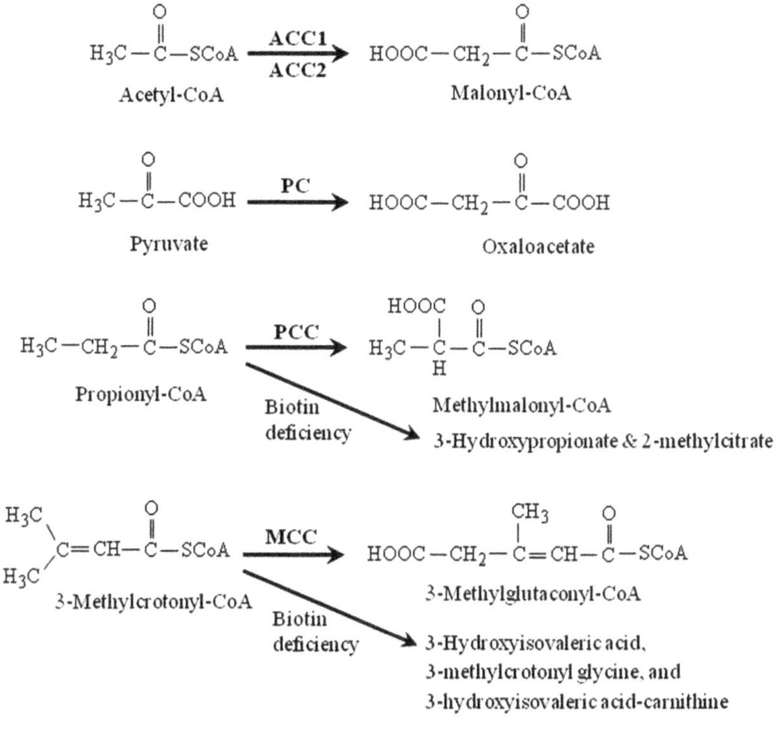

Figure 10.2 Biotin serves as a covalently bound coenzyme for carboxylases. (Data from Zempleni *et al.* 2009).

10.3.2 Biotinylation of Histones

Chromatin is made of repetitive nucleoprotein complexes, the nucleosomes (Wolffe 1998). The *N*-terminal tails of core histones protrude from the nucleosomal surface. Covalent modifications of these tails affect the structure of chromatin and are crucial for gene regulation by epigenetic mechanisms. Some modifications also exist in the globular domain and *C*-terminal regions (Kouzarides and Berger 2007; Zempleni *et al.* 2009) (Figure 10.3).

To date, at least five biotinylation sites have been identified in histones H3 [lysine (K)-4, K9, K18 and probably K23] and H4 (K8, K12 and probably K16) (Camporeale *et al.* 2004; Kobza *et al.* 2005; Kobza *et al.* 2008). K9 and K13 in histone H2A might also be biotinylated (Chew *et al.* 2006), but the abundance of these two marks appears to be very low (Stanley *et al.* 2001). Studies with synthetic HLCS substrates provide unambiguous evidence that biotinylation of histones by HLCS is a substrate-specific process (Hassan *et al.* 2009a). Histone biotinylation is a comparably rare event (<0.1% of histones are biotinylated), but the abundance of an epigenetic mark is not necessarily a marker for its importance. For example, serine-14 phosphorylation in histone H2B and histone poly(ADP-ribosylation) are detectable only after induction of apoptosis

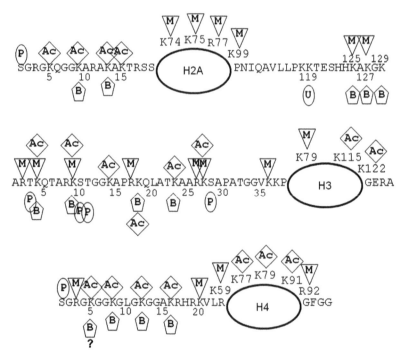

Figure 10.3 Histones H2A, H3 and H4 are posttranslationally modified. Abbreviations: Ac, acetate; B, biotin; M, methyl; P, phosphate; U, ubiquitin (data from Kouzarides and Berger 2007; Zempleni *et al.* 2009). Posttranslational modifications in histone H2B are not depicted.

and major DNA damage, respectively, but the role of these epigenetic marks in cell death is unambiguous (Zempleni *et al.* 2011).

HLCS-dependent biotinylation marks are enriched in telomeric repeats, long-terminal repeats, pericentromeric alpha satellite repeats and in repressed, yet transcriptionally poised, loci (Camporeale *et al.* 2007; Chew *et al.* 2008; Wijeratne *et al.* 2010). Repression of long terminal repeats and, therefore, genome stability depends on biotin and HLCS in human and murine cell cultures, *Drosophila melanogaster*, and humans (Chew *et al.* 2008). Due to the scarcity of histone biotinylation marks, it is uncertain whether the roles of biotin in gene repression at the chromatin level are caused by histone biotinylation or by physical interactions between HLCS and other chromatin proteins to form multiprotein repressor complexes. The two mechanisms are not mutually exclusive and we have evidence in support of both theories. First, we have shown that HLCS physically interacts with histone H3 and has biotinyl:histone ligase activity (Bao *et al.* 2011) and that biotinylation of histones mediates chromatin condensation (Filenko *et al.* 2011). Second, we have shown that HLCS-dependent histone biotinylation cross-talks with histone and cytosine methylation marks (Camporeale *et al.* 2007; Chew *et al.* 2008) and that HLCS interacts physically with distinct chromatin proteins (J. Jing *et al.*, and Y. Li *et al.*, unpublished observations).

10.3.3 HLCS

Consistent with the important roles of HLCS in intermediary metabolism and epigenetics, no living HLCS null individual has ever been reported, suggesting embryonic lethality. HLCS knockdown studies ($\sim 30\%$ residual activity) produce phenotypes such as decreased life span and heat resistance in *Drosophila melanogaster* (Camporeale *et al.* 2006) and aberrant gene regulation in human cell lines (Chew *et al.* 2008; Gralla *et al.* 2008). Mutations have been identified and characterized in the human *HLCS* gene; these mutations cause a substantial decrease in HLCS activity and metabolic abnormalities (Suzuki *et al.* 2005; NCBI 2008a). Unless diagnosed and treated early, HLCS deficiency appears to be uniformly fatal (Thuy *et al.* 1999). In addition to the approximately 30 mutations in the human *HLCS* gene that have been described, 2200 SNPs have been mapped in the *HLCS* locus (NCBI 2008b) but their biological significance is unknown.

HLCS is present in both nuclear and extranuclear structures (Narang *et al.* 2004; Chew *et al.* 2006). Nuclear HLCS is a chromatin protein (Camporeale *et al.* 2006); its binding to chromatin is mediated by physical interactions with histones H3 and H4 (Bao *et al.* 2011). Our knowledge of HLCS regulation is limited to the following observations:

(i) Both the abundance of HLCS mRNA and the nuclear translocation of HLCS depend on biotin (Gralla *et al.* 2008).
(ii) The human HLCS promoter has been tentatively identified (Warnatz *et al.* 2010) but not yet characterized in great detail.

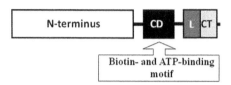

Figure 10.4 Four domains have been identified in holocarboxylase synthetase. Abbreviations: CD, central domain; L, linker domain; CT, C-terminus. (Data from Hassan et al. 2009b).

(iii) The expression of HLCS is repressed by miR-539 (Bao et al. 2010).
(iv) HLCS-dependent histone biotinylation cross-talks with cytosine methylation in gene regulation. Specifically, histone biotinylation is substantially impaired when cytosine methylation marks are erased by treating cells with 5-aza-2'-deoxycytidine (Chew et al. 2008). Partially, the effects of 5-aza-2'-deoxycytidine on HLCS expression are mediated by demethylation of the promoters in the two human *miR-153* genes, leading to high levels of miR-153 and, subsequently, miR-153 dependent degradation of HLCS mRNA (Bao et al. 2012).

Human HCLS contains four domains, namely *N*-terminus, central domain (which contains the biotin- and ATP-binding motifs), linker domain and *C*-terminus (Hassan et al. 2009b) (Figure 10.4). The *N*-terminus, central domain and *C*-terminus participate in substrate recognition.

10.3.4 Biotinidase

Breakdown of carboxylases leads to the release of biotinylated polypeptides. Biotinidase releases free biotin from these peptides for recycling in the synthesis of new holocarboxylases (cf. biotinidase deficiency by Wolf and Heard 1991). In the 1990s, biotinidase was considered the enzyme that might be responsible for mediating the binding of biotin to histones (Hymes et al. 1995). Clearly, biotinidase has catalytic activity to mediate biotinylation of histones *in vitro* (Camporeale et al. 2004). However, evidence suggests that HLCS is the enzyme that mediates biotinylation of histones *in vivo* (Camporeale et al. 2006) and that biotinidase might play a role in the enzymatic removal of biotin from histones (Ballard et al. 2002), which is consistent with their roles in carboxylase metabolism.

10.3.5 Cell Signalling

Pioneering studies by Dakshinamurti and co-workers suggested a role for biotin in the regulation of the *glucokinase* gene (Dakshinamurti and Cheah-Tan 1968). Subsequently, it was shown that biotin affects gene regulation by 'classical' signalling pathways such as cGMP, NF-κB, Sp1 and Sp3, nitric oxide and receptor tyrosine kinases, and by the intermediate biotinyl-5'-AMP (Solorzano-Vargas et al. 2002; Zempleni et al. 2009). Biotin also affects gene expression at the post-transcriptional level (Collins et al. 1988).

10.4 Biotin–Drug Interactions

The anticonvulsants primidone and carbamazepine inhibit biotin uptake into brush-border membrane vesicles from human intestine (Zempleni et al. 2009). Long-term therapy with anticonvulsants increases both biotin catabolism and urinary excretion of 3-hydroxyisovaleric acid. These effects might be due to displacement of biotin from biotinidase by anticonvulsants, thereby affecting plasma transport, renal handling or cellular uptake of biotin.

Lipoic acid may be administered to treat heavy metal intoxication, to reduce signs of diabetic neuropathy, and to enhance glucose disposal in patients with noninsulin-dependent diabetes mellitus. Lipoic acid is structurally similar and competes with biotin for cellular uptake by a sodium-dependent multivitamin transporter (SMVT) and, perhaps, displaces biotin from HLCS. Chronic administration of pharmacological doses of lipoic acid decreases the activities of PC and MCC in rat liver to 64–72% of controls (Zempleni et al. 2009).

Summary Points

- The sodium-dependent multivitamin transporter is the quantitatively most important biotin transporter in the majority of human tissues. The monocarboxylate transporter 1 may play a specialized role in biotin uptake in cells from the lymphoid lineage.
- Biotin is a covalently bound coenzyme for acetyl-CoA carboxylases 1 and 2, 3-methylcrotonyl-CoA carboxylase, propionyl-CoA carboxylase and pyruvate carboxylase, which play essential roles in macronutrient metabolism.
- Holocarboxylase synthetase catalyses the covalent binding of biotin to carboxylases, whereas biotinidase catalyses the release of free biotin from degraded carboxylases.
- Biotin is also covalently attached to histones, but the abundance of biotinylated histones is low. It is unclear whether the effects of histone biotinylation in gene repression and genome stability are mediated by histone biotinylation or by interactions of holocarboxylase synthetase with other chromatin proteins.
- Biotin modulates cell signals such as cGMP, NF-κB, Sp1 and Sp3, nitric oxide, receptor tyrosine kinases and biotinyl-5'-AMP, thereby altering gene expression.
- Biotin is catabolized by sulfur oxidation in its heterocyclic ring and by β-oxidation of its valeric acid side chain.

Key Facts about Histones

- Histones H1, H2A, H2B, H3 and H4 are the quantitatively most important families of histones in mammals.
- Histones are small proteins (102–225 amino acids) and carry a positive net charge due to a large number of lysine and arginine residues in these molecules.
- Two dimers of histones H2A and H2B and a tetramer of histone H3-H3-H4-H4 form an octamer, around which 147 basepairs of DNA are wrapped to form the nucleosomal core particle.
- The *N*-terminal tails of histones and some amino acids in hinge regions and the *C*-terminus are exposed at the nucleosomal surface and are subject to a large number of posttranslational modifications.
- Posttranslational modifications of histones play essential roles in gene regulation, DNA repair and replication.

Definitions of Words and Terms

Chromatin: Chromatin comprises of DNA and DNA-binding proteins, *i.e.* histones and non-histone proteins (see Key Facts of Histones).

Epigenetics: An epigenetic trait is a stably inherited phenotype resulting from changes in a chromosome without alterations in the DNA sequence. Epigenetic marks include cytosine (hydroxy)methylation and various histone modifications.

Single nucleotide polymorphism (SNP): SNPs are genetic variations in genomes that occur when a single nucleotide in the genome is altered. SNPs usually arose through point mutations that have been evolutionarily conserved in a significant proportion of the population.

List of Abbreviations

ACC	acetyl-CoA carboxylase
HLCS	holocarboxylase synthetase
K	lysine
MCC	3-methylcrotonyl-CoA carboxylase
miR	microRNA
PC	pyruvate carboxylase
PCC	propionyl-CoA carboxylase
SMVT	sodium-dependent multivitamin transporter

Acknowledgments

A contribution of the University of Nebraska Agricultural Research Division, supported in part by funds provided through the Hatch Act. Additional support was provided by NIH grants DK063945, DK077816 and DK082476.

References

Ballard, T.D., Wolff, J., Griffin, J.B., Stanley, J.S., Calcar, S.V., and Zempleni, J., 2002. Biotinidase catalyzes debiotinylation of histones. *European Journal of Nutrition.* 41: 78–84.

Bao, B., Rodriguez-Melendez, R., Wijeratne, S.S., and Zempleni, J., 2010. Biotin regulates the expression of holocarboxylase synthetase in the miR-539 pathway in HEK-293 cells, *Journal of Nutrition.* 140: 1546–1551.

Bao, B., Pestinger, V., Hassan, Y.I., Borgstahl, G.E.O., Kolar, C., and Zempleni, J., 2011. Holocarboxylase synthetase is a chromatin protein and interacts directly with histone H3 to mediate biotinylation of K9 and K18. *Journal of Nutritional Biochemistry.* 22: 470–475.

Bao, B., Rodriguez-Melendez, R., and Zempleni, J., 2012. Cytosine methylation in *miR-153* gene promoters increases the expression of holocarboxylase synthetase, thereby increasing the abundance of histone H4 biotinylation marks in HEK-293 human kidney cells. *The Journal of Nutritional Biochemistry.* 23: 635–639.

Boas, M.A., 1927. The effect of desiccation upon the nutritive properties of egg-white. *Biochemistry Journal.* 21: 712–724.

Camporeale, G., Shubert, E.E., Sarath, G., Cerny, R., and Zempleni, J., 2004. K8 and K12 are biotinylated in human histone H4. *European Journal of Biochemistry.* 271: 2257–2263.

Camporeale, G., Giordano, E., Rendina, R., Zempleni, J., and Eissenberg, J.C., 2006. *Drosophila* holocarboxylase synthetase is a chromosomal protein required for normal histone biotinylation, gene transcription patterns, lifespan and heat tolerance. *Journal of Nutrition.* 136: 2735–2742.

Camporeale, G., and Zempleni, J., 2006. Biotin. In: Bowman, B.A., and Russell, R.M. (ed.) *Present Knowledge in Nutrition*, Vol. 1. International Life Sciences Institute, Washington DC, USA, pp. 314–326.

Camporeale, G., Oommen, A.M., Griffin, J.B., Sarath, G., and Zempleni, J., 2007. K12-biotinylated histone H4 marks heterochromatin in human lymphoblastoma cells. *The Journal of Nutritional Biochemistry.* 18: 760–768.

Chew, Y.C., Camporeale, G., Kothapalli, N., Sarath, G., and Zempleni, J., 2006. Lysine residues in N- and C-terminal regions of human histone H2A are targets for biotinylation by biotinidase. *The Journal of Nutritional Biochemistry.* 17: 225–233.

Chew, Y.C., West, J.T., Kratzer, S.J., Ilvarsonn, A.M., Eissenberg, J.C., Dave, B.J., Klinkebiel, D., Christman, J.K., and Zempleni, J., 2008. Biotinylation of histones represses transposable elements in human and mouse cells and cell lines, and in Drosophila melanogaster. *Journal of Nutrition.* 138: 2316–2322.

Collins, J.C., Paietta, E., Green, R., Morell, A.G., and Stockert, R.J., 1988. Biotin-dependent expression of the asialoglycoprotein receptor in HepG2. *Journal of Biological Chemistry.* 263: 11280–11283.

Dakshinamurti, K., and Cheah-Tan, C., 1968. Liver glucokinase of the biotin deficient rat. *Canadian Journal of Biochemistry.* 46: 75–80.

du Vigneaud, V., Melville, D.B., Folkers, K., Wolf, D.E., Mozingo, D.E., Keresztesy, J.C., and Harris, S.A., 1942. The structure of biotin: A study of desthiobiotin. *Journal of Biological Chemistry*. 146: 475–485.

Filenko, N.A., Kolar, C., West, J.T., Hassan, Y.I., Borgstahl, G.E.O., Zempleni, J., and Lyubchenko, Y.L., 2011. The role of histone H4 biotinylation in the structure and dynamics of nucleosomes. *PLoS ONE*. 6: e16299.

Gralla, M., Camporeale, G., and Zempleni, J., 2008. Holocarboxylase synthetase regulates expression of biotin transporters by chromatin remodeling events at the SMVT locus. *The Journal of Nutritional Biochemistry*. 19: 400–408.

Hassan, Y.I., Moriyama, H., and Zempleni, J., 2009a. The polypeptide Syn67 interacts physically with human holocarboxylase synthetase, but is not a target for biotinylation. *Archives of Biochemistry and Biophysics*. 495: 35–41.

Hassan, Y.I., Moriyama, H., Olsen, L.J., Bi, X., and Zempleni, J., 2009b. N- and C-terminal domains in human holocarboxylase synthetase participate in substrate recognition. *Molecular Genetics and Metabolism*. 96: 183–188.

Hymes, J., Fleischhauer, K., and Wolf, B., 1995. Biotinylation of histones by human serum biotinidase: assessment of biotinyl-transferase activity in sera from normal individuals and children with biotinidase deficiency. *Biochemical and Molecular Medicine*. 56: 76–83.

Kim, K.-H., McCormick, D.B., Bier, D.M., and Goodridge, A.G., 1997. Regulation of mammalian acetyl-coenzyme A carboxylase. *Annual Reviews of Nutrition*. 17: 77–99.

Knowles, J.R., 1989. The mechanism of biotin-dependent enzymes. *Annual Reviews of Biochemistry*. 58: 195–221.

Kobza, K., Camporeale, G., Rueckert, B., Kueh, A., Griffin, J.B., Sarath, G., and Zempleni, J., 2005. K4, K9, and K18 in human histone H3 are targets for biotinylation by biotinidase. *FEBS Journal*. 272: 4249–4259.

Kobza, K., Sarath, G., and Zempleni, J., 2008. Prokaryotic BirA ligase biotinylates K4, K9, K18 and K23 in histone H., *Biochemistry and Molecular Biology Reports*. 41: 310–315.

Kögl, F., and Tönnis, B., 1932. Über das Bios-Problem. Darstellung von krystallisiertem Biotin aus Eigelb. *Zeitschrift für Physiologische Chemie*. 242: 43–73.

Kouzarides, T., and Berger, S.L., 2007. Chromatin modifications and their mechanism of action. In: Allis, C.D., Jenuwein, T., and Reinberg, D. (ed.) *Epigenetics*. Cold Spring Harbor Press, Cold Spring Harbor, NY, USA, pp. 191–209.

Lee, H.M., Wright, L.D., and McCormick, D.B., 1972. Metabolism of carbonyl-labeled [^{14}C] biotin in the rat. *Journal of Nutrition*. 102: 1453–1464.

McCormick, D.B., and Wright, L.D., 1971. The metabolism of biotin and analogues. In: Florkin, M., and Stotz, E.H. (ed.) *Metabolism of Vitamins and Trace Elements*. Elsevier Publishing Company, Amsterdam, The Netherlands, pp. 81–110.

Narang, M.A., Dumas, R., Ayer, L.M., and Gravel, R.A., 2004. Reduced histone biotinylation in multiple carboxylase deficiency patients: a nuclear role for holocarboxylase synthetase. *Human Molecular Genetics*. 13: 15–23.

National Center for Biotechnology Information (NCBI), 2008a. *Entrez SNP*. Available at: http://www.ncbi.nlm.nih.gov/snp. Accessed 28 November 2008.

National Center for Biotechnology Information (NCBI), 2008b. *OMIM®— Online Mendelian inheritance in man.* Available at: http://www.ncbi.nlm.nih.gov/sites/entrez?db=omim. Accessed 18 July 2008.

Rolfe, B., and Eisenberg, M.A., 1968. Genetic and Biochemical Analysis of the Biotin Loci of Escherichia coli K-12. *Journal of Bacteriology.* 96: 515–524.

Solorzano-Vargas, R.S., Pacheco-Alvarez, D., and Leon-Del-Rio, A., 2002. Holocarboxylase synthetase is an obligate participant in biotin-mediated regulation of its own expression and of biotin-dependent carboxylases mRNA levels in human cells. *Proceedings of the National Academy of Sciences of the United States of America.* 99: 5325–5330.

Stanley, J.S., Griffin, J.B., and Zempleni, J., 2001. Biotinylation of histones in human cells: effects of cell proliferation. *European Journal of Biochemistry.* 268: 5424–5429.

Suzuki, Y., Yang, X., Aoki, Y., Kure, S., and Matsubara, Y., 2005. Mutations in the holocarboxylase synthetase gene HLCS. *Human Mutation.* 26: 285–290.

Thuy, L.P., Belmont, J., and Nyhan, W.L., 1999. Prenatal diagnosis and treatment of holocarboxylase synthetase deficiency. *Prenatal Diagnosis.* 19: 108–112.

Warnatz, H.J., Querfurth, R., Guerasimova, A., Cheng, X., Haas, S.A., Hufton, A.L., Manke, T., Vanhecke, D., Nietfeld, W., Vingron, M., Janitz, M., Lehrach, H., and Yaspo, M.L., 2010. Functional analysis and identification of cis-regulatory elements of human chromosome 21 gene promoters. *Nucleic Acids Research.* 38: 6112–6123.

Wijeratne, S.S., Camporeale, G., and Zempleni, J., 2010. K12-biotinylated histone H4 is enriched in telomeric repeats from human lung IMR-90 fibroblasts. *The Journal of Nutritional Biochemistry.* 21: 310–316.

Wolf, B., Heard, G.S., McVoy, J.R.S., and Grier, R.E., 1985. Biotinidase deficiency. *Annals of the New York Academy of Sciences.* 447: 252–262.

Wolf, B., and Heard, G.S., 1991. Biotinidase deficiency. In: Barness, L., and Oski, F. (ed.) *Advances in Pediatrics.* Medical Book Publishers, Chicago, IL, USA, pp. 1–21.

Wolffe, A., 1998. *Chromatin*, 3th edition. Academic Press, San Diego, CA, USA, pp. 1–447.

Zempleni, J., Wijeratne, S.S., and Hassan, Y.I., 2009. Biotin. *BioFactors.* 35: 36–46.

Zempleni, J., Li, Y., Xue, J., and Cordonier, E.L., 2011. The role of holocarboxylase synthetase in genome stability is mediated partly by epigenomic synergies between methylation and biotinylation events. *Epigenetics.* 6: 892–894.

CHAPTER 11
The Chemistry of Folate

ABALO CHANGO

Institut Polytechnique Lasalle Beauvais, 19 rue Pierre Waguet, F-60026 Beauvais Cedex, France
Email: abalo.chango@lasalle-beauvais.fr

11.1 Introduction

A nutritional factor in yeast that both prevented and cured macrocytic anaemia in pregnant women was first described by Lucy Wills in 1931. Folic acid received its name in 1941 when isolated from spinach as a growth factor for *Streptococcus lactis* (Mitchell *et al.* 1941). Various investigators investigating this growth factor also referred to it as vitamin M, vitamin B_9, vitamin Bc, folic acid or folate (Whiteley 1971). The word 'folate' comes from the Latin word *folium*, which means leaf. Folate refers to all pteroylglutamates possessing vitamin activity. 'Folic acid' and 'folate' are the preferred synonyms for pteroylglutamic acid (PteGlu) and pteroylglutamate, respectively.

The biochemical functions of folate were first determined with bacteria. Folate plays a major role in carrying one-carbon units within cells. In the brush border of mucosal cells, the polyglutamyl chain is removed by the enzyme folate conjugase and folate monoglutamate is subsequently absorbed. Folate acts as both a donor and receiver of one-carbon moieties in a variety of reactions.

The chemistry and biochemistry of folate have been extensively discussed in several excellent reviews. Here we briefly summarize these aspects.

11.2 Chemistry

Folate (Figure 11.1) is used to designate a member of the family of pteroylglutamates, each having a different level of reduction of the pteridine ring,

The Chemistry of Folate

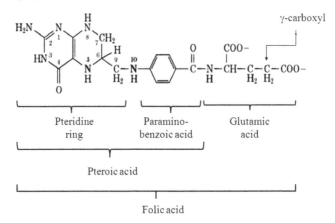

Figure 11.1 Chemical structure of folic acid (pteroylglutamic acid).

Table 11.1 One-carbon substituents and their positions in folate coenzymes.

Name[a]	Abbreviation	Position of substituent groups[b]	
		N5	N10
Pteroylglutamic acid	Folic acid		-H
7,8-Dihydrofolate		-H	-H
5-Methyl-5,6-dihydrofolate	5-CH$_3$-THF	-CH3	-H
5,6,7,8-Tetrahydrofolate (THF)	THF	-H	-H
5-Methyltetrahydrofolate	5-CH$_3$-THF	-CH3	-H
5-Formyltrtrahydrofolate	5-CHO-THF	-CHO	-H
10-Formyltetrahydrofolate	10-CHO-THF	-H	-CHO
5,10-Methenyltetrahydrofolate	5,10-CH$^+$=THF	=CH- ± bridge	—
5,10-Methylenetetrahydrofolate	5,10-CH$_2$THF	5 -CH2-bridge	—
5-Formiminotetrahydrofolate	5-CHNH-THF	-CHNH	–H

[a]The one-carbon units attached to the reduced tetrahydrofolate molecule.
[b]The one-carbon substituent can be at the N5 and/or N10 positions.

one-carbon unit [methyl (CH$_3$-), formyl (CHO-), methylene (CH$_2$=), methenyl (CH$_4$-)] substitution (Table 11.1) and a number of glutamate residues.

Folic acid is the parent compound of the group of folate vitamers including tetrahydrofolate (THF) and its derivatives. Folic acid consists of a pteridine ring, p-aminobenzoic acid and glutamic acid. Naturally occurring folates are generally reduced to THF with hydrogen at the 5, 6, 7, and 8 positions or to dihydrofolate with hydrogen at the 7 and 8 positions; they have a one-carbon unit at the N5 or N10 positions, or both (Blakley and Benkovic 1984; Wagner 1995; Brody and Shane 2001).

11.2.1 Extraction and Isolation

Folate in natural sources can be extracted and isolated under conditions that preserve the polyglutamyl chain and the reductive state of the folate. Optimal conditions for the recovery of intact folate are to heat a minced biological sample for 5–10 min in a boiling water bath (Brody and Shane 2001). The presence of antioxidants (for instance 1.0% sodium ascorbate or 0.2M 2-mercaptoethanol) is required to denature folate-metabolizing enzymes such as γ-glutamyl hydrolase (folate conjugase) and folate-binding proteins under conditions that protect the labile folate derivatives.

11.2.2 Structure

Folic acid has three main components: the pterin (2-amino-4-oxopteridine) moiety linked by a methylene bridge at the C6 position to a *para*-aminobenzoic acid moiety, which is joined by a peptide linkage to glutamic acid (Figure 11.1). Folate is used to describe a group of compounds derived from 5,6,7,8-tetrahydropteroyl-glutamate referred to as THF (or $H_4PteGlu$).

Folate polyglutamates are the preferred substrates for folate-dependent enzymes; they are more active metabolically in cells and also better retained in cells than folate monoglutamates. Natural folates found in foods are all conjugated to a polyglutamyl chain containing different numbers of glutamic acids depending on the type of food.

In the nomenclature of polyglutamate forms, $H_4PteGlu_n$ states the number (n) of glutamate residues (Glu). Derivatives of THF contain different one-carbon units attached to the N5 or N10 atom of the pteridine ring. Natural folates occur in the reduced, 7,8-dihydro- and 5,6,7,8-tetrahydro forms. Each L-glutamate is linked by amide bonds to the preceding molecule of glutamate through the γ-carboxyl group.

11.2.3 Physicochemical Properties

The molecular weight of folic acid is 441.40 g/mol. The molecule is very slightly soluble in water, quite soluble in the salt form and insoluble in alcohol. It is sensitive to light and ultraviolet (UV) radiation. Folic acid is readily soluble in dilute solutions of alkali hydroxides and carbonates. Aqueous solutions of folic acid are heat sensitive and decompose rapidly in the presence of light. Solutions should be stored in a cool place protected from light.

Folic acid exhibits substantially greater stability than most reduced folates and is consequently the form of the vitamin normally used in supplements and to fortify foods. The order of stability of these latter forms is 5-formyl-THF > 5-methyl-THF > 10-formyl-THF > THF. However, it is important to note that the stability is pH-dependent, with the reduced folates being most stable at pH > 8 and pH > 2 and least stable between pH 4–6.

11.3 Analysis

The multiplicity of forms of folate, their low stability and low levels in blood or foods make quantitative analysis of folate difficult. Different methods are available for measuring the status of folate in the population and in clinical

Table 11.2 Principles of most frequent methods used for folate analysis.

Method	Principle
Enzyme protein binding assay	Based on competitive reaction between folate isomers and folate binding protein.
Microbiological assay	Based on the microbial growth obtained an amount of folate in the medium.
Chromatography (HPLC) assay	Based on chromatographic separation of folate isomers followed by quantification with a detector.

settings, or folate analysis in foods (Table 11.2) (Arcot and Shrestha 2005; DeVries *et al.* 2005; Shane 2011).

The folate assay can be divided in three steps:

(i) Liberation of folates from the cellular matrix
(ii) Deconjugation of polyglutamates to generate the mono and di-glutamate forms
(iii) Detection of the biological activity or chemical concentration of the resulting folates.

The method used to identify folates depends on whether one needs to know the identities of the one-carbon unit and reductive state are the cofactor, or the length of the polyglutamyl chain of the folate, or both (Brody and Shane 2001). For identification of polyglutamate distributions, folates can be cleaved at the C9–N10 bond to yield *p*-aminobenzoylpolyglutamate derivatives, which can be separated and identified by high-performance liquid chromatography (HPLC) analysis. The detection methods most commonly used are a microbiological assay that relies on the turbidimetric measurement of the bacterial growth of *Lactobacillus casei* (renamed *Lactobacillus rhamnosus*, ATCC 7469). The folate content of natural sources, as well as that of the fractions recovered after chromatographic procedures, can be measured by microbiological assay.

A 96-well microtitre plate assay with microcomputer analysis has simplified the microbiological assay making it less laborious, less time-consuming and more reproducible. Recently, an automated 96-well plate isotope–dilution tandem mass spectrometry method and an ultra-performance liquid chromatography-tandem mass spectrometry method have been developed and validated for the quantification of synthetic folic acid in folate-fortified breads (Fazili and Pfeiffer 2004; Chandra-Hioe *et al.* 2011).

Folate can be measured in plasma or serum by microbiological assay using *L. casei* as the test organism, but this test can be confounded if the subject is on antibiotic treatment. Serum folate values reflect recent dietary intake and a vitamin deficiency is ascribed only where serum folate remains low over a period of time. Plasma folate levels are thought to reflect the day-to-day variations in dietary folate levels while red blood cell folate is a better indicator of long-term tissue storage levels.

Radioassay kits are used extensively in clinical laboratories. Folate values obtained by using the different commercially available kits fluctuate considerably and therefore no absolute values can be given to indicate folate deficiency. Each laboratory has to define its own lower limits based on a large number of sample assays from a representative population of normal subjects (Shane 2011). At present, the microbiological assay using *L. casei* is one of the best and most versatile methods for the determination of food folate.

11.4 Concluding Remarks

To be optimal, extraction and detection of folate need to be conducted carefully due to its chemical proprieties, multiple forms and low levels of derivatives in biological samples. The optimal conditions and more reliable methods for folate assay are currently under development.

Summary Points

- Folate designates a member of the family of pteroylglutamates, each having a different level of reduction of the pteridine ring, one-carbon unit.
- Folic acid is the preferred synonym for pteroylglutamic acid and folate is the preferred synonym for pteroylglutamate.
- Folic acid exhibits substantially greater stability than most reduced folates.
- Different methods are available for folate status measurement and for folate analysis in foods.

Key Facts

- Folic acid was first isolated from spinach.
- The biochemical functions of folate were first determined as a growth factor for *Streptococcus lactis*.
- Folic acid received its name in 1941.
- Folate has received different names: vitamin M, vitamin B_9, vitamin Bc and folic acid.

List of Abbreviations

Glu	glutamate
HPLC	high-performance liquid chromatography
PteGlu	pteroylglutamic acid
THF	tetrahydrofolate

Acknowledgements

The author thanks Professor Barry Shane (Department of Nutritional Science and Toxicology, University of California, Berkeley, CA) for his review of the manuscript.

References

Arcot, J., and Shrestha, A.K., 2005. Folates: methods of analysis in foods. *Trends in Food Science and Technology*. 16: 253–266.

Blakley R. L. 1984. Folates and Pterins. In: Blakley R. L., Benkovic S. J, (eds) Wiley-Interscience, New York, USA. Vol. 1, pp. 191–253.

Brody, T., and Shane, B., 2001. Folic Acid. In: Rucker, R. (ed.) Handbook of Vitamins. Marcel Dekker, New York, USA. pp. 427–462.

Chandra-Hioe, M.V., Bucknall, M.P., and Arcot, J., 2011. Folate analysis in foods by UPLC-MS/MS: development and validation of a novel, high throughput quantitative assay; folate levels determined in Australian fortified breads. *Analytical and Bioanalytical Chemistry*. 401: 1035–1042.

DeVries, J.W., Rader, J.I., Keagy, P.M., Hudson, C.A., Angyal, G., Arcot, J., Castelli, M., Doreanu, N., Hudson, C., Lawrence, P., Martin, J., Peace, R., Rosner, L., Strandler, H.S., Szpylka, J., van den Berg, H., Wo, C., and Wurz, C., 2005. Microbiological assay–trienzyme procedure for total folates in cereals and cereal foods: collaborative study. *Journal of AOAC International*. 88: 5–15.

Fazili, Z., and Pfeiffer, M., 2004. Measurement of folates in serum and conventionally prepared whole blood lysates: application of an automated 96-well plate isotope-dilution tandem mass spectrometry method. *Clinical Chemistry*. 50: 2378–2381.

Mitchell, H.K., Snell, E.E., and Williams, R.J., 1941. The concentration of 'folic acid'. *Journal of the American Chemistry Society*. 63: 2284.

Shane, B., 2011. Folate status assessment history: implications for measurement of biomarkers in NHANES. *The American Journal of Clinical Nutrition*. 94: 337S–342S.

Wagner, C., 1995. Biochemical role of folate in cellular metabolism. In: Bailey, L.B. (ed) Folate in Health and Disease. Marcel Dekker, New York, USA, pp. 23–42.

Whiteley, J.M., 1971. Some aspects of the chemistry of the folate molecule. *Annals of the New York Academy of Sciences*. 186: 29–42.

CHAPTER 12
The Chemistry of Cobalamins

ALEXIOS S. ANTONOPOULOS AND CHARALAMBOS ANTONIADES*

Department of Cardiovascular Medicine, University of Oxford, John Radcliffe Hospital, West Wing Level 6, OX3 9DU, Oxford, UK
*Email: charalambos.antoniades@cardiov.ox.ac.uk

12.1 Introduction

Cobalamins are a group of water-soluble molecules that have structural similarities with natural tetrapyrroles such as coenzyme F430 and chlorophyll. Cobalamins are organometallic compounds belonging to the corrin family and have the characteristic feature of a tetrapyrrole ring (corrin) in their moiety (Banerjee 1997; Jensen and Ryde 2009; Lodowski et al. 2011). However, what makes cobalamins different is the cobalt atom they have in their molecule. This rare in nature cobalt–carbon (Co–C) bond gives cobalamins unique chemical properties, rendering them suitable molecules to act as cofactors (see Key Facts) in various enzymatic systems in mammals (Lodowski et al. 2011). Recent advances in existing knowledge regarding the absorption and transport of cobalamins in human body are reviewed in details elsewhere (Quadros 2009). In this short chapter we discuss important aspects of the chemistry of cobalamins as well as their role in the biochemistry of humans.

12.2 Chemical Structure of Cobalamins

In the tetrapyrrole ring of cobalamins, the central cobalt ion has four nitrogen ligands provided by the corrin ring. The lower axial ligand or the so-called 'nucleotide tail' is a 5,6-dimethylbenzimidazole (DMB) group which offers a fifth nitrogen ligand to the central cobalt ion (Seetharam and Alpers 1982). The upper axial ligand is a variable R group. When the R group is cyanide, cyanocobalamin (the most known form of cobalamins) is formed. When R group is substituted by groups such as OH, H_2O, CH_3 and 5'-deoxyadenosyl, the other members of the cobalamins family are formed—namely hydroxycobalamin, aquacobalamin, methylcobalamin (MeCbl) and adenosylcobalamin (AdoCbl), respectively. Peripheral, side chains of the corrin ring may include acetamide and proprionamide groups (Seetharam and Alpers 1982). The chemical structure of cobalamins is presented in Figure 12.1. The corrin ring of cobalamins is flexible, with an upward folding towards the R ligand like a butterfly (Pett *et al.* 2002). The folding angle is determined by the effects of the R group and the axial *N*-donor ligand. As Co–N_{ax} distance shortens, the upward folding of the corrin ring increases. Co–N_{ax} distance and the folding

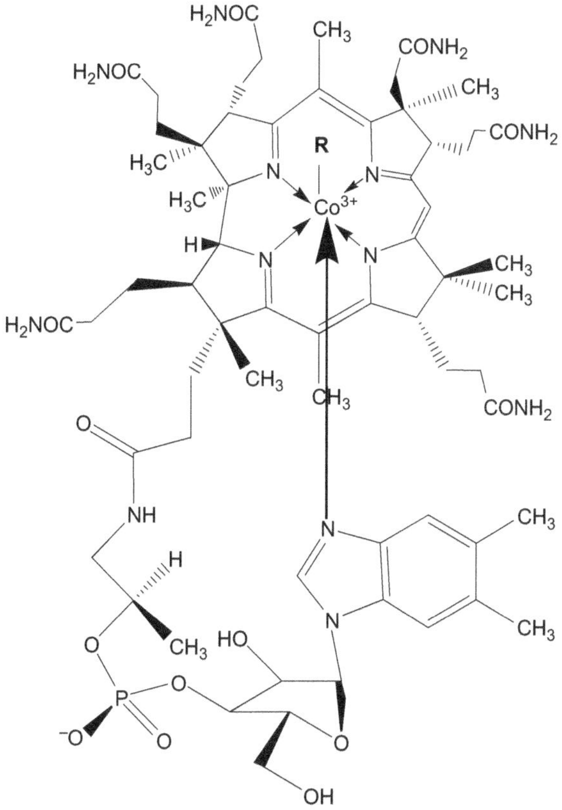

Figure 12.1 Chemical structure of cobalamins. R group may be CN (cyanocobalamin), OH (hydroxycobalamin), H_2O (aquacobalamin), CH_3 (methylcobalamin) and 5'-deoxyadenosyl (adenosylcobalamin).

angle have been highlighted as factors affecting the stability of the Co–C bond (Pett et al. 2002).

12.3 Natural Forms of Cobalamins

Cobalamin is biosynthesized *via* two distinct routes, an aerobe and an anaerobic pathway, that differ in the timing of cobalt ion insertion into precorrin macrocycle. The oxygen pathway, which was delineated in *Pseudomonas denitrificans*, incorporates molecular oxygen in the precorrin-3 molecule and cobalt is inserted to the macrocycle late in the pathway. In the anaerobic pathway found in *Salmonella typhimurium*, the cobalt ion is chelated into precorrin-2 in the absence of oxygen. The biosynthetic pathways of cobalamin are reviewed in detail by Roper et al. (2000).

Humans receive cobalamins from food—mainly from meat and meat products, and to a lesser extent from milk and milk products. Cobalamins are found in food in five forms: hydroxycobalamin, AdoCbl, MeCbl, suphitocobalamin and cyanocobalamin. AdoCbl and hydroxycobalamin are the most frequent types, followed by MeCbl (Farquharson and Adams 1976). Nevertheless both AdoCbl and MeCbl are light-sensitive, with light evoking a photodissociation of the Co–C bond. The naturally found hydroxycobalamin is the product of photolysis of the light-sensitive cobalamins (Farquharson and Adams 1976; Seetharam and Alpers 1982).

Even though cyanocobalamin is the least frequent form in the natural resources of cobalamins (found only in small amounts in certain foods), it is the most stable compound. Unlike the alkylcobalamins (AdoCbl and MeCbl), the -CN group in cyanocobalamin grants stability to the molecule protecting it from photodissociation after simple photon excitation (Lodowski et al. 2011). X-ray crystallographic studies (see Key Facts) have demonstrated that the length of the Co–C bond is shorter in cyanocobalamin (1.868 or 1.886 Å) than in AdoCbl (2.033 Å) or MeCbl (1.979 Å) (Lodowski et al. 2011). Furthermore, the importance of inverse *trans* influence on the Co–C bond has been highlighted by time-dependent density functional theory (DFT) studies (see Key Facts). Computation of electronically excited states of MeCbl and CNCbl molecules has demonstrated that photolysis in MeCbl is mediated by the dissociative $^3(\sigma_{Co-C} \rightarrow \sigma^*_{Co-C})$ triplet state. Other factors such as N_{axial}-Co–C bending might also influence the stability of the molecules (Lodowski et al. 2011). This property of cyanocobalamin renders its moiety a suitable compound for oral administration in medical practice in cases of vitamin B_{12} deficiency. Nevertheless it must be converted to the other naturally occurring forms inside the human body to become biologically active.

12.4 Biochemistry of Cobalamins

MeCbl and AdoCbl are the two active forms of cobalamins found *in vivo* in humans and mammals. MeCbl is found principally in the cytoplasm, while

AdoCbl is found mainly in the mitochondria (Banerjee 1997). In humans, cobalamins serve as cofactors only for two enzymes—the cytoplasmic methionine synthase and the mitochondrial methylmalonyl-CoA mutase.

12.4.1 Methionine Synthase

The first enzyme, methionine synthase, is a critical component of the homocysteine metabolic pathway. Methionine synthase requires cobalamin as a cofactor in order to metabolize homocysteine to methionine. The binding site of cobalamin in methionine synthase is a pocket site consisting of α-helices and β-sheets (Pett et al. 2002). The transmethylation reaction catalysed by methionine synthase involves conversion of 5-methyltetrahydrofolate to tetrahydrofolate and addition of the methyl group to the homocysteine molecule to form methionine. Methionine can then be adenylated to form S-adenosylmethionine, which serves as a methyl donor in a variety of metabolic pathways (Pett et al. 2002). 5-Methyltetrahydrofolate is the circulating form of folic acid in humans and conversion to tetrahydrofolate is an important step towards its utilization for DNA synthesis (Banerjee 1997). The overall reaction involves two nucleophilic substitution reactions with Cob(I)alamin as an intermediate. The methyl group is attached to cob(I)alamin and then heterolytic cleavage of the Co–C bond follows, donating the methyl group to the thiolate of homocysteine for the formation of methionine. Heterolytic cleavage of the Co–C bond is enhanced 10^5-fold by methionine synthase (Jarrett et al. 1996). Reactivation of the inactive cofactor cob(II)alamin, occasionally formed by the oxidation of cob(I)alamin, requires the presence of S-adenosylmethionine (SAM) (Banerjee 1997). Single nucleotide polymorphisms in the methionine synthase gene have been held responsible for some rare causes of severe homocysteinemia, due to reduced methionine synthase activity. Interestingly these genetic defects are clustered in the B_{12}-binding site of the enzyme (Banerjee 1997).

12.4.2 Methylmalonyl-CoA Mutase

In humans, methylmalonyl-CoA mutase is required for the metabolism of proprionate, derived from branched-chain amino acids, odd-chain fatty acids and cholesterol, into succinyl-CoA (Banerjee 1997; Pett et al. 2002). Methylmalonyl-CoA mutase requires AdoCbl as a cofactor. The mechanism involves the homolytic cleavage of the Co–C bond, forming cob(II)alamin and a 5′-deoxyadenosyl radical (Pett et al. 2002). The homolytic cleavage of the Co–C bond is increased 10^{12}-fold in the presence of the enzyme. The 5′-deoxyadenosyl radical first abstracts a hydrogen atom from the substrate methylmalonyl-CoA, which donates after a rearrangement reaction back to form succinyl-CoA (Pett et al. 2002). In humans, deficiency in methylmalonyl-CoA mutase causes an inherited metabolic disorder and is one of causes of methylmalonylacidemia. According to the severity of methylmalonyl-CoA mutase reduced activity, the deficiency is characterized as mut⁻ (detectable enzyme activity) or mut (non detectable enzyme activity) (Banerjee 1997). Most of the mutations in the methylmalonyl-CoA mutase gene are clustered in the B_{12}-binding domain of the enzyme (Banerjee 1997).

12.5 Concluding Remarks

Cobalamins are essential enzymatic cofactors in human biochemistry. Cobalamins' chemical structure is based on the tetrapyrrole ring, while the chemical properties of the Co–C bond located in the centre of their moiety have been the focus of extensive research. Cyanocobalamin, the most known form of cobalamins, is rarely found in nature. Methylcobalamin and adenosylcobalamin are the two active forms of cobalamins *in vivo* in humans. Weakening of the Co–C bond and its homolytic or heterolytic cleavage have been uncovered as an essential mechanism in the biochemical role of cobalamins as cofactors in humans. Recent studies using modern computational methods and application of quantum chemistry models have widened our knowledge of cobalamins' biochemistry and are expected to contribute to our further understanding of cobalamin-dependent enzymes.

Summary Points

- This chapter focuses on the chemistry of cobalamins.
- The cobalt–carbon (Co–C) bond in the chemical structure of cobalamins offers them unique chemical properties.
- There are five naturally occurring forms of cobalamins with methylcobalamin, adenosylcobalamin and hydroxycobalamin being the most frequently found forms.
- Methylcobalamin and adenosylcobalamin are light-sensitive, with light inducing dissociation of the Co–C bond.
- Cyanocobalamin is the most stable form of cobalamins, frequently also used for medical purposes.
- The active forms of cobalamins in humans are methylcobalamin and adenosylcobalamin.
- In humans, cobalamins serve as cofactors mainly for two enzymes—cytoplasmic methionine synthase and the mitochondrial methylmalonyl-CoA mutase.

Key Facts about the Chemistry of Cobalamins

- Cofactor is a non-protein chemical compound that facilitates the activity of an enzyme and is required for the biochemical reaction to occur. A further biochemical classification exists for the enzyme and its cofactor. The enzyme without its cofactor is an apoenzyme; the enzyme together with its cofactor is called an oloenzyme.
- The cellular substance, the gel-like fluid that exists between the nucleus of a cell and the cellular membrane and contains all cellular organelles, is

called cytoplasm. Such cellular organelles are the endoplasmic reticulum, the Golgi apparatus, the mitochondria and the lysosomes. The cytoplasm is separated from the nucleus by the nucleoplasm. Most of cellular biochemical reactions take place in the cytoplasm.
- *S*-adenosylmethionine (SAM) is present in every tissue of the human body, but mainly in the liver. SAM participates in biochemical reactions involving transmethylation, transsulfurylation and aminopropylation steps. It is important in the homocysteine metabolic pathway where it acts as a substrate for the conversion of homocysteine to methionine.
- X-ray crystallographic studies are an experimental technique that allows a three-dimensional reconstruction of the structure of a chemical compound. It is based on the property of crystals to diffract X-rays. The diffraction pattern is dependent on the atoms present in the molecule and can be read by an X-ray crystallographer who computationally constructs the structure and shape of a molecule.
- One of the most promising methods used in quantum physics and chemistry today is density functional theory (DFT). With application of DFT-based methods, the electronic structure of the matter can be predicted. DFT is based on the principle that the total energy for a system described by the laws of quantum mechanics can be theoretically calculated if the electrons' spatial distribution (electron density) is known. In modern chemistry, application of DFT models can predict the atomization and ionization energy, molecular structures, enzymatic reactions and more.

Definitions of Words and Terms

Heterolysis. Also known as heterolytic fission, it is the cleavage of a single bond in which the electron pair of the cleaved bond remains to the same atom.

Homolysis. Also known as homolytic fission, it is the cleavage of a single bond in which the two electrons of the cleaved bond are split between two atoms.

Organometallic compounds. The chemical compounds that contain bonds between carbon and a metal.

Photodissociation. Also known as photolysis or photodecomposition. Is the chemical breakdown of a chemical compound by photons. Chemical compounds that undergo photodissociation are called light-sensitive.

List of Abbreviations

AdoCbl	adenosylcobalamin
Co–C	cobalt–carbon
DFT	density functional theory
DMB	5,6-dimethylbenzimidazole
MeCbl	methylcobalamin
SAM	*S*-adenosylmethionine

References

Banerjee, R., 1997. The Yin-Yang of cobalamin biochemistry. *Chemistry & Biology.* 4: 175–186.

Farquharson, J., and Adams, J.F., 1976. The forms of vitamin B_{12} in foods. *British Journal of Nutrition.* 36: 127–136.

Jarrett, J.T., Amaratunga, M., Drennan, C.L., Scholten, J.D., Sands, R.H., Ludwig, M.L., and Matthews, R.G., 1996. Mutations in the B_{12}-binding region of methionine synthase: how the protein controls methylcobalamin reactivity. *Biochemistry.* 35: 2464–2475.

Jensen, K.P., and Ryde, U., 2009. Cobalamins uncovered by modern electronic structure calculations. *Coordination Chemistry Reviews.* 253: 769–778.

Lodowski, P., Jaworska, M., Kornobis, K., Andruniow, T., and Kozlowski, P.M., 2011. Electronic and structural properties of low-lying excited states of vitamin B_{12}. *The Journal of Physical Chemistry B.* 115: 13304–13319.

Pett, V.B., Fischer, A.E., Dudley, G.K., and Zacharias, D.E., 2002. Probing the mechanism of coenzyme B_{12}: synthesis, crystal structures, and molecular modeling of coenzyme B_{12} model compounds. *Comments on Inorganic Chemistry.* 23: 385–400.

Quadros, E.V., 2009. Advances in the understanding of cobalamin assimilation and metabolism. *British Journal of Haematology.* 148: 195–204.

Roper, J.M., Raux, E., Brindley, A.A., Schubert, H.L., Gharbia, S.E., Shah, H.N., and Warren, M.J., 2000. The enigma of cobalamin (vitamin B_{12}) biosynthesis in *Porphyromonas gingivalis*. Identification and characterization of a functional corrin pathway. *The Journal of Biological Chemistry.* 275: 40316–40323.

Seetharam, B., and Alpers, D.H., 1982. Absorption and transport of cobalamin (vitamin B_{12}). *Annual Review of Nutrition.* 2: 343–369.

Analysis

CHAPTER 13

Assay of B Vitamins and other Water-soluble Vitamins in Honey

MARCO CIULU,[a] NADIA SPANO,[a] SEVERYN SALIS,[a] MARIA I. PILO,[a] IGNAZIO FLORIS,[b] LUCA PIREDDU[c] AND GAVINO SANNA*[a]

[a] Dipartimento di Chimica e Farmacia, Università degli Studi di Sassari, *Via* Vienna, 2, I-07100 Sassari, Italy; [b] Dipartimento di Agraria, Università degli Studi di Sassari, *Via* De Nicola 9, 07100 Sassari, Italy; [c] CRS4, Parco Scientifico e Tecnologico, Edificio 1, 09010 Pula (CA), Italy
*Email: sanna@uniss.it

13.1 Honey

Honey is a natural food produced by bees (*Apis mellifera*) from nectar and honeydew. Its economic and cultural relevance make it the most important product of beekeeping (Krell 1996). It exhibits functional properties (Viuda-Martos *et al.* 2008) and its significance in traditional medicine has been recognized in various cultures (Gómez-Caravaca *et al.* 2006). Although the chemical composition of honey is extremely variable, it could be defined as an aqueous solution supersaturated in sugars. Table 13.1 summarizes the average composition and the ranges of concentration of the most important constituents of nectar honey.

While fructose and glucose are the quantitatively predominant compounds, many other mono- and oligosaccharides have been quantified in honey (Doner

Table 13.1 Average composition and respective range of the major constituents of nectar honey.

Constituent	Average (%)	Range (min–max) (%)
Fructose	38.2	30–45
Glucose	31.3	24–40
Water	17.2	15–20
Reducing disaccharides (expressed as maltose)	7.3	2.7–16.0
Others (*e.g.* phenolic compounds, vitamins, enzymes, waxes)	2	1.5–3.0
Other sugars	1.2	0.1–8.5
Sucrose	1	0.1–7.6
Erlose	0.8	0.5–6.0
Free acid (expressed as gluconic acid)	0.4	0.1–0.9
Amino acids and proteins	0.3	0.2–0.4
Ash	0.2	0.0–1.0
Lactones (expressed as glucolactone)	0.1	0.0–0.4

1977; Siddiqui 1970) such as sucrose, maltose, isomaltose and erlose. These minority sugars can provide useful information about the botanical origin of the product and evidence of adulteration (Krell 1996). Water represents the third most abundant compound in honey and its concentration is critical in the production and post-production phases (Krell 1996). Table 13.1 reveals that other minority compounds are also quantifiable and Table 13.2 reports the ranges of the concentrations of such trace analytes in nectar honeys of different floral origins.

The concentrations of minority analytes in a sample of honey form a sort of fingerprint that is a powerful instrument for identifying the floral and geographical origin of honey and detecting potential frauds and adulterations (Anklam 1998). Some of the most important classes of analytes are the non-aromatic organic acids (Mato *et al.* 2006; Suarez-Luque *et al.* 2002, White 1978), the free amino acids (Anklam 1998 and references therein; Spano *et al.* 2009 and references therein), the inorganic elements (Bogdanov *et al.* 2008), the polyphenols (Andrade *et al.* 1997; Dimitrova *et al.* 2007; Scanu *et al.* 2005; Stephens *et al.* 2010; Tuberoso *et al.* 2009) and the flavonoids (Andrade *et al.* 1997; Anklam 1998; Ferreres *et al.* 1996; Tomas-Barberan *et al.* 2001)—all present in concentrations between a few µg/kg and a few g/kg.

13.2 Vitamins in Honey

The vitamin content of honey is very low and its contribution to the human daily intake is generally negligible. Pollen is considered the main source of vitamins in honey and commercially clarified samples often show low concentrations of these analytes (Haydak *et al.* 1943). Table 13.3 reports the ranges of vitamin concentrations found in nectar honeys of different floral origins.

Table 13.2 Concentration range of trace constituents of nectar honey of different floral origin.

Class/Analyte	Range (mg/kg)	Floral origin of honey[c]	References
Organic acids			
Citric acid[a]	20–394	CHS, EUC, TRI, MUL	Suarez-Luque et al. 2002
Formic acid	46–908	CHS, EUC, MUL	Mato et al. 2006
Fumaric acid	0.04–7.29	CHS, EUC, TRI, MUL	Suarez-Luque et al. 2002
Gluconic acid	3.7–14.4	CHS, EUC, MUL	Mato et al. 2006
Lactic acid	29–632	CHS, EUC, MUL	Mato et al. 2006
Malic acid	13–434	CHS, EUC, MUL	Suarez-Luque et al. 2002
Oxalic acid	14–114	CHS, EUC, MUL	Mato et al. 2006
Succinic acid	12–759	CHS, EUC, MUL	Suarez-Luque et al. 2002
Amino acids			Spano et al. 2009 and references therein
Alanine	3.5–68.6	ACA, CHS, CIT, THY, ROS, LAV, ORA, EUC, RHO, LIM, SAV, BUC, BLL, TAR, TIL, RAP, FIR, LIN, SUN, HEA, STT	
Arginine	3.5–23	ACA, CHS, THY, ROS, LAV, ORA, EUC, RAP, FIR, LIN, SUN, STT	
Aspartic acid	2.3–205.9	ACA, CHS, CIT, THY, ROS, LAV, ORA, EUC, RHO, LIM, RAP, FIR, LIN, SUN, HEA, STT	
Glutamic acid	8.3–191.9	ACA, CHS, CIT, THY, ROS, LAV, ORA, EUC, RHO, LIM, RAP, FIR, LIN, SUN, HEA, STT	
Glycine	2.3–29.1	ACA, CHS, CIT, THY, ROS, LAV, ORA, EUC, RHO, LIM, SAV, BUC, BLL, TAR, TIL, RAP, FIR, LIN, SUN, HEA, STT	
Lysine	2.8–41.9	ACA, CHS, CIT, THY, ROS, LAV, ORA, EUC, RHO, LIM, SAV, BUC, BLL, TAR, TIL, RAP, FIR, LIN, SUN, HEA, STT	
Methionine	0.2–312	ACA, CHS, CIT, THY, ROS, LAV, ORA, EUC, RHO, LIM, SAV, BUC, BLL, TAR, TIL, RAP, FIR, LIN, SUN, HEA, STT	

Table 13.2 (*Continued*)

Class/Analyte	Range (mg/kg)	Floral origin of honey[c]	References
Phenylalanine	5.5–1152.8	ACA, CHS, CIT, THY, ROS, LAV, ORA, EUC, RHO, LIM, SAV, BUC, BLL, TAR, TIL, RAP, FIR, LIN, SUN, HEA, STT	
Proline	137–958	ACA, CHS, CIT, THY, ROS, LAV, ORA, EUC, RHO, LIM, SAV, BUC, BLL, TAR, TIL, RAP, FIR, LIN, SUN, HEA, STT	
Tyrosine	4.1–299	ACA, CHS, CIT, THY, ROS, LAV, ORA, EUC, RHO, LIM, SAV, BUC, BLL, TAR, TIL, HEA	
Valine	3.3–287	ACA, CHS, CIT, THY, ROS, LAV, ORA, EUC, RHO, LIM, SAV, BUC, BLL, TAR, TIL, RAP, FIR, LIN, SUN, HEA, STT	
Elements			Bogdanov 2008 and references therein
Aluminium	0.1–24		
Arsenic	0.14–0.26		
Bromide	4–13		
Calcium	30–310		
Chloride	4–56		
Chromium	0.1–3		
Copper	0.2–6		
Fluoride	4–13.4		
Iodide	100–1000		
Iron	0.3–40		
Magnesium	7–130		
Manganese	0.2–2		
Potassium	400–35000		
Selenium	0.02–0.1		
Sodium	16–170		
Zinc	0.5–20		
Phenolic compounds and flavonoids			
3-Hydroxybenzoic acid[b]	3.39–4.71	LIM	Dimitrova *et al.* 2007
Ellagic acid	0.93–6.13	HEA	Ferreres *et al.* 1996
Hesperitin	0.499–0.973	CIT	Ferreres *et al.* 1993
Homogentisic acid	73.13–604.4	STT	Scanu *et al.* 2005
Methoxyphenylacetic acid	8.8–161	KAN	Stephens *et al.* 2010

Table 13.2 (*Continued*)

Class/Analyte	Range (mg/kg)	Floral origin of honey[c]	References
Methyl syringate	122.6–343.8	ASP	Tuberoso *et al.* 2009
Myricetin 3-methyl ether	0.01–0.493	HEA	Ferreres *et al.* 1996
Naringenin	Not reported	LAV	Andrade *et al.* 1997
Quercetine	1.25–2.90	SUN	Tomas-Barberan *et al.* 2001
Rosmarinic acid	Not reported	THY	Andrade *et al.* 1997

[a] Reliable parameter for the differentiation between nectar honey and honeydew honey.
[b] Candidate markers of floral origin.
[c] ACA, acacia; ASP, asphodel; BLL, black locust; BUC, buckwheat; CHS, chestnut; CIT, citrus; EUC, eucalyptus; FIR, fir; HEA, heather; KAN, kanuka; LAV, lavender; LIM, limetree; LIN, linden; MUL, multiflora; ORA, orange; RAP, rape; RHO, rhododendron; ROS, rosemary; SAV, savory; STT, strawberry tree; SUN, sunflower; TAR, taraxacum; THY, thyme; TIL, tilia; TRI, trifolium.

The data reported in the literature almost exclusively concern water-soluble vitamins (B group vitamins and vitamin C). For this reason, this chapter limits its treatment of the analytical aspects of honey to these compounds.

13.2.1 B Group Vitamins in Honey

13.2.1.1 Vitamin B_1: Thiamine

Only traces of thiamine can be found in honey: its concentration rarely exceeds 0.1 mg/kg (Bogdanov *et al.* 2008; Haydak *et al.* 1942; Kitzes *et al.* 1943). Although thiamine is very unstable in aqueous solutions, it appears to be stable in honey. This fact is confirmed by a detailed examination of the data provided by Haydak *et al.* (1942), which show that thiamine has been found in aged honeys in the same concentration range as in fresh ones.

13.2.1.2 Vitamin B_2: Riboflavin

Whereas previous studies found very low amounts of riboflavin in honey samples—between 0.005 and 1.45 mg/kg (Bogdanov *et al.* 2008; Haydak *et al.* 1942; Kitzes *et al.* 1943; Viñas *et al.* 2004a), very recent results (Ciulu *et al.* 2011) provide evidence that this vitamin can also reach higher concentrations (*i.e.* close to 10 mg/kg).

13.2.1.3 Vitamin B_3: Niacin

Data from early studies are contradictory. Haydak *et al.* (1942) measured very high concentrations of niacin, *i.e.* between 40 and 940 mg/kg. However, only one year later Kitzes *et al.* (1943) measured concentrations that were two orders of magnitude lower, *i.e.* between 0.63 and 5.9 mg/kg. More than 60 years later

Table 13.3 Concentration ranges of vitamins in nectar honey of different floral origin.

Vitamin	Range (mg/kg)	Floral origin of honey[a]	References
Ascorbic acid (C)	<0.02–2960	EUC, SUL, CIT, ACA, ASP, THY, STT, LAV, HEA, ROS, LIN, MUL, COT, HBO, ALF, MES, SAG, SPW, TAM, WSA, WBW, TUP, TIT, HMI, FWE, MZT, RAT, POP, ALG, MIL, MEL, LOG, CAM, COF	Haydak et al. 1942; Bogdanov 2008 and references therein; White 1975; Abu-Tarboush et al. 1993; Castro et al. 2001, Casella and Gatta 2001, Gheldof et al. 2001; Matei et al. 2004; Ferreira et al. 2009; Alvarez-Suarez et al. 2010a, 2010b; Leon-Ruiz et al. 2011; Ciulu et al. 2011.
Folic acid (B_9)	<0.15–6.92	EUC, SUL, CIT, ACA, ASP, THI, STT, LAV, HEA, ROS, LIN, MUL	Ciulu et al. 2011
Niacin (B_3)	<0.75–940	EUC, SUL, CIT, ACA, ASP, THI, STT, LAV, HEA, ROS, LIN, MUL, COT, HBO, ALF, MES, SAG, SPW, TAM, WSA, WBW, TUP, TIT, HMI, FWE, MZT, RAT, POP, ALG, MIL, MEL, LOG, CAM, COF	Haydak et al. 1942; Kitzes et al. 1943; Ciulu et al. 2011.
Panthothenic acid (B_5)	0.009–28	EUC, SUL, CIT, ACA, ASP, THI, STT, LAV, HEA, ROS, LIN, MUL, COT, HBO, ALF, MES, SAG, SPW, TAM, WSA, WBW, TUP, TIT, HMI, FWE, MZT, RAT, POP, ALG, MIL, MEL, LOG, CAM, COF	Haydak et al. 1942; Kitzes et al. 1943; Bogdanov 2008 and references therein; Ciulu et al. 2011.
Pyridoxin (B_6)	0.039–4.8	ORA, EUC, ROS, MUL, COT, HBO, ALF, MES, SAG, SPW, TAM, WSA, WBW, TUP, TIT, HMI, FWE, MZT, RAT, POP, ALG, MIL, MEL, LOG, CAM, COF	Haydak et al. 1942; Kitzes et al. 1943; Bogdanov 2008 and references therein; Viňas et al. 2004.
Riboflavin (B_2)	0.009–9.2	EUC, SUL, CIT, ACA, ASP, THI, STT, LAV, HEA, ROS, LIN, MUL, SCA, CHS, COT, HBO, ALF, MES, SAG, SPW, TAM, WSA, WBW, TUP, TIT, HMI, FWE, MZT, RAT, POP, ALG, MIL, MEL, LOG, CAM, COF	Haydak et al. 1942; Kitzes et al. 1943; Bogdanov 2008 and references therein; Viňas et al. 2004; Ciulu et al. 2011.

Table 13.3 (*Continued*)

Vitamin	Range (mg/kg)	Floral origin of honey[a]	References
Thiamin (B_1)	0.014–0.12	MUL, TAR, ORA, THI, COT, HBO, ALF, MES, SAG, SPW, TAM, WSA, WBW, TUP, TIT, HMI, FWE, MZT, RAT, POP, ALG, MIL, MEL, LOG, CAM, COF, BTN, MAR, SFE	Haydak *et al.* 1942; Kitzes *et al.* 1943; Bogdanov 2008 and references therein.

[a]ACA, acacia; ALF, alfalfa; ALG, algaroba; ASP, asphodel; BLL, black locust; BUC, buckwheat; CAM, campanulla; CHS, chestnut; CIT, citrus; COF, coffee; COT, cotton; EUC, eucalyptus; FIR, fir; FWE, fireweed; HBO, holly blossom; HEA, heather; HMI, horsemint; KAN, kanuka; LAV, lavender; LIM, limetree; LIN, linden; LOG, logwood; MAR, marigold; MEL, melon; MES, mesquite; MIL, milkwort; MUL, multiflora; MZT, manzanita; ORA, orange; POP, poplar; RAP, rape; RAT, rattan; RHO, rhododendron; ROS, rosemary; SAG, sage; SAV, savory; SFE, sugar-feed; SPW, spikeweed; STT, strawberry tree; SUL, sulla; SUN, sunflower; TAM, tamarisk; TAR, taraxacum; THY, thyme; TIL, tilia; TIT, titi; TRI, trifolium; TUP, tupelo; WBW, wild buckwheat; WSA, white sage.

Ciulu *et al.* (2011) measured the concentration of niacin in 28 honey samples from Sardinia, obtaining data ranged from less than the limit of detection (LoD) to 12.4 mg/kg.

13.2.1.4 Vitamin B_5: Pantothenic acid

The presence of pantothenic acid in honey has been known since the 1940s (Haydak *et al.* 1942; Kitzes *et al.* 1943). These research groups measured this analyte's concentration in the range between 0.009 and 3.6 mg/kg. A review by Bogdanov *et al.* (2008) more recently reported concentrations between 0.2 and 1.1 mg/kg. Higher concentrations (between 4.4 and 28 mg/kg) were later found by Ciulu *et al.* (2011) in honeys from asphodel, sulla, citrus, strawberry tree and eucalyptus; the concentration of the analyte has always been under its quantification limit in acacia, thistle and lavender honeys.

13.2.1.5 Vitamin B_6: Pyridoxine

Vitamin B_6 is also a minority analyte in honey: Haydak *et al.* (1942) and Kitzes *et al.* (1943) measured concentration ranges between 0.039 and 4.8 mg/kg using colorimetric and microbiological methods, respectively. Furthermore, Bogdanov *et al.* (2008) reported concentrations of vitamin B_6 in honey of between 0.1 and 3.2 mg/kg, whereas Viñas *et al.* (2004b) placed the total concentrations of pyridoxal and pyridoxine at between 0.141 and 0.826 mg/kg.

13.2.1.6 Vitamin B_9: Folic Acid

Until the study released by Ciulu *et al.* (2011), no data were available in the literature regarding the presence of vitamin B_9 in honey. However, folic acid appears to be a relatively common vitamin in different honeys, although it is not present at high concentration values. Its presence has been ascertained in more than 75% of honey samples and the concentrations ranged from values less than the limit of quantification (LoQ) to 6.92 mg/kg.

13.2.2 Vitamin C: Ascorbic Acid

The attempt to establish a 'typical' concentration of vitamin C in honey may be a 'mission impossible'. Factors such as the age, handling and storage conditions of the honey particularly affect the concentration of this vitamin which is very easily decomposed. Moreover, the different floral origin of the honey and the performance of the analytical method can also influence the measurements. Consequently, the concentration data reported in the literature are spread over four orders of magnitude, going from less than 0.1 mg/kg to values very close to 3 g/kg!!!

A number of research groups have reported concentrations of ascorbic acid in honey between 5.4 and 54 mg/kg (Abu-Tarboush *et al.* 1993; Bogdanov *et al.* 2008; Castro *et al.* 2001; Ciulu *et al.* 2011; Haydak *et al.* 1942; White 1975). Other studies (Alvarez-Suarez *et al.* 2010a, 2010b; Casella and Gatta 2001; Gheldof *et al.* 2002) reported the substantial absence of vitamin C in honey using different analytical methods.

On the other hand, Matei *et al.* (2004) published extremely high concentrations of ascorbic acid (between 2.26 and 2.96 g/kg) in four Romanians honeys, measured using a titrimetric (oxidimetric) method. Also, Ferreira *et al.* (2009) determined vitamin C concentrations of between 140.01 and 145.80 mg/kg in three Portuguese honeys using a spectrophotometric method. Finally, Leon-Ruiz *et al.* (2011) measured concentrations of ascorbic acid between 21.8 and 1700 mg/kg in 82 honeys from Spain using both a redox titration with 2,6-dichloroindophenol and an high-performance liquid chromatography–ultraviolet (HPLC-UV) method.

In summary, the wide variability in the published data makes it very difficult to establish a typical concentration range for ascorbic acid in honey.

13.3 Assay of Water-soluble Vitamins in Honey

Very little work has been done until now on the determination of water-soluble vitamins in honey. Many factors contribute to this situation. Among them, the low amount of these analytes in honey has probably played a key role.

13.3.1 Assay of B Group Vitamins in Honey

Due to the low concentration of B group vitamins in honey, the measurement of the analyte requires a very sensitive method. Colorimetric, spectrophotometric, spectrofluorimetric, microbiological and reversed-phase HPLC (RP-HPLC) analytical approaches have been used until now.

Haydak et al. (1942) measured the concentration of vitamins B_1, B_3 and B_6 in honey using spectroscopic methods previously developed by Hennessey and Cerecedo (1939), Melnick and Field (1940) and Scudi (1941), respectively, for determinations in animal organs, biological fluids and a number of foodstuffs. The first is a spectrofluorimetric method, while the last two are both colorimetric methods.

Some microbial strains have the ability to produce, in presence of a specific target vitamin, a quantifiable reaction corresponding to the amount of the specific vitamin present. For instance, some will produce a corresponding quantity of an easily determinable molecule (e.g. lactic acid). Others will produce a measurable linear analytical response dependent on an intrinsic property of the growing microbial strains (e.g. turbidity). Microbes with these abilities have been advantageously used in the past to measure the amount of a specific type of vitamin in a sample. Specifically, microorganisms of the *Lactobacillus* genus or yeasts such as *Saccharomyces carlsbergensis* (Atkin et al. 1943) have been used to accurately determine riboflavin, niacin, pantothenic acid and pyridoxine in non-saccharidic matrices. Pioneering groups (Haydak et al. 1942; Kitzes et al. 1943) adapted these methods to honey with good results. Table 13.4 summarizes the microbiological methods used for to determine B group vitamins in honey.

In recent years, RP-HPLC with UV (or fluorescence) detection methods have also been successfully used for the determination of B vitamins in honey. Viñas et al. (2004a, 2004b) proposed quantifying vitamins B_2 and B_6 with chromatographic methods using fluorescence detection. In their protocol for B_2, the stationary phase is an end-capped amide-based column and the mobile phase is a 10 : 90 v/v acetonitrile/phosphate buffer (pH = 5). The protocol for B_6 is

Table 13.4 Microbiological methods used for the determination of B group vitamins in honey.

Vitamin	Microorganism	Metabolite	References Method	Application to honey
Niacin $(B_3)^a$	*Lactobacillus arabinosus*	Lactic acid	Snell and Wright 1941	Kitzes et al. 1943
Panthothenic acid $(B_5)^b$	*Lactobacillus casei*	Lactic acid	Strong et al. 1941	Haydak et al. 1942; Kitzes et al. 1943
Pyridoxin $(B_6)^{b,c}$	*Saccharomyces carlsbergensis*		Atkin et al. 1943	Kitzes et al. 1943
Riboflavin (B_2)	*Lactobacillus casei*	Lactic acid	Snell and Strong 1939	Haydak et al. 1942; Kitzes et al. 1943

[a]In addition to niacin, niacine adenine dinucleotide and nicotinuric acid can also be metabolized by *L. arabinosus*, producing corresponding amounts of lactic acid.
[b]A proper choice of basal medium provides specificity towards the vitamin.
[c]Quantification accomplished by means of turbidimetric measurements.

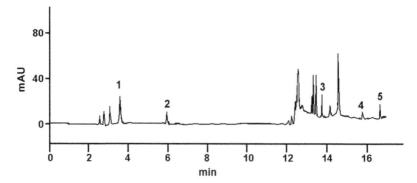

Figure 13.1 Chromatogram of an asphodel honey from Sardinia, Italy. 1, vitamin C; 2, vitamin B_3; 3, vitamin B_5; 4, vitamin B_9; 5, vitamin B_2. From Ciulu et al. (2011).

identical except for the mobile phase which becomes an acetonitrile solution containing a 5×10^{-2} mol/dm^3 phosphate buffer (pH = 7). In addition to measuring the vitamin content, these methods are able to simultaneously also determine riboflavin coenzymes or pyridoxal, pyridoxine and pyridoxamine, two of their phosphorylated forms (pyridoxal-5-phosphate and pyridoxamine-5-phosphate) and one excretion product (4-pyridoxic acid).

Finally, Ciulu et al. (2011) developed a RP-HPLC method with UV revelation able to simultaneously determine four B group vitamins and ascorbic acid. This method does not require an extraction phase and uses a conventional C18 column for separation. A gradient elution programme was adopted using a trifluoroacetic acid aqueous solution (0.025%, v/v) and acetonitrile. Figure 13.1 shows a typical chromatogram of an asphodel honey obtained by applying this method.

13.3.2 Assay of Vitamin C in Honey

The need to quantify ascorbic acid in honey has led to the development of different analytical approaches, sometimes with questionable results. The assay of ascorbic acid in honey has mainly been accomplished using HPLC methods (UV, fluorescence or electrochemical detection) and less frequently with UV–visible and titrimetric methods.

The extraction of ascorbic acid with metaphosphoric acid and the reaction of a citrate-buffered solution of analyte with a known amount of 2,6-dichlorophenol indophenol is the base for the colorimetric method developed by Bessey (1938). This method, which was later applied by Haydak et al. (1942) to honey, takes advantage of the fact that the absorbance decrease at 520 nm is proportional to the concentration of ascorbic acid. More recently, Ferreira et al. (2009) measured the concentration of vitamin C in three Portuguese unifloral honeys using the same principle, but unfortunately no validation parameters were provided in this study.

HPLC methods form an important class of techniques for the quantification of vitamin C in honey and the extraction of vitamin C with metaphosphoric acid is a key step in a number of these. We list some of the works using this methodology. Abu-Tarboush et al. (1993) reported the use of an AOAC International official method to measure the concentration of vitamin C in six commercial Arabian honeys. Their study used a PLRP-S column and the mobile phase was a H_3PO_4/NaH_2PO_4 buffer solution at pH 2.14. In contrast, the following three research groups did not find any ascorbic acid in any honey sample. First, Castro et al. (2001) added metaphosphoric acid in the mobile phase (a $MeOH/H_2O$ solution, 15/85 v/v) while using a C18-ODS column. The revelation was accomplished at 254 nm. On the other hand, Gheldof et al. (2002) pretreated the sample with dithiothreitol and metaphosphoric acid in a $0.1\,mol/dm^3$ aqueous solution of KH_2PO_4 (pH = 7.0). They then used a C18 column, a gradient elution with a $50\,mmol/dm^3$ aqueous solution of KH_2PO_4 (pH = 4.5) and methanol, and UV revelation. Alvarez-Suarez et al. (2010a, 2010b) used a very similar method, with the exception of a diode array detector in spite of a single wavelength UV detector. Casella and Gatta (2001) used an ion-chromatographic method combined with electrochemical detection at a copper-based chemically modified glassy carbon electrode. The stationary phase was a CarboPac PA1 column and the mobile phase was a $0.1\,mol/dm^3$ solution of NaOH.

Ciulu et al. (2011) and Leon-Ruiz et al. (2011) recently assessed and fully validated two RP-HPLC methods, but they reached very different results after analysing tens of honey samples of different botanical origins. The details of the method used by Ciulu and colleagues is described in Section; Leon-Ruiz and colleagues used a C18 column as the stationary phase and a mobile phase consisting of a 0.01% (v/v) aqueous solution of H_2SO_4, and performed the UV detection at 245 nm. While Ciulu et al. did not measure concentrations of vitamin C higher than $4 \pm 1\,mg/kg$ on fresh and well-stored honey samples directly provided by local beekeepers, Leon-Ruiz et al. found extremely high concentrations in thyme honey, sometimes close to 1700 mg/kg! Moreover, in lavender and rosemary honeys they found a concentration of ascorbic acid more than ten times higher than that measured by Ciulu et al. in similar samples.

Finally Matei et al. (2004) arbitrarily extended a titrimetric method (Skoog et al. 1998) for determination of the analyte in vitamin C tablets. The very high amounts of ascorbic acid reported are probably distorted due to the interferences of easily oxidizable species found naturally in honey.

In conclusion, while we pass over data we deem unreliable due to the use of bias-affected methods, the data provided by Leon-Ruiz et al. and Ciulu et al. demonstrate that the floral and geographical origins of the honey are important parameters influencing the concentration of ascorbic acid in this matrix.

13.4 Validation

Validation is the first of the universally recognized key steps for quality assurance in analytical methods (Thompson et al. 2002). It is important to remember that the term 'method' indicates not only the 'analytical protocol', but also the concentration level of the analyte and the type of test material. At the very least, a validation protocol should consider the following performance parameters: limit of detection, linearity, precision and trueness.

13.4.1 LoD and LoQ

The limit of detection (LoD) represents the minimum concentration at which the presence of the analyte can be distinguished from its absence with a high statistical probability. There are different possible ways to define the LoD pivoting about the interpretation of 'significantly different' from the blank. The choice of definition will influence which type of error is more likely: Type I, false positives; or Type II, false negatives. In addition to these considerations, the suitability of a specific LoD is of course affected by the chosen instrumentation and analytical method. Conversely, at higher levels the concentration of an analyte can be quantified. The lowest concentration that can be reliably quantified is known as the limit of quantification (LoQ). The LoQ is often set at a fixed multiple (usually between 2 and 3.3) of the LoD.

Given the low concentrations of vitamins in honey, the choice of LoD and the method used to validate are of particular importance. Table 13.5 summarizes the most representative validation parameters for LoD and LoQ reported in the literature. Notwithstanding its importance, the LoD is rarely reported in published methods. Moreover, when a LoD is indicated all too often the approach used for its evaluation remains unknown. As a consequence, it is not uncommon to see similar published methods exhibiting LoD values differing by several orders of magnitude!

As an example of how LoD levels can differ between studies, consider the work by Ciulu *et al.* (2011) and Viñas *et al.* (2004a) in the determination of vitamin B_2. Beyond the obvious different sensitivity between the methods used, the difference in the LoD values reported by Ciulu *et al.* (2011) and Viñas *et al.* (2004a) in the determination of vitamin B_2 is due to the decision to adopt different calculation methods. Viñas *et al.* used the 3σ method (Long and Winefordner 1983), whereas Ciulu *et al.* opted for the upper limit approach 1 (ULA1) (Mocak *et al.* 1997). The 3σ method allows one to obtain low LoD values, but raises the risk of Type II errors when measuring analyte concentrations close to the LoD. Conversely, the ULA1 approach minimizes the risk of making any decision error approach but tends to result in high LoD values.

13.4.2 Linearity

High linearity is a very desirable feature in a method because it allows it to operate in a wide analyte concentration range without additional sample treatments (*e.g.* dilutions). Linearity is usually evaluated by determining the correlation coefficient, R^2 in a calibration plot. The closer the value of R^2 is to 1, the better the linearity within the tested concentration range. In all cases, a graphical analysis of the distribution of the regression line residuals is strongly

Table 13.5 Validation data reported in literature for methods used for the determination of water-soluble vitamins in honey.

Vitamin	LoD (mg/kg)[a]	LoQ (mg/kg)[a]	Linearity (mg/L)[a]	Precision Repeatability (RSD %)	Reproducibility (RSD %)	Trueness Recovery tests (%)	References
B_1					36.3		Haydak et al. 1942
						85–105	Kitzes et al. 1943
B_2					32.4		Haydak et al. 1942
	0.3 µg/L	0.3 µg/L	1–150 µg/L	2.3	19.8	98.3–99.1	Viñas et al. 2004a
	0.25	0.75	0.01–100	2.6	77.3	100	Ciulu et al. 2011
B_3						85–108.7	Haydak et al. 1942
						90–110	Kitzes et al. 1943
B_5	0.25	0.75	0.05–500	6.9	18	99	Ciulu et al. 2011
					42.12		Haydak et al. 1942
						90–110	Kitzes et al. 1943
B_6	0.58	1.75	0.5–1000	7.6	7.6	103	Ciulu et al. 2011
					30.6		Haydak et al. 1942
						95–105	Kitzes et al. 1943
	3 µg/L	11 µg/L	30–200 µg/L	4			Viñas et al. 2004b
B_9	0.15	0.50	0.05–100	2.9	12.8	98	Ciulu et al. 2011
C					51.5		Haydak et al. 1942
	20 µg/L		0.5–800 µM	1.6		92.1–94.8	Castro et al. 2001
				4			Casella and Gatta 2001
	0.10	0.30	0.05–500	7.3	3.3	104	Ciulu et al. 2011
	0.68 mg/L	2.3 mg/L	0.11–276.70		2.5	[b]	Leon-Ruiz et al. 2011

[a] If not otherwise specified.
[b] The trueness has been successfully evaluated by means of the comparison of two independent methods.

recommended (Thompson *et al.* 2002): any curved pattern suggests the lack of fit due to a 'hidden' non-linear calibration function.

13.4.3 Precision

In a single-laboratory validation protocol, precision is evaluated in terms of repeatability and reproducibility. From the 'Definitions of words and terms', it is evident that uncertainty of reproducibility (expressed as RSD_R) is expected to be higher than that of repeatability (RSD_r). Horwitz's studies (1982) show that precision depends only on the concentration range of the analyte—the lower it is, the higher the relevant relative standard deviation (RSD) value. Horwitz's theory also describes a fitness-for-purpose method to evaluate the precision of the data in terms of the maximum RSD acceptable for a stated level of concentration—used by Ciulu *et al.* in their recent paper. Due to the fact that the water-soluble vitamins are normally present in honey in concentrations between tenths and tens of mg/kg, the 'acceptable' values of RSD_r and RSD_R must not to be higher than 22.6% and 11.3% and 34% and 17%, respectively. The precision RSD values reported in Table 13.5 substantiate that, except for the early results reported by Haydak and coworkers, the studies are well within Horwitz's acceptability range.

13.4.4 Trueness

The trueness definition suggests that a very critical role is bestowed on the 'accepted' reference value. This quantity can be provided by a number of different approaches:

(i) The analysis of a certified reference material (CRM)
(ii) The analysis of a reference material (RM)
(iii) The comparison of data obtained by two independent methods
(iv) The use of a spiking/recovery technique.

A CRM for vitamins in honey does not exist at present and it is unlikely that one could prepare a RM due to the known instability of analytes. Therefore, to evaluate the bias, one must analyse the material by independent methods (Leon-Ruiz *et al.* 2011) or, more commonly, evaluate the recovery of known amounts of analytes spiked to honey. The first method is based on the fact that there is a negligible statistical probability that two independent methods may provide data affected by same bias, whereas in the spiking/recovery technique, recovery values statistically indistinguishable from 100% (*e.g.* for a two-tailed *t*-test) indicate the absence of a bias. Table 13.5 shows recovery values of between 85 and 108.7%. Although a number of these recoveries different from 100% probably indicate some bias, it has been previously observed (AOAC 1998) that for low concentration levels recoveries can normally differ from 100%. AOAC guidelines (AOAC 1998) consider recovery levels between 80%

and 110% to be 'acceptable' for concentration levels between tenths and tens of mg/kg, *i.e.* those of the water-soluble vitamins in honey.

13.5 Concluding Remarks

Water-soluble vitamins represent a remarkable feature of the minority chemical composition of honey. Although their low concentrations preclude any specific interest from a nutritional viewpoint, it has been shown that their concentration levels provide information sufficient to determine the origin of the honey and verify its freshness.

The development of specific analytical methods for determining water-soluble vitamins in such matrix has not seen major activity in past decades. In the early years, microbiological and colorimetric assays were normally used for this task, while in more recent times, analytical procedures assessed for measuring vitamins in other matrices have been extended to honey, sometimes with questionable results. Only in the last decade has the attention of a number of research groups resulted in specialized and validated HPLC methods.

Summary Points

- Honey is far from being a vitamin-rich food.
- Only low concentrations of water-soluble vitamins are normally present in honey.
- The amount of water-soluble vitamins in honey and their concentrations can be a strong indicator of the geographical origin and the freshness of the honey.
- Analytical methods used to determine vitamins until now were based on spectroscopic, chromatographic, titrimetric and microbiological approaches.
- Water-soluble vitamins in honey have been often determined using methods not specifically developed for this matrix.
- While data published in the literature are quite homogeneous regarding the amount of B group vitamins in honey, the reported concentration of vitamin C in honey varies widely over several orders of magnitude.
- It is possible that data relative to high concentrations in honey of vitamin C are affected by some bias.
- Only a proper validation of the whole analytical protocol offers guarantees on the reliability of the data produced.

Key Facts

Key Facts about Honey

- Honey is the first natural sweetener used by humans.
- Honey exhibits many functional properties and is still widely used as part of the popular medicinal practice in many countries.

Key Facts about the Vitamins in Honey

- The amount of vitamins in honey is low and only water-soluble vitamins are usually found in honey.
- The amount of water-soluble vitamins in honey varies as a function of its floral and geographical origin.

Key Facts about Assaying Water-soluble Vitamins in Honey

- There is a general lack of analytical methods capable of determining contemporary water-soluble vitamins in honey.

Key Facts about Assaying B Group Vitamins in Honey

- B group vitamins have been measured starting in the early decades of the 20th century, mainly using microbiological methods.
- Vitamin B_8 and vitamin B_{12} have never been detected in honey.

Key Facts about Assaying Vitamin C in Honey

- Unlike what is observed with B group vitamins, the concentration of vitamin C in honey is spread over many orders of magnitude.

Key Facts about Validation Protocols

- Analytical protocols used to measure water-soluble vitamins in honey have rarely been validated.
- The application to honey of methods developed to measure water-soluble vitamins in different matrices is a questionable analytical approach.

Definitions of Words and Terms

Chromatographic methods of analysis. They are a large group of analytical techniques devoted to the separation, identification and quantification of the constituents of a mixture of compounds.

Floral origin of honey. It is the botanical variety that gives origin to the honey and its specific set of chemical and physical parameters.

Geographical origin of honey. It is the association of chemical or physical parameters of honey and its geographical area of production.

Honey. It is a natural food produced by bees (*Apis mellifera*) from the nectar and the honeydew of plants.

Limit of detection, LoD. The concentration which produces a measured signal that, statistically, is significantly different from the blank signal.

Limit of quantification, LoQ. The lowest concentration that can be quantified with a stated statistical confidence.

Linearity. It measures the extension of the linear relationship between analyte concentration and the corresponding instrument signal.

Precision. The level of agreement between independent measurements obtained under defined conditions (repeatability and reproducibility). Precision improves when the number of repeated measurements increases, and it is usually measured as a standard deviation (SD) or a relative standard deviation (RSD).

Repeatability. Describes any variations observed during a single analytical run, *i.e.* n replicates of the same analysis performed on the same sample, in the same analytical session, using the same reagents, instruments and operators.

Reproducibility. Describes the degree of variation in the analytical data observed during several different analytical sessions, *i.e.* n replicates of the same analysis performed on the same sample in different analytical sessions, using different reagents, instruments and operators.

Spectroscopic methods of analysis. They are a wide group of instrumental analytical methods all based on the interaction between energy (an electromagnetic radiation) and matter (usually the analyte).

Trueness. The level of agreement between an analytical value and an 'accepted' reference value. Bias is the parameter that quantitatively measures trueness. A small bias indicates high trueness.

Validation. A validation procedure is a set of tests able to evaluate the performance of a method according to selected performance criteria. Such a procedure is generally designed to test whether the method is fit for a particular analytical purpose.

Water-soluble vitamins. This is a group of nine compounds, all structurally different from each other. Eight are B group vitamins (B_1, B_2, B_3, B_5, B_6, B_8, B_9 and B_{12}) and the ninth is vitamin C. They mainly act as coenzymes and, due the fact that they cannot be accumulated in the human body, they have to be assumed regularly.

List of Abbreviations

ACA	acacia
ALF	alfalfa
ALG	algaroba
ASP	asphodel
BLL	black locust
BUC	buckwheat
BTN	buckthorn
CAM	campanulla
CHS	chestnut
CIT	citrus
COF	coffee
COT	cotton

CRM	certified reference material
EUC	eucalyptus
FIR	fir
FWE	fireweed
HBO	holly blossom
HEA	heather
HMI	horsemint
HPLC	high-performance liquid chromatography
KAN	kanuka
LAV	lavender
LIM	limetree
LIN	linden
MAR	marigold
MEL	melon
LoD	limit of detection
LOG	logwood
LoQ	limit of quantification
MES	mesquite
MIL	milkwort
MUL	multiflora
MZT	manzanita
POP	poplar
ORA	orange
RAP	rape
RAT	rattan
RHO	rhododendron
RM	reference material
ROS	rosemary
RP	reversed-phase
RSD	relative standard deviation
SAG	sage
SAV	savory
SD	standard deviation
SFE	sugar-feed
SPW	spikeweed
STT	strawberry tree
SUL	sulla
SUN	sunflower
TAM	tamarisk
TAR	taraxacum
THY	thyme
TIL	tilia
TIT	titi
TRI	trifolium
TUP	tupelo
ULA	upper limit approach

UV   ultraviolet
WBW   wild buckwheat
WSA   white sage

References

Abu-Tarboush, H.M., Al-Kahtani, H.A., and El-Sarrage, M.S., 1993. Floral-type identification and quality evaluation of some honey types. *Food Chemistry*. 46: 13–17.

Alvarez-Suarez, J.M., González-Paramás, A.M., Santos-Buelga, C., and Battino, M., 2010a. Antioxidant characterization of native monofloral Cuban honeys. *Journal of Agricultural and Food Chemistry*. 58: 9817–9824.

Alvarez-Suarez, J.M., Tulipani, S., Dìaz, D., Estevez, Y., Romandini, S., Giampieri, F., Damiani, E., Astolfi, P., Bompadre, S., and Battino, M., 2010b. Antioxidant and antimicrobial capacity of several monofloral Cuban honeys and their correlation with color, polyphenol content and other chemical compounds. *Food and Chemical Toxicology*. 48: 2490–2499.

Andrade, P., Ferreres, F., Gil, M.I., and Tomas-Barberan, F.A., 1997. Determination of phenolic compounds in honeys with different floral origin by capillary zone electrophoresis. *Food Chemistry*. 60: 79–84.

Anklam, E., 1998. A review of the analytical methods to determine the geographical and botanical origin of honey. *Food Chemistry*. 63: 549–562.

AOAC, 1998. AOAC Peer-Verified Methods Program. Manual on Policies and Procedures. 35 pp, AOAC International, Arlington, VA, USA.

Atkin, L., Schultz, A.S., Williams, W.L., and Frey C.N., 1943. Yeast microbiological methods for determination of vitamins. Pyridoxine. *Industrial & Engineering Chemistry Analytical Edition*. 15: 141–144.

Bessey, O.A., 1938. A method for the determination of small quantities of ascorbic acid and dehydroascorbic acid in turbid and colored solutions in the presence of other reducing substances. *Journal of Biological Chemistry*. 126: 771–784.

Bogdanov, S., Jurendic, T., Sieber, S., and Gallmann, P., 2008. Honey for nutrition and health: a review. *Journal of the American College of Nutrition*. 27: 677–689.

Casella, I.G., and Gatta, M., 2001. Determination of electroactive organic acids by anion-exchange chromatography using a copper modified electrode. *Journal of Chromatography A*. 912: 223–233.

Castro, R.N., Azeredo, L.C., Azeredo M.A.A., and de Sampaio, C.S.T., 2001. HPLC assay for the determination of ascorbic acid in honey samples. *Journal of Liquid Chromatography and Related Technologies*. 24: 1015–1020.

Ciulu, M., Solinas, S., Floris, I., Panzanelli, A., Pilo, M.I., Piu, P.C., Spano, N., and Sanna, G., 2011. RP-HPLC determination of water-soluble vitamins in honey. *Talanta*. 83: 924–929.

Dimitrova, B., Gevrenova, R., and Anklam, E., 2007. Analysis of phenolic acids in honeys of different floral origin by solid-phase extraction and high-performance liquid chromatography. *Phytochemical Analysis.* 18: 24–32.

Doner, L.W., 1977. The sugars of honey—a review. *Journal of the Science of Food and Agriculture.* 28: 443–456.

Ferreira, I.C.F.R., Aires, E., Barreira, J.C.M., and Estevinho, L.M., 2009. Antioxidant activity of Portuguese honey samples: different contributions of the entire honey and phenolic extract. *Food Chemistry.* 114: 1438–1443.

Ferreres, F., Andrade, P., Gil, M.I., and Tomas-Barberan, F.A., 1996. Floral nectar phenolics as biochemical markers for the botanical origin of heather honey. *Zeitschrift für ebensmitteluntersuchung und -Forschung A.* 202: 40–44.

Gheldof, N., Wang, X.H., and Engeseth, N.J., 2002. Identification and quantification of antioxidant components of honeys from various floral sources. *Journal of Agricultural and Food Chemistry.* 50: 5870–5877.

Gómez-Caravaca, A.M., Gómez-Romero, M., Arráez-Roman, D., Segura-Carretero, A., and Fernández-Gutiérrez, A., 2006. Advances in the analysis of phenolic compounds in products derived from bees. *Journal of Pharmaceutical and Biomedical Analysis.* 41: 1220–1234.

Haydak, M.H., Palmer, L.S., Tanqdary, M.C., and Vivino, A.E., 1942. Vitamin content of honeys. *The Journal of Nutrition.* 23: 581–588.

Haydak, M.H., Palmer, L.S., Tanquary, M.C., and Vivino, A.E., 1943. The effect of commercial clarification on the vitamin content of honey. *The Journal of Nutrition.* 319–321.

Hennessy, D.J., and Cerecedo, L.R., 1939. The determination of free and phosphorylated thiamin by a modified thiochrome assay. *Journal of the American Chemical Society.* 61: 179–183.

Horwitz, W., 1982. Evaluation of analytical methods used for regulation of foods and drugs. *Analytical Chemistry.* 54: 67A–76A.

Kitzes, G., Schuette, H.A., and Elvehjem, C.A., 1943. The B vitamins in honey. *The Journal of Nutrition.* 26: 241–250.

Krell, R., 1996. Value-added products from beekeeping. FAO Agricultural Services Bulletin No. 124. Food and Agriculture Organization of the United Nations, Rome. Available at: http://www.fao.org/docrep/w0076E/w0076E00.htm.

Leon-Ruiz, V., Vera, S., Gonzalez-Porto, A.V., and San Andres, M.P., 2011. Vitamin C and sugar levels as simple markers for discriminating Spanish honey sources. *Journal of Food Science.* C356–C361.

Long, G.L., and Winefordner, J.D., 1983. Limit of detection: a closer look at the IUPAC definition. *Analytical Chemistry.* 55: 713A–724A.

Matei, N., Birghila, S., Dobrinas, S., and Capota, P., 2004. Determination of C vitamin and some essential trace elements (Ni, Mn, Fe, Cr) in bee products. *Acta Chimica Slovenica.* 51: 169–175.

Mato, I., Huidobro, J.F., Simal-Lozano, J., and Sancho, M.T., 2006. Rapid determination of nonaromatic organic acids in honey by capillary zone electrophoresis with direct ultraviolet detection. *Journal of Agricultural and Food Chemistry.* 54: 1541–1550.

Melnick, D., and Field, Jr., H., 1940. Chemical determination of nicotinic acid: inhibitory effect of cyanogen bromide upon the aniline side reactions. *Journal of Biological Chemistry*. 135: 53–58.

Mocak, J., Bond, A.M., Mitchell, S., and Scholary, G., 1997. A statistical overview of standard (IUPAC and ACS) and new procedures for determining the limits of detection and quantification: application to voltammetric and stripping techniques (Technical Report). *Pure and Applied Chemistry*. 69: 297–328 (IUPAC recommendation, document 550/35/87).

Scanu, R., Spano, N., Panzanelli, A., Pilo M.I., Piu, P.C., Sanna, G., and Tapparo, A., 2005. Direct chromatographic methods for the rapid determination of homogentisic acid in strawberry tree (*Arbutus unedo* L.) honey. *Journal of Chromatography A*. 1090: 76–80.

Scudi, J.V., 1941. On the colorimetric determination of vitamin B_6. *Journal of Biological Chemistry*. 139: 707–720.

Siddiqui, I.R., 1970. The sugars of honey. *Advances in Carbohydrate Chemistry*. 25: 285–309.

Skoog, D.A., West, D.M., and Holler, F.J., 1998. Fundamentals of Analytical Chemistry, 7th ed. pp. 870, Saunders College Publishing, Philadelphia, PA, USA.

Snell, E.E., and Strong, F.M., 1939. A microbiological assay for riboflavin. *Industrial & Engineering Chemistry Analytical Edition*. 11: 346–350.

Snell, E.E., and Wright, L.D., 1941. A microbiological method for the determination of nicotinic acid. *Journal of Biological Chemistry*. 139: 675–686.

Spano, N., Piras, I., Ciulu, M., Floris, I., Panzanelli, A., Pilo, M.I., Piu, P.C., and Sanna, G., 2009. Reversed-phase liquid chromatographic profile of free amino acids in strawberry-tree (*Arbutus unedo* L.) honey. *Journal of AOAC International*. 92: S1145–S1156.

Stephens, J.M., Schlothauer, R.C., Morris, B.D., Yang, D., Fearnley, L., Greenwood, D.R., and Loomes, K.M., 2010. Phenolic compounds and methylglyoxal in some New Zealand manuka and kanuka honeys. *Food Chemistry*. 120: 78–86.

Strong, F.M., Feeney, R.E., and Earle, A., 1941. Microbiological assay for pantothenic acid. *Industrial & Engineering Chemistry Analytical Edition*. 13: 566–570.

Suarez-Luque, S., Mato, I., Huidobro, J.F., Simal-Lozano, J., and Sancho, M.T., 2002. Rapid determination of minority organic acids in honey by high-performance liquid chromatography. *Journal of Chromatography A*. 955: 207–214.

Thompson, M., Ellison, S.E.R., and Wood, R., 2002. Quality assurance schemes for analytical laboratories. Harmonized guidelines for single-laboratory validation of methods of analysis. (IUPAC Technical Report). *Pure and Applied Chemistry*. 74: 835–855.

Tomas-Barberan, F.A., Martos, I., Ferreres, F., Radovic, B.S., and Anklam, E., 2001. HPLC flavonoid profiles as markers for the botanical origin of European unifloral honeys. *Journal of the Science of Food and Agriculture*. 81: 485–496.

Tuberoso, C.I.G., Bifulco, E., Jerkovic, I., Caboni, P., Cabras, P., and Floris, I., 2009. Methyl syringate: a chemical marker of asphodel (*Asphodelus microcarpus* Salzm. et Viv.) monofloral honey. *Journal of Agricultural and Food Chemistry*. 57: 3895–3900.

Viñas, P., Balsalobre, N., López-Erroz, C., and Hernández-Córdoba, M., 2004a. Liquid chromatographic analysis of riboflavin vitamers in foods using fluorescence detection. *Journal of Agricultural and Food Chemistry*. 52: 1789–1794.

Viñas, P., Balsalobre, N., López-Erroz, C., and Hernández-Córdoba, M., 2004b. Determination of vitamin B_6 compounds in foods using liquid chromatography with post-column derivatization fluorescence detection. *Chromatographia*. 59: 381–386.

Viuda-Martos, M., Ruiz-Navajas, Y., Fernández-López, J., and Pérez-Alvarez, J.A., 2008. Functional properties of honey, propolis, and royal jelly. *Journal of Food Science*. 73: 117–124.

White, J.W. Jr., 1978. Honey. In: Advances in Food Research, Vol. 24. pp. 287–374, Academic Press, New York, USA.

CHAPTER 14

Analytical Trends in the Simultaneous Determination of Vitamins B_1, B_6 and B_{12} in Food Products and Dietary Supplements

ANNA LEBIEDZIŃSKA*[a] AND MARCIN L. MARSZAŁŁ[b]

[a] Department of Food Sciences, Medical University of Gdańsk, Al. Gen. J. Hallera 107, Gdańsk, 80–416, Poland; [b] Department of Toxicology, Medical University of Gdańsk, Al. Gen. J. Hallera 107, Gdańsk, 80–416, Poland
*Email: aleb@gumed.edu.pl

14.1 Introduction

The term 'vitamin' was derived from 'vitamine' a word made up by the Polish scientist Casimir Funk from 'vital' and 'amine', with the meaning 'amine of life'. It was suggested in 1912 that the organic micronutrient food factors that prevent beriberi and perhaps other similar dietary-deficiency diseases might be chemical amines. This proved incorrect for the micronutrient class and the word was shortened to vitamin. Vitamins have been produced as commodity chemicals since the middle of the 20th century and are widely available as inexpensive synthetic multivitamin dietary supplements. They are classified by their biological and chemical activity, not their structure (Ball 2006). For several of the vitamins, their biological activity is attributed to a number of structurally related compounds known as vitamers. In most cases, the vitamers pertaining to a particular vitamin display similar qualitative biological properties, but because of subtle differences in their chemical structures, they exhibit varying degrees of potency. Vitamers are by definition convertible to the active form of the vitamin in the body and are sometimes inter-convertible to one another as well.

Vitamins have diverse biochemical and physiological functions. Most vitamins (*e.g.* the B complex vitamins) function as precursors for enzyme cofactors, which help enzymes in their work as catalysts in the metabolism. In this role, vitamins may be tightly bound to enzymes as part of prosthetic groups. Vitamins may also be less tightly bound to enzyme catalysts as coenzymes (*i.e.* detachable molecules that carry chemical groups or electrons between molecules) such as folic acid, which carries various forms of carbon groups—methyl, formyl and methylene—in the cell (Fenech 2001).

The B vitamins help to maintain healthy nerves, skin, eyes, hair, liver and mouth, as well as muscle tone in the digestive tract. They are coenzymes involved in energy transfers necessary for the metabolism of carbohydrates, fats and proteins (Ball 2006).

Vitamins are natural constituents of food and a well-balanced diet supplies all the required vitamins. The correlation between diet and health has led consumers to intentionally ingest foods containing certain vitamins. Food is the main source of vitamins for both humans and animals, as most of the vitamins cannot be manufactured by the body (Ball 2006).

The status of B vitamins in a healthy population is generally satisfactory, but in high-risk populations, with decreased intake or increased needs, the status may be worse. It has been emphasized that making healthy food choices is an integral part of total health risk management. Health authorities have recommended an increased consumption of food-products rich in vitamins B (Lukaski 2004). Nutrition is one of the major factors affecting human health, fitness and well-being. An optimal diet is described as one fulfilling the consumers' energetic and nourishment needs (Morris *et al.* 2007).

At present, nutrient deficiencies, including vitamins, are particularly observed among pregnant women, children and elderly people. The causes of nutritional vitamin deficiencies are a combination of inadequate ingestion, poor absorption, and increased excretion (Morris *et al.* 2007). For some vitamins in the B group, a clinical deficiency results in a biochemical defect that is manifested as a disease with characteristic symptoms. A deficiency in vitamins B_{12} and B_6 rarely occurs alone, but is most commonly seen among people who are deficient in several vitamins in the B complex (Ball 2006).

Vitamin B_{12} deficiency is rarely caused by a lack of B_{12} in the diet, but is rather attributed to various absorption and transport disorders. Howard *et al.* (1998) investigated dietary cobalamin intake among elderly people with an abnormal cobalamin status and concluded that the high frequency of mildly abnormal cobalamin levels could not be attributed to poor intake of the vitamin. Instead, it has been found that a decreased gastric secretion due to chronic atrophic gastritis is the major cause of vitamin B_{12} deficiency among older persons (Ball 2006; Howard et al. 1998).

Vitamins B_6, B_{12} and folate have specific and vital functions in the metabolism; a deficiency (or excess) causes specific diseases such as hyperlipidemia, hypertension, obesity or cardiovascular diseases, which are related to a modern lifestyle and common in industrialized countries. As observed in several studies, cobalamin (B_{12}) deficiency in most cases coincides with an insufficient folate status (Ball 2006; Howard et al. 1998; Morris et al. 2007).

An improperly balanced diet may cause mutations and chromosomal aberrations. For the development of methods to counteract such changes, it is vital to determine the possible correlations between the dietary components and DNA changes. Folic acid and vitamin B_{12} are crucial elements in the methylation of DNA, proteins and neurotransmitters. The development of neural tube defects associated with folic acid deficiency is related to methylation dysfunction and a malfunctioning expression of the genes responsible for the tube formation. Furthermore, it is believed that DNA hypermethylation causes mutations related to tumor initiation or development. Previous observations have shown that diets supplemented with the B group vitamins for 15 years were correlated with a significant reduction in the development of tumour diseases (Fenech 2001).

Vitamin supplementation has grown in popularity among consumers as a means of preventing or retarding illnesses such as hyperlipidemia, hypertension, obesity and cardiovascular diseases, which are common in industrialized countries (Lukaski 2004). An adequate supply of vitamins B_1, B_6 and B_{12} is also essential for normal blood formation and neurological functions (Ball 2006). When clinical or subclinical deficiency symptoms occur in a particular population, it indicates that the food needs to be enriched with the lacking vitamins (Morris et al. 2007).

The stability of most vitamins is particularly susceptible to a large number of factors affected by food processing, including temperature, moisture, oxygen, light, pro-oxidants, reducing agents and pH (Ball 2006). Most countries promote the enrichment of food in order to correct potential nutrient deficiencies in the population; fortification of food with vitamins intends to compensate for the loss of these compounds in the manufacturing process.

Due to the nutritional importance of vitamins, new analytical methodologies have been developed for determination of these substances in food, pharmaceutical formulations, dietary supplements and biological fluids (Ball 2006; Eitenmiller et al. 2008; Rychlik 2011; Vidović et al. 2008).

14.2 Analysis of B Vitamins in Food Products and Dietary Supplements

The relatively high detection limits in combination with low natural concentrations in a complex sample matrix make it difficult to determine the content of B vitamins in food. The methods available for vitamin analyses require an individual approach for each vitamin with respect to the physical, chemical and microbiological analytical procedures (AOAC 2006; Ball 2006; Blake 2007).

Traditional methods for vitamin determination in food require the analysis of each vitamin by microbiological assays (AOAC 2006). The growth of microorganisms with a water-soluble vitamin requirement is dependent on the supply of the vitamin. Following the addition of the vitamin as a standard or as a compound of the sample, microbial growth continues until the vitamins are consumed. Microbiological assays (AOAC 2006) are still used for the determination of vitamin B_6, B_{12}, biotin, niacin and folic acid. For example, Tuskatani *et al.* (2011) described a method for determination of B vitamins (with the exception of B_{12}) using a colorimetric microbial viability assay based on the reduction of a tetrazolinum salt in a 96-well microtitre plate. A wide variety of other physical, chemical and biological methods are also currently used including colorimetric, fluorometric, spectrometric and electrometric techniques (Blake 2007; Eitenmiller *et al.* 2008). In comparison with other methods for chemical analyses, electroanalytical techniques are relatively easy to miniaturize as far as sensors or detectors are concerned. A small number of papers have already been published on the determination of vitamins B_6 and cyanocobalamin in pharmaceutical formulations and fortified food using polarographic and voltammetry techniques (Trojanowicz 2011).

A selective analysis can also be achieved using an element-selective detection approach such as atomic absorption spectrometry (AAS) for the determination of Co, which is the central atom of vitamin B_{12} (Kumar *et al.* 2009). A better detection capacity can be attained using inductively coupled plasma–mass spectrometry (ICP-MS) (Chen and Jiang 2008). However, the best detection limit for cobalamin has been reported for a combination of high-performance liquid chromatography (HPLC) or capillary electrophoresis with ICP-MS and electrochemical detection (ED). The latter is based on electrical charge transfer that occurs when electrons are given up by a molecule during oxidation or absorbed by a molecule during reduction. Although electrochemical detection methods can only detect substances that can be electrolysed, this limitation is actually an advantage when applied to complicated food matrices because it improves the selectivity. Electrochemical detection is used for the sensitive detection and measurement of electro-active analytes in many areas of analytical chemistry and biochemistry (Flanagan *et al.* 2005; Trojanowicz 2011).

When working with complicated matrices (biological samples or food products), the selectivity of electrochemical methods (ED) can usually be increased if they are combined with effective separation techniques, *e.g.* capillary electrophoresis and liquid chromatography (Rychlik 2011; Trojanowicz 2011). In separation science, electrochemical detection is used to detect and measure response analytes in flowing streams after separation by HPLC or capillary electrophoresis (Trojanowicz 2011). The former was introduced in the mid-20th century and is still an actively developing analytical technique. Employment of a new generation of columns, new detector types, new software and the possibility of working on-line with other analytical devices means that the importance of HPLC in food analysis is constantly growing. The use of HPLC has also improved the accuracy of quantitative measurements by eliminating interfering substances and permitted determination of different forms of each vitamin. Additional advantages in comparison with biological and manual chemical methods are a more rapid analysis, improved precision, reasonable accuracy and increased specificity with simultaneous separation of multiple vitamins, vitamers and metabolites (Kall 2003; Rychlik 2011; Trojanowicz 2011).

An HPLC procedure with an appropriate detection system is considered the most convenient method for the determination of B vitamins. Recently, numerous methods have been developed for the determination of vitamins B such as HPLC with fluorescence and ultraviolet (UV) detection (Ball 2006; Blake 2007; Eitenmiller et al. 2008).

One of the latest techniques that facilitates the analysis of ultra trace amounts of compounds is the combination of HPLC with mass spectrometry (HPLC-MS). This allows the separation of very small amounts of compounds in a multi-component mixture, and analysis of their quality and quantity (Heudi et al. 2006; Rychlik 2011).

14.3 Simultaneous Analysis of Vitamins B in Food Supplements and Food Products by HPLC

During the last decade there has been an increasing interest in the simultaneous determination of vitamins. When working with complicated matrices (biological samples or food products), the selectivity of electrochemical methods can usually be increased if they are combined with effective separation techniques, e.g. capillary electrophoresis and liquid chromatography (Rychlik 2011; Trojanowicz 2011). As mentioned above, HPLC analysis can conveniently be applied to the assay of high concentration of vitamins B in food products, fortified food or supplements, and pharmaceutical preparations (Ciulu et al. 2011; Esteve et al. 1998; Guggisberg et al. 2012; Luo et al. 2006; Pakin et al. 2005; Rychlik 2011; Vidović et al. 2008). For the analysis of fortified foods, Heudi et al. (2005) reported separation of nine water-soluble vitamins with HPLC with UV analysis in premixes.

A small number of papers on the determination of vitamins B_6 and B_{12} in fortified food using ED techniques have been published. A wide variety of compounds is capable of being monitored with a coulometric detector. In general, electroactivity is dependent on presence of functional group of molecules (Flanagan et al. 2005; Trojanowicz 2011).

The combined HPLC with electrochemical detection is a sensitive and selective method for simultaneous determination of different redox-active compounds. The coulometric detection is very similar to the amperometric and voltammetric ones in terms of electrochemical principles and, hence, has wide applications in liquid chromatography (Trojanowicz 2011; Zhang and Wang 2011).

We have developed a method for simultaneous analysis of thiamine hydrochloride, pyridoxine hydrochloride and cyanocobalamin in pharmaceuticals and dietary supplements (Marszałł et al. 2005) and in fortified food (Lebiedzińska and Marszałł 2006) using HPLC-ED. Vitamins were determined in their free forms, so an extraction step from fortified fruit juice was performed prior to the chromatographic isolations. The extraction procedure was based on a study by Ndaw et al. (2000). The enzymatic digestion prior to the separation and quantification step made it possible to release the vitamins bound to proteins or sugars and converted vitamin esters to free forms; thus we were able to obtain the total vitamin contents of the fruit juices. The supernatants were adjusted to pH 4.5 with 2.5 M sodium acetate and a single extraction procedure for all vitamins was carried out using mixture of the enzymes, papain and diastase (Lebiedzińska and Marszałł 2006).

The detection method is based on a two-step detection system, *i.e.* a combination of a coulometric detector and an ultraviolet–visible (UV-Vis) detector. In our study, the UV detector was used for determination of thiamine, while quantification of pyridoxine and cyanocobalamin was performed using the electrochemical detector. The ESA Inc. coulometric detector cells are designed so that the eluent flows through a porous graphite electrode rather than flowing the electrode as in traditional electrochemical detection. With this design, all the electroactive compound will be oxidized or reduced. Since a larger amount of the electroactive species is oxidized (or reduced) without a corresponding increase in noise, this detector can provide enhanced sensitivity (Flanagan et al. 2005; Trojanowicz 2011). This method has been proven to have good sensitivity and selectivity, linearity and accuracy, and has enlarged the range of B_1, B_6, and B_{12} quantification in a single run.

Figure 14.1 presents chromatograms of vitamins analysed in fruit drinks. The detection limit for thiamine hydrochloride detection was 9.2 ng/ml, whereas the limits for pyridoxine and cyanocobalamin were 2.7 and 0.08 ng/ml, respectively. The proposed separation and detection procedure was applied successfully for quantitative evaluation of the studied B vitamins in pharmaceutical preparations and dietary supplements, and for routine control of multivitamin enriched foods. Based on those successful results, we have developed also a method for analysis of vitamins B_6, B_{12} and B_1 in seafood products (Lebiedzińska et al. 2007).

14.4 Simultaneous Analysis of Vitamins B_6 and B_{12} in Seafood Products

One of the greatest challenges in vitamin analysis is to extract them from the food matrix; the problems are mainly due to their different physical and chemical properties. Vitamins release from proteins is generally obtained by autoclaving the food sample (121 °C) or heating at 100 °C (AOAC 2006).

The extraction procedure is considered to have the greatest impact on analytical results. Numerous extraction techniques have been developed to extend the

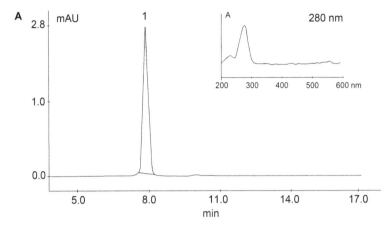
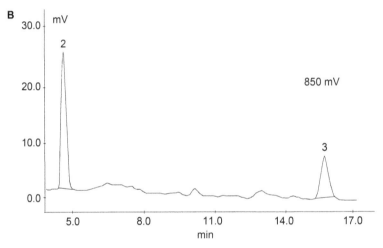

Figure 14.1 Chromatograms of a fruit drink sample: (A) 1 = thiamine (t_R 7.9 min, 15.5 ng ml^{-1}) at 280 nm; (B) 2 = PN (t_R 4.8 min, 35 ng ml^{-1}), 3 = cyanocobalamin (t_R 16.0 min, 9.5 ng ml^{-1}). LC 18 column 5 μm (4.6 mm × 25 cm) Supelco Inc. Mobile phase: 0.05 M phosphate buffer (pH = 3.55) with 10% methanol containing 0.018 M trimethylamine, flow rate 1 ml min^{-1}.

application of HPLC, including pre- or post-column labelling and solid phase extraction, reversed-phase solid phase extraction, and immunoaffinity solid–phase extraction (Guggisberg *et al.* 2012; Heudi *et al.* 2006; Pakin *et al.* 2005).

Vitamin B_6 extraction and analysis is complicated by several factors including its occurrence as six vitamers of differing structures. The total vitamin B_6 content of food has traditionally been determined by extraction at elevated temperatures (autoclaving) in the presence of mineral acids, followed by enzymatic hydrolysis (Rabinowitz and Snell 1947) with phosphatases and glucosidases to ensure complete release and conversion of all B_6 vitamers into pyridoxine (PN), pyridoxamine (PM) and pyridoxal (PL) (AOAC 2006).

The determination of total B_6 as PN was collaboratively studied by 12 laboratories in a comparative study (Bergaentzle *et al.* 1995); all the B_6 vitamers in food were dephosphorylated with acid phosphatase, after which PM was deaminated to PL in the presence of glyoxylic acid and ferrous sulfate. The PL was reduced to PN by sodium borohydride and the total vitamin B_6 content was determined as one vitamer (PN) by HPLC. Another study reported that the most suitable extraction procedure for determining endogenous total vitamin B_6 in food (meat, fish, milk) was treatment with 0.1 M hydrochloric acid at 120 °C for 30 min, followed by enzymatic hydrolysis with β-glucosidase alone or in combination with acid phosphatase. Some authors (Ndaw *et al.* 2000) have recommended the extraction procedure consisting of incubation with papain, α-amylase and phosphatase for 18 h at 37 °C, and compared this protocol with four other extraction protocols involving combinations with acid and enzymatic hydrolysis.

The cobalamin 'vitamers' refers to a group of vitamin B_{12} substances depending on the upper axial ligand of the cobalt ion (Figure 14.2). Food may contain a variety of cobalamins, and their release from proteins is generally obtained by autoclaving the food sample, though protease treatment has also been recommended. The extraction procedures of AOAC microbiological methods (AOAC 2006) are usually used for determination of the total vitamin B_{12} content in food samples. Many methods incorporate sodium cyanide to ensure that the native vitamin forms are converted into dicyanocobalamin (AOAC 2006; Eitenmiller *et al.* 2008). Those extraction procedures liberate cobalamins from proteins and convert the labile naturally occurring forms to a single, stable form, which is cyanocobalamin.

In our study (Lebiedzińska *et al.* 2007), the extraction procedure used for determination of the vitamers B_6 and B_{12} was based on a combination of acid digestion and enzymatic hydrolysis according to a study by Esteve *et al.* (1998) and recommended by AOAC's *Official Methods of Analysis* to release vitamins from food followed by HPLC analysis. Vitamins B_6 and B_{12} were detected following HPLC separation with coulometric detection. Our experiments performed with standards (*i.e.* methyl-, hydroxyl-, adenosyl- and cyanocobalamin) clearly indicated the same peak area response at the same potentials and retention times.

Computational chemistry analysis appears to be helpful as an additional tool to explain the lack of separation between different forms of cobalamins. Recent theoretical studies of the group deal with the electronic and vibrational spectra of some corrins, obtained using density functional theory (DFT) calculations (Andriuniow *et al.* 2005). The results indicated that, for all the structures used in the calculations, the same electrostatic potential on the molecule was obtained, which supports the observed simultaneous detection of cobalamins, irrespective of the axial ligand (cyano, adenosyl and methyl group). The lack of separation between different forms of cobalamins can be explained by analysis of molecular electrostatic potential (Andriuniow *et al.* 2005; Lebiedzińska *et al.* 2007).

The principle of our method is that electroactive functional groups with a free electron pair (hydroxyl and amino groups) are oxidized at a positive potential. The current peak at the positive potential is directly related to the

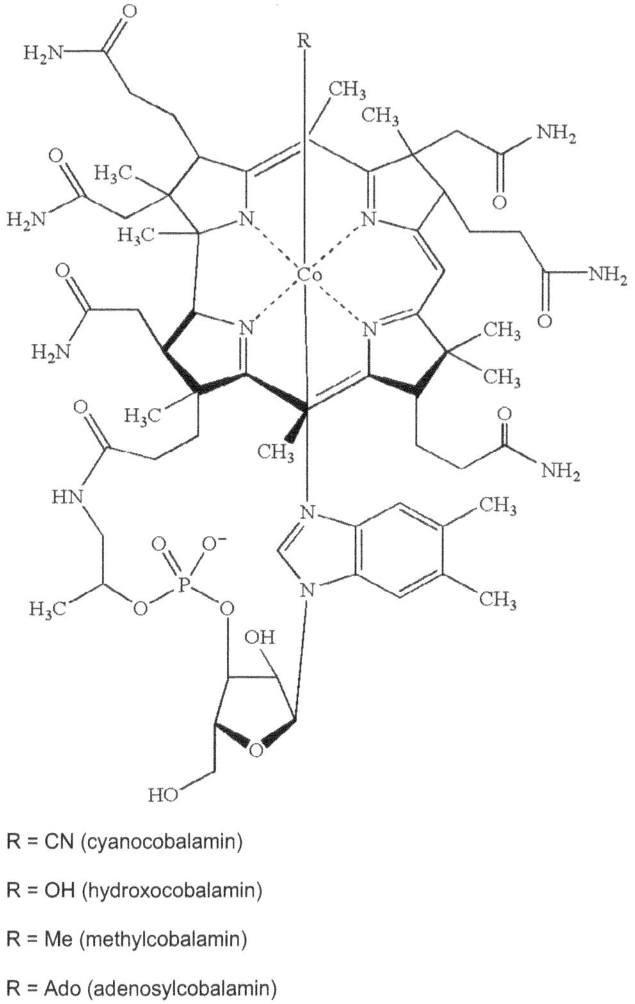

Figure 14.2 Structure of vitamin B12. Burtis, C.A., Ashwood, E.R., Bruns, D.E., 2012. *Tietz Textbook of Clinical Chemistry and Molecular Diagnostics*, 5 ed., chapter 3, p. 920, Elsevier, USA.

concentration of pyridoxamine, pyridoxal, pyridoxine and vitamin B_{12}, showing that the response is caused by these electroactive vitamins. The isocratic mobile phase that we used enabled satisfactory separation of the B_{12} and B_6 vitamins. The electroactivity of compounds monitored with a coulometric detector depends on the presence of a functional group of molecules. The detector response was set to give a full-scale detection for 1 µA and 50 µA current outputs from the analytical cell, and the peak area of the electrochemical signal at the porous graphite electrode was used for the quantitative analysis. The current peak at the positive potential is directly proportional to

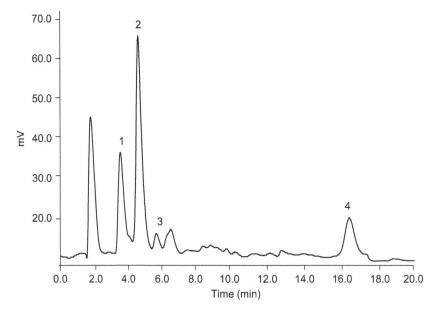

Figure 14.3 Chromatogram of vitamin B in a mussel sample: 1 = PM (t_R 3.7 min, 20.5 ng ml^{-1}); 2 = PL (t_R 4.7 min, 29.5 ng ml^{-1}); 3 = PN (t_R 5.6 min, 2.0 ng ml^{-1}); vitamin B_{12} (t_R 16.4 min, 18.6 ng ml^{-1}). The ED sensitivity is 1 µA. LC 18 column 5 µm (4.6 mm×25 cm) Supelco Inc. Mobile phase: methanol–phosphate buffer (10:90) phosphate buffer (pH = 3.55) containing 0.018 M trimethylamine, flow rate 1 ml min^{-1}.

the concentration of pyridoxamine, pyridoxal, pyridoxine and vitamin B_{12}, which shows that the response is caused by these electroactive vitamins. Figure 14.3 illustrates chromatograms of the vitamins in mollusk samples.

In a coulometric detection, all the co-eluting compounds with an oxidation potential lower than that of the compounds of interest will be oxidized at the screen electrode. This use of the screen mode is a powerful tool that enhances selectivity (Flanagan *et al.* 2005). The advantage of our HPLC method with coulometric detection is that is offers a possibility to quantify vitamins B_6 and B_{12} in food samples without prior transformation of cobalamins to cyanocobalamin.

Analysts are cautioned against applying a method to a food substrate unless it has been validated for that type of food; the food matrix can significantly affect the type of extraction and handling required to achieve a complete extraction without vitamin degradation or loss. Reference materials provide one way of assessing the validity of newly developed methods. Ideally, reference materials should be prepared from a matrix that closely reflects real samples. They should have a homogeneous composition and cover the range of concentrations expected in the samples. Food-based certified standard reference materials have been collaboratively studied, and are available from organizations such as the European Commission's Community Bureau of reference and the National Institute of Standards and Technology (USA).

The accuracy of our method (Lebiedzińska et al. 2007) was estimated by analysing a certified reference material (CRM 487 pig's liver), and the recoveries ranged from 96.5–100.5%. The method offers excellent linearity, good repeatability and reproducibility, and a relatively short analysis time (17 min). The limits of quantification (LOQ) for PM, PL, PN and vitamin B_{12} in seafood (in a 2 g sample) were 2.1, 2.01, 0.99 and 0.11 ng ml^{-1}, respectively. The proposed high sensitivity and selectivity of the HPLC-ED method upgraded by a coulometric detector with isocratic elution enabled accurate determination of vitamins B_6 and B_{12} in fish, mussels, oysters and other seafood products.

Food analytical studies, including novel techniques for the determination of the B group vitamins in food products, are important not only for food and dietary science but also for clinical practices as well as for the pharmaceutical and food industries.

Summary Points

- This chapter focuses on analytical methods for the simultaneous determination of vitamin B_1, B_6 and B_{12} in various food matrices.
- The hydrochloric acid and enzymatic mixture (papain and diastase) proposed for the extraction of vitamin B_{12} and B_6 (pyridoxamine, pyridoxal and pyridoxine) which made it possible to obtain vitamin contents in seafood products, and is in agreement with the certified values in reference materials, thus constitutes an enzymatic system for the analysis of these vitamins by our HPLC-ED method.
- The HPLC-ED and UV-Vis detections method has been developed for the simultaneous analysis of vitamins B_1, B_6 and B_{12} in a single run.
- The fusion of electrochemical detectors with liquid chromatographic systems represents the more successful and identifiable incursions of electrochemistry into conventional analytical methodologies.
- The most advantage of the presented method is a quantification of vitamin B_{12} in natural foodstuff samples without prior transformation of cobalamins to cyanocobalamin.

Key Facts about Electrochemical Detection

- An electrochemical detector responds to substances that are either oxidizable or reducible.
- The electrical output results from an electron flow caused by the chemical reaction that takes place at the surface of the electrodes.
- The coulometric electrode provides for essentially 100% conversion of the electroactive species.
- The potential is held constant and the area under the current/tome curve is integrated to obtain the number of coulombs.
- The current is directly related to the concentration according to Faraday's law.
- Coulometric detectors provide extremely high sensitivity and a low limit of detection.

Definitions of Words and Terms

Computational chemistry: A branch of chemistry that uses the results of theoretical chemistry, incorporated into efficient computer programs, to calculate the structures and properties of molecules.

Coulometry: An electrochemical process where the total quantity of electricity required to electrolyse a specific electroactive species is measured in a stirred solution under controlled potential or constant current conditions.

Density functional theory: A quantum mechanical modelling method used in physics and chemistry to investigate the electronic structure of atoms, molecules.

Dietary supplements: Is a preparation intended to supply nutrients (*e.g.* vitamins, minerals, fatty acids or amino acids) that are missing or not consumed in sufficient quantity in a person's diet.

Electrochemical detection (ED): An electrochemical detector responds to substances that are either oxidizable or reducible, and the electrical output results from an electron flow caused by the chemical reaction that takes place at the surface of the electrodes.

High-performance liquid chromatography (HPLC): A type of liquid chromatography that uses an efficient column containing small particles of a stationary phase.

Liquid chromatography: A physical method of separation in which the components to be separated are distributed between two phases: stationary phase and the mobile liquid phase.

Vitamins: An essential organic micronutrient that must be supplied exogenously and in many cases in the precursor to a metabolically derived coenzyme.

Vitamer: A term used to describe any of a number of compounds that possess a given vitamin activity.

List of Abbreviations

AAS	atomic absorption spectrometry
DFT	density functional theory
ED	electrochemical detection
HPLC	high performance liquid chromatography
HPLC-ED	high performance liquid chromatography with electrochemical detection
ICP-MS	inductively coupled plasma–mass spectrometry
LOQ	limit of quantification
MS	mass spectrometry
PL	pyridoxal
PM	pyridoxamine
PN	pyridoxine
RSD	relative standard deviation
UV	ultraviolet
UV-Vis	ultraviolet–visible

References

Andruniow, T., Kuta, J., Zgierski, M.Z., and Kozłowski, P.M., 2005. Molecular orbital analysis of anomalous *trans* effect in cobalamins. *Chemical Physics Letters*. 410: 410–416.

AOAC, 2006. Vitamins and Other Nutrients. Official Methods of Analysis of AOAC International, 18th ed., Current through Revision 1. AOAC International, Gaithersburg, MD, USA, 1–70.

Ball, G.F.M., 2006. Vitamins in Foods. Analysis, Bioavailability, and Stability, CRC Press, Taylor & Francis Group, Boca Raton, Fl, USA, 785 p.

Bergaentzle, M., Arella, F., Bourguignon, J.B., and Hasselmann C., 1995. Determination of vitamin B_6 in foods by HPLC: a collaborative study. *Food Chemistry*. 52: 8181–8186.

Blake, C.J,. 2007. Analytical procedures for water-soluble vitamins in foods and dietary supplements: a review. *Analytical Bioanalytical Chemistry*. 398: 63–76.

Chen, J.H., and Jiang, S.J. 2008. Determination of cobalamin in nutritive supplements and chlorella foods by capillary electrophoresis-inductively coupled plasma mass spectrometry. *Journal of Agricultural and Food Chemistry*. 56(4): 1210–1215.

Ciulu, M., Solinas, S., Floris, I., Panzanelli, A., Pilo, M.I., Piu, P.C., Spano, N., and Sanna, G., 2011. RP-HPLC determination of water-soluble vitamins in honey. *Talanta*. 83: 925–929.

Eitenmiller, R.R., Ye, L., and Landen, W.O., 2008. *Vitamin Analysis for the Health and Food Sciences*, 2nd ed. CRC Press, New York, USA. 401 p.

Esteve, M.J., Farré, R., Frígola, A., and García-Cantabella, J.M.J., 1998. Determination of vitamin B_6 (pyridoxamine, pyridoxal and pyridoxine) in pork meat products by liquid chromatography. *Journal of Chromatography A*. 795: 383–387.

Fenech, M., 2001. The role of folic acid and vitamin B_{12} in genomic stability of human cells. *Mutation Research*. 475: 57–67.

Flanagan, R.J., Perrett, D., and Whelpton, R., 2005. *Electrochemical Detection in HPLC: Analysis of Drugs and Poisons*. RSC Chromatography Monographs. RSC Publishing, Cambridge, UK. 56 p.

Guggisberg, D., Risse, M., C., and Hadorn, R., 2012. Determination of vitamin B_{12} in meat products by RP-HPLC after enrichment and purification on an immunoaffinity column. *Meat Science*. 90: 279–283.

Heudi, O., Kilinc, T., and Fontannaz, P., 2005. Separation of water-soluble vitamins by reversed-phase high performance liquid chromatography with ultra-violet detection: Application to polivitamined premixes. *Journal of Chromatography A*. 1070: 49–55.

Heudi, O., Kilinc, T., Fontannaz, P., and Marley, E., 2006. Determination of vitamin B_{12} products and premixes by reversed-phase high performance liquid chromatography and immunoaffinity extraction. *Journal of Chromatography A*. 1101: 63–68.

Howard, J.M., Azen, C., Jacobsen, D.W., Green R., and Carmel, R., 1998. Dietary intake of cobalamin in elderly people who have abnormal serum cobalamin, methylmalonic acid and homocysteine levels. *European Journal Clinical Nutrition.* 52: 582–587.

Kall, M.A., 2003. Determination of total vitamin B_6 in foods by isocratic HPLC: a comparison with microbiological analysis. *Food Chemistry.* 82: 315–327.

Kumar, S.S., Chouhan, R.S., and Thakur, M.S., 2009. Enhancement of chemiluminescence for vitamin B_{12} analysis. *Analytical Biochemistry.* 388: 312–316.

Lebiedzińska, A., and Marszałł, M., 2006. Fruit juices and fruit drinks as a source of vitamin B: HPLC determination of water soluble vitamins in fortified juices and drinks. *Polish Journal of Environmental Studies.* 15: 1318–1321.

Lebiedzińska, A., Marszałł, M., Kuta, J., and Szefer, P., 2007. Reversed-phase high-performance liquid chromatography method with coulometric electrochemical and ultraviolet detection for the quantification of vitamins B_1 (thiamine), B_6 (pyridoxamine, pyridoxal and pyridoxine) and B_{12} in animal and plant foods. *Journal of Chromatography A.* 1173: 71–80.

Lukaski, H., 2004. Vitamin and mineral status: effect on physical performance. *Nutrition.* 20: 632–644.

Luo, X., Chen, B., Ding, L., Tang, F., and Shouzhuo,Y., 2006. HPLC-ESI-MS analysis of vitamin B_{12} in food products and in multivitamins–multimineral tablets. *Analytica Chimica Acta.* 562: 185–189.

Marszałł, M., Lebiedzińska, A., Czarnowski, W., and Szefer, P., 2005. High-performance liquid chromatography method for the simultaneous determination of thiamine hydrochloride, pyridoxine hydrochloride and cyanocobalamin in pharmaceutical formulations using coulometric electrochemical and ultraviolet detection. *Journal of Chromatography A.* 1094: 91–98.

Morris, M.S., Jacques, P.F., Rosenberg, I.H., and Selhub, J., 2007. Folate and vitamin B_{12} status in relation to anemia, macrocytosis, and cognitive impairment in older Americans in the age of folic acid fortification. *The American Journal of Clinical Nutrition.* 85:193–200.

Ndaw, S., Bergentzle, M., Aoude-Werner, D., and Hasselmann, C., 2000. Extraction procedures for the liquid chromatographic determination of thiamin, riboflavin and vitamin B_6 in foodstuffs. *Food Chemistry.* 71: 129–138.

Pakin, C., Bergaentzle, M., Arella, F., Bourguignon, J.B., and Hasselmann, C., 2005. α-Ribazole, a fluorescent marker for the liquid chromatographic determination of vitamin B_{12} in foodstuffs. *Journal of Chromatography A.* 1081: 182–189.

Rabinowitz, J.C., and Snell, E.E., 1947. Vitamin B_6 group. Extraction procedures for the microbiological determination of vitamin B_6. *Analytical Chemistry.* 19: 277–280.

Rychlik, M. (ed.), 2011. *Fortified Foods with Vitamins. Analytical Concepts to Assure Better and Safer Products.* Wiley-Verlag, Weinheim, Germany. 159 p.

Trojanowicz, M., 2011. Recent developments in electrochemical flow detection: a review. Part II. Liquid chromatography. *Analytica Chimica Acta*. 688: 8–35.

Tsukatani, T., Suenaga, H., Ishiyama, M., Ezoe, T., and Matsumo, K., 2011. Determination of water soluble vitamins using a colorimetric microbial viability assay based on the reduction of water-soluble tetrazolium salts. *Food Chemistry*. 127: 711–715.

Vidović, S., Stojanović, B., Veljković, J., and Prazić-Arsić, L., 2008. Simultaneous determination of some water-soluble vitamins and preservatives in multivitamin syrup by validated stability-indicating high-performance liquid chromatography method. *Journal of Chromatography A*. 1202: 155–162.

Zhang, Y., and Wang, Y., 2011. Voltammetric determination of vitamin B_6 at glassy carbon electrode modified with gold nanoparticles and multi-walled carbon nanotubes. *American Journal of Analytical Chemistry*. 2: 194–199.

CHAPTER 15

Spectrofluorimetric Analysis of Vitamin B_1 in Pharmaceutical Preparations, Bio-fluid and Food Samples

SANG HAK LEE,* MOHAMMAD KAMRUZZAMAN AND AL-MAHMNUR ALAM

Department of Chemistry, School of Natural Science, Kyungpook National University, 1370, Sankyuck-Dong, Puk-Gu, Daegu, 702-701, South Korea
*Email: shlee@knu.ac.kr

15.1 Introduction

Vitamin B_1 (also called thiamine) is a water-soluble vitamin of B complex present in many foods as a natural nutrient. It is a biologically and pharmaceutically important compound containing a pyrimidine and a thiazole moiety such as 4-amino-5-hydroxymethyl-2-methyl-pyrimidine and 5-(2-hydroxyethyl)-4-methylthiazole linked by a methylene bridge (Figure 15.1). It is necessary for carbohydrate metabolism and for the maintenance of neural activity because most of the humans and mammals cannot synthesize vitamin B_1. Nerve cells need vitamin B1 for their normal function because vitamin B_1 has diphosphate-active sites which serve as a cofactor for several enzymes (Leopold *et al.* 2005). Vitamin B_1 is employed for the prevention and treatment of beriberi, neuralgia, *etc.* and played a vital role in enzymatic mitochondrial

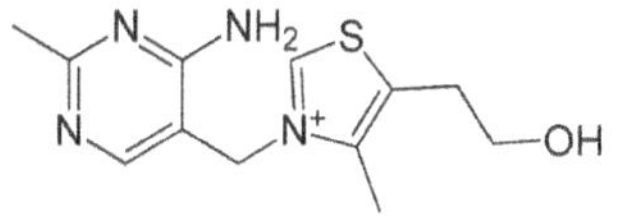

Figure 15.1 Structure of vitamin B_1 (Khan *et al.* 2009).
First appeared in Luminescence: The Journal of Biological and Chemical Luminescence, 24: 73–78, published by John Wiley & Sons.

energy liberation reactions and NADPH production for maintaining cellular redox, glutathione (GSH) levels, protein sulphydryl groups, nucleic acid and fatty acid synthesis (Depeint *et al.* 2006). Due to the several uses of vitamin B_1 for normal body functions and stability factors, the determination of vitamin B_1 in clinical analysis, food processing, pharmaceutical and biotechnological processes is very important. Therefore, much effort has been put into developing effective and sensitive analytical techniques for the determination of vitamin B_1.

A number of analytical methods have been reported for the determination of vitamin B_1 including gravimetric (Vannatta and Harris 1959), adsorptive polarographic assay (Wang *et al.* 2004), adsorptive voltammetric (Aboul-Kasim 2000), cathodic stripping voltammetry (Ciszewski and Wang 1992), ion square wave voltammetric analysis (Jiang and Sun 2007), potentiometric determination (Chang and Wang 2006; Jing *et al.* 2003), electrochemical analysis (Wan *et al.* 2002), thin-layer chromatography (Diaz *et al.* 1993; Ponder *et al.* 2004), gas chromatography (Echols *et al.* 1980; Velisek *et al.* 1986), capillary electrophoresis (Mrestani and Neubert 2000; Su *et al.* 2001), high performance liquid chromatography (Bohrer *et al.* 2004; He *et al.* 2005; Marszałł *et al.* 2005; Sanchez-Machado *et al.* 2004; Tang *et al.* 2006), liquid chromatography (Mancinelli *et al.* 2003; Zafra-Gomez *et al.* 2006), spectrofluorimetric analysis (Feng *et al.* 2004; Ni and Cai 2005; Tabrizi 2006; Zou and Chen 2007), spectrophotometric determination (Ghasemi and Abbasi 2005; Rocha *et al.* 2003; Tian and Zhang 2005), ultraviolet spectrophotometry (Lopez-de-Alba *et al.* 2006; Ozdemir and Dinc 2004).

Among these analytical methods, some methods are used to determine vitamin B_1 at macro level and also require expensive and sophisticated instruments. On the other hand, a spectrofluorimetric method promises high sensitivity and selectivity, does not need expensive instrumentation and has been widely used in pharmaceutical, food processing and biological analysis. In this chapter, developments of the spectrofluorimetric method are discussed as advanced analytical approaches for the determination of vitamin B_1.

15.2 Fluorescence Determination of Vitamin B_1 using Various Catalysts

Spectrofluorimetric methods are more sensitive and selective and have been used extensively to analyse vitamin B1 (thiamin). Thiamin is a non-fluorescent compound which can be converted to a fluorescent thiochrome (TC) derivative upon oxidation using an appropriate oxidant as catalyst. Thiamin is oxidized to TC by the formation of thiamin disulfide (TDS) from two thiamin molecules in the presence of oxidants.

Several spectrofluorimetric methods have been reported for the determination of thiamin based on the oxidation of thiamin using several oxidants including cyanogen bromide (Fujiwara and Matsui 1953), copper (Pérez-Ruiz et al. 1992), mercury(II) (Martinez-Lozano et al. 1990; Perez-Ruiz et al. 2004), cobalt(II) (Jie et al. 1993), iron(III) tetrasulfonatophthalocyanine (Chen et al. 1999), peroxidate (Amjadi et al. 2007), horseradish peroxidase (Khan et al. 2009).

Fujiwara and Matsui (1953) reported a thiochrome reaction to produce thiochrome from thiamin for the determination of thiamin. The reaction is based on the formation of a blue light emitting fluorescent compound, thiochrome, by the reaction of thiamin and cyanogen bromide in alkali medium at room temperature. The formed thiochrome was extracted using butyl alcohol, which exhibited intense fluorescent intensity.

A flow injection fluorimetric method has been reported for the determination of thiamin using copper as catalyst (Pérez-Ruiz et al. 1992). Thiamin is oxidized to produce fluorescent thiochrome by copper(II) in basic medium. The method is successfully applied to determine thiamin in pharmaceutical samples.

Thiamin has also been determined using mercury(II) (Martinez-Lozano et al. 1990). Mercury(II) is used to oxidize thiamin to the fluorescent thiochrome compound. Based on this reaction, a flow injection method has been developed to analyze thiamin in commercial preparations. A fluorimetric method coupled with flow injection has been developed for determining thiamine based on the oxidation of thiamin to thiochrome by mercury(II) (Pérez-Ruiz et al. 2004). A high mercury(II) concentration is utilized in this reaction for the largest yield of thiochrome and it produces an enhanced fluorescence emission signal at 440 nm when excited at 356 nm. The fluorescence intensity increases with the concentration of thiamin added, which exhibits outstanding results to determine vitamin B_1.

Cobalt(II) also possesses advantages suitable for utilization in the thiochrome reaction converting thiamin to thiochrome. A rapid fluorimetric method is reported for the determination of thiamin based on the oxidation of thiamin to thiochrome by Co(II) in the presence of alkali and Triton X100, a nonionic surfactant (Jie et al. 1993). During the oxidation reaction, the produced thiochrome is entrapped by the surfactants which enhance the fluorescence intensity with thiamin concentration.

Iron(III) tetrasulfonatophthalocyanine (FeTSPc), a mimetic enzyme, is used to catalyse the oxidation reaction of thiamin to produce thiochrome in the presence of H_2O_2 (Chen et al. 1999). Thiamin is oxidized to thiochrome by the catalytic activity of FeTSPc in the presence of H_2O_2 in alkaline medium, which exhibits an enhanced fluorescence emission peak at 440 nm when excited at 375 nm with the fluorogenic substrate, L-tyrosine (Chen et al. 1999). The fluorescence intensity is increased linearly with the concentration of thiamin in the range of 1.0×10^{-8} to 1.0×10^{-4} mol L^{-1}. The limit of detection is 4.3×10^{-9} mol L^{-1} and the relative standard deviation is 2.2% for six repeated measurement of 2.010^{-7} mol L^{-1} thiamin. Based on these results, Chen *et al.* (1999) developed a sensitive spectrofluorimetric method for the determination of thiamin using FeTSPc as catalyst.

Peroxidase as an enzyme can be applied for the oxidation of thiamin to thiochrome in the presence of hydrogen peroxide. Horseradish peroxidase is the main source of peroxidase, which utilizes hydrogen peroxide to oxidize a wide variety of organic and inorganic compounds in various analytical applications. A crude extract of kohlrabi, *Brassica oleracea* Gongylodes, rich in peroxidase is used in the thiochrome reaction (Amjadi *et al.* 2007). This extract with enzymatic activities is used as catalyst in the oxidation of thiamin by H_2O_2 in the presence of a stabilizer, polyvinylpyrrolidone (Amjadi *et al.* 2007). Thiochrome, the oxidation product, shows strong fluorescence intensity at 436 nm with excitation at 370 nm. The fluorescence intensity increases with increasing thiamin concentration in the linear range of 2.0×10^{-7} to 1.0×10^{-4} mol L^{-1}. The limit of detection (LOD) is 6.2×10^{-8} mol L^{-1} with the relative standard deviation (RSD) of 1.2% for 5×10^{-6} mol L^{-1} thiamin.

Based on the oxidation of thiamin by kohlrabi extract, Amjadi *et al.* (2007) presented a sensitive and effective fluorimetric method for determining vitamin B_1 (thiamin) in vitamin B complex tablets and syrup with good accuracy. Based on the catalytic activity of horseradish peroxidase (HRP), Khan *et al.* (2009) developed a new spectrofluorimetric method for the determination of vitamin B_1. HRP can oxidize the non-fluorescent compound to a fluorescent compound in the presence of H_2O_2:

$$AH_2 + H_2O_2 \rightarrow A + H_2O \qquad (1)$$

The non-fluorescent substrate (AH_2) can be determined by measuring the fluorescence intensity of the oxidation product (A) of the reaction shown in eqn (1) (Khan *et al.* 2009). The resulting product from the oxidation reaction of vitamin B_1 and H_2O_2 exhibits a strong fluorescence emission peak at 382 nm when excited at 368 (shown in Figure 15.2) in the presence of HRP; in the absence of HRP, it shows almost no fluorescence peak (Khan *et al.* 2009). Figure 15.2 implies that the non-fluorescent vitamin B_1 is converted to fluorescent thiochrome in the presence of H_2O_2 by the catalytic activity of HRP, which enhances the fluorescence intensity significantly. At the optimized conditions of HRP (Figure 15.3), the fluorescence intensity increases linearly with the logarithm of vitamin B_1 concentration (shown in Figure 15.4, Khan *et al.* 2009).

Figure 15.4 demonstrates that the fluorescence intensity is linearly correlated with vitamin B_1 concentrations in the range 0.026–16.83 µg mL^{-1}. The relative standard deviation is 1.75% for six replicate measurements of 10 µg mL^{-1}. The detection limit of vitamin B_1 is 0.015 µg mL^{-1}. The method is applied to determine vitamin in commercially available vitamin B_1 tablets (Biphane, Yuhan Co., Korea) to check the performance of the system (Table 15.1). The effect of the foreign substances that may influence the FL intensity was investigated to evaluate the selectivity of this method. The results indicated that most of the potentially interfering substances had very little or no effect on the

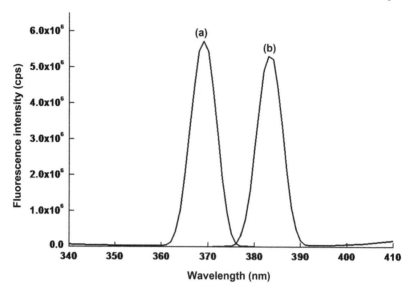

Figure 15.2 Excitation and emission spectra of a fluorescent compound produced from vitamin B_1 with H_2O_2 in the presence of HRP. A = excitation peak. B = emission peak.
First appeared in Luminescence: The Journal of Biological and Chemical Luminescence, 24: 73–78, published by John Wiley & Sons.

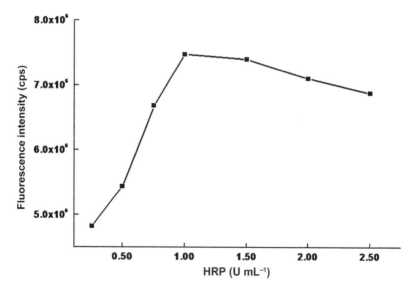

Figure 15.3 Effect of HRP on fluorescence response of vitamin B_1-HRP-NaOH-H_2O_2 system (Khan *et al.* 2009).
First appeared in Luminescence: The Journal of Biological and Chemical Luminescence, 24: 73–78, published by John Wiley & Sons.

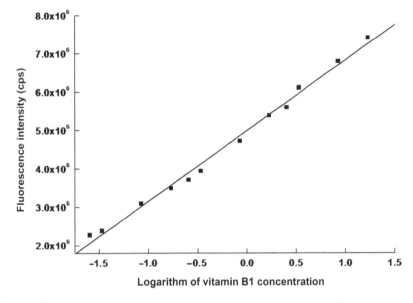

Figure 15.4 Calibration curve for the determination of vitamin B_1. The concentration of vitamin B_1 was in the range 0.026–16.83 μg mL^{-1}.
First appeared in Luminescence: The Journal of Biological and Chemical Luminescence, 24: 73–78, published by John Wiley & Sons.

Table 15.1 Assay of vitamin B_1 in a commercial vitamin B_1 and vitamin B containing dietary supplement (Khan *et al*. 2009). First appeared in Luminescence: The Journal of Biological and Chemical Luminescence, 24: 73–78, published by John Wiley & Sons.

Sample	Claimed (mg/tablet)	Proposed method	Added (μg/mL)	Found (μg/mL)	Recovery (%)
Centrum (vitamin B complex)	1.7	1.68	2.0	1.96	98
Biphane (vitamin B_1 tablet)	8.0	7.68	2.0	1.98	99

determination of vitamin B_1 (Table 15.2). The results demonstrate the feasibility of using HRP to oxidize vitamin B_1 for the sensitive and selective determination of vitamin B_1 in commercial vitamin B_1 tablets and vitamin B complex tablets with good reproducibility.

15.3 Extraction-based Determination

A solid-phase extraction technique coupled with a chip-based flow injection system was introduced to determine vitamin B_1 by spectrofluorimetry (Zhu *et al*.

Table 15.2 Effect of common pharmaceutical excipients on the determination of vitamin B_1 (Khan et al. 2009). First appeared in Luminescence: The Journal of Biological and Chemical Luminescence, 24: 73–78, published by John Wiley & Sons.

Excipients	Concentration ($\mu g/mL$)	Change in fluorescence intensity (%)
Sucrose	100	+1.52
Cellulose	100	+2.22
Magnesium stearate	50	+3.20
Lactose	100	+1.30
Starch	100	+1.73
Riboflavin	10	−3.52
Pyridoxine hydrochloride	10	+0.85
Cyanocobalamin	1	−1.05

2003). The method involved the oxidation of thiamin to fluorescent thiochrome by potassium hexaferricyanide. Thiochrome was isolated by solvent extraction. The solvent was prepared using a suspension of octadecyl-alkylated polystyrene/divinylbenzene micro beads. Thus thiochrome was adsorbed onto the surface of the micro beads during the extraction process. The obtained emission intensity of the thiochrome-adsorbed micro beads was linearly related to the concentration of vitamin B_1. The LOD was achieved at micro-molar scale.

Liquid phase extraction of the analyte by cloud point extraction (CPE) is a versatile technique in separation chemistry for the purpose of extraction, purification and preconcentration of numerous organic compounds in different samples (Saitoh and Hinze 1991; Shi et al. 2004). The cloud point refers to a certain temperature above which an aqueous homogeneous solution of non-ionic and zwitterionic surfactant becomes turbid and separates into two phases: (i) a surfactant-rich phase which contains most of the surfactant and (ii) a diluted aqueous phase which contains water and surfactant at a concentration close to the critical micelle concentration (Quina and Hinze 1999; Stalikas 2002).

CPE followed by spectrofluorimetry was applied to analyse the concentration of vitamin B_1 in samples of urine (Tabrizi 2006). In this method, Triton-X114 surfactant was used for CPE. The procedure was accomplished by adding an aqueous solution of thiamin, ferricyanide and Triton-X114 in an alkaline medium. The surfactant-rich turbid phase and diluted aqueous phase were attained after centrifugation of the mixture. The surfactant-rich phase was collected and diluted in an ethanol : water mixture prior to measurement of the fluorescence excitation and emission intensity. Thiamin was oxidized by ferricyanide under alkaline conditions to form thiochrome, which is a fluorescent species. During the extraction procedure, thiochrome was entrapped in surfactant micelles. Thus, thiamin was separated from the biological matrix after derivatization. The fluorescence intensity responded linearly with the concentration of thiamin under optimized conditions.

15.4 Nanomaterial-based Determination of Vitamin B_1

Nanomaterials in the form of a colloidal solution or quantum dots are attributed to have a tremendous impact in analytical chemistry for their unique physical and chemical properties (Alivisatos 2004; Katz and Willner 2004). A different methodology has been adopted to analyse vitamin B_1 spectrofluorimetrically by using cadmium selenide quantum dots (CdSe QDs), cadmium telluride (CdTe) nanorods, and silica and gold nanoparticles. A fluorescence resonance Rayleigh scattering (RRS) method was applied for determining vitamin B_1 at sub-nanomolar level (Liu et al. 2006). In this technique, vitamin B_1 was mixed with acidic buffer and prepared gold nanoparticles. After incubation, the solution mixture was excited in synchronous mode to obtain RRS spectra. The RRS spectral intensity correlated with the concentration of vitamin B_1.

Tetra-substituted carboxyl iron phthalocyanine entrapped silica (TCFePc-SiO_2) nanoparticles have been shown to analyse the concentration of vitamin B_1 (Zou and Chen 2007) at nanomolar scale. In this procedure, TCFePc-SiO_2 nanoparticles were used to catalyze the oxidation reaction of thiamine by hydrogen peroxide to form the highly fluorescent species, thiochrome. It produces a strong emission at 440 nm upon excitation at a wavelength of 370 nm. The fluorescence emission intensity and concentration of vitamin B_1 were linearly related. The method was applied to determine the vitamin B_1 content in tablets and a vitamin B complex.

CdSe QDs have been used as an optical probe for the analysis of vitamin B_1 concentration (Sun et al. 2008). The technique comprises the fluorescence quenching of light emitted by the CdSe QDs in presence of vitamin B_1 under alkaline conditions. Vitamin B_1 is an efficient quencher and can interrupt the radiative recombination process by adsorbing on the surface of CdSe QDs. The QDs produced an emission at 591 nm with excitation at 380 nm. Thus, vitamin B_1 was determined from the linear relationship between concentration of vitamin B_1 and fluorescence quenching of QDs. The obtained LOD was achieved at nanogram scale. The concentration of vitamin B_1 in tablet and injection samples was determined by this method.

Water-soluble CdTe nanorods modified with thioglycolic acid and cysteine have been used as a fluorescence probe for the determination of vitamin B_1 (Li et al. 2010). The size-dependent luminescence of the nanorods was studied in this work. A synthesized nanorod of short length was not luminous, even in the presence of vitamin B_1, but longer nanorods emitted strong light at a wavelength of 665 nm in the presence of vitamin B_1. The degree of enhancement of the luminescence intensity was proportional to the concentration of vitamin B_1. The LOD was achieved at micromolar scale. The method was applied to determine vitamin B_1 in a commercial tablet and vitamin B complex.

15.5 Flow Injection and Other Techniques

Flow injection Analysis (FIA) is a popular approach for automation of chemical analysis and the technique has been adapted successfully in various fields of analytical chemistry. Flow injection was first coupled with fluorescence to determine vitamin B_1 in 1980 (Karlberg and Thelandera 1980). In this technique, the thiochrome method was applied to a continuous flow system to analyse the concentration of vitamin B_1 in a pharmaceutical preparation.

Drop-based optical sensing coupled with a flow injection technique was first introduced to analyse vitamin B_1 from a mixture of vitamins B_1, B_2 and B_6 (Feng et al. 2004). Generally, an optically transparent cell is used in conventional fluorescence detection of a liquid sample in which a light beam is passed through the cell window. In this case, the possibility of light scattering and a background fluorescence signal cannot be avoided due to the adsorption or deposition of reagents inside the cell walls (Feng et al. 2004). Using a drop by itself has been reported to minimize the problems because it can serve as a windowless optical cell for fluorescence detection (Liu and Dasgupta 1996). In the reported method, vitamin B_1, B_2 and B_6 were detected and determined simultaneously in a mixture based on adsorption and desorption using the selective adsorption property of Sephadex CMC-25, a cation exchange resin. The mixture containing the vitamins was passed through a prepared ion-exchange column containing the resin. McIlvaine buffer or potassium ferricyanide was pumped through the flow channel. The vitamin mixture and pumped solution were mixed and formed continuous drops at the tip of the polytetrafluoroethylene (PTFE) tube. During the analysis procedure, vitamins B_1 and B_6 were trapped in the resin and vitamin B_2 was eluted as drops. Thus, when the system was excited the characteristic fluorescence peak of vitamin B_2 was recorded. Next, NaOH solution was pumped through the column to elute vitamin B_1 and B_2 from the resin. In this case, excitation on the falling drops provided fluorescence emitted only by vitamin B_6 because vitamin B_1 is a non-fluorescent compound in NaOH solution or water. Finally, potassium ferricyanide was passed through the other channel instead of McIlvaine buffer. Thus, vitamin B_1 was oxidized by potassium ferricyanide to produce fluorescent thiochrome and the vitamin B_6 present in the mixture was quenched in presence of ferricyanide. Therefore concentration of vitamin B_1, B_2 and B_6 were analysed sequentially from their individual fluorescence intensity.

In a FI photochemical technique coupled to spectrofluorimetry for the determination of vitamin B_1, thiamin was converted to thiochrome by a photochemical reaction (Chen et al. 1998). The photochemical reaction was conducted in continuous flow mood. Acetone was used as s sensitizer to enhance the fluorescence intensity. The obtained LOD was at microgram scale. Another photochemical procedure coupled with synchronous fluorimetry has been reported for the selective determination of vitamin B_1, B_2 and B_6 (Guo et al. 1993). In case of vitamin B_1, the LOD was obtained in nanogram scale.

A derivatization procedure of thiamin with *ortho*-phthalaldehyde (OPA) in the presence of 2-mercaptoethanol was applied to determine vitamin B_1 by FI-spectrofluorimetry (Viñas et al. 2000). The obtained LOD was at nanogram level. Spectrofluorimetric determination of vitamin B_1 was performed by a flow injection–solvent extraction (FI-SE) technique in which two different modes of FI-SE were conducted (Alonso et al. 2006) without phase separation. Phase segmentation was carried out in one mode and a single batch of organic solvent such as chloroform was used in another mode prior to measuring the fluorescence intensity of thiochrome oxidation product of thiamin by ferricyanide. Vitamin B_1 was determined in this study in pharmaceutical samples using both modes.

A spectrofluorimetric technique coupled with flow injection has been reported in which immobilized potassium hexaferricyanide on an anion exchange resin was used to oxidize thiamin to thiochrome (Calatayud et al. 1990). Vitamin B_1 was analysed from the correlation between the emission intensity of thiochrome and the concentration of vitamin B_1.

15.6 Concluding Remarks

Various analytical methods have been developed to determine vitamin B_1. However, they lack selectivity and sensitivity and most of them are also time-consuming. The determination of vitamin B_1 therefore requires critical consideration of the desired sensitivity, selectivity and specificity, the available time, and the process of preparation of the sample, as well as cost. Spectrofluorimetric methods have a high sensitivity and selectivity for vitamin B_1 determination and a short review of these methods is presented in this chapter. The most appropriate utilization of each spectrofluorimetric method based on its sensitivity, selectivity and accuracy is indicated. The methods for determining vitamin B_1 may be helpful for physicians, patients and food specialists.

Summary Points

- This chapter focuses on the spectrofluorimetric determination of vitamin B_1 (thiamin).
- Vitamin B_1 is an indispensable nutrient in the body in order to maintain normal neural action and to prevent beriberi disease.
- Most of the analytical techniques based on the spectrofluorimetry involve the generation of fluorescent thiochrome from thiamin *via* an oxidative or photochemical reaction in the absence or presence of catalyst.
- Thiochrome produces a strong emission band at a wavelength of 440 nm upon excitation and vitamin B_1 has been determined from the correlation between the emission intensity and the concentration of thiamin.
- Different strategies have been coupled with spectrofluorimetry (*e.g.* extraction and flow injection) to analyse vitamin B_1.
- In recent years, nanomaterials such as nanoparticles, quantum dots and nanorods have been introduced into the methodology to enhance the performance of the detection methods.

Key Facts

Key Facts about Spectrofluorimetry

- When an external energy is applied to a molecule as a photon, the molecule emits light by absorbing the photon corresponding to the energy of the applied photon.
- This phenomenon is known as fluorescence. The intensity of fluorescence produced from a sample molecule can be measured and termed in a technique known as spectrofluorimetry.
- Fluorescent molecules emit a characteristic fluorescence band.
- Fluorescence intensity is correlated with the amount of molecules present.
- A spectrofluorimetric procedure is applied to detect an analyte based on its characteristic emission band and to analyse the concentration of a target analyte from its emission intensity.

Key Facts about Nanomaterials

- Materials that are measured in nanoscale are classified as nanomaterials.
- Nanoscale is defined as smaller than one billionth of a metre at least in one dimension.
- Nanomaterials can be subdivided as nanoparticles, nanotubes, nanorods, quantum dots, *etc.* according to their size and properties.
- Nanomaterials exhibit unique chemical and physical properties due to their large surface area.
- Nanomaterials can participate as catalysts or fluorescent probes in the spectrofluorimetric determination of different analytes.

Key Facts about Oxidation

- Oxidation is a process that involves the loss of electron from chemical species such as atoms or molecules.
- Oxidation reactions always occur simultaneously with reduction, which involves the gain of an electron, and is called a redox process.
- The species which loses an electron is called the oxidant and the species which gains an electron is called the reductant.
- Strong oxidants include hydrogen peroxide, potassium ferricyanide, potassium permanganate and potassium dichromate.
- As an example, the strong oxidant potassium ferricyanide or hydrogen peroxide oxidizes thiamin to fluorescent thiochrome.

Definitions of Words and Terms

Adsorption. The physical or chemical process of accumulating of atoms, molecules, ions or biomolecules (known as 'adsorbed') on any surface is called adsorption.

Catalysts. Catalysts are reagents that change the rate of a chemical reaction without being consumed in the reaction. They are capable of speeding up or slowing down the reaction rate.

Desorption. This process is opposite of adsorption, *i.e.* accumulated atoms, molecules, ions or biomolecules are released from the surface.

Enzymes. Enzymes are mainly proteins (*i.e.* they are composed of amino acids) and are capable of catalysing a chemical reaction. Some enzymes are used for analytical purposes, *e.g.* horseradish peroxidase and glucose oxidase are used for the determination of vitamin B_1 and glucose, respectively.

Extraction. Extraction is a separation technique by which the required substance(s) is separated from a matrix. Different extraction techniques are available such as liquid–liquid extraction, solid–phase extraction (SPE) or cloud point extraction (CPE).

Flow injection analysis (FIA). FIA is an automated technique in which continuous flow of solution into flow channel is maintained using a pump such as peristaltic or syringe pump. Two or more flow streams are mixed in the mixing zone of the channel.

Fluorescence. When any atom or molecule absorbs energy, an electron from an orbital shifts to a higher energy level (called excitation) and emits corresponding light during the relaxation to ground state (called emission). This phenomenon is known as fluorescence.

Limit of detection (LOD). LOD stands for the lowest amount of analyte that can be measured in absence of the analyte within a given confidence limit. It can be estimated from the standard deviation of the blank signal and slope of the obtained calibration curve.

Photochemical reaction. A photochemical reaction is a special type of chemical reaction that is accomplished with absorption of light by atoms or molecules. Activation energy is attained by the collision of molecules in typical chemical reaction. On the other hand, the light source provides the required activation energy for chemical reactions to occur in a photochemical process.

Sensitized fluorescence. This is used for enhancing the weak fluorescence of a fluorophore. In this process, the excitation energy absorbed by the donor species is transferred to an acceptor species (fluorophores). A dramatic increase in the fluorescence of the acceptor molecule is observed through this indirect excitation process.

Synchronous fluorescence. Synchronous fluorescence stands for a type of fluorescence measurement technique in which the excitation and emission monochromator are scanned simultaneously and the constant wavelength difference between excitation and emission is maintained throughout the process. Synchronous spectra are a characteristic property of a molecule or compound and so the analyte can be detected selectively from a mixture or solution.

Zwitterions. A molecule which contains positive and negative charge in the same structure is called a zwitterion.

List of Abbreviations

CdSe QDs	cadmium selenide quantum dots
CPE	cloud point extraction
FI	flow injection
FI-SE	Flow Injection Solvent Extraction
GSH	glutathione
FeTSPc	iron(III) tetrasulfonatophthalocyanine
HRP	horseradish peroxidase
LOD	limit of detection
OPA	*ortho*-phthalaldehyde
PTFE	polytetrafluoroethylene
RRS	resonance Rayleigh scattering
RSD	relative standard deviation
TC	thiochrome
TCFePc-SiO$_2$	tetra-substituted carboxyl iron phthalocyanine entrapped silica
TDS	thiamin disulfide

References

Aboul-Kasim, E., 2000. Anodic adsorptive voltammetric determination of vitamin B$_1$ (thiamine). *Journal of Pharmaceutical and Biomedical Analysis.* 22: 1047–1054.

Alivisatos, P., 2004. The use of nanocrystals in biological detection. *Nature Biotechnology.* 22: 47–52.

Alonso, A., Almendral, M.J., Porras, M.J., and Curto Y., 2006. Flow-injection solvent extraction without phase separation fluorimetric determination of thiamine by the thiochrome method. *Journal of Pharmaceutical and Biomedical Analysis.* 42: 171–177.

Amjadi, M., Manzoori, J.L., and Orooj, M., 2007. The use of crude extract of kohlrabi (*Brassica oleracea* Gongylodes) as a source of peroxidase in the spectrofluorimetric determination of thiamine. *Bulletin of Korean Chemical Society.* 28: 246–250.

Bohrer, D., do Nascimento, P.C., Ramirez, A.G., Mendonca, J.K.A., de Carvalho, L.M., and Pomblum, S.C.G., 2004. Determination of thiamine in blood serum and urine by high-performance liquid chromatography with direct injection and post-column derivatization. *Microchemical Journal.* 78: 71–76.

Calatayud, J.M., Benito, C.G., and Gimenez, D.G., 1990. FIA-fluorimetric determination of thiamine. *Journal of Pharmaceutical and Biomedical Analysis.* 8: 667–670.

Chang, W., and Wang, L., 2006. Determination of concentration of vitamin B1 in unpolished rice by automatic potentiometric titration. *Huaxue Shijie.* 47: 589–592.

Chen, H., Zhu, J., Cao, X., and Fang, Q., 1998. Flow injection on-line photochemical reaction coupled to spectrofluorimetry for the determination of thiamine in pharmaceuticals and serum. *Analyst.* 123: 1017–1021.

Chen, Q.Y., Li, D.H., Yang, H.H., Zhu, Q.Z., Zheng, H., and Xu, J.G., 1999. Novel spectrofluorimetric method for the determination of thiamine with iron(iii) tetrasulfonatophthalocyanine as a catalyst. *Analyst.* 124: 771–775.

Ciszewski, A., and Wang, J., 1992. Determination of thiamine by cathodic stripping voltammetry. *Analyst.* 117: 985–988.

Depeint, F., Bruce, W.R., Shangari, N., Mehta, R., and O'Brien, P.J., 2006. Mitochondrial function and toxicity: Role of the B vitamin family on mitochondrial energy metabolism. *Chemico-Biological Interactions.* 163: 94–112.

Diaz, A.N., Paniagua, A.G., and Sanchez, F.G., 1993. Thin-layer chromatography and fiber-optic fluorometric quantitation of thiamine, riboflavin and niacin. *Journal of Chromatography.* 655: 39–43.

Echols, R.E., Harris, J., and Miller, R.H.J., 1980. Modified procedure for determining vitamin B_1 by gas chromatography. *Journal of Chromatography.* 193: 470–475.

Feng, F., Wang, K., Chen, Z., Chen, Q., Lin, J., and Huang, S., 2004. Flow injection renewable drops spectrofluorimetry for sequential determinations of vitamins B_1, B_2 and B_6. *Analytical Chimica Acta.* 527: 187–193.

Fujiwara, M., and Matsui, K., 1953. Determination of thiamine by the thiochrome reaction. *Analytical Chemistry.* 25: 810–812.

Ghasemi, J., and Abbasi, B., 2005. Simultaneous spectrophotometric determination of group B vitamins using parallel factor analysis. *Journal of the Chinese Chemical Society.* 52: 1123–1129.

Guo, X.-Q., Xu, J.-G., Wu, Y.-Z., Zhao, Y.B., Huang, X.-Z., and Chen G.-J., 1993. Determination of thiamine (vitamin B_1) by in situ sensitized photochemical spectrofluorimetry. *Analytical Chimica Acta.* 276: 151–160.

He, H.Z., Li, H.B., and Chen, F., 2005. Determination of vitamin B_1 in seawater and microalgal fermentation media by high-performance liquid chromatography with fluorescence detection. *Analytical Bioanalytical Chemistry.* 383: 875–879.

Jiang, X., and Sun, T., 2007. Indication ion square wave voltammetric determination of thiamine and ascorbic acid. *Analytical Letters.* 40: 2589–2596.

Jie, N., Yang, J., and Zhan, Z., 1993. Fluorimetric determination of thiamine with cobalt(II). *Analytical Letters.* 26: 2283–2289.

Jing, L., Yu, Y., Song, C., Mei, H., and Li, J., 2003. Determination of vitamin B_1 injection by potentiometric titration with chloride ion selective electrode. *Zhongguo Yiyao Gongye Zazhi.* 34: 521–522.

Karlberg, B.O., and Thelandera S., 1980. Extraction based on the flow-injection principle: Part 3. Fluorimetric determination of vitamin b1 (thiamine) by the thiochrome method. *Analytica Chimica Acta.* 114: 129–136.

Katz, E., and Willner, I., 2004. Integrated nanoparticle-biomolecule hybrid systems: synthesis, properties and applications. *Angewandte Chemie, International Edition.* 43: 6042–6108.

Khan, M.A., Jin, S.O., Lee, S.H., and Chung, H.Y., 2009. Spectrofluorimetric determination of vitamin B1 using horseradish peroxidase as catalyst in the presence of hydrogen peroxide. *Luminescence.* 24: 73–78.

Leopold, N., Cintă-Pínzaru, S.C., Baia, M., Antonescu, E., Cozar, O., Kiefer, W., and Popp, J., 2005. Raman and surface-enhanced Raman study of thiamine at different pH values. *Vibrational Spectroscopy.* 39: 169–176.

Li, Y., Wang, P., Wang, X., Cao, M., Xia, Y.S., Cao, C., Liu, M., and Zhu, C., 2010. An immediate luminescence enhancement method for determination of vitamin B_1 using long-wavelength emitting water-soluble CdTe nanorods. *Microchimica Acta.* 169: 65–71.

Liu, H., and Dasgupta, P.K., 1996. Analytical chemistry in a drop. *Trends in Analytical Chemistry.* 15: 468–475.

Liu, S., Chen, Y., Liu, Z., Hu, X., and Wang F., 2006. A highly sensitive resonance Rayleigh scattering method for the determination of vitamin B_1 with gold nanoparticles probe. *Microchimica Acta.* 154: 87–93.

Lopez-de-Alba, P.L., Lopez-Martinez, L., Cerda, V., and Amador-Hernandez, J., 2006. Simultaneous determination and classification of riboflavin, thiamine, nicotinamide, and pyridoxine in pharmaceutical formulations, by UV-visible spectrophotometry and multivariate analysis. *Journal of the Brazilian Chemical Society.* 17: 715–722.

Mancinelli, R., Ceccanti, M., Guiducci, M.S., Sasso, G.F., Sebastiani, G., Attilia, M.L., and Allen, J.P., 2003. Simultaneous liquid chromatographic assessment of thiamine, thiamine monophosphate and thiamine diphosphate in human erythrocytes: a study on alcoholics. *Journal of Chromatography B.* 789: 355–363.

Marszałł, M.L., Lebiedzinska, A., Czarnowski, W., and Szefer, P., 2005. High-performance liquid chromatography method for the simultaneous determination of thiamine hydrochloride, pyridoxine hydrochloride, and cyanocobalamin in pharmaceutical formulations using coulometric electrochemical and ultraviolet detection. *Journal of Chromatography A.* 1094: 91–98.

Martinez-Lozano, C., Pérez-Ruiz, T., Tomás, V., and Abellán, C., 1990. Flow injection determination of thiamine based on its oxidation to thiochrome by mercury(II). *Analyst.* 115: 217–220.

Mrestani, Y., and Neubert, R.H.H., 2000. Thiamine analysis in biological media by capillary zone electrophoresis with a high-sensitivity cell. *Journal of Chromatography A.* 871: 351–356.

Ni, Y.N., and Cai, Y.J., 2005. Simultaneous synchronous spectrofluorimetric determination of vitamin B_1, B_2 and B_6 by PARAFAC. *Guangpuxue Yu Guangpu Fenxi.* 25: 1641–1644.

Ozdemir, D., and Dinc, E., 2004. Determination of thiamine HCl and pyridoxine HCl in pharmaceutical preparations using UV-visible spectrophotometry and genetic algorithm based multivariate calibration methods. *Chemical and Pharmaceutical Bulletin.* 52: 810–817.

Pérez-Ruiz, T., Martínez-Lozano, C., Sanz, A., and Guillén, A., 2004. Successive determination of thiamine and ascorbic acid in pharmaceuticals by flow injection analysis. *Journal of Pharmaceutical and Biomedical Analysis.* 34: 551–557.

Pérez-Ruiz, T., Martinez-Lozano, C., Tomas, V., and Ibarra, I., 1992. Flow injection fluorimetric determination of thiamine and copper based on the formation of thiochrome. *Talanta*. 39: 907–911.

Ponder, E.L., Fried, B., and Sherma, J., 2004. Thin-layer chromatographic analysis of hydrophilic vitamins in standards and from *Helisoma trivolvis* snails. *Acta Chromatographica*. 14: 70–81.

Quina, F.H., and Hinze, L.W., 1999. Surfactant-mediated cloud point extractions: an environmentally benign alternative separation approach. *Industrial and Engineering Chemical Research*. 38: 4150–4168.

Rocha, F.R.P., Filho. O.F., and Reis, B.F., 2003. A multicommuted flow system for sequential spectrophotometric determination of hydrosoluble vitamins in pharmaceutical preparations. *Talanta*. 59: 191–200.

Saitoh, T., and Hinze, L.W., 1991. Concentration of hydrophobic organic compounds and extraction of protein using alkylammoniosulfate zwitterionic surfactant mediated phase separations (cloud point extractions). *Analytical Chemistry*. 63: 2520–2525.

Sanchez-Machado, D.I., Lopez-Cervantes, J., Lopez-Hernandez, J., and Paseiro-Losada, P., 2004. Simultaneous determination of thiamine and riboflavin in edible marine seaweeds by high-performance liquid chromatography. *Journal of Chromatographic Science*. 42: 117–120.

Shi, Z., He, J., and Chang, W., 2004. Micelle-mediated extraction of tanshinones from *Salvia miltiorrhiza* Bunge with analysis by high-performance liquid chromatography. *Talanta*. 64: 401–407.

Stalikas, C.D., 2002. Micelle-mediated extraction as a tool for separation and preconcentration in metal analysis. *Trends in Analytical Chemistry*. 21: 343–355.

Su, S.C., Chou, S.S., Hwang, D.F., Chang, P.C., and Liu, C.H., 2001. Capillary zone electrophoresis and micellar electrokinetic capillary chromatography for determining water-soluble vitamins in commercial capsules and tablets. *Journal of Food Science*. 66: 10–14.

Sun, J., Liu, L., Ren, C., Chen, X., and Hu, Z., 2008. A feasible method for the sensitive and selective determination of vitamin B_1 with CdSe quantum dots. *Microchimica Acta*. 163: 271–276.

Tabrizi, A.B., 2006. A cloud point extraction-spectrofluorimetric method for determination of thiamine in urine. *Bulletin of the Korean Chemical Society*. 27: 1604–1608.

Tang, X., Cronin, D.A., and Brunton, N.P., 2006. A simplified approach to the determination of thiamine and riboflavin in meats using reverse phase HPLC. *Journal of Food Composition and Analysis*. 19: 831–837.

Tian, L., and Zhang, A., 2005. Study on kinetic spectrophotometric determination of vitamin B_1. *Lihua Jianyan, Huaxue Fence*. 41: 255–258.

Vannatta, E.E., and Harris, L.E., 1959. The gravimetric silicotungstate method of assay for thiamin in the presence of interfering substances. *Journal of the American Pharmaceutical Association*. 48: 34–36.

Velisek, J., Davidek, J., Mnukova, J., and Pistek, T., 1986. Gas chromatographic determination of thiamin in foods. *Journal of Micronutrient Analysis*. 2: 73–80.

Viñas, P., López-Erroz C., Cerdán F.Z, Campillo, N., and Hernández-Córdoba, M., 2000. Flow-injection fluorimetric determination of thiamine in pharmaceutical preparations. *Michrochimica Acta*. 134: 83–87.

Wan, Q., Yang, N., and Ye, Y., 2002. Electrochemical behavior of thiamine on a self-assembled gold electrode and its square-wave voltammetric determination in pharmaceutical preparations. *Analytical Science*. 18: 413–416.

Wang, H., Luo, D., and Lan, J., 2004. Adsorptive polarographic wave of vitamin B1 in phosphate buffer and its analytical application. *Zhongnan Minzu Daxue Xuebao, Ziran Kexueban*. 23: 22–24.

Zafra-Gomez, A., Garballo, A., Morales, J.C., and Garcia-Ayuso. L.E., 2006. Simultaneous determination of eight water-soluble vitamins in supplemented foods by liquid chromatography. *Journal of Agricultural and Food Chemistry*. 54: 4531–4536.

Zhu, H., Chen, H., and Zhou, Y., 2003. Determination of thiamin in pharmaceutical preparation by sequential injection renewable surface solid-phase spectrofluorimetry. *Analytical Sciences*. 19: 289–294.

Zou, J.L., and Chen, X.L., 2007. Using silica nanoparticles as a catalyst carrier to the highly sensitive determination of thiamine. *Microchemical Journal*. 86: 42–47.

CHAPTER 16
Measurement of Thiamine Levels in Human Tissue

NATALIE M. ZAHR,*[a] MARY E. LOUGH,[b] YOUNG-CHUL JUNG[a] AND EDITH V. SULLIVAN[a]

[a] Stanford University School of Medicine, Department of Psychiatry and Behavioral Sciences, 401 Quarry Rd, Stanford, CA 94305, USA and SRI International, 333 Ravenswood Ave, Menlo Park, CA 94025, USA;
[b] Stanford Hospital and Clinics, 300 Pasteur Dr., Stanford, CA 94305-2295, USA and Department of Physiological Nursing, University of California at San Francisco, 505 Parnassus Ave, San Francisco, CA 94143, USA
*Email: nzahr@stanford.edu

16.1 Introduction

Thiamine (vitamin B_1) is considered an essential vitamin because it is required but not synthesized by the body and thus must be obtained from dietary sources at $\sim 1.4\,\text{mg/day}$ ($0.5\,\text{mg}/1000\,\text{kcal/day}$) (Lonsdale 2006). Natural sources of thiamine include yeast (*e.g.* Marmite), whole grains, brown rice, flax, asparagus, green peas, oranges, liver and pork (Anderson *et al.* 1986; Harper 2006; Jiang 2006). After ingestion, thiamine undergoes several transport steps to become available biochemically: (i) into intestine; (ii) into bloodstream; (iii) into cells of heart, liver and other tissue; (iv) within cells to access mitochondria and nuclei (Singleton and Martin 2001). The brain requires a fifth step, which requires thiamine to cross the blood–brain barrier.

Vitamin B_1 is transported as free thiamine (T^+), but the biochemically active forms are the phosphorylated derivatives, including thiamine monophosphate

(TMP), thiamine diphosphate (TDP; also known as thiamine pyrophosphate, TPP), and thiamine triphosphate (TTP). TDP, the best characterized form, in its role as a coenzyme for molecules involved in carbohydrate metabolism (*e.g.* transketolase and pyruvate dehydrogenase) is important for energy production and numerous metabolic functions (Lonsdale 2006).

Thiamine stores in the human body ($\sim$ 25–30 mg) are small and subject to rapid turnover (Martin *et al.* 2003). Thus, symptoms of mild thiamine deficiency, including poor sleep, malaise, weight loss, irritability and confusion, can be induced within 14 days if vitamin B_1 is absent from the diet. The heart and the nervous system, where thiamine levels are higher than in other organs, are dependent on high levels of oxidative metabolism and are especially sensitive to thiamine deficiency (Singleton and Martin 2001). Indeed, syndromes associated with severe thiamine deficiency are characterized by congestive heart failure (*e.g.* wet Beriberi), peripheral neuropathy (*e.g.* dry Beriberi) and damage to the nervous system (*e.g.* Wernicke–Korsakoff syndrome) (Zahr *et al.* 2011); untreated thiamine deficiency is fatal. Individuals susceptible to thiamine deficiency include: the elderly, children and the critically ill (particularly cancer and AIDS patients); pregnant women; chronic alcoholics; individuals with gastrointestinal problems including gastric bypass patients; and those malnourished from poverty or anorexia, among other reasons (Galvin *et al.* 2010; Lonsdale 2006).

Wernicke's encephalopathy, the traditional diagnosis of the reversible, thiamine deficiency-related complication, is based on the presence of the clinical triad of ocular motor abnormalities, lack of coordinated muscle control and altered mental state (Sechi and Serra 2007; Victor *et al.* 1989). However, Wernicke's encephalopathy is missed by routine clinical examination in 75–80% of cases and only $\sim$ 20% of patients diagnosed at autopsy presented with the full triad of clinical features (Galvin *et al.* 2010; Harper *et al.* 1986). Furthermore, thiamine levels are rarely measured in clinical settings (Galvin *et al.* 2010) because, in cases of suspected thiamine deficiency, vitamin B_1 can be administered with impunity (Harper 2006). Commercial laboratories typically perform thiamine quantification for clinical cases (see Table 16.1 for commercial reference standards). Thus, a caveat to this overview of healthy *versus* deficient thiamine levels is that rarely is the biological relevance (*e.g.* neuropsychological and motor characteristics) of measured thiamine levels described. Quantification is primarily accomplished in research studies and, in general, these refer to a 'thiamine deficiency' if quantified levels were two standard deviations from the mean.

16.2 Thiamine Measurements in Human Urine

Levels of excreted thiamine in urine have been measured using various techniques, including the trichrome method (based on the oxidation of thiamine) (Bessey *et al.* 1952), microbiological assay using *Lactobacillus viridescens* (Sauberlich *et al.* 1979), and high performance liquid chromatography (HPLC) (Roser *et al.* 1978). Urinary assessment of thiamine status presents several

Table 16.1 Thiamine reference standards in blood.

Laboratory test	Normal thiamine	Moderate thiamine deficiency	Severe thiamine deficiency
Erythrocyte Transketolase Activity Coefficient	1.00–1.19	1.20–1.25	>1.25
TDP Effect (%)	0–15%	16–24% (16–19%)	>25% (>20%)
Whole Blood Thiamine–EFNS Guidelines	60–220 nmol/L	36–60 nmol/L	<35 nmol/L
		Thiamine deficiency	
Whole Blood Thiamine–Quest Diagnosists, Madison NJ	87–280 nmol/L	<86 nmol/L	
Whole Blood TDP–ARUP Laboratories, Salt Lake City, UT	70–180 nmol/L	<69 nmol/L	
Whole Blood TDP–Mayo Clinics	80–150 nmol/L	<46 nmol/L	
Erythrocyte Thiamine– Bio-Center Laboratory, Wichita, KS	33–110 ng/mL	<32 ng/mL	

TDP = thiamine diphosphate; EFNS = European Federation of Neurological Societies.

challenges. Specifically, thiamine levels in urine vary within and between individuals (Clarke et al. 1966; Ihara et al. 2008; van Dokkum et al. 1990). Urine collections across 24 hours are difficult to obtain and random urine samples reveal within subject variability of thiamine levels with contributions from recent food consumption and the diurnal cycle (Clarke et al. 1966; Ihara et al. 2008). A morning void urine sample contains thiamine levels comparable to 24-hour urine collection results and to high-performance liquid chromatography (HPLC) blood measure results (Ihara et al. 2008); thus, a morning urine sample may be the most appropriate for measuring thiamine excretion. In general, any thiamine not utilized or stored by the body is excreted in urine in proportion to intake: when total daily thiamine intake is 0.50 mg/1000 kcal, ∼100 µg is excreted; when intake is ∼0.35 mg/1000 kcal, ∼40–90 µg is excreted; when intake is ∼0.2 mg/1000 kcal, 5–25 µg of thiamine is excreted (Horwitt et al. 1946; Melnick 1942; Oldham et al. 1946; Shimazono and Katasura 1965; Tsuji et al. 2010). In cases of beriberi, excretions as low as 0–15 µg have been reported, indicating a natural reserve mechanism when the body is nearing thiamine deficiency or depletion (Shimazono and Katasura 1965). Further to this point, healthy volunteers given a thiamine deficient diet for six weeks had reduced urinary thiamine excretion while those given an oral supplement of thiamine had increased urinary thiamine excretion (Brin 1963). Diuretic use can also increase urinary thiamine excretion (Hanninen et al. 2006; Suter et al. 2000), rendering individuals on diuretics, especially furosemide, at risk for subclinical thiamine deficiency (Rieck et al. 1999). More robust methods for quantification of thiamine have now replaced measurements in urine.

16.3 Thiamine Measurements in Human Blood

16.3.1 Erythrocyte Transketolase Activation and the TDP Effect

The erythrocyte transketolase (ETK) activation assay (also known as the B_1 saturation test) measures the functional capacity of the enzyme transketolase in red blood cells (*i.e.* erythrocytes). Transketolase is a thiamine-dependent enzyme in the non-oxidative branch of the pentose phosphate pathway (PPP), a process of glucose turnover that produces nicotinamide adenine dinucleotide phosphates (NADPH) as reducing equivalents and pentose sugars as essential components of nucleotides. In the absence of adequate thiamine, the PPP output is compromised.

There are multiple variations of the standard ETK method (Brin 1962). Two identical erythrocyte aliquots are prepared. A substrate (*e.g.* ribose-5-phosphate, erythrose-4-phosphate) is added to both samples. One aliquot is saturated with excess TDP while the other serves as the control sample. The amount of substrate remaining and the amount of product (*e.g.* sedoheptulose-7-phosphate, another sugar produced by the PPP) formed through the PPP pathway are measured after an incubation period of 30–60 minutes. The control aliquot gives the 'basal' or 'unsaturated' measure while the sample incubated with excess thiamine gives the 'stimulated' or 'saturated' measure. When erythrocyte thiamine status is normal, the addition of exogenous thiamine has a minimal effect on transketolase activity. By contrast, when thiamine levels are low, the addition of TDP can enhance ETK activity and PPP output (Sauberlich 1967). The 'saturated' value is divided by the 'unsaturated' value to give an activity coefficient (AC) also referred to as the 'saturation ratio.' A 'normal' ETK AC value is close to 1.0; a case is considered 'thiamine deficient' if the ratio is more than 1.25 (Table 16.1). Healthy controls from can have average ETK AC values of 1.09 ± 0.09 (range: 0.91–1.27, non-fasting, $n=54$) (Hoorn *et al.* 1975) to 1.13 ± 0.08 (fasting, $n=360$) (Herbeth *et al.* 1986). In adolescents, ETK AC was reported at 1.06 ± 0.03 (range: 0.69–1.45, fasting, $n=54$) (Bailey *et al.* 1994). In elderly individuals (>60 years old), average ETK AC can be above 1.2; thiamine deficiency, described by ETK AC values above 1.25, is seen in 3–23% of this population (Hoorn *et al.* 1975; Vir and Love 1979). ETK AC average 1.2 ± 0.18 in critically ill patients (Bradley *et al.* 1978) (Table 16.2). ETK based methods appear not to be influenced by age (below 60 years) or sex and fasting samples are not essential (Markkanen *et al.* 1969).

Percentage stimulation [*i.e.* $100 \times$ (saturated/unsaturated)] is referred to as the TDP effect. Based on the majority of the literature, normal thiamine is associated with 0–15% stimulation, moderate thiamine deficiency with 16–24% stimulation, and severe thiamine deficiency with more than 25% stimulation (Table 16.1). Studies of healthy control individuals report average TDP effects as low as 3.0 ± 4.7 (Kawai *et al.* 1980) to as high as $17 \pm 12.3\%$ (Lu and Frank 2008) with a range of intermediate results (Boni *et al.* 1980; Kuriyama *et al.* 1980; O'Rourke *et al.* 1990).

Table 16.2 Thiamine measures in blood (ETKAC and TDP effect).

Reference	Country	Participants	Fasting?	Sample	Status	Measure Mean and SD (Range) ETKAC
Bailey et al. 1994	England	adolescents (n = 54)	fasting	whole blood	normal	1.06 ± .03 (0.69–1.45)
Herberth et al. 1986	France	healthy controls (n = 360)	fasting	whole blood	normal	1.13 ± .08
Hoorn et al. 1975	Netherlands	healthy controls (n = 54)	nonfasting	whole blood	normal	1.09 ± .09 (0.91–1.27)
Hoorn et al. 1975	Netherlands	older adults in hospital (n = 153)	nonfasting	whole blood	22.9% TD	>1.27
Vir and Love 1979	Ireland	older adults, various locations (n = 196)	fasting	whole blood	13% MTD; 3% TD	>1.2
Bradley et al. 1978	England	patients, critically ill (n = 15)	N.S.	whole blood	100% MTD to TD	1.2 ± 0.18
Wetherilt et al. 1992	Turkey	children (n = 960)	N.S.	whole blood	20.1% TD	>1.18

Reference	Country	Participants	Fasting?	Sample	Status	TDP effect
Kawai et al. 1980	Japan	healthy controls (n = 12)	N.S.	whole blood	normal	3.0 ± 4.7%
O'Rourtke et al. 1990	England	healthy controls (n = 14)	nonfasting	whole blood	normal	(2.6–8.3%)
Kuriyama et al. 1980	Japan	healthy controls (n = 674)	N.S.	whole blood	normal	11.6 ± 11.5%
Boni et al. 1980	Japan	healthy controls (n = 54)	N.S.	whole blood	normal	13.5% (8.3–18.5%)
Chong and Ho 1970	Malaysia	healthy controls (n = 197)	N.S.	whole blood	normal	17 ± 12.3%
Pongpanich et al. 1974	Thailand	infants and children (n = 87)	N.S.	whole blood	10% MTD	>15%
Seear et al. 1992	Canada	children in ICU (n = 80)	N.S.	whole blood	12.5% MTD	>16%
Fattal-Valevski et al. 2005	Israel	infants with WE (n = 9)	N.S.	whole blood	100% MTD to TD	(17.6 to 37.8%)
Nichols and Basu 1994	Canada	older adults, community dwelling (n = 60)	N.S.	whole blood	50% MTD	>14%
Andrade Juguan et al. 1999	Indonesia	older adults, community dwelling (n = 204)	fasting	whole blood	36.6% MTD to TD	21.3 ± 17.7

Table 16.2 (Continued)

Study	Country	Subjects	Fasting	Sample		TDP effect
O'Keefe et al. 1994	Ireland	older adults in hospital ($n=36$)	N.S.	whole blood	31% MTD; 17% TD	MTD: 16–24%, TD: >25%
Pepersack et al. 1999	Belgium	older adults in hospital ($n=118$)	N.S.	whole blood	39% MTD; 5% TD	MTD: 16–21%, TD: >22%
Barbato and Rodriguez 1994	Australia	patients, critically ill ($n=50$)	N.S.	whole blood	36% MTD; 28% TD	>24%
Talwar et al. 2000	England	patients, critically ill ($n=63$)	N.S.	whole blood	13% TD	>25%
Saito et al. 1987	Japan	diabetic outpatients ($n=46$)	fasting	whole blood	21% TD	$16.6 \pm 13.2\%$
Chong and Ho 1970	Malaysia	pregnant women ($n=103$)	N.S.	whole blood	N.S.	$23.0 \pm 14.8\%$
McLaren et al. 1981	Scotland	patients with alcohol use disorders ($n=20$)	N.S.	whole blood	N.S.	$15.9 \pm 12.2\%$
Levy et al. 2002	France	patients with alcohol use disorders[a] ($n=40$)	N.S.	whole blood	N.S.	15.9%
Kawai et al. 1980	Japan	patients with Beriberi ($n=23$)	N.S.	whole blood	100% TD	$56.2 \pm 39.1\%$
Kuriyama et al. 1980	Japan	patients with Beriberi ($n=21$)	N.S.	whole blood	100% TD	$34.6 \pm 18.4\%$
Chong and Ho 1970	Malaysia	patients with Beriberi ($n=5$)	N.S.	whole blood	100% TD	$37.0 \pm 11.5\%$
Boni et al. 1980	Japan	patients with WE ($n=18$)	N.S.	whole blood	100% TD	(28–67%)

ETK AC = Erythrocyte Transketolase Activity Coefficient; TDP = thiamine diphosphate; N.S. = not specified; MTD = moderate thiamine deficiency; TD = thiamine deficiency.
a = with cirrhosis.

In Thailand, a small proportion (10%) of ostensibly healthy infants and children ($n = 87$) were shown to have moderate thiamine deficiency (TDP effect >15%) (Pongpanich et al. 1974), 12.5% of critically ill children ($n = 80$) in Canada showed moderate thiamine deficiency (TDP effect >16%) and Israeli infants ($n = 9$) presenting with signs of Wernicke's encephalopathy because of a soy-based formula deficient in thiamine showed TDP effects ranging from 17.6–37.8% (Fattal-Valevski et al. 2005). Elderly individuals can also have high TDP effects (>22): studies indicate that up to 50% of individuals over 60 years can be moderately thiamine deficient and 5–17% can be severely thiamine deficient (Andrade Juguan et al. 1999; Nichols and Basu 1994; O'Keeffe et al. 1994; Pepersack et al. 1999). Studies indicate TDP effects of ∼16% in individuals with alcohol use disorders (Levy et al. 2002; McLaren et al. 1981). Effects are higher in pregnant women (e.g. TDP effect $23.0 \pm 14.8\%$) (Chong and Ho 1970) and critically ill patients (TDP effect >24%) (Barbato and Rodriguez 1994; Talwar et al. 2000) but highest in patients with beriberi (TDP effect >35%) (Chong and Ho 1970; Kawai et al. 1980; Kuriyama et al. 1980) and Wernicke's encephalopathy (TDP effect 28–67%) (Boni et al. 1980) (Table 16.2).

ETK based methods, once considered the most reliable means of assessing thiamine status, are now considered inadequate because they only provide an indirect measure. Because transketolase activity requires thiamine, decreased transketolase activity is presumed to be due to a decrease in thiamine. However, other factors may decrease transketolase activity including decreased enzymatic binding and decreased enzyme synthesis as has been demonstrated in diseases such as diabetes (Friedrich 1988) and liver dysfunction (Fennelly et al. 1967). ETK based methods have also been criticized as unreliable, insensitive, and subject to poor precision (Bailey et al. 1994).

16.3.2 High Performance Liquid Chromatography Methods

Techniques using HPLC and fluorescent detection are now the standard for direct quantification of thiamine in blood (Lynch and Young 2000). Plasma contains less than 10% of whole blood thiamine (in the form of T^+ and TMP) and levels are strongly influenced by recent food consumption (Tallaksen et al. 1993). Following thiamine intake, plasma thiamine levels peak at 50 minutes (range: 20–120 minutes) and return to baseline within 12 hours (Tallaksen et al. 1993). Thus, a fasting state is often recommended to increase accuracy of plasma thiamine measures. However, measurements in whole blood or erythrocytes are preferred as plasma measures suffer from poor sensitivity and specificity.

Because 80% of total thiamine in whole blood is present in erythrocytes (Schrijver et al. 1982), whole blood or erythrocyte samples are preferred for thiamine quantification. Total thiamine in whole blood from healthy adults ranges from 45–70 ng/mL (Chong and Ho 1970; Kawai et al. 1980; Kuriyama et al. 1980) or from 104.7 ± 24.0 to 191.0 ± 32.0 nmol/L (Lu and Frank 2008; Poupon et al. 1990; Wyatt et al. 1989). Total thiamine in erythrocytes from healthy adults ranges from 73–220 ng/mL (Gold et al. 1998; O'Rourke et al. 1990) or averages 186.0 ± 30.0 nmol/L (Herve et al. 1994). Whole blood thiamine levels are more likely to be low in vulnerable populations such as critically ill children (e.g. 15.2–31.6 ng/mL) (Lima et al. 2011), the elderly (e.g. <50 ng/mL) (Lee et al. 2000; Vognar and Stoukides 2009) and individuals with alcohol use disorders (e.g. 48.1 ± 13.9 ng/mL or 95.2 ± 26.4 nmol/L) (Maschke et al. 2005; Poupon et al. 1990). Levels are typically lowest in individuals with beriberi (e.g. 22.9 ± 10.6 ng/mL) (Kawai et al. 1980) (Table 16.3).

HPLC analysis of TDP in whole blood or erythrocytes is the most sensitive, specific and precise method for determining the nutritional status of thiamine because TDP, the biologically active form of thiamine, is a reliable indicator of total body stores (Warnock et al. 1978). TDP is nearly undetectable in plasma or serum but is present in erythrocytes (Tallaksen et al. 1991). Although it has been suggested that thiamine measures are most accurate when performed in a sample of washed erythrocytes (Mancinelli et al. 2003), TDP levels in whole blood correlate strongly with TDP levels in erythrocytes (Talwar et al. 2000) and sample preparation is more efficient and reliable. TDP levels in whole blood samples from healthy individuals range from averages of 84.4 ± 19.0 to 194.7 ± 27.5 nmol/L (Lu and Frank 2008; Wyatt et al. 1989). In erythrocyte samples, TDP levels from healthy controls are relatively higher and average 223.5 ± 30.3 but can range from 165–286 nmol/L (Baines 1985; Herve et al. 1994; Lynch et al. 1997; Mancinelli et al. 2003; Warnock 1982). In the elderly, erythrocyte TDP levels can be below 140 nmol/L and thus indicate that ∼16% of the elderly are thiamine deficient (Wilkinson et al. 1997). Individuals with alcohol use disorders have erythrocyte TDP levels averaging as low as 101.6 ± 27.5 (Pitel et al. 2010) to as high as 218.0 ± 29.0 nmol/L (Baines et al. 1988) with various results in between (Levy et al. 2002; Mancinelli et al. 2003) (Table 16.4).

16.4 Thiamine Measurements in Other Human Tissue

The measurement of thiamine levels in human tissue other than blood is limited to biopsy or autopsy samples. This limitation translates into high sample variability due to factors such as age and nutritional, disease and medication status. Another limitation associated with measurement of thiamine in human tissue is the biological instability of thiamine derivatives. Because the turnover time of TTP is approximately one hour, significant hydrolysis can occur during extended delays between death and sample retrieval, preparation, and storage (Bettendorff et al. 1996a).

16.4.1 Thiamine Measurements in Human Heart, Liver, and Other Organs

TDP, the biologically active form of thiamine, is highest in the heart, followed by skin, liver, kidney and colon (Gangolf et al. 2010) (Table 16.5). Indeed, TDP

Table 16.3 Thiamine measures in blood (total thiamine in ng/mL and nmol/L).

Reference	Country	Participants	Fasting?	Sample	Status	Measure Mean and SD (Range)
						Total thiamine (ng/L)
Kimura et al. 1982	Japan	healthy controls (n = 20)	N.S.	whole blood	normal	46.2 ± 2.3 ng/mL
Kuriyama et al. 1980	Japan	healthy controls (n = 674)	N.S.	whole blood	normal	68.1 ± 31.2 ng/mL
Kawai et al. 1980	Japan	healthy controls (n = 12)	N.S.	whole blood	normal	69.6 ± 12.6 ng/mL
O'Rourke et al. 1990	England	healthy controls (n = 14)	nonfasting	erythrocytes	normal	(73–131 ng/mL)
Gold et al. 1998	USA	healthy controls (n = 24)	nonfasting	erythrocytes	normal	(134–220 ng/mL)
Lee et al. 2000	USA	older adults in nursing home (n = 75)	N.S.	plasma	14% TD	<10 ng/mL
Vognar & Stoukides 2009	Egypt	older adults in nursing home[a] (n = 150)	N.S.	plasma	13.3% TD	<50 ng/mL
Maschke et al. 2005	Germany	patients with alcohol use disorders (n = 45)	N.S.	whole blood	N.S.	48.1 ± 13.9 ng/mL (20–96 ng/mL)
Lima et al. 2011	Brazil	children in ICU (n = 202)	N.S.	whole blood	28.2% TD	25.8 ng/mL (15.2–31.6 ng/mL)
Saito et al. 1987	Japan	diabetic outpatients (n = 46)	fasting	whole blood	21% TD	46.9 ± 28.5 ng/mL
Kawai et al. 1980	Japan	patients with Beriberi (n = 23)	N.S.	whole blood	N.S.	22.9 ± 10.6 ng/mL
Kuriyama et al. 1980	Japan	patients with Beriberi (n = 21)	N.S.	whole blood	N.S.	39.2 ± 16.6 ng/mL
						Total thiamine (nmol/L)
Bailey et al. 1994	England	adolescents (n = 54)	fasting	erythrocytes	normal	216.5 ± 21.0 nmol/L (101.9–949.9)
Herve et al. 1994	France	healthy controls (n = 52)	fasting	erythrocytes	normal	186.0 ± 30.0 nmol/L
Poupon et al. 1990	France	healthy controls (n = 68)	fasting	whole blood	normal	104.7 ± 24.0 nmol/L
Lu and Frank et al. 2008	USA	healthy controls (n = 43)	nonfasting	whole blood	normal	124.8 nmol/L (75.2–193.8)
Wyatt et al. 1989	USA	healthy controls (n = 140)	N.S.	whole blood	normal	191.0 ± 32.0 nmol/L
Wyatt et al. 1991	USA	infants (n = 159)	N.S.	whole blood	normal	187.0 ± 39.0 nmol/L

Table 16.3 (Continued)

						Total thiamine (nmol/L)
Cromer et al. 1989	USA	adolescents ($n=38$)	N.S.	whole blood	normal	181.0 ± 23.0 nmol/L
Donnino et al. 2010	USA	patients, critically ill with sepsis ($n=30$)	N.S.	plasma	20% TD	≤ 9 nmol/L
Ortega et al. 2009	Spain	patients, obese ($n=57$)	N.S.	whole blood	21.8% TD	≤ 150 nmol/L
Li et al. 2008	USA	patients intoxicated in ER ($n=39$)	N.S.	whole blood	15% TD	129.69 ± 69.15 nmol/L (52–494 nmol/L)
Popupon et al. 1990	France	patients with alcohol use disorders ($n=37$)	fasting	whole blood	N.S.	95.2 ± 26.4 nmol/L

ng/mL × 2.69 = nmol/L.
nmol/L × 0.3373 = ng/mL.
molecular weight (thiamine hydrochloride): 337.28
N.S. = not specified; TD = thiamine deficiency; a = with cognitive deficits.

Table 16.4 Thiamine measures in blood (free thiamine and its phosphate derivative).

Reference	Country	Participants	Fasting?	Sample	Status	T*	TMP	TDP
						Measure Mean and SD (Range)		
Wyatt et al. 1991	USA	infants (n=159)	N.S.	whole blood	normal	67.0 ± 21.0 nmol/L		
Wyatt et al. 1989	USA	healthy controls (n=140)	N.S.	whole blood	normal	62.7 ± 17.0 nmol/L		
Poupon et al. 1990	France	healthy controls (n=68)	fasting	whole blood	normal	19.0 ± 8.6 nmol/L		
Herve et al. 1994	France	healthy controls (n=52)	fasting	erythrocytes	normal	4.0 ± 2.0 nmol	<2 nmol/L	176.0 ± 28.0 nmol/L
Lu and Frank 2008	USA	healthy controls (n=43)	nonfasting	whole blood	normal	7.4 nmol/L (3.3–12.4 nmol/L)	3.5 nmol/L (1.6–6.5 nmol/L)	114.0 nmol/L (70.3–178.6 nmol/L)
Tallaksen et al. 1992a	Norway	healthy controls (n=30)	fasting	whole blood	normal	24.8 ± 11.0 nmol/L	5.73 ± 2.46 nmol/L	84.4 ± 19.0 nmol/L
Tallaksen et al. 1997	Norway	healthy controls (n=34)	fasting	whole blood	normal	24.0 ± 1.0 nmol/L	5.7 ± 2.2 nmol/L	85.0 ± 18.1 nmol/L
Tallaksen et al. 1991	Norway	healthy controls (n=30)	fasting	whole blood	normal	(19.6–43.8 nmol/L)	(7.4–16.0 nmol/L)	(91.4–204.4 nmol/L)
Tallaksen et al. 1991	Norway	healthy controls (n=30)	fasting	serum	normal	(13.8–20.2 nmol/L)	<2 nmol/L	<2 nmol/L
Tallaksen et al. 1991	Norway	healthy controls (n=30)	fasting	erythrocytes	normal	(13.2–35.3 nmol/L)	(2.1–13.9 nmol/L)	(229.3–435.2 nmol/L)
Mancinelli et al. 2003	Italy	healthy controls (n=103)	fasting	erythrocytes	normal	89.6 ± 22.7 nmol/L	4.4 ± 6.6 nmol/L	222.2 ± 56.3 nmol/L
Poupon et al. 1990	France	patients with alcohol use disorders (n=37)	fasting	whole blood	N.S.	21.6 ± 8.3 nmol/L		
Levy et al. 2002	France	patients with alcohol use disorders[a] (n=40)	N.S.	erythrocytes	N.S.	1.15 nmol/L (0–18 nmol/L)	0	186.0 nmol/L (69–542 nmol/L)
Mancinelli et al. 2003	Italy	patients with alcohol use disorders (n=35)	fasting	erythrocytes	N.S.	69.4 ± 35.9 nmol/L	4.8 ± 5.2 nmol/L	127.4 ± 62.5 nmol/L

Table 16.4 (*Continued*)

Reference	Country	Participants	Fasting?	Sample	Status	T*	TMP	TDP
Tallaksen et al. 1992a	Norway	patients with alcohol use disorders ($n=22$)	fasting	whole blood	N.S.	19.3 ± 7.12 nmol/L	6.51 ± 6.47 nmol/L	63.9 ± 23.6 nmol/L
Warnock 1982	USA	healthy controls ($n=17$)	N.S.	erythrocytes	normal			123.0 ± 27.0 ng/mL (80–166 ng/mL)
Lynch et al. 1997	England	healthy controls ($n=45$)	nonfasting	erythrocytes	normal			174.0 ± 34.0 nmol/L
Baines 1985	England	healthy controls ($n=48$)	N.S.	erythrocytes	normal			223.5 ± 30.3 nmol/L (165–286 nmol/L)
Korner et al. 2009	Germany	healthy controls ($n=30$)	N.S.	whole blood	normal			194.7 ± 27.5 nmol/L
McLaren et al. 1981	Scotland	patients with alcohol use disorders ($n=20$)	N.S.	plasma	N.S.			56.0 ± 38.0 nmol/L
Wilkinson et al. 1997	New Zealand	older adults, community dwelling[b] ($n=222$)	N.S.	erthrocytes	15.8% TD			<140 nmol/L
Pitel et al. 2010	USA	patients with alcohol use disorders[c] ($n=21$)	nonfasting	whole blood	N.S.			101.6 ± 27.5 nmol/L
Pitel et al. 2010	USA	patients with alcohol use disorders ($n=7$)	nonfasting	whole blood	N.S.			123.3 ± 59.1 nmol/L
Baines et al. 1988	England	patients with alcohol use disorders ($n=9$)	N.S.	erthrocytes	N.S.			218.0 ± 29.0 nmol/L

N.S. = not specified; a = cirrhosis; b = on diuretics; c = with history of dietary deficiency.

Table 16.5 Thiamine measurements in human tissue[a,b] (pmol/mg of protein).

Organ	n	Thiamine	TMP	TDP	TTP
Heart auricle	5	0.63 [1]	1.4 ± 0.7	66 ± 41	0.4 ± 0.5 [4]
Skin	3	2.1 ± 0.6	3.6 ± 0.8	47 ± 12	0.44 ± 0.09
Liver	3	0.26 ± 0.07	3 ± 2	45 ± 29	1.7 ± 1.8
Kidney	1	3.5	80	33	0.19
Colon	2	0.07 ± 0.06	2.1 ± 2.4	30 ± 22	0.3 ± 0.2
Lung	7	2.2 ± 1.1	2.0 ± 0.9	30 ± 12	0.49 [1]
Adipose tissue	11	3 ± 4 [4]	3 ± 3	27 ± 23	2 ± 2 [5]
Skeletal muscle	11	0.6 ± 0.7 [7]	0.7 ± 0.4	17 ± 12	1 ± 1 [10]
Arteries	6	1.0 ± 0.8	2.4 ± 2.0	10 ± 7	0.8 ± 0.3 [4]
Veins	10	1.4 ± 1.0 [6]	1.7 ± 1.4	9 ± 6	3 ± 4 [6]
Thymus	6	0.23 ± 0.16 [2]	1.1 ± 0.4	7.1 ± 0.9	1.1 [1]
Pericardium	6	1.3 ± 0.6 [3]	0.8 ± 0.5	4.8 ± 2.3	0.25 [1]
Valves	5	2.6 ± 3.6	0.9 ± 0.5	4.7 ± 2.2	1.8 ± 1.5 [2]

[a] Biopsied tissue from subjects aged 37–82. Twenty-nine out of 33 were hospitalized due to cardiovascular problems [from Gangolf 2010].
[b] Numbers in brackets indicate measured samples if different from the number of patients.

levels in the heart exceed those in the cerebral cortex by about three times. As with measures in blood, there is a negative correlation between age and TDP levels, at least in skeletal muscles and lung (Bettendorff et al. 1996a). The limited storage capacity for thiamine in human body is likely due to relatively low total thiamine levels in the liver (Gangolf et al. 2010).

16.4.2 Thiamine Measurements in Human Cerebrospinal Fluid

Thiamine measures in cerebrospinal fluid (CSF) are rarely reported. A study in healthy individuals ($n = 31$) from which CSF samples were collected during a myelography procedure for back pain revealed that CSF contains predominately T^+ (8.6 ± 3.9 nmol/L) and TMP (28.1 ± 7.9 nmol/L) (Tallaksen et al. 1992a) at concentrations significantly higher than in serum (Tallaksen et al. 1992a). In Alzheimer's disease, alterations in brain thiamine metabolism have been noted (Butterworth and Besnard 1990). Studies have also demonstrated abnormally low reduced plasma or erythrocyte thiamine levels in individuals with Alzheimer's disease (Gold et al. 1995, 1998; Molina et al. 2002; Pepersack et al. 1999) but see (Scileppi et al. 1984) with therapeutic benefits from thiamine treatment (Blass et al. 1988; Meador et al. 1993). By contrast, there is little evidence for compromised blood thiamine levels in Parkinson's disease (Gold et al. 1998).

Studies of CSF thiamine measures corroborate these findings. Compared with healthy controls ($n = 32$), individuals with Alzheimer's disease ($n = 33$) had lower, but not significantly different, CSF levels of total thiamine, T^+, TMP and TDP (Molina et al. 2002). By contrast, healthy controls ($n = 40$) and patients with Parkinson's disease ($n = 24$) were indistinguishable with respect to CSF thiamine levels (Jiménez-Jiménez et al. 1999).

16.4.3 Thiamine Measurements in Human Brain

The particular sensitivity of the human nervous system to thiamine deficiency might be due to the fact that thiamine levels in the human brain are lower than in other species. TDP is the most abundant derivative, comprising ~60% of total thiamine in the human brain (Gangolf et al. 2010). Whereas the distribution in cortex and subcortical regions is relatively uniform, total thiamine and TDP are highest in mammillary bodies and lowest in the hippocampus (Bettendorff et al. 1996a) (Table 16.6). Consistent with blood findings, whole brain total thiamine and TDP are lower by 20–25% in elderly (>77 years) compared to middle aged (40–55 years) individuals who died without evidence of neurological or psychiatric disease (Bettendorff et al. 1996a). Human brain TDP content is less variable than in peripheral tissue suggesting tight regulation of thiamine homeostasis in the brain (Bettendorff et al. 1996a). Nevertheless, autopsy of brains from Alzheimer's disease patients revealed low cortical TDP levels (by ~20%) (Heroux et al. 1996; Mastrogiacoma et al. 1996). Autopsy of brains from patients with frontal lobe degeneration of the non-Alzheimer's type demonstrated a 40–50% reduction in cortical TDP levels (Bettendorff et al. 1997). By contrast, cortical TDP levels from brains of patients with Friedreich's

Table 16.6 Thiamine measurements in human brain tissue (mol/mg of protein).

		T^*	TMP	TDP	TTP
Autopsied normal tissue[a]	Frontal cortex	8.4 ± 2.7	2.6 ± 0.7	21 ± 3	n.d.
	Temporal cortex	9.1 ± 1.5	2.7 ± 0.5	20 ± 4	n.d.
	Parietal cortex	8.8 ± 1.9	3.0 ± 1.2	21 ± 4	n.d.
	Occipital cortex	11.3 ± 2.7	3.3 ± 0.9	20 ± 4	n.d.
	Insular cortex	9.4 ± 3.5	2.2 ± 0.7	19 ± 4	n.d.
	Cingulate cortex	10.5 ± 2.6	2.1 ± 0.7	20 ± 5	n.d.
	Cerebellar cortex	8.6 ± 1.2	1.8 ± 0.6	19 ± 3	n.d.
	Caudate nucleus	11.4 ± 5.5	1.8 ± 0.5	23 ± 5	n.d.
	Hippocampus	9.3 ± 2.7	1.7 ± 0.6	15 ± 4	n.d.
	Globus pallidus	15.8 ± 5.4	1.7 ± 0.7	17 ± 5	n.d.
	Thalamus (midline)	12.7 ± 4.2	1.9 ± 0.7	17 ± 6	n.d.
	Thalamus (pulvinar medial)	13.8 ± 2.6	2.1 ± 0.7	20 ± 3	n.d.
	Mammillary bodies	13.0 ± 3.1	2.6 ± 1.0	24 ± 4	n.d.
Biopsied normal tissue[b]	Temporal cortex (3yo)	4.1	1.9	18.4	0.20
	Temporal cortex (18yo)	8.3	1.0	22.4	0.15
	Frontal cortex (58yo)	7.9	3.0	24.9	0.12
Biopsied peritumoral tissue[c]	Temporal cortex	0.2 ± 0.3	3.5 ± 2.6	21 ± 5	0.4 ± 0.3

[a] 63 subjects with no neurological disorders [from Bettendorff 1996].
[b] 3 subjects (3 year old (yo) with temporal lobe epilepsy. 18 yo with astrocytoma, and 58 yo with meningioma [from Bettendorff 1996].
[c] 5 subjects with temporal lobe glioblastoma [from Gangolf 2010].
n.d., not detected.

ataxia and spinocerebellar ataxia type 1 were not different from controls (Bettendorff et al. 1996b).

16.5 Concluding Remarks

The classic method to assess thiamine status indirectly is to measure erythrocyte transketolase activity. Two samples of blood are incubated with excess substrate for the PPP; to one is also added excess TDP while the other sample serves as the control. The amount of substrate remaining and the product formed are quantified, and any enhancement in activity resulting from the added TDP indicates that the sample was originally deficient in thiamine to some extent. Based on a series of studies in healthy, control samples, a 'normal' ETK AC value is close to 1.0; a 'normal' TDP effect is associated with 0–15% stimulation. Currently, the most accurate assessment of thiamine status is considered the quantification of TDP in whole blood or erythrocytes. In healthy adults, whole blood TDP ranges from 65–220 nmol/L, while in erythrocytes, TDP ranges from 165–286 nmol/L. Challenges for the future include normalizing the methods of quantification, units of measure and reference standards, and establishing 'biologically relevant' thiamine levels by correlating, in population studies, thiamine levels and clinical signs and symptoms of deficiency.

Summary Points

- This chapter focuses on thiamine (vitamin B_1).
- Thiamine is an essential vitamin because the human body requires it but cannot make it.
- Natural sources of thiamine include whole grains, asparagus and pork.
- The heart and nervous system are especially vulnerable to thiamine deficiency.
- Thiamine deficiency can result in beriberi, Wernicke's encephalopathy or Korsakoff syndrome.
- The most accurate way to measure thiamine status is to quantify the biologically active form of thiamine, thiamine diphosphate, in whole blood or isolated red blood cells.
- Future challenges include normalization of procedures, units of measure and reference standards.

Key Facts

Key Facts about High Performance Liquid Chromatography (HPLC)

- HPLC is a technique used in biochemistry and analytical chemistry to identify, purify and quantify individual compounds from a mixture.
- The classic configuration is a column with a filter at the bottom. A dry column is first filled with a dry stationary phase and then filled with a wet mobile phase.

- A sample is introduced and the time at which the sample comes out of the column is called the retention time. Retention time is often used to identify components of a mixture.
- Variations in types of solvents, additives and gradients are common.
- There are also different types of HPLC such as displacement chromatography, reversed-phase chromatography and ion-exchange chromatography.

Key Facts about Thiamine Diphosphate (TDP)

- TDP is also known as thiamine pyrophosphate (*i.e.* TPP).
- It is currently the best characterized thiamine derivative.
- TDP is a coenzyme for many enzymes including pyruvate dehydrogenase, α-ketoglutarate dehydrogenase and transketolase where it catalyses the reversible cleavage of a substrate compound at a carbon–carbon bond.
- TDP is involved in the catabolism (break down) of sugars and amino acids.
- Erythrocyte TDP is considered a reliable indicator of thiamine body stores.

Key Facts about Wernicke–Korsakoff Syndrome (WKS)

- Wernicke–Korsakoff syndrome denotes the constellation of brain abnormalities and behavioural impairments commonly associated with untreated or undertreated thiamine depletion.
- Wernicke's encephalopathy is an acute, potentially reversible neurological disorder caused by severe thiamine deficiency, commonly occurring with chronic alcoholism, poor nutrition, long-term parenteral feeding, hyperemesis gravidarum or bariatric surgery.
- Incidence rates of Wernicke's encephalopathy in the general population—on the basis of autopsy findings in Western countries—range from 0.1–2.8% but can be as high as 12.5% in patients with alcoholism.
- When Wernicke's encephalopathy is left undiagnosed and untreated, ≈80% of patients with this condition develop Korsakoff syndrome—a severe, typically permanent neurological disorder characterized by anterograde amnesia.
- Alcoholics are at a high risk of thiamine deficiency because of poor diet. Chronic alcoholism compromises thiamine absorption from the gastrointestinal tract, impairs thiamine storage and may reduce the phosphorylation of thiamine to TDP.
- Guidelines for the diagnosis, treatment and prevention of Wernicke's encephalopathy have recently been released by the European Federation of Neurological Societies and are based on three decades of research.
- When Wernicke's encephalopathy is recognized, treatment with thiamine can result in rapid clinical improvement.

- The prevalence of Wernicke's encephalopathy has been reduced in a number of countries (including the US, the UK and Australia) that have instituted nationwide thiamine supplementation in staple foods such as bread.

Definition of Words and Terms

Alcohol use disorders—generally known as alcohol abuse or alcoholism—are the result of heavy, chronic drinking that causes distress or harm. Diagnoses includes some of the following: drinking more and longer than intended; failed attempts to cut down or stop; getting into situations that increase chances of getting hurt; drinking more to get the desired effects; continued drinking even though feeling depressed or anxious as a result; having memory blackouts; and experiencing withdrawal symptoms such as trouble sleeping, shakiness, restlessness, nausea, sweating, heart palpitations or seizure.

Beriberi is a nervous system disorder caused by a thiamine deficiency. Wet beriberi primarily affects the cardiovascular system, while dry beriberi affects the nervous system. Wet beriberi can also be characterized by vasodilation, peripheral oedema, shortness of breath and swelling of the lower legs. Dry beriberi causes partial paralysis resulting from damaged peripheral nerves and can be characterized by difficulty walking, loss of muscle function, mental confusion and involuntary eye movements.

Diuretic—a class of pharmacological agents that increase the excretion of water (*i.e.* urine) and are used to treat hypertension, heart failure, liver cirrhosis and certain kidney diseases.

Erythrocytes—red blood cells are the most common type of blood cell and are used to deliver oxygen to tissue through the circulatory system.

Erythrocyte transketolase activity was the classic method to assess thiamine status. Two samples of blood are incubated with excess substrate for the pentose phosphate pathway; to one is also added excess thiamine diphosphate while the other serves as the control. The amount of substrate remaining and product formed are quantified, and any enhancement in activity resulting from the added thiamine diphosphate indicates that the sample was originally deficient in thiamine to some extent.

Korsakoff's syndrome is a severe, typically permanent neurological disorder characterized by anterograde amnesia that results from untreated thiamine deficiency.

Thiamine (vitamin B_1) is a water-soluble, essential vitamin of the B complex.

Thiamine diphosphate—also known as thiamine pyrophosphate, thiamine diphosphate is the biologically active form of thiamine and is a coenzyme in the catabolism of sugars and amino acids.

Wernicke's encephalopathy is an acute, potentially reversible neurological disorder caused by a deficiency in, or severe depletion of, thiamine. Ocular motor abnormalities occur in ~30% of patients and may include involuntary eye movement or paralysis of muscles responsible for eye movement, while motor dysfunction can be found in ~25% of patients and may manifest as loss of equilibrium, or incoordination of gait or limb muscles. Approximately 80% of patients exhibit an altered mental state, which may comprise mental sluggishness, apathy, impaired awareness of an immediate situation, an inability to concentrate, confusion or agitation, hallucinations, behavioural disturbances mimicking an acute psychotic disorder, or coma.

List of Abbreviations

AC	activity coefficient
CSF	cerebrospinal fluid
ETK	erythrocyte transketolase
HPLC	high performance liquid chromatography
NADPH	nicotinamide adenine dinucleotide phosphate
PPP	Pentose Phosphate Pathway
T^+	free thiamine
TDP	thiamine diphosphate
TMP	thiamine monophosphate
TTP	thiamine triphosphate
WKS	Wernicke–Korsakoff syndrome

References

Anderson, S.H., Vickery, C.A., and Nicol, A.D., 1986. Adult thiamine requirements and the continuing need to fortify processed cereals. *Lancet*. 2: 85–89.

Andrade Juguan, J., Lukito, W., and Schultink, W., 1999. Thiamine deficiency is prevalent in a selected group of urban Indonesian elderly people. *Journal of Nutrition*. 129: 366–371.

Bailey, A.L., Finglas, P.M., Wright, A.J., and Southon, S., 1994. Thiamin intake, erythrocyte transketolase (EC 2.2.1.1) activity and total erythrocyte thiamin in adolescents. *British Journal of Nutrition*. 72: 111–125.

Baines, M., 1985. Improved high performance liquid chromatographic determination of thiamin diphosphate in erythrocytes. *Clinica Chimica Acta*. 153: 43–48.

Baines, M., Bligh, J.G., and Madden, J.S., 1988. Tissue thiamin levels of hospitalised alcoholics before and after oral or parenteral vitamins. *Alcohol and Alcoholism*. 23: 49–52.

Barbato, M., and Rodriguez, P.J., 1994. Thiamine deficiency in patients admitted to a palliative care unit. *Palliative Medicine*. 8: 320–324.

Bessey, O.A., Lowry, O.H., and Davis, E.B., 1952. The measurement of thiamine in urine. *The Journal of Biological Chemistry*. 195: 453–458.

Bettendorff, L., Mastrogiacomo, F., Kish, S.J., and Grisar, T., 1996a. Thiamine, thiamine phosphates, and their metabolizing enzymes in human brain. *Journal of Neurochemistry.* 66: 250–258.

Bettendorff, L., Mastrogiacomo, F., LaMarche, J., Dozic, S., and Kish, S.J., 1996b. Brain levels of thiamine and its phosphate esters in Friedreich's ataxia and spinocerebellar ataxia type 1. *Movement Disorders.* 11: 437–439.

Bettendorff, L., Mastrogiacomo, F., Wins, P., Kish, S.J., Grisar, T., and Ball, M.J., 1997. Low thiamine diphosphate levels in brains of patients with frontal lobe degeneration of the non-Alzheimer's type. *Journal of Neurochemistry.* 69: 2005–2010.

Blass, J.P., Gleason, P., Brush, D., DiPonte, P., and Thaler, H., 1988. Thiamine and Alzheimer's disease. A pilot study. *Archives of Neurology.* 45: 833–835.

Boni, L., Kieckens, L., and Hendrikx, A., 1980. An evaluation of a modified erythrocyte transketolase assay for assessing thiamine nutritional adequacy. *Journal of Nutrition Science and Vitaminology.* 26: 507–514.

Bradley, J.A., King, R.F., Schorah, C.J., and Hill, G.L., 1978. Vitamins in intravenous feeding: a study of water-soluble vitamins and folate in critically ill patients receiving intravenous nutrition. *The British Journal of Surgery.* 65: 492–494.

Brin, M., 1962. Erythrocyte transketolase in early thiamine deficiency. *Annals of the New York Academy of Sciences.* 98: 528–541.

Brin, M., 1963. Thiamine deficiency and erythrocyte metabolism. *The American Journal of Clinical Nutrition.* 12: 107–116.

Butterworth, R.F., and Besnard, A.M., 1990. Thiamine-dependent enzyme changes in temporal cortex of patients with Alzheimer's disease. *Metabolic Brain Disorders.* 5: 179–184.

Chong, Y.H., and Ho, G.S., 1970. Erythrocyte transketolase activity. *The American Journal of Clinical Nutrition.* 23: 261–266.

Clarke, R.P., Cosgrove, L.D., and Morse, E.H., 1966. Vitamin to creatinine ratios. Variability in separate voidings of urine of adolescents during a 24 hour period. *The American Journal of Clinical Nutrition.* 19: 335–341.

Cromer, B.A., Wyatt, D.T., Brandstaetter, L.A., Spadone, S., and Sloan, H.R., 1989. Thiamine status in urban adolescents: effects of race. *Journal of Pediatric Gastroenterology and Nutrition.* 9: 502–506.

Donnino, M.W., Carney, E., Cocchi, M.N., Barbash, I., Chase, M., Joyce, N., Chou, P.P., and Ngo, L., 2010. Thiamine deficiency in critically ill patients with sepsis. *Journal of Critical Care.* 25: 576–581.

Fattal-Valevski, A., Kesler, A., Sela, B.A., Nitzan-Kaluski, D., Rotstein, M., Mesterman, R., Toledano-Alhadef, H., Stolovitch, C., Hoffmann, C., Globus, O., and Eshel, G., 2005. Outbreak of life-threatening thiamine deficiency in infants in Israel caused by a defective soy-based formula. *Pediatrics.* 115: e233–238.

Fennelly, J., Frank, O., Baker, H., and Leevy, C.M., 1967. Red blood cell-transketolase activity in malnourished alcoholics with cirrhosis. *The American Journal of Clinical Nutrition.* 20: 946–949.

Friedrich, W., 1988. In: *Vitamins*. Walter de Gruyter, Berlin, Germany, pp. 341–394.

Galvin, R., Brathen, G., Ivashynka, A., Hillbom, M., Tanasescu, R., and Leone, M.A., 2010. EFNS guidelines for diagnosis, therapy and prevention of Wernicke encephalopathy. *European Journal of Neurology*. 17: 1408–1418.

Gangolf, M., Czerniecki, J., Radermecker, M., Detry, O., Nisolle, M., Jouan, C., Martin, D., Chantraine, F., Lakaye, B., Wins, P., Grisar, T., and Bettendorff, L., 2010. Thiamine status in humans and content of phosphorylated thiamine derivatives in biopsies and cultured cells. *PLoS One*. 5: e13616.

Gold, M., Chen, M.F., and Johnson, K., 1995. Plasma and red blood cell thiamine deficiency in patients with dementia of the Alzheimer's type. *Archives of Neurology*. 52: 1081–1086.

Gold, M., Hauser, R.A., and Chen, M.F., 1998. Plasma thiamine deficiency associated with Alzheimer's disease but not Parkinson's disease. *Metabolic Brain Disease*. 13: 43–53.

Graziano, A., Leggio, M.G., Mandolesi, L., Neri, P., Molinari, M., and Petrosini, L., 2002. Learning power of single behavioral units in acquisition of a complex spatial behavior: an observational learning study in cerebellar-lesioned rats. *Behavioral Neuroscience*. 116: 116–125.

Hanninen, S.A., Darling, P.B., Sole, M.J., Barr, A., and Keith, M.E., 2006. The prevalence of thiamin deficiency in hospitalized patients with congestive heart failure. *The Journal of the American College of Cardiology*. 47: 354–361.

Harper, C., 2006. Thiamine (vitamin B1) deficiency and associated brain damage is still common throughout the world and prevention is simple and safe! *European Journal of Neurology*. 13: 1078–1082.

Harper, C.G., Giles, M., and Finlay-Jones, R., 1986. Clinical signs in the Wernicke–Korsakoff complex: a retrospective analysis of 131 cases diagnosed at necropsy. *Journal of Neurology Neurosurgery and Psychiatry*. 49: 341–345.

Herbeth, B., Zittoun, J., Miravet, L., Bourgeay-Causse, M., Carre-Guery, G., Delacoux, E., Le Devehat, C., Lemoine, A., Mareschi, J.P., Martin, J., Potier de Courcy, G., and Sancho, J., 1986. Reference intervals for vitamins B1, B2, E, D, retinol, beta-carotene, and folate in blood: usefulness of dietary selection criteria. *Clinical Chemistry*. 32: 1756–1759.

Héroux, M., Raghavendra Rao, V.L., Lavoie, J., Richardson, J.S., and Butterworth, R.F., 1996. Alterations of thiamine phosphorylation and of thiamine-dependent enzymes in Alzheimer's disease. *Metabolic Brain Disorders*. 11: 81–88.

Herve, C., Beyne, P., and Delacoux, E., 1994. Determination of thiamine and its phosphate esters in human erythrocytes by high-performance liquid chromatography with isocratic elution. *Journal of Chromatography B: Biomedical Sciences and Applications*. 653: 217–220.

Hoorn, R.K., Flikweert, J.P., and Westerink, D., 1975. Vitamin B-1, B-2 and B-6 deficiencies in geriatric patients, measured by coenzyme stimulation of enzyme activities. *Clinica Chimica Acta*. 61: 151–162.

Horwitt, M.K., Liebert, E., Kreisler, O., and Wittman, P., 1946. Studies of vitamin deficiency. *Science*. 104: 407–408.

Ihara, H., Matsumoto, T., Kakinoki, T., Shino, Y., Hashimoto, R., and Hashizume, N., 2008. Estimation of vitamin B_1 excretion in 24-hr urine by assay of first-morning urine. *Journal of Clinical Laboratory Analysis*. 22: 291–294.

Jiang, Y.Y., 2006. Effect of B vitamins-fortified foods on primary school children in Beijing. *Asia Pacific Journal of Public Health*. 18: 21–25.

Jiménez-Jiménez, F.J., Molina, J.A., Hernánz, A., Fernandez-Vivancos, E., de Bustos, F., Barcenilla, B., Gómez-Escalonilla, C., Zurdo, M., Berbel, A., and Villanueva, C., 1999. Cerebrospinal fluid levels of thiamine in patients with Parkinson's disease. *Neuroscience Letters*. 271: 33–36.

Kawai, C., Wakabayashi, A., Matsumura, T., and Yui, Y., 1980. Reappearance of beriberi heart disease in Japan. A study of 23 cases. *American Journal of Medicine*. 69: 383–386.

Kimura, M., Fujita, T., and Itokawa, Y., 1982. Liquid-chromatographic determination of the total thiamin content of blood. *Clinical Chemistry*. 28: 29–31.

Korner, R.W., Vierzig, A., Roth, B., and Muller, C., 2009. Determination of thiamin diphosphate in whole blood samples by high-performance liquid chromatography—a method suitable for pediatric diagnostics. *Journal of Chromatography B: Analytical Technologies in the Biomedical and Life Sciences*. 877: 1882–1886.

Kuriyama, M., Yokomine, R., Arima, H., Hamada, R., and Igata, A., 1980. Blood vitamin B1, transketolase and thiamine pyrophosphate (TPP) effect in beriberi patients, with studies employing discriminant analysis. *Clinica Chimica Acta*. 108: 159–168.

Lee, D.C., Chu, J., Satz, W., and Silbergleit, R., 2000. Low plasma thiamine levels in elder patients admitted through the emergency department. *Academy of Emergency Medicine*. 7: 1156–1159.

Levy, S., Herve, C., Delacoux, E., and Erlinger, S., 2002. Thiamine deficiency in hepatitis C virus and alcohol-related liver diseases. *Digestive Diseases and Sciences*. 47: 543–548.

Li, S.F., Jacob, J., Feng, J., and Kulkarni, M., 2008. Vitamin deficiencies in acutely intoxicated patients in the ED. *The American Journal of Emergency Medicine*. 26: 792–795.

Lima, L.F., Leite, H.P., and Taddei, J.A., 2011. Low blood thiamine concentrations in children upon admission to the intensive care unit: risk factors and prognostic significance. *The American Journal of Clinical Nutrition*. 93: 57–61.

Lonsdale, D., 2006. A review of the biochemistry, metabolism and clinical benefits of thiamin(e) and its derivatives. *Evidence Based Complementary and Alternative Medicine*. 3: 49–59.

Lu, J., and Frank, E.L., 2008. Rapid HPLC measurement of thiamine and its phosphate esters in whole blood. *Clinical Chemistry*. 54: 901–906.

Lynch, P.L., Trimble, E.R., and Young, I.S., 1997. High-performance liquid chromatographic determination of thiamine diphosphate in erythrocytes using internal standard methodology. *Journal of Chromatography B: Analytical Technologies in the Biomedical and Life Sciences*. 701: 120–123.

Lynch, P.L.M., and Young, I.S., 2000. Determination of thiamine by high-performance liquid chromatography. *Journal of Chromatography A.* 881: 267–284.

Mancinelli, R., Ceccanti, M., Guiducci, M.S., Sasso, G.F., Sebastiani, G., Attilia, M.L., and Allen, J.P., 2003. Simultaneous liquid chromatographic assessment of thiamine, thiamine monophosphate and thiamine diphosphate in human erythrocytes: a study on alcoholics. *Journal of Chromatography B: Analytical Technologies in the Biomedical and Life Sciences.* 789: 355–363.

Markkanen, T., Heikinheimo, R., and Dahl, M., 1969. Transketolase activity of red blood cells from infancy to old age. *Acta Haematologica.* 42: 148–153.

Martin, P.R., Singleton, C.K., and Hiller-Sturmhöfel, S., 2003. The role of thiamine deficiency in alcoholic brain disease. *Alcohol Research & Health.* 27: 134–142.

Maschke, M., Weber, J., Bonnet, U., Dimitrova, A., Bohrenkamper, J., Sturm, S., Muller, B.W., Gastpar, M., Diener, H.C., Forsting, M., and Timmann, D., 2005. Vermal atrophy of alcoholics correlate with serum thiamine levels but not with dentate iron concentrations as estimated by MRI. *Journal of Neurology.* 252: 704–711.

Mastrogiacoma, F., Bettendorff, L., Grisar, T., and Kish, S.J., 1996. Brain thiamine, its phosphate esters, and its metabolizing enzymes in Alzheimer's disease. *Annals of Neurology.* 39: 585–591.

McLaren, D.S., Docherty, M.A., and Boyd, D.H., 1981. Plasma thiamin pyrophosphate and erythrocyte transketolase in chronic alcoholism. *The American Journal of Clinincal Nutrition.* 34: 1031–1033.

Meador, K., Loring, D., Nichols, M., Zamrini, E., Rivner, M., Posas, H., Thompson, E., and Moore, E., 1993. Preliminary findings of high-dose thiamine in dementia of Alzheimer's type. *Journal of Geriatric Psychiatry and Neurology.* 6: 222–229.

Melnick, D., 1942. Vitamin B_1 (thiamine) requirement of man. *Journal of Nutrition.* 24: 139.

Molina, J.A., Jiménez-Jiménez, F.J., Hernánz, A., Fernández-Vivancos, E., Medina, S., de Bustos, F., Gómez-Escalonilla, C., and Sayed, Y., 2002. Cerebrospinal fluid levels of thiamine in patients with Alzheimer's disease. *Journal of Neurology Transmisson.* 109(7–8):1035–44.

Nichols, H.K., and Basu, T.K., 1994. Thiamin status of the elderly: dietary intake and thiamin pyrophosphate response. *Journal of the American College of Nutrition.* 13: 57–61.

O'Keeffe, S.T., Tormey, W.P., Glasgow, R., and Lavan, J.N., 1994. Thiamine deficiency in hospitalized elderly patients. *Gerontology.* 40: 18–24.

O'Rourke, N.P., Bunker, V.W., Thomas, A.J., Finglas, P.M., Bailey, A.L., and Clayton, B.E., 1990. Thiamine status of healthy and institutionalized elderly subjects: analysis of dietary intake and biochemical indices. *Age and Ageing.* 19: 325–329.

Oldham, H.G., Davis, M.V., and Roberts, L.J., 1946. Thiamine excretions and blood levels of young women on diets containing varying levels of the B vitamins, with some observations on niacin and pantothenic acid. *Journal of Nutrition.* 32: 163–180.

Ortega, R.M., Lopez-Sobaler, A.M., Andres, P., Rodriguez-Rodriguez, E., Aparicio, A., and Bermejo, L.M., 2009. Increasing consumption of breakfast cereal improves thiamine status in overweight/obese women following a hypocaloric diet. *International Journal of Food Science Nutrition.* 60: 69–79.

Pepersack, T., Garbusinski, J., Robberecht, J., Beyer, I., Willems, D., and Fuss, M., 1999. Clinical relevance of thiamine status amongst hospitalized elderly patients. *Gerontology.* 45: 96–101.

Pitel, A.L., Zahr, N.M., Jackson, K., Sassoon, S.A., Rosenbloom, M.J., Pfefferbaum, A., and Sullivan, E.V., 2010. Signs of preclinical Wernicke's encephalopathy and thiamine levels as predictors of neuropsychological deficits in alcoholism without Korsakoff's syndrome. *Neuropsychopharmacology.* 36: 580–588.

Pongpanich, B., Srikrikkrich, N., Dhanamitta, S., and Valyasevi, A., 1974. Biochemical detection of thiamin deficiency in infants and children in Thailand. *The American Journal of Clinincal Nutrition.* 27: 1399–1402.

Poupon, R.E., Gervaise, G., Riant, P., Houin, G., and Tillement, J.P., 1990. Blood thiamine and thiamine phosphate concentrations in excessive drinkers with or without peripheral neuropathy. *Alcohol and Alcoholism.* 25: 605–611.

Rieck, J., Halkin, H., Almog, S., Seligman, H., Lubetsky, A., Olchovsky, D., and Ezra, D., 1999. Urinary loss of thiamine is increased by low doses of furosemide in healthy volunteers. *The Journal of Laboratory and Clinical Medicine.* 134: 238–243.

Roser, R.L., Andrist, A.H., Harrington, W.H., Naito, H.K., and Lonsdale, D., 1978. Determination of urinary thiamine by high-pressure liquid chromatography utilizing the thiochrome fluorscent method. *Journal of Chromatography.* 146: 43–53.

Saito, N., Kimura, M., Kuchiba, A., and Itokawa, Y., 1987. Blood thiamine levels in outpatients with diabetes mellitus. *Journal of Nutritional Science Vitaminology.* 33: 421–430.

Sauberlich, H.E., 1967. Biochemical alterations in thiamine deficiency—their interpretation. *The American Journal of Clinical Nutrition.* 20: 528–546.

Sauberlich, H.E., Herman, Y.F., Stevens, C.O., and Herman, R.H., 1979. Thiamin requirement of the adult human. *The American Journal of Clinincal Nutrition.* 32: 2237–2248.

Schrijver, J., Speek, A.J., Klosse, J.A., van Rijn, H.J., and Schreurs, W.H., 1982. A reliable semiautomated method for the determination of total thiamine in whole blood by the thiochrome method with high-performance liquid chromatography. *Annals of Clinical Biochemistry.* 19: 52–56.

Scileppi, K.P., Blass, J.P., and Baker, H.G., 1984. Circulating vitamins in Alzheimer's dementia as compared with other dementias. *Journal of the American Geriatriatric Society.* 32: 709–711.

Sechi, G., and Serra, A., 2007. Wernicke's encephalopathy: new clinical settings and recent advances in diagnosis and management. *Lancet Neurology.* 6: 442–455.

Seear, M., Lockitch, G., Jacobson, B., Quigley, G., and MacNab, A., 1992. Thiamine, riboflavin, and pyridoxine deficiencies in a population of critically ill children. *Journal of Pediatrics.* 121: 533–538.

Shimazono, N., and Katasura, E., 1965. Review of Japanese Literature on Beriberi and Thiamine. Vitamin B Research Committee, Kyoto, Japan, 315pp.

Singleton, C.K., and Martin, P.R., 2001. Molecular mechanisms of thiamine utilization. *Current Molecular Medicine.* 1: 197–207.

Suter, P.M., Haller, J., Hany, A., and Vetter, W., 2000. Diuretic use: a risk for subclinical thiamine deficiency in elderly patients. *The Journal of Nutrition, Health & Aging.* 4: 69–71.

Tallaksen, C.M., Bohmer, T., Bell, H., and Karlsen, J., 1991. Concomitant determination of thiamin and its phosphate esters in human blood and serum by high-performance liquid chromatography. *Journal of Chromatography.* 564: 127–136.

Tallaksen, C.M., Bell, H., and Bohmer, T., 1992a. The concentration of thiamin and thiamin phosphate esters in patients with alcoholic liver cirrhosis. *Alcohol and Alcoholism.* 27: 523–530.

Tallaksen, C.M., Bohmer, T., and Bell, H., 1992b. Blood and serum thiamin and thiamin phosphate esters concentrations in patients with alcohol dependence syndrome before and after thiamin treatment. *Alcoholism: Clinical and Experimental Research.* 16: 320–325.

Tallaksen, C.M., Bøhmer, T., Karlsen, J., and Bell, H., 1997. Determination of thiamin and its phosphate esters in human blood, plasma, and urine. *Methods Enzymology.* 279: 67–74.

Tallaksen, C.M., Sande, A., Bøhmer, T., Bell, H., and Karlsen, J., 1993. Kinetics of thiamin and thiamin phosphate esters in human blood, plasma and urine after 50 mg intravenously or orally. *European Journal of Clinical Pharmacology.* 44: 73–78.

Talwar, D., Davidson, H., Cooney, J., and O'Reilly, D.S.J., 2000. Vitamin B(1) status assessed by direct measurement of thiamin pyrophosphate in erythrocytes or whole blood by HPLC: comparison with erythrocyte transketolase activation assay. *Clinical Chemistry.* 46: 704–710.

Tsuji, T., Fukuwatari, T., Sasaki, S., and Shibata, K., 2010. Urinary excretion of vitamin B1, B2, B6, niacin, pantothenic acid, folate, and vitamin C correlates with dietary intakes of free-living elderly, female Japanese. *Nutrition Research.* 30: 171–178.

van Dokkum, W., Schrijver, J., and Wesstra, J.A., 1990. Variability in man of the levels of some indices of nutritional status over a 60-d period on a constant diet. *European Journal of Clinical Nutrition.* 44: 665–674.

Victor, M., Adams, R.D., and Collins, G.H., 1989. *The Wernicke-Korsakoff Syndrome and Related Neurologic Disorders Due to Alcoholism and Malnutrition*, 2nd ed. F.A. Davis Co, Philadelphia, USA.

Vir, S.C., and Love, A.H., 1979. Nutritional status of institutionalized and noninstitutionalized aged in Belfast, Northern Ireland. *The American Journal of Clinical Nutrition.* 32: 1934–1947.

Vognar, L., and Stoukides, J., 2009. The role of low plasma thiamin levels in cognitively impaired elderly patients presenting with acute behavioral disturbances. *Journal of the American Geriatrics Society*. 57: 2166–2168.

Warnock, L.G., Prudhomme, C.R., and Wagner, C., 1978. The determination of thiamin pyrophosphate in blood and other tissues, and its correlation with erythrocyte transketolase activity. *The Journal of Nutrition*. 108: 421–427.

Warnock, L.G., 1982. The measurement of erythrocyte thiamin pyrophosphate by high-performance liquid chromatography. *Analytical Biochemistry*. 126: 394–397.

Wetherilt, H., Ackurt, F., Brubacher, G., Okan, B., Aktas, S., and Turdu, S., 1992. Blood vitamin and mineral levels in 7-17 years old Turkish children. *International Journal of Vitamin and Nutritional Research*. 62: 21–29.

Wilkinson, T.J., Hanger, H.C., Elmslie, J., George, P.M., and Sainsbury, R., 1997. The response to treatment of subclinical thiamine deficiency in the elderly. *The American Journal of Cinical Nutrition*. 66: 925–928.

Wyatt, D.T., Lee, M., and Hillman, R.E., 1989. Factors affecting a cyanogen bromide-based assay of thiamin. *Clinical Chemistry*. 35: 2173–2178.

Wyatt, D.T., Nelson, D., and Hillman, R.E., 1991. Age-dependent changes in thiamin concentrations in whole blood and cerebrospinal fluid in infants and children. *American Journal of Clinical Nutrition*. 53: 530–536.

Zahr, N.M., Kaufman, K.L., and Harper, C.G., 2011. Clinical and pathological features of alcohol-related brain damage. *Nature Review Neurology*. 7: 284–294.

CHAPTER 17
The Assay of Thiamine in Food

HENRYK ZIELIŃSKI*[a] AND JUANA FRIAS[b]

[a] Division of Food Sciences, Institute of Animal Reproduction and Food Research of the Polish Academy of Sciences, Olsztyn, Poland; [b] Instituto de Ciencia y Tecnología de Alimentos y Nutrición (ICTAN-CSIC), Madrid, Spain
*Email: haziel@pan.olsztyn.pl

17.1 Food as a Source of Thiamine

Thiamine is a relatively simple compound consisting of a pyrimidine and a thiazole ring. It exists naturally in most types of foods as free thiamine and phosphorylated forms including thiamine monophosphate (TMP), thiamine diphosphate or pyrophosphate (TPP), and thiamine triphosphate (TTP) (Tanphaichitr 2001) (Figure 17.1). Although all forms exist in animal and plant foods, thiamine as the free (non-phosphorylated) form is mainly found in plant-based foods whereas, in animal products, 80% of thiamine is represented by TPP and lesser amounts by TMP and triphosphate TTP.

The primary dietary sources of thiamine to man are unrefined cereal grains or starchy roots, enriched cereal grain products, breakfast cereals, legumes, nuts and pork. The whole-grain cereal products contain nutritionally significant amounts of thiamine (Bui and Small 2007). Peas and other legumes are also good sources of thiamine; its content in those crops increases with the maturity of the seed. Dried brewer's yeast and wheat germs are both rich in thiamine. However, much of the dietary thiamine in developed countries appears in fortified cereals and breads, as well as in any functional foods. Thiamine fortification compensates for the losses occurring during food processing and tries to avoid thiamine deficiencies. Synthetic thiamine

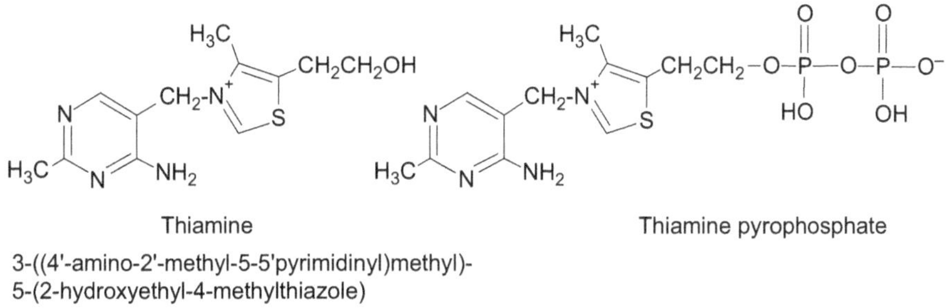

Figure 17.1 Structure of thiamine and thiamine pyrophosphate. Free thiamine is depicted as the free base form (Gregory 1997).

hydrochloride or mononitrate salts are the forms used in most nutritional supplements and in food fortification. Several synthetic disulfides are also commercially available in certain countries but are not approved for worldwide distribution (Gregory 1997). In general, the fortification of food products with vitamins is achieved with premixes that contain a high concentration of vitamins. Meat and meat products are also a good source of thiamine. Pork is the richest source, which is higher in thiamine than most of the other muscle foods. Fortified breakfast cereals contain twice as much thiamine as pork products (ca. 2.5 mg/100 g vs. 0.9–1.2 mg/100 g). The latter are considered the most concentrated non-enriched or non-fortified food source for thiamine. The highly refined cereal products, polished rice, fats, oils, refined sugar and non-enriched flours contain virtually no thiamine.

Thiamine is weakly stable when the pH value of the matrix approaches neutrality. The maximum stability in a solution is between pH 2.0 and pH 4.0. Thiamine is thermolabile and its destruction during thermal processing may be extensive, especially in low-acid foods. Thiamine is stable in slightly acidic water up to the boiling point, but can be leached out of food by boiling. Thiamine is destroyed by ultraviolet (UV) radiation and by thiaminases (thiamin-degrading enzymes) present in uncooked freshwater fish, shellfish and some bacteria. The loss of thiamine can be enhanced by using rapidly boiling water in cooking vegetables, as the amount of oxygen increases in contact with the food product. Thiamine stability in food would be greater if low-moisture products are formulated.

17.2 The Recommended Dietary Allowance for Thiamine

A varied diet should provide an adequate level of thiamine to prevent deficiency. The thiamine requirement of an individual is influenced by age, energy and carbohydrate intake, and body mass. On the basis of considerable evidence, the Food and Nutrition Board of the US Institute of Medicine recommends thiamine intake at a level of 0.5 mg/1000 kcal (4184 KJ) (National

Research Council 1989). Because there are some data indicating lower availability of thiamine in older people, it is recommended that they maintain an intake of 1 mg/day even if they consume less than 2000 kcal (8368 KJ daily). Since thiamine requirements increase during pregnancy and lactation, an additional intake of 0.4 mg/day is recommended during pregnancy and 0.5 mg/day during lactation.

Nutritional status assessment for thiamine is generally carried out by assaying the total thiamine in whole blood or erythrocytes, or by measuring the activity of erythrocyte transketolase before and after incubation with exogenous thiamine pyrophosphate. The latter serves as the sensitive index of thiamine nutritional status (Brin 1980). In addition to the enzymatic test, a measure of urinary thiamine in relation to dietary intake has been the basis for balance studies to assess the adequacy of intake. When thiamine excretion is low, a larger portion of the test dose is retained, indicating a tissue's need for thiamine. A high excretion indicates tissue saturation. In the deficient state, excretion drops to zero. Plasma pyruvate and lactate concentrations have also been used to assess thiamine status.

17.3 Overview of Analytical Methods

Established methods for the determination of B vitamins require the analysis of each vitamin individually by a wide number of different methods including colorimetric, fluorimetric, spectrophotometric and titrimetric techniques. These methods tend to be time-consuming and are based on older technology. The most sensitive are the high-performance liquid chromatography (HPLC) assays that are yielding excellent results. They have the advantages of sensitivity and reliability.

In the last decade increasing attention has been paid to the simultaneous determination of vitamins due to the interest in dietary supplements and fortified foods. In the case of such foods, the analytical procedures of vitamins determination should consider the fact that, apart from the added quantity of various vitamins, a food product may also contain its own so-called endogenous vitamins (Heudi et al. 2005). This means that the total content of vitamins in a food product represents the content of the endogenous free and bound forms together with the added quantity. According to the Regulation (EC) no 1925/2006 (European Commission 2006), this total sum of vitamins should be the one given on the label of food products. This Regulation stipulates that the presence of too low and insignificant amounts of added vitamins in fortified foods would not offer any benefit to consumers and would be misleading. Thus, enriched food products usually contain significant amounts of added vitamins. In addition, the Regulation specifies that in order to avoid any consumer confusion with the natural nutritional value of fresh foods, the addition of vitamins should not be allowed. It means that vitamin fortification can only be performed in those groups of foodstuffs that originally do not contain significant amounts of endogenous vitamins and thus the added significant quantities predominate in total thiamine content.

17.4 Analysis of Thiamine in Food

There are various microbiological, chemical and animal assays available for thiamine analysis. Method selection depends on the accuracy and sensitivity required, and on the interferences due to the sample matrix. Although microbiological assays do exist for thiamine, these are rarely used. The highly fluorescent products are the basis of the thiochrome procedure for thiamine quantification. Thiochrome analysis has been used to develop most of the analytical data available on the thiamine content of the food supply.

Since thiamine does not fluorescence, the development of the thiochrome analysis or closely related fluorescent compounds has provided a highly specific method for thiamine quantification from most matrices. The development of HPLC methods, often coupled with either pre- or post-column oxidation to thiochrome, permits highly sensitive measurement of total thiamine or its individual free and phosphorylated forms (Gregory 1997). Thiochrome-formed pre-column refers to procedures that form the thiochrome directly in the sample extract. Post-column procedures are based on the conversion of thiamine compounds to their respective thiochromes after chromatographic resolution of the thiamine forms. The oxidizing agent is usually 0.01% potassium hexacyanoferrate(III) [$K_3Fe(CN)_6$] solution in 15% sodium hydroxide.

17.4.1 Extraction Procedures

Extraction procedures depend on the investigator's need to quantitate total thiamine, or thiamine and the phosphate esters. To analyse total thiamine, the extraction steps usually include an acid disintegration followed by an enzymatic digestion. These steps are required in order to release the protein-bound and phosphorylated form of the vitamin. A solid-phase extraction (SPE) technique using disposable C18 columns is more often used for this end (Riccio *et al.* 2006), although an ion-exchange resin may be applied to remove some interferences. Quantification of the free thiamine and the individual phosphate esters requires conditions that do not hydrolyse the ester bond. Sample extraction is limited to acid hydrolysis to free matrix-bound thiamine. There is also typically no need for digestion methods to obtain thiamine from fortified food products, as the free vitamin can be successfully extracted from the bound form. In this case, determination of an endogenous vitamin content is not a strict requirement and the enzymatic sample preparation is not necessary (Engel *et al.* 2010). Thus, methods such as capillary electrophoresis (Fotsing *et al.* 1997, 2000), micellar electrokinetic chromatography (Dinelli and Bonetti 1994; Fujiwara *et al.* 1988), micellar liquid chromatography (Ghorbani *et al.* 2004; Monferrer-Pons *et al.* 2003) and liquid chromatography have been developed. Among these methods, liquid chromatography appears promising due to the improvements in both stationary phases and chromatography equipment but it is not validated as the method of the Association of Official Analytical Chemists (AOAC).

17.4.2 Microbiological Methods for Thiamine Analysis

At present, microbiological assays are the most frequently used for low-level, naturally occurring vitamins in food. The content of specific vitamins may be determined by monitoring the growth of certain microorganisms. Extraction procedures for the microbiological analysis of thiamine generally follow enzyme hydrolysis of the phosphate esters. This step is required to avoid differential growth response to TMP, TPP and TTP.

The species most commonly employed for thiamine determination are some Lactobacilli such as *L. fermentum* and *L. viridescens* (ATCC 12706). The latter is the microorganism most widely used to measure thiamine concentrations. It requires the intact thiamine molecule for growth. Other organisms such as *Phycomyces blakeslekanus, Kloeckera brevis* (ATCC 9774), *Ochromonas danica* and *Neurospora crassa* are available, but they are less useful.

In traditional microbiology, colonies of the target microorganisms should first be cultured and later maintained by regular inoculation. Before the actual assay procedure can begin, the cultures ought to be freshly prepared and the number of microorganisms should be regulated before the organisms are transferred to the medium. This requires a great deal of time and labour. There are commercially available microbiological kits for determining thiamine concentration in food such as Difco™ Thiamine Assay Medium and Difco™ Thiamine Assay medium LV. Thiamine Assay Medium is applied for determining thiamine concentration by the microbiological assay technique using *Lactobacillus fermentum* ATCC™ 9338 whereas Thiamine Assay Medium LV technique uses *Weissella* (*Lactobacillus*) *viridescens* ATCC™ 12706.

According to Blake (2007), the principles of this technique include:

(1) Extraction of thiamine from the food matrix either by autoclaving in the presence of acids, or by digesting with suitable enzymes such as takadiastase or protease
(2) Incubation of the food extract at several dilution levels with the growth medium and microbiological culture
(3) Read out of turbidity, which is related to the growth of the microorganism, with a spectrophotometer at 540–660 nm
(4) Calibration: semi-logarithmic plot of the absorbance against increasing concentrations of vitamin.

The assays have been a powerful tool for thiamine analysis in various foods. However, the analytical process may take up to 72 h and is often plagued by poor reproducibility. It allows for the detection of thiamine amounts between 5 and 50 ng. The aseptic technique should be used throughout the microbiological assay procedure under the same conditions for successful results.

The Assay of Thiamine in Food

However, these assays are no longer considered the gold standard in vitamin analysis (Blake 2007). In general the precision and accuracy of the microbiological assays is not high and a relative measurement uncertainty of about 20% is fairly common. A recent development is the use of precoated plates, which enables numerous problems to be avoided while maintaining stock cultures. The VitaFast® system, based on AOAC, EN and DIN reference methods, determines vitamin content microbiologically by utilizing microtitre plates coated with specific microorganisms. The test kit is marketed by R-Biopharm AG and it is ideal for routine analysis since the reagents are ready-to-use and the kit is user-friendly. Unlike other immunological assay systems, no washing step is required. The VitaFast® kit for thiamine was validated and has been used for thiamine analysis in some fruit juices (Niemeijer *et al.* 2009). Compared with traditional microbiological vitamin assays, the time required for analysis using the VitaFast® test kit is approximately around 48 h and the determination costs are reduced by approximately 70% (Niemeijer *et al.* 2009). The VitaFast® system has the potential to offer significant improvements for microbiological assays. It can be successfully used to determine thiamine as well as vitamin B_{12}, biotin, niacin, pantothenic acid, riboflavin and pyridoxine in fortified and natural foods. Besides the microbiological methods, there are other methods recognized as official methods.

17.4.3 Chemical Methods of Thiamine Analysis

Chemical methods are now the first choice for the analysis of thiamine. In these assays special care should be taken to protect samples and standards from light and heat at alkaline pH. The proposed methods are based on fluorimetry (FL) and HPLC (Blake 2007). Table 17.1 lists the main official methods for the chemical analysis of thiamine.

Table 17.1 Summary of official methods for the chemical analysis of thiamine.

Official method	Title	Field of application	Reference
AOAC 942.23	Thiamine (vitamin B_1) in foods–fluorometric method	Human and pet foods	AOAC 1990b
AOAC 953.17	Thiamine (vitamin B_1) in grain products—fluorometric (rapid) method	Grain products	AOAC 1990a
AOAC 957.17	Thiamine (vitamin B_1) in bread—fluorometric (rapid) method	Breads	AOAC 1990c
AOAC 986.27	Thiamine in milk-based infant formula—fluorometric method	Infant formula	AOAC 1990d
EN 14122	Foodstuffs—determination of vitamin B_1 by HPLC	All foods, supplements	CEN 2003

17.4.3.1 Fluorimetric Method

The analysis of thiamine levels in food has traditionally been based upon the oxidation of free thiamine to give thiochrome (Gregory 1996). Variations on the basic procedure have been developed and have generally involved the measurement of thiochrome with a spectrofluorimeter. The thiochrome method depends upon the alkaline oxidation of thiamine to thiochrome (Figure 17.2).

The initial steps include extraction and enzymatic hydrolysis of the phosphate esters of thiamine. The vitamin is then cleaned up by ion-exchange chromatography and finally oxidized to thiochrome. The thiochrome is finally extracted into an organic solvent such as isobutyl alcohol. Thiochrome exhibits an intense blue fluorescence which can be measured fluorimetrically. It is usually analysed by employing 360–365 nm as the excitation wavelength and 460–480 nm as the emission wavelength. Under standard conditions and in the absence of other fluorescing substances, the fluorescence is proportional to the thiochrome present and hence to the thiamine content (Gubler 1991). The procedures have been validated and accepted internationally as standard methods by the Association of Official Analytical Chemists (AOAC method 953.17; AOAC 1990a), the American Association of Cereal Chemists (AACC; AACC method 86-80) and the International Association of Cereal Science and Technology (ICC-Standard no. 117; ICC 1990).

The AOAC methods are accepted internationally by the analytical chemistry community. These standard methods offer the advantage of being well-established and reliable. They can be employed out rapidly and economically, and consequently are applicable to routine determination of thiamine (Ellefson 1985). Precision is known to be generally very good for these chemical procedures.

The extraction and assay of thiamine according to the fluorometric method (AOAC method 953.17; AOAC 1990a) can be summarized as follows: sample preparation, acidification (homogenized with 0.1 M HCl), autoclaving (20 min, 109 °C), centrifugation (20 min, 3500 rpm), oxidation using $K_3Fe(CN)_6$, extraction (isobutyl alcohol) and determination of thiochrome (fluorescence measurements using excitation wavelength set at 365 nm and emission wavelength at 435 nm). The measurement of the thiochrome is based on the fact that thiochrome is soluble in isobutyl alcohol (IUPAC nomenclature: 2-methyl-1-propanol)

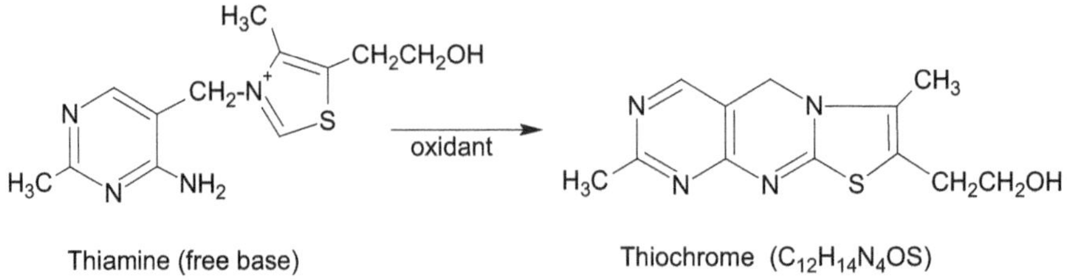

Figure 17.2 Oxidation of thiamine to thiochrome (Gregory 1996). The figure shows the structure of thiochrome formed after oxidation of thiamine. The following oxidant reagents are widely used: mainly $K_3Fe(CN)_6$ but also others, such as $KMnO_4$, MnO_2, $CNBr$, $HgCl_2$ and H_2O_2.

and the yield is very constant under standardized conditions (Ellefson 1985). Quinine sulfate solution is used to assess reproducibility.

17.4.3.2 HPLC Method with UV and Fluorescence Detection

HPLC with UV and fluorescence detection are the most common techniques used today. HPLC is the preferred technique for vitamin separation because of its high selectivity. Reverse-phase chromatography on a C18 column support with an isocratic mobile phase works efficiently for the resolution of thiamine.

In most food sample matrices, acid and/or enzymatic treatment is needed to release thiamine from phosphate and proteins. Different acids and enzymes have been proposed, including hydrochloric, trichloroacetic and sulfuric acids; single enzymes such as papain and pepsin, or enzyme mixtures such as takadiastase or claradiastase. Hydrolysis using claradiastase coupled with hydrochloric acid has proven to be the most efficient treatment.

Ultraviolet detection at 245–254 nm is not sensitive enough for naturally occurring levels of thiamine, TMP, TPP and TTP in foods, and therefore is more used for enriched foods. Although some methods enable the determination of thiamine directly by HPLC with UV detection, the low vitamin content, very low molar absorption of thiamine and the high quantity of interfering compounds in foodstuffs means that better thiamine analysis is obtained when it is preceded by oxidation to thiochrome by either pre- or post-column reaction with or without extraction into isobutyl alcohol with precise timings and when the final analysis is carried out using HPLC with FL detection.

Pre-column derivatization is normally used; although post-column derivatization should be more convenient for routine analyses, it is not often applied. In this case, the separation of thiamine requires the use of ion-pair liquid chromatography since it is not retained on a reverse-phase column (Blake 2007). This technique also increases the sensitivity and reproducibility of the thiamine assay.

The thiochrome determination involves two problems related to the instability of the thiochrome itself and to the chromatographic problems associated with the excess of oxidizing reagent. The former problem can be solved by the addition of orthophosphate acid, which minimizes the formation of thiamine disulfide and stabilizes the thiochrome. The second problem can be solved by solid-phase chromatography cleanup using C18 cartridges or by extracting the thiochrome into isobutyl alcohol as soon as it is formed. For the final analysis, reverse-phase HPLC offers the best results, although solutions have to be injected immediately after thiochrome formation. The most commonly employed chromatographic columns are C18 and C8. Mobile phases, including organic solvents, ion pair and organic–aqueous buffer mixtures, have also been used. Most of the methods are optimized for a single vitamin and may involve columns and equipment that are not common. An additional problem is a lack of suitable internal standards for thiamine and therefore external standard solutions that have undergone the same procedure as the sample are commonly used for thiamine quantification. The representative HPLC methods for thiamine analysis are listed in Table 17.2.

Table 17.2 Representative HPLC methods for thiamine analysis: a summary of progress in the analysis of thiamine over the past two decades.

Sample preparation	Stationary phase	Mobile phase	Detection[a]	Source	Reference
Acid hydrolysis with 0.1 M H_2SO_4, enzymatic hydrolysis, centrifugation, oxidation	Nucleosil NH2, 5 μm	25% K_2HPO_4 buffer pH 4.4 in acetonitrile	FL Ex λ = 370 nm Em λ = 425 nm	Dietetic food	Botticher and Botticher 1986
Acid hydrolysis with 0.1 and 6 M HCl, autoclaving, enzymatic hydrolysis, Amberlite GC-50 and Sep-pack C18 clean-up	μBondak C18 10 μm	Several mixtures of methanol, acetic acid, hexane sulfonate and heptane sulfonate	UV = 254 nm	Foodstuffs	Vidal-Valverde and Reche 1990
Acid hydrolysis with 0.2 N H_2SO_4, autoclaving, enzymatic hydrolysis, filtration, oxidation with potassium ferricyanide, extraction with isobutyl alcohol	Lichrosorb RP-8	Methanol/acetonitrile/isobutyl alcohol (80 : 10 : 10)	FL Ex λ = 370 nm Em λ = 425 nm	Meat and vegetables	Bognar 1992
Solid phase extraction with the Sep-Pak C18 cartridges, evaporation of the eluent to dryness, dissolving the residue in mobile phase, filtration using 0.45 μm filters	Discovery C18	(A) KH_2PO_4 pH 7 and (B) methanol	UV = 220 nm	Traditional Turkish cereal food	Ekinci and Kadakal 2005
Extraction at alkaline but finally with phosphate buffer at pH 5.5, sonication, final filtration through a 0.22 μm Millipore filter	YMC-Pack Pro C18 5 μm	(A) aqueous 0.025% solution of TFA pH 2.6 and (B) acetonitrile	UV = 275 nm	Polyvitaminated premixes	Heudi et al. 2005

Extraction	Column	Mobile phase	Detection	Food	Reference
Acid hydrolysis with 4% TCA, centrifugation and filtration through a 0.45 μm filter	Luna C8 5 μm	(A) octanesulfonic acid 5 mM, triethylamine 0.5%, acetic acid 2.4% and methanol 15%; (B) acetonitrile 100%	UV = 249 nm	Fortified meat products	Riccio et al. 2006
Acid hydrolysis with 0.3 M HCl, autoclaving, enzymatic hydrolysis with takadiastase, filtration, oxidation, and filtration through a 0.45 μm filter	μBondapak C18 column and a Porasil B Bondapak C18 guard column	methanol/water/acetic acid (31 : 68.5 : 0.5), containing 5 mM sodium hexasulfonate	FL Ex λ = 360 nm Em λ = 435 nm	Cruciferae sprouts	Zieliński et al. 2005
Acid hydrolysis with 0.1 N HCl, heating in water bath at 95–100 °C, enzymatic hydrolysis with takadiastase, filtration, oxidation	Novapak C18 4 μm	Aqueous 0.3% triethylamine (TEA) pH 7.4 with sulfuric acid/methanol	FL Ex λ = 365 nm Em λ = 436 nm	Cultivated mushrooms	Furlani and Godoy 2008
Single extraction by acid hydrolysis with 0.3 M HCl, autoclaving, enzymatic hydrolysis, filtration, post column derivatization	μBondapak C18 column and a Porasil B Bondapak C18 guard column	Methanol/water/acetic acid (31 : 68.5 : 0.5), containing 5 mM sodium hexasulfonate	FL Ex λ = 360 nm Em λ = 435 nm	Rye bread	Martínez-Villaluenga et al. 2009

[a] Ex λ, excitation wavelength; Em λ, emission wavelength.

17.4.3.3 Thiamine Analysis by Multivitamin Methods with Potential to Improve Performance in the Future

Although estimations of vitamin levels should be obtained more precisely by a 'one vitamin at a time' approach, the determination of not just a single vitamin in fortified foods and dietary supplements may also be of great interest. Since a range of water-soluble vitamins needs to be determined in fortified food products, the use of multivitamin methods would be advantageous particularly for compliance monitoring. In the US, the first approach is often reported as being suitable for product compliance monitoring whereas in Europe an approach is to determine total vitamin content, which involves a more complex and time-consuming extraction procedure. A simple precipitation of proteins with zinc acetate and phosphotungstate, perchloric acid or trichloric acid followed by centrifugation or filtration may be sufficient for matrices such as fortified infant formulas or beverages, but this approach is not sufficient for complex food products such as cereals or pet food (Blake 2007). Therefore for thiamine analysis performed together with B_2, B_3, nicotinamide and folic acid, a typical acid hydrolysis followed by enzymatic hydrolysis is usually required (Viñas *et al.* 2003). Several research groups have elaborated multivitamin methods for B vitamins assay in infant formulas and other fortified products using liquid chromatography with tandem mass spectrometry (LC-MS/MS) detection, micellar liquid chromatography (MLC), liquid chromatography/diode array detection–mass spectrometry (LC-DAD/MS), liquid chromatography/electrospray ionization–mass spectrometry (LC/ESI-MS) or liquid chromatography isotope dilution mass spectrometry (LC-IDMS) systems which offer better sensitivity and selectivity (Table 17.3).

The reversed-phase HPLC method with UV detection (RP-HPLC-UV) has been proven to be suitable for the simultaneous monitoring of vitamins B_1, B_2, B_3, B_6, B_9 and C without using an ion pairing reagent. This chromatographic method combined with the optimized sample preparation method, in contrast to an unspecified detection principle and insufficient sample clean-up, represents a simple and low-cost solution for screening several B group vitamins, including thiamine, at the same cost as of a single vitamin C measurement in enriched food products (Engel *et al.* 2010).

Reversed-phase HPLC with coulometric electrochemical and UV detection (RP-HPLC-ED-UV) has been successfully applied for the simultaneous determination of thiamine, vitamin B_6 (pyridoxamine, pyridoxal and pyridoxine) and vitamin B_{12} in a single run in fruit juices. The method offers linearity, good repeatability and reproducibility, and a relatively short analytical time. The determination of thiamine in fruit juices has been carried out in continuous flow in the UV region resulting in a lack of signals from interfering substances within the concentration range of 22–8200 ng/mL with a detection limit of 9.2 ng/mL (Lebiedzińska *et al.* 2007).

The LC-IDMS method is an excellent approach for establishing reference values and can be useful in investigating and validating methods. An isotopically labelled version of thiamine used in the study performed by

Table 17.3 Recent multivitamin analytical methods, including thiamine analysis: recent development of high quality instrumental methods for thiamine analysis when a food sample is subjected to the multivitamin assay. The application of the methods to pharmaceutical materials is indicated.

Method	Material	Vitamin B analysed	Reference
RP-LC-UV	Various enriched food products	Thiamine (B_1), riboflavin (B_2), niacinamide (B_3), pyridoxine (B_6),	Engel et al. 2010
LC-IDMS	SRM 1849 Infant/adult nutritional formula vitamin-fortified milk based powder	Thiamine (B_1), niacinamide (B_3), pyridoxine (B_6), pantothenic acid (B_5), folic acid (B_9), riboflavin (B_2), biotin (B_7)	Goldschmidt and Wolf 2010
LC-MS/MS	Infant formula	B_1, B_2, B_3, B_5 and B_6	Huang et al. 2009
LC-DAD/MS	Multimineral/ multivitamin dietary supplements	Thiamine (B_1), riboflavin (B_2), pyridoxine (B_6), pantothenic acid (B_5), biotin (B_7), folic acid (B_9)	Chen and Wolf 2007
LC/ESI-MS	Multivitamin tablets	All B vitamins except vitamin B_{12}, vitamin C, taurine	Chen et al. 2006
MLC	Capsules, pills and syrups	Thiamine (B_1), riboflavin (B_2), pyridoxine and pyridoxamine (B_6)	Monferrer-Pons et al. 2003
		Seven B vitamins and vitamin C	Ghorbani et al. 2004
HPLC-ED-UV	Fruit juices	Thiamine (B_1), pyridoxamine, pyridoxal and pyridoxine (B_6), cobalamins (B_{12})	Lebiedzińska et al. 2007

Goldschmidt and Wolf (2010) was 4,5,4-methyl-$^{13}C_3$-thiamine chloride. However, due to the cost and complexity, the IDMS methods may generally not be suitable for routine analysis.

The LC-MS/MS affords the possibility of performing multiple analyte measurements simultaneously. This method has been demonstrated to have high sensitivity and specificity, and has widespread applications in many areas including food and nutrient analysis. Some data derived by LC-MS/MS for typical infant formulas showed slightly lower values compared with those obtained with the current microbiological method. Given that the LC-MS/MS has higher specificity than the microbiological methods, the higher results obtained with the microbiological method are due to the cross-reactivity of the bacteria (Huang et al. 2009).

MLC offers a number of advantages compared with other chromatographic methods including low cost, low toxicity, low volatility of mobile phase constituents, the possibility of simultaneous separation of ionic and non-ionic compounds, and uncommon separation selectivity owing to the involvement of a large number of parameters (Ghorbani *et al.* 2004).

17.4.4 Electrochemical Methods

Another large group of practical methods used to determine thiamine together with vitamin B_2 and vitamin C is electrochemical methods. Voltammetric methods of analysis have attracted attention due to the high sensitivity, simplicity and relatively short analysis time, and the possibility of using inexpensive equipment. Thiamine has been simultaneously determined with vitamin B_2 and vitamin C on metallic electrodes such as gold, platinum, hanging mercury drop electrode (HMDE) and dropping mercury electrode (DME) (Aboul-Kasim 2000).

In respect of validation requirements in routine electroanalysis of vitamins, only renewable mercury electrodes are used. But despite their great analytical performance, the increased risk associated with the use, manipulation and disposal of metallic mercury has led to a search for an alternative sensor (Baś *et al.* 2011). The quantitative determination of thiamine together with B_2, C, E vitamins and quercetin has been shown to be effective using pulsed differential voltammetry with mercury film as the indicator electrode (Slepchenko *et al.* 2005).

Recently, a refreshable mercury film silver based electrode (Hg(Ag)FE) has been suggested to be applicable for vitamin B_1, B_2 and C determinations. Baś *et al.* (2011) successfully applied a novel renewable silver liquid amalgam film modified silver solid amalgam annular band electrode (AgLAF-AgSAE) for the determination of thiamine together with vitamins B_2 and C in commercially available pharmaceutical products and fruit juices using an adsorptive stripping voltammetry (AdSV) technique. Juice samples were analysed with no previous preparation and the measured thiamine concentration ranged from 0.01 to 0.1 mg/mL with a detection limit of 0.003 mg/mL. The use of AgLAF-AgSAE electrode does not pose any significant problem with the disposal and toxicity of mercury—unlike liquid mercury electrodes. Moreover, the AgLAF-AgSAE electrode was shown to be stable towards the joint determination of thiamine, vitamin B_2 and vitamin C since it was not damaged during the reduction/oxidation of vitamins.

17.4.5 Animal Assays (Biological Methods)

Thiamine can accumulate in all cells of the body but there is no single specific storage site *per se*. The body does not store the vitamin and thus a daily supply is needed. Animal assays are used for determining the availability of thiamine in a food source. They are based on the effect of thiamine on the growth and evolution of diseases related to thiamine-lacking effects. Thiamine deficiency is associated with anorexia, tissue wasting, impaired cardiac function, weakness and neurological abnormalities, all of which are classical symptoms of beriberi.

The animal assays are hardly ever used, as they are time consuming (6–8 weeks) and poorly reproducible. Experimental animals include rats, pigeons and chickens, although rats are preferable (Caster and Meadows 1980). The material being tested enables the measurement of the curative effect of a food source on rats that have been made thiamine-deficient and comparing it with the curative effect of pure synthetic thiamine hydrochloride.

Summary Points

- Microbiological methods for the analysis of thiamine are well-established procedures used for low-level naturally occurring vitamins in food. The use of microtitre plates coated with specific microorganisms has significantly improved thiamine analysis.
- The official methods of thiamine analysis are based on chemical procedures using fluorimetry and HPLC, which are now the first methods of choice.
- The animal assays for thiamine are hardly ever used, as they are time-consuming and poorly reproducible.
- The recently elaborated multivitamin methods for B vitamins assay using LC with MS/MS detection, micellar LC, LC-DAD/MS, LC/ESI-MS or LC-IDMS systems offer better sensitivity and selectivity, and can be potentially used for thiamine determination in fortified foods and infant formulas. These methods have potential to improve performance in the future.
- The voltammetric methods of analysis have attracted attention due to the high sensitivity, simplicity and relatively short analysis time, and the possibility of using inexpensive equipment. A refreshable mercury film silver based electrode [Hg(Ag)FE] has been suggested to be applicable for thiamine determination together with vitamin B_2 and vitamin C.

Key Facts

- Food is the primary sources of thiamine.
- Assays of thiamine in food, fortified food, infant formulas and premixes are based on microbiological and chemical methods.
- The chemical methods are widely accepted and recognized as official methods for thiamine analysis.
- The nutritional status for thiamine supported by current recommended dietary allowance for thiamine serves as a background to understanding animal assays of thiamine.
- Particular attention has been paid in recent years for thiamine analysis with multivitamin methods with potential to improve performance in the future. These methods include up-do-date techniques of liquid chromatography and electrochemical methods based mainly on the voltammetric methods of analysis.

Definitions of Words and Terms

C18 and C8 columns. The stationary phase of the column is generally made up of hydrophobic alkyl chains (-CH_2-CH_2-CH_2-CH_3) that interact with the analyte. The numbers 18 or 8 indicates the alkyl chain length.

Electrochemical detector. The electrochemical detector responds to substances that are either oxidizable or reducible, and the electrical output results from an electron flow caused by the chemical reaction that takes place at the surface of the electrodes. The detector normally has three electrodes—the working electrode (where the oxidation or reduction takes place), the auxiliary electrode and the reference electrode (which compensates for any change in the electrical conductivity of the mobile phase).

Fluorescence—the property of emitting radiation while irradiated. The radiation emitted is usually of longer wavelength than that incident or absorbed. A substance can be irradiated with invisible radiation and emit visible light.

High-performance liquid chromatography (HPLC)—a type of column chromatography in which the solvent is conveyed through the column under pressure, characterized by the high resolution.

Liquid chromatography (LC)—a form of chromatography employing a liquid as the moving phase, and a solid or a liquid on a solid support as the stationary phase.

Mass spectrometry (MS)—an analytical technique by which chemical substances are identified by sorting gaseous ions by mass using electric and magnetic fields. A mass spectrometer uses electrical means to detect the sorted ions.

Reversed-phase high performance liquid chromatography (RP-HPLC)—an HPLC technique where the mobile phase is more polar than the solid phase. In RP-HPLC compounds are separated based on their hydrophobic character.

Thiochrome. Tricyclic alcohol ($C_{12}H_{14}N_4OS$) formed by oxidation of thiamine and giving a blue fluorescence under ultraviolet light that serves as the basis of a method for thiamine assay.

Ultraviolet detector. This detector responds to those substances that absorb light in the range 180–350 nm, including those substances having one or more double bonds and substances having unshared electrons. The sensor of a UV detector consists of a short cylindrical cell through which passes the column eluent. UV light is arranged to pass through the cell and fall on a photoelectric cell. The output from the photocell passes to a modifying amplifier and then to a recorder or data acquisition system.

List of Abbreviations

AACC	American Association of Cereal Chemists
AdSV	adsorptive stripping voltammetry
AgLAF-AgSAE	silver liquid amalgam film-modified silver solid amalgam annular band electrode
AOAC methods	Official Methods of Analysis of the Association of Official Analytical Chemists

ATCC	American Type Culture Collection
CEN	European Committee for Standardization
CNBr	cyanogen bromide
DME	dropping mercury electrode
FL	fluorimetry
Hg(Ag)FE	mercury film silver based electrode
HMDE	hanging mercury drop electrode
HPLC	high-performance liquid chromatography
$K_3Fe(CN)_6$	potassium hexacyanoferrate(III)
ICC	International Association of Cereal Science and Technology
LC	liquid chromatography
LC-MS/MS	liquid chromatography–tandem mass spectrometry
LC/ESI-MS	liquid chromatography/electrospray ionization–mass spectrometry
LC-DAD/MS	liquid chromatography/diode array detection–mass spectrometry
LC-IDMS	liquid chromatography–isotope dilution mass spectrometry
MLC	micellar liquid chromatography
RP-HPLC-ED-UV	reversed-phase high-performance liquid chromatography with coulometric electrochemical and ultraviolet detection
RP-HPLC-UV	reversed-phase high-performance liquid chromatography with ultraviolet detection
SPE	solid-phase extraction
TMP	thiamine monophosphate
TPP	thiamine pyrophosphate
TTP	thiamine triphosphate
UV	ultraviolet

References

Aboul-Kasim, E., 2000. Anodic adsorptive voltammetric determination of the vitamin B_1 (thiamine). *Journal of Pharmaceutical and Biomedical Analysis.* 22: 1047–1054.

AOAC, 1990a. Method 953.17. Thiamine (vitamin B1) in grain products—fluorometric (rapid) method. Final action. In: *Official Methods of Analysis of the Association of Official Analytical Chemists*, 15th ed. Association of Official Analytical Chemists, Arlington VA, USA, p. 1051.

AOAC, 1990b. Method 942.23. Thiamine (vitamin B1) in foods—fluorometric method. Final action. In: *Official Methods of Analysis of the Association of Official Analytical Chemists*, 15th ed. Association of Official Analytical Chemists, Arlington VA, USA, pp. 1049–1051.

AOAC, 1990c. Method 957.17. Thiamine (vitamin B1) in bread—fluorometric method. Final action. In: *Official Methods of Analysis of the Association of*

Official Analytical Chemists, 15th ed. Association of Official Analytical Chemists, Arlington VA, USA, pp. 1051–1052.

AOAC, 1990d. Method 986.27. Thiamine in milk-based infant formula—fluorometric method Final action. In: *Official Methods of Analysis of the Association of Official Analytical Chemists*, 15th ed. Association of Official Analytical Chemists, Arlington VA, USA, pp. 1113–1114.

Baś, B., Jakubowska, M., and Górski, Ł., 2011. Application of renewable silver amalgam annular band electrode to voltammetric determination of vitamins C, B1 and B2. *Talanta*. 84: 1032–1037.

Blake, C.J., 2007. Analytical procedures for water-soluble vitamins in foods and dietary supplements: a review. *Analytical and Bioanalytical Chemistry*. 389: 63–76.

Bognar, A., 1992. Determination of vitamin B1 in food by high-performance liquid chromatography and postcolumn derivatization. *Fresenius' Journal of Analytical Chemistry*. 343: 155–156.

Botticher, B., and Botticher, D., 1986. Simple rapid determination of thiamin by a HPLC method in foods, body fluids, urine and faeces. *International Journal for Vitamin and Nutrition Research*. 56: 155–159.

Brin, M., 1980. Red cell transketolase as an indicator of nutritional deficiency. *The American Journal of Clinical Nutrition*. 33: 169–171.

Bui, L.T.T., and Small, D.M., 2007. The contribution of Asian noodles to dietary thiamine intakes. A study of commercial dried products. *Journal of Food Composition and Analysis*. 20: 575–583.

Caster, W.O., and Meadows, J.S., 1980. The three thiamin requirements of the rat. *International Journal for Vitamin and Nutrition Research*. 50: 125–130.

CEN, 2003. En 14122. *Foodstuffs*. Determination of vitamin B1 by HPLC. European Committee for Standardization, Brussels.

Chen, P., Chen, B., and Yao S., 2006. High-performance liquid chromatography/electrospray ionization-mass spectrometry for simultaneous determination of taurine and 10 water-soluble vitamins in multivitamin tablets. *Analytica Chimica Acta*. 569: 169–175.

Chen, P., and Wolf, W.R., 2007. LC/UV/MS-MRM for the simultaneous determination of water-soluble vitamins in multi-vitamin dietary supplements. *Analytical and Bioanalytical Chemistry*. 387: 2441–2448.

Dinelli, G., and Bonetti, A., 1994. Micellar electrokinetic capillary chromatography analysis of water-soluble vitamins and multi-vitamin integrators. *Electrophoresis*. 15: 1147–1150.

Ellefson, W.C., 1985. Thiamine. In: Augustin, J., Klein, B.P., Becker, D., and Venugopal, P.B. (ed). *Methods of Vitamin Assay*. Wiley, New York, USA, pp. 349–364.

Ekinci, R., and Kadakal, C., 2005. Determination of seven water-soluble vitamins in tarhana, a traditional Turkish cereal food, by high-performance liquid chromatography. *Acta Chromatography*. 15: 289–297.

Engel, R., Stefanovits-Banyai, E., and Abranko, L., 2010. LC simultaneous determination of the free forms of B group vitamins and vitamin C in various fortified products. *Chromatographia*. 11/12: 1069–1074.

European Commission, 2006. Regulation (EC) No 1925/2006 of the European Parliament and of the Council of 20 December 2006 on the addition of vitamins and minerals and of certain other substances to food. *Official Journal of the European Union*. L404: 26–38.

Fotsing, L., Fillet, M., Bechet, I., Hubert, P., and Crommen, J., 1997. Determination of six water-soluble vitamins in a pharmaceutical formulation by capillary electrophoresis. *Journal of Pharmaceutical and Biomedical Analysis*. 15: 1113–1123.

Fotsing, L., Boulanger, B., Chiap, P., Fillet, M., Hubert, Ph., and Crommen, J., 2000. Multivariate optimization approach for the separation of water-soluble vitamins and related compounds by capillary electrophoresis. *Biomedical Chromatography*. 14: 10–11.

Fujiwara, S., Iwase, S., and Honda, S., 1988. Analysis of water-soluble vitamins by micellar electrokinetic capillary chromatography. *Journal of Chromatography*. 447: 133–140.

Furlani, R.P.Z., and Godoy, H.T., 2008. Vitamins B1 and B2 contents in cultivated mushrooms. *Food Chemistry*. 106: 816–819.

Ghorbani, A.R., Momenbeik, F., Khorasani, J.H., and Amini, M.K., 2004. Simultaneous micellar liquid chromatographic analysis of seven water-soluble vitamins: optimization using super-modified simplex. *Analytical and Bioanalytical Chemistry*. 379: 439–444.

Goldschmidt, R.J., and Wolf, W.R., 2010. Simultaneous determination of water-soluble vitamins in SRM 1849 Infant/Adult Nutritional Formula powder by liquid chromatography-isotope dilution mass spectrometry. *Analytical and Bioanalytical Chemistry*. 397: 471–481.

Gregory, J.F., 1996. Vitamins. In: Fennema, O.R. (ed.) *Food Chemistry*. Marcel Dekker, New York, USA, pp. 532–610.

Gregory, J.F., 1997. Bioavailability of thiamin. *European Journal of Clinical Nutrition*. 51(SI 1): S34–S37.

Gubler, C.J., 1991. Thiamine. In: Machlin, L.J. (ed.). *Handbook of Vitamins*. Marcel Dekker, New York, USA, pp. 233–281.

Heudi, O., Kilinc, T., and Fontannaz, P., 2005. Separation of water-soluble vitamins by reversed-phase high performance chromatography with ultraviolet detection: application to polyvitaminated premixes. *Journal of Chromatography A*. 1070: 49–56.

Huang, M., Winters, D., Crowley, R., and Sullivan, D., 2009. Measurement of water-soluble B vitamins in infant formula by liquid chromatography/tandem mass spectrometry. *Journal of AOAC International*. 96: 1728–1738.

ICC, 1990. Chemical determination of thiamine in cereal products. *Standard Methods of the International Association of Cereal Science and Technology* No. 117. Verlag Moritz Schäfer, Detmold.

Lebiedzińska, A., Marszałł, M.L., Kuta, J., and Szefer P., 2007. Reversed-phase high-performance liquid chromatography method with coulometric electrochemical and ultraviolet detection for the quantification of vitamins B1 (thiamine), B6 (pyridoxamine, pyridoxal and pyridoxine) and B12 in animal and plant foods. *Journal of Chromatography A*. 1173: 71–80.

Martínez-Villaluenga, C., Michalska, A., Frias, J., Piskuła, M.K., Vidal-Valverde, C., and Zieliński, H., 2009. Effect of flour extraction rate and baking on thiamine and riboflavin content and antioxidant capacity of traditional rye bread. *Journal of Food Science.* 74: 49–55.

Monferrer-Pons, L., Capella-Peiró, M.E., Gil-Agustí, M., and Esteve-Romero, J., 2003. Micellar liquid chromatography determination of B vitamins with direct injection and ultraviolet absorbance detection. *Journal of Chromatography A.* 984: 223–231.

National Research Council, 1989. Water-Soluble Vitamins. In: *Recommended Dietary Allowance*, 19th ed. National Academy Press, Washington D.C., USA, pp. 125–132.

Niemeijer, R., Haas-Lauterbach, S., Stengi, S., and Weber, W., 2009. Determination of water-soluble B-vitamins with VitaFast® tests in fruits and fruit products. In: *Proceedings of 4th International Symposium on Recent Advances in Food Analysis.* November 4–6, 2009, Prague, Czech Republic, p. 566.

Riccio, F., Mennella, C., and Fogliano, V., 2006. Effect of cooking on the concentration of Vitamins B in fortified meat products. *Journal of Pharmaceutical and Biomedical Analysis.* 41: 1592–1595.

Slepchenko, G.B., Anisimova, L.S., Slipchenko, V.F., Mikheeva, E.V., and Pikula, N.P., 2005. Voltammetric quality control of bioactive additives: determination of B1, B2, C, E vitamins and quercetin. *Pharmaceutical Chemistry Journal.* 39: 166–168.

Tanphaichitr, V., 2001. Thiamine. In: Ruker, R.B., and Suttie. J. (ed.). *The Handbook of Vitamins*, 3rd ed., Marcel Dekker, New York, USA, pp. 275–310.

Vidal-Valverde, C., and Reche, A., 1990. An improved high performance liquid chromatographic method for thiamin analysis in foods. *Zeitschrift für Lebensmittel-Untersuchung und -Forschung.* 191: 313–318.

Viñas, P., López-Erroz, C., Balsalobre, N., and Hernández-Córdoba, M., 2003. Reversed-phase liquid chromatography on an amide stationary phase for the determination of the B group vitamins in baby foods. *Journal of Chromatography A.* 1007: 77–84.

Zieliński, H., Frias, J., Piskuła, M.K., Kozłowska, H., and Vidal-Valverde, C., 2005. Vitamin B1 and B2, dietary fiber and minerals content of Cruciferae sprouts. *European Food Research and Technology.* 221: 78–83.

CHAPTER 18
Assays of Riboflavin in Food using Solid-phase Extraction

MARCELA A. SEGUNDO,* MARCELO V. OSÓRIO,
HUGO M. OLIVEIRA, LUÍSA BARREIROS AND
LUÍS M. MAGALHÃES

REQUIMTE, Department of Chemistry, Faculty of Pharmacy, University of Porto, R. Jorge Viterbo Ferreira, 228, 4050-313 Porto, Portugal
*Email: msegundo@ff.up.pt

18.1 Introduction

Riboflavin is found at low levels in almost all biological tissues, playing a relevant role as a precursor of coenzymes and as a component of specific metabolic pathways. Its uptake is solely by food ingestion and therefore its quantification is necessary for evaluation of nutrient content, particularly in vitamin-fortified foods. The determination of riboflavin is also relevant for food quality assessment in different commodities, namely as a marker of degradative conditions in milk processing and storage or as a precursor of 'sunlight flavour' compounds in wines and beers (Andrés-Lacueva et al. 1998). Moreover, riboflavin is an EU authorized natural food colorant (Scotter 2011). Though no legal limit is currently imposed, knowledge about its content as a component of total colorants is necessary. Information about riboflavin content in food products is summarized in Table 18.1.

Several difficulties arise when analyzing riboflavin in food. Total riboflavin exists in food in several forms, from which the most representative are

Table 18.1 Riboflavin content in food. Interval range for riboflavin content in food, expressed in mg of riboflavin per 100 g of edible food except for beers, for which values are expressed in mg of riboflavin per 100 mL. The data was gathered from McCance and Widdowson's The Composition of Foods integrated dataset © Crown Copyright, which can be fully consulted at http://tna.europarchive.org/20110116113217/http://www.food.gov.uk/science/dietarysurveys/ (accessed May 2012).

Food	Riboflavin content (mg)
Beverages	
Beers	0.02–0.06
Fruit juices	0.01–0.13
Powdered drinks and essences	0.06–1.60
Cereal and cereal products	
Biscuits	0.02–0.26
Bread	0.03–0.32
Breakfast cereals	0.01–2.70
Flours, grains and starches	0.02–0.72
Infant cereal foods	0.90–1.70
Fish and fish products	
Crustacea	0.01–0.86
Fatty fish	0.06–0.52
Molluscs	0.02–0.48
White fish	0.02–0.32
Fruits	0.01–0.50
Herbs and spices	0.03–1.74
Meat	
Beef	0.13–0.32
Poultry	0.06–0.80
Lamb	0.09–0.38
Offal	0.08–5.65
Pork	0.08–0.41
Milk and milk products	
Cheeses	0.13–0.65
Ice creams	0.04–0.36
Cow's milk	0.14–0.24
Yogurts	0.02–0.37
Nuts and seeds	0.01–0.75
Chocolate confectionery	0.06–0.79
Vegetables	0.01–1.34

riboflavin itself (free), followed by riboflavin-5′-phosphate (FMN) and riboflavin-5′-adenosyldiphosphate (FAD). Generally, the determination of total riboflavin is accomplished after hydrolysis of the phosphorylated analogues and/or those bound to proteins. Hot (100–120 °C) acid hydrolysis followed by enzymatic digestion with takadiastase and clarase were recommended some years ago. Current methods resort to milder conditions for acid digestion, with concomitant application of acid phosphatase (Tang et al. 2006).

The major problems in riboflavin assessment are related to interference from other matrix components on the instrumental method of analysis employed (ultraviolet spectrophotometry or fluorimetry). Though fluorimetry is more sensitive and more selective than spectrophotometry, it still suffers from matrix interference—generally non-specific effects from components that elicit enhancement or quenching of fluorescence from riboflavin itself. Hence, chromatographic separation assisted by a previous sample treatment that isolates the riboflavin analyte and/or remove the interferences is often required.

In this context, solid-phase extraction (SPE) appears a suitable preparative technique (see Key Facts), fostering elimination of interferences from the matrix, selective extraction of analyte and even analyte pre-concentration. This chapter addresses some of the methods that adopt the use of SPE prior to chromatographic determination by high-performance liquid chromatography (HPLC), with particular emphasis on automated flow-based methods.

18.2 Batch Solid-phase Extraction Methods

Several methods have been described using conventional silica C18 cartridges for selective extraction of riboflavin in different types of food samples. The method proposed by Ollilainen and co-workers was one of the first to be applied (Ollilainen *et al.* 1990), coinciding with the boom of SPE applications in the beginning of 1990s. The main analytical conditions are described in Table 18.2. Selective extraction of riboflavin was performed on a silica C18 commercially available cartridge. A simple protocol was implemented, consisting of conditioning the sorbent with 10 mL of methanol and twice with 10 mL of water, loading of sample extract (2.5–10 mL), washing with water, and elution with 40–70% methanol.

Sample treatment prior to SPE was required to convert FMN and FAD to riboflavin using a hot acid extraction procedure followed by enzymatic digestion, as the HPLC method applied allows the determination of only riboflavin. In this case, the recovery of riboflavin from FAD in pork samples was 94–98% with 2–4 hours of incubation time; prolonging the incubation time to 24 hours did not improve the recovery. Indeed, the SPE procedure allowed a concentration factor of two to four fold, with recoveries for samples in the range 96–108%. The method was applied to 21 food samples, providing comparable results to the AOAC method (modified) apart from the crispbread sample. This was justified by the presence of a fluorescent contaminant in this sample that caused an overestimation of the amount of riboflavin in the fluorimetric, non-separative method.

Variants of this methodology have been applied to the determination of riboflavin in sea urchin (Quirós *et al.* 2004) and in a food extract (Rudenko and Kartsova 2010). In the first case (Quirós *et al.* 2004), the extraction protocol was changed slightly by conditioning and washing the sorbent column with a buffer solution (5 mM acetic acid/ammonium acetate, pH 5.0) and by eluting the retained riboflavin (and thiamine too, in this case) with the chromatographic mobile phase. In the other method (Rudenko and Kartsova 2010), besides silica C18, other sorbents were tested, namely XAD2 Amberlite, finely

Table 18.2 Characteristics and figures of merit for solid-phase extraction of riboflavin and determination by HPLC-FL. The procedure proposed by Ollilainen *et al.* (1990) utilizes a silica C18 cartridge for selective extraction of riboflavin after acid and enzymatic digestion of samples. Riboflavin in the extract is quantified by HPLC using a reversed-phase column.

SPE procedure	
Type of sorbent	Silica C18 (cartridge Sep-Pak®)
Sample volume	2.5–10 mL (sample extract)
Eluent	MeOH 40–70% (v/v)
Other sample treatment	Hot acid digestion, pH adjustment to 4.5, overnight enzymatic digestion with takadiastase (FMN conversion) and with claradiastase (FAD conversion) (prior to SPE)
HPLC separation	
Column	Silica C18 (Spherisorb ODS-2, 250 × 4.6 mm i.d., 5 µm)
Mobile phase	H_2O/MeOH (65:35, v/v)
Run time	10 min
Injection volume	50 µL
Detector wavelength	$\lambda_{excitation} = 445$ nm; $\lambda_{emission} = 525$ nm
Linear range	0.0038–3.8 mg L^{-1}
Limit of detection	0.4 µg L^{-1}
Repeatability	<6.2%
Application to samples	Dairy products (butter, cheese, milk, yoghurt), meat and fish (liver, pork, herring, trout, sausage), vegetables and fruits (cauliflower, orange, potato, tomato), beverages (beer, coffee)
Other analytes assayed	none

grained activated carbon (grain size 1.5 mm) and polystyrene. Nevertheless, the best results in purification efficiency and recovery rates were obtained on reversed phase adsorbent silica C18, as activated carbon and polystyrene were not suitable while XAD2 Amberlite did not retain pyridoxine, thiamine and ascorbic acid (other target analytes in this method).

18.3 Automatic Flow Based Solid-phase Extraction Methods

Solid-phase extraction for the determination of riboflavin has been automated using flow injection methods (see Key Facts). Three systems have been proposed, from which two are based on a flow through column containing the sorbent which was placed in the loop of a six-port injection valve as schematically depicted in Figure 18.1. The third and most recent flow system relies on a novel strategy for sorbent handling inside mesofluidic channels in an automated renewable fashion. More detail about the automated protocols is given in the following sections.

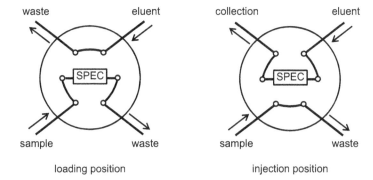

Figure 18.1 Schematic representation of solid-phase extraction performed in flow injection systems using a SPE column fitted in the injection loop of a six-port injection valve. In the loading position, sample is loaded into the SPE column, where riboflavin is retained. By switching to the injection position, eluent is sent through the SPE column, originating analyte desorption which is then collected in a vial placed on one of the ports. SPEC: solid-phase extraction column.

18.3.1 Off-line Method

Valcárcel and co-workers proposed in 2003 an automatic system for the determination of riboflavin in lyophilized food products using a solid-phase extraction process that included columns filled with cotton or silica C18 for the sequential retention of synthetic colorants and natural colorants, respectively (González et al. 2003). In this case riboflavin was assessed as an authorized natural colorant for total estimation of this class of compounds. The analytical characteristics of this methodology are summarized in Table 18.3.

Automated sample preparation was carried out in a continuous flow system containing two six-port injection valves, connected sequentially. Each sorbent column was placed in the injection loop of each valve as depicted in Figure 18.1 (for a single valve). In summary, the silica C18 column was conditioned with 1.5 mL of methanol and then with 3 mL of 1 M acetic acid/acetate buffer, pH 4.7. Next, the sample extract (3 mL) was passed first through one column filled with cotton (where synthetic colorants were retained) and then through the next column, containing silica C18 sorbent (where natural colorants, including riboflavin, were retained). After washing with 1.5 mL of water, riboflavin (and other natural colorants) was eluted by two aliquots (75 µL each) of methanol and only the second aliquot was processed for HPLC diode array detector (DAD) quantification.

This automated SPE flow system presents two major drawbacks. First, for an inexperienced operator, manual packing of the SPE column is not an easy task and prone to introduce low reproducibility into the methodology. Second, the same sorbent column is used for different samples, increasing the risk of cross-contamination and carryover. Therefore, sorbent washing with 3 mL of hexane is recommended for every 30 samples analyzed.

Table 18.3 Characteristics and figures of merit for automated solid-phase extraction of riboflavin and determination by HPLC-DAD. The procedure proposed by González et al. (2003) relies on automated SPE using a flow-through silica C18 column for extraction of several natural colorants, including riboflavin. The SPE extract is subsequently analyzed by reversed phase chromatography.

SPE procedure	
Type of sorbent	Silica C18 (lab packed microcolumn)
Sample volume	3 mL (treated sample)
Eluent	MeOH (2×75 µL)
Other sample treatment	Homogenization, lyophilization, extraction with 1 M acetic acid/sodium acetate buffer, pH 4.7 (prior to SPE)
HPLC separation	
Column	Silica C18 (Spherisorb ODS-2, 250×4.6 mm i.d., 5 µm)
Mobile phase	Component A: MeOH (85% v/v) + aqueous 0.07 g L^{-1} cetyltrimethylammonium bromide (15% v/v), pH 6.0; component B: MeOH (20%) + ethyl acetate (80%); gradient elution
Run time	15 min
Injection volume	20 µL
Detector wavelength	λ = 425 nm
Linear range	0.01–200 mg L^{-1} (treated extract)
	3–300 µg g^{-1} (lyophilized sample)
Limit of detection	0.8 µg g^{-1} (lyophilized sample)
Repeatability	<3.5%
Application to samples	Dairy (milkshakes, desserts, ice creams) and fatty (sauces, bouillon cubes, dehydrated soups) samples
Other analytes assayed	Synthetic (tartrazine, lissamine green B, indigo carmine, brilliant blue FCF, brilliant black BN) and natural (caramel, carminic acid, curcumin, trans-β-carotene) colorants

The procedure was tested with particularly difficult samples, with high protein (dairy samples) and fat contents (fatty foods, Table 18.3). Excellent recoveries were attained (94–95%) for seven different samples (yoghurt, milkshake, milk based dessert, ice cream, sauce, dehydrate soup and bouillon cubes) with figures of merit adequate for control of additives. It is also important to emphasize that sample lyophilization and reconstitution prior to SPE also contributed to the successful applicability of this methodology by preserving the sample for at least two months without changes in the concentrations of the colorants.

18.3.2 On-line Methods

Nowadays, online hyphenation of SPE to HPLC can be easily attained using expensive, high maintenance robotic stations. However, simpler alternatives based on continuous flow injection analysis can also be implemented. In this context, an online sample preparation method (Greenway and Kometa 1994) was described for the determination of riboflavin and FMN in milk and cereal samples by HPLC coupled to fluorimetric detection (FL) (Table 18.4).

A commercial silica C18 chromatographic precolumn was employed for SPE, which is a robust alternative compared with the laboratory-produced column proposed on the off-line flow system (González et al. 2003). In summary, 500 µL of sample dialysate was introduced in the flow system by means of an auxiliary injection valve and directed towards the flow through SPE column (as depicted for loading position in Figure 18.1). After analyte retention, the injection valve was switched and both FMN and riboflavin were eluted in a column switching approach, where the elution is carried out by the chromatographic mobile phase itself (as depicted for injection position in Figure 18.1). Initially, samples were minimally prepared and introduced as a solution (milk) or a suspension (cereals) of fine particles in a flow through reactor for microwave-assisted digestion. The digested sample was further dialysed before

Table 18.4 Characteristics and figures of merit for automated solid-phase extraction of riboflavin and determination by HPLC-FL. The procedure proposed by Greenway and Kometa (1994) is based on automatic SPE using a commercial flow-through column with silica C18. Samples are first digested by microwave irradiation and small molecules were separated by dialysis prior to SPE. The SPE extract is subsequently analyzed by reversed phase chromatography.

SPE procedure	
Type of sorbent	Silica C18 (chromatographic pre column)
Sample volume	1 mL (dialysate)
Eluent	Mobile phase (initial composition)
Other sample treatment	Microwave digestion and dialysis (prior to SPE)
HPLC separation	
Column	Silica C18 (Spherisorb ODS-2, 250×4.6 mm i.d., 5 µm)
Mobile phase	Component A: 0.1 M acetate/sodium acetate buffer (pH 4.8); component B: H_2O/ACN/MeOH (50 + 40 + 10); gradient elution with organic phase increased from 6 to 100% in linear gradient up to 30 min
Run time	30 min
Injection volume	Not defined; column switching scheme
Detector wavelength	$\lambda_{excitation} = 450$ nm; $\lambda_{emission} = 520$ nm
Linear range	0–3 mg L^{-1}
Limit of detection	Not given
Repeatability	<3.3%
Application to samples	Milk, fortified cereal products (cornflakes), non-fortified cereal samples (wheat and oats)
Other analytes assayed	FMN

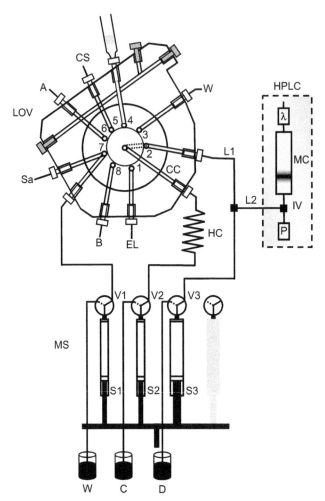

Figure 18.2 Schematic representation of online SPE performed in an automatic flow system for determination of riboflavin in foodstuff. LOV: lab-on-valve, MS: multisyringe, HPLC: liquid chromatograph, S_i: syringe, V_i: three way commutation valve (position off: dashed line, position on: solid line), A: air, CS: conditioning solvent (50% (v/v) MeOH/H$_2$O), BS: bead suspension in conditioning solvent, C: carrier solution (H$_2$O), D: diluent (H$_2$O), W: waste, CC: central channel, EL: eluent (50% (v/v) MeOH/H$_2$O + 1% (v/v) CH$_3$COOH), B: channel for bead discarding, Sa: sample/standard solution, HC: holding coil, L1: connection tubing (8 cm), L2: connection tubing (44 cm), P: chromatographic pump, IV: injection valve, MC: monolithic chromatographic column, λ: diode array detector.
(Reproduced from Oliveira *et al.* 2010 with kind permission from Springer Science + Business Media).

injection into the online SPE system for separating the analytes from the ground solid particles leaving the microwave oven.

Microwave extraction provided an efficient online method for the liberation of flavins from complex food matrices. Because of the acidic medium used and the microwave power, the conversion of flavins from one form into another occurred after liberation from the sample matrix and 100% conversion of FAD to FMN after extraction was observed. Some conversion of FMN into riboflavin was also observed (about 15.2%). This was taken into account when analyzing samples where the FMN content was determined by increasing the figure reported by 15%. For those products in which FMN was not detected no assumptions could be made. The method was compared with the AOAC fluorimetric method commonly applied for milk and cereal samples, with recoveries of 94–106% for spiked riboflavin.

A more recent online method was proposed by Segundo and co-workers, where SPE of riboflavin was performed on lab-on-valve (LOV) equipment coupled to HPLC-DAD (Figure 18.2) (Oliveira *et al.* 2010). In this case the sorbent applied was a commercial molecularly imprinted polymer (MIP) selective to riboflavin and previously applied to milk and beer samples

Table 18.5 Characteristics and figures of merit for automated solid-phase extraction of riboflavin and determination by HPLC-DAD. The procedure proposed by Oliveira *et al.* (2010) is based on automatic SPE using a renewable column of commercially available molecularly imprinted polymers selective to riboflavin. The solid-phase extraction procedure was coupled online to the reversed phase chromatographic determination, resorting to a monolithic silica C18 column.

SPE procedure	
Type of sorbent	Molecularly imprinted polymer (lab packed microcolumn)
Sample volume	1 mL
Eluent	50% (v/v) MeOH/H_2O + 1% (v/v) acetic acid (312.5 µL)
Other sample treatment	Acidic enzymatic digestion (acid phosphatase) for pig liver (prior to SPE)
HPLC separation	
Column	Monolithic silica C18 (chromolith RP-18e, 100 × 4.6 mm i.d.)
Mobile phase	5 mM of octanesulfonic acid, 0.5% (v/v) of triethylamine, 2.4% (v/v) of acetic acid and 15% (v/v) of methanol in water
Run time	7 min
Injection volume	200 µL
Detector wavelength	$\lambda = 268$ nm
Linear range	0.45–5.00 mg L^{-1}
Limit of detection	0.05 mg L^{-1}
Repeatability	<3.8%
Application to samples	Milk based infant formula, pig liver, energy drink
Other analytes assayed	none

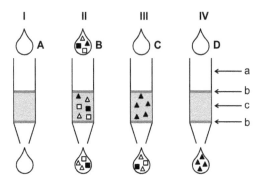

Figure 18.3 Schematic representation of solid-phase extraction protocol. a: cartridge, b: frits, c: sorbent. Protocol steps: I: sorbent conditioning, II: sample loading, III: washing operation, IV: analyte elution. Solutions employed: A: conditioning solution, B: sample, C: washing solution, D: eluent. ▲: analyte; △, ■, □: other matrix components.

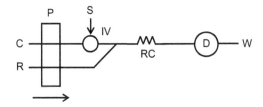

Figure 18.4 Schematic representation of a flow injection system. C: carrier; IV: injection valve; P: pump; D: detector; RC: reaction coil; S: sample; R: reagent; W: waste.

(Manesiotis et al. 2009). Furthermore, a novel strategy for sorbent handling, designated as bead injection, was implemented by inline sorbent column formation in one of the channels of the LOV piece followed by extraction, elution of riboflavin to the HPLC injection valve and discarding of the solid phase. Hence, contaminations from successive samples are avoided, as a fresh portion of sorbent is utilized for each analysis. All these operations were computer-controlled by dedicated software, requiring no operator intervention.

The analytical characteristics are presented in Table 18.5. The method was validated by application to CRM NIST 1846 (milk-based infant formula) and to CRM BCR 487 (pig liver), providing values that were statistically comparable to the certified values. Good precision and good recoveries (91–93%) were attained after addition to an energy drink. However, there is still a question about this method as no study has been undertaken regarding the MIP selectivity towards different forms of riboflavin, namely FMN and FAD. As pig liver was digested and treated with enzymes prior to analysis and the two other samples analyzed probably had mainly free riboflavin, no assumptions about the selectivity can be drawn from these results.

18.4 Concluding Remarks

Solid-phase extraction is an effective way of separating riboflavin from complex samples, providing an enriched extract that can be directly processed by HPLC systems. Until now two types of sorbents have been applied, where first applications resorted to the generally applied silica C18. More recently, MIP specific for riboflavin became commercially available, providing molecular recognition towards riboflavin for enhanced selectivity. Solid-phase extraction also fostered pre-concentration of riboflavin (factors of 2–5) which is not as relevant for food but it can be of significance when analyzing biological samples. Finally, automation of solid-phase extraction has been successfully implemented using flow injection based systems with application to real samples with complex matrices.

Summary Points

- The determination of riboflavin in food is important because of its role as a nutrient, as a marker of food processing quality, as a precursor of compounds with organoleptic relevance and as a natural food colorant.
- Methods for the determination of riboflavin in food samples should account for its different forms, namely flavin mononucleotide and flavin adenine mononucleotide.
- Solid-phase extraction is an effective technique for minimization or elimination of interferences for complex food matrices prior to determination of riboflavin using spectrophotometric or fluorimetric detection.
- Silica C18 or molecularly imprinted polymers selective to riboflavin have been employed as sorbents for solid-phase extraction of riboflavin in foods.
- Automation of solid-phase extraction protocols has been successfully implemented using continuous flow injection based systems.
- Automatic solid-phase extraction for the determination of riboflavin in foods was performed using flow through columns, commercially available or customized in the lab, and also using a bead injection approach for inline solid-phase extraction column formation and renewal.
- Automation of solid-phase extraction for riboflavin separation has been carried out in both offline and online modes, in the last case with direct coupling to the injection valve of HPLC equipment.

Key Facts

Key Facts about Solid-phase Extraction

- Solid-phase extraction (SPE) was proposed for sample treatment as an alternative to liquid–liquid extraction procedures employed for analyte extraction prior to HPLC separations.
- Generally, SPE protocols consist of four steps (Figure 18.3). The first step is the conditioning of the sorbent bed by a solution or solvent which will

activate the sorbent functional groups to retain the target analytes. The second step is sample loading, where analytes are selectively retained by covalent, ionic or van der Waals interactions with the sorbent functional moieties. The third step consists of removing possible interfering species, co-retained with the analyte, and is critical to eliminating interferences and for attaining high recovery rates by avoiding bleeding of analyte. The fourth and last step is analyte elution, using a solvent that elicits a more strong interaction with the analyte compared with those established between analyte and sorbent.
- In the first years following its development, a low number of sorbents were available, mainly silica or octadecyl (ODS) coated silica (also called reversed phases).
- In the last decade, a large number of sorbents were made commercially available, with a large diversity in composition (polymeric structures for instance, that enhance sorbent compatibility to samples with more extreme pH values) and in functional groups (for anion and cation exchange coupled with reversed phase characteristics, originating mix-mode sorbents).

Key Facts about Flow Injection Analysis

- Flow injection analysis (FIA) was defined by its inventors, Jaromir Ruzicka and Elo Hansen, as 'a means of information-gathering from a concentration gradient formed from an injected, well-defined zone of fluid, dispersed into a continuous unsegmented flow stream of a carrier'.
- FIA is based in three principles: (1) reproducible sample injection or insertion in a flowing carrier stream; (2) controlled dispersion of the sample zone; and (3) reproducible timing of its movement from the injector point to the detection system (schematic representation in Figure 18.4).
- The current definition of FIA expands its scope because, besides automating chemical reactions, flow injection systems foster online sample treatment, namely solvent extraction, gas diffusion, dialysis and solid-phase extraction, as found in many applications.
- Although no computer is required for simple manifold operation, FIA has evolved to novel strategies of fluid handling, using computer-controlled devices, but still based on the three principles mentioned above.
- Computer-controlled systems are applied in other techniques derived from FIA, such as sequential injection analysis (SIA, introduced in 1990), multicommutation flow analysis (MCFA, introduced in 1994), and multisyringe flow injection analysis (MSFIA, introduced in 1999).

Definitions of Words and Terms

Bead injection: flow-based technique employed for handling particle suspensions in mesofluidic channels.

Eluent: solvent employed in SPE protocols for elution of analyte(s) retained in sorbents.

Elution by column switching: elution mode in SPE where the sorbent column is placed in the loop of a HPLC injection valve and retained analyte(s) are eluted by the chromatographic mobile phase.

Lab-on-valve: micro-machined device, placed atop a multi-position valve, containing mesofluidic channels where solutions or suspensions are handled by means of a syringe pump.

Mobile phase (also called eluent): solution or solvent employed in chromatographic separations that flow through the analytical column.

Molecularly imprinted polymer: polymers with cavities, which were defined by a template molecule during its synthesis, with specific affinity for the target analyte or class of analytes.

Offline: situation where two procedures (or equipment) are not physically connected, requiring intervention by an operator (for transfer of solutions or reaction products, for instance) in order to proceed sequentially.

Online: situation where two procedures (or equipment) are physically connected, taking place sequentially, without intervention by an operator.

Silica C18: sorbent material composed by a silica core to which octadecyl moieties are bonded.

Sorbent: material employed as solid-phase for analyte retention in SPE protocols.

List of Abbreviations

ACN	acetonitrile
AOAC	Association of Official Analytical Chemists
DAD	diode array detector
FAD	riboflavin-5'-adenosyldiphosphate (or flavin adenine mononucleotide)
FIA	flow injection analysis
FL	fluorimetric detector
FMN	riboflavin-5'-phosphate (or flavin mononucleotide)
HPLC	high performance liquid chromatography
LOV	lab-on-valve
MCFA	multicommutation flow analysis
MSFIA	multisyringe flow injection analysis
MeOH	methanol
MIP	molecularly imprinted polymer
ODS	octadecylsilica
SIA	sequential injection analysis
SPE	solid-phase extraction
UV	ultraviolet

References

Andrés-Lacueva, C., Mattivi, F., and Tonon, D., 1998. Determination of riboflavin, flavin mononucleotide and flavin-adenine dinucleotide in wine and other beverages by high-performance liquid chromatography with fluorescence detection. *Journal of Chromatography A*. 823: 355–363.

González, M., Gallego, M., and Valcárcel, M., 2003. Liquid chromatographic determination of natural and synthetic colorants in lyophilized foods using an automatic solid-phase extraction system. *Journal of Agricultural and Food Chemistry*. 51: 2121–2129.

Greenway, G.M., and Kometa, N., 1994. Online sample preparation for the determination of riboflavin and flavin mononucleotides in foodstuffs. *Analyst*. 119: 929–935.

Manesiotis, P., Borrelli, C., Aureliano, C.S.A., Svensson, C., and Sellergren, B., 2009. Water-compatible imprinted polymers for selective depletion of riboflavine from beverages. *Journal of Materials Chemistry*. 19: 6185–6193.

Oliveira, H.M., Segundo, M.A., Lima, J.L.F.C., Miró, M., and Cerdà, V., 2010. Exploiting automatic on-line renewable molecularly imprinted solid-phase extraction in lab-on-valve format as front end to liquid chromatography: application to the determination of riboflavin in foodstuffs. *Analytical and Bioanalytical Chemistry*. 397: 77–86.

Ollilainen, V., Mattila, P., Varo, P., Koivistoinen, P., and Huttunen, J., 1990. The HPLC determination of total riboflavin in foods. *Journal of Micronutrient Analysis*. 8: 199–207.

Quirós, A.R.B., López-Hernández, J., and Simal-Lozano, J., 2004. Simultaneous determination of thiamin and riboflavin in the sea urchin, *Paracentrotus lividus*, by high-performance liquid chromatography. *International Journal of Food Sciences and Nutrition*. 55: 259–263.

Rudenko, A.O., and Kartsova, L.A., 2010. Determination of water-soluble vitamin B and vitamin C in combined feed, premixes, and biologically active supplements by reversed-phase HPLC. *Journal of Analytical Chemistry*. 65: 71–76.

Scotter, M.J., 2011. Methods for the determination of European Union-permitted added natural colours in foods: a review. *Food Additives and Contaminants Part A: Chemistry Analysis Control Exposure & Risk Assessment*. 28: 527–596.

Tang, X.Y., Cronin, D.A., and Brunton, N.P., 2006. A simplified approach to the determination of thiamine and riboflavin in meats using reverse phase HPLC. *Journal of Food Composition and Analysis*. 19: 831–837.

CHAPTER 19
Isotope Dilution Mass Spectrometry for Niacin in Food

ROBERT J. GOLDSCHMIDT*[a] AND WAYNE R. WOLF[b]

[a] USDA, ARS, BHNRC, Food Composition and Methods Development Laboratory, 10300 Baltimore Avenue, Building 161 Room 102 BARC-East, Beltsville, MD, 20705, USA; [b] Brookeville, MD, USA
*Email: Robert.Goldschmidt@ars.usda.gov

19.1 Introduction

The present methodology used to measure niacin (also called vitamin B_3) in foods consists mainly of three types: microbiological, colorimetric and liquid chromatography (LC) using ultraviolet (UV) detection. AOAC's Official Methods of Analysis for niacin are microbiological (Horwitz and Latimer 2005a) or colorimetric (Horwitz and Latimer 2005b). Microbiological methods are sensitive but laborious and time-consuming. Standard levels recommended in the AOAC methods are in the range 10–40 parts per billion (ppb) niacin. Reported relative standard deviations (RSD) are often near 5% (Hirayama 1998; Rose-Sallin *et al.* 2001; Tyler and Shrago 1980; van Niekerk *et al.* 1984). Colorimetric methods are faster but lack specificity and are not as sensitive.

The most recent reports on niacin measurement describe the use of LC with UV detection (Hirayama 1998; LaCroix *et al.* 1999, 2002; Lombardi-Boccia *et al.* 2005; Rizzolo *et al.* 1991; Saccani *et al.* 2005; Tyler and Shrago 1980; Tyler and Genzale 1990; van Niekerk *et al.* 1984). There can be significant interference from food matrices due to the low specificity at the UV region of interest ($\lambda_{max} = 260$ nm for nicotinic acid). Successful analysis usually requires either significant sample clean-up prior to chromatographic analysis (Hirayama 1998; Juraja et al. 2003; LaCroix et al. 1999, 2002; Tyler and Shrago 1980; Tyler and Genzale 1990; van Niekerk et al. 1984) or else extra chromatographic steps such as column switching (van Niekerk et al. 1984). As interferents may be matrix specific, published LC-UV methods may be limited in applicability. Reported limits of detection (LOD) are generally in the range of 200 ppb to 1 part per million (ppm). Better sensitivity and specificity can be achieved with post-column derivatization and fluorescence detection (Lahely et al. 1999; Rose-Sallin et al. 2001), with reported LOD as low as 5 ppb (Lahely et al. 1999). Most of the published LC methods have not been rigorously validated.

LC coupled with detection by mass spectrometry (MS) offers the potential for excellent sensitivity and specificity. Sample preparations suitable for niacin analysis by LC-UV and other methods should also be suitable for LC-MS. However, chromatographic methods developed for LC-UV or LC-fluorescence methods would in most cases have to be modified due to issues such as ion suppression and problems created by non-volatile species in commonly used mobile phases. Niacin has a relatively low molecular mass (123 Da), and thus in selected ion recording (SIR) mode, there may still be significant interference problems. With instruments that allow a multiple reaction monitoring mode (MRM), however, interference problems can potentially be avoided even without optimized chromatography or sample clean-up.

The method described here makes use of stable isotope dilution mass spectrometry (IDMS) for quantitative analysis of niacin. Isotopically labelled versions of both nicotinic acid and niacinamide are commercially available at a reasonable cost. The use of an isotopically labelled internal standard has distinct advantages in quantitative analysis, as it can correct for analyte losses and makes possible high levels of accuracy and precision (Fassett and Paulson 1989). Sample digestion and clean-up is based on a previously published LC-UV method (LaCroix et al. 1999, 2002; LaCroix and Wolf 2001). This chapter expands on a previous report of LC-IDMS analysis of niacin (Goldschmidt and Wolf 2007), with material from that report used with permission of the publisher.

Niacin occurs in foods in various forms and the interpretation of reported niacin values is not always straightforward. Endogenous niacin can be found in foods in the forms of nicotinic acid and niacinamide, and in free or bound forms (Ball 2006; Eitenmiller and Landen 1999; Friedrich 1988). Some bound forms of niacin, particularly in grains, are not readily bioavailable. Many processed foods are vitamin fortified, including many grain-based ones. In the case of niacin, fortification is normally done using either free nicotinic acid or free niacinamide, with the latter being more common. Sample preparation for niacin analysis typically involves either acid or alkaline digestion, although in cases where only free niacin is of interest, extraction may be sufficient. Alkaline hydrolysis is normally reported to give total niacin levels and acid hydrolysis normally to give a measure of available niacin, but the specific hydrolysis conditions (e.g. acid strength) can affect the completeness of the measurement. From a nutritional standpoint, levels of the amino acid tryptophan are also of interest, due to the *in vivo* conversion of tryptophan to niacin (Angyal 1985; Ball 1998). The conversion efficiency depends on various factors, but a reasonable approximation is given by the so-called 'niacin equivalent', according to which 60 mg of tryptophan is considered equivalent to 1 mg of available niacin. Tryptophan measurement is not considered in this chapter.

In the method discussed here, nicotinic acid is the form of niacin monitored, as the specified acid digestion converts any niacinamide present to nicotinic acid. The method calls for a relatively harsh acid digestion which should give a complete measure of niacin for most food samples; however, in some cases, such as grain sources, an alkaline digestion may be necessary for measurement of total niacin.

19.2 Methodological Considerations

19.2.1 Equilibration Requirement in IDMS

IDMS is recognized as a highly accurate quantitative technique as long as certain conditions are met (Fassett and Paulson 1989). Two crucial conditions are (1) equilibration of the labelled spike and the natural compound of interest and (2) accurate measurement of the relevant isotopic ratios. Equilibration of the spike and natural analogue means that they achieve chemically equivalent states, apart from any isotopic effects, and that they are free to exchange with one another.

Niacin in foods, as mentioned above, can exist in various chemical forms. In the IDMS analysis under consideration, the labelled niacin is added in a single form, which could be either free nicotinic acid or free niacinamide. The spike is best added during the sample weighing process at the start of the analysis before any losses of endogenous niacin can occur. The sample and spike are subjected to an acid digestion which frees chemically bound forms of niacin, subject to the constraints already mentioned, and which converts niacinamide into nicotinic acid. The digestion thus satisfies the equilibration condition, and any niacin losses which might occur thereafter in the analysis would affect the endogenous and labelled niacin proportionately. Thus the ratio of natural niacin to labelled niacin upon which the IDMS measurement is based is preserved throughout the sample treatment. With regard to measurement of the isotopic ratios, it should be noted that there is typically a mass dependence to the response of mass spectrometers though it is possible to correct for it such that such measurements are accurate. The correction is discussed in more detail below.

19.2.2 Sample Preparation and Analysis

Accurate IDMS measurements ultimately depend upon a standard of accurately known concentration. It is important, then, to use as a standard a material of known purity and of good stability. Niacin is not a primary standard material, but it is available in grades which have undergone testing and characterization of purity. The labelled niacin we have used is D_4-nicotinic acid (99.4 atom% D, labelled at the ring carbons), but other labelled forms, including labelled niacinamide, would also work. In our experience, solutions of the standard and labelled niacin prepared in 0.1 M HCl, when stored under refrigeration (4 °C), are stable for over a year, as determined by IDMS comparison to freshly made standards.

Sample amounts, usually in the range from about 0.5 g to 7 g, are chosen so that the expected niacin contents of the final sample solutions are appropriate for MS detection (roughly 50 ppb to 2 ppm). An accurately weighed portion of the isotopically labelled niacin spike solution is added to the sample. The amount of spike should be chosen to roughly match the expected amount of niacin in the sample on a molar basis. The sample container should be appropriate for autoclaving; we use 150 mL 'fleaker' flasks covered with a watch glass. About 10 mL of water and 2 mL of 9 M H_2SO_4 are added, and the samples are then autoclaved at 120 °C for 45 minutes. The original protocol upon which our sample preparation is based calls for a protein precipitation step after autoclaving. It may not be necessary for all samples, but we have generally used it in our work. It is accomplished, after cooling samples to room temperature, by adding 5 mL of 7.5 M NaOH, mixing and then re-acidifying by addition of 1.5 mL of 9 M H_2SO_4. Following the protein precipitation step, samples are gravity filtered using Whatman Grade 5V or similar filter papers, and diluted to a total volume of 35 mL with water. A solid-phase extraction (SPE) is then performed: aromatic sulfonic acid SPE columns are pretreated with one column volume of methanol followed by one column volume of 0.1 M pH 2.1 H_3PO_4/ NaH_2PO_4 buffer. The filtered samples are then applied to the columns (usually about one half of a column volume). Samples are washed with two column volumes of the pH 2.1 phosphate buffer and then niacin is eluted with two column volumes of 0.1 M pH 5.6 acetic acid/sodium acetate buffer. If necessary, sample volumes can be adjusted by evaporation or addition of water or buffer. Note that the SPE step is not absolutely necessary for operation in MRM mode but is essential when using SIR mode. Even for MRM mode, however, it does reduce the amount of potentially fouling or otherwise troublesome material produced during the digestion that is injected into the LC-MS system.

Samples are analysed by LC-MS. For the samples discussed below, the system consisted of an Agilent 1100 high-performance liquid chromatography (HPLC) system and a Waters Quattromicro triple quadrupole mass spectrometer. The HPLC was operated in reversed phase mode using a C18 column. In most cases gradient elution was used with a binary mobile phase consisting of (A) water with 0.1% formic acid and (B) acetonitrile with 0.1% formic acid. This results in a retention time for nicotinic acid of about 2.5 minutes. Nicotinic acid is fairly hydrophilic and its retention time in this case corresponds to about 1.5 column volumes. For some food samples, this chromatographic treatment would not isolate nicotinic acid sufficiently for a mass spectrometer operated in SIR mode but, due to the extra stage of selectivity, it works perfectly well for operation in MRM mode. Other chromatographic approaches that would

Table 19.1 Ions monitored for determination of niacin by isotope dilution mass spectrometry using positive electrospray ionization. Both natural and labelled nicotinic acid are protonated in positive ion electrospray ionization, giving quasi-molecular ions at mass-to-charge ratios (m/z) of 124 and 128, respectively, which can be monitored directly in selected ion recording (SIR) experiments and selected as the parent ions in multiple reaction monitoring (MRM) experiments. In MRM experiments protonated nicotinic acid can be induced to produce daughter ions at several other m/z values, but the given transitions are the ones with the highest signal intensity.

Mass spectrometer mode	m/z of ion or transition monitored for natural nicotinic acid	m/z of ion or transition monitored for labeled nicotinic acid
SIR	124	128
MRM	124 → 80	128 → 84

provide better retention for nicotinic acid, such as hydrophilic interaction liquid chromatography (HILIC), could also be used and might be a better match for systems without MRM capabilities. More recently developed approaches that offer improved resolution, such as ultra-performance liquid chromatography (UPLC), should also work well. Positive ion electrospray (ESI) is appropriate for ionization. Table 19.1 shows the mass-to-charge ratios (m/z) of the ions monitored for nicotinic acid in both SIR and MRM modes when using positive ESI.

19.3 IDMS Calculations

Calculations of niacin levels proceed from the ratio defined by:

$$R = \frac{AreaM_1}{AreaM_2} = \frac{(X_{nat})(A_{1nat}) + (X_{sp})(A_{1sp})}{(X_{nat})(A_{2nat}) + (X_{sp})(A_{2sp})} \quad (1)$$

where areas M_1 and M_2 are integrated peak areas for the selected, isotopically related ions (m/z 124 and m/z 128 for SIR, or 124 → 80 and 128 → 84 for MRM, respectively, in this work); X_{nat} and X_{sp} refer to the amounts of the natural and isotopic spike material present in the sample, respectively; and the A terms refer to the relevant isotopic abundances of the natural and spike materials. This equation is solved for X_{nat}, the ratio being experimentally determined, and all other terms being known. In the case of natural and D_4-niacin, there is no contribution of the natural to the signal monitored for the spike or of the spike to the signal monitored for the natural. Thus solving for X_{nat} gives:

$$X_{nat} = C \times R \times X_{sp} \quad (2)$$

where R is experimentally determined, X_{sp} is known, and C is a factor introduced to correct for instrumental and other bias. C is obtained by analysis of a standard containing known amounts of both natural niacin and labelled niacin. It is then given by:

$$C = \frac{\frac{X_{nat}}{X_{sp}}}{R} \qquad (3)$$

where all terms refer to the standard used. All niacin measurements are thus calibrated to the natural niacin standard.

19.4 Isotopic Effect in Fragmentation of Nicotinic Acid

For MRM mode analysis of niacin by IDMS one should be aware of an isotopic effect in the fragmentation of nicotinic acid. The natural nicotinic acid daughter spectrum (Figure 19.1b) suggests that the transitions 124 → 80 or 124 → 78 would be suitable for MRM analysis. The fragment at m/z 80 corresponds to a loss of CO_2 and that at m/z 78 to losses of H_2O and CO. For the latter, a corresponding transition in electron impact (EI) ionization is 123 → 78 and involves losses of hydroxyl and carbon monoxide, but it also involves an exchange between the hydroxylic and β-hydrogens (Neeter and Nibbering 1971). The presence of the ring deuteriums in D_4-nicotinic acid results in different relative intensities for the corresponding fragment peaks (m/z 84 and m/z 81, Figure 19.1a) compared with those of natural nicotinic acid. Use of a correction factor in IDMS calculations as discussed earlier can correct for this source of bias. In our work with niacin, correction factors used in MRM mode monitoring the transitions 124 → 80 (natural) and 128 → 84 (D4) have generally been within 5% of unity, where unity would be consistent with no bias.

19.5 Illustrative Results for Food Samples

Table 19.2 contains the results of niacin determinations performed in MRM mode for some grain-based food materials, including appropriate reference materials, and compares them to expected levels. For the commercial food materials, expected values are as listed on the packaging. The US National Institute of Standards and Technology (NIST) does not provide a reference value for niacin for RM 8437; however, there have been determinations of its niacin value reported in the literature (LaCroix *et al.* 1999; Tanner *et al.* 1988). The expected value for RM25C, an in-house reference material used by an analytical testing and consulting laboratory, was provided by the producer.

Precision obtained for the samples in Table 19.2, expressed as relative standard deviations (RSD), ranges from 0.5% to 2.7%. Mean levels of niacin obtained for the commercial flour and cereal samples are higher than the expected values, but this is not surprising. Due to labelling requirements for fortified foods (Bender *et al.* 1998), such foods are often fortified in excess of the stated amounts. Results for the flour (RM 8437) and cereal (RM25C) reference

Isotope Dilution Mass Spectrometry for Niacin in Food 291

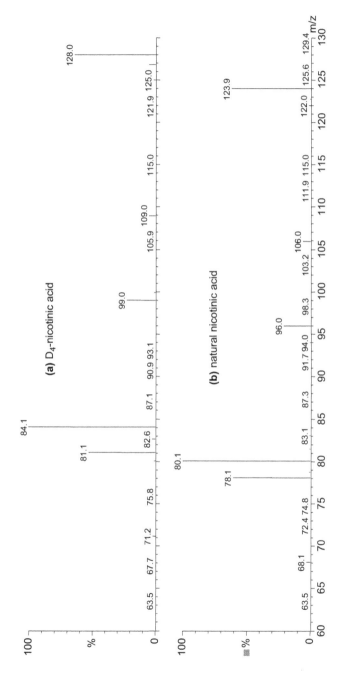

Figure 19.1 Daughter ion mass spectra of natural nicotinic acid and D_4-nicotinic acid obtained using electrospray ionization (ESI) in positive ion mode and collision gas argon. The daughter ion spectra of natural and D_4-nicotinic acid obtained under similar conditions are similar, allowing for the isotopic mass differences, but there is some difference in the relative abundances of the fragment peaks. Data are from Goldschmidt and Wolf (2007), with permission from the publisher.

Table 19.2 Results of niacin determinations for food samples. Niacin determinations by LC-IDMS are compared to expected values for six grain-derived food materials. Expected levels for the commercial flour and cereal samples (Brands A, B, C, and D) are according to the package labelling, although it should be noted that actual vitamin levels of vitamin fortified foods are often higher than the stated amounts. The determined niacin levels for these samples, listed in the third column with their 95% confidence limits, are from 30% to 70% higher than the claimed levels. Determined niacin levels for the two reference materials (RM) are in agreement with the expected levels. Data are from Goldschmidt and Wolf (2007), with permission from the publisher.

Sample	ppm niacin, expected (s)[a]	ppm niacin by LC-IDMS[b]
Brand A wheat flour	53	79.4 ± 2.4 $n=4$
Brand B corn meal	40	68.0 ± 1.1 $n=4$
Brand C brown rice flour	46	59.4 ± 2.7 $n=4$
RM 8437 (hard red springwheat)	39.4 $n=2$[c]	40.0 ± 0.3 $n=4$
	40.6 (5.4) $n=6$[d]	
Cereal Brand D	161	279.0 ± 11.9 $n=4$
Medallion Cereal RM25C	302.1 (31.1)	267.4 ± 2.9 $n=4$

[a] Values in parentheses (s) are sample standard deviations and n is the number of samples analyzed, where applicable.
[b] Given ranges are means with 95% confidence limits.
[c] From Tanner et al. 1988.
[d] From LaCroix et al. 1999.

materials are in agreement with the expected values. The value reported for RM 8437 in Table 19.2 is on a dry mass basis (measured moisture level 8.9%, following the procedure from the NIST Report of Investigation for RM 8437). All other values reported in Table 19.2 are on an 'as is' basis. Note that the levels as determined by IDMS for these fortified grain products include the combined added (fortified) niacin plus the endogenous niacin released by the digestion process. Endogenous levels of niacin in these materials are perhaps in the range 10–30 ppm, as estimated from levels given for similar, unfortified materials in the USDA Nutrient Database for Standard Reference, Release 23 (US Department of Agriculture 2010).

Table 19.3 contains results of niacin determinations of some milk samples, again in MRM mode, including RM 8435 (whole milk powder) from NIST. These samples are not fortified and contain niacin at considerably lower levels than the samples in Table 19.2. Precision obtained for these samples is in the range 1–5% RSD. An example of an extracted ion chromatogram for commercial milk Brand F is shown in Figure 19.2. The USDA Nutrient Database for Standard Reference, Release 23 (US Department of Agriculture 2010) lists a value of 0.89 ppm niacin for 'milk, whole, 3.25% milkfat' and a value of 0.84 ppm niacin for 'milk, producer, fluid, 3.7% milkfat'. We thus assume that expected values for niacin for whole milk samples are near 1 ppm. The levels

Isotope Dilution Mass Spectrometry for Niacin in Food 293

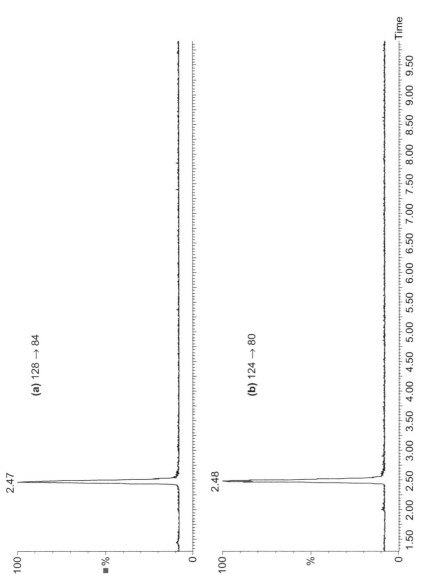

Figure 19.2 Extracted ion chromatograms obtained for commercial milk sample Brand F spiked with D_4-niacin. These extracted ion chromatograms illustrate that the retention times are essentially the same for natural and D_4-niacin, and that even at relatively low concentrations signal-to-noise ratios (S/N) are good and the chromatograms are essentially free of interfering signal. To obtain the chromatograms, 5 microlitres (μL) of a sample that was about 90 ppb each in natural niacin and D_4-niacin were injected. The milk sample from which the sample was prepared was determined to be about 1 ppm in niacin. Data are from Goldschmidt and Wolf (2007), with permission from the publisher.

Table 19.3 Results of niacin determinations for milk samples. Niacin determinations by liquid chromatography–isotope dilution mass spectrometry (LC-IDMS) are compared to expected values for four milk samples. Expected niacin levels for milk are roughly 1 ppm, according to the USDA Nutrient Database for Standard Reference (US Department of Agriculture 2010) and results obtained for two commercial milk samples (Brands F and G) are a little under 1 ppm. The result for sample NFY0409F6 is about 30% lower, but is consistent with results obtained for other milk samples from the same source. In addition, the niacin level for NFY0409F6 was estimated by a standard additions experiment, the result from which is in agreement with the estimate from the normal LC-IDMS procedure. The level obtained for the reference material (RM) RM 8435 whole milk powder, reported on a dry mass basis, is in agreement with the reference value. Data are from Goldschmidt and Wolf (2007), with permission from the publisher.

Sample	Expected niacin level, ppm	ppm niacin by LC-IDMS[a]
RM 8435 (whole milk powder, dry)	7.65 ± 0.93	7.28 ± 0.37 $n=4$
Whole Milk Brand F	1 ppm	0.911 ± 0.070 $n=4$
Whole Milk Brand G	1 ppm	0.968 ± 0.016 $n=4$
NFY0409F6	1 ppm	0.638 $n=1$
NFY0409F6	1 ppm	0.657 (by standard additions)

[a]Given ranges are means with 95% confidence limits.

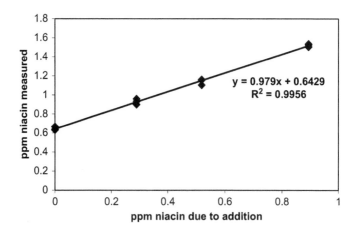

Figure 19.3 Standard additions curve for milk sample NFY0409F6. Four levels of a standard natural niacin and a fixed amount of D_4-niacin were added to milk sample NFY0409F6 in a standard additions experiment. The level of niacin determined for the sample, 0.657 ppm niacin, is in agreement with the level obtained by a normal IDMS determination. The linear relationship observed is consistent with complete recovery of added niacin and with an absence of matrix effects. Data are from Goldschmidt and Wolf (2007), with permission from the publisher.

obtained for the commercial samples are consistent with the expected level and that obtained for RM 8435 is in agreement with the reference value. The level reported for RM 8435 is on a dry mass basis (measured moisture level of 2.7%). The value obtained for sample NFY0409F6 is lower than the others, but is in the range of other milk samples we have tested from the same supplier. Sample NFY0409F6 was also determined by a standard additions experiment using four levels of added niacin and a constant amount of the D_4-niacin spike (see Figure 19.3). The standard additions curve exhibits good linearity and has a slope close to unity, consistent with complete recovery of added niacin and with an absence of matrix interference. The level obtained from the additions experiment, 0.657 ppm niacin, is close to the level obtained from the single LC-IDMS determination listed in Table 19.3. The solutions injected into the LC-MS system for the milk samples were 50–100 ppb in niacin. In separate experiments we have estimated a limit of detection (LOD) for the method of a little less than 1 ppb niacin when using 5 µL injections, a 200 or more times improvement in comparison to values typically reported for LC-UV determinations of niacin.

19.6 Concluding Remarks

For those with suitable MS capabilities, perhaps the chief obstacle to the adoption of IDMS for chemical measurements is the cost of isotopically labelled compounds. In the case of niacin, however, labelled analogues are available at moderate prices. IDMS should thus be considered an attractive option for the analysis of niacin in foods. Subject to certain conditions, it is capable of excellent accuracy, precision, selectivity and sensitivity. MS measurement techniques in general have the latter two characteristics, but MS signals are often subject to fluctuations which limit precision. Quantitative MS measurements are thus often made using an internal standard which, besides helping to correct for losses during sample processing, can also to some degree address changes in instrument response. The isotopically labelled analogues used in IDMS can be considered as ideal internal standards because they are virtually identical chemically and physically to their natural analogues. As long as both natural analyte and labelled spike are subject to the same treatment, any losses that might occur are experienced proportionately. Also, because the analogues behave virtually identically chromatographically, they are detected at very close to the same time, and so IDMS can correct for even rapid fluctuations in instrument response. Thus the measured ratio of natural analyte to labelled analogue should accurately reflect the ratio at the time of spiking.

The IDMS method using an acid digestion and reversed phase LC as described here should be applicable to niacin measurement in a wide variety of food matrices, but IDMS treatments should also be valid when using alkaline digestions, extractions and other chromatographic approaches. The key methodological requirement is that the labelled spike achieves equilibration with the analyte to be measured.

Summary Points

- This chapter is about the determination of niacin in foods using isotope dilution mass spectrometry (IDMS) and liquid chromatography (LC).
- Niacin, also called vitamin B_3, refers to two vitamer forms: nicotinic acid and niacinamide.
- Both free nicotinic acid and free niacinamide occur naturally in foods but, more commonly, niacin occurs in foods in various chemically bound forms.
- In order to extract niacin and to release the chemically bound forms, a digestion step is necessary as part of the sample preparation.
- Digestion can be accomplished by an acid or alkaline treatment; in this method we use an acid digestion that converts niacinamide to nicotinic acid.
- An isotopically enriched version of nicotinic acid is added early in the sample preparation procedure, prior to digestion.
- The added, isotopically labelled nicotinic acid and the nicotinic acid deriving from the endogenous niacin of the food sample have virtually identical physical and chemical properties and, by virtue of the digestion step, are assumed to be in a state of equivalence in regard to their chemical and physical environment thereafter in the analysis.
- Gravity filtration and solid-phase extraction (SPE) are used to separate nicotinic acid from many of the extraneous digestion products.
- Reversed phase LC using a C18 column is used for further isolation of nicotinic acid.
- Detection is accomplished using a triple quadrupole mass spectrometer operated in multiple reaction monitoring (MRM) mode.
- The accuracy of the method is verified by results obtained for reference materials.
- Other method performance criteria, such as precision, selectivity and sensitivity, compare very favourably with more common methods of niacin analysis in foods.

Key Facts about Isotope Dilution Mass Spectrometry

- IDMS is a technique for quantitative determinations by mass spectrometry.
- A known amount of an isotopically enriched analogue of the analyte to be determined is added to the sample as early in the sample treatment protocol as is practical; often it is added at the time of sample weighing.
- The isotopically enriched analogue has the same chemical formula and structure as the analyte of interest, but it has an isotopic distribution significantly different from that which is found naturally.
- For example, an isotopically enriched version of a compound containing C, H, N and O would be one that contained a high percentage of ^{13}C, ^{2}H, ^{15}N or ^{18}O at one or more positions in the molecule.

- The enriched analogue is almost identical to the natural compound in its chemical and physical properties, but because of the isotopic mass difference, it can be detected separately by mass spectrometry.
- Because of their similar properties, any losses undergone by the compound of interest during sample treatment are experienced to a proportional degree by the labelled analogue, as long as the two experience fully equivalent conditions, *i.e.* they must be free to exchange with one another in relation to the chemical and physical environment they experience throughout the sample treatment.
- As with an internal standard method of quantitation, of which IDMS can be considered a special case, one measures the ratio of two signals—in this case the signal corresponding to the labelled analogue and the signal corresponding to the natural compound.
- By the use of the ratio and because the labelled analogue and analyte of interest are detected at very close to the same time, IDMS also allows for correction of instrumental bias that can affect other kinds of quantitative approaches used in mass spectrometry.

Definitions of Words and Terms

Dalton. A dalton (Da) is a unit used for expressing atomic and molecular masses. It is equivalent to one-twelfth of the mass of an atom of carbon-12. The same unit is also referred to as a unified atomic mass unit, with abbreviation u, or, in older usage, an atomic mass unit (amu).

Electrospray ionization. A mass spectrometric ionization technique, commonly used when coupling liquid chromatography to mass spectrometry, in which a liquid is nebulized upon emergence from a metal capillary held at high voltage (a few kilovolts) relative to ground. Usually the chromatographic mobile phase or liquid contains a small amount of an organic acid (*e.g.* 0.1% formic acid) or other electrolytic additive. The droplets produced upon nebulization decay, leading to ejection of various ionic species. Electrospray is a 'soft' ionization technique in that it usually produces quasi-molecular ions that undergo little or no decomposition. For example, when formic acid is used as the additive, analyte ions observed in positive ion mode are typically protonated.

Isotope. Atoms having the same number of protons (*i.e.* of a particular element) having different numbers of neutrons.

Limit of detection. The limit of detection (LOD) of a chemical measurement technique is the lowest amount or concentration of an analyte that gives rise to a signal that can reliably be distinguished from a measurement made on a blank. It is usually expressed as the analyte level for which the observed signal corresponds to some number of standard deviations, often three, of the signal observed for the blank.

Mass-to-charge ratio. Mass-to-charge ratio (m/z) refers to the mass of an ion in Daltons (Da) divided by its net number of charges. Mass spectrometry, through the use of electric and/or magnetic fields, separates ions according to their mass-to-charge ratios.

MRM. MRM is an abbreviation for multiple reaction monitoring—a mode of mass spectrometric acquisition in which one or more parent ion-daughter ion decomposition reactions is monitored. As used with triple quadrupole mass spectrometers, parent ions of a particular mass-to-charge ratio selected using the first quadrupole interact with a collision gas, usually argon, in the second quadrupole. The purpose of the collision gas is to increase the energy of the selected ions and to induce them to undergo a characteristic decomposition into two or more fragments. Charged fragments (daughter ions) can be selected according to their mass-to-charge ratios with the third quadrupole, and can then be detected. Detection is thus limited to selected daughter ions that come only from selected parent ions. The additional level of selectivity in MRM means that the only signal collected is the desired signal; extraneous signal from compounds of the same mass or from other sources of chemical noise is virtually eliminated.

Reference material. As it pertains to chemical measurements, a reference material (RM) is a material with a reliably characterized level of one or more of its chemical components. Reference materials are used for verification of measurement results, instrument calibration and validation of measurement procedures. The material must have suitable stability and homogeneity, such that it can reliably be used for such purposes.

Reversed phase. Reversed phase refers to a type of liquid chromatography that uses a column packing characterized by its non-polar nature. For example, many reversed phase columns are based on octadecylsilyl, or C18, groups that are chemically bonded to silica particles. Non-polar analytes are more strongly attracted to these groups than are polar analytes. The mobile phase usually consists of water and an organic solvent such as acetonitrile or methanol and may also contain additives for various purposes, such as establishing a pH buffer. The mobile phase may be run at a fixed composition (isocratic) or in a gradient whereby the proportion of the organic part is increased as the run progresses. In either case, polar analytes are eluted before non-polar analytes in reversed phase chromatography.

Reverse isotope dilution. Reverse isotope dilution refers to the determination of the level of the isotopically labelled analogue used as a spike in an isotope dilution experiment by addition of the spike to a standard containing a known amount of the natural form of the analyte. Measurement is made as in normal isotope dilution mass spectrometry, but in this case it is the level of the labelled form that is treated as the unknown.

Selectivity. As it pertains to chemical measurements, selectivity refers to the ability of a measurement technique to distinguish analytes from potentially interfering compounds.

Sensitivity. As it pertains to chemical measurements, sensitivity strictly refers to the change in signal strength associated with a change in level of the analyte being measured. Thus in comparing two techniques of measurement, the one which exhibits a larger increase in signal strength for a given increase in level of the analyte would be considered the more sensitive for that analyte. In common usage sensitivity also refers to the ability of a measurement technique to measure low levels of an analyte. In this sense the technique which has the lower limit of detection for the analyte would be considered the more sensitive one.

SIR. SIR is an abbreviation for selected ion recording—a mode of mass spectrometric acquisition in which detection is limited to ions of one or more selected mass-to-charge ratios. In comparison to scanning modes, in which all ions over a selected range of mass-to-charge ratios are monitored, greater sensitivity is achieved for the ions of interest, since the bulk of the detection time is devoted to only those ions.

Vitamer. The term vitamer is used to signify one of two or more different chemical forms of a particular vitamin. Vitamers are structurally related and exhibit biological activity related to the vitamin but have some difference in their chemical structure.

List of Abbreviations

AOAC	Association of Official Analytical Chemists
D	deuterium (^{2}H)
Da	Dalton
EI	electron impact (a method of MS ionization)
ESI	electrospray ionization
HILIC	hydrophilic interaction liquid chromatography
HPLC	high performance liquid chromatography
IDMS	isotope dilution mass spectrometry
λ_{max}	lambda max (wavelength of maximum absorbance in UV spectrum)
LC	liquid chromatography
LOD	limit of detection
M	molar
mM	millimolar
MRM	multiple reaction monitoring
MS	mass spectrometry
m/z	mass-to-charge ratio
NIST	National Institute of Standards and Technology
RM	reference material
ppb	parts per billion
ppm	parts per million
RSD	relative standard deviation
SIR	selected ion recording
S/N	signal-to-noise ratio
SPE	solid phase extraction
µL	microlitre
UPLC	ultra-performance liquid chromatography
UV	ultraviolet

References

Angyal, G., 1985. Contribution of tryptophan to niacin equivalent of infant formula: microbiological method. In: *Production, Regulation, and Analysis of Infant Formula: A Topical Conference*. May 14–16, 1985. Association of Official Analytical Chemists, Virginia Beach VA, USA, pp. 152–159.

Ball, G. F. M., 2006. *Vitamins in Foods: Analysis, Bioavailability, and Stability*. CRC Press, Boca Raton, FL, USA, pp. 177–188.

Bender, M.M., Reder, J.J., and McClure, F.D., 1998. Guidelines for industry: nutrition labeling manual—a guide for developing and using data bases. Available at: http://www.fda.gov/Food/GuidanceComplianceRegulatoryInformation/GuidanceDocuments/FoodLabelingNutrition/ucm063113.htm. Accessed 15 May 2012.

Eitenmiller, R.R., and Landen, Jr. W.O., 1999. *Vitamin Analysis for the Health and Food Sciences*. CRC Press, Boca Raton, FL, USA, pp. 339–367.

Fassett, J. D., and Paulsen, P. J., 1989. Isotope dilution mass spectrometry for accurate elemental analysis. *Analytical Chemistry*. 61: 643A–649A.

Friedrich, W., 1988. *Vitamins*. Walter de Gruyter, Berlin, Germany, pp. 473–542.

Goldschmidt, R.J., and Wolf, W.R., 2007. Determination of niacin in food materials by liquid chromatography using isotope dilution mass spectrometry. *Journal of AOAC International*. 90: 1084–1089.

Hirayama, S., 1998. Determination of small amounts of niacin in vinegar: comparison of liquid chromatographic method with microbiological methods. *Journal of AOAC International*. 81: 1273–1276.

Horwitz, W., and Latimer, G.W., (ed.), 2005a. Methods 944.13 and 985.34. In: *Official Methods of Analysis of AOAC International*, 18th ed. AOAC International, Gaithersburg, MD, USA. Available at http://www.eoma.aoac.org/.

Horwitz, W., and Latimer, G.W., (ed.), 2005b. Methods 961.14, 975.41, 981.16 and 968.32. In: *Official Methods of Analysis of AOAC International*, 18th ed. AOAC International, Gaithersburg, MD, USA. Available at http://www.eoma.aoac.org/.

Juraja, S.M., Trenerry, V.C., Millar, R.G., Scheelings, P., and Buick, D.R., 2003. Asia Pacific Food Analysis Network (APNFAN) Training Exercise: The determination of niacin in cereals by alkaline extraction and high performance liquid chromatography. *Journal of Food Composition and Analysis*. 16: 93–106.

LaCroix, D.E., Wolf, W.R., and Vanderslice, J.T., 1999. Determination of niacin in infant formula and wheat flour by anion-exchange liquid chromatography with solid phase extraction cleanup. *Journal of AOAC International*. 82: 128–133.

LaCroix, D.E., and Wolf, W.R., 2001. Determination of niacin in infant formula by solid-phase extraction and anion exchange liquid chromatography. *Journal of AOAC International*. 84: 789–804.

LaCroix, D.E., Wolf, W.R., and Chase, Jr. G.W., 2002. Determination of niacin in infant formula by solid phase extraction/liquid chromatography: peer-verified method performance validation. *Journal of AOAC International*. 85(3): 654–664.

Lahely, S., Bergaentzle, M., and Hasselmann, C., 1999. Fluorometric determination of niacin in foods by high-performance liquid chromatography with post-column derivatization. *Food Chemistry*. 65: 129–133.

Lombardi-Boccia, G., Lanzi, S., and Aguzzi, A., 2005. Aspects of meat quality: trace elements and b vitamins in raw and cooked meats. *Journal of Food Composition and Analysis*. 18: 39–46.

Neeter, R., and Nibbering, N.M.M., 1971. Mass spectrometry of pyridine compounds ii: hydrogen exchange in the molecular ions of isonicotinic and nicotinic acid as a function of internal energy. *Organic Mass Spectrometry*. 5: 735–742.

Rizzolo, A., Baldo, C., and Polesello, A., 1991. Application of high-performance liquid chromatography to the analysis of niacin and biotin in italian almond cultivars. *Journal of Chromatography*. 553: 187–192.

Rose-Sallin, C., Blake, C.J., Genoud, D., and Tagliaferri, E.G., 2001. Comparison of microbiological and HPLC-fluorescence detection methods for determination of niacin in fortified food products. *Food Chemistry*. 73: 473–480.

Saccani, G., Tanzi, E., Mallozzi, S., and Cavalli, S., 2005. Determination of niacin in fresh and dry cured pork products by ion chromatography: experimental design approach for the optimisation of nicotinic acid separation. *Food Chemistry*. 92: 373–379.

Tanner, J.T., Angyal, G., Smith, J.S., Weaver, C., Bueno, M., Wolf, W.R., and Ihnat, M., 1988. Survey of selected materials for use as an organic nutrient standard. *Frezenius Zeitschrift fur Analytische Chemie*. 322: 701–703.

Tyler, T.A., and Shrago, R.R., 1980. Determination of niacin in cereal samples by HPLC. *Journal of Liquid Chromatography*. 3: 269–277.

Tyler, T.A., and Genzale, J.A., 1990. Liquid chromatographic determination of total niacin in beef, semolina, and cottage cheese. *Journal of the Association of Official Analytical Chemists*. 73: 467–469.

US Department of Agriculture, Agricultural Research Service, 2010. USDA Nutrient Database for Standard Reference, Release 23. Nutrient Data Laboratory Home Page. Available at: http://www.ars.usda.gov/nutrient-data. Accessed 6 June, 2011.

van Niekerk, P.J., Salomien, C.C., Smit, E., Strydom, S.P., and Armbruster, G., 1984. Comparison of a high-performance liquid chromatographic and microbiological method for the determination of niacin. *Journal of Agricultural and Food Chemistry*. 32: 304–307.

CHAPTER 20
Analysis of Pantothenic Acid (Vitamin B_5)

TSUTOMU FUKUWATARI* AND KATSUMI SHIBATA

Laboratories of Food Science and Nutrition, Department of Life Style Studies, School of Human Cultures, The University of Shiga Prefecture, 2500 Hassaka, Hikone, Japan
*Email: fukkie@shc.usp.ac.jp

20.1 Introduction

Biological, chemical and physiological methods are available for the measurement of pantothenic acid (vitamin B_5). However, the development of analytical methods for pantothenic acid lags far behind that for other vitamins. A classical microbiological assay remains the most practical method to measure pantothenic acid, mainly because there has been little interest in assessing pantothenic acid deficiency.

As indicated by its name, which is derived from the Greek word 'pantos' meaning everywhere, pantothenic acid is widely distributed among animals and plants, and therefore a low pantothenic acid diet is unlikely. Historically, only one report showed that burning feet syndrome was plausibly a pantothenic acid deficiency among malnourished populations imprisoned during World War II, and their conditions were improved by supplementation with pantothenic acid but not other B group vitamins (Gopalan 1946). Taking the pantothenic acid antagonist ω-methyl pantothenic acid induced pantothenic acid deficiency, while feeding a pantothenic acid-free diet for more than eight weeks barely induced the deficiency (Fry *et al.* 1976; Hodges *et al.* 1958). Because there is no

---

Food and Nutritional Components in Focus No. 4
B Vitamins and Folate: Chemistry, Analysis, Function and Effects
Edited by Victor R. Preedy
© The Royal Society of Chemistry 2013
Published by the Royal Society of Chemistry, www.rsc.org

Table 20.1 Pantothenic acid contents of foods listed in Standard Tables of Food Composition in Japan, 2010.

Food (description)	Content (mg/100 g edible portion)	Food (description)	Content (mg/100 g edible portion)
Wheat flour (soft flour)	0.53	Spinach (boiled)	0.13
White table bread	0.47	Orange (raw)	0.36
Spaghetti (boiled)	0.25	Apples (raw)	0.09
Cooked paddy rice	0.25	Mushrooms (boiled)	1.43
Corn (whole grain, raw)	0.57	Tuna (raw)	0.36
Sweet potatoes (baked)	1.30	Salmon (baked)	1.47
Potatoes (steamed)	0.52	Beef (Inside round, baked)	1.08
Peas (boiled)	0.39	Pork (loin, baked)	1.19
Soy beans (boiled)	0.29	Chicken (thigh with skin, baked)	1.74
Peanuts (roasted)	2.19	Liver (pork, raw)	7.19
Celery (raw)	0.26	Eggs (boiled)	1.35
Tomatoes (raw)	0.17	Milk	0.55
Carrots (without skin, boiled)	0.27	Cheese (natural, cheddar)	0.43

possibility of habitual consumption of a pantothenic acid-free diet for several months without suffering any other nutrient deficiency, a lack of interest in pantothenic acid deficiency hampers the development of analytical methods.

Pantothenic acid is measured in the blood, urine and food. Determination of blood and urinary pantothenic acid contents can be a biomarker for the evaluation of pantothenic acid status. Measurement of pantothenic acid in as many foodstuffs as possible gives important information for determining pantothenic acid status. For example, the Standard Tables of Food Composition in Japan 2010 records pantothenic acid contents for 1878 foodstuffs (Ministry of Education, Culture, Sports, Science and Technology 2010). The pantothenic acid content of a number of foodstuffs is shown in Table 20.1.

Because pantothenic acid must be measured in food, regardless of deficiency, analytical methods have been developed through measurement in foodstuffs. These techniques have then been applied to measurement of pantothenic acid in blood and urine. In this chapter, the development of analytical methods is first reviewed before the characteristics and issues are explained for each method.

20.2 Extraction for Measurement

Pantothenic acid exists in foodstuffs mainly in bound forms such as a component of coenzyme A (CoA), acyl-CoA, acyl-carrier protein or phosphopantetheine. These bound forms are digested to pantothenic acid or pantetheine in the digestive tract and these compounds are absorbed by the intestine. Because pantetheine is taken up rapidly in the liver, pantothenic acid is circulated in the plasma and excreted into the urine.

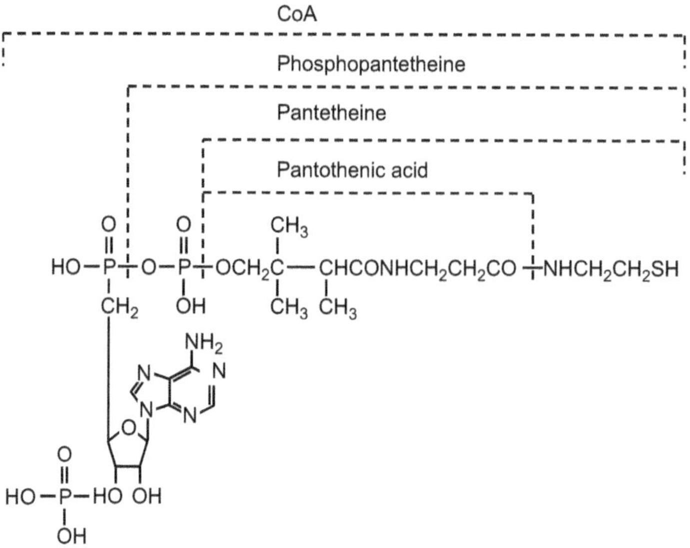

Figure 20.1 Structure of pantothenic acid and bound form coenzyme A (CoA), phosphopantetheine and pantetheine.

No analytical method has been developed that will simultaneously measure the bound forms in foodstuffs, so conversion of the bound forms to the free form is required to determine pantothenic acid (Figure 20.1). Several enzyme treatments have been proposed for liberation of the bound form from foodstuffs and a mixture of alkaline phosphatase and pantetheinase extracted from pigeon liver has been recommended (Gonthier *et al.* 1998a). However, endogenous enzymes contain pantothenic acid and this must be removed from the enzyme solution before use. In contrast, pantothenic acid can be determined directly in plasma and urine samples without enzyme treatment. Red blood cells contain abundant amounts of CoA and enzyme treatment is required to measure pantothenic acid levels in blood.

20.3 Analytical Methods

A summary of analytical development is shown in Table 20.2 and the characteristics of each method are summarized in Table 20.3.

20.3.1 Microbiological Assay

Following the discovery of pantothenic acid in 1933, several biological methods were reported based on the growth response to pantothenic acid of chicks, yeast and bacteria such as *Lactobacillus*, *Streptococcus* and *Proteus*. Microbiological assays have been used more commonly than animal tests because they are cheaper, faster and simpler methods. *Lactobacillus*, in particular, responds well

Table 20.2 Principal methods for pantothenic acid determination.

Researchers	Year	Method
Skeggs and Wright	1944	Microbiological assay
Wyse et al.	1979	RIA
Morris et al.	1980	ELISA
Song et al.	1990	ELISA
Akada	1986	HPLC-UV
Endo et al.	1991	HPLC-UV
Romera et al.	1996	HPLC-UV
Blanco et al.	1995	HPLC-fluorimetric
Pakin et al.	2004	HPLC-fluorimetric
Takahashi et al.	2009	HPLC-fluorimetric
Banno et al.	1990	GC-MS
Rychlik	2000	GC-MS
Rychlik	2003	LC-MS/MS[a]
Mittermayr et al.	2004	LC-MS
Haughey et al.	2005	Optical biosensor-based immunoassay

[a]Liquid chromatography with tandem mass spectrometry.

and is easily handled; thus the microbiological assay using *Lactobacillus plantarum* (ATCC 8014) has been the most popular method for pantothenic acid measurement in blood, urine and foodstuffs.

Lactobacillus plantarum (ATCC 8014) reacts to free pantothenic acid but not to pantetheine or bound forms. Using this microorganism can therefore determine free or total pantothenic acid content after liberation from bound forms. The advantages of this microbiological assay lie in the ease of maintenance of cultures, rapid growth and absence of the need for expensive equipment. Taken together with its approval as the official method for determination of pantothenic acid in foodstuffs by AOAC International, the microbiological method using *Lactobacillus plantarum* (ATCC 8014) has been the most practical method for pantothenic acid determination (AOAC 1996).

The method's disadvantage derives from the fact that it is a non-chemical determination. When a microorganism reacts to compounds that do not show pantothenic activity in humans, the measured values indicate pantothenic acid with pseudopantothenic acid. However, values determined by microbiological assay using *Lactobacillus plantarum* (ATCC 8014) are the same as those by other chemical analyses such as radioimmunoassay (RIA), high-performance liquid chromatography (HPLC) methods and liquid chromatography–mass spectrometry (LC-MS) methods in blood, urine and foodstuffs, and thus there seems to be a low risk of this method picking up pseudopantothenic acid (Rychlik and Roth-Maier 2005; Srinivasan et al. 1981; Takahashi et al. 2009). However, when samples contain compounds such as antibiotics which prevent growth of the microorganism in question, the microbiological assay fails.

It takes half to one day for sufficient microbial growth to be measured by a spectrophotometer, while two or three days are required from sample preparation to data analysis. The limit of quantification (LOQ) is 20 pmol/tube (*ca.* 4 ng/tube) in the microbiological assay using *Lactobacillus plantarum*

Table 20.3 Comparison of methods for pantothenic acid determination. A: excellent, B: good, C: below average, D: poor, –: not reported.

Method	Sensitivity	Precision	Handling	Time-consuming	Cost for equipment	Running cost	Measurement		
							Blood	Urine	Foods
Microbiological assay	C	D	A	D	A	A	C	B	B
RIA	C	C	B	C	C	B	–	B	B
ELISA	C	C	B	C	B	B	C	–	B
HPLC-UV	D	C	B	B	C	B	–	–	C
HPLC-fluorimetric	B	B	B	B	C	B	–	B	B
GC/MS	B	A	D	C	D	C	B	–	B
LC/MS	B	A	C	B	D	C	–	B	B
Optical biosensor-based immunoassay	C	C	B	B	D	C	–	–	B

(ATCC 8014). This assay shows sufficient sensitivity for determination of pantothenic acid in urine and total pantothenic acid in foodstuffs. However, concentrations in samples prepared from plasma are close to the limit of quantification.

20.3.2 Radioimmunoassay (RIA) and Enzyme-linked Immunosorbent Assay (ELISA)

The radioimmunoassay (RIA) was reported as the first immunoassay for pantothenic acid determination in blood and tissues in 1979 (Wyse et al. 1979). RIA can also determine pantothenic acid content in urine and foodstuffs (Srinivasan et al. 1981; Walsh et al. 1981). The enzyme-linked immunosorbent assay (ELISA) was reported in 1988 (Morris et al. 1988) and can measure pantothenic acid levels in plasma and food (Gontheir et al. 1998b; Morris et al. 1988; Song et al. 1990). The limits of quantification are 50 pmol/tube (ca. 10 ng/tube) and 10 pmol/well (ca. 2 ng/well) in RIA and ELISA, respectively (Gontheir et al. 1998b; Morris et al. 1988; Song et al. 1990; Wyse et al. 1979). These sensitivities are at the same level as the microbiological assay, that is, sensitive enough for measurement of pantothenic acid in urine and foodstuffs but less sensitive for blood and plasma samples. RIA and ELISA show specificity for pantothenic acid and require several hours or a day for measurement. Thus, RIA and ELISA are suitable for routine analysis for foodstuffs and clinical examination. However, the critical issue in RIA and ELISA is that the antibody is not commercially available.

20.3.3 High-performance Liquid Chromatography (HPLC)

20.3.3.1 HPLC–Ultraviolet (UV) Method

Analytical methods using HPLC have been reported since the 1980s. The first report showed that pantothenic acid was separated from multivitamin preparations by a reversed phase column and detected at 208 nm using an ultraviolet (UV) detector (Akada 1986). Several HPLC-UV methods were reported in the 1990s with improved separation conditions (Endo et al. 1991; Romera et al. 1996; Woollard et al. 2000). The HPLC method has superior specificity and is less time-consuming, and in particular, sample preparation is simpler than for the microbiological assay. Pantothenic acid can be analysed within 30 min in a single run. However, the disadvantage is that detection requires a low wavelength of approximately 200 nm. There are many compounds that absorb UV around 200 nm in plasma, urine and foodstuffs and the complete separation of pantothenic acid from miscellaneous compounds is very difficult prior to UV detection. Furthermore, pantothenic acid only weakly absorbs UV, so its absorption coefficient is not high at 200 nm. The limit of quantification is 200 pmol/injection (ca. 40 ng/injection) by the HPLC-UV method, which is less sensitive than the microbiological assay, RIA and ELISA.

Because of these disadvantages, the use of the HPLC-UV method is restricted only to measure pantothenic acid in multivitamin tablets and supplemented foods such as infant formula.

20.3.3.2 HPLC–Fluorimetric Method

Fluorescent compounds can be synthesized from pantothenic acid and the first HPLC-fluorimetric method was reported in 1995 (Blanco *et al.* 1995). In principle, pantothenic acid can be hydrolysed to β-alanine and pantoic acid by hot alkaline hydrolysis and the product β-alanine is reacted with *o*-phthaldialdehyde and 3-mercaptopropionic acid (3-MPA) to form a fluorescent compound, 1-alkylthio-2-alkylisoindole. This fluorescence is monitored at an excitation wavelength of 345 nm and emission wavelength of 455 nm (Figure 20.2). This method can determine the total pantothenic acid contents in foodstuffs and urinary pantothenic acid levels (Pakin *et al.* 2004; Takahashi *et al.* 2009).

Purification of pantothenic acid from an extract is required to measure the total pantothenic acid in foodstuffs (Pakin *et al.* 2004), while urine samples can

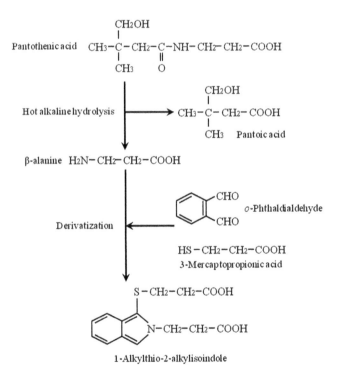

Figure 20.2 Derivatization of pantothenic acid to fluorescent compound 1-alkylthio-2-alkylisoindole.

Analysis of Pantothenic Acid (Vitamin B_5)

be directly injected into an HPLC system without any pre-cleanup treatment because urinary pantothenic acid is derivatized to 1-alkylthio-2-alkylisoindole after separation using a column (Takahashi et al. 2009). A schematic HPLC system for urinary pantothenic acid determination and its chromatogram for a urine sample are shown in Figures 20.3 and 20.4, respectively. The limit of

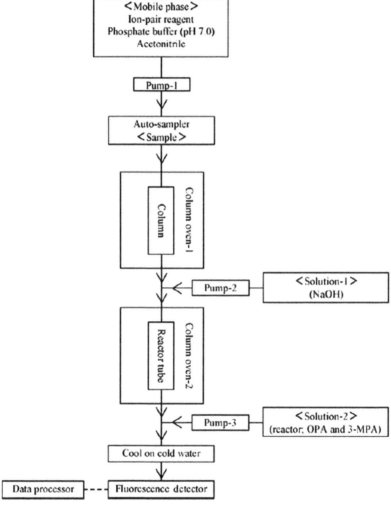

Figure 20.3 Schematic HPLC system for urinary pantothenic acid determination. Analytical conditions are shown in Table 20.4. A urine sample was directly injected into this system. Urinary pantothenic acid is separated in the reversed phase column and hydrolysed to β-alanine and pantoic acid. The product β-alanine reacts with o-phthaldialdehyde (OPA) and 3-mercaptopropionic acid (3-MPA), and is derivatized to 1-alkylthio-2-alkylisoindole which is detected by fluorescence (Takahashi et al. 2009).

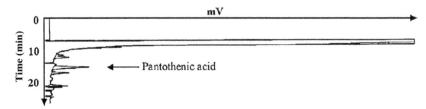

Figure 20.4 Determination of urinary pantothenic acid by HPLC-fluorimetric method. A 20 μL of urine sample was injected into HPLC system shown in Figure 20.3. Urinary pantothenic acid was detected as a sharp peak of fluorescent compound 1-alkylthio-2-alkylisoindole by fluorescence and its retention time was at 14 min (Takahashi *et al.* 2009).

quantification is 5 pmol/injecyion (*ca.* 1 ng/injection) in the HPLC-fluorimetric method, and this sensitivity is much higher than that of other methods (Takahashi *et al.* 2009). Other methods do not have sufficient sensitivity to precisely measure blood and plasma pantothenic acid levels. Although suitable conditions for determination of pantothenic acid in blood or plasma remain to be developed, the HPLC-fluorimetric method can resolve the issue surrounding sensitivity.

20.3.4 Mass Spectrometry (MS)

The recent dramatic development of mass spectrometry has developed gas chromatography–mass spectrometry (GC-MS) and LC-MS methods. Although pantothenic acid is not volatile enough for direct GC, pantothenic acid can be detected by MS after conversion to volatile compounds such as trimethylsilyl derivatives. Derivatization to volatile compounds from pantothenic acid requires an internal standard and the pantothenic acid homologue hopantothenic acid or radioisotope labelled [$^{13}C_3,^{15}N$]-pantothenic acid are used as internal standards for measurement by GC-MS (Banno *et al.* 1990; Rychlik 2000).

The GC-MS method can determine total pantothenic acid in foodstuffs and plasma (Banno *et al.* 1990, Rychlik 2000) and has been further developed and simplified to eliminate the derivatization step with an LC-MS method. LC-MS can determine total pantothenic acid in foodstuffs and urinary pantothenic acid (Heudi and Fontannaz 2005; Mittermayr *et al.* 2004; Rychlik 2003). These methods detect pantothenic acid by MS after separation by GC or LC, and MS detection is considered to have the highest selectivity. The limit of quantification is 5 pmol/injection. (*ca.* 1 ng/injection) by these methods and this sensitivity is as high as the HPLC-fluorimetric method. Pantothenic acid can be analysed within two hours by these methods. The disadvantage of these methods is the requirement for expensive equipment, specific techniques for maintenance and high running costs. Although the LC-MS method has high

Table 20.4 Basal data for the HPLC-fluorimetric method for pantothenic acid determination. DTMA: dodecyltrimethylammonium.

Auto-sampler	SIL-10AD$_{VP}$ (SHIMADZU)
Pump 1	L-2130 (HITACHI)
Pump 2	L-2130 (HITACHI)
Pump 3	L-2130 (HITACHI)
Column oven 1	L-2350 (HITACHI)
Column oven 2	655A-52 (HITACHI)
Fluorescence detector	RF-10A$_{XL}$ (SHIMADZU)
Data processor	C-R8A (SHIMADZU)
Column	Tosoh ODS-80Ts (4.6 i.d. × 250 mm)
Length of reactor tube	60 m (PTFE tube, 0.5 mm i.d.)
Temperature (column oven 1)	40 °C
Temperature (column oven 2)	100 °C
Flow-speed 1 (pump 1)	1.0 mL min^{-1}
Flow-speed 2 (pump 2)	0.5 mL min^{-1}
Flow-speed 3 (pump 3)	0.5 mL min^{-1}
Excitation/emission wavelength	345/455 nm
Mobile phase	60 mmol/L KH$_2$PO$_4$-NaOH buffer (pH7.0) containing 5 mmol/L DTMA and 5% acetonitrile (v/v)
Solution-1	600 mmol/L NaOH
Solution-2	10 mmol/L o-phthaldialdehyde and 16 mmol/L 3-mercaptopropionic acid

sensitivity and selectivity, only a few groups have used it for pantothenic acid measurement.

20.3.5 Optical Biosensor-based Immunoassay

Recent developments in affinity-based immunosensor techniques have exploited the potential for analyte detection through coupling of a biospecific binding protein–antigen interaction *via* optical, piezoelectric or electrochemical signal transducers. In 2005, a new method was reported in which an optical biosensor inhibition immunoassay was developed using a specific pantothenic acid binding protein for the quantification of pantothenic acid in foodstuffs (Haughey *et al.* 2005). In principal, a commercial biosensor system based on the optical phenomenon of surface plasmon resonance exploits the unlabelled interaction of a specific binding protein with a covalently immobilized ligand on a carboxymethyldextran-modified gold sensor surface. The optical detector monitors changes in resonance angle due to molecular interaction at the interface that is directly proportional to a change of mass at the surface. A commercially available kit can determine total pantothenic acid contents in foodstuffs and its analytical values are consistent with values determined by the microbiological assay and LC-MS methods (Haughey *et al.* 2005). The limit of quantification is 50 pmol/mL (*ca.* 10 ng/mL) by these methods and this sensitivity is at the same level as the microbiological assay and ELISA.

This method is suitable for routine measurements but the equipment and kit are expensive.

20.4 Concluding Remarks

Several new methods have been exploited during the last decade and these show high selectivity, sensitivity and rapid determination. In particular, HPLC-fluorimetric, LC-MS and optical biosensor-based immunoassay procedures are expected to be applied for routine measurement after further development.

Summary Points

- This chapter focuses on analytical methods for pantothenic acid.
- Free pantothenic acid must be liberated from bound forms such as coenzyme A (CoA) and pantetheine in foodstuffs prior to measurement.
- The classical microbiological assay using *Lactobacillus* is still the most frequently used procedure for determination of pantothenic acid in blood, urine and foodstuffs.
- Radioimmunoassay (RIA) and enzyme-linked immunosorbent assay (ELISA) procedures are no longer used for pantothenic acid determination because no commercially available antibody is available for these techniques.
- The high-performance liquid chromatography (HPLC)-fluorimetric method has been developed from HPLC-ultraviolet (UV) and this procedure shows high selectivity and sensitivity.
- The liquid chromatography–mass spectrometry (LC-MS) method has been developed from gas chromatography–mass spectrometry (GC-MS) and this procedure shows high selectivity and sensitivity.
- A commercially available kit can determine pantothenic acid contents in foodstuffs by recently developed optical biosensor-based immunoassay.
- Therefore, HPLC-fluor-metric, LC-MS and optical biosensor-based immunoassay procedures are expected to be applied for routine measurement after further development.

Key Facts about AOAC International

- AOAC International was formally called the Association of Official Analytical Chemists.
- It is a globally recognized, independent, not-for-profit association founded in 1884.
- Its vision is to attain 'worldwide confidence in analytical results'.
- Its website is www.aoac.org.
- AOAC serves communities of the analytical sciences by providing the tools and processes necessary to develop voluntary consensus standards or

technical standards through stakeholder consensus and working groups in which the fit-for-purpose and method performance criteria are established and fully documented.
- AOAC provides a science-based solution and its *Official Methods of Analysis* gives defensibility, credibility and confidence in decision-making.
- AOAC *Official Methods*^{SM} are accepted and recognized worldwide.

Definitions of Words and Terms

American Type Culture Collection (ATCC). ATCC is an independent, private, non-profit biological resource centre (BRC) and research organization established in 1925. ATCC's collections include a wide range of biological materials for research, including cell lines, molecular genomics tools, microorganisms and bioproducts.

Enzyme-linked immunosorbent assay (ELISA). ELISA is a procedure used to detect or determine an antigen or an antibody in a sample based on immunology. An antigen binds to its specific antibody linked to an enzyme and a subsequent enzyme reaction converts a substrate to a visible signal. ELISA is widely used to detect serum antibodies and serum proteins in clinical measurements.

Radioimmunoassay (RIA). RIA is a procedure used to detect or determine an antigen in a sample based on immunology. A radioisotope-labelled antigen is mixed into a sample, an antigen specific antibody is added to the mixture, and the amount of antibody bound with radiolabeled antigen is then measured by scintillation counter. The amount of antigen in the sample is calculated from the radioactivity. RIA is more sensitive than ELISA, but ELISA has replaced RIA because the latter requires the use of radioactive compounds.

High-performance liquid chromatography (HPLC). HPLC is a chromatographic method that separates compounds from other compounds through differences in molecular size or affinity to a stationary phase. In the HPLC system, a sample solution passes though a column with a mobile phase at high pressure and separated compounds are monitored by a detector. HPLC has advantages in selectivity and sensitivity, and is thus widely used for determination of small molecules.

Mass spectrometry (MS). MS is an analytical technique used to separate and detect an ionized molecule by its mass-to-charge ratio. LC-MS is an analytical system used to separate and detect an ionized molecule after separation by HPLC. When MS is combined with gas chromatography (GC), the system is called GC-MS. The advantages of LC-MS and GC-MS are very high selectivity and sensitivity, and thus these procedures are considered to offer the most precise analytical methods.

Microbiological assay. This bioassay technique utilizes the growth of an organism in the presence of a compound. It is popular method for detection or measurement of nutrients, antibiotics and mutagens. To measure vitamin contents in a sample by microbiological assay, a microorganism that requires the objective vitamin is incubated in a culture medium with the sample, and the turbidity of the medium is determined by using a spectrophotometer.

List of Abbreviations

CoA	coenzyme A
DTMA	dodecyltrimethylammonium
ELISA	enzyme-linked immunosorbent assay
GC-MS	gas chromatography–mass spectrometry
HPLC	high-performance liquid chromatography
LC-MS	liquid chromatography–mass spectrometry
LOQ	limit of quantification
3-MPA	3-mercaptopropionic acid
MS	mass spectrometry
OPA	*o*-phthaldialdehyde
RIA	radioimmunoassay
UV	ultraviolet

References

Akada, Y., 1986. Determination of calcium pantothenate and panthenol in pharmaceutical preparations by high performance liquid chromatography. *Bunseki Kagaku*. 35: 320–322.

AOAC, 1996. Method 945.74. Pantothenic acid in vitamin preparations-microbiological methods, In: Official Methods of Analysis of the Association of Official Analytical Chemists, 16th ed. Association of Official Analytical Chemists, Arlington VA, USA, pp. 45.2.05.

Banno, K., Matsuoka, M., Horimoto, S., and Kato, J., 1990. Simultaneous determination of pantothenic acid and hopantenic acid in biological samples and natural products by gas chromatography-mass fragmentography. *Journal of Chromatography*. 525: 255–264.

Blanco, M., Coello, J., Iturriaga, S., Maspoch, S., and Pagés, J., 1995. FIA fluorimetric determination of calcium pantothenate. Validation and quantification in multivitamin preparations. *Analytical Letters*. 28: 821–833.

Endo, M., Yao, B., Saotome, M., and Kamei, K., 1991. Determination of pantothenic acid in infant formulas by high performance liquid chromatography. *Nippon Shokuhin Kogyo Gakkaishi*. 38: 275–279.

Fry, P.C., Fox, H.M., and Tao, H.G., 1976. Metabolic response to a pantothenic acid deficient diet in humans. *Journal of Nutritional Science and Vitaminology*. 22: 339–346.

Gonthier, A., Fayol, V., Viollet, J., and Hartmann, D. J., 1998a. Determination of pantothenic acid in foods: influence of the extraction method. *Food Chemistry*. 63: 287–294.

Gontheir, A., Boullanger, P., Fayol, V., and Hartmann, D.J., 1998b. Development of an ELISA for pantothenic acid (vitamin B_5) for application in the nutrition and biological fields. *Journal of Immunoassay.* 19: 167–194.

Gopalan, C., 1946. The burning-feet syndrome. *The Indian Medical Gazette.* 81: 22–26.

Haughey, S.A., O'Kane, A.A., Baxter, G.A., Kalman, A., Trisconi, M.J., Indyk, H.E., and Watene, G.A., 2005. Determination of pantothenic acid in foods by optical biosensor immunoassay. *Journal of AOAC International.* 88: 1008–1014.

Heudi, O., and Fontannaz, P., 2005. Determination of vitamin B5 in human urine by high-performance liquid chromatography coupled with mass spectrometry. *Journal of Separation Science.* 28: 669–672.

Hodges, R.E., Ohlson, M.A., and Bean, W.B., 1958. Pantothenic acid deficiency in man. *Journal of Clinical Investigation.* 37: 1642–1657.

Ministry of Education, Culture, Sports, Science and Technology, 2010. Standard Tables of Food Composition in Japan, 2010. Ministry of Education, Culture, Sports, Science and Technology, Tokyo, Japan. p. 510.

Mittermayr, R., Kalman, A., Trisconi, M.J., and Heudi, O., 2004. Determination of vitamin B_5 in a range of fortified food products by reversed-phase liquid chromatography-mass spectrometry with electrospray ionization. *Journal of Chromatography A.* 1032: 1–6.

Morris, H.C., Finglas, P.M., Faulks, R.M., and Morgan, M.R.A., 1988. The development of an enzyme-linked immunosorbent assay (ELISA) for the analysis of pantothenic acid and analogues. Part I: production of antibodies and establishment of ELISA system. *Journal of Micronutrient Analysis.* 4: 33–45.

Pakin, C., Bergaentzlé, M., Hubscher, V., Aoudé-Werner, D., and Hasselmann, C., 2004. Fluorimetric determination of pantothenic acid in foods by liquid chromatography with post-column derivatization. *Journal of Chromatography A.* 1035: 87–95.

Romera, J.M., Ramirez, M., and Gil, A., 1996. Determination of pantothenic acid in infant milk formulas by high performance liquid chromatography. *Journal of Dairy Science.* 79: 523–526.

Rychlik, M., 2000. Quantification of free and bound pantothenic acid in foods and blood plasma by a stable isotope dilution assay. *Journal of Agricultural and Food Chemistry.* 48: 1175–1181.

Rychlik, M., 2003. Pantothenic acid quantification by a stable isotope dilution assay based on liquid chromatography-tandem mass spectrometry. *Analyst.* 128: 832–837.

Rychlik, M., and Roth-Maier, D., 2005. Pantothenic acid quantification: method comparison of a stable isotope dilution assay and a microbiological assay. *International Journal for Vitamin and Nutrition Research.* 75: 218–223.

Skeggs, H.R., and Wright, L.D., 1944. The use of *Lactobacillus arabinosus* in the microbiological determination of pantothenic acid. *Journal of Biological Chemistry.* 156: 21–26.

Song W.O., Smith A., Wittwer C., Wyse B.W., and Hansen R.G., 1990. Determination of plasma pantothenic acid by indirect enzyme linked immunosorbent assay. *Nutrition Research.* 10: 439–448.

Srinivasan, V., Christensen, N., Wyse, B.W., and Hansen, R.G., 1981. Pantothenic acid nutritional status in the elderly-institutionalized and -noninstitutionalized. *The American Journal of Clinical Nutrition.* 34: 1736–1742.

Takahashi, K., Fukuwatari, T., and Shibata, K., 2009. Fluorometric determination of pantothenic acid in human urine by isocratic reversed-phase ion-pair high-performance liquid chromatography with post-column derivatization. *Journal of Chromatography B.* 877: 2168–2172.

Walsh, J.H., Wyse, B.W., and Hansen, R.G., 1981. Pantothenic acid content of 75 processed and cooked foods. *Journal of The American Dietetic Association.* 78: 140–144.

Woollard, D.C., Indyk, H.E., and Christiansen, S.K., 2000. The analysis of pantothenic acid in milk and infant formulas by HPLC. *Food Chemistry.* 69: 201–208.

Wyse, B.W., Wittwer, C., and Hansen, R.G., 1979. Radioimmunoassay for pantothenic acid in blood and other tissues. *Clinical Chemistry.* 25: 108–111.

CHAPTER 21

High-performance Liquid Chromatography Mass Spectrometry Analysis of Pantothenic Acid (Vitamin B_5) in Multivitamin Dietary Supplements

PEI CHEN

Food Composition and Methods Development Laboratory, Beltsville Human Nutrition Research Center, Agricultural Research Service, US Department of Agriculture, 10300 Baltimore Ave. Beltsville, MD 21029, USA
Email: pei.chen@ars.usda.gov

21.1 Introduction

Pantothenic acid, vitamin B_5, is a B family water-soluble vitamin discovered in 1933 (Williams *et al.* 1933). The name was derived from the Greek word 'panthos' meaning everywhere. As its name indicates, pantothenic acid is widely distributed in foods, mostly incorporated into coenzyme A. Its richest sources are yeast and animal organs (liver, kidney, heart, brain) but eggs, milk, vegetables, legumes and whole-grain cereals are also common sources. Thus, human deficiency of pantothenic acid is rare and has been reported only in

Food and Nutritional Components in Focus No. 4
B Vitamins and Folate: Chemistry, Analysis, Function and Effects
Edited by Victor R. Preedy

Published by the Royal Society of Chemistry, www.rsc.org

severely malnourished patients (Smith and Song 1996). However, there is speculation that pantothenic acid may get lost during the manufacturing process of foods in modern society, creating a potential for pantothenic acid deficiency. Deficiency in pantothenic acid may lead to symptoms such as physical or muscle weakness, abdominal pains, susceptibility to infection, and depression (Bean and Hodges 1954).

21.1.1 Pantothenic Acid Supplementation: Aims and Problems

Pantothenic acid is important in the oxidation of fatty acids and carbohydrates and in the synthesis of amino acids, fatty acids, ketones, cholesterol, phospholipids and steroid hormones. As part of acetyl CoA or acyl carrier protein, it is essential in energy metabolic pathways (Lipmann et al. 1947; Pugh et al. 1966). It is a relatively safe vitamin supplement with the only documented side effects being diarrhoea and gastrointestinal disturbances when a large amount of pantothenic acid is taken (10 grams or a thousand times the minimum daily requirement) (Bean and Hodges 1954; Zucker and Zucker 1954; Bean et al. 1955). The amount of calcium pantothenate present in multivitamin dietary supplements is usually 10 mg/dose. In some B vitamin dietary supplements, the amount of calcium pantothenate can be 50 mg/dose. For pantothenic acid only dietary supplements, the calcium pantothenate content can be as high as 1000 mg/dose.

21.1.2 Chemical Properties of Pantothenic Acid

The structure of pantothenic acid is shown in Figure 21.1. Due to the chirality at the hydroxylated carbon atom of the pantothenic acid moiety, the vitamin is optically active. Only the $d(+)$-pantothenic acid enantiomer is biologically active and present in nature.

Pantothenic acid is stable at room temperature and is one of the most stable B family vitamins. It is not affected by storage at $-20\ °C$ for a month (Machlin 1991). The United States Pharmacopeia (USP) standard for pantothenic acid is the $d(+)$-enantiomer. Commercially available standards often are in the forms of sodium or calcium salts. The calcium salt of pantothenic acid (calcium pantothenate) is often used for fortification in the food and dietary supplement industries.

21.1.3 History of Analysis of Pantothenic Acid

Methods of pantothenic acid analysis vary widely in approach. Historically, pantothenic acid analysis has relied on chemical and physical methods, animal

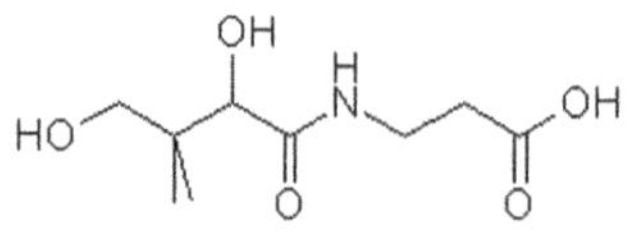

Figure 21.1 Chemical structure of pantothenic acid.

bioassay, microbiological methods (for a review see Wyse *et al.* 1985) and, more recently, chromatographic and immunological methods.

The AOAC official method for pantothenic acid analysis is a microbiological assay that was developed over 60 years ago (Crandall and Burr 1947; Ketchum and Conklin 1945). It is highly sensitive, but imprecise due to very detailed, laborious and time-consuming procedures that require multiple determinations to achieve the necessary precision for estimation of the mean value. This method is outdated in light of new technology (Chen and Wolf 2007). The USP official method uses high-performance liquid chromatography (HPLC) with ultraviolet (UV) detection (210 nm). The method is well suited for analysis of pantothenic acid by itself and in a vitamin premix (Heudi *et al.* 2005), but may or may not work well when applied to the analysis of pantothenic acid in multivitamin dietary supplements. The complex matrices of multivitamin dietary supplements may cause potential interferences with the particular vitamin being analysed.

21.2 Review of Analysis of Pantothenic Acid Using High-performance Liquid Chromatography

21.2.1 Introduction to High-performance Liquid Chromatography

There are a variety of HPLC columns and most of them are based on silica particles. The most popular column used today is the reversed-phase C18 column that uses a silica support with a bonded organic surface layer. Traditional HPLC analytical columns are usually 4.6 mm × 250 mm with a 5 µm particle size, as used in the USP official method. Modern columns for HPLC–mass spectrometry (LC-MS) applications are more efficient and offer the same or better resolving power in a much smaller package (2.1 mm × 150 mm with a 3.5 µm particle size or smaller). Ultra high-performance liquid chromatography (UHPLC) columns (sub 2µm particle size, 1–2 mm in diameter and 30–100 mm in length) are becoming more and more popular.

21.2.2 Earlier HPLC Method for Pantothenic Acid Analysis

Many early HPLC methods for water-soluble vitamin (WSV) analysis employed a reversed-phase ion-pair technique (Amin and Reusch 1987; Dong *et al.* 1988; Ivanovic *et al.* 1999; Kwok *et al.* 1981; Woollard 1984). The technique is called ion-pair chromatography (IPC) and uses ion-pair reagents to modulate retention time of analytes. Theoretically, IPC may provide reduced separation times and sharper peak shapes. In practice, IPC is more difficult to run and its method development usually takes longer compared with regular HPLC. The ion-pair reagents used may interfere with non-specific detectors such as UV (especially in the 190–220 nm range) and refractive index (RI) detector. Moreover, IPC may not be compatible with mass spectrometric detection. For WSV analysis, IPC should be avoided.

For pantothenic acid analysis, HPLC has advantages compared with the other methods mentioned in Section 21.1—it is easy to use, fast, accurate, precise and reproducible.

21.2.3 Common Detection Techniques for HPLC

The most common detector for HPLC and UHPLC is a diode array detector (DAD). Although there is an earlier report of determination of pantothenic acid using UV (210 nm in multivitamin pharmaceutical preparations (Hudson and Allen 1984), most HPLC-DAD methods used for the analysis of WSVs do not include pantothenic acid as part of the analysis (Kirchmeier and Upton 1978; Kwok *et al.* 1981; Li 2002; Markopoulou *et al.* 2002; Walker *et al.* 1981; Wills *et al.* 1977) as the interferences may undermine the quality of the quantitation.

Other widely used detectors for HPLC include refractive index (RI), fluorescence and evaporative light-scattering (ELS). The use of RI and ELS detectors for pantothenic acid analysis in multivitamin dietary supplements has not been reported. The main reason is that the two detectors are not selective and thus cannot resolve pantothenic acid from other components existing in a multivitamin dietary supplement. Although fluorescence detection can be highly selective depending on the application, pantothenic acid does not have fluorescence excitation and emission and so fluorescence detection cannot be used for pantothenic acid analysis unless derivatization methods are applied (Pakin *et al.* 2004; Takahashi *et al.* 2009). Derivatization adds more complexity to analytical method and should not be used unless necessary. For detection and quantitation of pantothenic acid in multivitamin dietary supplements with HPLC/UHPLC, a highly selective detector such as MS should be the instrument of choice.

21.3 Analysis of Pantothenic Acid Using Liquid Chromatography Mass Spectrometry

21.3.1 LC-MS Interfaces

The most common interfaces between LC and MS used today are electrospray ionization (ESI) and atmospheric pressure chemical ionization (APCI). Compared with ESI, APCI requires the analytes to be thermally stable and may yield multiple fragmentation ions. For these reasons, APCI is often used when a particular analyte is not ionized well with ESI. ESI should be used for LC-MS analysis of pantothenic acid since pantothenic acid is ionized very well under ESI.

21.3.2 Ion Suppression

An interference to which analytical chemists need to pay attention to during method development for LC-MS is ion suppression. Because of the specificity of the LC-MS and/or liquid chromatography–tandem mass spectrometry (LC-MS/MS) methods, researchers may have the misconception that good chromatographic separation is necessary. This is true in some cases, but is a compromise that must be dealt with very carefully as there is a fine line between minimizing analysis time and compromising chromatographic separation, especially when complex matrices are involved. Ion suppression occurs when a co-eluted component influences the ionization of the analyte and will lead to poor performance. It is important to realize that LC-MS/MS methods are just as susceptible to ion suppression effects as single LC-MS techniques because the advantages of MS/MS begin only after the ion formation.

21.3.3 Mass Analysers

There are a number of different methods for separating ions. The more popular ones used today are quadrupole MS (QMS), ion-trap MS (ITMS), time-of-flight MS (TOF-MS) and hybrids based on the combination of these mass analysers.

Quadrupoles can be used in either a scanning or a filtering mode. Quantitative studies are performed with the quadrupole working in filtering mode where only selected ions are allowed to pass and all other ions are 'filtered out'. This mode is called 'selected ion monitoring' (SIM). Quadrupole MS is highly selective compared with the optical detectors mentioned earlier. However, interferences may still occur if the analytes are from a complex matrix. In such cases, multiple reaction monitoring (MRM) with triple quadrupole tandem mass spectrometers may be used. Quadrupole MS in either SIM or MRM mode should be used as the instrument of choice for pantothenic acid analysis. In addition to quadrupole MS, other mass spectrometers can also be used for pantothenic acid analysis.

ITMS has the ability to trap and accumulate ions. A very useful feature is that with ITMS it is possible to isolate and accumulate one ion specie in the trap. The isolated ions can then be fragmented and the fragments detected (MS^2). The isolation and fragmentation process can be sequentially repeated multiple times (n) to perform multi-stage ion fragmentation analysis (MS^n). Therefore, ITMS is usually used for structure elucidation. Although ITMS can also be used for quantitative analysis, the quality of quantitation is highly dependent on the analytes studied and the ITMS used. For quantitative analysis, either SIM mode or selected reaction monitoring (SRM) mode can be used. The SRM mode for ITMS is similar to the MRM mode for triple quadrupole MS.

TOF-MS is not usually considered as the first choice for quantitative applications. However, there are many applications that successfully use TOF-MS for quantitative analysis, especially when the number of targeted analytes is large, as in the case of pesticides and drug residues analysis.

In summary, LC-MS or LC-MS/MS should be the instrument of choice for pantothenic acid analysis with quadrupole as the first choice. Both ITMS and TOF-MS can be used if a quadrupole MS is not available.

21.3.4 Stable Isotope Dilution Mass Spectrometry

Stable isotope dilution mass spectrometry (IDMS) can be a definitive analytical method for very accurate concentration determination of pantothenic acid (Cornelis 2003). However, definitive accuracy of the IDMS technique requires full equilibration between endogenous natural amounts of the compound and the isotopically labelled spike, which is usually accomplished by bringing both in the same chemical form in solution.

21.3.5 Mobile Phases for LC-MS

Phosphate buffers are often used as mobile phases in HPLC. Ion-pair agents are also widely used for various applications. For LC-MS, the use of non-volatile salts and/or ion-pair agents in the mobile phases is not advised. Earlier LC-MS applications often used trifluoroacetic acid (TFA) in mobile phases, but TFA has been replaced by formic acid due to ion-suppression and ion-adduction issues.

The key for LC-MS mobile phases is simplicity. Additives other than formic acid should not be used unless absolutely necessary. For pantothenic acid analysis, the use of 0.1% LC-MS grade formic acid in the mobile phase is advised (mobile phase A was 0.1% formic acid in water and mobile phase B was 0.1% formic acid in acetonitrile) (Chen, *et al.* 2007 and Chen *et al.* 2009).

21.4 Analysis of Pantothenic Acid Using LC/MS in Multivitamin Dietary Supplements

21.4.1 Reference Standard

Pure reference standards of B complex vitamins can be obtained from USP. Once a LC-MS method for pantothenic acid analysis from multivitamin dietary supplements is developed, the multi-element multivitamin dietary supplement Standard Reference Material 3280 (SRM 3280) from the National Institute of Standards and Technology (NIST, Gaithersburg, MD, USA) can be used for validation and quality control.

21.4.2 Sample Preparation

Because most B family vitamins are more stable in acidic conditions than in basic conditions, extraction of these vitamins in acidic conditions is preferred. Since pantothenic acid itself is very stable at room temperature, extraction can be kept as simple as possible. Sonication with water or buffer at pH 2–3 is sufficient, even with the extraction of other B family vitamins in mind (Chen and Wolf 2007; Chen *et al.* 2009). After sonication, centrifugation and/or filtration for sample clean-up is usually sufficient. More complicated sample clean-up procedures, *i.e.* solid-phase extraction (SPE), are not usually necessary.

21.4.3 Chromatography

Theoretically, if LC-MS or LC-MS/MS is used, a completely resolved pantothenic acid peak is not necessary since MS in the SIM or MRM mode can resolve the pantothenic acid peak from co-eluting peaks. However, a full chromatographic separation, at least from other WSVs in the sample, should be attempted to avoid ion suppression and to achieve the best possible results.

Another benefit of having completely resolved WSV peaks is the ability to simultaneously measure multiple WSVs in a single chromatographic run. To achieve this goal, an end-capped C-18 column that is compatible with a 100% aqueous mobile phase should be used for separation of multiple WSVs. A newer LC-MS/MS method reported by Chen and co-workers offers good separation of eight WSVs in multivitamin dietary supplement matrices (Chen and Wolf 2007; Chen et al. 2009).

LC-DAD and LC-MS/MS chromatograms for NIST SRM 3280 multivitamin dietary supplement are shown in Figures 21.2 and 21.3, respectively. There are no interferences for pantothenic acid in LC-MS/MS mode (Figure 21.3). Although well-separated from other WSVs, the pantothenic acid peak is small and subjected to interferences in the LC-UV chromatogram (UV 210 nm; Figure 21.4). However, with careful manual integration, quantitation of pantothenic acid using UV is possible, if not desirable. The method uses a Hydro-RP C18 (4 μm particle size, 2.0 mm × 250 mm) reversed-phase column (Phenomenex, Torrance, CA, USA) in combination with a Column-Saver™ pre-column filter (MAC-MOD Analytical, Inc., Chadds Ford, PA). A Unison UK-C18 column (3 μm particle size, 2.0 mm × 250 mm; Imtakt, Philadelphia, PA, USA) was also tried to test the method and the results are equally

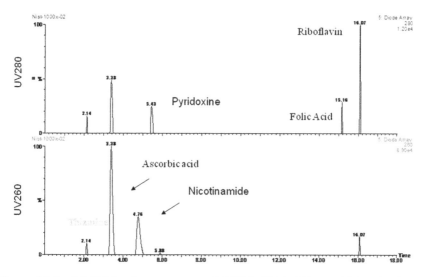

Figure 21.2 LC-DAD chromatogram of NIST SRM 3280 multivitamin dietary supplement.

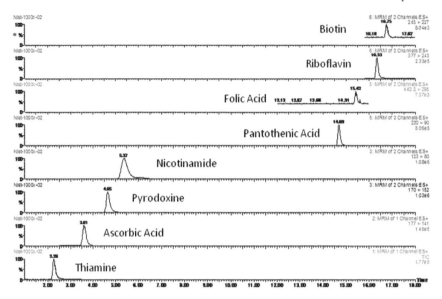

Figure 21.3 LC-MRM chromatogram of NIST SRM 3280 multivitamin dietary supplement.

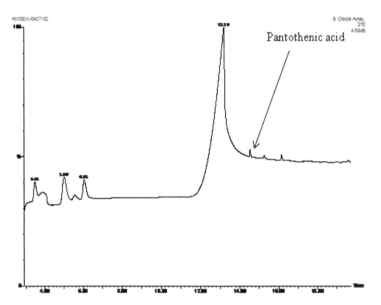

Figure 21.4 LC-UV (220 nm) chromatogram of NIST SRM 3280 multivitamin dietary supplement.

successful. Note that the chromatograms in Figures 21.2 and 21.3 are not necessarily applicable to other multivitamin dietary supplements.

The key to the successful separation of WSVs in multivitamin dietary supplements is the use of end-capped C18 columns that are compatible with

100% aqueous mobile phase. A 100% aqueous mobile phase was used for the first five minutes of the chromatographic run enabling the separation of thiamine, ascorbic acid, nicotinic acid and nicotinamide from each other, as well as from other components of the SRM 3280 (Figures 21.2 and 21.3). Pantothenic acid elutes at around 14.6 minutes. If a regular C18 reversed-phase column with a more aggressive HPLC elution programme were used, it is possible to elute pantothenic acid earlier, probably saving about five minutes in analysis time. However, this would sacrifice the benefit of measuring other WSVs in multivitamin dietary supplements in the same chromatographic run. Considering that a gradient HPLC profile must be used to wash out the fat-soluble vitamins and other potential hydrophobic components in multivitamin dietary supplements, the time saved with a faster elution is insignificant as washes and re-equilibration of column take about 7–10 minutes.

The detailed HPLC programme (mobile phase consisting of A: 0.1% formic acid in water and B: 0.1% formic acid in acetonitrile) is as follows:

0–5 min, isocratic at 100% A;
5–15 min, linear gradient from 100 : 0 A : B (v/v) to 50 : 50 A : B (v/v);
15–17 min, linear gradient from 50 : 50 A : B (v/v) to 5 : 95 A : B (v/v).

The post-run time is set to eight minutes (column equilibrating). The flow rate used was 0.25 mL/min. The total run time was 25 minutes. The retention times of the WSVs analysed are as follows:

1. thiamine, $t_R = 2.14$ (UV) and 2.26 min (MRM)
2. ascorbic acid, $t_R = 3.38$ (UV)
3. nicotinamide, $t_R = 4.76$ (UV) and 4.80 min (MRM)
4. pyridoxine, $t_R = 5.43$ (UV) and 5.56 min (MRM)
5. pantothenic acid, $t_R = 14.61$ min (MRM)
6. folic acid, $t_R = 15.16$ (UV) and 15.28 min (MRM)
7. riboflavin, $t_R = 16.07$ (UV) and 16.19 min (MRM)
8. biotin, $t_R = 16.65$ min (MRM)

21.4.4 LC-MS Analysis

If possible, ESI in MRM mode in the positive ion mode should be used for pantothenic acid analysis in multivitamin dietary supplements. MRM is the preferred mode for quantitation. The triple quadrupole MRM transition for pantothenic acid is m/z 220 → 90 (Figure 21.5). Excellent linearity can be achieved ($r^2 > 0.998$) (Chen et al. 2009). If a triple quadrupole MS is not available, single quadrupole MS in SIM mode (m/z 220, molecular ion) can be used.

For simultaneous determination of multiple WSVs in multivitamin dietary supplements, the main problem is the large variation in the WSV concentrations. In a typical multivitamin dietary supplement like NIST SRM 3280, the most concentrated WSV is ascorbic acid, at 42 mg/g, followed by

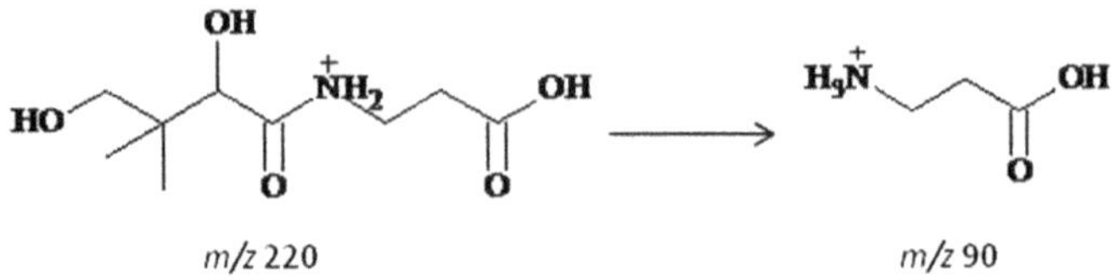

Figure 21.5 MRM transition for pantothenic acid m/z 220 > 90.

nicotinamide (14 mg/g), pantothenic acid (7.3 mg/g), pyridoxine hydrochloride (1.8 mg/mg), riboflavin (1.3 mg/g), thiamine (1.1 mg/g), folic acid (0.4 mg/g), biotin (0.024 mg/g) and cyanocobalamin (0.006 mg/g) (Sander et al. 2011).

Using a single LC-MS/MS method to quantify all WSVs at such different levels of concentrations is very difficult. However, quantitation of all the WSVs except cyanocobalamin in a single chromatographic run has been achieved (Chen and Wolf 2007; Chen et al. 2009). Compromises have to be made to accommodate WSVs at different concentration levels. For multi WSV analysis using LC-MS or LC-MS/MS in multivitamin dietary supplements, the instrument may need to be intentionally tuned to lower the sensitivity for pantothenic acid to accommodate higher concentrations if folic acid and biotin need to be analysed together (Chen and Wolf 2007).

ITMS can also be used for pantothenic acid analysis in multivitamin dietary supplements. Although the quality of the quantitative analysis using ITMS is highly dependent on the analytes, pantothenic acid displays good linearity and dynamic range when the LCQ series (Thermo Fisher Scientific Inc. Waltham, MA, USA) ITMS is used (unpublished data). Both the SIM mode and SRM mode (SRM transition 220→90) can be used.

Although TOF-MS can be used for pantothenic acid analysis in multivitamin dietary supplements, there is no report for such an application to date.

Isotope dilution mass spectrometry (IDMS0 has been successfully used for pantothenic acid analysis (Chen et al. 2007; Rychlik 2003). IDMS is an important method that can acquire the most accurate value possible for the pantothenic acid concentration in a sample. However, with the release of the NIST SRM 3280, the accuracy of a method used can be determined easily. Since IDMS is very expensive, routine use of the method is not recommended due to the associated cost.

Summary Points

- LC-MS or LC-MS/MS should be used as the instrument of choice for analysis of pantothenic acid in multivitamin dietary supplements.
- Quadrupole MS should be the first choice. ITMS and TOF-MS can be used as well.
- MRM or SIM modes should be used if available.
- The LC mobile phase should be as simple as possible. Water and acetonitrile with 0.1% formic acid work well.

- The HPLC column should be an end-capped C18 reversed-phase column that is compatible with 100% aqueous mobile phase for multi WSV analysis. A regular C18 column can be used if simultaneous determination of multi WSV is not required.

Key Facts

Key Facts about the National Institute of Standards and Technology

- Founded in 1901, NIST is a non-regulatory federal agency within the US Department of Commerce.
- NIST's mission is to promote US innovation and industrial competitiveness by advancing measurement science, standards and technology.
- NIST has four cooperative programmes: the NIST Laboratories, conducting scientific research; the Baldrige Performance Excellence Program promotes performance excellence among US manufacturers, service companies, educational institutions, health care providers and nonprofit organizations; the Hollings Manufacturing Extension Partnership offering technical and business assistance to smaller manufacturers; and the Technology Innovation Program provides cost-shared awards to industry, universities, and consortia for research purposes.

Key Facts about Standard Reference Material (SRM) 3280

- Standard Reference Material (SRM) 3280 for multivitamin/multimineral tablets was created by NIST in collaboration with the Office of Dietary Supplements (ODS) at the National Institutes of Health (NIH).
- SRM 3280 is a non-commercial batch of multi-element multi-vitamin tablets produced according to a contracting manufacture's normal procedures. NIST scientists tested and certified the concentrations of 24 elements and 17 vitamins and carotenoid compounds in the tablets.
- Researchers can use the SRM to benchmark their assays for vitamins and elements.
- SRMs are intended for analytical chemists to use to make sure their methods are working properly, not a benchmark for what a good product should be.

Definitions of Words and Terms

Atmospheric pressure chemical ionization (APCI): APCI is a common LC-MS interface. With APCI, a LC eluent is introduced as a spray using in a pneumatic (usually nitrogen powered) nebulizer. The emerging plume of liquid droplets is generated in the atmosphere and is directed towards a corona discharge electrode that is maintained typically at 1–3.5 kV. Analyte molecules react with the corona discharge plasma to generate ions. APCI is often used when a particular molecule is poorly ionized using ESI since APCI requires that the analytes must be thermally stable and abundant fragmentation may occur.

Diode array detector (DAD): DAD is used to measure molecular absorption of at a certain ultraviolet–visible (UV/Vis) wavelength. The amount of light absorbed will depend on the amount of a particular compound that is passing through the beam at the time and the absorbance of that compound at a particular UV wavelength. The difference between a DAD and a fixed wavelength UV detector is that a DAD can scan across a predetermined wavelength.

Electrospray ionization (ESI): ESI is the first true interface between LC and MS that could be used for highly polar and non-volatile compounds. In ESI, the solution with the analyte is sprayed into a mist of highly charged droplets after passing through a capillary which is held at high potential. As the droplets move towards the orifice of the mass spectrometer, the droplets reduce in size by evaporation of the solvent, resulting in a higher and higher charge density as the surface area of the droplets are reduced. This leads to 'coulomb explosion' (droplet subdivision resulting from the high charge density). Finally, fully desolvated ions resulting from complete evaporation of the solvent form gas phase ions that can be analysed by a mass analyser. ESI is one of the softest ionization methods available, yielding a molecular ion with minimal fragmentation.

Evaporative light-scattering (ELS) detection: With ELS detection, the solvent emerging from the end of the column is evaporated in a stream of air in a heated chamber. The solute does not evaporate, but is nebulized and passes in the form of minute droplets through a light beam, which is reflected and refracted. The amount of scattered light is measured and bears a relationship to the concentration of the material that is eluting. There are no special wavelength requirements for the light source. ELS detection will respond to any solute that does not evaporate before passing through the light beam.

Fluorescence detection: Fluorescence detectors are used almost exclusively in liquid chromatography and are a specific detector that senses only those substances that fluoresce. When a solute that fluoresces at the excitation wavelength passes through a flow cell, the fluorescent light it emits is detected by a photo cell. The excitation radiation is usually a wavelength selected from a deuterium lamp. The substance may then be selectively detected from other solutes which do not fluoresce at the excitation wave length or emit at a difference wavelength even if they co-elute. Fluorescence detection can be highly selective depending on the application.

High-performance liquid chromatography (HPLC): High-performance liquid chromatography is basically a highly improved form of column chromatography. Instead of a solvent (mobile phase) system being allowed to pass through a column (stationary phase) under gravity, it is forced through under high pressures of up to 400 bar and can exceed 1000 bar for modern ultra high-performance liquid chromatography (UHPLC). UHPLC allows a much smaller particle size for the column packing material (1.7–2 μm) to be used, which gives a much better separation of the components in the mixture.

Ion suppression: Ion suppression occurs in the early stages of the ionization process in the LC-MS interface, when a co-eluted component influences the ionization of the analyte. This is why proper chromatographic separation is always beneficial even with highly selective detectors such as MS or tandem mass spectrometry (MS/MS). It is important to realize that LC-MS/MS methods are just as susceptible to ion suppression effects as single LC-MS techniques because the advantages of MS/MS begin only after the ion formation.

Ion trap: In an ion trap, rather than passing through a quadrupole analyser with a superimposed radio frequency field, the ions are trapped in a radio frequency quadrupole field. The motion of the ions induced by the electric field on these electrodes allows ions to be trapped or ejected from the ion trap. The ion trap's ability to trap and accumulate ions make it possible to achieve high sensitivity. Another very useful feature of the ion trap is that it is possible to isolate one ion species by ejecting all others from the trap. It can be ejected and detected (SIM), or the isolated ions can subsequently be fragmented and the fragments detected (MS^2). The isolation and fragmentation process can be repeated multiple times (n) to perform multi-stage sequential MS analysis (MS^n). Therefore, ion traps are usually used for structure elucidation.

Ion-pair chromatography (IPC): Some HPLC applications use ion-pair reagents to modulate retention of the polar and ionic analytes, IPC reagents are usually large ionic molecules having a charge opposite to the analyte of interest, as well as a hydrophobic region to interact with the stationary phase. The counter-ion combines with the ions of the analytes, becoming ion pairs. This results in different retention, thus potentially facilitating separation of analytes. Theoretically, IPC may provide reduced separation times and sharper peak shapes. Some of the typical IP reagents are alkylsulfonates, alkylsulfates and tetra-alkylammonium ions.

Mass analysers: A mass analyser is used to separate ions in a mass spectrometer. The most popular ones used today are quadrupoles, ion traps, time-of-flight (TOF), Fourier transform MS (FTMS) and hybrids based on the combination of these.

Mass spectrometry (MS): MS can be considered as the smallest scale in the world since its function is to weigh molecules, *i.e.* to determine their molecular weight. To achieve the goal, a mass spectrometer converts sample molecules into ions in the gas phase, separates them according to their mass-to-charge ratio (m/z), and sequentially records the individual ion current intensities at each mass to generate a mass spectrum. The detector converts the ion energy into electrical signals, which can be recorded and calculated to determine the m/z of the ions and their ion counts.

Multiple reaction monitoring (MRM): MRM is the most common mode of using a triple quadrupole mass spectrometer for quantitative analysis,

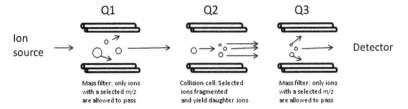

Figure 21.6 Schematics for MRM.

allowing enhanced sensitivity and selectivity. Triple quadrupole tandem mass spectrometers have three quadrupoles in series (Figure 21.6). In MRM mode, both Q1 and Q3 operate in SIM mode and only the selected product ions of a selected parent ion are allowed to reach the detector. All other product ions are filtered out. The second quadrupole (Q2), also known as the collision cell, focuses and transmits the ions while introducing a collision gas (argon or helium) into the flight path of the selected ion to induce collision-induced dissociation (CID), resulting in fragment ions of the parent ion. The MRM mode works like a double mass filter which drastically increases the signal-to-noise ratio (S/N).

Quadrupoles: The quadrupole acts as a mass filter. It consists of a linear array of four symmetrically arranged rods (Figure 21.6, Q1) to which a radio frequency (RF) generator and a DC potential are connected. The ions oscillate in a specific RF field inside the quadruple array with frequencies which depend on their m/z values. If the oscillations of an ion inside the array are stable, the ion will continue to pass through and reach the detector. Quadrupoles can be used in either a scanning or a filtering mode. Quantitative studies are performed with quadrupole working in filtering mode where only selected ions are allowed to pass according to a fixed set(s) of DC and RF voltages and all other ions 'filtered out'. This mode is called 'selected ion monitoring' (SIM).

Refractive index (RI) detection: Refractive index is one of the least sensitive LC detectors. It measures the change in refractive index of the column effluent passing through the flow-cell. Refractive index is very sensitive to changes in ambient temperature, pressure and flow rate, and cannot be used for gradient elution. Although it is useful for detecting those compounds that are non-ionic, do not absorb in the UV region and do not fluoresce, the many disadvantages make it less popular with the advancement of other detection techniques, such as ELS or MS.

Stable isotope dilution mass spectrometry (IDMS): IDMS is a technique (method) based on the addition to an analytical sample of a known amount of a stable isotopically labelled analogue of a desired analyte as an internal standard. The ratio of the amount of added isotopic analogue and the naturally occurring compound, as measured by mass spectrometry, provides highly accurate values. Definitive accuracy of the IDMS technique requires full equilibration between endogenous natural amounts of the compound and the isotopically labelled spike, which is usually considered to be accomplished by bringing both in the same chemical form in solution.

Time-of-flight (TOF): In TOF, all the ions generated in the source of the instrument are introduced into the mass analyser and routed to the detector. All the ions receive the same initial kinetic energy when they enter the fly tube–TOF mass analyser. As they pass along the field-free drift zone of the fly tube, ions are separated by their masses as the lighter ions travel faster. This enables the instrument to record all ions as they arrive at the detector and use their arrival times to calculate their m/z. TOF-MS gives rapid, full spectral acquisition that allows for high sensitivity across a wide mass range. Although it can provide very high resolution, TOF-MS cannot selectively monitor targeted analytes like other types of mass spectrometers.

List of Abbreviations

AOAC	Association of Official Analytical Chemists
APCI	atmospheric pressure chemical ionization
CID	collision-induced dissociation
DAD	diode array detector
ELS	evaporative light-scattering
ESI	electro-spray ionization
HPLC	high-performance liquid chromatography
IDMS	isotope dilution mass spectrometry
IPC	ion-pair chromatography
ITMS	ion-trap mass spectrometry
LC	liquid chromatography
MRM	multiple reaction monitoring
MS	mass spectrometry
MS/MS	tandem mass spectrometry
MS^n	multi-stage sequential MS analysis
m/z	mass-to-charge ratio
NIST	National Institute of Standards and Technology
QMS	quadrupole mass spectrometry
RF	radio frequency
RI	refractive index
SIM	selected ion monitoring
SPE	solid-phase extraction
SRM	selected reaction monitoring or Standard Reference Material
S/N	signal-to-noise ratio
TFA	trifluoroacetic acid
TOF	time-of-flight
UHPLC	ultra high performance liquid chromatography
USP	United States Pharmacopeia
UV	ultraviolet
WSV	water-soluble vitamins

References

Amin, M. and Reusch, J., 1987. High-performance liquid chromatography of water-soluble vitamins. Part 3. Simultaneous determination of vitamins B1, B2, B6, B12 and C, nicotinamide and folic acid in capsule preparations by ion-pair reversed-phase high-performance liquid chromatography. *Analyst.* 112: 989–991.

Bean, W.B. and Hodges, R.E., 1954. Pantothenic acid deficiency induced in human subjects. *Proceedings of the Society for Experimental Biology and Medicine.* 86: 693–698.

Bean, W.B., Hodges, R.E., Daum, K., Bradbury, J.T., Gunning, R., Manresa, J., Murray, W., Oliver, P., Routh, J.I., Schedl, H.P., Townsend, M. and Tung, I.C., 1955. Induced pantothenic acid deficiency in man. *Nutrition Reviews.* 13: 36–37.

Chen, P. and Wolf, W.R., 2007. LC/UV/MS-MRM for the simultaneous determination of water-soluble vitamins in multi-vitamin dietary supplements. *Analytical and Bioanalytical Chemistry.* 387: 2441–2448.

Chen, P., Ozcan, M., and Wolf, W.R., 2007. Contents of selected B vitamins in NIST SRM 3280 multivitamin/multielement tablets by liquid chromatography isotope dilution mass spectrometry. *Analytical and Bioanalytical Chemistry.* 389: 343–347.

Chen, P., Atkinson, R., and Wolf, W.R., 2009. Single-laboratory validation of a high-performance liquid chromatographic-diode array detector-fluorescence detector/mass spectrometric method for simultaneous determination of water-soluble vitamins in multivitamin dietary tablets. *Journal of AOAC International.* 92: 680–687.

Cornelis, R., 2003. Handbook of Elemental Speciation: Techniques and Methodology. Wiley, Chichester, UK and Hoboken, NJ, USA. 670 Pages.

Crandall, W.A. and Burr, M.M., 1947. Precision of microbiological assays for riboflavin, niacin and pantothenic acid. *Federation Proceedings.* 6(1 Pt 2): 246 pages.

Dong, M.W., Lepore, J., and Tarumoto T., 1988. Factors affecting the ion-pair chromatography of water-soluble vitamins. *Journal of Chromatography.* 442: 81–95.

Heudi, O., Kilinc, T., and Fontannaz, P., 2005. Separation of water-soluble vitamins by reversed-phase high performance liquid chromatography with ultra-violet detection: application to polyvitaminated premixes. *Journal of Chromatography A.* 1070: 49–56.

Hudson, T.J. and Allen, R.J., 1984. Determination of pantothenic acid in multivitamin pharmaceutical preparations by reverse-phase high-performance liquid chromatography. *Journal of Pharmaceutical Sciences.* 73: 113–115.

Ivanovic, D., Popovic, A., Radulovica, D., and Medenicac, M., 1999. Reversed-phase ion-pair HPLC determination of some water-soluble vitamins in pharmaceuticals. *Journal of Pharmaceutical and Biomedical Analysis.* 18: 999–1004.

Ketchum, H.M. and Conklin, R.L., 1945. A simplified medium for the microbiological assay for pantothenic acid. *Journal of Bacteriology*. 50: 717.

Kirchmeier, R.L. and Upton, R.P., 1978. Simultaneous determination of niacin, niacinamide, pyridoxine, thiamine, and riboflavin in multivitamin blends by ion-pair high-pressure liquid chromatography. *Journal of Pharmaceutical Sciences*. 67: 1444–1446.

Kwok, R.P., Rose, W.P., Tabor, R., and Pattison, T.S., 1981. Simultaneous determination of vitamins B1, B2, B6, and niacinamide in multivitamin pharmaceutical preparations by paired-ion reversed-phase high-pressure liquid chromatography. *Journal of Pharmaceutical Sciences*. 70: 1014–1017.

Li, K., 2002. Simultaneous determination of nicotinamide, pyridoxine hydrochloride, thiamine mononitrate and riboflavin in multivitamin with minerals tablets by reversed-phase ion-pair high performance liquid chromatography. *Biomedical Chromatography*. 16: 504–507.

Lipmann, F., Kaplan, N.O. Novelli, G.D, Tuttle, L.C., and Guirard, B.M., 1947. Coenzyme for acetylation, a pantothenic acid derivative. *Journal of Biological Chemistry*. 167: 869.

Machlin, L.J., 2001. Handbook of Vitamins. Marcel Dekker, New York, USA. 600 Pages.

Markopoulou, C.K., Kagkadis, K.A., and Koundourellis J.E., 2002. An optimized method for the simultaneous determination of vitamins B1, B6, B12 in multivitamin tablets by high performance liquid chromatography. *Journal of Pharmaceutical and Biomedical Analysis*. 30: 1403–1410.

Pakin, C., Bergaentzle, M., *et al.*, 2004. Fluorimetric determination of pantothenic acid in foods by liquid chromatography with post-column derivatization. *Journal of Chromatography A*. 1035: 87–95.

Pugh, E.L., Sauer, F., Waite, M, Toomey, R.E., and Wakil S.J., 1966. Studies on the mechanism of fatty acid synthesis. 13. The role of beta-hydroxy acids in the synthesis of palmitate and cis vaccenate by the *Escherichia coli* enzyme system. *Journal of Biological Chemistry*. 241: 2635–2643.

Rychlik, M., 2003. Pantothenic acid quantification by a stable isotope dilution assay based on liquid chromatography-tandem mass spectrometry. *Analyst*. 128: 832–837.

Sander, L.C., Sharpless, K.E., Wise, S. A., Nelson, B.C., Phinney, K.W., Porter, B.J., Rimmer, C.A., Thomas, J.B., Wood, L.J., Yen, J.H., Duewer, D. L., Atkinson, R., Chen, P., Goldschmidt, R., Wolf, W.R., Ho, I-P, and Betz, J. M., 2011. Certification of vitamins and carotenoids in SRM 3280 multivitamin/multielement tablets. *Analytical Chemistry*. 83: 99–108.

Smith, C.M. and Song, W.O., 1996. Comparative nutrition of pantothenic acid. *Journal of Nutritional Biochemistry*. 7: 312–321.

Takahashi, K., Fukuwatari, T., Shibata K., 2009. Fluorometric determination of pantothenic acid in human urine by isocratic reversed-phase ion-pair high-performance liquid chromatography with post-column derivatization. *Journal of Chromatography B: Analytical Technologies in the Biomedical and Life Sciences*. 877: 2168–2172.

Walker, M.C., Carpenter, B.E., Cooper, E.L., 1981. Simultaneous determination of niacinamide, pyridoxine, riboflavin, and thiamine in multivitamin products by high-pressure liquid chromatography. *Journal of Pharmaceutical Science*. 70: 99–101.

Williams, R.J., Lyman, C.M., Goodyear, G.H., Truesdail, J.H., and Holaday, D., 1933. 'Pantothenic acid,' a growth determinant of universal biological occurrence. *Journal of the American Chemical Society*. 55: 2912–2927.

Wills, R B., Shaw, C.G., and Day, W.R., 1977. Analysis of water soluble vitamins by high pressure liquid chromatography. *Journal of Chromatography Science*. 15: 262–266.

Woollard, D.C., 1984. New ion-pair reagent for the high-performance liquid chromatographic separation of B-group vitamins in pharmaceuticals. *Journal of Chromatography*. 301: 470–476.

Wyse, B.W., Song, W.O., Walsh, J.H. and Hansen, RG.. 1985. Pantothenic acid. In: Augustin, J., Klein, B.P., Becker, D.A. and Venugopal, P.B. (ed.) Methods of Vitamin Assay, 4th ed. Wiley, New York, USA, pp. 399–416.

Zucker, T.F. and Zucker, L.M.,1954. Pantothenic acid deficiency and loss of natural resistance to a bacterial infection in the rat. *Proceedings of the Society for Experimental Biology and Medicine*. 85(3): 517–521.

CHAPTER 22
Enzymatic HPLC Assay for all Six Vitamin B_6 Forms

TOSHIHARU YAGI

Faculty of Agriculture and Agricultural Science Program, Graduate School of Integral Arts and Sciences, Kochi University, Monobe Otsu 200, Nankoku, Kochi 783-8502, Japan
Email: yagito@kochi-u.ac.jp

22.1 Current Status and Challenges of Vitamin B_6 Analysis

Vitamin B_6 consists of six forms in nature, *i.e.* pyridoxine (PN), pyridoxal (PL), pyridoxamine (PM), pyridoxine 5′-phosphate (PNP), pyridoxal 5′-phosphate (PLP), and pyridoxamine 5′-phosphate (PMP) (Figure 22.1). PLP is a coenzyme for many enzymes involved in amino acid and carbohydrate metabolism, and plays a key role in the nutritional function of vitamin B_6. The other forms show the same nutritional efficiency because they are converted into PLP in cells. The free forms of vitamin B_6 are adsorbed through the intestinal mucosa, and then are phosphorylated and converted into PLP in the liver. Some PLP exits the liver and travels in the blood on albumin, being turned over slowly. PL is the form most actively transported to other cells from the liver. The cells adsorb PL and then phosphorylate it to yield PLP. The final metabolite derived from vitamin B_6 is 4-pyridoxic acid (4-PA, Figure 22.1), which is excreted into the urine. Plants contain a storage form of vitamin B_6, pyridoxine-β-glucoside

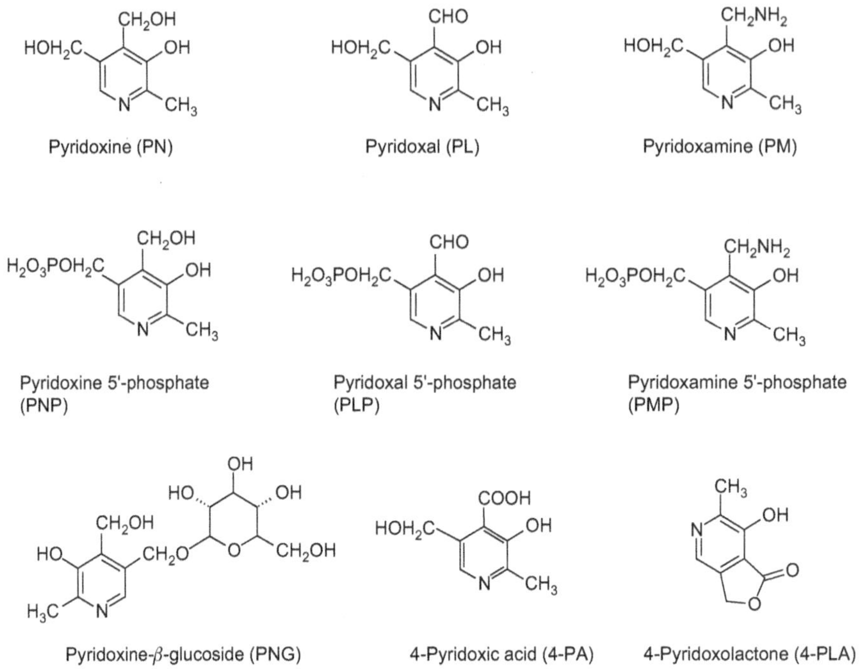

Figure 22.1 Structures of natural vitamin B_6 forms, pyridoxine-β-glucoside, 4-pyridoxic acid and 4-pyridoxolactone.

(PNG; Figure 22.1), which is not counted as a vitamin B_6 form because its nutritional contribution in the human body is controversial.

Recent studies have shown that the individual vitamin B_6 compounds have specific biochemical functions. PM detoxifies active carbonyl products, prevents the formation of advanced glycation end-products (AGEs) (Voziyan et al. 2003), and prevents diabetic complications (Jain 2007). PLP prevents the progression of diabetic nephropathy. PLP and PMP show stronger protection of yeast cells from oxidative death than vitamin C, which is a representative antioxidative vitamin (Chumnantana et al. 2005). Vitamin B_6 forms are strong quenchers of singlet oxygen (Bilski et al. 2000). Thus, the development of an assay method for all individual vitamin B_6 compounds in foods is required to estimate their biochemical functionality.

Many methods have been developed for the determination of vitamin B_6. They can be classified into two groups: gross and individual assays. The most familiar gross assay method is a microbioassay involving the yeast, *Saccharomyces cerevisiae* (former name, *S. carlsbergensis* and *S. uvarum*), which was developed 68 years ago and has been used to determine the total contents of vitamin B_6 in many foods described in nutrition tables.

The individual vitamin B_6 assay methods can be divided into two groups: ones that are used to determine all vitamin B_6 forms; and ones that are used to determine specific vitamin B_6 compounds. All vitamin B_6 forms can be determined by high-performance liquid chromatography (HPLC) with fluorescence detection (Coburn and Mahuren 1983), which are applicable to the analysis of samples containing high amounts of vitamin B_6 forms in comparison to the amounts of fluorescent contaminants. However, foods, tissue extracts and urine contain too many fluorescent contaminants to use a HPLC method to determine all vitamin B_6 forms in such samples. The specificity and sensitivity of the HPLC method needs to be improved to make it applicable to such samples.

One way to increase the specificity and sensitivity of the HPLC method is to specifically convert each vitamin B_6 forms into a highly fluorescent vitamin B_6 derivative prior to the HPLC analysis. Thus, a new method was developed in which the individual vitamin B_6 forms and PNG are specifically converted into 4-pyridoxolactone (Figure 22.1), a highly fluorescent derivative of vitamin B_6. The method was applicable to determination of the individual vitamin B_6 forms in the foods examined and human urine.

22.2 Enzymatic Reactions for Conversion of Vitamin B_6 Forms and PNG into 4-PLA

The sequence of enzymatic reactions is shown in Figure 22.2. The basic reaction is the irreversible conversion of PL into 4-PLA by pyridoxal 4-dehydrogenase (PLDH) from *Mesorhizobium loti* (Yokochi *et al.* 2006). PM is converted into 4-PLA through the coupled reaction involving PLDH and pyridoxamine-pyruvate aminotransferase (PPAT) from *M. loti* (Yoshikane *et al.* 2006). Although the PPAT reaction is reversible, PLDH makes the coupled reaction irreversible. PN is converted into 4-PLA through the coupled reaction involving pyridoxine oxidase (PNOX) from *M. loti* and PLDH. PNOX irreversibly oxidizes PN into PL (Yuan *et al.* 2004). PLP, PMP and PNP are irreversibly hydrolysed to PL,

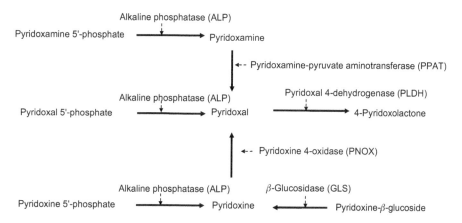

Figure 22.2 Enzymatic reactions for specific conversion of vitamin B_6 forms and pyridoxine-β-glucoside into 4-pyridoxolactone.

PM and PN, respectively, by alkaline phosphatase (ALP); the PL, PM and PN are then converted to 4-PLA as described above. PNG is irreversibly hydrolysed into PN by β-glucosidase (GLS) and the PN produced is converted into 4-PLA. Thus, all six forms of vitamin B_6 and PNG are specifically and completely converted into 4-PLA.

22.3 Materials and Methods

22.3.1 Materials

PN and PLP were purchased from Nacalai Tesque, Inc. (Kyoto, Japan), and PL and PM from Sigma Chemical Co. (St. Louis, MO, USA). PMP was a gift from Daiichi Fine Chemical Co. (Takaoka, Japan). PNP was prepared from PLP through reduction with sodium borohydride. PNG was prepared enzymatically from D-glucose and PN using the transglucosidase activity of a bacterial β-glucosidase. 4-PLA was biotechnologically prepared from PN as described previously (Tamura *et al.* 2008). The concentrations of standard solutions were spectrophotometrically calibrated. Homogeneous recombinant PLDH (Yokochi *et al.* 2006), PPAT (Yoshikane *et al.* 2006), and PNOX (Yuan 2004) were prepared as described previously. Homogenous recombinant *Escherichia coli* ALP with a His-tag was prepared according to the standard protocol. Almond GLS was purchased from Sigma Chemical Co.

22.3.2 Enzyme Activity Determination

PLDH, PNOX, PPAT, ALP and GLS were assayed as described below. One unit of enzyme was defined as the amount that catalysed the formation of 1 µmol of product per min. The activities of the enzymes were assayed every time just before usage for enzymatic conversion of the vitamin B_6 forms and PNG into 4-PLA. PLDH activity was determined by measuring the initial increase in A_{340} of NADH at 30 °C in 1 mL of a reaction mixture consisting of 50 mM CHES (pH 9.2), 1 mM pyridoxal, 3 mM NAD^+ and the enzyme.

The molecular extinction coefficients of NADH and 4-PLA at 340 nm are 6600 and 7500 $M^{-1}cm^{-1}$, respectively. Thus, a molecular extinction coefficient value of 14 100 was used for calculation of one unit of enzyme (Yokochi *et al.* 2006). PNOX activity was determined as follows. The reaction mixture (0.4 mL), consisting of 0.1 M Tris-acetate buffer (pH 7.0), 5 mM PN and the enzyme was incubated aerobically at 30 °C for 30 min, and then 66 µl of 9 M H_2SO_4 was added to it to stop the enzyme reaction. After centrifugation at 10 000g for 5 min, a portion (0.3 mL) of the supernatant was mixed with 0.6 mL of 1 M H_2SO_4 and 0.1 mL of 2% (w/v) phenylhydrazine dissolved in 5 M H_2SO_4, and the resultant mixture was incubated at 60 °C for 20 min. Then, A_{410} of the phenylhydrazone of PL was measured.

PPAT activity was also determined by the phenylhydrazine method, as follows. The reaction mixture (0.4 mL) consisted of 0.1 M borate-KOH, pH 9.0, 3.33 mM sodium pyruvate, 3.33 mM PM and the enzyme. The reaction was performed at 30 °C for 10 min and stopped by the addition of 66 μl of 9 M H_2SO_4. ALP activity was determined with *p*-nitrophenylphosphate as the substrate. The reaction mixture (1 mL) consisted of 50 mM Tris-HCl, pH 9.0, 0.5 mM *p*-nitrophenylphosphate and the enzyme. The reaction was performed at 30 °C for 10 min and stopped by the addition of 0.5 mL of 1 M NaOH. Then, A_{405} of the *p*-nitrophenol produced was measured. The molecular extinction coefficient of *p*-nitrophenol was 17.7 ($mM^{-1}cm^{-1}$). GLS activity was determined with nitrophenyl-β-D-glucose as the substrate. The reaction mixture consisted of 50 mM sodium acetate, pH 5.0, and 0.5 mM nitrophenyl-β-D-glucose. The reaction was performed at 30 °C for 10 min and stopped by the addition of 0.5 mL of 1 M NaOH. Then, A_{405} of the *p*-nitrophenol produced was measured.

22.3.3 HPLC System

4-PLA was determined by reversed-phase isocratic HPLC. The HPLC system consisted of a JASCO (Tokyo, Japan) AS-2055 autosampler, a PU-2080 pump, a CO-2060 column thermostat, an FP-2025 fluorescence detector and 807-IT integrator. The column used was of Cosmosil 5C18MS-II column (250 cm × 4.6 cm; Nacalai Tesque, Kyoto, Japan). The mobile phase buffer was 20 mM potassium phosphate, pH 7.0, containing 10% (v/v) methanol. The flow rate, sample volume and column temperature were 0.5 mL/min, 100 μL and 20 °C, respectively. The fluorescence intensity of the eluted 4-PLA was monitored at 430 nm (excitation at 360 nm).

22.3.4 Enzymatic Conversion into 4-PLA and Calculation

The vitamin B_6 forms and PNG in samples were converted into 4-PLA in seven reaction mixtures. The reaction mixtures and conditions are shown in Figure 22.3.

PL was determined under reaction conditions 1. A food sample (5 μL) was added to reaction mixture 1 to start the PLDH reaction. After 1 h, the reaction was stopped with HCl. Then, the mixture was filtered through a Dismic 13 syringe filter (pore size, 0.2 μm; Advantech, Tokyo, Japan). The filtrate (100 μL) was applied to the HPLC column. The amount of sample could be increased up to 200 μL depending on the PL content in the sample: the amount of water should be decreased to make the volume of the reaction mixture 380 μL. HCl was added before addition of the sample to obtain a control (no reaction) reaction mixture. For determination of the yield, a food sample containing a definite amount of standard PL was also analysed. The amount was calculated based on the results of preliminary analysis of the PL content of the food sample. For analysis of urine, 8 pmol of standard PL was added to reaction mixture 1. The yield was calculated as follows:

$$\text{Yield}(\%) = \frac{(\text{Amount of PL in sample with standard PL} - \text{Amount in the sample})(\text{mol}) \times 100}{\text{Amount of standard PL added (mol)}}$$

Pyridoxal (Reaction conditions 1)	Pyridoxal 5'-phosphate (Reaction conditions 4)	Pyridoxine 5'-phosphate (Reaction conditions 6)
250 mM Tris-HCl (pH 9.0), 40 µL	250 mM Tris-HCl (pH 9.0), 40 µL	250 mM Tris-HCl (pH 9.0), 40 µL
1 U/mL PLDH, 1 µL	5 mM MgCl₂, 40 µL	5 mM MgCl₂, 40 µL
100 mM NAD⁺, 2 µL	1.7 U/mL ALP, 6 µL	1.7 U/mL ALP, 6 µL
Sample, 5 µL	Sample, 5 µL	Sample, 5 µL
Water, 332 µL	Water, 286 µL	Water, 284 µL
→ 380 µL	→ 377 µL	→ 375 µL
30 °C, 1 h	30 °C, 2 h	30 °C, 2 h
→ 1.1 M HCl, 20 µL	→ 1 U/mL PLDH, 1 µL	→ 1 U/mL PLDH, 1 µL
Filtration	→ 100 mM NAD⁺, 2 µL	→ 100 mM NAD⁺, 2 µL
↓	30 °C, 1 h	→ 1 U/mL PNOX, 1 µL
HPLC (100 µL)	→ 1.1 M HCl, 20 µL	→ 1 mM FAD, 1 µL
	Filtration	30 °C, 1 h
	↓	→ 1.1 M HCl, 20 µL
	HPLC (100 µL)	Filtration
		↓
		HPLC (100 µL)
Pyridoxamine (Reaction conditions 2)	**Pyridoxamine 5'-phosphate (Reaction conditions 5)**	**Pyridoxine-β-glucoside (Reaction conditions 7)**
250 mM Tris-HCl (pH 9.0), 40 µL	250 mM Tris-HCl (pH 9.0), 40 µL	0.5 M sodium acetate (pH 5.0), 40 µL
1 U/mL PLDH, 1 µL	5 mM MgCl₂, 40 µL	1 M NaH₂PO₄, 20 µL
100 mM NAD⁺, 2 µL	1.7 U/mL ALP, 6 µL	25 U/mL β-glucoside, 2 µL
1 U/mL PPAT, 1 µL	Sample, 5 µL	Sample, 5 µL
100 mM Sodium pyruvate, 8 µL	Water, 277 µL	Water, 133 µL
Sample, 5 µL	→ 368 µL	30°C, 2 h → 200 µL
Water, 323 µL	30 °C, 2 h	→ 500 mM NaOH, 40 µL
→ 380 µL	→ 1 U/mL PLDH, 1 µL	→ 250 mM Tris-HCl (pH 9.0), 40 µL
30 °C, 1 h	→ 100 mM NAD⁺, 2 µL	→ 1 U/mL PLDH, 1 µL
→ 1.1 M HCl, 20 µL	→ 1 U/mL PPAT, 1 µL	→ 100 mM NAD⁺, 2 µL
Filtration	→ 100 mM sodium pyruvate, 8 µL	→ 1 U/mL PNOX, 1 µL
↓	30 °C, 1 h	→ 1 mM FAD, 1 µL
HPLC (100 µL)	→ 1.1 M HCl, 20 µL	Water, 95 µL
	Filtration	30 °C, 1 h
	↓	→ 2.0 M HCl, 20 µL
	HPLC (100 µL)	Filtration
		↓
		HPLC (100 µL)
Pyridoxine (Reaction conditions 3)		
250 mM Tris-HCl (pH 9.0), 40 µL		
1 U/mL PLDH, 1 µL		
100 mM NAD⁺, 2 µL		
1 U/mL PNOX, 1 µL		
1 mM FAD, 1 µL		
Sample, 5 µL		
Water, 330 µL		
→ 380 µL		
30°C, 1 h		
→ 1.1 M HCl, 20 µL		
Filtration		
↓		
HPLC (100 µL)		

Figure 22.3 Seven sets of reaction conditions for analysis of vitamin B_6 forms and pyridoxine-β-glucoside.

PM was determined under reaction conditions 2. Reaction mixture 2 comprised reaction mixture 1 additionally containing PPAT and pyruvate. Thus, 4-PLA produced in this reaction mixture is derived from PL and PM in the sample: subtraction of the amount of 4-PLA produced in reaction 1 from that produced in reaction 2 gives the amount of 4-PLA derived from PM.

PN was determined under reaction conditions 3. Reaction mixture 3 comprised reaction mixture 1 additionally containing PNOX and flavin adenine dinucleotide (FAD). 4-PLA produced in this reaction mixture is derived from PN and PL in the sample: subtraction of the amount of 4-PLA produced in reaction 1 from that produced in reaction 3 gives the amount of 4-PLA derived from PN.

PLP was determined under reaction conditions 4. The reaction was comprised two steps in one reaction tube. The first step was hydrolysis of PLP to PL by alkaline phosphatase (ALP) at 30 °C for 2 h in reaction mixture 4. The second step was the conversion of PL into 4-PLA. After the two-hour reaction, PLDH and nicotinamide adenine dinucleotide (NAD^+) were added to reaction mixture 4 and the resultant mixture was additionally incubated at 30 °C for 1 h. 4-PLA produced in this reaction mixture is derived from PLP and PL in the sample: subtraction of the amount of 4-PLA produced in reaction 1 from that produced in reaction 4 gives the amount of 4-PLA derived from PLP.

PMP was determined under reaction conditions 5. The reaction also comprised two steps in one reaction tube. The first reaction was conversion of PMP into PM in the same reaction mixture as that used for PLP. After the two-hour reaction, PLDH, NAD^+, PPAT and pyruvate were added and the reaction mixture was additionally incubated at 30 °C for 1 h. 4-PLA produced in this mixture is derived from PMP, PM, PLP and PL. Thus, subtraction of the amount of 4-PLA derived from PM, PLP and PL gives the amount of 4-PLA derived from PMP.

PNP was determined under reaction conditions 6. The first reaction was conversion of PNP into PN in the same reaction mixture as that used for PLP. After the two-hour reaction, PLDH, NAD^+, PNOX and FAD were added, and the reaction mixture was additionally incubated at 30 °C for 1 h. 4-PLA produced in this reaction mixture is derived from PNP, PN, PLP and PL. Thus, subtraction of the amount of 4-PLA derived from PN, PLP and PL gives the amount of 4-PLA derived from PNP.

PNG was determined under reaction conditions 7. The reaction comprised two steps in one reaction tube. The first step was conversion of PNG into PN at 30 °C for 2 h by β-glucosidase in reaction mixture 7, which contained a high concentration of sodium dihydrogen phosphate to inhibit phosphatase activity in the almond β-glucosidase preparation. After the two-hour reaction, NaOH and Tris-HCl buffer were added to the reaction mixture to make the pH 9.0. Then, PLDH, NAD^+, PNOX and FAD were added to the reaction mixture, which was additionally incubated at 30 °C for 1 h. 4-PLA in this reaction mixture is derived from PNG, PN and PL. Subtraction of the amount of 4-PLA derived from PN and PL gives the amount of 4-PLA derived from PNG.

22.3.5 Sample Preparation

22.3.5.1 Foods

The edible parts of foods were used. Dried foods were ground with a Mini Blender (Osaka Chemical Co., Osaka, Japan). The foods (0.1–1.0 g, dry or wet) were homogenized in 5 mL of 0.1 M HCl with a Polytron homogenizer, and then each homogenate was incubated at 100 °C for 30 min to free protein-bound PLP in a SCINICS ALB-121 aluminium dry block bath (Tokyo, Japan). After cooling, 0.1 mL of 50% (w/v) trichloroacetic acid was added to the homogenate, followed by additional incubation at 100 °C for 5 min in the bath. After cooling, the pH of the mixture was adjusted to 7.5 by the addition of 0.5 mL of 0.5 M Tris-HCl (pH 7.5) and 0.57 mL of 1.0 M NaOH. Then, water was added to make the volume of the reaction mixture 10 mL. The mixture was centrifuged at 8000 rpm for 5 min at 4 °C and the supernatant (5–50 µL) was used for the enzymatic reactions.

Some foods may need additional treatment before the homogenization in 0.1 M HCl. For example, meat, leafy vegetables and root crops should be ground, chopped and grated, respectively, with a food processor at 4 °C.

22.3.5.2 Human Urine

Trichloroacetic acid (50% w/v, 0.1 mL) was added to 5 mL of urine and the mixture was incubated at 100 °C for 5 min in the dry block bath. After cooling on ice, the mixture was centrifuged at 8000 rpm for 10 min at 4°C. The supernatant (5 µL) was used for analysis.

22.4 Chromatograms from Analyses of Standard 4-PLA and Pistachio Nuts, and Contents of Vitamin B_6 Forms and PNG in the Sample and Human Urine

22.4.1 Chromatographic Analyses of Standard 4-PLA, Vitamin B_6 Forms and PNG

Standard 4-PLA was eluted as a sharp single peak at around 11–14 min depending on the HPLC column lot number (Figure 22.4). The concentrations (0.1–100 pmol) of 4-PLA and the peak area or height fitted a straight line with $R^2 = 0.99$. Standard PL, PM, PN, PLP, PMP, PNP and PNG were quantitatively converted into 4-PLA under reaction conditions 1–7, respectively. Thus, they gave straight calibration lines with the same slope as that for standard 4-PLA and $R^2 = 0.99$ (Nishimura *et al.* 2008).

22.4.2 Analysis of all Vitamin B_6 Forms and PNG in Pistachio Nuts

Figure 22.5 shows HPLC chromatograms obtained when vitamin B_6 forms and PNG in a pistachio nut sample were analysed (Yagi *et al.* 2010a,

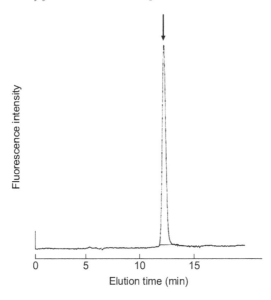

Figure 22.4 HPLC chromatogram of standard 4-pyridoxolactone. Standard 4-PLA (2 pmol) was applied. The chart attenuation was 8 mV/full scale. The fluorescence intensity is an arbitral unit.

Do et al. 2012). PL was determined under reaction conditions 1 (Figure 22.5A). The chromatogram of the reaction mixture (PL, R) contained a 4-PLA peak, which was not observed in the chromatogram of the control reaction mixture (PL, No-R). Thus, PL was contained in the pistachio nut sample, and the peak height or area gave the amount of 4-PLA derived from PL in the sample (Table 22.1). The sample containing 2 pmol (the amount in the 100 μL applied to the HPLC column) of standard PL gave a higher 4-PLA peak on chromatography (PL + Sd, R), showing that the added standard PL was converted into 4-PLA. Thus, PL in the pistachio nut sample was satisfactorily determined. PM in the sample was determined under reaction conditions 2 (Figure 22.5B).

The chromatogram of the reaction mixture (PM, R) contained a 4-PLA peak, which was not observed in that of the control reaction mixture (PM, No-R), and its height was higher than that of the reaction mixture (PL, R). Thus, PM was contained in the pistachio nut sample, and the peak height or area gave the amount of 4-PLA derived from PM and PL in the sample. The PM content in the pistachio nuts sample was determined by subtracting the amount of 4-PLA derived from PL (Table 22.1).

The sample containing 2 pmol (the amount in the 100 μL applied to the HPLC column) of standard PM gave a higher 4-PLA peak (PM + Sd, R), showing that added standard PM was converted into 4-PLA and that PM in the sample was satisfactorily determined. PN in the sample was determined under reaction conditions 3 (Figure 22.5C). The chromatogram of the reaction mixture (PN, R) contained a 4-PLA peak, which was not observed in that of the control reaction mixture (PN, No-R) and its height was higher than that of the reaction mixture (PL, R). Thus, PN was contained in the pistachio nut sample, and the peak height or area gave the amount of 4-PLA derived from PN and PL in the sample. The PN content in the pistachio nut sample was determined by subtracting the amount of 4-PLA derived from PL (Table 22.1). Interestingly, some fluorescent compounds eluted before 4-PLA are produced from unknown compounds contained in the pistachio nut sample in the reaction mixture through the PNOX-catalysed reaction.

The sample containing 2 pmol (the amount in the 100 μL applied to the HPLC column) of standard PN gave a higher 4-PLA peak (PN + Sd, R), showing that the added standard PN was converted into 4-PLA and that PN in the sample was satisfactorily determined. PLP in the sample was determined under reaction conditions 4 (Figure 22.5D). The chromatogram of the reaction mixture (PLP, R) contained a 4-PLA peak, which was not observed in that of the control reaction mixture (PLP, No-R) and its height was higher than that of the reaction mixture (PL, R). Thus, PLP was contained in the pistachio nut sample, and the peak height or area gave the amount of 4-PLA derived from PLP and PL in the sample. The PLP content in the pistachio nut sample was determined by subtracting the amount of 4-PLA derived from PL (Table 22.2).

The sample containing 2 pmol (the amount in the 100 μL applied to the HPLC column) of standard PLP gave a higher 4-PLA peak (PLP + Sd, R), showing that the added standard PLP was converted into 4-PLA and that PLP in the sample was satisfactorily determined. PMP in the sample was determined under reaction conditions 5 (Figure 22.5E). The chromatogram of the reaction mixture (PMP, R) contained a 4-PLA peak, which was not found in that of the control reaction mixture (PMP, No-R). The peak height or area gave the amount of 4-PLA derived from PMP, PM, PLP and PL in the sample. Thus, the PMP content in the pistachio nut sample was determined by subtracting the amount of 4-PLA derived from PM, PL, and PLP (Table 22.2).

The sample containing 2 pmol (the amount in the 100 μL applied to the HPLC column) of standard PMP gave a higher 4-PLA peak (PMP + Sd, R), showing that the added standard PMP was converted into 4-PLA and that PMP in the sample was satisfactorily determined.

PNP in the sample was determined under reaction conditions 6 (Figure 22.5F). The chromatogram of the reaction mixture (PNP, R) contained a 4-PLA peak, which was not observed in that of the control reaction mixture (PNP, No-R). PNP was contained in the pistachio nut sample and the peak height or area gave the amount of 4-PLA derived from PN, PLP and PL in the sample. Thus, the PNP content in the pistachio nut sample was determined by subtracting the amount of 4-PLA derived from PN, PL and PLP (Table 22.2). The sample containing 2 pmol (the amount in the 100 μL applied to the HPLC column) of standard PNP gave a higher 4-PLA peak (PNP + Sd, R), showing that the added standard PNP was converted into 4-PLA and that PNP in the sample was satisfactorily determined.

PNG in the sample was determined under reaction conditions 7 (Figure 22.5G). The chromatogram of the reaction mixture (PNG, R) contained a 4-PLA peak, which was not observed in that of the control reaction mixture

Enzymatic HPLC Assay for all Six Vitamin B_6 Forms 345

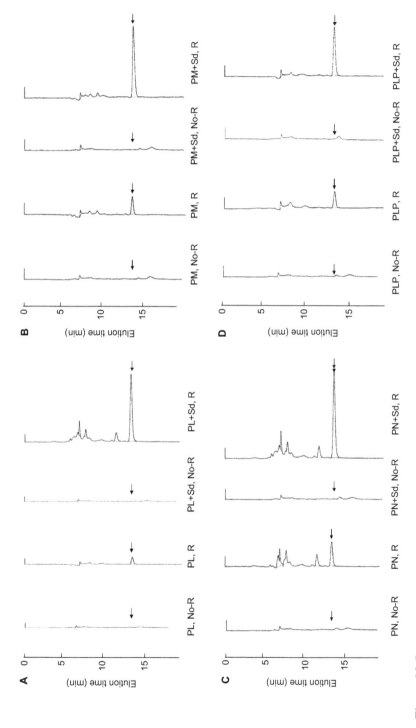

Figure 22.5

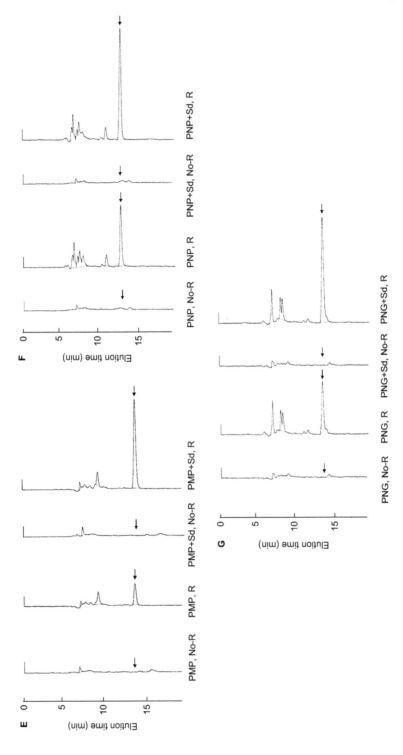

Figure 22.5 (Continued)

(PNG, No-R). PNG was contained in the pistachio nut sample, and the peak height or area gave the amount of 4-PLA derived from PN and PL in the sample. Thus, the PNG content in the pistachio nut sample was determined by subtracting the amount of 4-PLA derived from PN and PL (Table 22.2). The formation of fluorescent compounds eluted before 4-PLA by PNOX was augmented, suggesting that the β-glucosidase reaction increased the unknown compounds, which were converted into the fluorescent compounds. The sample containing 2 pmol (the amount in the 100 µL applied to the HPLC column) of standard PNG gave a higher 4-PLA peak (PNG + Sd, R), showing that the added standard PNG was converted into 4-PLA and that PNG in the sample was satisfactorily determined.

The yields of the vitamin B_6 forms and PNG added to the pistachio nuts prior to the sample preparation were successfully determined (Table 22.2). The yield of PLP was 38%, suggesting that PLP in the pistachio nuts was hard to dissociate as a free from. The tight binding of PLP in some foods is well known (Snell 1980).

Figure 22.5 HPLC chromatograms of analysis of all forms of vitamin B_6 and PNG in pistachio nuts. The chart attenuation was 8 mV/full scale in all cases. (A) PL was determined. The pistachio nut sample (5 µL) was subjected to enzymatic conversion (reaction conditions 1 in Figure 22.3) and then to HPLC: the elution pattern is shown above PL, R. The sample containing 2 pmol (the amount in the 100 µL applied to the HPLC column) of standard PL was subjected to enzymatic conversion (PL + Sd, R). The controls were PL, No-R and PL + Sd, No-R, respectively. (B) PM in the sample was determined. The sample (5 µL) was subjected to enzymatic conversion (reaction conditions 2) and then to HPLC (PM, R). The sample containing 2 pmol of standard PM was subjected to enzymatic conversion (PM + Sd, R). The controls were PM, No-R, and PM + Sd, No-R, respectively. (C) PN in the sample was determined. The sample (5 µL) was subjected to enzymatic conversion (reaction conditions 3) and then to HPLC (PN, R). The sample containing 2 pmol of standard PN was subjected to enzymatic conversion (PN + Sd, R). The controls were PN, No-R, and PN + Sd, No-R, respectively. (D) PLP in the sample was determined. The sample (5 µL) was subjected to enzymatic conversion (reaction conditions 4) and then to HPLC (PLP, R). The sample containing 2 pmol of standard PLP was subjected to enzymatic conversion (PLP + Sd, R). The controls were PLP, No-R, and PLP + Sd, No-R, respectively. (E) PMP in the sample was determined. The sample (5 µL) was subjected to enzymatic conversion (reaction conditions 5) and then to HPLC (PMP, R). The sample containing 2 pmol of standard PLP was subjected to enzymatic conversion (PMP + Sd, R). The controls were PMP, No-R, and PMP + Sd, No-R, respectively. (F) PNP in the sample was determined. The sample (5 µL) was subjected to enzymatic conversion (reaction conditions 6) and then to HPLC (PNP, R). The sample containing 2 pmol of standard PNP was subjected to enzymatic conversion (PNP + Sd, R). The controls were PNP, No-R, and PNP + Sd, No-R, respectively. (G) PNG in the sample was determined. The sample (5 µL) was subjected to enzymatic conversion (reaction conditions 7) and then to HPLC (PNG, R). The sample containing 2 pmol of standard PNG was subjected to enzymatic conversion (PNG + Sd, R). The controls were PNG, No-R, and PNG + Sd, No-R, respectively.

Table 22.1 Contents of free forms of vitamin B_6 forms and PNG in pistachio nuts and human urine.

	PL ($\mu mol/100\,g$ or mL)	PM ($\mu mol/100\,g$ or mL)	PN ($\mu mol/100\,g$ or mL)
	Yield (%)	Yield (%)	Yield (%)
Pistachio nuts (dry)	1.14 ± 0.02	0.43 ± 0.02	0.68 ± 0.05
	102	67	89
Urine (five young healthy men)[a]	0.038 ± 0.002	0.007 ± 0.000	ND
	89	78	89

[a]Adapted from Yagi et al. 2010b.
ND: not detected.

Table 22.2 Contents of phosphoester forms of vitamin B_6 and PNG in pistachio nuts and human urine.

	PLP ($\mu mol/100\,g$ or mL)	PMP ($\mu mol/100\,g$ or mL)	PNP ($\mu mol/100\,g$ or mL)	PNG ($\mu mol/100\,g$ or mL)
	Yield (%)	Yield (%)	Yield (%)	Yield (%)
Pistachio nuts (dry)	0.59 ± 0.01	0.48 ± 0.06	598 ± 0.12	6.31 ± 0.17
	38	83	81	57
Urine (five young healthy men)[a]	ND	ND	ND	0.023 ± 0.008
	79	92	93	85

[a]Adapted from Yagi et al. 2010b.
ND: not detected.

22.4.3 Analysis of Vitamin B_6 Forms and PNG in Human Urine

The contents of the vitamin B_6 forms in the urine of healthy young men consuming a usual diet were determined (Yagi et al. 2010b). The average values for five men are shown in Table 22.1. Although the contents were very low compared to those in foods, the urine contained PL, PM and PNG. The PNG content was similar to that of PL. The PM content was 18% of the PL content. The average content of 4-PA determined by the HPLC method (Schuster et al. 1984) was 0.12 ± 0.08 μg/100 mL. Thus, the PL content was 32% of the 4-PA content.

The results coincide with the general consensus that 4-PA is the major excretory metabolite of vitamin B_6 and that its content in urine is a direct measure of the nutritional state of vitamin B_6 in the human body (Leklem 1990). Interestingly, the PL and 4-PA contents showed a strong correlation with a correlation coefficient of 0.98 (Yagi et al. 2010b), suggesting that the PL content in urine also reflects the vitamin B_6 status in the human body. In contrast, no correlation (correlation coefficient of 0.07) was found between the PM and 4-PA contents. Analysis of individual vitamin B_6 forms in urine from various people consuming usual foods should provide us with novel aspects of vitamin B_6 metabolism.

22.5 Concluding Remarks

With the new method we were able to quantitate the six vitamin B_6 forms and PNG together in foods and human-derived samples. The contents in foods so far studied (Yagi *et al.* 2010a) (chicken liver and white meat, egg yolk and white, dried anchovy, tuna meat used as sashimi, milk, carrot, garlic, and pistachio nuts) were high enough to use this method. Although the human urine contained a very small amount of vitamin B_6 forms together with many fluorescent contaminants, this method was applicable.

The disadvantage of the method is the subtraction for calculation of the contents of the vitamin B_6 forms other than PL. In particular, for calculation of the PNP and PMP contents, the contents of three other vitamin B_6 forms have to be subtracted. Thus, the accuracy of their contents is low. If we could use an enzyme that specifically converts PNP and PMP into 4-PLA, the method would be much improved.

So far studies on the absorption, circulation, metabolism and excretion of vitamin B_6 have been performed by feeding or injecting high amounts of radiolabelled PN. However, the contents of PN in the foods examined were low compared with those of the other vitamin B_6 forms. Thus, information on the absorption, circulation, metabolism and excretion of vitamin B_6 when animals or human consume usual foods should be obtained using enzymatic HPLC assay.

Summary Points

- This chapter focuses on a determination method for all six vitamin B_6 forms and pyridoxine-β-glucoside (PNG).
- This is the first method applicable to analysis of foods and human-derived samples.
- The vitamin B_6 forms and PNG were specifically converted into highly fluorescent 4-pyridoxolactone, which was then measured by HPLC.
- The three enzymes involved in the vitamin B_6 degradation pathway found in *Mesorhizobium loti* play key roles in the conversion of free forms of vitamin B_6 in this method.
- Phosphate forms were hydrolysed to free forms with alkaline phosphatase.
- PNG was converted into PN with β-glucosidase.

Key Facts

Key Facts about Vitamin B_6

- Vitamin B_6 was identified as a preventive factor for dermatitis in rats by György in 1934.
- A deficiency of vitamin B_6 is rare in human consuming a usual diet. But in some cases deficiency symptoms including anaemia, dermatologic disorders, weakness, sleeplessness, nervous disorders, and appetite and growth depression occur.

- High doses (higher than 100 mg/day for adult males and females) of vitamin B_6 (pyridoxine) have adverse effects such as sensory neuropathy and dermatological lesions.
- Vitamin B_6 and PNG in foods are measured by means of a microbiological assay after HCl hydrolysis, and their contents are shown in a Standard Table of Food Composition in many countries.
- Recent studies clarified novel functions of vitamin B_6 forms. All forms are strong quenchers of singlet oxygen. Pyridoxamine reduces the incidence of diabetic complications. PLP prevents the progression of diabetic nephropathy.

Key Facts about Enzymes

- Enzymes are bio-catalysts necessary for the life of organisms. Almost all are proteins.
- Each enzyme catalyses a specific reaction at a moderate temperature and shows reactivity towards a specific substrate.
- The reactions catalysed by enzymes are reversible or irreversible. The reversible reactions become irreversible on coupling with an irreversible reaction.
- Many enzymes require cofactors for catalysis. Coenzymes are organic cofactors and derivatives of B vitamers. NAD^+ and FAD are well-known cofactors for dehydrogenases and oxidases, respectively. Many metal ions are inorganic cofactors.
- Enzyme reactions are performed at a constant temperature and stopped by inactivation of enzymes in the reaction mixtures. Enzymes are easily inactivated at a low pH by the addition of an acid such as HCl to the reaction mixture.

Definitions of Words and Terms

Aminotransferase. Aminotransferase is also called transaminase. Aminotransferase belongs to a group of enzymes that require pyridoxal 5′-phosphate as a coenzyme and catalyse amino group transfer between an amino acid and a 2-oxo acid. The PPAT described in this chapter exceptionally does not require a coenzyme.

Dehydrogenase. Dehydrogenase is an enzyme that catalyses the removal of hydrogen from a substrate (hydrogen donor) and the transfer of the hydrogen to a donor substrate. NAD^+ or NADH is generally used as a cosubstrate.

Fluorescence. The emission of light by a substance when it adsorbs light or electromagnetic waves with a shorter wavelength than visible light is called fluorescence.

Glucosidase. Glucosidase is an enzyme that hydrolyses a glucosidic bond. Glucosidase specifically hydrolyses α- or β-glucosidic bond between a sugar and a sugar or between a sugar and an aglycone (the non-sugar part of a glucoside).

Glycation end-products (AGEs). AGEs are heterogeneous chemical compounds produced through formation of a Schiff base and an Amadori intermediate from glucose and proteins. AGEs are causative factors for diabetic complications.

Microbioassay. A microbioassay is a microbiological determination method mainly used for biochemicals. The amounts of chemicals are measured based on the growth levels of microorganisms. Chemicals support or inhibit their growth.

Molar extinction coefficient. The molar extinction coefficient of a substance at a specific wavelength is usually represented as the unit $M^{-1}cm^{-1}$. A one molar solution of a substance in a cuvette of 1 cm in length absorbs light corresponding to the molar extinction coefficient at the specific wavelength.

Oxidase. Oxidase is an enzyme which catalyses the oxidation of a substrate with an oxygen molecule. The oxygen molecule is converted into H_2O or H_2O_2. Many oxidases require FAD or FMN as a coenzyme.

Phosphatase. Phosphatase is an enzyme that hydrolyses the phosphoester bond of a substrate to form phosphate and the product with an OH group. Alkaline and acid phosphatases have alkaline and acidic optimum pH values, respectively.

Singlet oxygen. $^1O_2$ is the symbol for singlet oxygen, which is a diamagnetic form of molecular oxygen and one of the reactive oxygen species. Singlet oxygen oxygenates organic compounds with unsaturated bonds and causes damage to organisms.

List of Abbreviations

ALP	alkaline phosphatase
FAD	flavin adenine dinucleotide
GLS	β-glucosidase
HPLC	high-performance liquid chromatography
M	molar
NAD^+	nicotinamide adenine dinucleotide
4-PA	4-pyridoxic acid
PL	pyridoxal
4-PLA	4-pyridoxolactone
PLDH	pyridoxal 4-dehydrogenase
PLP	pyridoxal 5′-phosphate
PM	pyridoxamine
PMP	pyridoxamine 5′-phosphate
PN	pyridoxine
PNG	pyridoxine-β-glucoside
PNOX	pyridoxine 4-oxidase
PNP	pyridoxine 5′-phosphate
PPAT	pyridoxamine-pyruvate aminotransferase

References

Bilski, P., Li, M.Y., Ehrenshaft, M., Daub, M.E., and Chignell, C.F., 2000. Vitamin B_6 (pyridoxine) and its derivatives are efficient singlet oxygen quenchers and potential fungal antioxidants. *Photochemistry and Photobiology*. 71: 129–134.

Chumnantana, R., Yokochi, N., and Yagi, T., 2005. Vitamin B_6 compounds prevent the death of yeast cells due to menadione, a reactive oxygen generator. *Biochimica et Biophysica Acta*. 1722: 84–91.

Coburn, S.P., and Mahuren, J.D., 1983. A versatile cation-exchange procedure for measuring the seven major forms of vitamin B-6 in biological samples. *Analytical Biochemistry.* 129: 310–317.

Do, H.T.V., Ide, Y., Mugo, A.N., Yagi, T., 2012. All-enzymatic HPLC method for determination of individual and total contents of vitamin B_6 in foods. *Foods & Nutrition Research.* 56: 5409–5416.

Jain, S.K., 2007. Vitamin B_6 (pyridoxamine) supplementation and complications of diabetes. *Metabolism: Clinical and Experimental.* 56: 168–171.

Leklem, J.E., 1990. Vitamin B-6. A status report. *Journal of Nutrition.* 120: 1503–7.

Nishimura, S., Nagano, S., Crai, C.A., Yokochi, N., Yoshikane, Y., Ge, F., and Yagi, T., 2008. Determination of individual vitamin B_6 compounds based on enzymatic conversion to 4-pyridoxolactone. *Journal of Nutritional Science and Vitaminology.* 54: 18–24.

Schuster, K., Bailey, L.B., Cerda, J.J., and Gregory, J.F. 3rd, 1984. Urinary 4-pyridoxic acid excretion in 24-hour versus random urine samples as a measurement of vitamin B_6 status in humans. *The American Journal of Clinical Nutrition.* 39: 466–470.

Snell, E.E., 1980. Vitamin B_6 analysis: some historical aspects. In: Leklem, J.E., and Reynolds, R.D. (ed.) *Methods in Vitamin B-6 Nutrition: Analysis and Status Assessment.* Plenum Press, New York, USA, pp. 1–20.

Tamura, A., Yoshikane, Y., Yokochi, N., and Yagi, T., 2008. Synthesis of 4-pyridoxolactone from pyridoxine using a combination of transformed *Escherichia coli* cells. *Journal of Bioscience and Bioengineering.* 106: 460–465.

Voziyan, P.A., Khalifah, R.G., Thibaudeau, C., Yildiz, A., Jacob, J., Serianni, A.S., and Hudson, B.G., 2003. Modification of proteins in vitro by physiological levels of glucose: pyridoxamine inhibits conversion of amadori intermediates to advanced glycation end-products through binding of redox metal ions. *Journal of Biological Chemistry.* 278: 46616–46624.

Yagi, T., Ide, Y., Do, T.V.H., and Murayama, R., 2010a. All-enzymatic HPLC method for determination of individual and total contents of vitamin B_6 in foods. In: *Proceedings of 1st International Vitamin Conference*, May 19–21, 2010, Copenhagen, Abstract 17.

Yagi, T., Murayama, R., Do, H.T.V., Ide, Y., Mugo, A.N., and Yoshikane, Y., 2010b. Development of a simultaneous enzymatic assay method for all six individual vitamin B_6 forms and pyridoxine-β-glucoside. *Journal of Nutritional Science and Vitaminology.* 56: 157–163.

Yokochi, N., Nishimura, S., Yoshikane, Y., Ohnishi, K., and Yagi, T., 2006. Identification of a new tetrameric pyridoxal 4-dehydrogenase as the second enzyme in the degradation pathway for pyridoxine in a nitrogen-fixing symbiotic bacterium, *Mesorhizobium loti. Archives of Biochemistry and Biophysics.* 452: 1–8.

Yoshikane, Y., Yokochi, N., Ohnishi, K., Hayashi, H., and Yagi, T., 2006. Molecular cloning, expression and characterization of pyridoxamine-pyruvate aminotransferase. *Biochemical Journal.* 396: 499–507.

Yuan, B., Yoshikane, Y., Yokochi, N., Ohnishi, K., and Yagi, T., 2004. The nitrogen-fixing symbiotic bacterium *Mesorhizobium loti* has and expresses the gene encoding pyridoxine 4-oxidase involved in the degradation of vitamin B_6. *FEMS Microbiology Letters.* 234: 225–230.

CHAPTER 23

Analysis of Biotin (Vitamin B_7) and Folic Acid (Vitamin B_9): A Focus on Immunosensor Development with Liposomal Amplification

JA-AN ANNIE HO,*[a] YU-HSUAN LAI,[a] LI-CHEN WU,[b] SHEN-HUAN LIANG,[a] SONG-LING WONG[a] AND JR-JIUN LIOU[a]

[a] BioAnalytical and Nanobiomedicinal Laboratory, Department of Biochemical Science and Technology, College of Life Science, National Taiwan University, No. 1, Sec. 4, Roosevelt Road, Taipei, 10617, Taiwan; [b] Department of Applied Chemistry, College of Science and Technology, National Chi Nan University, No. 1, University Road, Puli, Nantou, 54561, Taiwan
*Email: jaho@ntu.edu.tw

23.1 Introduction

Biotin and folic acid (FA) (Figure 23.1) are two members of the water-soluble B complex vitamins. Biotin plays an important role in gene expression, cell signalling and histone biotinylation, and functions as a coenzyme in the tricarboxylic acid (TCA) cycle. It also functions in the metabolism of fatty

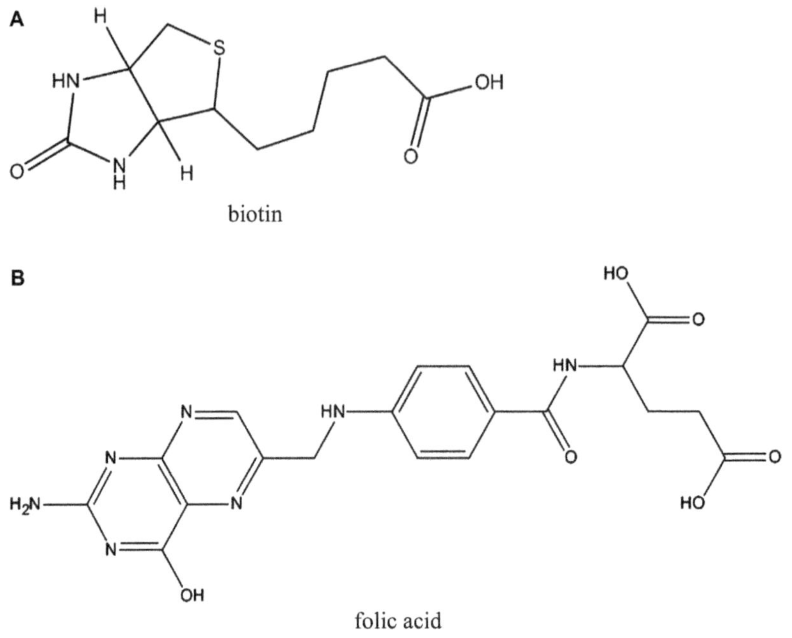

Figure 23.1 Structures of (A) biotin and (B) folic acid.

acid and amino acids, including the conversion of amino acids to glucose (gluconeogenesis), fatty acid synthesis, release of energy from fatty acids and DNA synthesis (Gropper *et al.* 2009; Insel *et al.* 2010).

Folic acid, also called folate, is essential for transferring one-carbon units involved in phospholipid, DNA/RNA, protein and neurotransmitter syntheses (Selhub and Rosenberg 1996). Furthermore folate is an essential dietary component for the formation of red and white blood cells and the epithelial cells in the digestive tract (Gregory *et al.* 1984).

Biotin deficiency is rare, but it can still occur. Inherent gene defects and consumption of excess raw egg white are two possible factors responsible for biotin deficiency. Symptoms progress from initial brittle hair, hair loss or baldness, skin rash, and development of convulsions and other neurological problems. Early diagnosis and daily biotin supplements can easily clear up symptoms. However, if untreated, biotin deficiency may induce depression, paresthesias, myalgias, or lead to coma and death (Insel *et al.* 2010; Pindolia *et al.* 2011). Sufficient folate nutriture is also understood to reduce the incidence of cardiovascular disease by lowering homocysteine concentration in blood, which has been implicated as a risk factor (Boushey *et al.* 1995; MRC Vitamin Study Research Group 1991; Hankey and Eikelboom 1999; Refsum *et al.* 1998). Given such health benefits, the US Food and Drug Administration in 1996 commanded the fortification of folate in cereal grain products at a level of 140 mg per 100 g (Food and Drug Administration 1996), anticipating that this would result in a reduction in incidences of folate-associated diseases in the US (Tucker *et al.* 1996).

Moreover, the correlation between folate intake and the occurrence of pregnancy neural tube defects has been previously studied and extensively discussed (Hibbard 1964; Mulinare *et al.* 1988). Based on the study by Johnson and Lund (2007), folate intake was also found to be inversely related to the risk of colorectal cancer.

23.2 Current Available Analytical Methods for Biotin and Folate

23.2.1 Analytical Methods for Biotin

23.2.1.1 Avidin-binding Assays

The proteins, avidin and streptavidin, are widely utilized in biotin analysis due to their outstanding affinity and specificity toward the binding of biotin (Zempleni *et al.* 2009). Generally an avidin-binding assay for the determination of biotin operates through the competition of sample biotin and labelled biotin (such as isotope-labelled biotin or biotinylated enzyme) with the limited number of avidin. Finally, the signals are acquired spectrophotometrically or electrochemically from the reaction of labelled enzyme with corresponding substrate or based on radioactivity counts.

23.2.1.1.1 Spectrophotometric Methods. On the binding of biotin with avidin, a shift (from 280 nm to 233 nm) can be observed on the absorption spectrum of the tryptophan residues of avidin. Biotin can therefore be determined by measuring spectral changes, especially the absorption at 233 nm (Livaniou *et al.* 2000). Another format of avidin-binding assay was described previously, in which sample biotin is capable of displacing *p*-hydroxy- azobenzene-2′-carboxylic acid (HABA), a dye molecule, from the avidin-HABA complex, leading to the dose-dependent decrease in the corresponding absorption at 500 nm (Zempleni *et al.* 2009).

23.2.1.1.2 Electrochemical Methods. Related methods make use of horseradish peroxidase (HRP), which is capable of reducing hydrogen peroxide electrocatalytically. Such a sensing system usually comprises a detecting electrode fabricated with avidin. The function of this amperometric sensing system is based on the competition of biotin–HRP complex and free biotin toward the immobilized avidin on the surface of electrode. As the final step, the current signals are obtained from the enzymatic reaction of biotin–HRP with a known concentration of hydrogen peroxide presented in the reaction mixture (Vreeke *et al.* 1995; Wright *et al.* 1995).

23.2.1.2 Bioluminescence-binding Assays

Aequorin (AEQ) is a photoprotein, originating from the jellyfish *Aequorea Victoria*, which offers exceptional sensitivity for being detected at levels as low as 10^{-21} mol. For this reason, it has been extensively used in the development of bioluminescence-binding assays for various targets, including biotin (Feltus *et al.* 2001). The bioluminescence-binding assays were constructed on the competition between a known amount of biotin–aequorin conjugate (AEQ-biotin) and a variable concentration of sample biotin for binding sites on protein avidin. The less free biotin present in the sample, the greater biotin-labelled photoprotein can bind with avidin, leading to higher bioluminescence signal emitted by the AEQ photoprotein (Feltus *et al.* 2001). Based on a similar principle, assays can be further developed by combining the bioluminescence resonance energy transfer (BRET) system and protein fusion technique (Vinokurov *et al.* 2003).

23.2.1.3 High-performance Liquid Chromatographic Analysis of Biotin

High-performance liquid chromatography (HPLC) methods have been comprehensively applied to the determination of biotin in pharmaceutical multivitamin preparations. Some of them have also been used to analyse more complicated matrices of samples, *e.g.* food material (Livaniou *et al.* 2000). Biotin detection by HPLC can be achieved with (either pre- or post-column) or without sample derivatization (Livaniou *et al.* 2000).

Biotin and its analogues can be discriminated by their structural difference using reversed phase or anion exchange HPLC (Bowers-Komro *et al.* 1986; Chastain *et al.* 1985; Livaniou *et al.* 2000) with no further derivitization. However, fluorescent derivatizations with 4-bromomethyl-7-methoxycumarin (Br-Mmc), 9-anthryldiazomethane (ADAM), 1-pyrenyldiazomethane (PDAM), thiamine, *o*-phthalaldehyde (OPA) or 3-mercaptopropionic acid (3-MPA) have usually been attempted to obtain better detection limit (Nojiri *et al.* 1998; Yokoyama and Kinoshita 1991).

In addition to the spectrometric detectors, HPLC has also been coupled with other detection systems, such as mass spectrometry (MS). Specificity can be highly improved using tandem mass spectrometry (HPLC-MS/MS) (Höller *et al.* 2006; Yomota and Ohnishi 2007). High sensitivity and selectivity can also be achieved by HPLC coupled with electrochemical detection (Kucera *et al.* 2007).

23.2.1.4 Other Biotin Assays

Biotin can also be determined by various other methods. In gas chromatographic analysis, the active hydrogen of the carboxyl group in biotin can be substituted with a silyl group using bis-(trimethylsilyl) acetamide (BSA) prior to the analysis to increase sensitivity and quantitative reproducibility (Viswanathan *et al.* 1970). A thin-layer chromatography method with a visualizing spraying system, such as *p*-dimethylaminocinnamaldehyde, used in the determination of biotin in multi-vitamin preparations has been described (Gröningsson and Jansson 1979). Microbiological assays (MBAs) using *Lactobacillus plantarum* (Baker 1985), capillary zone electrophoresis (Schiewe *et al.* 1996), paper chromatographic (Reio 1970), and polarographic methods (Serna *et al.* 1973) are also mentioned in previous literature on the determination of biotin.

23.2.2 Analytical Methods for Folic Acid

23.2.2.1 Binding Assays

Due to its high affinity toward folic acid, folate binding protein (FBP) is frequently exploited in the development of folate binding assays. Affinity chromatography, based on the utilization of FBP, has been used to extract an assortment of folate forms, including 5-methyltetrahydrofolic acid (5MT), from biological samples (Nelson *et al.* 2003). Such assay operates through the mixing of the sample folate and signal molecule-labelled folate conjugate prior to the introduction of the FBP. Upon the addition of FBP, unknown folate in the sample will compete with the folate conjugate; the signal detected will decrease when the quantity of folate in sample is higher. The signal molecules for labelling folate include radiolabels and enzymatic labels such as glucose 6-phosphate dehydrogenase (G6P-DH) (Bachas *et al.* 1984).

23.2.2.2 Gas Chromatography and Gas Chromatography–Mass Spectroscopy

Gas chromatography (GC) and GC–mass spectroscopy (GC-MS) methods often rely on the quantitative cleavage of folate to *para*-amino benzoic acid (pABA), in which molar concentration of HCl is used to hydrolyse folates to pABA, which is subsequently assayed by GC or GC-MS after derivatization (Dueker *et al.* 2000; Gregory *et al.* 2006; Lin *et al.* 2003) with an appropriate reagent.

23.2.2.3 HPLC and HPLC–mass spectroscopy

Ion-pair HPLC (IP-HPLC) and reversed-phase HPLC (RP-HPLC) have been used to determine folate (Wilson and Horne 1984, 1986), most commonly in conjugation with fluorescence, ultraviolet and UV-visible (UV-vis) spectrometry (Vahteristo *et al.* 1996) or electrochemical detection (ECD) systems (Kohashi *et al.* 1986; Lucock *et al.* 1989). The pteridine ring structure of the folate can be determined on the basis of its retention time determined by HPLC and spectral characteristics *via* the diode array detection (DAD) system (Selhub 1989).

It was revealed previously that ECD was well-suited for the analysis of 5-methyltetrahydrofolic acid (5MeTHF) in plasma, in which only a relatively low voltage was required to attain a peak height (thus being fairly independent of the applied potential), meaning that most of the electroactive species in sample remain unoxidized. This leads to a sensitive and highly selective form of measurement (Lucock *et al.* 1989). Since many reduced folates are fluorescent under acidic conditions (Gregory *et al.* 2006), such a unique characteristic is apt for fluorescence detection.

In addition, HPLC coupled with mass spectrometry (HPLC-MS) has been widely utilized for the analysis of folate. Liquid chromatography–tandem mass spectroscopic (LC-MS/MS) determination of pABA released by the acid hydrolysis of erythrocyte folates was previously described (Clifford *et al.* 2005), in which [$^{13}C_6$]pABA was used as internal standard and derivatization of pABA with diazomethane was performed prior to analysis.

23.2.2.4 Trienzyme Extraction Method

The trienzyme extraction method is often used to determine the folate content in foods. The method involves a simultaneous incubation of foods with folate conjugase and α-amylase, followed by the incubation with protease (Tamura and Hyun 2005). Folate conjugase treatment is intended to hydrolyse polyglutamyl folate (the primary food folate form) to monoglutamyl forms, which can be assayed subsequently. α-Amylase and protease treatments, however, permit enzymatic digestion of carbohydrate and protein matrices of foods in order to liberate trapped folate (Tamura *et al.* 2002). More efficient extraction of food folates is provided by these treatments compared with the classical combination of heating and folate conjugase treatment.

23.3 Application of Liposome in the Development of New Immunosensing Systems for Biotin and Folic Acid

Liposomes, the microscopic spherical vesicles that form through self-assembly of phospholipid molecules dispersed in water, were discovered in the mid-1960s (Bangham *et al.* 1965) and initially studied as cell membrane models. The resulting closed sphere may encapsulate water-soluble molecules within the central aqueous compartment, which leads to a variety of applications such as gene and drug delivery (Brandl *et al.* 1990; Lurquin 1992), adjuvant formulation (Heath *et al.* 1976), and immunoassays (Ho and Huang 2005; Ho and Hung 2008; Ho *et al.* 2009a, 2009b, 2010).

23.3.1 Immunosensors for Biotin

23.3.1.1 Flow Injection Liposome Immunoanalytical System (FILIA)

The majority of conventional immunosensors uses the 96-well microtitre plate format, which is somewhat cumbersome for use in automated operation, and

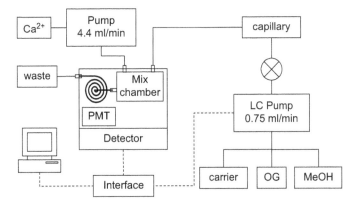

Figure 23.2 Schematic diagram of flow injection liposome immunoanalytical (FILIA) system.
(Reprinted with permission from *Analytical Chemistry*, 77, Ho, J. A. et al., Application of a liposomal bioluminescent label in the development of a flow injection immunoanalytical system, 3431–3436. Copyright (2005) American Chemical Society.)

often suffers from well-to-well, plate-to-plate and batch-to-batch variations. To overcome these problems, our group has previously developed a flow injection liposome immunoanalytical system (the arrangement of the FILIA in current use in our laboratory is shown in Figure 23.2) using biotin as model analyte and liposomal aequorin as signal amplification label (Ho and Huang 2005).

Aequorin, as mentioned previously in Section 23.2.1.2, is able to emit blue light in the presence of a trace of Ca^{2+} or some other divalent ions, such as Mg^{2+}. For this reason, a bioluminescent label was designed by encapsulating aequorin inside the liposome, whose outer surface was sensitized with biotin. The biotin-tagged liposomal aequorin was subsequently engaged in the development of a heterogeneous bioluminescence immunoassay for biotin. This immunoassay was based upon the competition between the sample biotin and aequorin-encapsulating, biotin-tagged liposomes for a limited number of anti-biotin antibody-binding sites immobilized on the inner wall of silica-fused capillary immunoreactor *via* protein A. Thirty percent methanol was found to be workable for the regeneration of antibody-binding sites after each measurement, allowing the re-use of immunoreactor for up to 50 sample injections. The dose–response curve (as indicated in Figure 23.3) for biotin in Tris-buffered saline (TBS) solution had a linear range of 1×10^{-11} to 1×10^{-3} M. The detection limit of the assay was calculated as 50 pg (Ho and Huang 2005).

23.3.1.2 Immunoaffinity Chromatographic (IAC) Biosensing System

Another format of immunosensor was also developed by our group. The combination of an immunoaffinity chromatography (IAC) assay and a

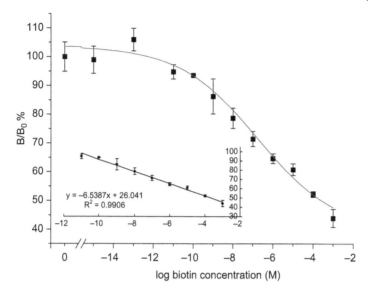

Figure 23.3 Dose–response curve for biotin measured by FILIA-bioluminescence (BL) system. (Each point represents the mean of three measurements; error bars represent ±1 standard deviation.) The inset shows a linear fit to the central data.
(Reprinted with permission from *Analytical Chemistry*, 77, Ho, J. A. et al., Application of a liposomal bioluminescent label in the development of a flow injection immunoanalytical system, 3431–3436. Copyright (2005) American Chemical Society.)

liposomal fluorescent biolabel turns into an alternative analytical method for the detection of biotin (Ho and Hung 2008). This method employs liposomal biolabels, biotin-derivatized liposomes encapsulating carboxyfluorescein (CF) as signal amplifiers that compete with biotin for a limited number of immobilized anti-biotin antibody-binding sites on the stationary phase, which is synthesized by covalently bonding anti-biotin monoclonal antibodies (Mab) onto 90 µm, *N*-hydroxysuccinimide (NHS) activated Sepharose beads. Subsequently the beads were packed into 1.9 cm diameter plastic tubes to form a column having a volume of 3.0 mL (Ho and Hung 2008).

IAC columns are often used for sample concentration. However, when working with liposomal biolabels, the IAC columns hold great promise for being further developed as sensitive immunoaffinity sensors for the determination of biotin; the schematic representation of the operation is illustrated in Figure 23.4. The number of liposomes bound to the anti-biotin antibody-binding sites was found to be inversely proportional to the amount of biotin present in the sample. The CF molecules were released from the lysed bound liposomes and concurrently antibody-binding activity was restored after flushing 35% methanol through the column. Finally, a buffer was passed through the system prior to performing the next measurement. A fluorimeter

Analysis of Biotin (Vitamin B_7) and Folic Acid (Vitamin B_9)

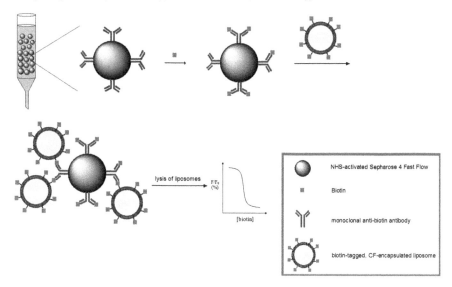

Figure 23.4 Schematic representation of the operation of an immunosensor for the detection of biotin (drawing not to scale).
(Reprinted with permission from *Analytical Chemistry*, 80, Ho, J. A. et al., Using liposomal fluorescent biolabels to develop an immunoaffinity chromatographic biosensing system for biotin, 6405–6409. Copyright (2008) American Chemical Society.)

was used to measure the fluorescence intensity of the released markers. The calibration curve for biotin was linear over eight orders of magnitude, from 10^{-12} to 10^{-4} M. The limit of detection of this immunoaffinity chromatographic biosensing system reached as low as 5.0 pg of biotin (equivalent to 500 μL of 4.10×10^{-11} M biotin) (Ho and Hung 2008).

23.3.1.3 Gold-Nanostructured Electrochemical Immunosensor

Two optical sensor designs for biotin detection are described above. In this section, we present a liposome-amplified signalling strategy for electrochemical immunosensors, which converts the bio-specific affinity binding reactions into an electrochemical signal (Ho et al. 2009a).

A cost-effective, easy-to-use, portable immunoanalytical platform with sufficient sensitivity for use in the detection of biotin was built. The detecting system comprises biotin-tagged, potassium ferrocyanide-encapsulated liposomes serving as the signal amplifier and a poly allylamine hydrochloride (PAH) modified, gold nanoparticle (AuNP) assembled screen-printed electrode (AuNP-SPE) as the sensing surface (working electrode). The diagnostic procedures, which are outlined in Figure 23.5, are based on specific immunorecognition and sensitive electrochemical detection.

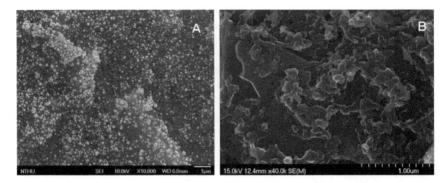

Figure 23.5 SEM images of surfaces of: (A) nanoAu-structured screen-printed electrode following electrodeposition treatment; and (B) bare SPE (before electrodeposition treatment).
(Reprinted from *Journal of Nanoscience and Nanotechnology*, 9, Ho, J. A. et al., Gold-nanostructured immunosensor for the electrochemical sensing of biotin based on liposomal competitive assay, 2324–2329, Copyright (2009), with permission from American Scientific Publishers.)

The target biotin was measured by relying on a 'competitive-type' immunoassay in which competition occurs between the analyte biotin and potassium ferrocyanide-encapsulated, biotin-tagged liposomes for a limited number of anti-biotin antibody binding sites, which were immobilized on the PAH/AuNP/SPE surface. The nanostructured gold SPE (scanning electron microscope image is shown in Figure 23.6) surface was covalently bonded to the PAH layer, which subsequently interacted with anti-biotin antibodies. The ferrocyanide released from ruptured bound-liposomes was oxidized at a detection potential of $+0.2$ V *versus* Ag/AgCl in neutral pH and finally measured using square-wave voltammetry. The dose-dependent square-wave voltammetric response is depicted in Figure 23.7. This immunodetection system offers a calibration curve for biotin with a linear range of 10^{-11} to 10^{-2} M, covering nine orders of magnitude. The detection limit of it was obtained as low as 9.1 pg of biotin (equivalent to 4.5 µL of 8.3×10^{-9} M) (Ho *et al.* 2009a).

23.3.1.4 Ultrasensitive Electrochemical Detection of Biotin using an Electrically Addressable Site-oriented Antibody Immobilization Approach

This section describes a modified electrochemical immunosensor where a boronic acid approach is employed for attaching antibodies onto the surfaces of assay devices. It is commonly observed that antibodies often adopt random orientations and therefore fail to display their original immunoaffinity toward their corresponding antigen when they are directly adsorbed onto sensing surfaces.

Analysis of Biotin (Vitamin B_7) and Folic Acid (Vitamin B_9)

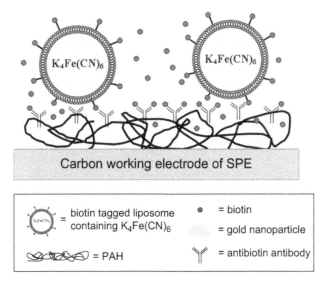

Figure 23.6 Outlines of immunosensor for the detection of biotin.
(Reprinted from *Journal of Nanoscience and Nanotechnology*, 9, Ho, J. A. et al., Gold-nanostructured immunosensor for the electrochemical sensing of biotin based on liposomal competitive assay, 2324–2329, Copyright (2009), with permission from American Scientific Publishers.)

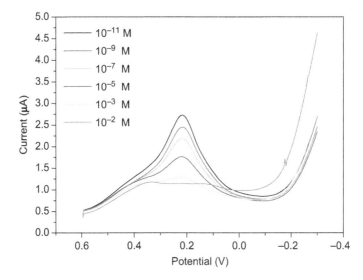

Figure 23.7 Dose-dependent square-wave voltammetric responses after immobilization of anti-biotin antibody on PAH/gold nano-structured SPE in 0.1 M phosphate-buffered saline (PBS) buffer at frequency 15 Hz, with scanning from 0.6 V to -0.3 V.
(Reprinted from *Journal of Nanoscience and Nanotechnology*, 9, Ho, J. A. et al., Gold-nanostructured immunosensor for the electrochemical sensing of biotin based on liposomal competitive assay, 2324–2329, Copyright (2009), with permission from American Scientific Publishers.)

Many coupling strategies are available for immobilizing antibodies onto solid surfaces, including glutaraldehyde, periodate, succinimide ester and carbodiimide linkages, with which a loss of biological activity is often associated (Lu *et al.* 1996). Moreover, in such coupling schemes, the linkers are not able to distinguish between attachment points close to or remote from the paratopes, resulting in spatial orientations of the antibodies on the sensing surfaces that might prohibit formation of antibody–antigen immunocomplexes. For this reason, there is an urgent need to develop specific and novel linking chemistries for attaching the antibodies to the sensing surfaces in an oriented manner.

Two approaches have been demonstrated in our group work published in 2010 (Ho *et al.* 2010). The first approach involves the deposition of gold nanoparticles (AuNPs) onto the SPGE (screen-printed graphite electrode) and subsequent adsorption of monovalent half-antibody (monoAb) fragments of the anti-biotin antibody *via* Au–thiol bonds, leading to a detection limit of 20.5 pg of biotin. For the second technique, we took advantage of the affinity of boronic acid towards sugar moieties by preparing a boronic acid-presenting SPGE surface to interact with the carbohydrate unit of this anti-biotin antibody.

It was found that the boronic acid approach enables us to prepare an electrochemical immunosensor with ultra-sensitivity, possessing a maximized paratope density, for the detection of biotin at concentrations as low as 0.19 pg. This notable improvement in detection sensitivity for biotin is believed to be attributed to the controlled orientation of the antibody when using the boronic acid/saccharide binding strategy for immobilization of antibodies, resulting in less denatured protein and more preserved immunobinding affinity on the SPGE surface. Moreover, we trust that a minimal, but sufficient, number of antibody binding sites were presented on the antibody/APBA/SPGE biosensor, leading to a more noticeable competition and an improved limit of detection (LOD). Results of differences in protein densities and signal intensities after using various approaches to immobilize antibody units onto the surfaces of electrodes are illustrated in Figure 23.8. We also believe that the improved orientation and large density of paratopes on the immunosurface lead to the distinct improvement in detectability of trace amounts of biotin. This antibody/APBA/SPGE electrochemical immunosensor allows the detection of biotin in a single assay that can be completed within 5 min at ambient temperature (Ho *et al.* 2010).

23.3.2 Immunosensors for Folic Acid

The immunoaffinity chromatography (IAC) approach mentioned in Section 3.1.1.2 was extended to the development of a simple, yet sensitive, immunodiagnostic column assay for the determination of an important water-soluble vitamin, folic acid, *via* liposomal amplification (Ho *et al.* 2009b). The schematic diagram of the operation of a liposomal immunodiagnostic assay for the detection of folic acid is shown in Figure 23.9.

The synthesis of distearoylphosphatidylethanolamine-poly(ethylene glycol) $_{2000}$-folic acid bioconjugate (DSPE-PEG$_{2000}$-FA), which is depicted in

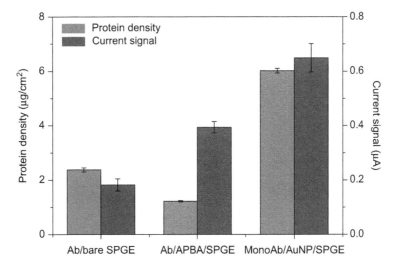

Figure 23.8 Differences in protein densities and signal intensities after using various approaches to immobilize antibody units onto the surfaces of electrodes. Ab/bare SPGE was prepared through physical adsorption; Ab/APBA/SPGE was prepared using the boronic acid approach; half-Ab/Au/SPGE was prepared using the thiolated half-Ab.
(Reprinted from *Biosensors & Bioelectronics*, 26, Ho, J. A. et al., Ultrasensitive electrochemical detection of biotin using electrically addressable site-oriented antibody immobilization approach *via* aminophenyl boronic acid, 1021–7, Copyright (2010), with permission from Elsevier.)

Figure 23.10, was subsequently used for preparation of FA-anchored, PEGylated liposomes. The utilization of CF-encapsulated, FA-anchored, PEGylated liposomes in conjugation with an IAC column packed with anti-FA Mab-immobilized Sepharose beads enables the construction of an IAC for determining trace levels of folic acid in real samples such as multivitamin preparations; the results of the recovery study are summarized in Table 23.1.

This IAC approach offers satisfactory sensitivity (LOD for FA: 6.8 ng), comparable with those reported previously, with a wide dynamic range, four orders of magnitude (dose–response curves obtained from IAC columns are exhibited in Figure 23.11). With its acceptably sensitive LOD, reusability, portability of the assay procedure, high accuracy and precision [coefficient of variation (CV) < 10%], cost-effectiveness and high assay speed (analysis times < 30 min), this optimized liposome-based IAC assay should allow the

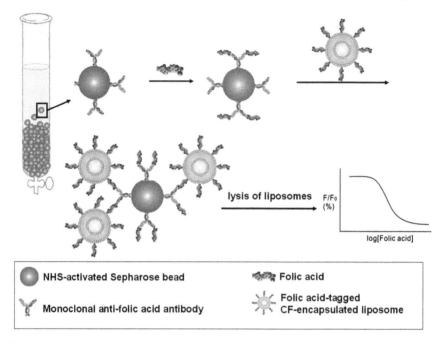

Figure 23.9 Schematic representation of the operation of a liposomal immunodiagnostic assay for the detection of folic acid.
(Reprinted with permission from *Analytical Chemistry*, 81, Ho, J. A. A. *et al.*, Folic acid-anchored PEGgylated phospholipid bioconjugate and its application in a liposomal immunodiagnostic assay for folic acid, 5671–5677, Copyright (2009) American Chemical Society.)

user-friendly monitoring and surveillance of folic acid levels in foodstuffs, infant formulas, feeds and pharmaceutical multivitamin preparations (Ho *et al.* 2009b).

Several advantageous features can be noted for our IAC assay for folic acid determination compared with conventional chromatographic or MBA methods, including:

- more rapid sample preparation (sample pre-concentration and analysis can be performed almost concurrently);
- savings in labour and time (no pre-incubation or repetitive washing are needed);
- less consumption of hazardous solvents (commonly used in chromatographic assays);
- increased analytical turnover rate due to the simple sampling and clean-up of samples prior to each analysis.

A plot of the binding activity of each folate derivative on the IAC column with respect to that of folic acid is shown in Figure 23.12. This demonstrates that our IAC method for the determination of folic acid presents a high specificity toward its detection and thus very low cross-reactivity was observed (Ho *et al.* 2009b).

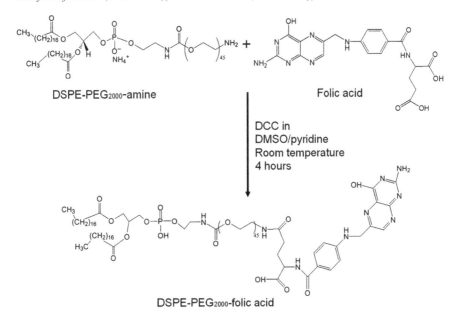

Figure 23.10 Synthesis of distearoylphosphatidylethanolamine–poly(ethylene glycol)$_{2000}$-folic acid bioconjugate (DSPE-PEG$_{2000}$-FA).
(Reprinted with permission from *Analytical Chemistry*, 81, Ho, J. A. A. et al., Folic acid-anchored PEGgylated phospholipid bioconjugate and its application in a liposomal immunodiagnostic assay for folic acid, 5671–5677, Copyright (2009) American Chemical Society.)

Table 23.1 Analysis of over-the-counter multivitamin tablets for their folate content. Recoveries of folic acid from the IAC analyses of real samples. (Reprinted with permission from *Analytical Chemistry*, 81, Ho, J.A.A. et al., Folic acid-anchored PEGgylated phospholipid bioconjugate and its application in a liposomal immunodiagnostic assay for folic acid, 5671–5677, Copyright (2009) American Chemical Society.)

Vitamin brand	Labelled amount (μg/tablet)	Found amount (μg)	Recovery (found/labelled; %)
Stresstabs	400	449 ± 62	112
Health Diary	800	764 ± 72	95.5
Jamieson	400	362 ± 36	90.5

23.4 Concluding Remarks

This chapter has described a number of liposome-amplified competitive immunosensing systems for two members of the B family of vitamins, biotin and folic acid, in flow injection analysis (FIA), immunoaffinity column (IAC) and electrochemical formats. The use of the immunoaffinity sensors is

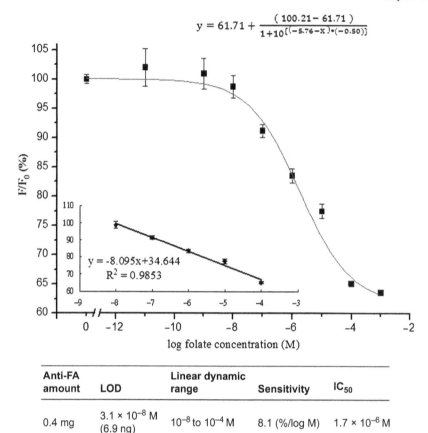

Figure 23.11 Dose–response curves obtained from IAC columns coated with 0.4 mg anti-FA; error bars: ±1 standard deviation; sensitivity: defined as the capability of responding reliably and measurably to changes in folic acid concentration, and represented as slope of calibration curve. Inset: linear portion of the main curve.
(Reprinted with permission from *Analytical Chemistry*, 81, Ho, J. A. A. *et al.*, Folic acid-anchored PEGgylated phospholipid bioconjugate and its application in a liposomal immunodiagnostic assay for folic acid, 5671–5677, Copyright (2009) American Chemical Society.)

demonstrated to be an easier, faster and more user-friendly analytical alternative for the quantitative detection of biotin and folic acid in serum, plasma, urine, pharmaceutical multivitamins preparations, and food or feeds than assays performed through currently common microbiological procedures or isotope dilution tests. The high sensitivity, low detection limit, cost-effectiveness and high portability of operation suggest that our proposed immunosensing platforms with liposomal amplification will be extendable as a technology for monitoring and determining a range of water-soluble vitamins in biological fluids, foods, feeds and drugs.

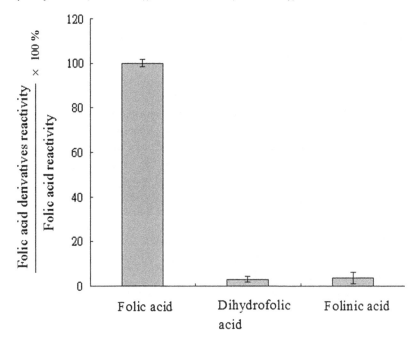

Figure 23.12 Plot of the binding activity of each folate derivative on the IAC column with respect to that of folic acid.
(Reprinted with permission from *Analytical Chemistry*, 81, Ho, J. A. A. *et al.*, Folic acid-anchored PEGgylated phospholipid bioconjugate and its application in a liposomal immunodiagnostic assay for folic acid, 5671–5677, Copyright (2009) American Chemical Society.)

Summary Points

- This chapter focuses on traditional and modern analytical methods of biotin and folic acid.
- Immunosensors offer high sensitivity, rapidity and simplicity of operation for screening of biologically and environmentally important target molecules.
- Applications of liposomes were demonstrated in the development of various formats of immunosensors for biotin and folic acid.
- The use of liposomes instead of the more common enzyme-produced signal as seen in enzyme-linked immunosorbent assay (ELISA) has a number of advantages.
- The signal enhancement produced by liposomal biolabels is instantaneous, eliminating the timed enzymatic incubation step required in ELISA.

Key Facts about Immunosensors

- An immunosensor is a biochemical tool that qualitatively or quantitatively measures the presence or concentration of a substance in solutions that frequently contain a complex mixture of substances.

- Such sensors are based on the unique ability of an antibody to bind with high specificity to one group of molecules. A molecule that binds to an antibody is called an antigen.
- The use of immunosensors for monitoring biologically and environmentally important target molecules is well-established, and a variety of laboratory sensors have been developed.
- Immunosensors are mostly in the form of ELISA, where measurements are made of the colour produced from a chromogenic substrate by the action of an enzyme conjugated to either an antibody or an analyte molecule.
- Flow injection liposome immunoanalytical system (FILIA) and immunoaffinity chromatographic assay (IAC) are two new formats for immunosensors, possessing several advantageous features.

Definitions of Words and Terms

Competitive liposome immunosensor. Liposomal biolabels and the sample analyte are passed over a solid surface where antibody to the analyte of interest has been immobilized. Competition occurs between the free analyte molecules and the analyte molecules conjugated to the liposomes. The number of liposomes that bind to the antibodies is inversely proportional to the amount of free analyte presented in the sample.

Flow injection analysis (FIA). Flow injection analysis is a chemical analysis approach accomplished by injecting a plug of sample into a flowing carrier stream.

Flow injection liposome immunoanalysis (FILIA). Flow injection liposome immunoanalysis is an approach to bioanalytical analysis that is a combination of FIA and immunoassay in conjugation of the utilization of liposome biolabels.

High-performance liquid chromatography (HPLC). HPLC is a form of liquid chromatography commonly used to separate a mixture of compounds utilizing shorter column and smaller beads, which are densely packed in the column, and with higher mobile phase pressures. This allows for a better separation compared with standard column chromatography.

Immunoaffinity chromatography (IAC). Immunoaffinity chromatography, based upon a highly specific biological interaction such as that between antigen and antibody, is a bioanalytical method for separating biochemical mixtures.

Immunosensor. An immunosensor is a biochemical tool that qualitatively or quantitatively measures the presence or concentration of a substance in solutions that frequently contain a complex mixture of substances.

Liposomes. Liposomes are lipid bilayer vesicles that are formed spontaneously when lipids are dispersed in water.

Mass spectrometry (MS). MS is an analytical technique based on the principle of ionizing chemical compounds to generate charged molecules or molecule fragments and measuring their mass-to-charge ratios.

List of Abbreviations

3-MPA	3-mercaptopropionic acid
5MeTHF	5-methyltetrahydrofolic acid
5MT	5-methyltetrahydrofolic acid
ADAM	9-anthryldiazomethane
AEQ	aequorin
AEQ-biotin	aequorin-biotin conjugate
AuNP	gold nanoparticle
BL	bioluminescence
BRET	bioluminescence resonance energy transfer
Br-Mmc	4-bromomethyl-7-methoxycumarin
BSA	bis-(trimethylsilyl) acetamide
CF	carboxyfluorescein
CV	coefficient of variation
DAD	diode array detection
DSPE	distearoylphosphatidylethanolamine
ECD	electrochemical detection
ELISHA	enzyme-linked Immunosorbent assay
FA	folic acid
FBP	folate binding protein
FIA	flow injection analysis
FILIA	flow Injection liposome immunoanalytical system
G6P-DH	glucose 6-phosphate dehydrogenase
GC	gas chromatography
GC-MS	GC–mass spectroscopy
HABA	*p*-hydroxy-azobenzene-2'-carboxylic acid
HPLC	high-performance liquid chromatography
HRP	horseradish peroxidase
IAC	immunoaffinity chromatography
IP-HPLC	ion-pair HPLC
LC-MS/MS	liquid chromatography–tandem mass spectrometry
LOD	limit of detection
pABA	*para*-aminobenzoic acid
PBS	phosphate-buffered saline
PEG_{2000}	poly(ethylene glycol)$_{2000}$
Mab	monoclonal antibody
MBA	microbiological assay
monoAb	monovalent half-antibody
MS	mass spectrometry
MS/MS	tandem mass spectrometry
NHS	*N*-hydroxysuccinimide
OPA	*o*-phthalaldehyde
PAH	poly allylamine hydrochloride
PDAM	1-pyrenyldiazomethane
RP-HPLC	reversed-phase HPLC

SPE   screen-printed electrode
SPGE   screen-printed graphite electrode
UV/Vis   ultraviolet–visible
TBS   Tris-buffered saline
TCA   tricarboxylic acid

References

Bachas, L.G., Lewis, P.F., and Meyerhoff, M.E., 1984. Cooperative interaction of immobilized folate binding-protein with enzyme folate conjugates—an enzyme-linked assay for folate. *Analytical Chemistry*. 56: 1723–1726.

Bangham, A.D., Standish, M.M., and Weissman. G., 1965. Action of steroids and streptolysin S on permeability of phospholipid structures to cations. *Journal of Molecular Biology*. 13: 253–259.

Baker, H., 1985. Assessment of biotin status—clinical implications. *Annals of the New York Academy of Sciences*. 447: 129–132.

Boushey, C.J., Beresford, S.A., Omenn, G.S., and Motulsky, A.G., 1995. A quantitative assessment of plasma homocysteine as a risk factor for vascular disease. Probable benefits of increasing folic acid intakes. *The Journal of the American Medical Association*. 274: 1049–1057.

Bowers-Komro, D.M., Chastain, J.L., and McCormick, D.B., 1986. Separation of biotin and analogs by high-performance liquid chromatography. *Methods in Enzymology*. 122: 63–67.

Brandl, M., Bachmann, D., Drechsler, M., and Bauer, K.H., 1990. Liposome preparation by a new high-pressure homogenizer Gaulin Micron Lab-40. *Drug Development and Industrial Pharmacy*. 16: 2167–2191.

Chastain, J.L., Bowers-Komro, D.M., and McCormick, D.B., 1985. High-performance liquid chromatography of biotin and analogues. *Journal of Chromatography A*. 330: 153–158.

Clifford, A.J., Owens, J.E., and Holstege, 2005. Quantitation of total folate in whole blood using LC-MS/MS. *Journal of Agricultural and Food Chemistry*. 53: 7390–7394.

Dueker, S.R., Lin, Y.M., Jones, A.D., Mercer, R., Fabbro, E., Miller, J.W., Green, R., and Clifford, A.J., 2000. Determination of blood folate using acid extraction and internally standardized gas chromatography-mass spectrometry detection. *Analytical Biochemistry*. 283: 266–275.

Feltus, A., Grosvenor, A.L., Conover, R.C., Anderson, K.W., and Daunert, S., 2001. Detection of biotin in individual sea urchin oocytes using a bioluminescence binding assay. *Analytical Chemistry*. 73: 1403–1407.

Food and Drug Administration, 1996. Food Standards: Amendment of standards of identity for enriched grain products to require addition of folic acid. Final Rule (21 CFR Parts 136, 137, and 139). *The Federal Register*. 61: 8781–8797.

Gregory, J.F., Ristow, K.A., Sartain, D.B., and Damron, B.L.1984. Biological activity of the folacin oxidation products 10–formylfolic acid and 5-methyl-5,6-dihyrofolic acid. *Journal of Agricultural and Food Chemistry*. 32: 1337–1342.

Gregory, J.F., Quinlivan, E.P., and Hanson, A.D., 2006. The analysis of folate and its metabolic precursors in biological samples. *Analytical Biochemistry*. 348: 163–184.

Gröningsson, K., and Jansson, L., 1979. TLC determination of biotin in a lyophilized multivitamin preparation. *Journal of Pharmaceutical Sciences*. 68: 364–366.

Gropper, S.A.S., Smith, J.L., and Groff, J.L., 2009. *Advanced Nutrition and Human Metabolism*. 5th ed. Wadsworth Cengage Learning, Belmont, CA, USA. p. 624.

Hankey, G.J., and Eikelboom, J.W., 1999. Homocysteine and vascular disease. *Lancet*. 354: 407–413.

Heath, T.D., Edwards, D.C., and Ryman, B.E., 1976. The adjuvant properties of liposomes. *Biochemical Society Transactions*. 4: 129–133.

Hibbard, B.M., 1964. The Role of Folic Acid in Pregnancy; with Particular Reference to Anaemia, Abruption and Abortion. *Journal of Obstetrics and Gynaecology of the British Commonwealth*. 71: 529–542.

Ho, J.A., and Huang, M.R., 2005. Application of a liposomal bioluminescent label in the development of a flow injection immunoanalytical system. *Analytical Chemistry*. 77: 3431–3436.

Ho, J.A.A., and Hung, C.H., 2008. Using liposomal fluorescent biolabels to develop an immunoaffinity chromatographic biosensing system for biotin. *Analytical Chemistry*. 80: 6405–6409.

Ho, J.A.A., Chiu, J.K., Hong, J.C., Lin, C.C., Hwang, K.C., and Hwu, J.R.R., 2009a. Gold-Nanostructured Immunosensor for the Electrochemical Sensing of Biotin Based on Liposomal Competitive Assay. *Journal of Nanoscience and Nanotechnology*. 9: 2324–2329.

Ho, J.A.A., Hung, C.H., Wu, L.C., and Liao, M.Y., 2009b. Folic acid-anchored PEGgylated phospholipid bioconjugate and its application in a liposomal immunodiagnostic assay for folic acid. *Analytical Chemistry*. 81: 5671–5677.

Ho, J.A., Hsu, W.L., Liao, W.C., Chiu, J.K., Chen, M.L., Chang, H.C., and Li, C.C., 2010. Ultrasensitive electrochemical detection of biotin using electrically addressable site-oriented antibody immobilization approach *via* aminophenyl boronic acid. *Biosensors and Bioelectronics*. 26: 1021–1027.

Höller, U., Wachter, F., Wehrli, C., and Fizet, C., 2006. Quantification of biotin in feed, food, tablets, and premixes using HPLC–MS/MS. *Journal of Chromatography B*. 831: 8–16.

Insel, P.M., Turner, R.E and., Ross, D., 2010. *Discovering Nnutrition*, 3rd ed. Jones and Bartlett Publishers, Sudbury, MA, USA.

Johnson, I.T., and Lund, E.K., 2007. Review article: nutrition, obesity and colorectal cancer. *Alimentary Pharmacology & Therapeutics*. 26: 161–181.

Kohashi, M., Inoue, K., Sotobayashi, H., and Iwai, K., 1986. Micro-determination of folate monoglutamates in serum by liquid-chromatography with electrochemical detection. *Journal of Chromatography*. 382: 303–307.

Kucera, R., Zerzanova, A., Zizkovsky, V., Klimes, J., Jesensky, I., Dohnal, J., and Barron, D., 2007. Using of HPLC coupled with coulometric detector for the determination of biotin in pharmaceuticals. *Journal of Pharmaceutical and Biomedical Analysis*. 45: 730–735.

Lin, Y., Dueker, S.R., and Clifford, A.J., 2003. Human whole blood folate analysis using a selected ion monitoring gas chromatography with mass selective detection protocol. *Analytical Biochemistry*. 312: 255–257.

Livaniou, E., Costopoulou, D., Vassiliadou, I., Leondiadis, L., Nyalala, J.O., Ithakissios, D.S., and Evangelatos, G.P., 2000. Analytical techniques for determining biotin. *Journal of Chromatography A*. 881: 331–343.

Lu, B., Smyth, M.R., and O'Kennedy, R., 1996. Oriented immobilization of antibodies and its applications in immunoassays and immunosensors. *Analyst*. 121: R29–R32.

Lucock, M.D., Hartley, R., and Smithells, R.W., 1989. A rapid and specific hplc-electrochemical method for the determination of endogenous 5-methyltetrahydrofolic acid in plasma using solid-phase sample preparation with internal standardization. *Biomedical Chromatography*. 3: 58–63.

Lurquin, P.F., 1992. Incorporation of genetic material into liposomes and transfer to cells. In: Greroriadis, G. (ed.), *Liposome Technology*: Entrapment of Drugs and Other Materials, Vol. II., CRC Press, Boca Raton, Florida, USA, pp. 129–139.

MRC Vitamin Study Research Group, 1991. Prevention of neural tube defects: results of the Medical Research Council vitamin study. *Lancet*. 338: 131–137.

Mulinare, J., Cordero, J.F., Erickson, J.D., and Berry, R.J., 1988. Periconceptional use of multivitamins and the occurrence of neural tube defects. *The Journal of American Medical Association*. 260: 3141–3145.

Nelson, B.C., Pfeiffer, C.M., Margolis, S.A., and Nelson, C.P., 2003. Affinity extraction combined with stable isotope dilution LC/MS for the determination of 5-methyltetrahydrofolate in human plasma. *Analytical Biochemistry*. 313: 117–127.

Nojiri, S., Kamata, K., and Nishijima, M., 1998. Fluorescence detection of biotin using post-column derivatization with OPA in high performance liquid chromatography. *Journal of Pharmaceutical and Biomedical Analysis*. 16: 1357–1362.

Pindolia, K., Jordan, M., Guo, C., Matthews, N., Mock, D.M., Strovel, E., Blitzer, M., and Wolf, B., 2011. Development and characterization of a mouse with profound biotinidase deficiency: a biotin-responsive neurocutaneous disorder. *Molecular Genetics and Metabolism*. 102: 161–169.

Refsum, H., Ueland, P.M., Nygard, O., and Vollset, S.E., 1998. Homocysteine and cardiovascular disease. *Annual Review of Medicine*. 49: 31–62.

Reio, L., 1970. Third supplement for paper chromatographic separation and identification of phenol derivatives and related compounds of biochemical interest using a reference-system. *Journal of Chromatography*. 47: 60–85.

Schiewe, J., Gobel, S., Schwarz, M., and Neubert, R., 1996. Application of capillary zone electrophoresis for analyzing biotin in pharmaceutical formulations – a comparative study. *Journal of Pharmaceutical and Biomedical Analysis*. 14: 435–439.

Selhub, J., 1989. Determination of tissue folate composition by affinity-chromatography followed by high-pressure ion-pair liquid-chromatography. *Analytical Biochemistry*. 182: 84–93.

Selhub, J., and Rosenberg, I. H., 1996. Folic acid. In: Ziegler, E. E., and Filer, L.J., Jr., et al (ed.), *Present Knowledge in Nutrition*. ILSI Press, Washington D.C., USA, pp. 206–219.

Serna, A., Vera, J., and Marin, D., 1973. Polarographic Behavior of Biotin. *Journal of Electroanalytical Chemistry*. 45: 156–159.

Tamura, T., Johnston, K.E., and Lofgren, P.A., 2002. Folate concentrations of fast foods measured by trienzyme extraction method. *Food Research International*. 35: 565–569.

Tamura, T., and Hyun, T.H., 2005. Trienzyme extraction in combination with microbiologic assay in food folate analysis: An updated review. *Experimental Biology and Medicine*. 230: 444–454.

Tucker, K.L., Mahnken, B., Wilson, P.W., Jacques, P., and Selhub, J., 1996. Folic acid fortification of the food supply. Potential benefits and risks for the elderly population. *The Journal of American Medical Association*. 276: 1879–1885.

Vahteristo, L.T., Ollilainen, V., Koivistoinen, P.E., and Varo, P., 1996. Improvements in the analysis of reduced folate monoglutamates and folic acid in food by high-performance liquid chromatography. *Journal of Agricultural and Food Chemistry*. 44: 477–482.

Vinokurov, L.M., Gorokhovatsky, A.Y., Rudenko, N.V., Marchenkov, V.V., Skosyrev, V.S., Arzhanov, M.A., Burkhardt, N., Zakharov, M.V., Semisotnov, G.V., and Alakhov, Y.B., 2003. Homogeneous assay for biotin based on Aequorea victoria bioluminescence resonance energy transfer system. *Analytical Biochemistry*. 313: 68–75.

Viswanathan F.P., Mahn, V., Venturella, V.S., and Senkowski, B.Z., 1970. Gas-liquid chromatography of d-biotin. *Journal of Pharmaceutical Sciences*. 59: 400–402.

Vreeke, M., Rocca, P., and Heller, A., 1995. Direct electrical detection of dissolved biotinylated horseradish-peroxidase, biotin, and avidin. *Analytical Chemistry*. 67: 303–306.

Wilson, S.D., and Horne, D.W., 1984. High-performance liquid-chromatographic determination of the distribution of naturally-occurring folic-acid derivatives in rat-liver. *Analytical Biochemistry*. 142: 529–535.

Wilson, S.D., and Horne, D.W., 1986. High-performance liquid-chromatographic separation of the naturally-occurring folic-acid derivatives. *Methods in Enzymology*. 122: 269–273.

Wright, J.D., Rawson, K.M., Ho, W.O., Athey, D., and McNeil, C.J., 1995. Specific binding assay for biotin based on enzyme channelling with direct electron transfer electrochemical detection using horseradish peroxidase. *Biosensors and Bioelectronics*. 10: 495–500.

Yokoyama, T., and Kinoshita, T., 1991. High-performance liquid chromatographic determination of biotin in pharmaceutical preparations by post-column fluorescence reaction with thiamine reagent. *Journal of Chromatography A*. 542: 365–372.

Yomota, C., and Ohnishi, Y., 2007. Determination of biotin following derivatization with 2-nitrophenylhydrazine by high-performance liquid chromatography with on-line UV detection and electrospray-ionization mass spectrometry. *Journal of Chromatography A*. 1142: 231–235.

Zempleni, J., Wijeratne, S.S., and Hassan, Y.I., 2009. Biotin. *Biofactors*. 35: 36–46.

CHAPTER 24
Biotin Analysis in Dairy Products

DAVID C. WOOLLARD*[a] AND HARVEY E. INDYK[b]

[a] New Zealand Laboratory Services, 35 O'Rorke Rd, Penrose, Auckland 1642, New Zealand; [b] Fonterra Co-operative Group Ltd, PO Box 7, Waitoa 3341, New Zealand
*Email: david.woollard@nz.bureauveritas.com

24.1 Introduction

Biotin is an essential micronutrient for all living mammals because of its role as a cofactor of carboxylation and decarboxylation enzymes. Human cells have four biotin-dependent carboxylases that catalyse key reactions in gluconeogenesis, amino acid catabolism and fatty acid synthesis. Mammalian cells cannot biosynthesize biotin, and thus it becomes an essential component of the diet and is therefore classified as a vitamin. Biotin is widely distributed in the human diet, either mostly free in the case of milk, vegetables and fruit, or partly bound in animal tissues and plant seeds. Protein-bound biotin is generally formed *via* covalent linkage to lysine residues, and gastrointestinal (GI) tract proteases can release biotin in the form of biocytin (ε-*N*-biotinyl-L-lysine). The biotin content of cow's milk is considered to be modest at approximately 2 µg/100 g. In the context of this discussion it is also relevant to note that biotin is central to several metabolic pathways involved with mammalian milk biosynthesis.

The structure of biotin has three asymmetric carbons and can therefore exist as eight potential stereoisomers, but only the *d*-biotin form is both biologically

Table 24.1 Daily intake of biotin to meet nutritional needs.[a]

Life stage and gender	Amount (µg)
Babies: birth to 6 months	5
Infants: 7–12 months	6
Children: 1–3 years	8
Children: 4–8 years	12
Children: 9–13 years	20
Teens: 14–18 years	25
Adults	30
Pregnant women	30
Breast-feeding women	35

[a]Estimated recommended human daily intake required as a function of age (source: Institute of Medicine).

active and found in nature, and is the only form of the parent vitamin that requires consideration. Biotin rarely exists in the free form in animal tissues, the majority being bound as biocytin within apoenzyme protein structures *via* an amide link between its terminal carboxyl group and protein lysine residues.

Clinical deficiencies of biotin in humans are not common in view of the low dietary intake levels required (see Table 24.1), but have been reported when prolonged and excessive amounts of raw egg white are consumed. However, in infants, biotin deficiency can occur as a consequence of insufficient production of biotinidase, the enzyme required to release bioavailable biotin from biocytin. Rare genetic deficiencies of biotinidase are also known. Obvious cutaneous symptoms include hair loss and skin rashes, but many neurological problems can develop. Deficiency during pregnancy is considered to be tetragenic in the newborn. Biotin-producing microorganisms within the intestine can contribute to the human biotin pool and complicate the evaluation of nutritional need.

As with all nutritional studies, reliable analytical methods are a pre-requisite for population-based policies related to Recommended Daily Allowances (RDA), total diet surveys and label claims, as well as supporting clinical and food compliance testing of biotin and its biomarkers. In general, for an estimation of total biotin in food, a proteolytic step is required to release the bioactive biocytin, which may then be included with free biotin in the analysis. The specific extraction procedures employed for the release of matrix-bound biotin remain a major analytical challenge and in fact may influence results even more than the end-point analytical measurement technique. Several extraction strategies developed for the analysis of bound biotin have, depending on sample type, most commonly employed high temperature mineral acid digestion and/or ambient temperature enzymatic hydrolysis techniques.

The first biotin tests for foods relied on microbiological principles, although some non-specific colorimetric tests were developed for biotin in pharmaceutical preparations. This chapter summarizes the important analytical techniques that have been employed, or are in use today, with emphasis on dairy products. The analytical techniques reported for the determination of biotin content can be categorized under (i) microbiological, (ii) biological,

Table 24.2 Overview of the attributes and limitations of the principal analytical strategies available for the analysis of biotin in foods.

Principle	Key steps	Advantages	Limitations
Microbiological assay (MBA). Growth of biotin-dependent microorganism proportional to biotin in sample extract.	Extraction of biotin and biocytin from sample. Incubate with microorganism and measure turbidity.	Low equipment costs. Versatile, sensitive, reference Infant Formula Council method.	Test microorganism vulnerable to growth inhibition or stimulation from matrix components. Highly manipulative, time-consuming.
High performance liquid chromatography (HPLC). Chromatographic separation of biotin species with quantitation based on detector response.	Extraction of biotin forms, purification, chromatographic separation, post-column derivatisation - fluorescence, or MS detection.	High specificity for biotin. Proven and commonly available technology.	High equipment costs
Ligand-binding assays, relying on biospecific molecular recognition of biotin. Labelled or non-labelled platforms available.	Extraction of biotin species. Direct or indirect detection by biotin-specific antibody or binding protein in either a labelled or non- labelled heterogenous format.	Rapid, specific inexpensive equipment costs if labelled technique. Proprietary kit-based assays available.	High equipment costs if non-labelled platform utilised.

(iii) chromatographic and physicochemical or (iv) biospecific ligand-binding principles. The three principal analytical techniques are briefly compared in Table 24.2. Recommended comprehensive literature reviews are provided in the reference list (Eitenmiller *et al.* 2007; Frappier 1993; Livaniou *et al.* 2000; Ploux 2000).

24.2 Microbiological Methods

These highly sensitive methods for the determination of biotin in complex matrices were among the first to be developed and are based on the principle that growth of specific microorganisms is dependent on the presence of limiting amounts of biotin in the culture media (Livaniou *et al.* 2000). Although these methods have the advantage of low cost, they are sensitive to many factors that can compromise results and require two days for completion.

Since the microorganisms respond only to free biotin, bound forms must be treated prior to analysis to release biotin in its free form. Following extraction of biotin from the sample by high-temperature acidic and/or ambient enzymatic strategies, the methodology relies on the incubation of the food extract at multiple dilution levels with the microbiological culture in a biotin-deficient growth medium. Quantification of culture growth, typically after 18–24 h, is generally achieved by measurement of turbidity at 540–660 nm, with response calibrated against added biotin solutions of known concentration and interpolation of the growth responses from unknown sample extracts.

Several microorganisms have been applied to the development of microbiological assays for biotin, with *Lactobacillus plantarum* employed predominantly, and less commonly, *Saccharomyces cerevisiae*. *L. plantarum* has been mainly used for the analysis of dairy-based food products (Angyal 1996; Bell 1974; Blake 2007; Eitenmiller *et al.* 2007; Hoppner and Lampi 1992). However, fatty acids are growth stimulatory and lipids therefore need to be solvent extracted prior to acid hydrolysis. Although AOAC International does not provide a collaboratively validated method for biotin, the International Formula Council does recommend a method based on *L. plantarum*. Internationally accepted compendia methods generally describe tube-based assays which can be automated using a programmable dispenser. However, these methods have more recently been successfully deployed in a 96-well microtitre plate format.

In general, microbiological assay (MBA) is acknowledged to be limited by poor precision, with measurement uncertainties of up to 20% reported. Analytical failures can originate from either poor growth or contamination issues. The questionable accuracy of microbiological methods and the view that the method tends to overestimate biotin content are significant concerns given that existing biotin data in food composition tables are predominantly based on these methods. Despite poorly defined standardization of methods between laboratories and issues related to different growth responses, the microbiological assay for biotin nonetheless remains the 'gold standard' reference method, since the method yields a single biological response to biotin activity.

A recent development is the commercial availability of proprietary microbiological assay kits that incorporate microtitration plates pre-coated with lyophilized microorganism, thereby avoiding the problems of maintaining stock cultures (VitaFast®, R-Biopharm). Incubation at 37 °C for 44–48 h is followed by measurement of growth using a microplate reader and construction of a dose–response calibration curve. Dairy products must first be treated to remove fat and protein that would otherwise interfere with the turbidimetric measurements. This is typically achieved by the addition of Carrez I (150 g/L potassium hexacyanoferrate(II)-3-hydrate) and Carrez II (300 g/L zinc sulfate-7-hydrate) reagents, followed by centrifugation. This method has recently received official AOAC Research Institute status based on a comprehensive single-laboratory validation report. It has the potential to offer significant advantages for routine testing laboratories, despite the costs associated with the kits.

24.3 Biological Methods

These earliest developed methods were mainly applied to the determination of biotin in foods and are generally considered to be less sensitive and far less technically practical than microbiological assays (Frappier 1993). Biotin bioassays were based on using rats or chicks as test subjects. The rat method exploited the inclusion of raw egg white (or avidin) in the diet to induce biotin deficiency over several weeks. The gain in animal weight was then followed after feeding of the diet with calibration against pure biotin, with weight gain related to the logarithm of the biotin dosage.

Bioassays also include methods that determine biotin indirectly through its biological function in the test animal. Thus, measurement of the activity of biotin-dependent pyruvate carboxylase in the blood has been utilized as an indicator of biotin content.

24.4 Chromatographic and Physicochemical Methods

Simple chemical laboratory procedures for biotin are restricted to pure materials as described in pharmacopeia monographs using alkaline titration. Similarly, spectral methods for determining native biotin are unknown in most samples except for highly purified materials as it has no useful absorption above 210 nm. To obtain suitable sensitivity and selectivity, biotin must first be derivatized. For example, the disruption of a *p*-hydroxy-azobenzene-2'-carboxylic acid (HABA) avidin complex by biotin, which binds to avidin with higher affinity, can be followed titrimetrically or spectroscopically to give useful determinations of biotin.

As recently as 2008, a reliable colorimetric method has been reported for biotin in pharmaceuticals based on its catalytic impact on the azide–triiodide reaction (Walash *et al.* 2008). However, biotin is best chromatographically separated from other compounds to allow practical determinations in all but the simplest matrices. This is particularly true for biological and food materials. Reviews such as Livaniou *et al.* (2000) summarize many of the historical physico-chemical techniques.

Following chromatographic separation, spectrometry has been used because the vitamin can be visualized in isolation from potential interferences. For example, planar chromatographic techniques, followed by spraying with an acidic solution of *p*-dimethylaminocinnamaldehyde has found significant practical use, since the intense red colour (533 nm) is specific to the ureido ring of biotin. Thin-layer chromatography (TLC) on silica-gel plates has been applied to biotin detection in multivitamin premixes and capsules, although this technique is not used greatly in the modern era. Ploux (2000) has reviewed some of the past applications with respect to TLC developing solutions and other visualization sprays such as *o*-toluidine-potassium iodide. TLC still has a place in some investigational laboratories as a tool for biochemical studies.

Ion chromatography (IC) using Dowex 1X2 has played an important role in biotin determinations of the past, being an essential component of biotin discovery and purification. Nowadays, high-performance IC platforms with preparative columns have replaced earlier techniques and remain an important tool for investigators of biotin biochemistry. The various fractions can be collected into tubes or wells, and subsequently studied for composition.

Gas chromatographic platforms have had a minor role in detection of the non-volatile and labile biotin vitamin. However, a biotin-silyl ester has been reported though it has not been used widely outside the pharmaceutical industries. The carboxylic acid and amido groups of biotin can be exploited to produce volatile silyl esters or N-silanization compounds, with bis-trimethylsilyl acetamide the preferred reagent. Such derivatives are readily separated from inferences on OV-17 capillary columns, although reports state that oxidation products of biotin can co-elute. The silyl-ester is sensitive enough for flame ionization detection (FID) in simple applications, but is challenged for sensitivity in biological studies. Some researchers question the repeatability of this reaction and have proposed other options. Nevertheless, gas chromatography (GC) and gas chromatography–mass spectrometry (GC-MS) applications have not become a commonplace analytical technique for biotin and certainly not within food industries. Similarly, capillary electrophoresis has been used for high concentration samples, but has not been applied to foods.

The first step for the chromatographic determination of biotin in foods is to release the analyte from its protein-bound environment where it may be covalently linked to lysine residues (biocytin, ε-N-biotinyl-L-lysine). The vitamin is stable enough to survive, at least in part, the elevated temperatures associated with autoclaving required for protein cleavage, in order to release biocytin. The degree of acid hydrolysis varies with the food matrix, with meat requiring 2 h at 121 °C in 6 M sulphuric acid. Such harsh conditions will promote some oxidation of biotin to its sulfone and sulfoxide, so compromises are often made in acid strength and heating regimen to achieve optimum recoveries. Hot hydrochloric acid is very damaging to biotin loss and so is rarely used. Plant materials are more easily hydrolysed and therefore are typically subjected to milder conditions (*e.g.* 2 M sulphuric acid), while some dairy products require minimal extraction other than mild heating to affect dissolution. Papain or trypsin enzyme digestion will break peptide bonds without risking damage to biotin and has become a preferred method for food product digestion. Coupled with takadiastase, the food matrix can be dispersed and solubilized, thereby releasing biotin. Although biocytin itself is not hydrolysed, this is unimportant provided that the method can uniquely separate biocytin from biotin (Gimenez *et al.* 2010; Lahely *et al.* 1999).

Although high-performance liquid chromatography (HPLC) separations of biotin and its analogues have been reported using many chromatographic modes, reversed-phase chromatography remains at the forefront, irrespective of which detection method is employed. Anion-exchange chromatography with electrochemical detection has not gained the same prominence, despite its apparent usefulness. Biotin, its metabolites and oxidation products are readily separated using aqueous–organic gradients.

Although a water-soluble vitamin, biotin is sufficiently non-polar to exhibit good retention on reversed-phase columns. Depending on the pH of the mobile phase, biotin elutes from C8 and C18 columns well separated from other

B vitamins, but typically between folic acid and riboflavin. For direct low-wavelength ultraviolet (UV) detection, the organic solvent must be acetonitrile and acidification should be made with trifluoroacetic acid to avoid high background absorption by the mobile phase. In recent times, ultra-HPLC equipment has been available, thereby improving the sensitivity and speed of analysis achievable with sub-3 µm particles. This important advance provides an improved strategy towards overcoming the risks of chromatographic interferences. For conventional HPLC platforms, chromatographic columns are recently available with solid particle cores that mimic ultra-HPLC, without the associated high system back-pressures.

Since it is unpractical to use direct UV detection, biotin is best quantitated following derivatization to a suitable chromophore or fluorophore. One of the early advances was an off-line coupling with the horseradish peroxidase–avidin binding assay, whereby fractions were collected and subjected to fluorimetic determination. This method was used to survey a wide range of foods and subsequently reveal that original MBA-derived data for the biotin content of many foods was questionable (Staggs *et al.* 2004). Pre-column derivatization has been used with equal success, yielding the nanogram sensitivities to analyse this vitamin. 9-Anthryldiazomethane (ADAM) forms a fluorescent ester that is subsequently separated by reversed-phase HPLC ($\lambda_{ex} = 365$ nm, $\lambda_{em} = 425$ nm). Similarly, biotin esters with panacyl bromide in the presence of crown ether are fluorescent ($\lambda_{ex} = 380$ nm, $\lambda_{em} = 470$ nm). 4-Bromomethylmethoxycoumarin derivatives of biotin are UV absorbing and provide another useful analytical method for biotin. One of the complications of pre-column chemistries can be the need for significant clean-up of the extract prior to injection.

From an operational viewpoint, the post-column reaction of biotin with avidin labelled with fluorescein isothiocyanate (avidin–FITC) has become the simplest HPLC method for the quantitation of biotin and biocytin, and has allowed routine testing laboratories to generate quality control (QC) data (Lahely *et al.* 1999). Typical chromatograms are shown in Figure 24.1 of a standard and a milk powder extract. The analytical time is only a few minutes because of the absence of interfering fluorescent substances. Biocytin, if present, elutes in front of biotin and can be concurrently detected.

An interesting feature of the reaction is its non-linearity which can confuse investigators, but is easily overcome using a quadratic fit to multiple (five or more) calibration points as shown in Figure 24.2 This method has been validated among ten laboratories to become a European Norm (EN15607; European Committee for Standardization 2009) and has also been further developed by others with modified sample preparation and HPLC configurations to gain greater robustness (Campos-Gimenez *et al.* 2010; Thompson *et al.* 2006). Such facile methodology should catalyse accumulation of reliable biotin data and promote enhanced nutritional and food compositional studies. Streptavidin-FTIC can replace avidin-FTIC and exhibits enhanced reagent stability, although at greater cost. Nonetheless, this HPLC technique is no more expensive in time and consumables compared with alternative testing strategies.

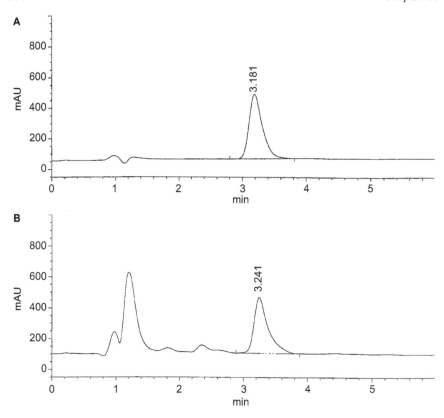

Figure 24.1 HPLC chromatogram of biotin standard with post-column avidin reaction. Note the fast chromatography permitted with this sensitive and specific reaction. Detection is using fluorescence. A: biotin standard, 0.04 ug/mL. B: supplemented milk powder extract with 12 ug/100 g.

High-performance affinity chromatography has recently been reported with trypsin-modified avidin supported on 5 μm silica. While the separations were successful and a wide range of foods were studied, elution times were 80 minutes and ADAM post-column reactions were still required (Hayakawa et al. 2009). However, such affinity columns within a solid-phase extraction (SPE) platform make realistic choices for sample preparation, whereby the biotin can be purified and concentrated prior to reversed-phase HPLC. R-Biopharm has recently developed a commercially available antibody-based immunoaffinity column to bind biotin from aqueous extracts, providing an excellent technique to clean up complex samples.

The next generation of HPLC detectors, namely mass spectrometers, are now available and being deployed for water-soluble vitamin determinations in dietary supplements with electrospray ionization interface (ESI) (Höller et al. 2006). There remain obstacles to overcome for multivitamin analyses in complex food matrices, as co-elution of vitamins or excipients can compromise

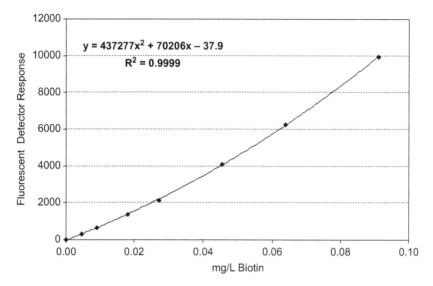

Figure 24.2 Non-linear calibration curve for biotin by post-column avidin reaction. The curve fits very well to a quadratic equation. It is important to have many data points to describe the curvature unless a narrow concentration range is selected and sample extracts diluted into this range.

ionization efficiency. These issues can be resolved using isotope dilution techniques when deuterated internal standards become available. Nevertheless, the use of positive ESI and [^{2}H$_2$]-biotin has successfully allowed biotin levels to be characterized in NIST 3280 multivitamin tablets (Nelson et al. 2006). Although for food samples, sample preparation restrictions will limit LC-MS ability to quantitatively analyse all total B vitamins in a single run, gains are currently being made to detect free (non-bound) vitamins in supplemented infant formula.

24.5 Ligand-binding Methods

24.5.1 Labelled Techniques

During recent decades, many biospecific ligand-binding assay variants have been applied to the determination of biotin in biological matrices, including foods (Bitsch et al. 1989; Eitenmiller and Landon, 2007; Finglas et al. 1986; Finglass and Morgan 1994; Livaniou et al. 2000; Mock et al. 1992; Ploux 2000). Such methods include protein-binding assays and immunoassays, which both share similar procedural principles, but differ in that they utilize either a biotin-specific protein or antibody. In the case of antibody-based immunoassays, the small molecular mass of biotin requires its coupling as a hapten in order to stimulate the production of antibodies in the host animal that can then be utilized in a concentration assay. Despite being more prevalent in the clinical diagnostic field, labelled biospecific protein-binding and immunoassay methods have also provided an alternative analytical strategy for food analysis and offer unique advantages. In general, these methods have the attributes of high sensitivity, specificity and speed, and are relatively simple to implement. They can therefore be easily adopted by the food industry for routine compliance testing of product and represent potentially useful alternatives to traditional microbiological assays and contemporary HPLC strategies.

To-date, all reported protein binding assays utilize the glycoprotein avidin, or its structurally related streptavidin, which both exhibit an extremely high affinity and specificity for biotin. The unique features of the avidin/streptavidin–biotin interaction system have been widely exploited, including application to the quantitation of biotin in foods. Both antibody-based immunoassays and avidin-based protein binding assays for biotin have been reported utilizing the 96-well microtitration plate format and applied to foods, including milk. Each offers different analytical attributes as a consequence of the different affinities and specificities for biotin and related structures. The majority of such assays are formatted utilizing a heterogeneous solid-phase platform and require radioisotopes, enzymes, and chemiluminescent and fluorescent substances as labelling systems to signal binding levels. Although radioisotope-based assays are commonly applied for clinical diagnostics, they are less desirable in the food analysis environment, where enzyme-linked formats are generally preferred. Although avidin-based protein binding assays are highly specific for biotin, they cannot discriminate between biotin and biotin metabolites that bind with typically lower affinity. In such circumstances, a chromatographic separation prior to a binding assay end-point has overcome this limitation and has been applied to the analysis of human milk.

There are a number of commercially available and proprietary protein-binding kits utilizing the enzyme protein binding assay (EPBA) format for measurement of biotin in biological samples and foods (*e.g.* Ridascreen®, R-Biopharm). As with microbiological assays, dairy samples must be depleted of fat and protein interferences prior to analysis. While kit assays are most commonly applied to the measurement of free, rather than total biotin, information on cross-reactivity towards biocytin and biotin-4-amidobenzoate is frequently provided. In one kit, avidin is bound to the surface of microtitre wells into which sample extracts or standards and a biotin-alkaline phosphatase conjugate are dispensed. The conjugate competes with endogenous free biotin for avidin binding sites. Incubation is allowed for one hour and the wells washed with Tris buffer to remove unbound material. A self-indicating phosphatase substrate is added and incubated at 37 °C, resulting in the development of a yellow colour. The reaction is stopped after 30 minutes and the spectral absorbance of the wells read at 405 nm. The concentration of biotin is indirectly related to the colour intensity and a non-linear calibration determines the unknown concentrations. The samples, typically 5–10 g are initially warmed in water (40 °C) to solubilize biotin and adjusted to pH 6–7 if necessary, followed by equal volumes of the two Carrez solutions. The extract is diluted to a defined volume and centrifuged to produce a clear supernatant. 100 µL is required for most EPBA tests. Further clarification with a 0.45 µm filter is sometimes necessary.

Several alternative assay format variants have been reported, including an EPBA whereby sample extracts and standards are incubated with a streptavidin–peroxidase conjugate, and dispensed in microtitre wells coated with biotin–albumin conjugate. After 30 minutes incubation, 3,3′,5,5′-tetramethyl benzidine (TMB) in acidic solution both stop the reaction and create a yellow colour that is measureable at 450 nm.

24.5.2 Non-Labelled Techniques

More recent developments in the biosensor field have provided novel platforms which have been increasingly applied to both protein-binding and immunoassays for the quantitation of vitamins, including biotin, in a range of foods. They represent an alternative to conventional ligand-binding assays and other analytical techniques, providing high-throughput, rapid and cost-effective strategies that can meet the increasing demands within the food industry with respect to compliance testing. As with traditional binding assays, the specific biorecognition of analyte can dramatically reduce the need for extensive sample preparation. Biosensors are based on the intimate contact of the biorecognition element with a physicochemical transducer, which functions to convert the specific interaction with analyte into a measurable signal, thereby facilitating real-time measurement. Several biorecognition molecular species have been utilized, including enzymes, antibodies and binding proteins, while electrochemical, optical and thermal devices have been used as the transducer.

The most commonly applied biosensor system applied for the analysis of a wide range of target analytes, including vitamins, is based on the optical transducer phenomenon of surface plasmon resonance (SPR), although other detection principles have been exploited (Blake, 2007; Kalman et al. 2006; Reyes et al. 2001). Such SPR-based optical biosensors are providing an innovative platform for the application of non-labelled biospecific techniques to aspects of food composition, safety and compliance, and specifically for the analysis of water-soluble vitamins in foods including folic acid, vitamin B_{12}, pantothenic acid and biotin. SPR has been successfully integrated into a commercially available instrument platform and provides for the real-time and label-free detection of ligand-binding events from which target analyte concentration can be derived. In simple terms, detection is based on real-time binding events between two biospecifically interacting molecular species at the surface of a sensor that moderate an evanescent field established within the SPR optical detector under conditions of total internal reflection. The change in SPR angle caused by association or dissociation at the sensor surface is proportional to the mass of bound material and is continuously recorded in the form of a sensorgram. The advantages of this platform compared with conventional ligand-binding methods include real-time measurement, freedom from enzyme or radioisotope labelling requirements, rapid analysis times, minimal sample preparation requirements, sensitivity, specificity and enhanced precision. The sensor chips used provide low non-specific binding, stability over many analysis cycles and excellent precision.

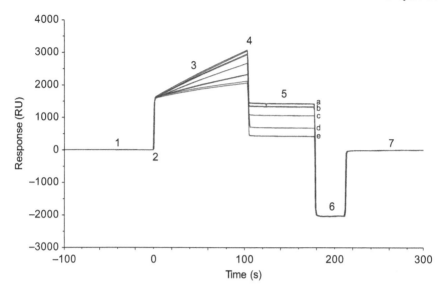

Figure 24.3 Sensorgrams derived from an SPR-biosensor analysis of biotin. Superimposed sensorgrams of biotin calibration standards: (a) 1.23 ng/mL; (b) 3.7 ng/mL; (c) 11.1 ng/mL; (d) 33.3 ng/mL; (e) 100 ng/mL. SPR-based biosensor assay utilizing a biotin specific antibody under inhibition conditions: (1) baseline with buffer flow; (2) sample injection; (3) binding association phase; (4) end of injection; (5) stable binding response acquired; (6) surface regeneration; (7) baseline for subsequent cycle.

In view of the low molecular mass of the vitamins, biosensor assays employ the principle of competitive inhibition. Prior to the analysis for biotin, a biotin derivative is covalently tethered to the sensor surface. Sample extract or biotin standard is mixed under equilibrium conditions with an excess of a biotin-specific antibody, and unbound antibody binds to the surface generating an SPR binding response in real-time, as illustrated in Figure 24.3.

The relative binding response is interpolated from a calibration curve in order to compute concentration. As an inhibition assay, the response is inversely related to biotin concentration and exhibits a sigmoidal dose–response relationship that is typical of most ligand-binding assays. With respect to specificity, the routine compliance assay is targeted to the quantitation of free biotin only in nutritional dairy products, and therefore does not include biocytin (Indyk *et al.* 2000). However, in milk and supplemented infant formulas, the overwhelming majority of biotin is present in the free form.

It remains speculative as to the potential impact on biotin analysis that may result from current innovations in the field of biosensor technologies. New alternatives to classical biorecognition elements will include artificial receptors such as molecularly imprinted polymers and aptamers, which promise increased stability. Although they are currently not widely utilized in commercial biosensors, such biosensor innovations are likely to offer future advantages.

24.6 Biotin Forms and Concentration in Milk and Dairy Products

A large number of clinical studies, spearheaded by Donald Mock and his associates (Mock *et al.* 1992, 1997; Staggs *et al.* 2004), have been conducted concerning the distribution of biotin in tissues and biological fluids. These have in turn attempted to assess the extent to which human nutritional requirements vary with age and gender. Biotin is required in such low quantity that average diets, whether of animal or vegetable origin, supply sufficient vitamin to prevent widespread deficiency states, making it problematical to determine recommended dietary intake (RDI) in humans without deliberate dietary intervention. It is also suggested that microbiological fauna in the intestinal tract contribute to the host biotin pool. Other measures, such as comparison of consumed *versus* excreted biotin, homeostasis of circulatory biotin and use of radiolabelled (tritiated) biotin have all helped to determine the current RDI for adults at 20–30 µg/day, increasing during pregnancy and lactation to 30–35 µg/day, while other estimates are 1.5 µg/kg body weight across all age groups.

Human milk is the primary agent for infant nurture and thereby guides the composition of manufactured infant formula and milk substitutes. The reported concentration of biotin in human milk is variable with lactation (and unfortunately between analytical methods), but is more than sufficient to supply the newborn infant with the RDI of 5–6 µg/day, as evidenced by the absence of reported deficiency syndromes in breast-fed babies. Interestingly, most biotin in milk is present in a free form and therefore unbound with any macromolecules. As expected, when milk is separated into its fat and aqueous fractions, the water-soluble biotin is found predominantly in the skim-milk phase. Biotin has some lipophilicity and so a small percentage is carried into the cream as part of the fat-globule membrane. The total concentration of human milk is not large and somewhat similar to bovine milk. With respect to breast milk substitutes, it is necessary to ensure the biotin status remains comparable, thus international guidelines recommend 0.4–2.4 µg/100 kJ of reconstituted or ready-to-feed infant formula.

Table 24.3 shows biotin levels in selected consumer dairy products and compares the milk of the cow with that of the goat and human. When commercial milk is converted by spray-drying into dehydrated powders, the biotin contents increase proportionately. Skim-milk powder thus becomes a reasonable source of the vitamin at 13–15 µg/100 g. Purified casein contains essentially

Table 24.3 Typical biotin content of species milks and cow's milk consumer products (µg/100 g).[a]

Cow					Goat Milk	Human Milk
Milk	Yoghurt	Cheese	Butter	Ice-cream		
2.0–4.3	0.9–4.0	0.8–5.9	0.5	1.1	3.2–3.9	0.4–0.8

[a] Data from cited references and Food Standards Australia New Zealand (2010).

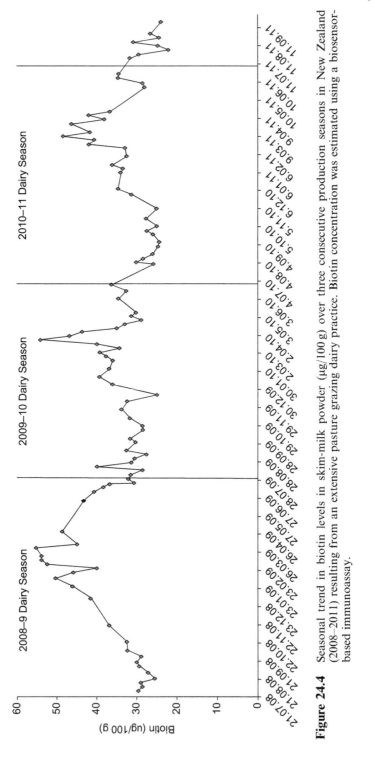

Figure 24.4 Seasonal trend in biotin levels in skim-milk powder (μg/100 g) over three consecutive production seasons in New Zealand (2008–2011) resulting from an extensive pasture grazing dairy practice. Biotin concentration was estimated using a biosensor-based immunoassay.

no biotin, as it is removed during the precipitation and washing processing stages. Although the whey fraction of processed milk contains the majority of the endogenous milk biotin, purified whey proteins lose most of their associated biotin during ultrafiltration, since in milk, biotin is not covalently bound to this fraction.

Countries with high biotin intake, as shown in total diet surveys, generally consume more meat, eggs and fish, which represent the predominant sources of the vitamin (Murakami *et al.* 2008). However, cheese contains measurable biotin across a range of concentrations (approximately 1–6 µg/100 g) and can contribute significant dietary biotin in some European societies where cheese consumption is high. On a solids basis, cheese has less biotin than its raw material (milk), so speculation that starter microorganisms can contribute additional biotin during production is probably false and, in fact, may consume biotin. However, hard cheeses such as Cheddar and Gruyère contain lesser amounts of biotin than soft cheeses (*e.g.* Brie). Similarly, yoghurt and other cultured dairy products have diminished biotin levels as compared to the parent milk on a solids-non-fat basis.

Recent work in New Zealand has shown a seasonality of biotin levels in bovine milk, as illustrated in Figure 24.4. Extensive pasture grazing and animal husbandry practices based on herd calving scheduled to maximize lactating cows at peak grass growth result in a minor but consistent seasonal trend.

Summary Points

- This chapter focuses on the analysis of biotin in dairy products.
- Biotin is a member of the B group of water-soluble vitamins.
- As a vitamin, biotin is an essential component of the human diet.
- Levels of biotin in dairy products formulated for infant nutrition are rigorously regulated.
- Reliable analytical methods are essential to support the formulation and compliance of manufactured dairy products.
- The attributes and limitations of different analytical platforms for the measurement of biotin are reviewed.

Key Facts

- Biotin is a vitamin required in very small (microgram) quantities as a cofactor to essential enzyme reactions.
- Biotin is hard to detect in dairy and food samples, different methods often giving different answers.
- Detection using microbiological organisms has traditionally been used, but is currently being replaced by modern chromatographic and immunochemical techniques.
- Biotin levels in dairy products are only moderate, and insufficient to supply human daily needs. However, since it is ubiquitous in many foods, deficiency states are uncommon.
- Infants fed on dairy-based formulae need additional biotin added to their food.

Definitions of Words and Terms

Affinity chromatography. A separation technique that exploits the high biological specificity and affinity of specific naturally occurring proteins for the analyte.

Apoenzyme. An enzyme devoid of its cofactor.

Avidin, streptavidin. Avidin is a protein expressed in the white of the reptilian, amphibian and avian egg with very high binding affinity for biotin. Streptavidin is a structurally related protein expressed in certain bacteria that exhibits comparable affinity for biotin.

Biosensor. An analytical device for the detection of an analyte that combines a biospecific detection element with a physicochemical detector that may be based on a number of principles.

Chromatography. A group of analytical techniques whereby complex mixtures of compounds are both separated into individual forms and then sequentially detected by a range of techniques. The techniques described in this chapter are predominantly based on a high-pressure liquid chromatographic (HPLC) platform.

Cofactor. A non-protein compound that is essential for facilitating the biological activity of an enzyme.

Enzyme. A protein that catalyses the conversion of one biologically active compound to another.

Ligand-binding. The biospecific binding interaction between an analyte and a biological macromolecule, usually a protein. Such an interaction is non-covalent, usually reversible and either crucial for biological activity, or can be exploited for analytical purposes.

Mass spectrometry. An analytical detection technique that measures the mass-to-charge ratio of charged particles, and provides confirmatory identification of target analytes separated by chromatography.

Stereoisomers. Compounds that have the same molecular formula and sequence of bonded atoms, but differ in the three-dimensional orientations of those atoms.

Vitamin. An organic compound that the body cannot biosynthesize yet is essential for normal growth and development and therefore must be derived from the diet.

List of Abbreviations

ADAM	9-anthryldiazomethane
EPBA	enzyme protein binding assay
ESI	electrospray ionisation interface
FID	flame ionization detection
FITC	fluorescein isothiocyanate
GC	gas chromatography
GC-MS	gas chromatography–mass spectrometry
GI	gastrointestinal
HABA	p-hydroxy- azobenene-2′-carboxylic acid

HPLC	high-performance liquid chromatography
IC	ion chromatography
MBA	microbiological assay
QC	quality control
RDA	Recommended Daily Allowance
RDI	recommended dietary intake
SPE	solid-phase extraction
SPR	surface plasmon resonance
TLC	thin-layer chromatography
TMB	3,3',5,5'-tetramethyl benzidine
UV	ultraviolet

References

Angyal, G. (ed.), 1996. Biotin. In: *US Food and Drug Administration. Methods for the Microbiological Analysis of Selected Nutrients*. Association of Official Analytical Chemists, Arlington, VA, USA, pp. 9–10.

Bell, J.S., 1974. Microbiological assay of vitamins of the B-group in foodstuffs. *Laboratory Practice*. 23: 235–241.

Bitsch, R., Salz, I., and Hotzel, D., 1989. Biotin assessment in foods and body fluids by a protein binding assay (PBA). *International Journal of Vitamins and Nutritional Research*. 59: 59–64.

Blake, C.J., 2007. Analytical procedures for water-soluble vitamins in foods and dietary supplements: a review. *Analytical and Bioanalytical Chemistry*. 389: 63–76.

Campos-Gimenez, E.C., Trisconi, M-J., Kilinc, T., and Andrieux, P., 2010. Optimization and validation of an LC-FLD method for biotin in infant formula, infant cereals, cocoa-malt beverages and clinical nutrition products. *Journal of AOAC International*. 93: 1494–1502.

Eitenmiller R.R., Landen Jr., W.O., and Ye, L., 2007. Biotin. In: *Vitamin Analysis for the Health and Food Sciences*, 2nd ed. CRC Press, Boca Raton, FL, USA, pp. 535–560.

European Committee for Standardization, 2009. *EN 15607: Foodstuffs. Determination of d-biotin by HPLC*. CEN, Brussels, Belgium. 14 pages.

Finglas, P.M., Faulks, R.M., and Morgan, M.R.A., 1986. The analysis of biotin in liver using a protein-binding assay. *Journal of Micronutrient Analysis*. 2: 247–257.

Finglas, P.M., and Morgan, M.R.A., 1994. Application of biospecific methods to the determination of B-group vitamins in food: a review. *Food Chemistry*. 49: 191–201.

Frappier, F., 1993. Biotin: properties and determination. In: Macrae, R., Robinson, R.K., and Sadler M.J. (ed.) Encyclopedia of Food Science, *Food Technology and Nutrition.*, Academic Press, London, UK, Vol. 1, pp. 395–399.

Hayakawa, K., Katsumata, N., Abe, K., Hirano, M., Yoshikawa, K., Ogata, T., Horikawa, R., and Nagamine, T., 2009. Wide range of biotin (vitamin H) content in foodstuffs and powdered milks as assessed by high-performance affinity chromatography. *Clinical Pediatric Endocrinology*. 18: 41–49.

Höller, U., Wachter, F., Wehrli, C., and Fizet, C., 2006. Quantification of biotin in feed, food, tablets and premixes using HPLC-MS/MS. *Journal of Chromatography B*. 831: 8–16.

Hoppner, K., and Lampi, B., 1992. Biotin content of cheese products. *Food Research International*. 25: 41–43.

Indyk, H.E., Evans, E.A., Bostrom-Caselunghe, M.C., Persson, B.S., Finglas, P.M., and Woollard, D.C., 2000. Determination of biotin and folate in infant formula and milk by optical biosensor-based immunoassay. *Journal of AOAC International*. 83: 1141–1148.

Kalman, A., Caelen, I., and Svorc, J. 2006. Vitamin and psuedovitamin analysis with biosensors in food products: a review. *Journal of AOAC International*. 89: 819–825.

Lahely, S., Ndaw, S., Arella, F., and Hasselmann, C., 1999. Determination of biotin in foods by high-performance liquid chromatography with post-column derivatization and fluorimetric detection. *Food Chemistry*. 65: 253–258.

Livaniou, E., Costopoulou, D., Vassiliadou, I., Leondiadis, L., Nyalala, J.O., Ithakissios, D. S., and Evangelatos, G.P., 2000. Analytical techniques for determining biotin. *Journal of Chromatography A* 881: 331–343.

Mock, D.M., Mock, N.I., and Langbehn, S.E., 1992. Biotin in human milk: methods, location and chemical form. *Journal of Nutrition*. 122: 535–545.

Mock, D.M., Mock, N.I., and Stratton, S., 1997. Concentrations of biotin metabolites in human milk. *Journal of Pediatrics*. 131: 456–458.

Murakami, T., Yamano, T., Nakama, A., and Mori, Y., 2008. Estimation of dietary intake of biotin and its measurement uncertainty using total diet samples in Osaka, Japan. *Journal of AOAC International*. 91: 1402–1408.

Nelson, B.C., Sharpless, K.E., and Sander, L.C,. 2006. Improved liquid chromatography methods for the separation and quantitation of biotin in NIST standard reference material 3280: multivitamin/multielement tablets. *Journal of Agricultural and Food Chemistry*. 54: 8710–8716.

Ploux, O., 2000. Chapter 11 'Biotin'. In: De Leenheer, A.P., Lambert, W.E., and Van Bocxlaer, J.F. (ed.) *Modern Chromatographic Analysis of Vitamins*, 3rd ed. Marcel Dekker, New York, USA, pp. 479–509.

Reyes, F.D., Romero, J.M.F., and de Castro, M.D.L., 2001. Determination of biotin in foodstuffs and pharmaceutical preparations using a biosensing system based on the streptavidin-biotin interaction. *Analytica Chimica Acta*. 436: 109–117.

Ridascreen®. Enzyme binding assay for the quantitative analysis of biotin. Cat. No. R2201, R-Biopharm GmbH, Darmstadt, Germany.

Staggs, C.G., Sealey, W.M., McCabe, B.J., Teague, A.M., and Mock, D.M., 2004. Determination of the biotin content of select foods using accurate and sensitive HPLC/avidin binding. *Journal of Food Composition and Analysis*. 17: 767–776.

Thompson, L.B., Schmitz, D.J., and Pan, S-J., 2006. Determination of biotin by high-performance liquid chromatography in infant formula, medical nutritional products and vitamin premixes. *Journal of AOAC International.* 89: 1515–1518.

Walash, M.I., Rizk, M., Sheribah, Z.A., and Salim, M.M., 2008. Kinetic spectrophotometric determination of biotin in pharmaceutical preparations. *International Journal of Biomedical Science.* 4: 238–244.

VitaFast®. Quantitative determination of biotin by microbiological assay. Cat. No. P1003. R-Biopharm GmbH, Darmstadt, Germany.

CHAPTER 25

Quantitation of Folates by Stable Isotope Dilution Assays

MICHAEL RYCHLIK

BIOANALYTIK Weihenstephan, Research Centre for Nutrition and Food Sciences, Technische Universität München, Alte Akademie 10, D-85354 Freising, Germany
Email: michael.rychlik@tum.de

25.1 Folates

The vitamins of the folate group are essential coenzymes in the metabolism of one-carbon groups (Selhub 2002) and are involved in DNA synthesis, amino acid metabolism and methylations, in general. This vitamin group is characterized by its multitude of derivatives, the so-called vitamers, which are distinguished by the oxidation status of the pterin ring, the attachment of one-carbon groups to N5 or N10 or both, and the number of glutamates bound via γ-peptide linkages (Figure 25.1).

For methylations, the vitamer 5-methyltetrahydrofolate transfers its methyl group to homocysteine thus yielding methionine, which can act as a methyl donor via S-adenosylmethionine to diverse methyl acceptors such as DNA, phospholipids and various amino acids. However, formyl-tetrahydrofolates are involved in the generation of purines as part of DNA synthesis.

Generally, consumers' intake of folates from natural sources is considered to be below the recommendations to meet physiological requirements. In consequence, folate deficiency is known to increase the risk of neural tube

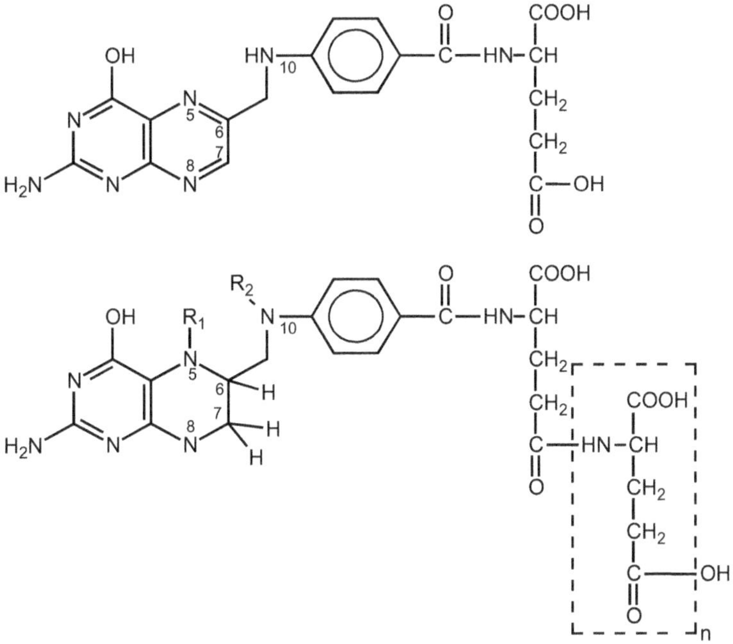

Figure 25.1 Structures of folic acid and naturally occurring folate vitamers. The formulae and the table show all possible variations of folates that have been detected in foods.

Folate	R_1	R_2	bridge R1–R2
Tetrahydrofolic acid	H	H	
5-Methyltetrahydrofolic acid	–CH$_3$	H	-
5-Formyltetrahydrofolic acid	–CH=O	H	-
10-Formyltetrahydrofolic acid	H	–CH=O	-
5-Formiminotetrahydrofolic acid	–CH=NH	H	-
5,10-Methylenetetrahydrofolic acid	-	-	–CH$_2$–
5,10-Methenyltetrahydrofolic acid	-	-	–CH$^+$=

n: number of glutamate residues (n = 0–7)

defects and is suspected to promote the development of certain forms of cancers, Alzheimer's disease and cardiovascular disease. Therefore, over 50 countries all over the world have introduced mandatory fortification with synthetic folic acid. Administration of the latter was implemented in 1998 in the USA and Canada and most recently in Australia in September 2009.

However, a dispute about folate fortification arouse in the last years as the decreasing trend of colon cancer inverted in some countries with mandatory folate fortification since its implementation (Mason *et al.* 2007). Therefore, many countries in the EU refuse mandatory fortification and favour the consumption of foods endogenously high in folates or increasing folate content in foods, generally.

25.2 Current Methods for the Analysis of Folates

25.2.1 Microbiological Assays

The most widely used method to quantitate folates is a microbiological assay (MBA), which provides a total figure for all folate vitamers. During this bioassay, the growth of *Lactobacillus casei* ssp. rhamnosus, for which folate is essential, is measured in a specifically deficient nutrient solution. The medium is then supplemented with a sample or a folate standard, and the resulting growth is measured at intervals by turbidity, which conveniently can be measured in an ultraviolet–visible (UV-vis) spectrophotometer.

However, MBAs are difficult to carry out routinely in the general setting of a modern laboratory for food analysis. In particular, the need to maintain inoculum cultures, sterile conditions and the suited water supply is a decisive restriction. Additionally, the measurement of a total vitamin figure no longer meets actual requirements in many cases. In modern food and nutrition sciences, differentiation of vitamers, conjugates or pro-vitamins is important due to different bioavailabilities or due to different responses of the microorganisms.

Despite the permanently increasing importance of chromatographic methods, MBAs still keep their specific place in vitamin analysis. As they are generally very sensitive, MBAs are commonly used for vitamin B_{12}, folic acid and biotin. For vitamin groups such as the folates, when detection, separation and quantitation of all vitamers is difficult to achieve, MBAs offer the advantage of obtaining a total value without the need for extensive method development. Because of these considerations, MBA is the 'gold' standard for folates and is still in use as reference method in many countries (AOAC 2006).

The microbiological determination of vitamins is quite labour-intensive and time-consuming besides requiring considerable laboratory organization. In contrast to maintaining active inoculum cultures, cryopreservation was introduced by Horne and Patterson (1988) and later the use of a chloramphenicol resistant strain of *Lactobacillus casei* allowed it to be run completely open on the laboratory bench (O'Broin and Kelleher 1992).

Advances in microtitre plate technology and the plates' improved optical qualities enabled MBAs for folates to be performed on 96-well plates (Newman and Tsai 1986). The obvious advantages of such miniaturisation of the assays include not only speed of reading and reduced reagent costs but also compatibility with a modern clinical or research laboratory.

25.2.2 Binding Assays

A further analytical approach is based on the reaction of folates with folate binding protein (Finglas *et al.* 1988) in protein-binding assays (PBAs) or with monoclonal antibodies in biosensor assays (Indyk *et al.* 2000). The latter have particular advantages with regard to time efficiency and are mainly based on surface plasmon resonance (SPR) measurement, a mass-sensitive equipment which transduces with high sensitivity the amount of bound molecules to surfaces by measuring the intensity of reflected polarized light at the resonance angle (O'Kane and Wahlström 2011). The specific binding of vitamins to the layer can be conveyed by interaction with folate-binding protein (FBP). This technology is commercially available as the BioCore® platform and has been reported for several other vitamins along with folic acid. In analogy to the microbiological assay, these methods also (i) cannot distinguish between the single vitamers and (ii) show different responses to them.

25.2.3 Chromatography

If differentiation between folates is needed, high-performance liquid chromatography (HPLC) or capillary electrophoresis coupled to fluorescence detection (FD), electrochemical detection (ED) or mass spectrometry (MS) has to be applied. As HPLC-FD and HPLC-ED are subject to matrix interferences, sophisticated clean-up procedures such as affinity chromatography using folate-binding protein (Pfeiffer *et al.* 1997a) are applied. To improve specificity, Stokes and Webb (1999) developed the first coupling with mass spectrometry, which enabled quantitation of folic acid and 5-/10-formyltetrahydrofolic acid. Because this method did not use an internal standard to correct for losses during sample clean-up and for variations in ionization efficiency, Garbis *et al.* (2001) introduced methothrexat as internal standard. A consequent development was the use of labelled analogues as internal standards in stable isotope dilution assays.

25.3 Stable Isotope Dilution Assays

Stable isotope dilution assays (SIDAs) can be traced back to the beginning of the 20th century and coincide with the discovery of isotopes by Soddy (1913) and the application of radioactive isotopes to determine the content of lead in rocks and the solubility of lead salts in water (Hevesy and Paneth 1913). Most elements consist of both stable and radioactive isotopes, and examples of natural distributions are listed in Table 25.1.

Although isotopes show different masses, their physical and chemical properties are quite similar. This is because the nuclei contain the same number of protons and hence the isotopes possess the same number of electrons that determine their chemical behaviour. Therefore, chemical differentiation is difficult. If the isotopic distribution of the elements of a

Table 25.1 Abundance of naturally occurring stable isotopes of important elements in stable isotope dilution assays (SIDAs). The occurrence of naturally occurring isotopes has to be considered as these can cause spectral overlaps between the analytes and the labelled standards.

Element	Isotope	Natural abundance
Hydrogen	H-1	99.988%
	H-2	0.012%
Carbon	C-12	98.89%
	C-13	1.11%
Nitrogen	N-14	99.63%
	N-15	0.37%
Silicon	Si-29	92.23%
	Si-30	4.67%
	Si-31	3.10%
Chlorine	Cl-35	75.77%
	Cl-37	24.23%
Sulfur	S-32	95.02%
	S-33	0.75%
	S-34	4.21%

chemical compound is altered, different kind of 'isomers' have to be distinguished: 'isotopomers' can be regarded as 'isotopic isomers', *i.e.* they have the same sum formula and same mass but the positions of the isotopes in the molecule are different. In contrast, 'isotopologues' or 'isotopologs' are 'isotopic homologues', *i.e.* they differ in their isotopic composition and masses. Generally, in stable isotope dilution assays, isotopologues are used as internal standard.

Although their properties are very similar, isotopes can be enriched or depleted due to their different masses leading to an alteration of the natural distribution. If an element or compound at natural isotopic distribution is mixed with such an isotopically different material, the naturally abundant isotopes are diluted in the resulting material. This is the origin of the term 'dilution' in SIDA.

During SIDAs, isotopic dilution takes place after addition of the labelled standard and its equilibration with the analyte. Due to their almost identical chemical and physical properties, the ratio of the isotopologues is stable throughout all subsequent analytical steps. Final mass spectrometry enables the determination of the isotopologues. From this ratio, the content of the analyte in the sample can be calculated with the known amount of the internal standard added in the beginning. In contrast, a structurally different internal standard may be discriminated against and, thus, cause systematic errors and imprecision. Therefore, in SIDA losses of the analyte are completely compensated for by identical losses of the isotopologue, whereas a structurally different internal standard may show different losses.

25.4 Benefits and Limitations of Folate SIDAs

25.4.1 Benefits

The ideal compensation for losses renders SIDA a perfect tool for analysing trace compounds such as folates. Besides the analytes occurring only in traces, folate analysis is challenging because folates are labile against oxygen, light and elevated temperatures, and occur endogenously as over 50 different vitamers. Therefore, tedious clean-up procedures due to matrix interferences are essential, which typically evoke losses of the analyte. The use of structurally different internal standards requires additional recovery and spiking experiments, which often result in imprecise data. In all these cases, SIDA offers significant benefits.

Another important feature of an isotopologic standard is the enhanced specificity of detection. Along with the detection of the analyte in its specific mass trace, the internal standard shows an almost identical retention time at a distinct mass shift. Therefore, the identity of the analyte unambiguously is visible in the chromatogram of a SIDA with the co-eluting peak of the internal standard (Figure 25.2).

Besides enhancing specificity, sensitivity of detection is improved by the so-called 'carrier effect'. Due to adsorption phenomena on glassware or chromatographic columns, an absolute amount of the analyte is likely to be lost during sample clean-up. If the total amount of the analyte in an extract falls below this loss, the compound will no longer be detectable. However, if an isotopologic standard is added in an amount exceeding this loss, the total sum of standard and analyte is higher than the loss and, therefore, the isotopologues will be detected. For drug analysis there are some applications showing a significant enhancement of sensitivity (Haskins *et al.* 1978).

With its specificity and ideal compensation for losses, SIDAs have the potential to become official reference methods for folates. A SIDA can be traced back to a gravimetric (*i.e.* primary) measurement and, therefore, is considered a primary method, which is a 'method having the highest metrological qualities, for which a complete uncertainty statement can be written down in terms of SI units, and whose results are, therefore, accepted without reference to a standard of the quantity being measured' (Quinn 1997). In case of being validated intensively for the absence of systematic errors, it has the potential to be accurate, *i.e.* producing the 'true' value.

25.4.2 Limitations

A principal limitation of methods applying extraction of the analyte also affects SIDAs. Although complete extraction is not necessary, at least a complete equilibration of standard and analyte has to be ensured.

As the labelled standard is added to the sample as a solution either in the extracting solvent or directly to the matrix itself, it is likely be recovered to a high extent during the extraction procedure. For the analyte, this might not be true, as it can be trapped in compartments of the matrix and might be less

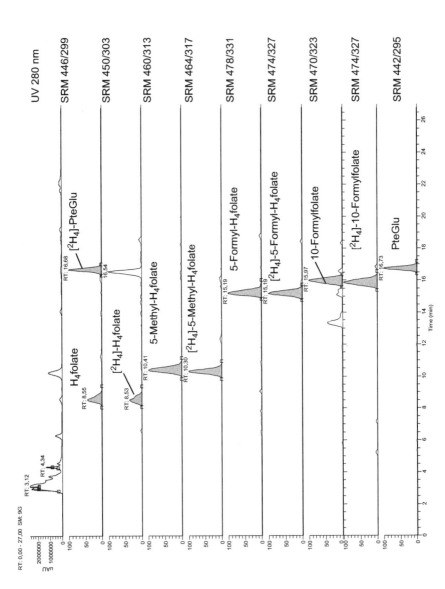

Figure 25.2 SIDA of dried white beans: LC-MS/MS chromatogram of the purified extract. Upper trace: UV-signal. Selected reaction monitoring (SRM) traces: m/z precursor ion/m/z product ion.

extractable by the solvent. Therefore, sufficient time has to be given to enable equilibration of standard and analyte in all parts of the sample and to assure that analyte and standard show the same concentration ratio in all compartments as far as possible.

Particularly for folate SIDAs, the high number of vitamers also imposes possible errors. Until now, of all the naturally occurring folates, only the monoglutamates are available as labelled isotopologues. Therefore, polyglutamate vitamers have to be converted to monoglutamate forms by the use of suitable deconjugases. As natural polyglutamates are not available as labelled standards, their conversion to monoglutamates cannot be monitored during SIDAs and possible incompletion of the reaction cannot be seen during analysis.

Furthermore, an important roadblock for propagation of SIDAs is the price of labelled standards. However, this is not a convincing argument as illustrated by the following example: 20 mg of commercially available [^{13}C]-labelled folate may cost around $1500. As less than 200 ng of the labelled standard is required for a SIDA, 20 mg of the standard enables the performance of at least 10 000 analyses. Hence the material costs for using a labelled standard accounts for $0.10 per sample, which is negligible compared with the cost of labour and equipment.

25.5 Application of SIDAs to Folate Analysis

25.5.1 Stable Isotopologues of Folates

Several folate isotopologues have been synthesized and some have been used or are still in use as an internal standard (Figure 25.3). At the end of the 1980s, [^{2}H$_2$]-folic acid as the first isotopologue was prepared by Gregory and Toth (1988a) and used as tracer in bioavailability studies. As a second tracer isotopologue, [^{2}H$_4$]-folic acid labelled in the glutamate moiety was also prepared by Gregory and Toth (1988b), and it was not until 1999 that an isotopologue, namely [^{13}C$_5$]-folic acid, was commercially available and applied as internal standard (Pawlosky and Flanagan, 2001).

The first deuterated folate to be used as internal standard was [^{2}H$_4$]-folic acid labelled in the 4-aminobenzoate moiety; the latter was the base for the first folate analogues for naturally occurring folates (Freisleben *et al.* 2002). The latter were the first to be used as internal standards for folates in foods not fortified with folic acid (Freisleben *et al.* 2003b). Further folate isotopologues were synthesized thereafter, and an extensive synthetic study was presented by Maunder *et al.* (1999) with routes to [^{2}H$_4$]-folic acid and [^{13}C$_6$]-folic acid labelled in the glutamate moiety and the benzene moiety, respectively.

At present, tetrahydrofolate (H$_4$folate), 5-methyl-H$_4$folate (5-CH$_3$-H$_4$folate), 5-formyl-H$_4$folate, 5,10-methenyl-H$_4$folate and folic acid (PteGlu) are commercially available all labelled as [^{13}C$_5$]-glutamates and are used by several groups in SIDAs of clinical samples and foods (see below). As tracer

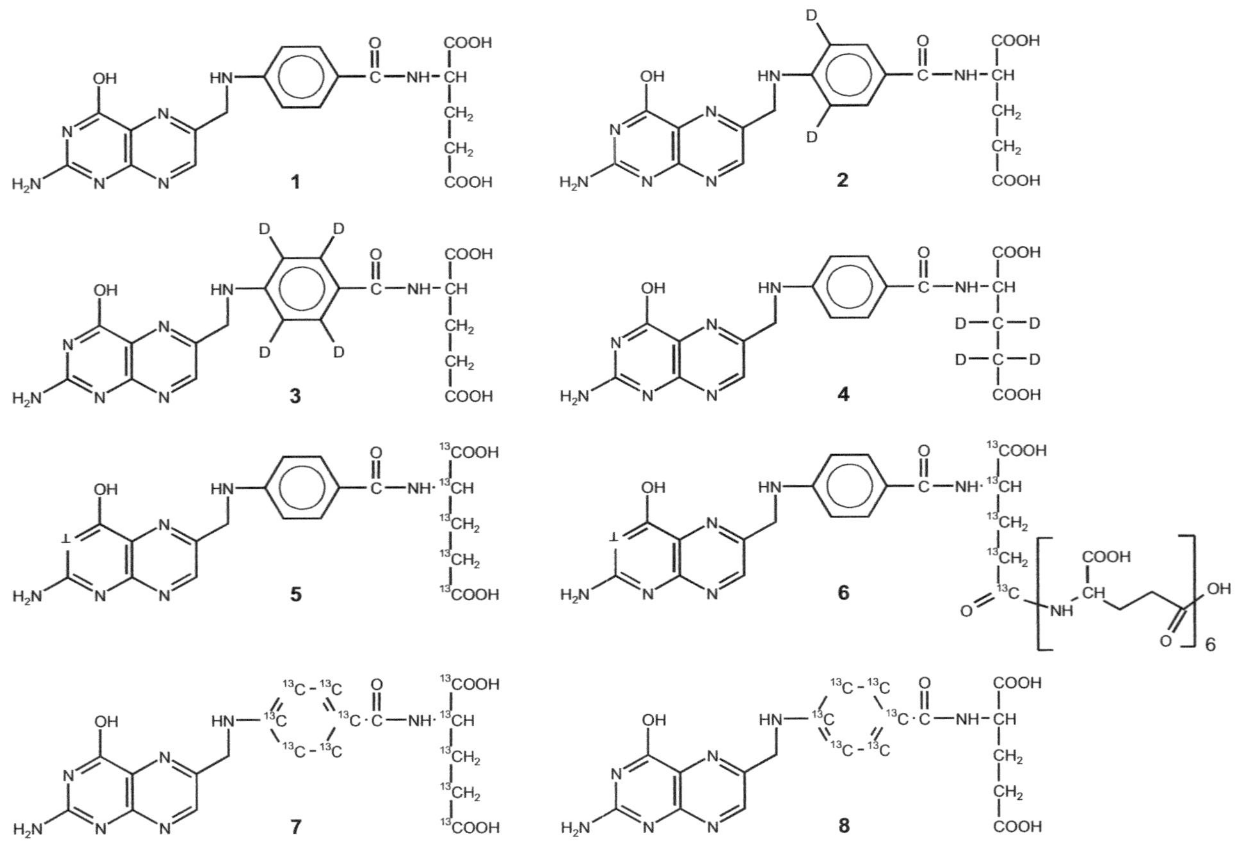

Figure 25.3 Isotopologues of folic acid (1) used as internal standards or tracers. Several isotopologues of folic acid have been used in SIDAs or tracer studies. Labelling with deuterium (2, 3, 4) can be differentiated from labelling with [^{13}C] (6, 7, 8).

isotopologues, $[^2H_2]$-5,10-methylene-H_4folate (Hong and Kohen 2005), $[^{15}N_{1-7}]$-5-CH_3-H_4folate (Hart et al. 2006) and $[^{13}C_{11}]$-PteGlu (Melse-Boonstra et al. 2006) have also been applied, mainly in studies on folate bioavailability and methylation physiology. Further isotopologues related to folates were $[^{13}C_5]$-4-aminobenzoylglutamate, acetyl $[^{13}C_5]$-4-aminobenzoylglutamate as catabolite internal standard and $[^2H_4]$-carboxyethylfolate for quantification of folate glycation products. The most recent folate isotopologue and the first to be used as tracer isotopologue of deconjugation for folate analysis was $[^{13}C_5]$-pteroylheptaglutamate, which was produced *via* a *Merrifield*-like solid phase synthesis (Mönch and Rychlik 2011).

25.5.2 SIDAs in Food Folate Analysis

The first SIDA for folic acid in fortified foods was applied by Pawlosky and Flanagan (2001) using commercially available $[^{13}C_5]$-PteGlu as the internal standard in single-stage MS. Thereafter, the first SIDA for quantitation of endogenous food folates using fourfold deuterated folic acid along with the most abundant folate monoglutamates was reported by Freisleben et al. (2003a). Further developments of this SIDA included the use of chicken pancreas in addition to rat plasma along with applying liquid chromatography–tandem mass spectrometry (LC-MS/MS) for enhancing specificity and sensitivity (Rychlik et al. 2007) and the use of 4-morpholineethanesulfonic acid (MES) buffer for improved deconjugation (Mönch and Rychlik 2011).

With $[^{13}C]$-labelled H_4folate, 5-CH_3-H_4folate, 5-formyl-H_4folate, 5,10-methenyl-H_4folate and folic acid being offered commercially, SIDA for folates in foods is currently used by several groups all over the world (Table 25.2). However, the analysis of endogenous folates remains challenging. In particular, the polyglutamates and their complete conversion to the monoglutamates still have to be ensured and monitored. The use of $[^{13}C_5]$-pteroylheptaglutamate for confirmation and quantitation of the degree of deconjugation is the most recent development. Moreover, the task of quantifying labile folates such as 10-formyl-H_4folate by LC-MS/MS still is not solved.

25.5.3 Tracer Folates and SIDAs in Folate Analysis of Clinical Samples

The first tracer folate in bioavailability studies was $[^2H_2]$-PteGlu (Gregory and Toth 1988b) followed by the first dual label study using $[^2H_2]$-PteGlu and $[^2H_4]$-PteGlu by Pfeiffer et al. (1997b). These early investigations were restricted to the sole measurement of total urinary folate isotopologues by gas chromatography–mass spectrometry (GC-MS) and were often hampered by spectral overlap due to insufficient mass increments interfering with naturally occurring isotopologues.

Wright et al. (2005) measured $[^{13}C_6]$-labelled, $[^{15}N_{1-7}]$-labelled and unlabelled 5-CH_3-H_4folate in plasma using $[^2H_2]$-folic acid as the internal standard in the

Table 25.2 Applications of stable isotope dilution assays (SIDAs) to food folates. Reported SIDAs for foods with and without fortification with folic acid are listed along with the isotopologic standard used.

Sample type	Sample	Internal standards	Reference
Fortified foods			
	Diverse	[^{13}C$_5$]-PteGlu	Pawlosky and Flanagan 2001
	Diverse	[^{2}H$_4$]-PteGlu	Rychlik 2003
	Multivitamin tablets	[^{13}C$_5$]-PteGlu	Nelson et al. 2006
	Infant formula	[^{13}C$_5$]-PteGlu	Jung et al. 2007
	Infant formula	[^{13}C$_5$]-PteGlu	Goldschmidt and Wolf 2010
	Ready-to-eat-cereals	[^{13}C$_5$]-PteGlu	Phillips et al. 2010
	Ready-to-eat-cereals	[^{2}H$_4$]-PteGlu	Krishnan et al. 2011
Foods			
	Diverse	[^{2}H$_4$]-folates	Freisleben et al. 2003a
	Vegetables	[^{13}C$_5$]-5-MTHF	Wang et al. 2010
	Diverse	[^{13}C$_5$]-folates	Vishnumohan et al. 2011
	Diverse	[^{2}H$_4$]-folates, [^{13}C$_5$]-pteroylheptaglutamate	Mönch and Rychlik 2012

single ion monitoring mode. However, the use of a structurally different internal standard such as [^{2}H$_2$]-PteGlu may decrease accuracy by ion suppression and, moreover, quantitation was hampered by spectral overlap of [^{15}N$_{1-7}$]-5-CH$_3$-H$_4$folate with, on the one hand, [^{13}C$_6$]-5-CH$_3$-H$_4$folate and, on the other hand, unlabelled 5-CH$_3$-H$_4$folate in single stage LC-MS. The first to use a better suited internal standard for serum analysis were Pawlosky et al. (2001) followed by Hart et al. (2002) in plasma and urine with [^{13}C$_5$]-5-CH$_3$-H$_4$folate, still in one dimensional mass spectrometry (Table 25.3). The first LC-MS/MS application for plasma folate analysis was reported by Freisleben et al. (2003b) using [^{2}H$_4$]-labelled folates.

The handicap of spectral overlaps of differently labelled isotopologues was overcome by Melse-Boonstra et al. (2006) who measured [^{13}C$_6$]-labelled along with [^{13}C$_{11}$]-labelled 5-CH$_3$-H$_4$folate as tracer isotopologues and simultaneously quantified unlabelled 5-CH$_3$-H$_4$folate by using [^{13}C$_5$]-5-CH$_3$-H$_4$folate as the internal standard. In the latter study, spectral overlaps of 5-CH$_3$-H$_4$folate isotopologues were avoided by labelling of different moieties of the target molecule and their differentiation by LC-MS/MS.

A further simultaneous use of tracer isotopologues and internal standard isotopologues was reported for a bioavailability study using an ileostomy model (Büttner et al. 2011). The absence of spectral overlaps was verified for

Table 25.3 Applications of stable isotope dilution assays (SIDAs) to folates in clinical samples. Reported SIDAs for clinical samples are listed along with the isotopologic standard used.

Sample type	Sample	Internal standards	Reference
Clinical samples			
	Serum	[$^{13}C_5$]-5-CH_3-H_4folate	Pawlosky et al. 2001
	Plasma, urine	[$^{13}C_5$]-5-CH_3-H_4folate	Hart et al. 2002
	Plasma	[$^2H_4$]-folates	Rychlik et al. 2003
	Plasma, ileostomy effluents	[$^2H_4$]-folates	Büttner et al. 2011
	Red blood cells	[$^{13}C_5$]-folates	Fazili et al. 2005 Huang et al. 2008
	Plasma	[$^2H_2$]-PteGlu	Wright et al. 2005
	Plasma	[$^{13}C_5$]-5-CH_3-H_4folate	Melse-Boonstra et al. 2006
	Plasma, Red blood cells, urine	[$^2H_4$]-folates	Mönch et al. 2010
	Plasma and serum	[$^{13}C_5$]-folates	Nelson et al. 2004
	Serum	[$^{13}C_5$]-folates	Kirsch et al. 2010
	Tissue	[$^{13}C_5$]-folates	Liu et al. 2011

the [$^{13}C_5$]-, the [$^2H_4$]-labelled and the unlabelled isotopologues of H_4folate, 5-CH_3-H_4folate, 10-HCO-PteGlu, 5-HCO-H_4folate and PteGlu. An example of the structures and MS/MS transitions of 5-CH_3-H_4folate isotopologues, respectively, is depicted in Figure 25.4. None of the abundant product ions of [$^{13}C_5$]-5-CH_3-H_4folate was identical with those of the internal standard, [$^2H_4$]-5-CH_3-H_4folate. This observation is in good agreement with the loss of the glutamate group during MS/MS, which generated different product ions for labelling in the glutamate and the 4-aminobenzoic acid moiety. In accordance with these considerations, unlabelled 5-CH_3-H_4folate shows the same abundant product ion as the tracer [$^{13}C_5$]-5-CH_3-H_4folate, since the label of the latter is lost during MS/MS. However, both can be differentiated by the mass increment of the precursor ions.

Response curve linearity and good agreement of response factors for [$^{13}C_5$]-labelled and unlabelled folate relative to the [$^2H_4$]-labelled internal standard confirmed the absence of any 'cross-talk' effects in the applied mass spectrometer.

25.6 Method Comparisons

As folate analysis is highly challenging, the accuracy of the single quantitation methods is controversial. Different methods often yield conflicting results and the 'true' folate content remains open. Several method comparisons have been performed to shed light on this issue: the more complex the sample matrix, the more disagreement has to be expected.

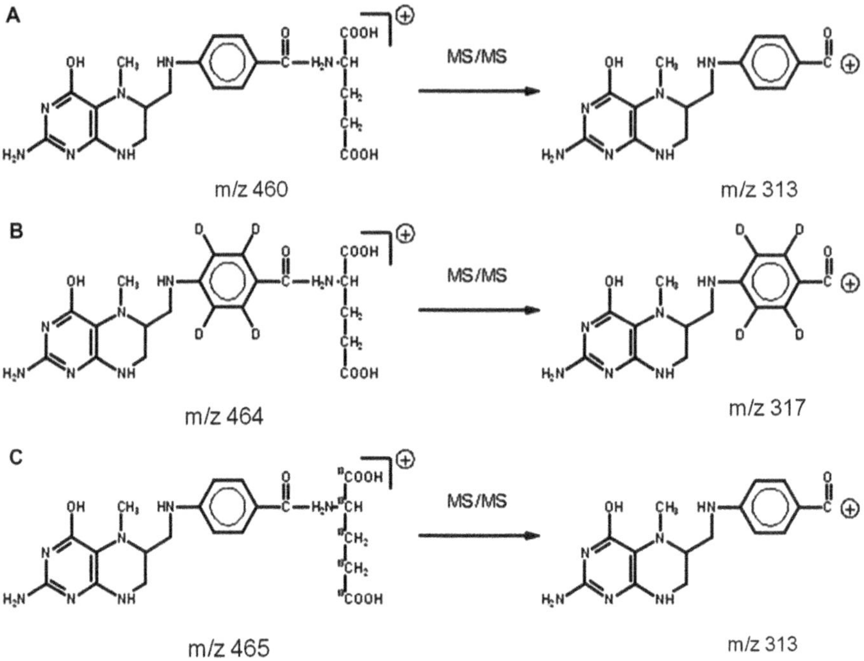

Figure 25.4 Fragmentation of 5-methyltetrahydrofolate isotopologues in tandem mass spectrometry. The fragmentation enables differentiation of unlabelled 5-methyltetrahydrofolate (A) from its [^{2}H$_4$]-labelled (B) and the [^{13}C$_5$]-labelled isotopologue (C) due to different precursor ions or product ions.

25.6.1 Folates in Blood Plasma

Regarding its vitamer distribution, blood plasma is rather simple as it contains mainly 5-methyl-H$_4$folate. Therefore, MBAs and protein-binding assays are likely to yield accurate results.

The first comparison between SIDA, MBA and a radiobinding assay was performed by Pfeiffer *et al.* (2004). As internal standards, [^{13}C$_5$]-labelled 5-CH$_3$-H$_4$folate, 5-HCO-H$_4$folate and PteGlu were used. The authors analysed 50 serum samples using all three methods and correlated the results by least-squares regression (Figure 25.5). Good correlations were found between all methods with the radiobinding assay giving quite close results to SIDA, showing a small but significant mean difference of − 2.8 nmol/L.

In contrast, the microbiological assay gave significant higher results than SIDA at a difference of 30.2% over the whole concentration range. To unravel the cause of this difference, the authors tested different calibrators for the assay and found a significant influence on the results. When using a different folic acid source, the relative difference between MBA and SIDA decreased to 15%, although the purity of both calibrators was confirmed by spectrophotometry. If 5-methyltetrahydrofolate as the mainly occurring vitamer in plasma was used

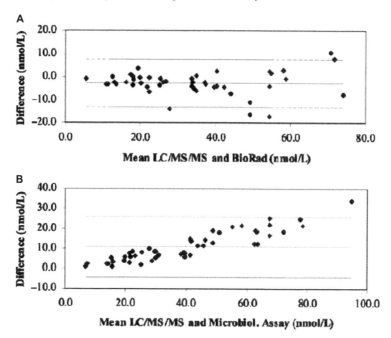

Figure 25.5 Method comparison between SIDA, radiobinding assay (A) or microbiological assay (B) for serum folates. The results are presented as Bland–Altman difference plots: *solid line* represents the mean difference between the two methods. The SIDA method is used as the comparison method. *Dashed lines* represent the limits of agreement or 0.95 central intervals of the differences between the two methods (mean difference = 2 SD). A total of 46 serum samples were used in the analysis in A and 49 for the analysis in B.
(Reproduced with permission from Pfeiffer *et al.* 2004).

as the calibrant and the results were recalculated, no significant difference between SIDA and MBA was found. This result points to the need to use the convenient vitamer for calibration of the MBA.

Another comparison of plasma samples was performed using SIDA and HPLC coupled to fluorescence detection (LC-FD) by Freisleben *et al.* (2003b). The authors applied [^{2}H$_4$]-5-CH$_3$-H$_4$folate as the internal standard for SIDA and found 5.6 µg/L and 6.8 µg/L for 5-CH$_3$-H$_4$folate by SIDA and LC-FD, respectively. Although the results were only marginally different, the coefficient of variation of LC-FD determination was significantly higher than that of SIDA. This drawback is attributable to interferences that were present in the fluorescent chromatogram even though the samples were purified by solid-phase extraction (SPE) on strong anion exchange cartridges. On the contrary, the SIDA mass chromatograms of 5-methyl-H$_4$folate and of the internal standard were both devoid of any matrix interferences and allowed an unequivocal quantification.

25.6.2 Comparisons of Folates Analyses in Foods

A comparison between HPLC with UV detection and SIDA of folic acid in fortified foods was reported recently (Krishnan *et al.* 2011). Cereal-based foods from the US market were analysed by SIDA using [^{2}H$_4$]-folic acid as the internal standard as well as by HPLC-UV in an inter-laboratory study by seven laboratories.

In general, the SIDA data were well in accordance with the labelled contents of folic acid (Table 25.4). Label values are usually target enrichment levels intended by food manufacturers but often include overages to ensure that minimum levels are maintained over the whole shelf life. In contrast, SIDA and HPLC interlaboratory study results were different in folic acid values for several samples. In particular for corn flakes, noodles and bread, the high signals of HPLC-UV were attributable to interfering compounds.

In a further study on ready-to-eat cereals fortified with folic acid, Phillips *et al.* (2010) compared SIDA using [^{13}C$_5$]-PteGlu as internal standard to a MBA using *Lactobacillus casei* ssp. rhamnosus. Interestingly, MBA gave significantly higher results than SIDA with the excess of MBA data ranging between 10 and 67%. The content of endogenous folates in the samples was negligible and, with folic acid, the correct calibrating vitamer was used for MBA. Thus, the reason for the discrepancy remained open.

A further study compared endogenously occurring folates in different foods by SIDA and by LC-FD (Freisleben *et al.* 2003b). By applying SIDA, a 67%

Table 25.4 Comparison of labelled folic acid data with those obtained by stable isotope dilution assay (SIDA) and by HPLC-UV in an inter-laboratory study (internal standard). SIDA results generally are in good accordance with the labels. Differences for corn flakes, noodles and bread to HPLC-UV were attributable to interfering compounds.

µg/g	Sample	Labelled folate content	SIDA	Mean HPLC internal standard data
Breakfast cereals	Krispies No. 1	3.03	2.82	3.32
	Krispies No. 2	9.66	9.65	7.87
	Krispies No. 3	8.48	9.84	4.80
	Corn flakes	3.57	7.42	5.92
	Wheat cereals No. 1	6.66	6.30	5.88
	Wheat cereals No. 2	6.66	6.55	5.61
	Wheat cereals No. 3	1.00	1.23	1.32
Noodles		1.79	1.62	2.16
Cookies		1.33	0.81	0.94
Flour	Wheat flour	1.33	1.63	1.22
Bread	Wheat bread	0.88	1.38	1.62
Infant formula	NIST Reference Material	1.29[a]	1.18	0.73

[a]Total folate value as measured by microbiological analysis.

Table 25.5 Comparison between stable isotope dilution assay (SIDA) and Liquid chromatography-fluorescence detection (LC-FD) for food folates. Folate contents are calculated as folic acid in µg/100 g. SIDA generally yields higher folate contents as 5-formyl-H$_4$folate was not detectable in LC-FD due to weak fluorescence.

µg folic acid /100 g	SIDA	LC-FD
Spinach	159.2	95.5
Wheat bread	19.8	16.2
Beef	1.2	0.7

higher folate content than by LC-FD was quantified in an identical aliquot of spinach, although the LC-FD values were corrected for recovery (Table 25.5). One reason for the lower LC-FD values was that 5-formyl-H$_4$folate could not be detected by LC-FD due to its weak fluorescence and amounted to 12.0 µg/100 g 5-formyl-H$_4$folate according to SIDA. However, even if this amount of 5-formyl-H$_4$folate was included into the total folate content gathered by the LC-FD method, a significant lower value than that of SIDA was obtained.

For bread, SIDA after affinity chromatography clean-up revealed the samples to contain 5-methyl-H$_4$folate, 5-formyl-H$_4$folate, H$_4$folate, 10-formylfolic acid and folic acid, of which the latter was below quantifiable levels. By LC-FD, however, neither 5-formyl-H$_4$folate nor folic acid was detected, which was again attributable to the poor fluorescence of both compounds. Although the vitamer pattern measured by LC-FD was so different from that found by SIDA, the total folate figures in wheat bread were quite similar.

For meat samples, SIDA required affinity clean-up prior to LC-MS/MS as the folate levels in this matrix are known to be low. In analogy to other foods, 5-formyl-H$_4$folate and folic acid could not be detected by LC-FD, whereas the more selective SIDA confirmed the presence of both compounds although the latter were below limit of quantitation. The total folate values calculated as folic acid for both methods were quite close to another. This comparison demonstrated that the LC-FD assay only produces reliable results if the matrix is not too demanding.

A further comparison was carried out between SIDA using [^{2}H$_4$]-folates and MBA in a commercial microtitre format. Several vegetables were tested for different variations of the usual SIDA and various calibrators used in MBA. Generally, after some experience, MBA yielded consistent results for different dilutions of the extracts, which rendered the presence of inhibiting compounds unlikely. Due to the high variance of MBA, the difference to SIDA was only significant for broccoli (Table 25.6). Some but not differences could be attributed to losses during sample preparation as addition of the labelled standards and measuring by LC-MS/MS yielded lower results in comparison with the usual SIDA with addition of the internal standard at the beginning of extraction. To some degree, the difference may be attributed to incomplete deconjugation of extracts for MBA when measured by LC-MS/MS, although *Lactobacillus casei* ssp. rhamnosus is known also to respond to polyglutamate

Table 25.6 Comparison between stable isotope dilution assay (SIDA) and microbiological assay (MBA) for food folates. Considerable differences between SIDA and MBA were obtained, which were attributable to losses during sample preparation and unsuitable calibrators in MBA.

µg folate/100 g	SIDA	MBA
Spinach	81	98 (76)[a]
Cauliflower	101	89
Savoy cabbage	155	>115[b]
Broccoli	180	127

[a] Result in parentheses refers to calibration with 5-methyltetrahydrofolate.
[b] Estimation as the used dilution was out of calibration range.

forms. Considering relative responses of vitamers and the vitamer distribution determined by SIDA, better agreement was obtained and confirmed the finding of Pfeiffer *et al.* (2004) that correct calibration is crucial for MBA.

25.7 Perspectives

Microbiological assay has been the standard method in folate analysis for about 50 years. With the new 96-well assay formats, which render its use as simple as enzyme-linked immunosorbent assay (ELISA), this method will retain its particular place for samples showing simple folate distributions. However, with the advance of LC-MS/MS in routine laboratories and the need to distinguish single vitamers, LC-MS and in particular SIDAs will gain importance in folate analysis. All important monoglutamates are commercially available as labelled isotopologues and are used by several groups analysing folates. However, important issues remain to be solved in folate SIDAs. To ensure complete deconjugation of polyglutamates, labelled isotopologues of naturally occurring vitamers have to be generated, which after deconjugation also enable differentiation from the isotopologues used as internal standards for monoglutamates. Recent synthesis of [$^{13}C_5$]-pteroylheptaglutamate is an important step in this direction.

In order to establish SIDAs as official reference methods, standard operation protocols have to be developed. In particular, the highly variable mass spectrometric settings depending on the respective instruments retarded collaborative trials. However, in the area of mycotoxins, isotope dilution assays have been elaborated as reference methods and point to a way to implement SIDAs as official methods also in folate analysis (Rychlik and Asam 2008). This could lead to the establishment of reference materials with certified values obtained by SIDA. Lability of folates is still a roadblock but will be essential for quality assurance in accredited laboratories.

Summary Points

- Folates are a group of water-soluble vitamins involved in DNA synthesis, amino acid metabolism and methylations.

- The most common analytical method for folates is a microbiological assay, which measures the growth of *Lactobacillus casei* ssp. rhamnosus in a specifically deficient nutrient solution.
- The microbiological assay is not able to differentiate between single folate vitamers.
- Stable isotope dilution assays are based on the use of stable isotopologues as internal standards that are detected by mass spectrometry.
- Stable isotope dilution assays are becoming increasingly used as they enable to differentiate between the vitamers and to correct for losses during sample preparation.
- As internal standards in folate stable isotope dilution assays, [^{13}C]-labelled and [^{2}H$_4$]-labelled folates are mainly used.
- Applications of folate stable isotope dilution assays are reported for foods with or without fortification by folic acid and for clinical samples such as blood serum, plasma, red blood cells, urine and ileostomy effluents.
- In method comparisons of stable isotope dilution assays with microbiological assays, good agreement is obtained if 5-methyltetrahydrofolate is used as the calibrator for the microbiological assay.
- Method comparisons of stable isotope dilution assay with MBA for food folates revealed good agreement for fortified foods and, for endogenous food folates, the vitamer distribution has to be considered for microbiological assays to yield accurate results.

Key Facts

- In a SIDA, stable isotopic analogues of the analytes, *i.e.* stable isotopologues, are used as internal standards.
- Internal standards are added at the beginning of the extraction in definite amounts to equilibrate with the analytes.
- Due to their almost identical chemical and physical properties, the ratio of the isotopologues is stable throughout all subsequent analytical steps.
- Final mass spectrometry enables to determine the isotopologic ratio.
- From the isotopologic ratio, the content of the analyte in the sample can be calculated with the known amount of the internal standard added in the beginning.
- In SIDA, losses of the analyte are completely compensated for by identical losses of the isotopologic internal standard.

Definitions of Words and Terms

Deconjugation of folates. Enzymatic reaction to convert polyglutamate vitamers of folates to the respective monoglutamates.

Isotopologues, isotopologs. Isomers that differ in their isotopic composition and their masses. Generally, in stable isotope dilution assays, isotopologues are used as internal standards.

Isotopomers. Isomers that have the same chemical formula and same mass but the positions of the isotopes in the molecule are different.

Stable isotope dilution assay. Analytical method that uses isotope-labelled analogues of the analytes as internal standards to compensate for losses and discrimination during analysis.

Vitamer. A defined species within a vitamin group.

List of Abbreviations

5-CH$_3$-H$_4$folate	5-methyltetrahydrofolate
ED	electrochemical detection
ELISA	enzyme-linked immunosorbent assay
FBP	folate-binding protein
FD	fluorescence detection
GC-MS	gas chromatography–mass spectrometry
HPLC	high-performance liquid chromatography
LC-FD	liquid chromatography–fluorescence detection
LC-MS	liquid chromatography–mass spectrometry
LC-MS/MS	liquid chromatography–tandem mass spectrometry
H$_4$folate	tetrahydrofolate
MBA	microbiological assay
MES	4-morpholineethanesulfonic acid
MS	mass spectrometry
PBA	protein-binding assay
PteGlu	folic acid
SIDA	stable isotope dilution assay
SPE	solid-phase extraction
SPR	surface plasmon resonance

Acknowledgment

Appreciation is due to Maria Grübner, Technische Universität München, for performing the method comparison between SIDA and MBA for endogenous food folates.

References

AOAC, 2006. AOAC Official Method 944.12 Folic Acid in Vitamin Preparations–Microbiological Methods. In: Official Methods of Analysis of AOAC International, 18th ed., Current through Revision 1. AOAC International, Gaithersburg, MD, USA.

Büettner, B., Oehrvik, V., Witthoeft, C.M., and Rychlik, M., 2011. Quantitation of isotope-labelled and unlabelled folates in plasma, ileostomy and food samples, *Analytical and Bioanalytical Chemistry*. 399: 429-439.

Fazili, Z., Pfeiffer, C.M., Zhang, M., and Jain, R., 2005. Erythrocyte folate extraction and quantitative determination by liquid chromatography-tandem mass spectrometry: Comparison of results with microbiologic assay. *Clinical Chemistry.* 51: 2318–2325.

Finglas, P.M., Faulks, R.M., and Morgan, M.R.A., 1988. The development and characterization of a protein-binding assay for the determination of folate—potential use in food analysis. *Journal of Micronutrient Analysis.* 4: 295–308.

Freisleben, A., Schieberle, P., and Rychlik, M., 2002. Syntheses of labelled vitamers of folic acid to be used as internal standards in stable isotope dilution assays. *Journal of Agricultural and Food Chemistry.* 50: 4760–4768.

Freisleben, A., Schieberle, P., and Rychlik, M., 2003a. Comparison of folate quantification in foods by high-performance liquid chromatography-fluorescence detection to that by stable isotope dilution assays using high-performance liquid chromatography-tandem mass spectrometry. *Analytical Biochemistry.* 315: 247–255.

Freisleben, A., Schieberle, P., and Rychlik, M., 2003b. Specific and sensitive quantification of folate vitamers in foods by stable isotope dilution assays using high-performance liquid chromatography-tandem mass spectrometry. *Analytical and Bioanalytical Chemistry.* 376: 149–156.

Garbis, S.D., Melse-Boonstra, A., West, C.E., and van Breemen, R.B., 2001. Determination of Folates in Human Plasma Using Hydrophilic Interaction Chromatography-Tandem Mass Spectrometry. *Analytical Chemistry.* 73: 5358–5364.

Goldschmidt, R.J., and Wolf, W.R., 2010. Simultaneous determination of water-soluble vitamins in SRM 1849 Infant/Adult Nutritional Formula powder by liquid chromatography-isotope dilution mass spectrometry. *Analytical and Bioanalytical Chemistry.* 397: 471–481.

Gregory, J.F., and Toth, J.P., 1988a. Preparation of folic acid specifically labeled with deuterium at the 3',5'-positions. *Journal of Labelled Compounds and Radiopharmaceuticals.* 15: 1349–1359.

Gregory, J.F.I., and Toth, T.G., 1988b. Chemical synthesis of deuterated folate monoglutamate and in vivo assessment of urinary excretion of deuterated folates in man. *Analytical Biochemistry.* 170: 94–104.

Hart, D.J., Finglas, P.M., Wolfe, C.A., Mellon, F., Wright, A.J.A., and Southon, S., 2002. Determination of 5-methyltetrahydrofolate (13C-labeled and unlabeled) in human plasma and urine by combined liquid chromatography mass spectrometry. *Analytical Biochemistry.* 305: 206–213.

Hart, D.J., Wright, A.J.A., Wolfe, C.A., Dainty,J., Perkins, L.R., and Finglas, P.M., 2006. Production of intrinsically labeled spinach using stable isotopes (13C or 15N) for the study of folate absorption. *Innovative Food Science & Emerging Technologies.* 7: 147–151.

Haskins, N.J., Ford, G.C., Grigson, S.J.W., and Waddell, K.A., 1978. A carrier effect observed in assays for antidiarrheal drug compounds. *Biomedical Mass Spectrometry.* 5: 423–4.

Hevesy, G., and Paneth, F., 1913. The solubility of lead sulfide and lead chromate. *Zeitschrift für anorganische und allgemeine Chemie.* 82: 323–328.

Hong, B., and Kohen, A., 2005. Microscale synthesis of isotopically labeled (6R)-N5,N10-methylene-5,6,7,8-tetrahydrofolate. *Journal of Labelled Compounds & Radiopharmaceuticals.* 48: 759–769.

Horne, D.W., and Patterson, D., 1988. Lactobacillus casei microbiological assay of folic acid derivatives in 96-well microtiter plates. *Clinical Chemistry.* 34: 2357–2359.

Indyk, H.E., Evans, E.A., Caselunghe, M.C.B., Persson, B.S., Finglas, P.M., Woollard, D.C., and Filonzi, E.L., 2000. Determination of biotin and folate in infant formula and milk by optical biosensor-based immunoassay. *Journal of AOAC International.* 83: 1141–1148.

Jung, M., Kim, B., Boo, D.W., and So, H.-Y., 2007. Development of isotope dilution-liquid chromatography/tandem mass spectrometry as a candidate reference method for the determination of folic acid in infant milk formula. *Bulletin of the Korean Chemical Society.* 2007: 745–750.

Kirsch, S.H., Knapp, J.-P., Herrmann, W., and Obeid, R., 2010. Quantification of key folate forms in serum using stable-isotope dilution ultra performance liquid chromatography-tandem mass spectrometry. *Journal of Chromatography B.* 878: 68–75.

Krishnan P.G., Musukula, S.R., Rychlik, M., Nelson, D.R., DeVries, J.W., and MacDonald, J.L., 2011, Optimization of HPLC methods for analyzing added folic acid in fortified foods. In: Rychlik, M. (ed.) *Fortified Foods with Vitamins*, Wiley-VCH, Weinheim, Germany, pp. 135–142.

Liu, J., Pickford, R., Meagher, A. P., Ward, R. L., 2011. Quantitative analysis of tissue folate using ultra high-performance liquid chromatography tandem mass spectrometry. *Analytical Biochemistry.* 411: 210–217.

Lu, W., Kwon, Y.K., and Rabinowitz, J.D., 2007. Isotope ratio-based profiling of microbial folates. *Journal of the American Society of Mass Spectrometry.* 18: 898–909.

Mason, J.B., Dickstein, A., Jacques, P.F., Haggarty, P., Selhub, J., Dallal, G., and Rosenberg, I.H., 2007. A temporal association between folic acid fortification and an increase in colorectal cancer rates may be illuminating important biological principles: a hypothesis. *Cancer Epidemiology, Biomarkers & Prevention.* 16: 1325–1329.

Maunder, P., Finglas, P.M., Mallet, A.I., Mellon, F.A., Razzaque, M.A., Ridge, B., Vahteristo, L., and Witthoft, C., 1999. The synthesis of folic acid, multiply labelled with stable isotopes, for bio-availability studies in human nutrition. *Journal of the Chemical Society, Perkin Transactions.* 1: 1311–1323.

Melse-Boonstra, A., Verhoef, P., West, C.E., van Rhijn, J.A., van Breemen, R.B., Lasaroms, J.J.P., Garbis, S.D., Katan, M.B., and Kok, F.J., 2006. A dual-isotope-labeling method of studying the bioavailability of hexaglutamyl folic acid relative to that of monoglutamyl folic acid in humans by using multiple orally administered low doses. *The American Journal of Clinical Nutrition.* 84: 1128–1133.

Mönch, S., and Rychlik, M., 2012. Improved Folate Extraction and Tracing Deconjugation Efficiency by Dual Label Isotope Dilution Assays in Foods. *Journal of Agricultural and Food Chemistry.* 60: 1363–1372.

Mönch, S., Netzel, M., Netzel, G., and Rychlik, M., 2010. Quantitation of folates and their catabolites in blood plasma, erythrocytes, and urine by stable isotope dilution assays. *Analytical Biochemistry.* 398: 150–160.

Nelson, B.C., Pfeiffer, C.M., Margolis, S.A., and Nelson, C.P., 2004. Solid-phase extraction-electrospray ionization mass spectrometry for the quantification of folate in human plasma or serum. *Analytical Biochemistry.* 325: 41–51.

Nelson, B.C., Sharpless, K.E., and Sander, L.C., 2006. Quantitative determination of folic acid in multivitamin/multielement tablets using liquid chromatography/tandem mass spectrometry. *Journal of Chromatography A.* 1135: 203–211.

Newman, E.M., and Tsai, J.F., 1986. Microbiological analysis of 5-formyltetrahydrofolic acid and other folates using an automatic 96-well plate reader. *Analytical Biochemistry.* 154: 509–515.

O'Broin, S., and Kelleher, B., 1992. Microbiological assay on microtitre plates of folate in serum and red cells. *Journal of Clinical Pathology.* 45: 344–347.

O'Kane, A., and Wahlström, L. 2011, Biosensors in vitamin analysis of foods. In: Rychlik, M. (ed.) *Fortified Foods with Vitamins*, Wiley-VCH, Weinheim, Germany, pp. 65–75.

Pawlosky, R.J., and Flanagan, V.P., 2001. A quantitative stable-isotope LC-MS method for the determination of folic acid in fortified foods. *Journal of Agricultural and Food Chemistry.* 49: 1282–1286.

Pawlosky, R.J., Flanagan, V.P., and Pfeiffer, C.M., 2001. Determination of 5-methyltetrahydrofolic acid in human serum by stable-isotope dilution high-performance liquid chromatography-mass spectrometry. *Analytical Biochemistry.* 298: 299–305.

Pfeiffer, C.M., Rogers, L.M., and Gregory, J.F., 1997a. Determination of folate in cereal-grain food products using trienzyme extraction and combined affinity and reversed-phase liquid chromatography. *Journal of Agricultural and Food Chemistry.* 45: 407–413.

Pfeiffer, C.M., Rogers, L.M., Bailey, L.B., and Gregory, 3rd, J.F., 1997b. Absorption of folate from fortified cereal-grain products and of supplemental folate consumed with or without food determined by using a dual-label stable-isotope protocol. *The American Journal of Clinical Nutrition.* 66: 1388–1397.

Pfeiffer, C.M., Fazili, Z., McCoy, L., Zhang, M., and Gunter, E.W., 2004. Determination of folate vitamers in human serum by stable-isotope-dilution tandem mass spectrometry and comparison with radioassay and microbiologic assay. *Clinical Chemistry.* 50: 423–432.

Phillips, K.M., Ruggio, D.M., Ashraf-Khorassani, M., Eitenmiller, R.R., Cho, S., Lemar, L.E., Perry, C.R., Pehrsson, P.R., and Holden, J.M., 2010. Folic acid content of ready-to-eat cereals determined by liquid chromatography-mass spectrometry: comparison to product label and to values determined by microbiological assay. *Cereal Chemistry.* 87: 42–49.

Quinn, T.J., 1997. Primary methods of measurement and primary standards. *Metrologia*. 34: 61.

Rychlik, M., 2003. Simultaneous analysis of folic acid and pantothenic acid in foods enriched with vitamins by stable isotope dilution assays. *Analytica Chimica Acta*. 495: 133–141.

Rychlik, M., Netzel, M., Pfannebecker, I., Frank, T., and Bitsch, I., 2003. Application of stable isotope dilution assays based on liquid chromatography-tandem mass spectrometry for the assessment of folate bioavailability. *Journal of Chromatography B*. 792: 167–176.

Rychlik, M., Englert, K., Kapfer, S., and Kirchhoff, E., 2007. Folate contents of legumes determined by optimized enzyme treatment and stable isotope dilution assays. *Journal of Food Composition and Analysis*. 20: 411–419.

Rychlik, M., and Asam, S., 2008. Stable isotope dilution assays in mycotoxin analysis. *Analytical and Bioanalytical Chemistry*. 390: 617–628.

Selhub, J., 2002. Folate, vitamin B12 and vitamin B6 and one carbon metabolism. *Journal of Nutrition, Health and Aging*. 6: 39–42.

Soddy, F., 1913. Intra-atomic charge. Nature. 92: 399–4.

Stokes, P., and Webb, K., 1999. Analysis of some folate monoglutamates by high-performance liquid chromatography-mass spectrometry. I. *Journal of Chromatography*. 864: 59–67.

Vishnumohan, S., Arcot, J., and Pickford, R., 2011. Naturally-occurring folates in foods: method development and analysis using liquid chromatography-tandem mass spectrometry (LC-MS/MS). *Food Chemistry*. 125: 736–742.

Wang, C., Riedl, K.M., and Schwartz, S.J. 2010. A liquid chromatography-tandem mass spectrometric method for quantitative determination of native 5-methyltetrahydrofolate and its polyglutamyl derivatives in raw vegetables. *Journal of Chromatography*. 878: 2949–2958.

Wright, A.J.A., Finglas, P.M., Dainty, J.R., Wolfe, C.A., Hart, D.J., Wright, D.M., and Gregory, J.F., 2005. Differential kinetic behavior and distribution for pteroylglutamic acid and reduced folates: a revised hypothesis of the primary site of PteGlu metabolism in humans. *Journal of Nutrition*. 135: 619–623.

CHAPTER 26
Analysis of Cobalamins (Vitamin B_{12}) in Human Samples: An Overview of Methodology

DORTE L. LILDBALLE* AND EBBA NEXO

Department of Clinical Biochemistry, Aarhus University Hospital, Norrebrogade 44 Building 9, DK-8000 Aarhus, Denmark
*Email: dolild@rm.dk

26.1 Introduction

In 1926, intake of high amounts of liver was shown to cure pernicious anaemia (Minot and Murphy 1926). During the mid-1940s and following decade, the factor present in liver was isolated and crystallized, and finally its structure was resolved (for historical review, see (Lanska 2010). The identified compound was named vitamin B_{12} or cyanocobalamin.

In humans, cobalamins are essential coenzymes for two cellular enzymes. One is the mitochondrial enzyme, methyl malonyl-coA-mutase, which converts methylmalonic acid (MMA) to succinyl-CoA. The other is the cytoplasmic methionine syntase, which converts homocysteine (Hcy) to methionine (reviewed by Banerjee 2006). An insufficient amount of cobalamin in the cells, therefore, leads to increase levels of MMA and total homocysteine (tHcy).

Since the discovery and purification of cobalamin, methods suitable for its identification and quantification has been explored. The first analytical methods used microorganisms which were dependent on presence of cobalamin to grow (Shorb 1947). Several other methods have since been developed. Spectrophotometry is useful for the measurement of micromolar concentrations of cobalamins in solution (Pratt 1972), whereas a number of other methods have been developed for measurement of the minute amounts of the vitamin present in biological samples such as human serum. This review focuses on the methods used for analysis of cobalamins in human samples. The general considerations are, however, equally relevant for the analysis in food items and samples of veterinarian origin.

26.2 Cobalamins in Humans

Cobalamin, like other vitamins, is an essential micronutrient that humans obtain from food sources. Cobalamins are synthesized by microorganisms and enter the food chain through animals eating such organisms. Thus, the main sources of cobalamin are animal-derived food items including meat, fish, milk and eggs (Chanarin 1969). Plants have cobalamin-independent enzymes and do not contain the vitamin. Therefore, vegans do not receive a sufficient intake of cobalamin.

In the digestive tract, cobalamins bind to the gastric protein intrinsic factor and are subsequently absorbed in the final part of the small intestine. After absorption, cobalamin is distributed *via* the blood (reviewed by Quadros 2010). Symptoms of poor cobalamin status in humans are neurological and haematological changes such as numbness and megaloblastic anaemia (reviewed by Carmel 2008). If left untreated, cobalamin deficiency can lead to irreversible damage of the nervous system. As a means to decide whether the observed symptoms are due to cobalamin deficiency or not, there is a need to determine the cobalamin status of patients. This has for decades been done by measuring cobalamin in human serum samples. The next section describes the nature of cobalamins in serum and the following sections discuss different methods for measuring cobalamin.

26.2.1 Cobalamins in Human Serum

The total amount of cobalamin present in serum is in the magnitude of 400 pmol/L. Circulating cobalamins are bound to two proteins. Transcobalamin (TC) carries approximately 20% of the cobalamins in serum and the protein haptocorrin (HC) carries the remaining 80% (Nexo and Andersen 1977). In contrast, the amount of unsaturated (apo) transcobalamin is high compared with that of haptocorrin (Figure 26.1).

Only cobalamin present on transcobalamin is available for the cells (for a review, see Quadros 2010), while the function of cobalamin bound to haptocorrin remains to be clarified (for an overview, see Morkbak *et al.* 2007). Transcobalamin carries both of the two co-enzymes, 5′-deoxy adenosyl-cobalamin and methyl-cobalamin, and also other forms of the vitamin that can be converted into the coenzymes such as cyanocobalamin and hydroxo-cobalamin

Analysis of Cobalamins (Vitamin B_{12}) in Human Samples

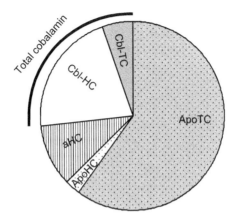

Figure 26.1 Distribution of cobalamins in human serum. A graphic overview of the distribution of the average amounts of cobalamins (Cbl) (400 pmol/L) and analogues (a) (150 pmol/L) on the two transport proteins, transcobalamin (TC, total of 1000 pmol/L) and haptocorrin (HC, total of 500 pmol/L). The fraction of total transport protein without cobalamins or analogues (apo-protein) [grey (TC) and white (HC) areas] adds up to 60% of the total amounts of the cobalamin-binding proteins, showing that there is high capacity to bind any cobalamins released into the blood. Of the total amount of active cobalamins in serum, only approximately 20% is bound to TC (holoTC). This is the cobalamin available for the cells.

(Kerwar et al. 1971; Peirce et al. 1975). The chemical structure of cobalamin and its derivatives in human serum is shown in Figure 26.2.

Haptocorrin carries both the active and inactive forms of the vitamin (Hardlei and Nexo 2009). Methyl-cobalamin is the most abundant active form of the cobalamins bound to haptocorrin (Nexo 1977). The nature of the inactive forms, the so-called analogues, remains to be clarified. Cobalamin-analogues cannot act as coenzymes in the cells. Therefore, care is usually taken to choose a method that measures the active forms of cobalamin without interference from the analogues. The distribution of cobalamins and analogues associated with the two binding proteins is shown in Figure 26.1.

To summarize, human serum contains low concentration of various active and inactive forms of cobalamins bound to carrier proteins. Thus, in the development of any method for the measurement of cobalamin, these aspects must be considered.

26.3 Analysis of Cobalamins in Human Serum

Serum isolated from a blood sample allowed to coagulate is a convenient human sample to collect in a reproducible manner. Therefore, the majority of routine analyses of cobalamins are designed to be used on serum samples. Alternatively, plasma can be used.

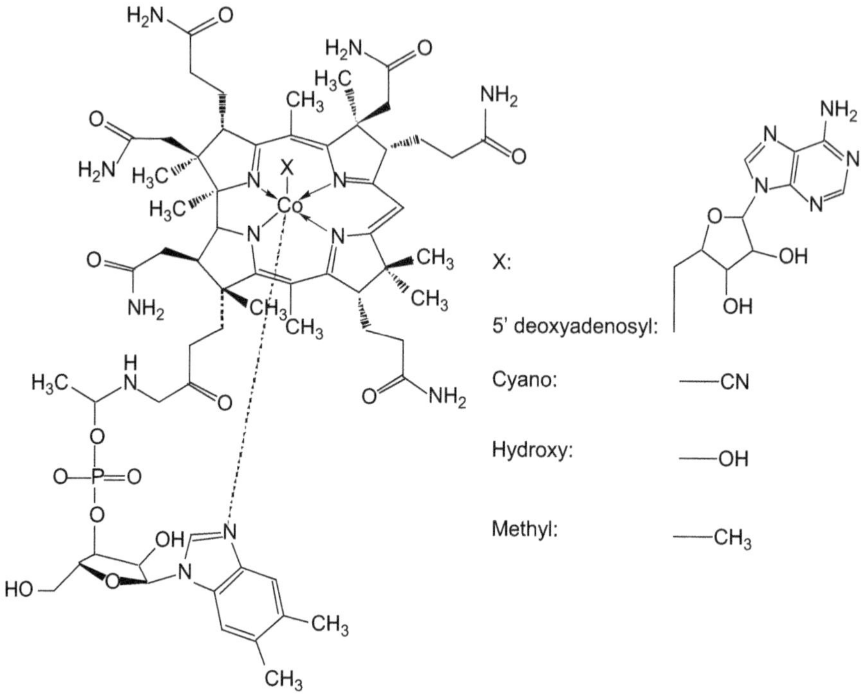

Figure 26.2 Chemical structure of cobalamin. X denotes the groups coupled to the central cobalt atom in cobalamins isolated from human serum. The crystal structure was first solved by Hodgkin and co-workers (Hodgkin et al. 1956).

The general considerations for choosing a method to analyse cobalamin concentration in any sample can be split into different steps as shown in Figure 26.3:

(I) Sample preparation
(II) Considerations concerning calibrators
(III) Choice of analytical method
(IV) Assay validation.

These issues are discussed below.

26.3.1 Sample Preparation

As mentioned above, cobalamins present in serum are bound to two different transport proteins and are present in many molecular forms. To avoid the influence of these in the downstream analytical steps, it is crucial to release cobalamins from the proteins and to transform all cobalamins into a single subtype.

Analysis of Cobalamins (Vitamin B_{12}) in Human Samples

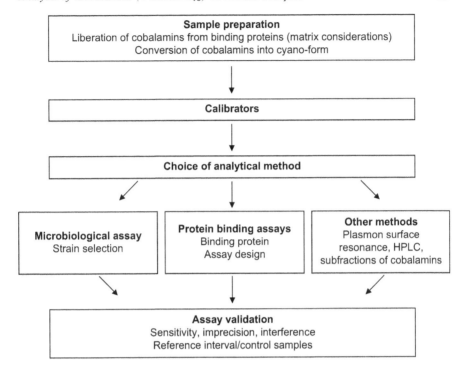

Figure 26.3 Flow chart for measurement of cobalamin. These steps are necessary to establish a cobalamin analysis. See the main text for details on each aspect.

26.3.1.1 *Extraction of Cobalamins from Protein Matrix*

Protein denaturation was initially accomplished by autoclaving the sample (Raven *et al.* 1968) or by boiling in acetate buffer (Tibbling 1969). In both cases, denaturation is followed by a centrifugation step to remove the denatured protein debris. Autoclaving has been the method of choice for microbiological determination of cobalamin, while acid extraction was the preferred method in relation to protein binding assays until the introduction of the alternative no-boil alkaline extraction procedure.

In the 1980s, analysis of cobalamin on automatic platforms was introduced in clinical biochemistry. In relation to this, manufacturers introduced the no-boil methods involving alkaline protein denaturation in combination with addition of KCN (Higgins and Wu 1983). The advantages compared to the boiling principle are no evaporation, no need to cool, and ease in automation. The no-boil concept is now universally used on automated platforms available today.

The use of extreme pH and/or temperature for extraction of cobalamins has been extensively debated in the literature. An important matter of debate has been whether these rather harsh conditions can be employed without destroying the cobalamin to be analysed. For some time, it was challenged whether the so-called analogues were indeed formed during the extraction procedure (Gimsing and Beck 1989). A more gentle extraction of cobalamins can be achieved by proteolytic degradation and this method has proven very useful in combination with immunoprecipitation of the cobalamin binding proteins (Hardlei and Nexo 2009). Using this method we finally showed that analogues do not represent an artefact produced during the extraction procedure, but that they are naturally occurring and are bound to the protein haptocorrin. We do not recommend the use of proteolytic degradation occurring protease inhibitors.

26.3.1.2 Conversion of Serum Cobalamins into one Molecular Form

As mentioned above, the pool of cobalamins in a serum sample consists of several variants. The methyl- and 5′-adenosyl-cobalamins are light-sensitive and are converted into aquo-cobalamin when exposed to light. Aquo-cobalamin readily forms complexes with the side chain of the histidine amino acid; the aquo-group is replaced from the cobalt atom by the histidine side chain (Pratt 1972) and the cobalamin is associated with the denatured protein pellet. Addition of cyanide (*e.g.* 20 µmol/L KCN; Raven *et al.* 1968) prevents this phenomenon as the cyanide ion rapidly displaces other ligands on the cobalt atom (X on Figure 26.2). Therefore, cyanide is added during protein denaturation to prevent precipitation of cobalamin with the protein debris. Simultaneously, addition of cyanide converts all cobalamins into cyanocobalamin. This ensures that the sample contains the same form of the vitamin as the form present in the calibrator of the assay.

26.3.2 Calibrators

For most analytical processes, a measured signal of the sample is compared with the signals of calibrators in order to calculate the concentration present in the sample. An ideal calibrator contains a known concentration of the analyte and is by all means comparable to the sample to be analysed. As serum contains various protein bound cobalamins and cobalamin analogues in a very complex matrix, it is not possible to use an ideal calibrator when analysing cobalamin in serum. A suitable approximation is a solution of cyanocobalamin prepared as described in Box 26.1. For the final dilution of the calibrator, a buffer comparable to the one used for dilution of the samples to be analysed should be employed.

Box 26.1 Stock Solution of Cobalamin

The stock solution is to be used for preparation of calibrators of cyanocobalamin. A similar principle can be used for other forms of cobalamin.

- Dissolve approximately 3.5 mg cyanocobalamin in 100 mL H_2O. Mix well.
- Prepare 18 tubes each containing exactly 1 mL and add exactly 1 mL 0.2 M KCN. Mix well and incubate for 1 h at room temperature to

> ensure that only dicyanocobalamin is present. Leave the samples at room temperature until analysed.
> - Measure the extinction at 369 nm (E_{max}) in a 1 cm cuvette. Measure six samples on three consecutive days. Discard the samples once they have been analysed.
> - Calculate the concentration of dicyanocobalamin in the solution employing the equation (Pratt 1972): conc $\frac{E_{369}}{30400}$ mol/L and calculate the mean and coefficient of variation (CV) for the 18 determinations. CV should be below 3%.
> - The concentration to be assigned to your stock solution of cyanocobalamin is twice that of the one calculated for dicyanocobalamin.
> - Aliquot your stock solution and store at 4 °C. It is stable for years.
> - To prepare your calibrators, dilute accordingly. If the final concentration is below 10 µmol/L, use of a buffer including cobalamin-free albumin (0.1 mg/mL) is recommended.

26.3.3 Analytical Principles

Microbiological assays taking advantage of cobalamin-depending microorganisms were the first assays developed for analysis of cobalamin in human serum samples (Shorb 1947). Today, most assays employed in clinical settings are based on competition between sample cobalamin and a labelled compound for binding to a cobalamin binding protein. These two assays are discussed below. Finally in this chapter more specialized methods are briefly described.

26.3.3.1 Microbiological Assay

Microbiological assays were developed in the late 1940s for use on human samples (Shorb 1947). Several microorganisms have been used including *Lactobacillus leichmannii*, *Lactobacillus delbrueckii*, *Euglena gracilis*, *Ochromonas malhamensis*, and *Escherichia coli* (summarized in Nexo and Olesen 1982). The growth is proportional to the amount of cobalamin present. The growth of the microorganism is most commonly measured by turbidimetry (Girdwood 1954) but measurement of CO_2-production has also been explored (Chen *et al.* 1977). The choice of bacterial strain is crucial for the specificity of the assay. The assay must measure all active cobalamins but not measure the inactive cobalamin-analogues. This was the case for some of the strains used initially, thereby giving rise to falsely too high values. Another important issue is that antibiotics or cytotoxic drugs present in a human serum sample to be analysed can inhibit the growth. The result is an underestimation of the cobalamin concentrations.

Today, a commercial kit based on the bacteria *L. delbrueckii subsp. Lactis* is available (ImmunDiagnostik 2010; Obeid and Herrmann 2006); see also Table 26.1. The bacteria are pre-coated onto microtitre plates and a set of calibrators are incubated on individual wells the plate as are the extracted samples to be analysed. The growth of the microorganism is measured by turbidimetry using an enzyme-linked immunosorbent assay (ELISA) reader. We judge the assay to be suitable especially in a laboratory accustomed to microbiological analysis, but obviously also to be hampered by the limitations of any microbiological assay (see above). It should also be noted that in-house microbiological assays have been employed in recent years in relation to various large clinical studies (see, for example, (Valente *et al.* 2011).

26.3.3.2 Protein Binding Assay

Protein binding assays for cobalamin analysis were first suggested in the 1960s (Rothenberg 1961). The principle is based on two main steps:

(1) Cobalamins extracted from the sample matrix (see Section 26.3.1) are allowed to bind to a cobalamin binding protein in competition with labelled cobalamin.
(2) The fraction of protein-bound cobalamin is measured.

Critical points are the choice of the cobalamin binding protein and the choice of method to separate the non-bound cobalamin from the protein-bound fraction.

26.3.3.2.1 Binding Protein. Over the years, three soluble cobalamin binding proteins (transcobalamin, haptocorrin and intrinsic factor) have been used and the origin of the protein has spanned human, pig, dog, chicken and fish (Kolhouse *et al.* 1978). In the late 1970s, Kolhouse and coworkers convincingly demonstrated the importance of using a binding protein with a well-defined specificity for cobalamin not cross-reacting with cobalamin analogues (Kolhouse *et al.* 1978).

If the binding protein is of the haptocorrin type, both cobalamin and cobalamin analogues are recognized. Before this discovery, no special attention was paid to the type of binding protein employed and numerous binders of unknown specificity were used both for research purposes and in routine assays of cobalamin. The preparations of porcine intrinsic factor employed in those days contained high amounts of haptocorrin. Current assays employing a haptocorrin-like protein are warranted only if the goal is to measure the sum of cobalamin and cobalamin analogues (Hardlei and Nexo 2009).

To our knowledge, all commercially available assays today employ intrinsic factor as the binding protein. The challenge with this is that the most commonly used source of intrinsic factor is the porcine stomach, a source that contains a substantial amount of haptocorrin in addition to intrinsic factor. Because of this, care is needed to remove haptocorrin prior to using porcine gastric intrinsic factor. It seems obvious that, in order to standardize assays for cobalamin across the analytical platforms, the use of recombinant intrinsic factor is relevant. Recombinant human intrinsic factor is available and has

Analysis of Cobalamins (Vitamin B_{12}) in Human Samples

been shown suitable for measurement of cobalamin (Bor *et al.* 2003), but has yet to be introduced in commercial assays.

26.3.3.2.2 Labelled Cobalamin or Labelled Binding Protein. The first protein binding assays for cobalamin employed radiolabelled cobalamin and measured the amount of radioactive cobalamin bound when this label was added alone or in combination with calibrators or the sample to be tested (Rothenberg 1961). The amount of bound label decreased as the concentration of cobalamin in the calibrator or sample increased. As an alternative to radioactive labels, the commercial companies have developed derivatives of cobalamin still able to bind to the binding protein but at the same time coupled to a property such as fluorescence (Abbott 2007; Bayer Diagnostics 2008). Another approach is to in-solubilize cobalamin and measure the amount of labelled binding protein not trapped on the in-solubilized cobalamin (Roche, 2008).

26.3.3.2.3 Separation of Free and Bound Label. Once the labelled and sample cobalamin has reacted with the binding protein, a method to separate free and protein bound cobalamin is required. At first, the separation step used activated charcoal in order to trap surplus free cobalamin (Lau *et al.* 1965), but today's method employs solid-phase separation (Figure 26.4) (Abbott 2007;

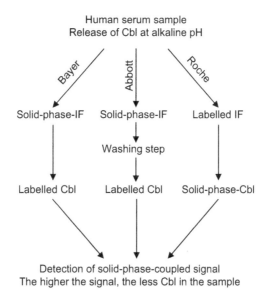

Figure 26.4 Automatic measurement of cobalamin (Cbl) in human serum samples on platforms: design of the methods. All three methods employ protein binding assays using intrinsic factor (IF) but differ in the detailed assay design. See the main text for details. The figure is published with permission from *Clinical Chemistry Laboratory Medicine* (Lildballe *et al.* 2009).

Bayer Diagnostics 2008; Roche 2008). One way to do this is to couple the binding protein covalently to solid-phase particles in the form of paramagnetic beads. Another approach is to use biotinylated cobalamin as the competitive agent; after binding to the protein, biotinylated cobalamin is precipitated onto streptavidin linked to solid-phase particles. In either case, the protein–cobalamin equilibrium may be disturbed in washing steps, and leakage from the solid phase may also be an issue to consider especially if these analytical principles are used in in-house assays.

26.3.3.2.4 Three Automated Platforms Reacts Differently Towards Changes in Sample Type. This section describes the design of assays for cobalamin from the three main players in automated cobalamin analysis on human serum (Abbott, Bayer Diagnostics and Roche) and demonstrates how these assays react if employed for the measurement of cobalamin in human milk. Human milk contains up to 200-fold more haptocorrin than serum and most of it is unsaturated with cobalamin. If haptocorrin is insufficiently denatured, it can interfere in the assays (Lildballe *et al.* 2009).

All three assays use extraction at high pH and include cyanide. All assays use chemiluminescence for the final detection and all three assays employ intrinsic factor as the binding protein. As outlined in Figure 26.4, the designs of the assays differ with regard to the washing procedures and whether the chemiluminescence is linked to cobalamin or to the binding protein.

In the Bayer Diagnostics analyser, cobalamin in the sample mixed with labelled cobalamin competes for binding with in-solubilized intrinsic factor (Bayer Diagnostics 2008). The presence of active haptocorrin in the sample will give rise to a spurious result that is too high for cobalamin since part of the labelled cobalamin will bind to haptocorrin rather than to the intrinsic factor.

In the Abbott analyser, solid-phase intrinsic factor and non-denatured haptocorrin in the sample to be analysed compete for sample cobalamin (Abbott 2007). Prior to the addition of labelled cobalamin, the sample is washed and haptocorrin—as well as the part of sample cobalamin bound to the haptocorrin—is removed. Consequently, the content of cobalamin present in the sample will be underestimated.

In the Roche analyser, the label is on intrinsic factor and the solid-phase is linked to cobalamin (Roche 2008). Unsaturated sample haptocorrin and intrinsic factor compete for binding to sample cobalamin and thereby less samples cobalamin is bound to intrinsic factor. Following this, solid-phase cobalamin is added and the amount of solid-phase–cobalamin–intrinsic factor complex is measured. Provided that enough solid-phase cobalamin is present, the amount of cobalamin present in the sample will also be underestimated by this method.

We discovered these analytical problems when seeking to analyse the content of cobalamin in human milk (Lildballe *et al.* 2009). However, we encountered the same problem for serum samples with a high content of unsaturated haptocorrin (Lildballe *et al.* 2011). Once such a problem is realized, it is possible to circumvent it. We did so by developing a method whereby apo-haptocorrin is removed prior to further analysis of the content of cobalamin in human milk (Lildballe *et al.* 2009).

26.3.3.3 Other Methods for Cobalamin Analysis

Besides the two major analytical principles described above, other methods have been applied. Some of these methods also aim at measuring specific subgroups of the serum cobalamins.

26.3.3.3.1 Plasmon Surface Resonance. Based on the principle of plasmon surface resonance, Biacore Life Sciences has developed a kit assay optimized for cobalamin analysis in food items (Biacore 2005). The recommended extraction method is autoclaving in acid followed by filtration or centrifugation. As the extraction is carried out manually before analysis on the instrument, we expect that other extraction methods may work equally well (see Section 26.3.1).

The extracted cobalamins are mixed with excess cobalamin-binding protein of unknown origin and are allowed to bind. The reaction mixture is applied to a cobalamin-precoated microchip. The amount of unsaturated cobalamin-binding protein that binds to the chip is detected and is inversely proportional to the cobalamin concentration in the sample. The specificity of the method depends on the binding protein employed. If intrinsic factor is used as binding protein, only cobalamin will be quantified. If a haptocorrin-like protein is employed, the sum of cobalamin and analogues is measured.

26.3.3.3.2 HoloTC: the Cell-available Fraction of Serum Cobalamins. In humans, only the fraction of cobalamin on the transport protein transcobalamin is available for cellular uptake. Therefore, recent diagnostic tools have been designed to analyse this component, which is named holoTC. Increasing evidence suggests that holoTC is a slightly better diagnostic marker of cobalamin status than total serum cobalamin (Fedosov 2010). When transcobalamin binds cobalamin, it induces a conformational change in the protein structure. It has therefore been possible to produce antibodies specific for holoTC without cross-reactivity towards unsaturated transcobalamin or towards the other cobalamin binding protein in serum, haptocorrin (Orning *et al.* 2006). Measurement of holoTC has been applied in commercially available ELISA and radioimmunoassays (RIA) (Loikas *et al.* 2003; Orning *et al.* 2006).

An alternative strategy has been applied in our laboratory (Nexo *et al.* 2002). The principle is to initially remove all unsaturated cobalamin binding proteins in the sample by absorption to a solid-phase cobalamin surface. Measurement of transcobalamin in this pre-treated sample by ELISA reflects the amount of holoTC in the sample.

26.3.3.3.3 Determining the Subtypes of Cobalamin. Based on differences in retention times between different forms of cobalamins, reversed-phase high-performance liquid chromatography (HPLC) has been used to separate the active subtypes of cobalamin and also to separate cobalamin from cobalamin analogues (Jacobsen et al. 1982).

The detection system in HPLC is often based on ultraviolet (UV) light measurement of the effluent. Since the concentration of cobalamins in human samples is very low, our laboratory has developed a sophisticated method to determine low-level concentrations of HPLC-separated cobalamins (Hardlei and Nexo 2009). After immunoprecipitation of haptocorrin or transcobalamin followed by a proteolytic extraction and reversed-phase HPLC in darkness (to avoid conversion of methyl- and adenosyl-cobalamin into hydroxo-cobalamin), the fractions are vacuum-dried, re-dissolved in buffer and the cobalamin in the fractions is measured either by employing a sensitive cobalamin specific assay or an assay recognizing also the cobalamin analogues capable of measuring levels as low as 4 pmol/L.

26.3.3.3.4 Mass Spectrometry. Different varieties of mass spectrometry have for several years been used to identify the chemical nature of cobalamins produced from microorganisms (Kumudha et al. 2010). The sensitivity of mass spectrometry is still insufficient for measurement of cobalamins in the picomolar range and this method is therefore of limited interest for analysis of human serum. If the sensitivity is improved, mass spectrometry is predicted to be a very strong tool for identification of the varieties of cobalamins and cobalamin analogues present in human serum.

26.3.4 Choice of Method

As apparent from the description above, a number of different assays are available for the measurement of cobalamin. Some are commercially available and some are in-house methods. In order to make an educated choice concerning the method to be used, a number of issues need to be considered. Many of these are linked to the standard characteristics of the assays concerning sensitivity, specificity imprecision, practicability and costs. The number of samples to be analysed per year as well as the availability of specialized equipment will also highly influence the choice of method.

We have compared the assay validation information available for four commercial kits based on three different analytical principles:

- microbiological assays by ImmunDiagnostik (ImmunDiagnostik 2010);
- automatic and manual protein binding assay by Roche and MP Biomedicals, respectively (MP Biomedicals 2007; Roche 2008);
- plasmon surface resonance by Biacore (Biacore 2005).

The comparative data are shown in Table 26.1.

Table 26.1 Comparison of four different commercial available methods for analysis of total serum cobalamin. The four methods represent different principles for analysis of cobalamin in human serum.

	Microbiological assay (immunDiagnostik 2010)	Automatic protein binding assay (Abbott 2007)[a]	Manual protein binding assay (MP Biomedicals 2007)	Surface plasmon resonance (Biacore 2005)
Calibration range (pM)	0–40[b]	0–1476	0–1480	60–2950
Sample pre-treatment	Manual Boiling in acidic buffer, cooling, centrifugation, dilution	Automatic No-boil denaturation in alkaline/cyanide solution	Manual No-boil denaturation in alkaline solution, manual addition of tracer and binding protein	Manual Autoclaving in acidic buffer, cooling, filtration
Lower detection limit (pM)	4.4	62	55	44
Designed for, sample type	Serum (human and veterinarian)	Serum, EDTA–plasma	Serum, EDTA–plasma, whole blood hemolysate	Food items (e.g. milk powder, vitamin tablets)
Imprecision	5–8%	2.7–5.6%	4.2–12.3%	not reported
Reported reference interval, human serum (pmol/L)	18–500	138–652	118–716	141–489
Special equipment requested	ELISA reader at 610–630 nm	Architect system	Gamma counter with possibility to distinguish the isotopes Co^{57} and I^{125}	Plasmon surface resonance instrument

[a] We chose the method developed by Abbott as an example of automatic protein binding assays for analysis of serum cobalamin. Minor variations occur for other automatic platforms.
[b] The calibration range requires that the final dilution of human serum samples is 25 times.

26.3.4.1 Use of Control Samples

In order to follow the imprecision and the accuracy of the analytical procedure over time, control samples must be included in each analysis. The control samples should be representative of the samples to be analysed and have concentrations comparable to those of the samples. Thus, for measurement of cobalamin in serum one should use a pool of human serum with concentrations of cobalamin in the low, intermediate and high level of the measuring range. Several programmes exist for monitoring the continuous quality of analysis (Thorpe *et al.* 2007) and are not discussed here.

26.4 Concluding Remarks

Measurement of cobalamin in serum was introduced more than 60 years ago and, for clinical use, a number of assays are currently available on automatic platforms. All these assays have been well-validated but, due to minor differences in the assays, the results obtained by different methods are not directly comparable. Because of this, it is not possible to use a common reference interval across methods. In addition, comparison of the results from the measurement of cobalamin in food items and on human samples is currently hampered by a lack of a 'golden standard' for the measurement of cobalamin both in biological samples and in food items. This issue should be solved in order to relate an accurate estimate of the intake of cobalamin to the level of cobalamin in human serum and in order to compare the results obtained by different laboratories.

Today, measurement of cobalamin in serum is the most widely used blood test to judge the cobalamin status in humans. Currently, it is debated whether measurement of the fraction of cobalamin available for the cells, holoTC, will prove a more useful marker. The first assays suitable for measurement of this active fraction of circulating cobalamin have already been introduced in the clinical setting.

In the years to come it is to be expected that definitive methods for measurement of cobalamin will be developed, most likely employing mass spectrometry. It is also expected that measurement of holoTC may partly replace measurement of cobalamin as a means to determine cobalamin deficiency. At the same time there will remain a need for new methods in order to understand the flow of cobalamin from food items to the end station inside the cells of the body.

Summary Points

- This chapter gives an overview of cobalamin analysis in human serum samples.
- Measurement of cobalamin (also known as vitamin B_{12}) is used in relation to nutritional issues and as a research tool.
- Measurement of total serum cobalamin is used to explore the presence of cobalamin deficiency.

- Serum cobalamin consists of picomolar amounts of several forms of the vitamin bound to transcobalamin or haptocorrin.
- Measurement of serum cobalamin includes release of cobalamin from its binding proteins transcobalamin and haptocorrin, conversion of the various forms of cobalamin into one form of the vitamin, and quantification employing microbiological or protein binding assays.
- False levels of cobalamin may be encountered if the concentration of cobalamin unsaturated haptocorrin is high as observed in human milk.
- Measurement of holoTC, the part of serum cobalamin available for the cells, may be a better marker of cobalamin deficiency than total serum cobalamin.
- Measurement of cobalamin and analogues bound to haptocorrin is mainly used for research purposes.
- Standardized protein binding assays have been developed for measurement of total cobalamin in serum. Further studies are required to optimize assays of cobalamin in other matrixes than serum and in order to explore the use of measurement of sub-fractions of cobalamin such as holoTC.

Key Facts

Key Facts about Assay Validation

- Measurement of an analyte concentration in human blood samples is generally used to diagnose disease and to distinguish between disease and health.
- An analytical procedure is characterized by its measurement range, imprecision, accuracy, linearity, quality control programme, reference interval and ease of performance.
- A quality control programme is the procedure employed to ensure that the analysis continuously meets the goal for imprecision and accuracy. Several external quality control programmes exist for measurement of cobalamin, *e.g.* United Kingdom National External Quality Assessment Service (www.ukneqas.org.uk).
- The measurement range indicates the concentrations that can be accurately measured. For cobalamin in serum this covers 25–1400 pmol/L.
- Imprecision indicates the variation observed by repeat measurement of the same sample. For cobalamin in serum, the imprecision is typically below 10%.
- Accuracy indicates the ability to measure the true value. The use of a set of calibrators with a traceable concentration is important to ensure accuracy. For cobalamin in serum, calibrators can be prepared from stock solutions of cyanocobalamin where the concentration is determined by spectrophotometry.
- Linearity ensures that samples diluted, *e.g.* 1 + 1 gives half the value of the undiluted sample.

- A reference interval, denoting the 95% central interval obtained for healthy individuals, must be determined.
- The final choice among analyses meeting the validation criteria depends on practical issues including the costs of the assay.

Key Facts about Cobalamin Analysis in Human Samples

- Analysis of cobalamin (also known as vitamin B_{12}) in human serum samples is very common worldwide as a mean to determine the cobalamin status in patients suspected to be cobalamin deficient.
- Currently, no reference method ('gold standard') exists for this analysis.
- Well-validated automated assays have been developed and are currently available.
- When analysing cobalamin in human serum or developing new method a number of issues need to be considered: cobalamin in serum is bound to two different proteins; and it is present in various chemical forms, including inactive forms (the so-called cobalamin analogues).
- Methods for analysis of cobalamin in other sample types other than human serum are less well-validated and it is important to validate them before employing such assays.
- In order to be able to make an accurate comparison between cobalamin intake and cobalamin status of a patient in the future, there is a need to analyse cobalamin present in food items and in human serum employing comparable methodologies.

Definitions of Words and Terms

Anticoagulant: An anticoagulant is a compound that prevents the blood from clotting (coagulating). It is added in the sample tube when collecting plasma as in contrast to serum. In this respect, citrate, ethylenediaminetetraacetic acid (EDTA) and heparin are frequently used chemicals.

Apo and holo-protein: Apo-protein denotes the amount of protein not saturated with its ligand (*e.g.* cobalamin) while holo-protein denotes the part of the protein saturated with its ligand.

Assay validation: Assay validation is the procedures carried out prior to the introduction of a new analysis. It should include information on the sensitivity (the lowest concentration that can be detected), the specificity (the degree of cross-reactivity with other components than the analyte in question), the reproducibility (how close are repeat measurement to each other), linearity (can dilution of a sample be performed without a change in the final result obtained) and special precautions concerning sample collection.

Cobalamin analogues: Cobalamin analogues are cobalamin-related molecules not able to act as coenzymes for the cobalamin dependent human enzymes. In serum, both cobalamin and cobalamin analogues are present. The protein transcobalamin can only bind active forms of cobalamin, whereas the protein haptocorrin can bind both cobalamins and analogues.

Cobalamin-binding proteins: The term 'cobalamin-binding proteins' most often refers to the soluble-binding proteins involved in the absorption and distribution of cobalamin. Intrinsic factor ensures the intestinal uptake of the vitamin and recognizes only the biologically active forms of the vitamin. The same holds for transcobalamin that transports cobalamin after absorption in the intestine and to all the cells of the body. The third binding protein, haptocorrin, has a widespread distribution and is present both in the circulation and in most human secretions such as milk. Haptocorrin binds both cobalamin and cobalamin analogues, but the physiological function of the protein remains to be clarified. Over time, the protein has carried a variety of names including transcobalamin I and III, and R-proteins. All three binding proteins have been employed for measurement of cobalamin, but today all commercially available assays use intrinsic factor.

Cobalamin deficiency: Cobalamin deficiency is a common condition caused by an insufficient intake of cobalamin or an inability to absorb the vitamin. If left untreated, the condition may result in irreversible neurological damage and eventually death caused by a severe anaemia. Measurement of serum cobalamin is the most commonly employed test in order to confirm or rule out the presence of cobalamin deficiency.

Holotranscobalamin (holoTC): HoloTC is cobalamin bound to transcobalamin and this is the available cobalamin for the cells. There is now a debate as to whether measurement of holoTC is a better marker for identifying individuals with cobalamin deficiency than total serum cobalamin.

Microbiological assay: A microbiological assay depends on a microorganism that will grow only if the analyte to be measured is added. Most often such assays are very sensitive (picomolar range) but are hampered by the difficulty in standardization and the influence of interfering substances such as bacteriostatic agents.

Plasma and serum: Plasma and serum are obtained from a blood sample. Serum remains if the blood is allowed to clot, while plasma is obtained if coagulation is avoided by addition of an anticoagulation factor such as EDTA or heparin. Both plasma and serum can be employed for the analysis of cobalamin.

Protein binding assay: A protein binding assay is based on the competition between the analyte in the sample and an added labelled analyte for the binding to a limited amount of binding protein. In a traditional radioimmunoassay, the binding protein is an antibody. In assays employed for measurement of cobalamin, the binding protein is by and large intrinsic factor. The advantage of a protein binding assay is that it can detect picomolar concentrations. The limitations are the specificity of the binding protein and the existence of a suitable method for separation of free and bound analyte.

Reference interval: Reference interval, also known as reference range, is calculated as the central 95% fraction of values measured for a healthy population. This means that 2.5% of the reference population has cobalamin levels below the reference interval and 2.5% has cobalamin levels above the reference interval. The reference interval for cobalamin depends on the method employed, but is approximately 125–600 pmol/L.

List of Abbreviations

Apo	Cbl unsaturated protein
Cbl	cobalamin
CV	coefficient of variation
EDTA	ethylenediaminetetraacetic acid
ELISA	enzyme-linked immunosorbent assay
HC	haptocorrin
HCy	homocysteine
Holo	Cbl saturated protein
HPLC	high-performance liquid chromatography
MMA	methylmalonic acid
RIA	radioimmunoassay
TC	transcobalamin
tHCy	total homocysteine
UV	ultraviolet

References

Abbott Diagnostic Division, 2007. B_{12} analysis using Architect system. Abbott.

Banerjee, R., 2006. B_{12} trafficking in mammals: A for coenzyme escort service. *ACS Chemical Biology.* 1: 149–159.

Bayer Diagnostics, 2008. ADVIA Centaur reference and assay manual.

Biacore, 2005. Qflex kit vitamin B_{12} PI handbook.

Bor, M.V., Fedosov, S.N., Laursen, N.B., and Nexo, E., 2003. Recombinant human intrinsic factor expressed in plants is suitable for use in measurement of vitamin B_{12}. *Clinical Chemistry.* 49: 2081–2083.

Carmel, R., 2008. How I treat cobalamin (vitamin B_{12}) deficiency. *Blood.* 112: 2214–2221.

Chanarin, I., 1969. *The Megaloblastic Anaemias.* Blackwell Scientific, Oxford, UK, pp. 1007.

Chen, M.F., McIntyre, P.A., and Wagner, H.N., Jr., 1977. A radiometric microbiologic method for vitamin B_{12} assay. *Journal of Nuclear Medicine.* 18: 388–393.

Fedosov, S.N., 2010. Metabolic signs of vitamin B(12) deficiency in humans: computational model and its implications for diagnostics. *Metabolism.* 59: 1124–1138.

Gimsing, P., and Beck, W.S., 1989, Cobalamin analogues in plasma. An *in vitro* phenomenon?. *Scandinavian Journal of Clinical Laboratory Investigation.* 194(Suppl): 37–40.

Girdwood, R.H., 1954, Rapid estimation of the serum vitamin B_{12} level by a microbiological method. *British Medical Journal.* 2(4894): 954–956.

Hardlei, T.F., and Nexo, E., 2009. A new principle for measurement of cobalamin and corrinoids, used for studies of cobalamin analogs on serum haptocorrin. *Clinical Chemistry.* 55: 1002–1010.

Higgins, T., and Wu, A., 1983. Differences in vitamin B_{12} results as measured with boil and no-boil kits. *Clinical Chemistry*. 29: 587–588.

Hodgkin, D.C., Kamper, J. Mackay, M., Pickworth, J., Trueblood, K.N., and White, J.G., 1956. Structure of vitamin B_{12}. *Nature*. 178(4524): 289–198.

ImmunDiagnostik, 2010. ID-Vit Vitamin B12 Manual. Microbiological test kit for the determination of vitamin B_{12} in serum using a *Lactobacillus delbrueckii* subsp. lactis coated microtitre plate.

Jacobsen, D.W., Green, R., Quadros, E.V., and Montejano, Y.D., 1982. Rapid analysis of cobalamin coenzymes and related corrinoid analogs by high-performance liquid chromatography: *Analytical Biochemistry*. 120: 394–403.

Kerwar, S.S., Spears, C., McAuslan, B., and Weissbach, H., 1971. Studies on vitamin B_{12} metabolism in HeLa cells. *Archives of Biochemistry and Biophysics*. 142: 231–237.

Kolhouse, J.F., Kondo, H., Allen, N.C., Podell, E., and Allen, R.H., 1978. Cobalamin analogues are present in human plasma and can mask cobalamin deficiency because current radioisotope dilution assays are not specific for true cobalamin. *The New England Journal of Medicine*. 299: 785–792.

Kumudha, A., Kumar, S.S., Thakur, M.S., Ravishankar, G.A., and Sarada, R., 2010. Purification, identification, and characterization of methylcobalamin from *Spirulina platensis*: *Journal of Agricultural Food Chemistry*. 58: 9925–9930.

Lanska, D.J., 2010. Chapter 30: historical aspects of the major neurological vitamin deficiency disorders: the water-soluble B vitamins. *Handbook of Clinical Neurology*. 95: 445–476.

Lau, K.S., Gottlieb, C., Wasserman, L.R., and Herbert, V., 1965. Measurement of serum vitamin B_{12} level using radioisotope dilution and coated charcoal. *Blood*. 26: 202–214.

Lildballe, D.L., Hardlei, T.F., Allen, L.H., and Nexo, E., 2009. High concentrations of haptocorrin interfere with routine measurement of cobalamins in human serum and milk. A problem and its solution. *Clinical Chemistry and Laboratory Medicine*. 47: 182–187.

Lildballe, D.L., Nguyen, K.Q., Poulsen, S.S., Nielsen, H.O., and Nexo, E., 2011. Haptocorrin as marker of disease progression in fibrolamellar hepatocellular carcinoma. *European Journal of Surgical Oncology*. 37: 72–79.

Loikas, S., Lopponen, M., Suominen, P., Moller, J., Irjala, K., Isoaho, R., Kivela, S.L., Koskinen, P., and Pelliniemi, T.T., 2003. RIA for serum holotranscobalamin: method evaluation in the clinical laboratory and reference interval. *Clinical Chemistry*. 49: 455–462.

Minot, G.R., and Murphy, W.P., 1926. Treatment of pernicious anemia by a special diet. *The Journal of the American Medical Association*. 250: 3328–3335.

Morkbak, A.L., Poulsen, S.S., and Nexo, E., 2007. Haptocorrin in humans. *Clinical Chemistry and Laboratory Medicine*. 45: 1751–1759.

MP Biomedicals, 2007. SimulTRAC-SNB radioassay kit vitamin B_{12} [^{57}Co]/folate [^{125}I].

Nexo, E., 1977. Characterization of the cobalamins attached to transcobalamin I and transcobalamin II in human plasma. *Scandinavian Journal of Haematology*. 18: 358–360.

Nexo, E., and Andersen, J., 1977. Unsaturated and cobalamin saturated transcobalamin I and II in normal human plasma. *Scandinavian Journal of Clinical Laboratory Investigation.* 37: 723–728.

Nexo, E., and Olesen, H., 1982. Quantification of cobalamins in human serum. In: Dolphin, D. (ed.). *B12.* John Wiley, Chichester, UK, and New York, USA, pp. 88–100.

Nexo, E., Christensen, A.L., Hvas, A.M., Petersen, T.E., and Fedosov, S.N., 2002. Quantification of holo-transcobalamin, a marker of vitamin B_{12} deficiency. *Clinical Chemistry.* 48: 561–562.

Obeid, R., and Herrmann, W., 2006. Mechanisms of homocysteine neurotoxicity in neurodegenerative diseases with special reference to dementia. *FEBS Letters.* 580: 2994–3005.

Orning, L., Rian, A., Campbell, A., Brady, J., Fedosov, S.N., Bramlage, B., Thompson, K., and Quadros, E.V., 2006. Characterization of a monoclonal antibody with specificity for holo-transcobalamin. *Nutrition & Metabolism.* 3: 3.

Peirce, K., Abe, T., and Cooper, B.A., 1975. Incorporation and metabolic conversion of cyanocobalamin by Ehrlich ascites carcinoma cells *in vitro* and *in vivo*. *Biochimica et Biophysica Acta.* 381: 348–358.

Pratt, J.M., 1972. *Inorganic Chemistry of Vitamin B_{12}.* Academic Press, London, UK. 348 pp.

Quadros, E.V., 2010. Advances in the understanding of cobalamin assimilation and metabolism. *British Journal of Haematology.* 148: 195–204.

Raven, J.L., Robson, M.B., Walker, P.L., and Barkhan, P., 1968. The effect of cyanide, serum and other factors on the assay of vitamin B_{12} by a radioisotope method using $^{57}CoB12$, intrinsic factor and coated charcoal. *Guy's Hospital Reports.* 117: 89–109.

Roche, 2008. Vitamin B_{12} analysis using Elecys and Cobas analytical instruments. Roche instrument guide.

Rothenberg, S.P., 1961. Assay of serum vitamin B12 concentration using Co57-B12 and intrinsic factor: *Proceedings of the Society for Experimental and Biological Medicine.* 108: 45–48.

Shorb, M.S., 1947. Unidentified essential growth factors for *Lactobacillus lactis* found in refined liver extracts and in certain natural materials. *Journal of Bacteriology.* 53: 669.

Thorpe, S.J., Heath, A., Blackmore, S., Lee, A., Hamilton, M., O'Broin, S., Nelson, B.C., and Pfeiffer, C., 2007, International standard for serum vitamin B(12) and serum folate: international collaborative study to evaluate a batch of lyophilised serum for B(12) and folate content. *Clinical Chemistry and Laboratory Medicine.* 45: 380–386.

Tibbling, G., 1969. A method for determination of vitamin B_{12} in serum by radioassay. *Clinical Chimica Acta.* 23: 209–218.

Valente, E., Scott, J.M., Ueland, P.M., Cunningham, C., Casey, M., and Molloy, A.M., 2011. Diagnostic accuracy of holotranscobalamin, methylmalonic acid, serum cobalamin, and other indicators of tissue vitamin B_{12} status in the elderly. *Clinical Chemistry.* 57: 856–863.

CHAPTER 27

Assay by Biosensor and Chemiluminescence for Vitamin B_{12}

M.S. THAKUR* AND L. SAGAYA SELVA KUMAR

Fermentation Technology and Bioengineering Department, Central Food Technological Research Institute, Mysore-570020, Karnataka, India
*Email: msthakur@cftri.res.in

27.1 Introduction to Biosensors

In brief, a biosensor is an instrument or device containing a sensing element of biological origin, which is either integrated within or is in intimate contact with a physicochemical transducer designed for quantitative or semiquantitative analysis of analyte (Thakur and Karanth 2003) as shown in Figure 27.1.

The biocomponent is the responsive element which provides specificity and selectivity to the biosensor system. The biological components used in the biosensor construction can be divided into two categories:

- those where the primary sensing event results from catalysis (*e.g.* enzymes, microorganisms, cells, and tissues);
- those that depend on an essentially irreversible binding of the target molecule (*e.g.* antibodies, receptors, nucleic acids).

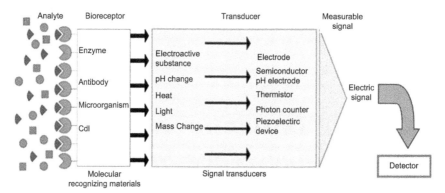

Figure 27.1 Schematic diagram of a biosensor unit. The unit comprises molecular recognizing materials with signal transducers which generate signals amplified and detected using a detector.

Depending upon the physical/chemical change brought about by the biological element, different physicochemical transducers can be used as shown in Figure 27.2. An ideal biosensor should meet the following criteria:

1. *Selectivity*: The device should be highly selective for the target analyte and have no cross-reactivity with moieties having similar chemical structure.
2. *Sensitivity*: The device should be able to measure in the range of interest for a given target analyte with minimum additional steps such as pre-cleaning and pre-concentration of the samples.
3. *Linearity and reproducibility of signal response*: The linear response range of the system should cover the concentration range over which the target analyte is to be measured.
4. *Response time and recovery time*: The sensor response should be rapid so that real-time monitoring of the target analyte can be done efficiently. The recovery time should be short for reusability of the system.
5. *Stability and operating life*: The biological element used for sensing should be active enough for a long time in order to obtain a reproducible response.

27.2 Biosensor-based Assay for Vitamin B_{12}

The analytical technology based on biosensors is an extremely wide field, which impacts on all the major industrial sectors such as pharmaceutical, healthcare, food, agriculture, environment and water. Biosensors offer advantages as an alternative to conventional methods as ideally they overcome issues in the latter in order to analyse vitamins simultaneously. Specific biorecognition of the analyte by a biological system or biologically derived elements used in a biosensor can potentially reduce extensive and time-consuming sample preparation.

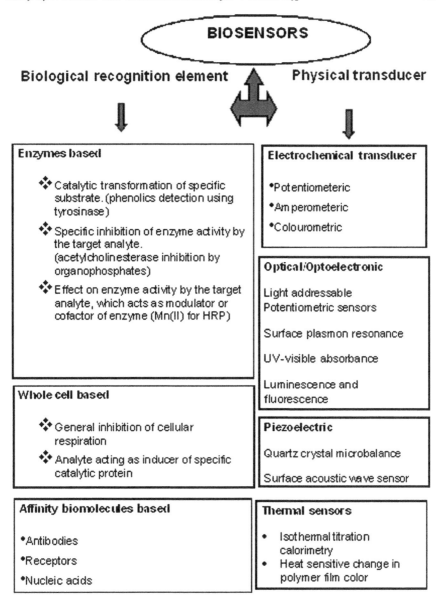

Figure 27.2 Classifications of biosensors: schematic representation with various combinations of physical and biological elements.

The range of biosensors for the analysis of vitamins that use enzymes as the biorecognition elements is rather restricted. However, affinity biosensors have opened a new horizon in the field of vitamin analysis with surface plasmon resonance (SPR) biosensors at the forefront. Commercialization of optical SPR analysers has led to development and routine applications for analysis of vitamin B_{12} using antibodies and binding protein (Indyk et al. 2002).

A rapid electrochemical microbioassay for vitamin B_{12} was reported in 1987 using *Escherichia coli* 215 and an oxygen electrode (Karube et al. 1987). The sample solution is injected into the microbial electrode system; increased consumption of oxygen by the microorganisms leads to a decrease in current which is directly proportional to amount of vitamin B_{12}.

In 2002, Indyk and co-workers evaluated vitamin B_{12} in a range of foods such as milks, infant formulas, meat and liver using SPR technology with performance parameters including a quantitation range of 0.08–2.4 ng/mL with recoveries of 89–106% (Indyk et al. 2002). The analytical technique was biosensor based utilizing bimolecular interaction.

A non-labelled inhibition protein binding assay using non-intrinsic R-protein is exploited under SPR technology. A wide selection of milk, infant formulas, meat and liver were evaluated for their vitamin B_{12} content with studied interactions between various cobalamins and vitamin B_{12} binding protein; the results indicated the importance of biomolecular interaction for biosensor design (Cannon et al. 2002). The same assay format was used in the SPR Biocore platform: the first assay consisted of a competitive binding assay format where cyanocobalamin-β-(5-aminopentylamide) was immobilized on an activated carboxymethyl (CM) 5 chip. A cobalamin analogue was mixed with the cobalamin binding protein solution, and apparent solution equilibrium dissociation constants were determined. The second assay was established to study the comparative kinetic rate constants of cyanocobalamin (CN-Cbl) and CN-Cbl-β-(5-aminopentylamide) binding to recombinant human transcobalamin II immobilized on an activated CM 5 chip.

Similar SPR technology has been widely used for quantification of water-soluble vitamins in milk-based infant formulas with the key component of Biacore biosensor technology patented on sensor chip surfaces coated with CM groups (Yaling et al. 2008). Thus, when molecules in the test solution bind to a target immobilized on the chip surface the molecular mass increases, resulting in an increase in the local refractive index. Thus, real-time changes in mass on the sensor chip surface are optically detected and reported in the form of sensor gram with a detection limit of $2\,\mu g/100\,g$ of infant food formulas.

Commercial biosensors are available in several forms such as autoanalysers, manual laboratory instruments and portable (handheld) devices. One biosensor system, Biacore Q (Biacore, Uppsala, Sweden), is based on optical detection— the SPR technique for vitamin B_{12} analysis.

27.3 Chemiluminescence

Chemiluminescence (CL) is the generation of electromagnetic radiation as light through the release of energy from a chemical reaction. While the light, in principle, is emitted in the ultraviolet, visible or infrared region, those reactions emitting visible light are the most common and useful for several applications in biosensor development. Various types of CL are mentioned below.

27.3.1 Enzyme-based Enhanced CL

The enzyme, horseradish peroxidase (HRP), has grown as a potential enzyme for enhanced chemiluminescence reaction (usually through labelling an immunoglobulin that specifically recognizes the molecule). The enzyme catalyses the oxidation of luminol to 3-aminophthalate via several intermediates. which on further oxidation by hydrogen peroxide (H_2O_2) enhances the light up to 1000-fold. The enhancement of light emission is called enhanced chemiluminescence (ECL). Enhanced CL allows detection of minute quantities of a biomolecule and, using this method, it is possible to detect the pesticides at the picogram range (Chouhan et al. 2006).

CL-based analytical methods have several advantages such as rapidity, specificity and cost-effectiveness; one minor disadvantage is that they require semi-skilled personnel. There is no requirement for an excitation source as in fluorescence and phosphorescence, a monochromator (often not even a filter), or the use of radioactive or hazardous chemicals. Thus, CL is advantageous for routine analysis.

27.3.2 Chemiluminescence of Luminol

The CL of luminol (5-amino-2,3-dihydrophthalazine-1,4-dione) was first described by Albrecht in 1928 (Albrecht 1928). Since then, luminol and its derivatives have been extensively studied and applied for analytical applications (Yamaguchi et al. 2002).

In an aqueous medium, luminol reacts with a potent oxidizing agent in the presence of a catalyst (generally a metal, a metal-containing compound or an enzyme) in alkaline solution to yield 3-aminophthalate in an excited electronic state, which returns to ground state with the production of light. Metal ions such as Co(II), Cu(II) and Fe(II) or enzymes like HRP and microperoxidase catalyse the luminol reaction efficiently (Figure 27.3). The CL reaction can be summarized as follows:

$$HRP + H_2O \rightarrow Complex\ 1 + H_2O$$

$$Complex\ 1 + luminol\ (LH_2) \rightarrow luminol\ radical\ (LH^-)$$

$$Complex\ 2 + luminol\ radical\ (LH^-) \rightarrow HRP + Oxidized\ luminol + light$$

Luminol solutions are sensitive to light and the presence of metal cations; typically they are only stable for 8–12 h. Luminol has been shown to be thermally unstable, and so luminol and its solutions should be protected from high temperatures (Stott and Kricka 1987).

27.3.3 Chemiluminescence of the Dioxetane, CDP-*Star*

The dioxetane, disodium 2-chloro-5-(4-methoxyspiro {1,2-dioxetane-3,2′-(5′-chloro)tricyclo[3.3.1.13,7]decan}-4-yl) phenyl phosphate (CDP-*Star*), is a

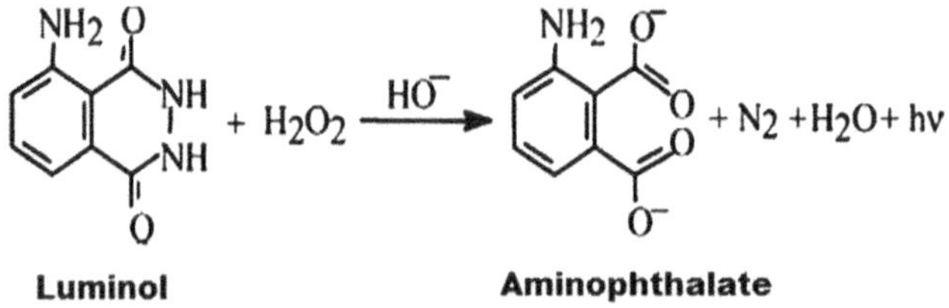

Figure 27.3 Chemiluminescence of luminol. The chemiluminescence technique is based on generating free oxygen radicals (HO$^{\bullet}$, $O_2^{\bullet-}$, O_2, ROO$^{\bullet}$) in a luminescence system. The luminol (LH$_2$)–hydrogen peroxide system in alkaline solution leads to aminophtalated anion in an excited electronic state, a compound that when de-excited emits a quantum of light (a band with maximum at 430 nm).

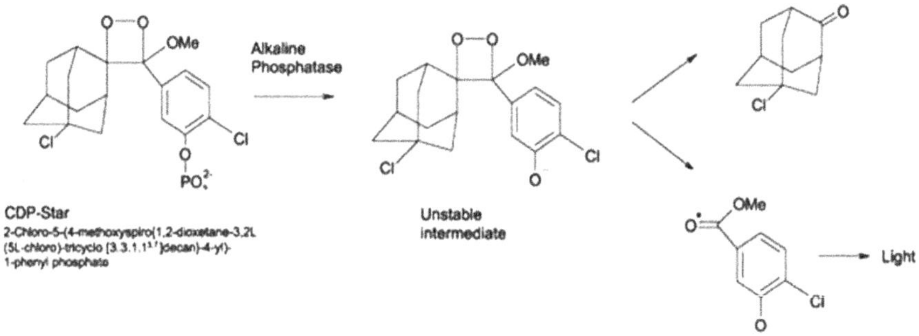

Figure 27.4 Chemiluminescence of CDP-*Star*. Enzymatic dephosphorylation of dioxetane by alkaline phosphatase leads to the formation of the metastable dioxetanephenolate anion which decomposes and emits light at 466 nm (adapted from CDP-*Star* product sheet available at https://e-labdoc.roche.com/LFR_PublicDocs/ras/11759051001_en_08.pdf).

chemiluminescent substrate for alkaline phosphatase (ALP). In the presence of ALP, CDP-*Star* is dephosphorylated and produces an unstable intermediate which further decomposes, emitting a glow of light in direct proportion to the amount of ALP (Figure 27.4).

27.3.4 Chemiluminescence of Acridans

Another substrate for HRP is a stable trifluorophenyl substituted acridan known as Lumigen PS-3, which emits photons at 430 nm. The detection of HRP activity by PS-3 is based on a CL process involving the enzymatic generation of an intermediate acridinium ester. In the presence of hydrogen peroxide and *p*-iodophenol as enhancer, the action of the acridan substrate with the HRP label generates thousands of acridinium ester intermediates ([AE]+) per minute, which react with hydrogen peroxide, at alkaline pH, to produce

Assay by Biosensor and Chemiluminescence for Vitamin B_{12}

N-methylacridone in the singlet excited state. This compound comes back to the ground state, producing a sustained high-intensity CL, according to the scheme:

$$PS\text{-}3 \xrightarrow{HRP,\,phenol,\,H_2O_2} [AE]^+$$

$$[AE]^+ \xrightarrow{Buffer,\,H_2O_2} [N\text{-methylacridone}]^* + trifluorophenol$$

$$[N\text{-methylacridone}]^* \longrightarrow N\text{-methylacridone} + light$$

27.3.5 Electrochemiluminescence

Electrochemiluminescence (ECL) is a form of CL produced during electrochemical reactions (Knight *et al.* 1999). The advantages of CL are retained, but the ECL reaction allows the timing and position of the light-emitting reaction to be controlled. By controlling the time, light emission can be delayed until events such as immuno- or enzyme-catalysed reactions have taken place.

Luminol is one of the earliest known synthetic compounds exhibiting CL. In alkaline solution, the luminol ECL reaction is thought to proceed *via* electrooxidation of the luminol anion, followed by reaction with either H_2O_2 (the primary reaction) or dissolved oxygen, leading to formation of the excited 3-aminophthalate species, which can subsequently emit light (Haapakka and Kankare 1982).

27.4 Chemiluminescence Analysis of Vitamin B_{12}

CL is a process that results in the emission of light due to chemical reactions. Traditionally, analysis used only a small number of chemicals, such as luminol–hydrogen peroxide and acridinium ester. The luminol system was most used due to its sensitivity and ease of handling.

27.4.1 Acridium Ester Based Chemiluminescence

Acridium ester based chemiluminescence system was applied in food analysis with an acridium ester labelled vitamin B_{12} derivative and the most specific vitamin B_{12} binding protein, intrinsic factor (IF) (Emi *et al.* 2006; Watanabe *et al.* 1998). The system was used in food analysis for shellfish and spirulina, and compared well with the microbiological method.

27.4.2 Luminol Based Chemiluminescence

The luminol system was first used for the analysis of vitamin B_{12} in 1977 (Terry *et al.* 1977) when the method was applied to pharmaceutical tablets containing vitamin B_{12}, though vitamin B_{12} as such did not enhance the light due to its cobalt being in an oxidation state of 3+ (Sheehan and Hercules 1977). As it was tightly bound by the porphyrin linkages, harsh treatment was necessary to remove the cobalt from the vitamin B_{12} structure; Terry *et al.* utilized an in-line Jones reductor, which reduced vitamin B_{12} to Co^{2+}, which in turn acted as a strong catalyst. Oxygen constitutes interference and was removed from all solutions. Diffusion of oxygen into flow lines was eliminated by use of a dry box filled with nitrogen. The limit of detection was 2×10^{-9}M vitamin B_{12} and the linear range was 10^{-9} to 10^{-6} M.

Qin *et al.* (1997) proposed the same protocol with a slight modification including a flow injection analysis (FIA) system. The protocol was based on the catalytic effect of cobalt(II), liberated from vitamin B_{12} by acidification, on the CL reaction between luminol, immobilized electrostatically on an anion exchange column, and hydrogen peroxide generated electrochemically in-line via a negatively biased electrode from dissolved oxygen in the flow cell. The sensor responded linearly to vitamin B_{12} concentration in the 1.0×10^{-3} to 10 mg/l range, and the detection limit was 3.5×10^{-4} mg/l vitamin B_{12}. The system was stable for over 500 determinations and has been applied successfully to the determination of vitamin B_{12} in pharmaceutical preparations. This CL system was commercialized by Chiron Diagnostics (Erst Walpole, MA, USA) into a fully automated CL analyser for the determination of vitamin B_{12} in serum and has been used in clinical applications.

In 2002, Sato and co-workers replaced IF with vitamin B_{12}-targeting *Lactobacillus helveticus* B-1 in the vitamin B_{12} assay by CL method due to its cost and non-availability (Sato *et al.* 2002). *Lactobacillus helveticus* B-1 is assumed to have a vitamin B_{12} targeting (or binding) site on its cells and binds vitamin B_{12} instantly and quantitatively. This reaction is specific to complete vitamin B_{12} compounds, cobalamins, and was used for a vitamin B_{12} assay method by CL. The calibration graph was linear from 0.1 to 10 ng/mL.

In 2008, Akbay and co-workers produced a modified protocol for the catalytic effect of cobalt from vitamin B_{12} in basic medium using flow injection analysis (Akbay and Gok 2008), in which the increment in CL intensity is proportional to the concentration of vitamin B_{12} in the range 8.68–86.9 ng/mL, with a detection limit of 0.89 ng/mL. Their paper presents a CL flow system for the determination of vitamin B_{12}. The method was based on the catalytic effect of Co(II), which was released from vitamin B_{12} by an acidification process, on the CL reaction between luminol and hydrogen peroxide. As noted above, cobalt in vitamin B_{12} is at an oxidation state of 3+ but it is Co(II) that catalyses the luminol CL reaction. Thus, the reduction of vitamin B_{12} was necessary and the effectiveness of vitamin B_{12} acidification was optimized to produce Co(II) in the medium. Akbay and Gok (2008) investigated the complex formation between Co(II) and triethanolamine, which was immobilized on a strongly acidic cation exchanger resin. The colours of the immobilization samples on the resin showed that acidified vitamin B_{12} samples contained Co(II) which was then followed by CL analysis.The method developed by Akbay and Gok has been successfully applied to the determination of vitamin B_{12} in pharmaceutical injections.

27.4.3 Chemiluminescence Work at CFTRI

Even after many modifications on chemiluminescence methods, vitamin B_{12} sensitivity was lacking. Therefore in CFTRI, a highly sensitive CL method was proposed for vitamin B_{12} using carbonate enhancement effect. Vitamin B_{12} was detected up to 5 pg/ml. Experimental parameters were optimized, including luminol concentration, urea–H_2O_2 concentration, effect of pH, and sequence of addition of reactants for obtaining maximum CL, which was not explored previously. The limit of detection was 5 pg/mL and the linear range was 10 pg/mL to 1 mg/mL with a regression coefficient of $R^2 = 0.9998$ (Sagaya *et al.* 2009).

The importance of these experimental parameters and the carbonate enhancement effect is discussed below based on the knowledge of the mechanism of oxidation of luminol and the decomposition of urea–H_2O_2 in the presence of vitamin B_{12}. The crucial step in analysis of vitamin B_{12} is the acidification procedure to remove cobalt from vitamin B_{12}. In the present work, 100 mg/mL vitamin B_{12} was acidified with 2mL of 9N nitric acid and heating until the solution was evaporated completely, followed by the addition of 2 mL of 5.5 N hydrochloric acid to remove any traces of nitric acid (Figure 27.5). The reaction mixture was further heated at 95 °C for 2min and cooled; the residue was re-dissolved with a dilution of 5 pg/mL to 1 mg/mL in bicarbonate buffer (pH 10.7).

The overall novelty of CFTRI work was on the pH factor of the analysis which enhanced the CL strongly and increased the sensitivity. The effect of pH was studied on CL. Maximum CL signals occurred above pH 10 and were relatively the same over a pH range of 10.5–12. Therefore, pH 10.7 was chosen for further experimental purposes. It has been reported that, at neutral pH, hydroxyl radicals were very active and attacked the carbon sites on the aromatic ring of luminol, producing various species that do not lead to sufficient production of hydroperoxide (Buxton *et al.* 1988; Neta *et al.* 1988), and thus fewer CL signals are observed. In the presence of bicarbonate buffer (pH 10.7, 0.2M), the hydroxyl radicals are converted into carbonate radicals that react almost selectively with luminol radicals, leading to the formation of luminol hydroperoxide (Merenyi *et al.* 1984, 1990) and the production of more CL signals. Thus, bicarbonate enhances CL intensity by selectively increasing the steady state concentration of the luminol hydroperoxide. The carbonate buffer catalyses the decomposition of hydrogen peroxide to yield hydroxyl radicals, which are converted to carbonate radicals in the presence of carbonate. When compared with hydroxyl radicals, bicarbonate radicals react slowly with luminol radicals and there is a delay in the signals that can be recorded and reproduced.The crucial factor in CL was the addition sequence, which had not been explored previously. It was found that the addition sequence of the reactants during the CL reaction had a considerable impact on CL when comparing a control and a sample containing vitamin B_{12}.

Various addition sequences of reactants were performed out to achieve maximum CL. In sequence 1 (luminol + urea–H_2O_2 + vitamin B_{12}) and sequence 2 (urea–H_2O_2 + luminol + vitamin B_{12}), the reaction between luminol

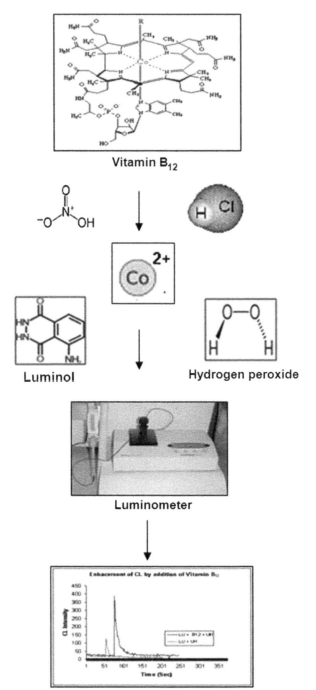

Figure 27.5 Flow chart of chemiluminescence response for vitamin B_{12} detection. Acid hydrolysis of vitamin B_{12} resulting in free of cobalt helps in the enhancement of CL.

and urea–H_2O_2 yielded instantaneous light emission. However, the catalytic role of vitamin B_{12} in the reaction was not significant, making it difficult to distinguish between the signals obtained from the two reactions. The unstable nature of H_2O_2 in sequence 4 (urea–H_2O_2 + vitamin B_{12} + luminol) made it unsuitable for performing the reaction due to self-oxidation resulting in less CL signal production.

Based on these observations, the addition of urea–H_2O_2 as the final reactant was deemed more appropriate than adding it at the initial stage of the reaction. It was also observed that, with the addition of luminol as a second or third reactant in sequence 6 (vitamin B_{12} + luminol + urea–H_2O_2) and sequence 3 (vitamin B_{12} + urea–H_2O_2 + luminol), the added urea–H_2O_2 would react instantly with luminol, emitting signals that were difficult to distinguish, whereas in sequence 5 (luminol + vitamin B_{12} + urea–H_2O_2), the catalytic effect of vitamin B_{12} and complete reaction of free radicals with luminol is prominent. It can be concluded that the addition of reactants as in sequence 5 (luminol + vitamin B_{12} + urea–H_2O_2) facilitated the reaction to occur in a controlled manner and maximum CL was observed. Therefore, further experiments were carried out using the luminol, vitamin B_{12} and urea–H_2O_2 sequence. The experimental set up is as shown in Scheme 27.1.

The influence of luminol on CL was examined with varying concentrations of luminol from 1 mM to 1 mM; 5 mM was found to be the optimal concentration

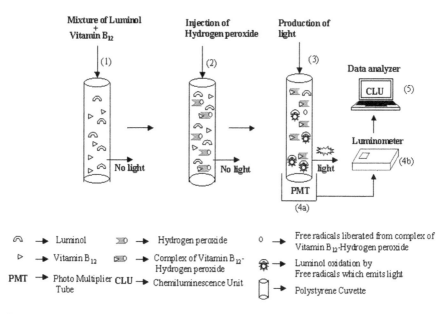

Scheme 27.1 Design of CL-based detection systems for vitamin B_{12} analysis. The addition of sequential addition of luminol (5 µM, 50 µL), vitamin B_{12} (10 pg/mL, 100 µL) and urea–H_2O_2 (0.1 mM, 100 µL) at 60-s intervals produces CL signals that were analysed using a photomultiplier tube-based luminometer (Sagaya et al. 2009).
Reprinted with permission from Elsevier.

due to higher CL signal production. It also offered higher sensitivity than the other concentrations (Sagaya et al. 2009). It was observed that, with increasing concentrations, the decrease in light intensity may be due to vitamin B_{12} complexing with luminol or the aminophthalate product of luminol oxidation, thereby reducing the availability of vitamin B_{12} for catalysis with H_2O_2 to form carbonate radicals. Therefore, the use of 50 mL of 5 mM luminol was chosen for further experiments.

Among different concentrations of urea–H_2O_2 from 0.025 to 0.1 mM, 0.1 mM was found to be optimal due to a higher production of CL signals and offered higher sensitivity than any other concentration (Sagaya et al. 2009). At higher concentrations, undesirable background CL signals were produced due to greater formation of free radicals that interfered in the enhancement of the CL signal due to the presence of vitamin B_{12}. Therefore, 100 mL of 0.1 mM urea–H_2O_2 was chosen for further experiments. The proposed CL method was successfully applied to the determination of vitamin B_{12} in vitamin B_{12} tablets, multivitamin capsules, and vitamin B_{12} injections as shown in Table 27.1.

A possible mechanism for vitamin B_{12} detection using CL is proposed (Scheme 27.2). The mechanism involves the formation of peroxobicarbonate

Table 27.1 Analysis of vitamin B_{12} in pharmaceutical preparations using proposed CL method in comparison with atomic absorption spectroscopy (AAS). Values are the averages of three measurements ($\pm$ SD) [amount of labelled vitamin B_{12} in injection (mg/mL) and in tablets or capsules (μg/tablet or capsule)] (Sagaya et al. 2009). Reprinted with permission from the Elsevier.

Pharmaceutical samples	Amount labelled	Amount found	
		Proposed CL method	AAS
Injection	0.334	0.323 ($\pm$ 0.01)	0.322 ($\pm$ 0.04)
Tablet	15	14.8 ($\pm$ 0.01)	14.8 ($\pm$ 0.1)
Capsule	15	14.9 ($\pm$ 0.01)	14.8 ($\pm$ 0.1)

$$H_2O_2 \longrightarrow HO_2^- + H^+ \quad (1)$$
$$H_2O_2 + HCO_3^- \longrightarrow HOOCO_3^- + H^+ \quad (2)$$
$$VitB_{12}CO\,(II) + HOOCO_3^- \longrightarrow [Vit\,B_{12}CO\,(II).HOOCO_3] \quad (3)$$
$$[VitB_{12}CO\,(II).HOOCO_3^-] \longrightarrow Vit\,B_{12}CO\,(III) + HOOCO_3^{*-} \quad (4)$$
$$Vit\,B_{12}CO\,(II) + HOOCO_3^{*-} \longrightarrow Vit\,B_{12}CO\,(III) + CO_3^{*-} + HO_2^- \quad (5)$$
$$\text{Carbonate radicals } (CO_3^{*-}) + \text{Luminol Radicals} \longrightarrow \text{Luminol Hydroperoxide (fully Oxidized)} + N_2 + H_2O + \text{light } (h\nu) \quad (6)$$

Scheme 27.2 Possible mechanism for HCO_3-H_2O_2-CO_2^+ chemiluminescence involving carbon dioxide dimer. Details of the schematic principle are described in the text (Sagaya et al. 2009).
Reprinted with permission from Elsevier.

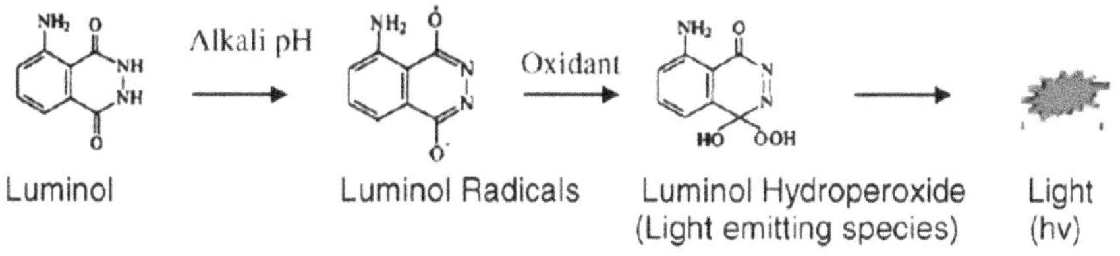

Luminol → (Alkali pH) Luminol Radicals → (Oxidant) Luminol Hydroperoxide (Light emitting species) → Light (hv)

Scheme 27.3 Formation of light-emitting species from luminol using carbonate free radicals generated at alkaline pH leads to the formation of luminol hydroperoxide which becomes excited and decay results in emission of light (Sagaya et al. 2009).
Reprinted with permission from Elsevier.

Table 27.2 Different types of CL for vitamin B_{12} analysis in pharmaceutical samples. Sensitivity, applications and references are given for various types of CL for vitamin B_{12}.

SL no.	Types of vitamin B_{12} CL method	Sensitivity	Application in pharmaceutical samples	References
1.	Luminol CL	2 ng/mL	Vitamin pills	Terry et al. 1977
2.	Luminol FIA of CL	20 ng/mL	Vitamin B_{12} injections	Zhou et al. 1991
3.	CL receptor assay	20 ng/mL	Serum	Styia et al. 1994
4.	Electrochemically generated CL using Flow.	350 pg/mL	Pharmaceutical preparations	Qin et al. 1997
5.	Flow injection analysis CL	8.68 ng/mL	Pharmaceutical injections	Akbay and Gok 2008
6.	Luminol CL	5 pg/mL	Pharmaceutical products	Sagaya et al. 2009
7.	FIA CL using charge-coupled device (CCD) photodetector	9.3 ng/mL	Pharmaceutical products	Jose et al. 2011

radicals when H_2O_2 comes into contact with carbonate (steps 1 and 2), leading to complex formation with cobalt as the important intermediate leading to luminescence. After formation of the complex, peroxocarbonate needs to lose two electrons to be converted to carbonate radicals (steps 3–6). These free radicals (CO_3) help in the oxidation of luminol radicals to yield luminol hydroperoxide, the excited state for emission of light (Scheme 27.3).

A similar kind of observation was made in the catalysis of H_2O_2 with cobalt(III) (Wangila and Jordan 2006). The luminol CL reaction system most often used is luminal + H_2O_2 + OH, which has been widely used for luminal based biochemical analysis. A novel method for activating H_2O_2 by using bicarbonate ion was described by Drago and co-workers (Drago et al. 1998), involving the formation of peroxymonocarbonate ion HCO_4, a previously unrecognized reactive oxygen species seldom used for analytical purposes.

Table 27.3 Vitamin B_{12} contents determined by chemiluminescence methods in foods. All values obtained represent mean ± SEM ($n = 3$).

SL no.	Types of CL for vitamin B_{12}	Applications in foods	Sensitivity ($\mu g/100 g$)	References
1.	Automated CL analyser (acridinium ester-labelled B_{12} derivative)	Beef muscle (raw) Beef liver (raw) Chicken muscle (raw) Chicken egg yolk (fresh) Tuna lean meat (raw) Short-necked clam (raw) Ordinary liquid milk	1.53 ± 0.04 62.51 ± 1.28 1.73 ± 0.02 2.71 ± 0.04 3.69 ± 0.08 17.57 ± 0.19 0.22 ± 0.01	Watanabe et al. 1998
2.	Intrinsic factor-CL method	Edible shellfish, oyster, clam and mussel	46.77 ± 2.38 31.12 ± 1.26 57.76 ± 0.36 13.02 ± 0.06	Watanabe et al. 2001
3.	Intrinsic factor-CL method	Spirulina tablets	6.2–17.4	Watanabe et al. 1999
4.	Intrinsic factor-CL method	Chlorella tablets	200.9–211.6	Kittaka et al. 2002
5.	Vitamin B_{12}-targeting site on *Lactobacillus helveticus B-1* to vitamin B_{12} assay by CL method	Oyster and sardine	75.9 ± 2.9 39.4 ± 2.4	Sato et al. 2002
6.	Intrinsic factor-CL method	Edible Cyanobacterium *Aphanizomenonflos-aquae*	32.3 ± 2.0	Emi et al. 2006
7.	Luminol CL	*Spirulina plantesis*	35.7 ± 2.0	Kumudha et al. 2010

More recently, a similar kind of mechanism was reported in the CL reaction of cobalt using carbonate/bicarbonate (Liang et al. 2008).

27.5 Concluding Remarks

Various CL-based analyses of vitamin B_{12} are listed in Tables 27.2 and 27.3 for pharmaceutical and food samples. Compared with conventional analytical techniques, such as microbiological assay, high-performance liquid chromatography (HPLC), atomic absorption spectroscopy (AAS), radioisotope isotope assay, and fluorimetric detection, the current proposed CL method was more sensitive with a detection limit of 5 pg/mL.

Summary Points

- This chapter focuses on the analysis of vitamin B_{12} using biosensors and chemiluminescence (CL).
- Luminol, CDP-*Star* and acridan esters are commonly used CL reagents.
- A sensitive and simple CL method has been proposed for vitamin B_{12} detection at a detection limit of 5 pg/mL level by using the carbonate

enhancement effect and optimum sequence of addition of reactants, alkaline pH, and luminol and H_2O_2 concentration.
- This method takes prominent advantage of simplicity, reduced reagent consumption, improved sensitivity and analytical efficiency, and it is easy to perform. It has been successfully applied to the determination of vitamin B_{12} in pharmaceutical preparations.
- Conventional methods were time-consuming, tedious, less safe, less sensitive and expensive. Drawbacks that need to be overcome are the stability and analysis in mass as well as the practicality in handling. Nevertheless, biosensors and CL offer a unique solution to vitamin B_{12} analysis in terms of specificity and time saving.
- Biosensors offer a unique solution to vitamin B_{12} analysis in terms of specificity and time-saving commercial biosensors are available in several forms such as autoanalysers, manual laboratory instruments and portable (handheld) devices. One biosensor system, Biacore Q, is based on optical detection—the surface plasmon resonance (SPR) technique for vitamin B_{12} analysis.

Key Facts

- **Methods to assess activity of vitamin B_{12}.** These include macrocytic anaemia improvements, homocysteine (Hcy) reduction and methyl malonic acid (MMA) reduction.
- **Macrocytic anaemia improvements.** Vitamin B_{12} deficiency can cause macrocytic anaemia (large red blood cells). However, folate deficiency can do the same. If an individual with macrocytic anaemia is known to have adequate folate status and is fed food that is thought to have vitamin B_{12}, and their anaemia improves, it is fairly safe to say that the food has some vitamin B_{12} activity for red blood cells. Unfortunately, it is not known for sure whether vitamin B_{12} that is active for blood cells is always active for nerve cells.
- **Homocysteine (Hcy) reduction.** Vitamin B_{12} deficiency can cause elevated Hcy levels in the blood. However, folate and vitamin B_6 deficiency can do the same. Reducing Hcy levels might give a good idea about the vitamin B_{12} activity of a food, but because folate and vitamin B_6 can confound the results, it might not be the safest test for determining vitamin B_{12} activity.
- **Methyl malonic acid (MMA) reduction.** Because the biochemical pathway that reduces MMA levels in the blood uses only vitamin B_{12}, lowering MMA levels is a test specific for vitamin B_{12} activity. Although it is not known for certain, it is likely that this biochemical pathway is an integral part of the function of vitamin B_{12} in nerve tissue. Thus, if food lowers MMA levels, it can be assumed to provide full vitamin B_{12} activity. Similarly, the bioavailability of vitamin B_{12} in lyophilized purple liver was assessed by MMA excretion to find total vitamin B_{12} and vitamin B_{12} analogue contents in the liver.

- **Biosensor market.** A biosensor as defined by IUPAC is a device able to provide quantitative analytical information by employing a biological recognition element linked to a transduction system. The first biosensor was designed by Clark in 1962 based on an amperometric biosensor that immobilized glucose oxidase on an oxygen electrode. A global strategic business report published by Global Industry Analysts Inc. stated that the global market for biosensors grew from $6.1 billion in 2004 to $8.2 billion in 2009, and would continue at an average annual growth rate of 6.3% in 2010 and market is expected to reach US$12 Billion by 2015, according to a new report by Global Industry Analysts, Inc. (Global Industry Analysts, 2012) (Global Industry Analysts Inc. new report, 2012, http://www.prweb.com/releases/biosensors/medical_biosensors/prweb8067456.htm)
- **Reports on biosensors.** From 1984 to 1990, there were about 3000 scientific publications and 200 patents on biosensors. The same number of publications (~3300 articles) but almost double the patent activity (400 patents) appeared between 1991 and 1997. However, the explosion of nanobiotechnology between 1998 and 2004 generated over 6000 articles and 1100 patents issued/pending. Indications are that the number of publications and patents on biosensors is steadily increasing.

Definition of Words and Terms

Chemiluminescence. Defined as the emission of light from a molecule as a result of the transition from an electron in an electronically excited state (produced by the energy liberated from a chemical reaction) to the ground state.

Fluorescence. Fluorescence occurs if the photon emission (loss of energy) occurs between states of the same spin state (*e.g.* S_1 to S_0).

Jones reductor. A Jones reductor is a device which can be used to reduce a metal ion in aqueous solution to a very low oxidation state.

Phosphorescence. Phosphorescence occurs if the photon emission (loss of energy) occurs between states of the different spin state (*e.g.* T_1 to S_0).

Surface plasmon resonance. When a metal absorbs light of a resonant wavelength, it causes the electron cloud to vibrate, dissipating the energy. This process usually occurs at the surface of a material (as metals are not usually transparent to light) and is therefore called surface plasmon resonance. Plasmon is the name for the oscillation of the electron cloud.

List of Abbreviations

AAS	atomic absorption spectroscopy
ALP	alkaline phosphatase
CDP-star	disodium 2-chloro-5-(4-methoxyspiro {1,2-dioxetane-3,2'-(5'-chloro) tricycle [3.3.1.13,7]decan}-4-yl) phenyl phosphate
CM	carboxymethyl
CL	chemiluminescence

CN-Cbl   cyanocobalamin
CFTRI   Central Food Technological Research Institute
ECL   electrochemiluminescence
FIA   flow injection analysis
Hcy   homocysteine
HPLC   high-performance liquid chromatography
HRP   horseradish peroxidase
IF   intrinsic factor
MMA   methyl malonic acid
SPR   surface plasmon resonance
urea–H_2O_2 urea–hydrogen peroxide

References

Albrecht, H.O., 1928. Chemiluminescence of aminophthalichydrazide. *Zeitschrift fur Physikalische Chemie*. 136: 321–330.

Akbay, N., and Gok, E., 2008. Determination of vitamin B_{12} using a chemiluminescence flow system. *Journal of Analytical Chemistry*. 63: 1073–1077.

Buxton, G.V., Greenstock, C.L., Helman, W.P., and Ross, A.B., 1988. Critical review of rate constants for reactions of hydrated electrons, hydrogen atoms, and hydroxyl radicals in aqueous solution. *Journal of Physical Chemistry*. 17: 513–886.

Cannon, M.J., Myszka, D.G., Bagnato, J.D., Alpers, F.G., and Grisom, C.B., 2002. Equilibrium and kinetic analyses of the interactions between vitamin B_{12} binding proteins and cobalamins by surface plasmon resonance. *Analytical Biochemistry*. 305: 1–9.

Chouhan, R.S., VivekBabu, K., Kumar, M.A., Neeta, N.S., Thakur, M.S., Amitha Rani, B.E., Pasha, A., Karanth, N.G., and Karanth, N.G.K., 2006. Detection of methyl parathion using immuno-chemiluminescence based image analysis using charge coupled device. *Biosensor and Bioelectronics*. 21: 1264–1271.

Drago, R.S., Frank, K.M., Yang, Y.C., and Wagner, G.W., 1998. Catalytic activation of hydrogen peroxide - A green oxidant system. In: Dorothy, A. Berg (ed.), *Proceedings of the 1997 ERDEC Scientific Conference on Chemical and Biological Defense Research*, US Army Edgewood Research, Development, and Engineering Center, Aberdeen Proving Ground, MD, USA. UNCLASSIFIED Report (AD-A356-165).

Emi, M., Yuri, T., Tomoyuki, N., Florin, B., Hiroshi, I., Tomoyuki, F., Watanabe, F., and Yoshihisa, N., 2006. Purification and characterization of a corrinoid-compound in an edible cyanobacterium *Aphanizomenonflosaquae* as a nutritional supplementary food. *Journal of Agriculture Food Chemistry*. 54: 9604–9607.

Haapakka, K., and Kankare, J.J., 1982. The mechanism of the electrogenerated chemiluminescence of luminol in aqueous alkaline solution. *Analytica Chimica Acta*. 138: 263–275.

Indyk, H.E., Persson, B.S., Caselunghe, M.C.B., Moberg, A., Filonzi, E.L., and Woolard, D.C., 2002. Determination of vitamin B_{12} in milk products and selected foods by optical biosensors protein binding assay: method comparisons. *Journal of AOAC International.* 85: 72–81.

Jose, A.M.P., Luisa, F.G.B., and Nieves, M.S.G., 2011. Flow injection chemiluminescence determination of vitamin B_{12} using on-line UV-persulfatephotooxidation and charge coupled device detection. *Luminescence.* 26: 536–542.

Karube, I., Wang, Y., Tamiya, E., and Kawarai, M., 1987. Microbial electrode sensor for vitamin B_{12}. *Analytica Chimica Acta.* 199: 93–97.

Kittaka-Katsura, H., Fujita, T., Watanabe, F., and Nakano, Y., 2002. Purification and characterization of a corrinoid-compound from chlorella tablets as an algal health food. *Journal of Agriculture Food Chemistry.* 50: 4994–4997.

Knight, A.W., 1999. A review of recent trends in analytical applications of electro generated chemiluminescence. *Trends in Analytical Chemistry.* 18: 47–62.

Kumudha, A., Selva Kumar, L.S., Thakur, M.S., Ravishankar, G.A., and Sarada, R., 2010. Purification, identification, and characterization of methylcobalamin from *Spirulina platensis*. *Journal of Agriculture Food Chemistry.* 58: 9925–9930.

Liang, S., Zhao, L., Zhang, B., and Lin, J., 2008. Experimental studies on the chemiluminescence reaction mechanism of carbonate/bicarbonate and hydrogen peroxide in the presence of cobalt(II). *Journal of Physical Chemistry A.* 112: 618–623.

Merenyi, G., Lind, J., and Eriksen, T.E., 1984. The equilibrium reaction of the luminol radical with oxygen and the one electron reduction potential of 5-aminophthalazine-1,4-dione. *Journal of Physical Chemistry.* 88: 2320–2323.

Merenyi, G., Lind, J., Shen, X., and Eriksen, T.E., 1990. Oxidation potential of luminol: is the autooxidation of singlet organic molecule an outer sphere electron transfer? *Journal of Physical Chemistry.* 94: 748–752.

Neta, P., Huie, R.E., and Ross, A.B., 1988. Rate constants for reactions of inorganic radicals in aqueous solutions. *Journal of Physical Chemistry.* 17: 1027–1284.

Qin, W., Zhang, Z., and Lu, H., 1997. Chemiluminescence flow sensor for the determination of vitamin B_{12}. *Analytica Chimica Acta.* 357: 127–132.

Sagaya Selva Kumar, L., Chouhan, R.S., and Thakur, M.S., 2009. Enhancement of chemiluminescence for vitamin B_{12} analysis. *Analytical Biochemistry.* 388: 312–316.

Sato, K., Muramatsu, K., and Amano, S., 2002. Application of vitamin B_{12}-targeting site on *lactobacillus helveticus* B-1 to vitamin B_{12} assay by chemiluminescence method. *Analytical Biochemistry.* 308: 1–4.

Sheehan, T.L., and Hercules, D.M., 1977. Analytical study of chemiluminescence from the vitamin B_{12}–luminol system. *Analytical Chemistry.* 49: 446–450.

Stott, R.A.W., and Kricka, L.J., 1987. In: Schoelmerich, J. (ed.), *Proceedings of 4th International Symposium on Bioluminescence and Chemiluminescence*, September 8–18 1986. John Wiley & Sons, Chichester, UK, pp. 237–240.

Styia, W., McBride, J.A., and Walker, W.H., 1994. Chemiluminescence receptor assay for measuring vitamin B_{12} in serum evaluated. *Clinical Chemistry*. 40: 537–540.

Terry, L., Sheehan, H., and David M., 1977. Analytical study of chemiluminescence from the vitamin B_{12}-luminol system. *Analytical Chemistry*. 446: 3–49.

Thakur, M.S., and Karanth, N.G., 2003. Research and development on biosensors for food analysis in India. In: Malhotra, B.D., and Turner, A.P.F. (ed.) *Advances in Biosensors: 5*. Elsevier, Amsterdam, The Netherlands, pp. 131–160.

Wangila, W.G., and Jordan, B.R., 2006. Reaction of hydrogen peroxide with hexacobalt(III) perchlorate. *Inorganic Chimica Acta*. 359: 3177–3182.

Watanabe, F., Takenaka, S., Abe, K., Tamura, Y., and Nakano, Y., 1998. Comparison of a microbiological assay and a fully automated chemiluminescent system for the determination of vitamin B_{12} in food. *Journal of Agriculture Food Chemistry*. 46: 1433–1436.

Watanabe, F., Katsura, H., Takenaka, S., Fujita, T., Abe, K., Tamura, Y., Nakatsuka, T., and Nakano, Y., 1999. Pseudovitamin B_{12} is the predominate cobamide of an algal health food, Spirulina tablets. *Journal of Agriculture Food Chemistry*. 47: 4736–4741.

Watanabe, F., Katsura, H., Takenaka, S., Enomoto, T., Miyamoto, E., Nakatsuka, T., and Nakano, Y., 2001. Characterization of vitamin B_{12} compounds from edible shellfish, clam, oyster, and mussel. *International Journal of Food Sciences and Nutrition*. 52: 263–268.

Yaling, G., Fei, G., Sumangala, G., Andrew, C., Qinghai, S., and Mingruo, G., 2008. Quantification of water soluble vitamins in milk based infant formulae using biosensor-based assays. *Food Chemistry*. 110: 769–777.

Yamaguchi, M., Yoshida, H., and Nohta, H., 2002. Luminol-type chemiluminescence derivatization reagents for liquid chromatography and capillary electrophoresis. *Journal of Chromatography A*. 950: 1–19.

Zhou, Y.K., Li. H., and Liu.Y., 1991. Chemiluminescence determination of vitamin B_{12} by a flow-injection method. *Analytica Chimica Acta*. 243: 127–130.

CHAPTER 28

The Diagnostic Value of Measuring Holotranscobalamin (Active Vitamin B_{12}) in Human Serum: A Clinical Biochemistry Viewpoint

FABRIZIA BAMONTI* AND CRISTINA NOVEMBRINO

Dipartimento di Scienze Mediche, Università degli Studi di Milano, Fondazione IRCCS, Ca' Granda Ospedale Maggiore Policlinico, via F. Sforza, 35, 20122 Milan, Italy
*Email: fabrizia.bamonti@unimi.it

28.1 Introduction

Vitamin B_{12} (B_{12}) or cobalamin (Cbl), a micronutrient supplied by meat and dairy products, is essential for mammalian intracellular metabolism, particularly one-carbon groups, and for cell proliferation and differentiation (Herrmann *et al.* 2003a, 2005). Vitamin B_{12} absorption from animal food is a complex process (Herrmann *et al.* 2003a).

Holotranscobalamin (holoTC, or active B_{12}), the cobalamin–transcobalamin II complex released into the portal circulation and recognized by ubiquitous specific receptors, is considered the specific Cbl-saturated transporter, the biologically active form of vitamin B_{12}. Only 6–20% of total circulating vitamin

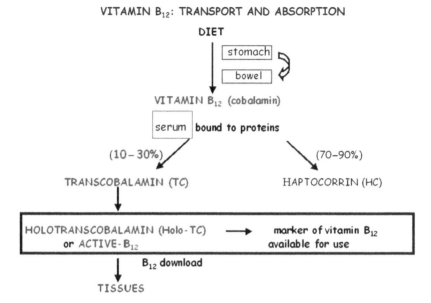

Figure 28.1 Scheme of cobalamin metabolism, transport and absorption after dietary intake in humans. The active vitamin B_{12} fraction is highlighted.

B_{12} is present in holoTC, the remaining cobalamin is bound to transcobalamin (TC) I and III. Figure 28.1 summarizes this complex process.

Final vitamin B_{12} absorption is limited ($<3\,\mu g$ per meal) except when the diet is rich in animal food (Baik and Russel 1999). Vitamin B_{12} low intake of from food may lead to a negative balance and finally to functional deficiency when tissue stores of vitamin B_{12} are depleted. In addition, several factors causing intestinal malabsorption may lead to cobalamin deficiency (Carmel 2011):

- achlorhydria resulting in inability to release cobalamin from animal protein;
- lack of intrinsic factor (IF) due to total gastrectomy or pernicious anaemia (idiopathic atrophy of gastric mucosa in association with antibodies against parietal cells and IF);
- exocrine failure leading to cobalamin malabsorption;
- inability to degrade cobalamin–R-protein complexes;
- high concentrations of bacteria or certain parasites in the small intestine, which can absorb cobalamin;
- resection of distal ileum.

Deficiency from IF–related malabsorption usually progresses inexorably (and was once lethal), whereas non-malabsorptive or slightly malabsorptive deficiency progresses very slowly or not at all (Carmel 2011). Transcobalamin genetic absence or disorders involving circulating transport as well as genetic disorders involving intracellular conversion into coenzyme forms manifest themselves as the typical haematological and neurological diseases caused by vitamin B_{12} deficiency. Haptocorrin (HC) genetic absence is rare but not a serious condition and is usually discovered accidentally (Quadros 2010).

Cobalamin deficiency is relatively common, but the great majority of cases in epidemiologic surveys have not an overt but only a subclinical cobalamin deficiency. Figure 28.2 summarizes the different stages leading to overt cobalamin deficiency.

Clinicians are uncertain about what to do with asymptomatic patients with mild biochemical abnormalities and tend to treat patients with oral vitamin B_{12} without a specific diagnosis; clear cobalamin deficiency, usually discovered shortly after birth as a failure to thrive, requires an aggressive therapy with vitamin B_{12} if long-term and irreversible neurologic damage is to be avoided (Carmel 2011).

The current standard clinical test for vitamin B_{12} deficiency is the measurement of serum total B_{12} (tB_{12}) concentrations. However, a number of subjects, whose tB_{12} concentrations are considered low, show no clinical or biochemical evidence of deficiency. Conversely, there can be neuropsychiatric and metabolic abnormalities when circulating levels of tB_{12} are within the reference interval (Brady *et al.* 2008).

As suggested several years ago by Herbert (1994), and confirmed by studies on vegetarian/vegan populations (Herrmann *et al.* 2003b), and more recently on elderly people (Valente *et al.* 2011), holoTC can be considered the very first marker of vitamin B_{12} deficiency (see Figure 28.2).

For nearly 20 years, holoTC concentrations have been thought to be a better indicator of vitamin B_{12} status, but until a few years ago, holoTC could only be

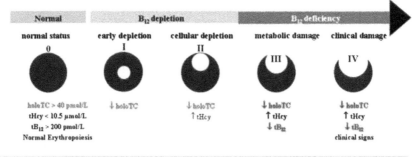

Modified from V. Herbert et al. Am J Clin Nutr 1994

holoTC = serum holotranscobalamin; tHcy = plasma total homocysteine; tB_{12} = serum total vitamin B_{12}

Figure 28.2 Illustration of the different stages leading to overt cobalamin deficiency and the role of the most important cobalamin-related analytes (adapted from Herbert *et al.* 1994).

measured by calculating the difference between tB_{12} (*e.g.* 500 pmol/L) and the B_{12} fraction not attached to transcobalamin (*e.g.* 420 pmol/L). This method of evaluating holoTC levels was clinically unreliable. There are now new reliable and sensitive methods for measuring holoTC, as reported by Ulleland *et al.* (2002).

28.2 HoloTC Analytical Performance

28.2.1 Pre-analytical Phase

28.2.1.1 Blood Sampling and Storage

In general, both plasma and serum can be used to analyse holoTC concentrations. HoloTC, like tB_{12}, is a stable analyte and no special precautions need to be taken when drawing blood samples to measure holoTC concentrations. HoloTC is also stable in storage at $-20\,°C$ or $-70\,°C$, probably for years (Loikas *et al.* 2003).

28.2.1.2 Modulation and Daytime Variation

Serum holoTC concentrations can be influenced by several factors, *i.e.* the amount of cobalamin absorbed, the rates of holoTC hepatic and renal uptake, the production and release of ileal and possibly renal holoTC, tissue cobalamin requirement and other unknown factors. Serum holoTC daytime variations after meals were slight and blood could be drawn from both fasting and non-fasting subjects. However, repeated intakes of B_{12} at high pharmacologic doses (*e.g.* 9 g three times daily) were found to increase holoTC and, to a lesser degree, tB_{12} concentrations (Nexø and Hoffmann-Lücke 2011).

28.2.1.3 Sex, Race and Life Cycle

HoloTC, like tB_{12}, varies to a certain degree according to sex and/or age. However, while Refsum *et al.* (2006) reported only slightly lower concentrations in young women, other authors observed marked age- and sex-related alterations in holoTC and other metabolic analytes (*e.g.* tHcy) (Nexø and Hoffmann-Lücke 2011). During pregnancy, both haptocorrins and tB_{12} decrease while holoTC concentrations remain unchanged. HoloTC is, therefore, a better marker than tB_{12} for monitoring vitamin B_{12} status in population-based studies including pregnant women. There are very few studies on holoTC concentrations and other vitamin B_{12} markers in other countries and ethnic groups (Herrmann *et al.* 2003a; Nexø and Hoffmann-Lücke 2011; Vander Jagt *et al.* 2011).

28.2.1.4 Transcobalamin Genotypes

Different genotypes for transcobalamin have been observed. Several authors have clearly shown that the most common genotype, P259R (TCN2 776C/G), influences the total concentration of this protein: 776CC individuals have total transcobalamin concentrations 20% lower than 776GG subjects, while 776GC individuals have intermediate values. Genotype distribution varies somewhat according to the literature, but generally 30% of subjects have P259P (776CC), 50% P259R (776CG), and 20% R259R (776GG). HoloTC concentrations are much less influenced by transcobalamin genotypes, though some studies have reached different conclusions and the issue is still controversial (Nexø and Hoffmann-Lücke 2011).

28.2.2 Analytic Phase

28.2.2.1 HoloTC Measurement

Table 28.1 summarizes several methods for measuring holoTC, leading to different analytical and clinical aspects.

Recently Nexø and Hoffmann-Lücke (2011) reported three different methods devised for measuring holoTC, with reasonably similar results. The first method combined ionic precipitation of transcobalamin and measurement of the amount of vitamin B_{12} trapped in the precipitate (Begley and Hall 1975). Lindemans *et al.* (1983) improved this method by using antibodies against transcobalamin rather than ionic separation. The second method, proposed by Ulleland *et al.* (2002), consisted of measuring trapped vitamin B_{12} by an isotope dilution assay (holoTC radioimmunoassay, RIA). Refsum *et al.* (2006) used a third technique, a microbiological method, to measure vitamin B12 levels (Nexø and Hoffmann-Lücke 2011).

HoloTC RIA and holoTC enzyme immunoassay (EIA) were evaluated in a multicentre study involving four European laboratories (Nexø and Hoffmann-Lücke 2011) with similar results: calibration curve 1.36–24.5 pmol/L; imprecision 6% for EIA and 10% for RIA. Moreover, RIA results were also similar to those of the microbiological assay (calibration curve: 1.36–24.5 pmol/L; imprecision: 4–7%) (Refsum *et al.* 2006).

Table 28.1 Methods for measuring circulating holotranscobalamin concentrations: a summary of the evolution of innovative analytical technologies for routine holotranscobalamin measurement (Nexø and Hoffmann-Lücke 2011).

Method[a]	Reference
TC ionic precipitation + measure of trapped vitamin B_{12}	Begley and Hall 1975
antiTC Ab + measure of trapped vitamin B_{12}	Lindemans *et al.* 1983
RIA	Ulleland *et al.* 2002
Microbiological	Refsum *et al.* 2006
EIA	Ornig *et al.* 2006
EIA on fully automated platforms	Brady *et al.* 2008

[a]antiTC Ab = antibody anti-transcobalamin; EIA = enzyme immunoassay; RIA = radioimmunoassay; TC = transcobalamin.

Recently, holoTC-specific monoclonal antibodies, adopted for developing a new EIA (Axis-Shield Diagnostics, Dundee, UK), were found to be a considerable improvement on previous manual RIA and EIA assays. This new immunoassay was utilized for developing two new immunoassays on fully automated platforms (AxSYM or ARCHITECT, Abbott Labs, North Chicago, IL). These methods have a wider measurement range (3–128 pmol/L) and total imprecision (3–9%) similar to the methods described above (Brady et al. 2008).

28.2.2.2 Cut-off Point

There is enough evidence to suggest that holoTC is an early marker of changes in cobalamin homeostasis. The determination of a suitable holoTC cut-off point is essential for this purpose, but the cut-off thresholds reported by different studies range between 35 and 45 pmol/L. In a previous study (Bamonti et al. 2010), we evaluated the analytical performance of the automated EIA assessment of holoTC concentrations (active vitamin B_{12}, Abbott Diagnostics, Wiesbaden, Germany) using the AxSYM analyser. In addition, we aimed to calculate an appropriate cut-off threshold for holoTC in order to identify cobalamin deficiency in an Italian population.

The analytical part of this study confirmed the adequate imprecision of holoTC determination, with mean intra- and inter-assay coefficients of variation (CVs) ranging from 2.9% to 4.1% on assay controls and from 6.0% to 7.7% on our local pooled control serum, good holoTC linearity from 8.8 to 143.3 pmol/L ($r = 0.99$), good recovery in spiked specimens (mean 95%, interval: 90–100%), mean recovery determined by dilution (100%, range 93–111%) and detection limit (0.07 pmol/L).

The evaluation of an optimal diagnostic threshold for active vitamin B_{12} was carried out by testing holoTC concentration on 250 selected serum specimens. Figure 28.3 shows weak correlation between holoTC and tB_{12} levels on all specimens: ($r = 0.420$). No correlation was found between holoTC and other metabolically correlated parameters (folate, homocysteine and creatinine).

The optimal threshold for holoTC was estimated according to these data: the maximum phi correlation coefficient was at 40 pmol/L of holoTC and, at this cut-off, the sensitivity and specificity by receiver operating characteristic (ROC) curve analysis were 0.86 and 0.66, respectively. HoloTC values and estimated cut-off were not affected by gender and age (p-value 0.54 and 0.298, respectively); the 40 pmol/L cut-off value we selected was also reported by other authors (Morkbak et al. 2005). HoloTC predictive ability at the tB_{12} cut-off was also verified and area under curve (AUC) data were more predictive than other cobalamin deficiency predictors (folate and tHcy). The poor correlation between active B_{12} and tB_{12} values, and lack of correlation between holoTC and the other parameters related to vitamin B_{12} status were also key findings of this study. Thus, a suitable threshold has been established for assessing cobalamin deficiency in populations with reduced tB_{12} values. HoloTC concentrations, rather than circulating tB_{12}, highlight metabolic cobalamin status.

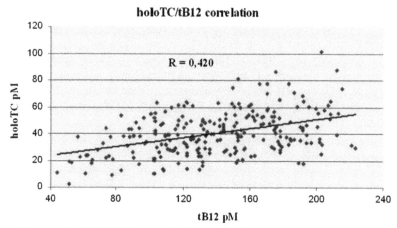

Figure 28.3 Correlation between tB_{12} and holoTC concentrations evaluated in all the 250 selected serum specimens evaluated at our laboratory.
Data are from Bamonti et al. 2010.

In fact, according to Chen et al. (2005), holoTC concentrations do not depend on recent vitamin B_{12} absorption.

28.2.3 Post-analytical Phase: Clinical Studies

Early identification, especially in cases of asymptomatic subjects with normal tB_{12} concentrations, is crucial due to the high prevalence of subclinical cobalamin deficiency. In 2001, Refsum et al. showed a widespread prevalence of impaired cobalamin status in Asian Indians: 47% of subjects had cobalamin deficiency (<150 pmol/L) and 73% had low holoTC levels (<35 pmol/L). Metabolic signs of vitamin B_{12} deficiency were partly explained by their low dietary intake of animal food and/or by other health conditions (e.g. cardiovascular diseases, diabetes) (Refsum et al. 2001). As subsequently reported by Refsum and Smith (2003), patients diagnosed with Alzheimer's disease often had impaired cobalamin status, in particular low levels of holoTC, but not of tB_{12}.

28.2.3.1 Vitamin B_{12} Status of Elderly People

Elderly people's vitamin B_{12} deficiency is often associated with cognitive deficiency. According to recent studies by some authors, holoTC is an important marker with which to assess cognitively impaired elderly patients (Valente et al. 2011). In particular, a significant association was found between cobalamin and holoTC levels and the decrease in brain volume: lower vitamin concentrations, greater rate of brain volume loss (Smith and Refsum 2009). de Lau et al. (2009) evaluated the association of markers of vitamin B_{12} status with cerebral white matter lesions, infarcts and cognition; the study showed that lack of vitamins status was significantly associated with greater severity of white-matter lesions in a concentration-related manner.

28.2.3.2 Vitamin B_{12} Status of Vegetarians and Vegans

Further studies on vegetarians have confirmed that tB_{12} measurement is not an effective tool for diagnosing vitamin B_{12} deficiency (Herrmann *et al.* 2003a). According to this assay, vitamin B_{12} deficiency ranged between 25% and 52% depending on how strict the vegetarian diet was, whereas, according to the holoTC assay, the number of vegetarians with very low vitamin B_{12} levels was much higher (73–90%). This study on vegetarians showed good correlation between holoTC and tB_{12} within the reference interval ($r = 0.75$) but poor correlation in the very low interval ($r = 0.37$). Most of tB_{12} is bound to haptocorrins with six days' half-life, and only 6–20% is bound to serum TC II with only six minutes' half-life. This could explain the diagnostically reduced efficiency of the serum tB_{12} assay as well as the difference in correlation between tB_{12} and active vitamin B_{12} reported in different studies (Herrmann *et al.* 2003a).

Vitamin deficiency (low cobalamin and/or folate levels) due to a physiopathological condition and/or lifestyle (*e.g.* drugs, alcohol) or insufficient dietary intake can cause mild hyperhomocysteinemia (HHcy). Another study by Hermann *et al.* (2003a) on vegetarians showed that HHcy incidence and severe vitamin B_{12} deficiency were related to the degree of animal food restriction as confirmed by vegans' B_{12} deficiency. Subjects with combined elevated methylmalonic acid (MMA) and reduced holoTC concentrations most probably had hyperhomocysteinemia, a risk factor for cardiovascular and neurodegenerative diseases (Herrmann *et al.* 2003a).

28.2.3.3 Vitamin B_{12} Status in Different Countries and Ethnic Groups

Syrian subjects, like Indian populations, showed combined low holoTC and elevated MMA (Herrmann *et al.* 2003a), confirming how diagnostically important MMA and holoTC are in the evaluation of cobalamin status. A widespread prevalence of low serum vitamin B_{12} has also been reported in a multi-ethnic Israeli population. It is unknown how significant the prevalence of functional vitamin B_{12} deficiency was in Syrian subjects even if recent evidence has suggested that B_{12} deficiency may induce secondary folate deficiency and, therefore, HHcy. Studies in Mexico, Venezuela and Guatemala (Herrmann *et al.* 2003a) and Nigeria (Vander Jagt *et al.* 2011) have shown a high rate of cobalamin deficiency.

28.2.3.4 HoloTC in Subjects with Renal Failure

The kidney plays an important role in regulating cobalamin flux. The lower the glomerular filtration rate (GFR), the more circulating holoTC is required to achieve sufficient cellular active vitamin B_{12} uptake and maintain a balanced, intracellular **B_{12}**-dependent metabolism. Serum tB_{12} concentrations within the reference interval are not likely to ensure vitamin delivery into the cells (Obeid et al. 2005). Supra-physiological doses of vitamin B_{12} may be necessary to deliver a sufficient amount of vitamins to the cells via mechanisms largely independent of holoTC receptors (Obeid et al. 2005). However, the mechanisms responsible for high cobalamin levels in renal failure are still unknown (Herrmann et al. 2003a, 2005).

28.2.3.5 Vitamin B_{12} Status in Subjects at Risk of B_{12} Deficiency

A study by our group aimed to determine whether it might be clinically and diagnostically useful to evaluate holoTC levels together with homocysteine metabolism in asymptomatic subjects belonging to the following groups:

a. Elderly people 40 subjects (19M) frequently suffering from vitamin B_{12} deficiency (>20%), often neither identified nor investigated because of its subclinical manifestations, or mainly caused by reduced production or lack of intrinsic factor (15–20%), or altered cobalamin absorption, or possibly associated with insufficient dietary intake.
b. Smokers, 61 subjects (35M) whose serum cobalamin concentrations are significantly lower than those of non-smokers. Organic nitrites, nitric oxide, cyanates and isocyanates inhaled with cigarette smoke interact with vitamin B_{12}, neutralizing it.
c. Obese people 85 subjects (25M) with low levels of micronutrients due to altered dietary intake and/or altered absorption.
d. Women (21) undergoing estroprogestinic therapy often suffering from reduced tB_{12} serum levels.
e. Vegans 20 subjects (11M) at risk of animal food deficiency because of a solely vegetarian diet.

Serum tB_{12} and holoTC, serum and erythrocyte folate (s-Fol and ery-Fol) and plasma tHcy concentrations were determined by the AxSYM analyser. The findings are reported in Table 28.2 and Figure 28.4.

Subjects with holoTC concentrations below the 40 pmol/L cut-off were about twice as many as those with total B_{12} concentrations below the 164 pmol/L cut-off. Interestingly, among vegans, holoTC and tB_{12} deficiency percentages were both higher than other groups' (77%) due to their strictly vegetarian diet. Moreover, holoTC deficiency did not always correspond to tB_{12} deficiency because of a weak correlation between the two analytes ($r=0.39$). Elderly people (70%), smokers (56%), obese (45%), women in estroprogestinic therapy (29%) and vegans (71%) had mild HHcy, partly due to cobalamin rather than folate deficiency.

Table 28.2 Homocysteine metabolism parameters evaluated in asymptomatic subjects at risk. Data expressed as median (interquartile range). Analyte concentrations correlated with homocysteine metabolism and percentages of cases with abnormal values. Data are from a study by our group (Novembrino et al. 2008).

Group	tB12 (pM)	tB12 <164 pM[a]	Holo-TC (pM)	Holo-TC <40 pM[a]	tHcy (μM)	tHcy >10.0 μM[a,b]	s-Fol (nM)	s-Fol <7 nM[a]	ery-Fol (nM)	ery-Fol <421 nM[a]
a	237 (202–330)	20%	51.1 (33.9–64.8)	43%	13.4 (10.7–17.9)	70%	15.6 (12.5–23.5)	0%	704 (552–904)	0%
b	272 (211–351)	11%	58.0 (43.7–77.7)	28%	10.8 (8.1–13.2)	56%	12.9 (10–17.6)	15%	441 (356–598)	49%
c	275 (225–350)	7%	55.5 (33.9–64.8)	21%	10.4 (8.4–12.4)	45%	16.5 (11.7–22.6)	4%	668 (547–847)	7%
d	217 (178–283)	24%	51.1 (43.9–64.6)	43%	8.1 (7.5–11.2)	29%	23.3 (14–28.5)	5%	679 (617–786)	5%
e	118 (72–164)	77%	15.4 (7.9–33.8)	77%	14.2 (10.1–18.7)	71%	34.8 (28.7–37.3)	0%	1070 (731–1660)	0%

[a] Reference cut-off.
[b] cut-off = 12.5 μM for group a.

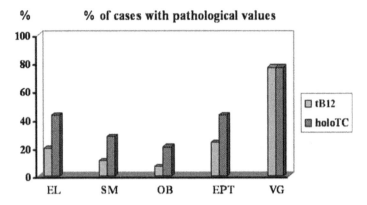

EL=elderly; SM=smokers; OB=obese subjects; EPT=women on estroprogestinic therapy; VG=vegans; tB12=serum total B12; holoTC=serum holotranscobalamin

Figure 28.4 Percentage of subjects with holoTC and tB_{12} concentrations below cut-off. Histograms represent the percentage of subjects with pathological values. Data are from a study by our group (Novembrino *et al.* 2008).

28.2.3.6 Vitamin B_{12} Status in Pregnancy

Maternal vitamin B_{12} deficiency during pregnancy is associated with possible risks of adverse pregnancy outcomes both for mother (*i.e.* preeclampsia, early miscarriage) and foetus (*i.e.* neural tube defects, intrauterine growth retardation). As recently reported by Vander Jagt *et al.* (2011), global prevalence of vitamin B_{12} deficiency, particularly in developing countries, is underestimated. In fact, they found that 36% of pregnant Nigerian women were classified as vitamin B_{12}-deficient using plasma holoTC levels as an indicator of physiologically available vitamin B_{12}; moreover, holoTC concentrations correlated negatively with Hcy and positively with ery-Fol concentrations. The study underlines the importance of giving adequate vitamin B_{12} supplementation to pregnant women in Nigeria in order to prevent anaemia and decrease the risk of neural tube defects; women attending antenatal clinics are given routinely iron and folic acid as dietary supplements, but not vitamin B_{12} (Vander Jagt *et al.* 2011).

Ours is the first ever study carried out on preeclampsia, a major cause of maternal and infant mortality and morbidity. This preliminary study aimed to verify a possible involvement of Hcy metabolism in preeclampsia. At present, ten cases of preeclampsia (aged 28–38; median age 35) are being evaluated and eight cases of normal pregnancy (aged 29–43; median age 36), all within 20–26 weeks' pregnancy. Our preliminary data on preeclampsia (Table 28.3) showed about 75% of cases with Hcy levels significantly higher, vitamin B_{12} and holoTC levels significantly lower than in controls and folate status similar to that of normal pregnant women; moreover, notably, subjects with holoTC under cut-off were about twice as many as those with tB_{12} under cut-off (78% *vs.* 45%).

Table 28.3 Homocysteine metabolism parameters in preeclampsia. Data expressed as median (interquartile range). Serum and erythrocyte folate levels not reported because these were within relevant reference interval in both groups. Data are from a study by our group.

	$tHcy$ (μM)	$tB12$ (pM)	$holoTC$ (pM)
Preeclampsia (ten women)	8.5 (4.9–20)[a] *75%*	200.2 (116–379)[a] *45%*	33.9 (26.4–74.7)[a] *78%*
Controls (eight women)	5.6 (4.2–5.1) *0%*	292 (261–324) *0%*	49 (46.5–51.3) *0%*

[a] $p<0.05$ vs. controls.

28.2.3.7 Vitamin B_{12} Status in Infancy

The study by Hay et al. (2011) on the cobalamin status of healthy infants (breastfeeding and weaning up to two-years-old) showed that vitamin B_{12} status (i.e. holoTC levels) remained significantly lower in breastfed than in non-breastfed infants. Maternal holoTC concentrations were the best predictor of a newborn's cobalamin status. Moreover, cobalamin intake correlated with serum holoTC levels in healthy toddlers (Hay et al. 2011).

28.3 Clinical Significance and Utility

The assessment of serum holoTC concentrations is clinically and diagnostically useful. In fact, the biologically active form of vitamin B_{12} represents the best predictor of cobalamin status from infancy to old age: maternal holoTC concentrations can predict newborns' vitamin status, serum holoTC levels correlate with infants' dietary cobalamin intake, holoTC measurement is useful to diagnose cognitively impaired patients (Refsum et al. 2006).

28.3.1 Vitamin B_{12} Status of Elderly People

de Lau et al. (2009) highlighted the association between vitamin B_{12} status and the severity of white matter lesions (especially periventricular lesions) caused by cobalamin deficiency on myelin integrity rather than by vascular risk factors. Elderly people's low cobalamin status should be further investigated as a cause of brain atrophy and of likely subsequent cognitive impairment (Smith and Refsum 2009). Elderly people's cobalamin deficiency can cause brain atrophy and subsequent cognitive impairment, even in the absence of classical haematological diseases. Folic acid treatment can improve the shape of megaloblastic erythrocytes, but cannot prevent neurological damage in the case of cobalamin-deficient subjects (Herrmann *et al.* 2003a). Herrmann *et al.* (2003a) have suggested that even slight changes in the circulating vitamin B_{12} pool most probably correspond to changes in serum holoTC levels. In addition to vitamin B_{12} deficiency, malabsorption is a common cause of decreasing serum holoTC levels; consequently, cobalamin deficiency caused by gastro-intestinal malabsorption must be differentiated from deficiency caused by diet (Herrmann *et al.* 2003a).

Elderly people's HHcy and vitamin B_{12} deficiency can be due to the following causes:

- malabsorption of cobalamin due to the lack of intrinsic factor, pernicious anaemia and achlorhydria;
- gastrointestinal disorders causing inadequate cobalamin uptake (surgery, atrophy, drugs that suppress acid secretion, bacterial overgrowth);
- inadequate dietary intake of B vitamins and folic acid;
- low serum TC concentrations;
- renal failure;
- drugs interacting with vitamins (*e.g.* theophylline inhibiting pyridoxal kinase);
- alcohol interference with cobalamin and folate/folic acid absorption and metabolism.

An additional cause is an age-related reduced activity of cystathionine-β-synthase (and possibly of other enzymes involved) and a substantial decrease in the amount of vitamin B_{12} bound to TC II, responsible for trans-cellular B_{12} in- and out-transportation (Herrmann *et al.* 2003a). According to Obeid *et al.* (2005), vitamin B_{12} supra-physiological doses are probably necessary to deliver a sufficient amount of vitamins to cells *via* mechanisms largely independent of holoTC receptors but still unknown.

28.3.2 Vitamin B_{12} Status of Vegetarians and Vegans

While a vegan lifestyle is most probably related to vitamin B_{12} deficiency, there are different stages in B_{12} deficiency among vegetarians (Herbert 1994). In stage I/II, a good number of vegetarians have vitamin B_{12} deficiency with low holoTC levels, but no signs of metabolic abnormalities (normal Hcy and MMA). A transient decrease in active B_{12} but not in tB_{12} caused by diet can also be due to transient malabsorption (*e.g.* alcohol abuse or temporary exposure to drugs such as colchicines or omeprazole); therefore, repeated holoTC assays and an interview with the patient may help clarify the situation and enhance the diagnostic value of the result. In stage III, there is an increase in vitamin B_{12} deficiency (high MMA, HHcy and low holoTC) and cells no longer have the proper vitamin B_{12} catalysed reactions; this stage is, therefore, of clinical relevance leading to organ failure (Herbert 1994).

28.3.3 Vitamin B_{12} Status in Different Countries and Ethnic Groups

Vegetarian people from developing countries and from certain ethnic groups can have functional cobalamin deficiency. The significance of a high rate of cobalamin deficiency in Mexico, Venezuela and Guatemala remains unknown; poor growth, anaemia and neurological manifestations (*i.e.* mood changes, altered sensory, motor and cognitive functions) are possible functional consequences (Herrmann *et al.* 2003b). According to a psycho-educational evaluation, short-term memory and perception in Guatemalan children are more frequently due to low rather than to adequate cobalamin concentrations (Herrmann *et al.* 2003a).

28.3.4 HoloTC in Subjects with Renal Failure

Renal patients' increased metabolite levels could be due to holoTC abnormal distribution, altered receptor activity for renal TC II uptake and altered TC II function. Treatment with folic acid and vitamin B_{12} reduces MMA and Hcy levels, but only during the treatment, which suggests a chronic alteration in homocysteine remethylation (Herrmann *et al.* 2003a). Elderly subjects with renal failure require more circulating holoTC to deliver sufficient amounts of vitamin B_{12} into the cells and maintain normal cellular vitamin B_{12} status (Herrmann *et al.* 2003a).

28.3.5 Vitamin B_{12} Status in Pregnancy

Our preliminary findings on preeclampsia confirmed that symptoms of cobalamin deficiency during pregnancy, leading to HHcy can be undervalued, especially if women's serum tB_{12} concentrations are within the reference interval. As shown in Table 28.3, serum holoTC concentrations represent an important diagnostic result. Therefore, holoTC determination could be considered as a useful additional diagnostic factor in assessing homocysteine status, accurately evaluated by the five 'Hcy panel' parameters: tHcy, s- and ery- folate, tB_{12} and holoTC. Our results are in line with the study by Vander Jagt *et al.* (2011) on Nigerian pregnant women.

28.4 Discussion

The symptoms of vitamin B_{12} deficiency manifest themselves at a later stage as megaloblastic anaemia and/or neuropsychiatric disorders; vitamin B_{12} deficiency, if untreated, leads to irreversible degeneration of the nervous system. To prevent irreversible neurological damage, early diagnosis of vitamin B_{12} deficiency can serve as a warning.

We can now measure holoTC levels to highlight in advance vitamin B_{12} deficiency. Vitamin B_{12} and folate are involved in homocysteine metabolism as cofactor and cosubstrate of enzymatic reactions; deficiency of either or both vitamins can lead to HHcy.

Elevated serum MMA and tHcy concentrations can be considered alternative specific metabolic parameters of cobalamin deficiency. Measurement of functional metabolite MMA requires sophisticated equipment and is, therefore, unsuitable for routine use. Total homocysteine is a more sensitive analyte than tB_{12} in diagnosing subclinical vitamin B_{12} deficiency because its plasma levels increase before clinical symptoms appear. However, the lack of specificity of this analyte represents a serious limit to its use. Total homocysteinemia depends on genetic or physiological factors, life style, diseases in progress, and drugs. HHCY is caused by folate or vitamin B_6 deficiency and renal failure.

28.4.1 Asymptomatic Subjects' HoloTC Concentrations

In order to evaluate holoTC reliability as an early index of cobalamin status, we analysed holoTC levels of asymptomatic subjects at risk of developing HHcy because of age, life style (*e.g.* smoking, drugs); our results seemed to confirm that a considerable percentage of subjects suffer from vitamin B_{12} deficiency. According to our study, quite a high percentage of subjects suffer from mild HHcy, presumably due to vitamin B_{12} rather than to folate deficiency. In agreement with literature on elderly people (Herrmann *et al.* 2003a; Refsum and Smith 2003), we found that 70% of elderly subjects suffer from vitamin B_{12} deficiency and mild HHcy.

Smokers' reduced vitamin B_{12} levels are usually attributed to interaction between methylcobalamin and cyanide inhaled with cigarette smoke, which transforms cobalamin into an inactive compound. According to our study, more than 50% of the smokers suffered from mild HHcy and 49% from low ery-Fol levels, confirming the concealed toxicity of tobacco smoke caused directly by noxious substances inhaled and also by food lacking in vitamins. According to literature, cigarette smoking belongs to a life style closely and independently associated with an increased risk of cardiovascular disease, as demonstrated by a slight increase in tHcy (Okumura and Tsukamoto 2011).

Overweight/obese subjects show lower concentrations of nutrients than normal weight subjects due to insufficient dietary intake or altered absorption, distribution and metabolism of micronutrients. In our study, obese subjects' vitamin B_{12} values were below the reference interval (only 7%) and holoTC levels below cut-off (only 21%), perhaps thanks to the Mediterranean diet which redress the imbalance caused by weight, particularly in young subjects.

According to the literature, there are contrasting data about the connection between the use of estroprogestinic therapy and vitamin B_{12} concentrations. According to some studies, tB_{12} levels were significantly lower in women in progestinic therapy than in controls, while other authors did not observe any difference (Lussana *et al.* 2003). According to our results quite a number of women (30%) had tB_{12} and, above all, holoTC deficiency; given their young age, even mild HHcy must not be ignored.

A vegan diet promotes health and longevity and reduces the risk of cardiovascular diseases, but increases the risk of developing essential nutrients deficiency (above all vitamins from animal food). Group B vitamins (and specifically vitamin B_{12}) deficiency can influence the functioning of vegans' correlated metabolisms (Herrmann et al. 2003b). According to us and other authors, a high percentage of vegans (77%) suffer from vitamin B_{12} and holoTC deficiency (Herrmann et al. 2003b). It is important to highlight that many, even young, subjects suffer from mild HHcy with possible clinical damage due to cobalamin deficiency despite adequate folate status (as shown by serum and erythrocyte determination). We must not forget the close correlation between the two vitamins and the 'folate trap': the only known direct metabolic link between vitamins consists in the vitamin B_{12} dependent remethylation of Hcy to methionine in order to regenerate tetrahydrofolate and maintain the folate cycle.

Summing up, except for vegan population where cobalamin deficiency was so severe that even tB_{12} was below normal, we must once again underline that low holoTC levels were not always correlated with low vitamin B_{12} concentrations ($r^2 = 0.39$). Moreover, in our routine experience, only a small percentage of subjects had both vitamin B_{12} and holoTC deficiency (Figure 28.5).

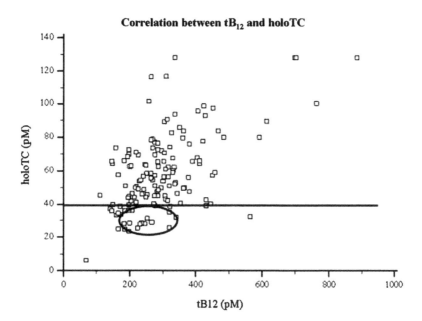

$tB12$ = serum total B12; holoTC = serum holotranscobalamin

Figure 28.5 Correlation between tB_{12} and holoTC concentrations in routine serum specimens evaluated in our laboratory. Horizontal line indicates holoTC cut-off. Good correlation seen between vitamin B_{12} and holoTC in very few cases (see circle). Data are from an unpublished study by our group.

28.4.2 Asymptomatic Subjects' 'Hcy Panel'

Our preliminary study on preeclampsia also showed that subjects with holoTC under cut-off were about twice as many as those with tB_{12} under cut-off, but all of them had HHcy and therefore their pregnancy was at risk. The study by Vander Jagt et al. (2011) on pregnant women in Nigeria totally confirmed our findings.

In agreement with other authors (de Lau et al. 2008; Gonzalez-Gross et al. 2007; Herrmann et al. 2033a,b; Miller et al. 2006; Refsum et al. 2006; Nexø and Hoffmann-Lücke 2011; Smith and Refsum 2009; Vander Jagt et al. 2011), our preliminary findings highlight the clinical importance of measuring the 'Hcy panel' for an early identification of cobalamin deficiency, especially in asymptomatic subjects with normal folate levels.

In our opinion, diagnosing even slight cobalamin deficiency can be clinically useful to start an adequate treatment as soon as possible. Indeed, according to Miller et al. (2006) and Gonzalez-Gross et al. (2007), measuring both analytes together, active B_{12} and tB_{12}, guarantees a better screening for cobalamin deficiency.

28.5 Concluding Remarks

HoloTC assay is reliable as an early diagnostic tool but not yet sufficiently utilized by clinicians: whenever there is subclinical or suspected vitamin deficiency, the measurement of both tB_{12} and its active fraction holoTC provides different and complementary information, allowing us to identify latent cobalamin deficiency and treat seemingly healthy subjects. Additionally, as shown by several authors (de Lau et al. 2008; Gonzalez-Gross et al. 2007; Herrmann et al. 2033a,b; Miller et al. 2006; Refsum et al. 2006; Smith and Refsum 2009; Nexø and Hoffmann-Lücke 2011, Valente et al. 2011; Vander Jagt et al. 2011) and also by us, measuring all the analytes related to Hcy metabolism can be clinically and diagnostically useful to highlight metabolic impairment from pregnancy to old age. Finally, we hope all clinicians, including Italian ones, will realize the importance of an early diagnosis and adequate treatment of cobalamin deficiency in asymptomatic subjects.

Summary Points

- This chapter focuses on the clinical–diagnostic value of measuring serum holoTC concentrations.
- HoloTC: the most important cobalamin fraction (10–30%), biologically available for ubiquitous cellular uptake.
- HoloTC: an early marker of vitaminB_{12} deficiency, more reliable than tB_{12}.
- Serum tB_{12} assay: routinely used standard method, but not always predictive of cobalamin deficiency especially where with tB_{12} levels are within the reference interval.
- HoloTC assay: allows us to identify subjects with tB_{12} concentrations within the reference interval but with latent cobalamin deficiency.
- HoloTC assay: not yet sufficiently utilized, particularly, by Italian clinicians.

- Evaluation of cobalamin status (tB$_{12}$, holoTC, tHcy) by automated immunoassays: a new approach for routine assessment and an important diagnostic procedure.

Key Facts about Cobalamin Deficiency

- Overt cobalamin deficiency is not as common as subclinical cobalamin deficiency.
- Prevalence of subclinical cobalamin deficiency is higher than previously expected, especially where in asymptomatic subjects are apparently healthy but at risk of developing vitamin deficiency due to physiopathological conditions and/or lifestyle.
- Global prevalence of vitamin B$_{12}$ deficiency, particularly in developing countries.
- Clinical studies highlight the diagnostic importance of evaluating vitamin 'status' and homocysteine metabolism where populations are at risk.
- Diagnostic utility for clinicians: to prevent irreversible neurologic damage, even if folate status is normal, and treat patients adequately from pregnancy to old age.
- Further studies on a larger number of subjects should be promoted.

Definitions of Key Terms

Cobalamin (vitamin B$_{12}$): a water-soluble B vitamin, normally involved in the human body metabolism, affecting DNA synthesis and regulation, fatty acid synthesis and energy production. Supplied by animal food, its deficiency leads to macrocytic anaemia, decreased bone marrow cell production, neurological problems, as well as metabolic issues (methylmalonyl-CoA acidosis, hyperhomocysteinemia).

Estroprogestinic therapy (hormone replacement therapy): estrogen and progestin given to women to relieve menopause symptoms.

Folate (vitamin B$_9$): a natural, water-soluble B vitamin. Tetrahydrofolate is its biologically active form converted to dihydrofolate in the liver. Folate, involved in DNA synthesis, repair, and methylation, acts also as cofactor in several biological reactions together with vitamin B$_{12}$. Folate is an essential factor for rapid cell division and growth (*e.g.* infancy and pregnancy). Leafy green vegetables (like spinach), fruits (like orange fruits and juices), dried beans and peas, yeast are all natural sources of folate.

Folic acid (pteroyl-L-glutamic acid): the synthetic oxidized form of folate, found in supplements and added to fortified foods.

Hyperhomocysteinemia: high homocysteine concentrations mostly due to genetic mutation, deficiency in B group vitamins and/or renal insufficiency; an independent risk factor for cardiovascular disease.

Holotranscobalamin (transcobalamin II–cobalamin complex): a minor fraction (about 20%) of circulating vitamin B$_{12}$ and, thanks to specific ubiquitous receptors, its active form (active B$_{12}$).

Homocysteine: aminothiol biosynthesized from methionine by removal of its terminal C methyl group; can be recycled into methionine or converted into cysteine with B vitamin co-factors. Deficiencies of vitamin B_6, B_9 or B_{12} and/or genetic mutations can lead to hyperhomocysteinemia.

Obesity: abnormal or excessive fat accumulation due to an imbalance between energy intake and consumption. This condition represents a risk factor for a number of chronic diseases, including diabetes, cardiovascular diseases and cancer. A body mass index $\geq 30 \, \text{kg/m}^2$ is generally considered obese.

Preeclampsia: a serious condition developing in late pregnancy characterized by a sudden rise in blood pressure, excessive weight gain, generalized oedema, proteinuria, severe headache, and visual disturbances; if untreated may result in eclampsia.

Vegans: vegetarians avoiding the use of all animal products including dairy and eggs.

List of Abbreviations

AUC	area under curve
B_{12}	vitamin B_{12}
Cbl	cobalamin
CV	coefficient of variation
EIA	enzyme immunoassay
ery-Fol	erythrocyte folate
GFR	glomerular filtration rate
HC	haptocorrin
holoTC	holotranscobalamin
HHcy	hyperhomocysteinemia
IF	intrinsic factor
MMA	methylmalonic acid
RIA	radioimmunoassay
ROC	receiver operating characteristic
s-Fol	serum folate
tB_{12}	total vitamin B_{12}
tHcy	total homocysteine
TC	transcobalamin

References

Baik, H.W., and Russel, R.M., 1999. Vitamin B12 Deficiency in the elderly. *Annual Review of Nutrition*. 19: 357–377.

Bamonti, F., Moscato, G.A., Novembrino, C., Gregori, D., Novi, C., De Giuseppe, R., Galli, C., Uva, V., Lonati, S., and Maiavacca, R., 2010. Determination of serum holotranscobalamin concentrations with the AxSYM active B_{12} assay: cut-off point evaluation in the clinical laboratory. *Clinical Chemistry and Laboratory Medicine*. 48: 249–253.

Begley, J.A., and Hall, C.A., 1975. Measurement of vitamin B12-binding proteins of plasma. I. Technique. *Blood*. 45: 281–6.

Brady, J., Wilson, L., McGregory, L., Valente, E., and Orning, L., 2008. Active B12: a rapid, automated assay for holotranscobalamin on the Abbott AxSYM analyzer. *Clinical Chemistry*. 54: 567–573.

Carmel, R., 2011. Biomarkers of cobalamin (vitamin B-12) status in the epidemiologic setting: a critical overview of context, applications, and performance characteristics of cobalamin, methylmalonic acid, and holotranscobalamin II. *The American Journal of Clinical Nutrition*. 94: 348S–358S.

Chen, X., Remacha, A.F., Sarda, M.P., and Carmel, R., 2005. Influence of cobalamin deficiency compared with that of cobalamin absorption on serum holo-transcobalamin II. *The American Journal of Clinical Nutrition*. 81: 110–114.

de Lau, L.M., Smith, A.D., Refsum, H., Johnston, C., and Breteler, M.M., 2009. Plasma vitamin B12 status and cerebral white-matter lesions. *Journal of Neurology, Neurosurgery and Psychiatry*. 80(2): 149–157.

Gonzalez-Gross, M., Sola, R., Albers, U., Barrios, L., Alder, M., Castillo, M.J., and Pietrzik, K., 2007. B-vitamins and homocysteine in Spanish institutionalised elderly. *International Journal for Vitamin and Nutrition Research*. 77: 22–33.

Hay, G., Trygg, K., Whitelaw, A., Johnston, C., and Refsum, H., 2011. Folate and cobalmin status in relation to diet in healthy 2-y-old children. *The American Journal of Clinical Nutrition*. 93(4): 727–735.

Herbert, V., 1994. Staging vitamin B-12 (cobalamin) status in vegetarians. *The American Journal of Clinical Nutrition*. 59(suppl): 1213S–1222S.

Herrmann, W., Obeid, R., Schorr, H., and Geisel, J., 2003a. Functional vitamin B12 deficiency and determination of holotranscobalamin in populations at risk. *Clinical Chemistry and Laboratory Medicine*. 41: 1478–1488.

Herrmann, W., Schorr, H., Obeid, R., and Geisel, J., 2003b. Vitamin B-12 status, particularly holotranscobalamin II and methylmalonic acid concentrations, and hyperhomocysteinemia in vegetarians. *The American Journal of Clinical Nutrition*. 78: 131–136.

Herrmann, W., Obeid, R., Schorr, H., and Geisel, J., 2005. The usefulness of holotranscobalamin in predicting vitamin B12 status in different clinical settings. *Current Drug Metabolism*. 6: 47–53.

Lindemans, J., Schoester, M., and van Kapel, J., 1983. Application of a simple immunoadsorption assay for the measurement of saturated and unsaturated transcobalamin II and R-binders. *Clinica Chimica Acta*. 132: 53–61.

Loikas, S., Lopponen, M., Suominen, P., Møller, J., Irjala, K., Isoaho, R., Kivelä, S.L., Koskinen, P., and Pelliniemi, T.T., 2003. RIA for serum holotranscobalamin: method evaluation in the clinical laboratory and reference interval. *Clinical Chemistry*. 49: 455–462.

Lussana, F., Zighetti, M.L., and Bucciarelli, P., 2003. Blood levels homocysteine, folate, vitamin B6 and B12 in women using oral contraceptives compared to non-users. *Thrombosis Research*. 112: 37–41.

Miller, J.W., Garrod, M.G., Rockwood, A.L., Kushnir, M.M., Allen, L.H., Haan, M.N., and Green, R. 2006. Measurement of total vitamin B12 and holotranscobalamin, singly and in combination, in screening for metabolic vitamin B12 deficiency. *Clinical Chemistry*. 52: 278–285.

Morkbak, A.L., Heimdal, R.M., Emmens, K., Molloy, A., Hvas, A.M., Schneede, J., Clarke, R., Scott, J.M., Ueland, P.M., and Nexo, E., 2005. Evaluation of the technical performance of novel holotranscobalamin (HoloTC) assays in a multi center European demonstration project. *Clinical Chemistry and Laboratory Medicine*. 43: 1058–1064.

Nexø, E., and Hoffmannn-Lücke, E., 2011. Holotranscobalamin, a marker of vitamin B-12 status: analytical aspects and clinical utility. *The American Journal of Clinical Nutrition*. 94: 359S–365S.

Novembrino, C., De Giuseppe, R., Uva, V., Bonara, P., Moscato, G., Galli, C., Maiavacca, R., and Bamonti, F., 2008. Subclinical vitamin B_{12} deficiency in asymptomatic subjects: the importance of holotranscobalamin (Holo-TC i.e. active B_{12}) assay. *Ligand Assay*. 13: 243–249.

Obeid, R., Kuhlmann, M., Kirsch, C.M., and Herrmann, W., 2005. Cellular uptake of vitamin B12 in patients with chronic renal failure. *Nephron Clinical Practice*. 99(2): c42–48.

Okumura, K., and Tsukamoto, H., 2011. Folate in smokers. *Clinica Chimica Acta*. 412(7–8): 521–526.

Orning, L., Rian, A., Campbell, A., Brady, J., Fedosov, S.N., Bramlage, B., Thompson, K., Quadros, E.V., 2006. Characterization of a monoclonal antibody with specificity for holo-transcobalamin. *Nutrition & Metabolism* (Lond). 3:3.

Quadros, E.V., 2010. Advances in the understanding of cobalamin assimilation and metabolism. *British Journal of Haematology*. 148(2): 195–204.

Refsum, H., Yajnik, C.S., Gadkari, M., Schneede, J., Vollset, S.E., Orning, L., Guttormsen, A.B., Joglekar, A., Sayyad, M.G., Ulvik, A., and Ueland, P.M., 2001. Hyperhomocysteinemia and elevated methylmalonic acid indicate a high prevalence of cobalamin deficiency in Asian Indians. *The American Journal of Clinical Nutrition*. 74(2): 233–241.

Refsum, H., and Smith, A.D., 2003. Low vitamin B-12 status in confirmed Alzheimer's diseases as revealed by serum holotranscobalamin. *Journal of Neurology, Neurosurgery and Psychiatry*. 74: 959–61.

Refsum, H., Johnston, C., Guttormsen, A.B., and Nexo, E., 2006. Holotranscobalamin and total transcobalamin in human plasma: determination, determinants, and reference values in healthy adults. *Clinical Chemistry*. 52: 129–137.

Smith, A.D., and Refsum, H., 2009. Vitamin B-12 and cognition in the elderly. *The American Journal of Clinical Nutrition*. 89(2): 707S–711S.

Ulleland, M., Eilertsen, I., Quadros, E.V., Rothenberg, S.P., Fedosov, S.N., Sundrehagen, E., and Orning, L., 2002. Direct assay for cobalamin bound to transcobalamin (holo-transcobalamin) in serum. *Clinical Chemistry*. 48: 526–532.

Valente, E., Scott, J.M., Ueland, P.M., Cunningham, C., Casey, M., and Molloy, A.M., 2011. Diagnostic accuracy of holotranscobalamin, methylmalonic acid, serum cobalamin, and other indicators of tissue vitamin B12 status in the elderly. *Clinical Chemistry*. 57(6): 856–863.

Vander Jagt, D.J., Ujah, I.A., Ikeh, E.I., Bryant, J., Pam, V., Hilgart, A., Crossey, M.J., and Glew, R.H., 2011. Assessment of the vitamin B12 status of pregnant women in Nigeria using plasma holotranscobalamin. *Obstetric and Gynecolology*. 2011:365894. Epub 2011 Jul 14.

Function and Effects

CHAPTER 29

B Vitamins (Folate, B_6 and B_{12}) in Relation to Stroke and its Cognitive Decline

CONCEPCIÓN SÁNCHEZ-MORENO* AND
ANTONIO JIMÉNEZ-ESCRIG

Institute of Food Science, Technology and Nutrition, Spanish National Research Council (ICTAN-CSIC), José Antonio Novais 10, Ciudad Universitaria, ES-28040 Madrid, Spain
*Email: csanchezm@ictan.csic.es

29.1 Introduction

Stroke is one of the leading causes of death in developed countries. Nearly one third of stroke survivors have some degree of dementia after stroke: 32% of patients who had suffered ischemic stroke had dementia based on comprehensive neurological and psychological testing, clinical mental status interviews, magnetic resonance imaging scans, and detailed histories collected (Henon et al. 1999; Kase et al. 1998; Pohjasvaara et al. 1998). Importantly, dementia is more common in stroke patients who are older, smoke and have lower levels of education (Pohjasvaara et al. 1998). In addition, vascular dementia often coexists with Alzheimer's disease (AD) and the presence of AD may predispose one to the development of vascular dementia. In fact, the 5% prevalence in cognitive impairment that occurs in the elderly over the age of 65 increases sharply after ischemic stroke up to 38%. Cognitive impairment is associated with death or disability at four years after a stroke (Patel et al. 2002).

However, vascular dementia, which is the second most important cause of cognitive impairment and dementia associated with aging in the US, is the most preventable form affecting the elderly.

The realization that brain ischemia elicits more robust brain damage when nutritional status is poorer provides a fertile ground for the discovery of novel therapeutic agents and nutritional intervention for stroke. Deficiency of B vitamins appears to increase cognitive impairment in stroke patients (Gonzalez-Gross *et al.* 2001). Better understanding of the role that specific nutrients play on vasculature and brain cell response in stroke patients may be relevant to reduce the incidence of cognitive impairment and dementia associated with stroke.

Interestingly, B vitamins play critical roles in cell function (Singleton and Martin 2001). For example, folate (Figure 29.1) in the 5-methyltetrahydrofolate (MTHF) form is a cosubstrate required by methionine synthase to convert homocysteine (Hcy) to methionine; consequently, Hcy accumulates when folate is low (Mesnard *et al.* 2002; Nilsson *et al.* 1999). High Hcy is strongly associated with atherosclerotic vascular disease and stroke (Sarkar and Lambert 2001). Vitamin B_6 (Figures 29.2, 29.3 and 29.4) is needed for the synthesis of neurotransmitters such as serotonin and dopamine. A deficiency of vitamin B_6 may also contribute to increase levels of Hcy. Vitamin B_{12} (Figure 29.5) is important in maintaining the nervous system where it plays a vital role

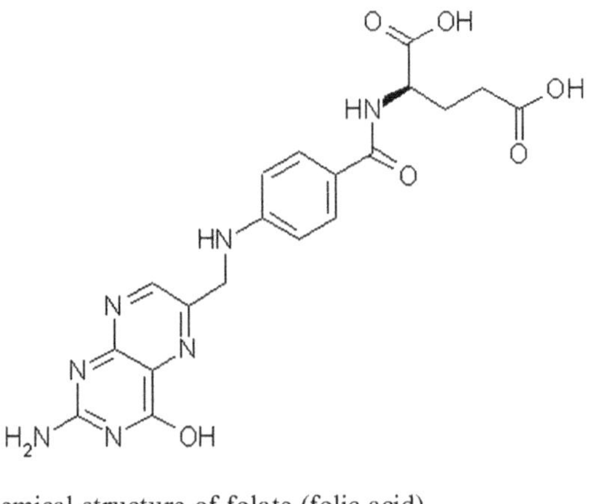

Figure 29.1 Chemical structure of folate (folic acid).

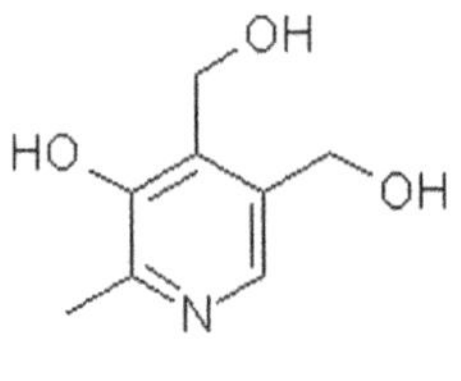

Figure 29.2 Chemical structure of vitamin B_6 (pyridoxine).

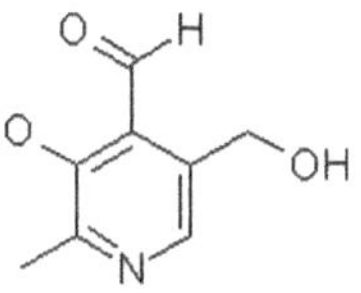

Figure 29.3 Chemical structure of vitamin B_6 (pyridoxal).

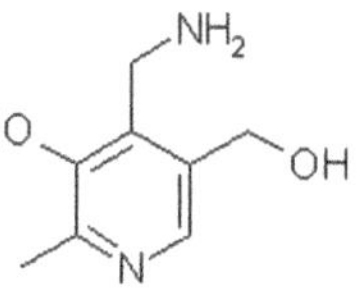

Figure 29.4 Chemical structure of vitamin B_6 (pyridoxamine).

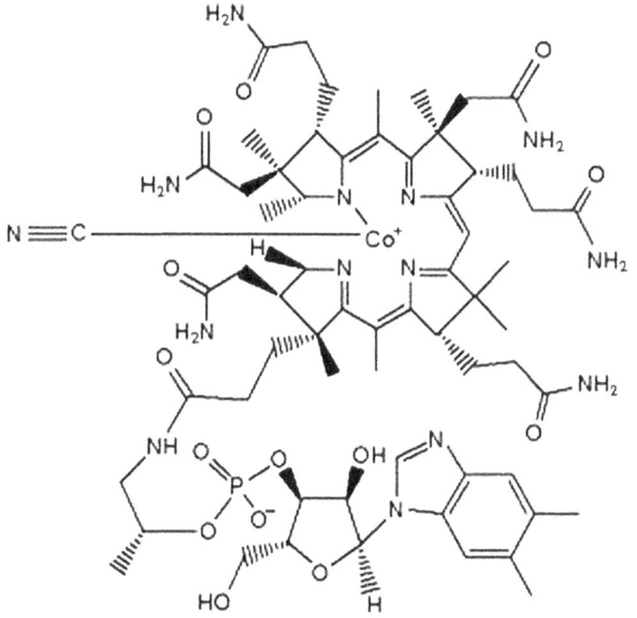

Figure 29.5 Chemical structure of vitamin B_{12} (cyanocobalamin).

in the metabolism of fatty acids essential to maintain myelin (Rogers 2001). Vitamin B_{12} is also required for methionine synthesis from Hcy (Robinson 2000). Figure 29.6 shows the biochemical pathways of Hcy metabolism with the roles of folate, B_6 and B_{12} (Robinson 2000).

Table 29.1 provides a summary of some of the observational and intervention studies in the literature regarding B vitamins, homocysteine and stroke.

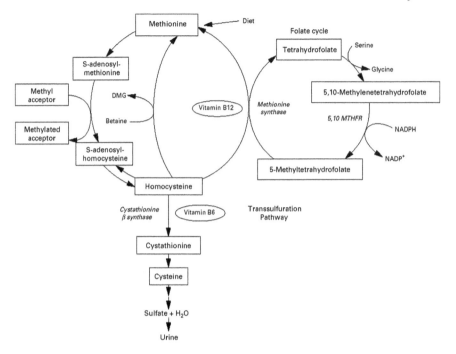

Figure 29.6 Pathways for the metabolism of homocysteine. Normal transsulfuration requires cystathionine β-synthase with vitamin B_6 as cofactor. Remethylation requires 5,10-methylenetetrahydrofolate reductase and methionine synthase. The latter requires folate as cosubstrate and vitamin B_{12} (cobalamin) as cofactor. An alternative remethylation pathway also exists using the cobalamin independent betaine–homocysteine methyltransferase (Robinson 2000).

29.2 B Vitamins: Folate, B_{12}, B_6

The major manifestation of folate deficiency is megaloblastic anaemia. Gastrointestinal disturbances may also accompany folate deficiency. Numerous animal studies have also shown that folate deficiency during pregnancy can impair proper development of foetal central nervous system structures. Folate deficiency in pregnant women is associated with an increased incidence of spina bifida and other neural defects (Molloy et al. 1999). Vitamin B_{12} deficiency in adults is usually not the result of reduced dietary intake but rather reflects reduced intestinal absorption of the vitamin. A number of disease conditions can alter production of the intrinsic factor that is essential for absorption of vitamin B_{12} from the intestine. Atrophic gastritis is an important cause of vitamin B_{12} malabsorption. This condition primarily affects the ability to extract vitamin B_{12} from foods. Deficiency of vitamin B_{12} impairs the ability of the bone marrow to produce red blood cells and thus leads to anaemia, similar to what is observed with folate deficiency. Vitamin B_{12} deficiency can also result in irreversible damage to the nervous system causing swelling, demyelination and death of neurons.

Table 29.1 B vitamins, homocysteine and stroke.

Study	Study focus	Number of subjects	Results
Observational studies			
Verhoef et al. (1994)	Levels of Hcy and risk of ischemic stroke	14 916 male physicians, 40–84 years	The data were compatible with a small but non-significant association between elevated plasma Hcy and risk of ischemic stroke.
Perry et al. (1995)	Hcy and risk of stroke in middle-aged British men	107 stroke patients 118 controls	Hcy was significantly higher among stroke patients than controls.
Giles et al. (1998)	Hcy and the likelihood of non-fatal stroke	4534 US adults, ≥ 35 years	Hcy concentration was independently associated with an increased likelihood of nonfatal stroke. This association was present in both black and white adults.
Robinson et al. (1998)	Hcy, B vitamins and vascular diseases	750 patients with vascular disease 800 control subjects	High Hcy is associated with an elevated risk of vascular disease independent of all traditional risk factors. Low folate and B_6 confer an increased risk of atherosclerosis.
Beamer et al. (1999)	Vitamin use in patients with ischemic stroke	231 stroke patients	Use of vitamin supplements may be associated with lower levels of Hcy in elderly persons whether or not stroke or stroke-related risk factors are present.
Bots et al. (1999)	Hcy, myocardial infarction and stroke in the elderly: the Rotterdam Study	104 myocardial infarct patients 120 stroke patients 533 control subject	The risk of stroke and myocardial infarction increased with total Hcy.
Fallon et al. (2001)	Risk of ischemic stroke and Hcy: the Caerphilly study	2254 men, 50–64 years	No significant relation between Hcy and ischemic stroke was observed in this cohort. However, its etiological importance may be greater for premature ischemic strokes (<65 years).

Table 29.1 (Continued)

Study	Study focus	Number of subjects	Results
Yoo and Lee (2001)	Elevated Hcy in elderly patients with stroke	50 stroke patients 35 controls	Elevated asymmetric dimethylarginine concentrations are at increased risk for ischemic stroke in the elderly, and may account for increased risk of stroke in patients with hyper-Hcy.
Bazzano et al. (2002)	Dietary intake of folate and risk of stroke and CVD	9764 men and women, 25–74 years	Increasing dietary intake of folate from food sources may be an important approach to the prevention of CVD in the US population.
Howard et al. (2002)	Hcy in the acute phase after stroke	76 stroke patients, 40–85 years	Hcy levels at baseline of 11.3 ± 0.5 μmol/L increased consistently to a 12.0 ± 0.05, 12.4 ± 0.5, 13.3 ± 0.5 and 13.7 ± 0.7 μmol/L at days 3, 5, 7 and 10 to 14, respectively after stroke. These data suggest that the clinical interpretation of Hcy after stroke and the eligibility for clinical trials assessing treatment for elevated Hcy levels require an adjustment in time since stroke.
He et al. (2004)	Folate, vitamin B_6 and B_{12} intakes and risk of stroke	43 732 men free of cardiovascular diseases and diabetes at baseline, 40–75 years	Intake of vitamin B_{12}, but not B_6, was inversely associated with risk of ischemic stroke. Increased folate intake was associated with decreased risk of ischemic stroke in men.
Iso et al. (2004)	Elevated Hcy and risk of total stroke and its subtypes	11 846 Japanese subjects	High serum Hcy level (>11.0 μmol/L) was associated with a threefold higher risk of total stroke. The excess risk was confined to ischemic stroke, more specifically lacunar infarction.

Zhang et al. (2010)	Hcy levels and the risk of stroke recurrence and all-cause mortality in a large prospective stroke population	1823 Chinese stroke patients	After adjustment for age, gender and other cardiovascular risk factors, a high Hcy concentration was associated with an increased risk of 1.74-fold for stroke recurrence and 1.75-fold for all-cause mortality when highest and lowest categories were compared. This study suggest that elevated Hcy concentrations can predict the risk of stroke recurrence and mortality in patients with stroke.
Intervention studies			
Clarke and Armitage (2000)	Vitamins supplements and CVD risk	Meta-analysis of 12 randomised trials	Dietary folate reduced Hcy levels by 25%. Vitamin B_{12} produced and additional reduction in blood Hcy of 7%, whereas vitamin B_6 did not have any significant effect.
Spence et al. (2001)	Vitamins to prevent stroke	3688 stroke patients, $\geq$ 35 years	The Vitamin Intervention for Stroke Prevention (VISP) study is a double-masked, randomised, multicentre clinical trial designed to determine if, in addition to best medical/surgical management, high-dose folic acid, vitamin B_6 and vitamin B_{12} supplements will reduce recurrent stroke compared to lower doses of these vitamins.
VITATOPS Trial Study Group (2002)	Vitamins to prevent stroke	8000 stroke patients, any age	VITATOPS study follows up 8000 patients between 2000 and 2009 provided a reliable estimate of the safety and effectiveness of dietary supplementation with folic acid, vitamin B_{12} and vitamin B_6 in reducing recurrent serious vascular events and transient ischemic attacks stroke.

Table 29.1 (Continued)

Study	Study focus	Number of subjects	Results
Lonn et al. (2006)	B vitamins in vascular disease	5522 patients with a history of vascular disease, ≥ 55 years	Supplements combining folic acid (2.5 mg) and vitamins B_6 (50 mg) and B_{12} (1 mg) did not reduce the risk of major cardiovascular events in patients with vascular disease.
Albert et al. (2008)	Daily intake of folic acid and B vitamins and risk of cardiovascular events and total mortality	5442 women, ≥ 42 years	After 7.3 years of treatment and follow-up, a combination pill of folic acid (2.5 mg), vitamin B_6 (50 mg) and vitamin B_{12} (1 mg) did not reduce a combined end-point of total cardiovascular events among high-risk women, despite significant homocysteine lowering.
Saposnik et al. (2009)	B vitamins and stroke risk, severity, and disability	5522 patients with known cardiovascular disease, ≥ 55 years	Lowering of Hcy with folic acid (2.5 mg) and vitamins B_6 (50 mg) and B_{12} (1 mg) did reduce the risk of overall stroke, but not stroke severity or disability.
VITATOPS Trial Study Group (2010)	Hcy, B vitamins and patients with previous stroke or transient ischemic attack	8164 patients	Patients were followed up for a median duration of 3.4 years; 616 (15%) patients assigned to B vitamins and 678 (17%) assigned to placebo reached the primary end-point. There were no unexpected serious adverse reactions and no significant differences in common adverse effects between the treatment groups.
Almeida et al. (2010)	B vitamins and long-term prevalence of depression in stroke survivors	563 patients	Long-term treatment (1–10.5 years) of post-stroke survivors with folic acid (2 mg), B_6 (25 mg) and B_{12} (0.5 mg) was associated with a reduction in the hazard of major depression in the patient population.

Editorials and reviews

Kuller and Evans (1998)	Hcy, vitamins and CVD	Editorial	Possible associations between Hcy, B vitamins and vascular disease.
Perry (1999a) Perry (1999b)	Hcy and risk of stroke	Review	High Hcy in apparently well-nourished populations and the tendency of high Hcy with aging, and the effects of Hcy on stroke, reduction of Hcy will have profound implications for public health.
Goldstein (2000)	Hcy and risk of stroke	Review	Association between moderately elevated levels of Hcy and stroke patients with chronic inflammation, as well as those with chronic or acute infection, are at elevated stroke risk.
Robinson (2000)	Hcy, B vitamins and risk of CVD	Editorial	Routine measurement of Hcy still remains speculative until the results of the some intervention trials became known.
Rosenberg (2001)	B vitamins and cognitive function	Review	Importance of B vitamins (B_6 and B_{12}) and folate in neurocognitive and other neurologic functions.
Hankey (2002)	Hcy and risk of stroke	Editorial	Plasma Hcy and routine treatment of high Hcy with vitamins to prevent symptomatic cerebrovascular disease remains insufficient.

Evidence of the importance of the B vitamins folic acid, vitamin B_6 and vitamin B_{12} for the well-being and normal function of the brain derives from data showing neurological and psychological dysfunction in vitamin deficiency states and in cases of congenital defects of one-carbon metabolism. A review by Selhub et al. (2000) indicated that the status of these vitamins is frequently inadequate in the elderly and recent studies have shown associations between loss of cognitive function or AD and inadequate B vitamin status. The authors suggest that low B vitamin intake may affect methylation reactions, which are crucial for normal brain function. Poor B vitamin status can also result in high Hcy, a risk factor for occlusive vascular disease, stroke and thrombosis, and thus may contribute to brain ischemia.

Evidence of the importance of B vitamins (B_{12} and B_6, and folate) in neurocognitive and other neurological functions is derived from reported cases of severe vitamin deficiencies (particularly pernicious anaemia) or homozygous defects in genes that encode for enzymes of one-carbon metabolism (Rosenberg 2001). However, the data from a recent systematic review of randomized trials does not yet provide adequate evidence of an effect of vitamin B_6 or B_{12} or folic acid supplementation, alone or in combination, on cognitive function testing in people with either normal or impaired cognitive function (Balk et al. 2007). Low levels of folate and vitamin B_6 have been regarded to confer an increased risk of atherosclerosis (Robinson et al. 1998). High plasma Hcy concentration is a risk factor for atherosclerosis and circulating concentrations of Hcy are related to vitamin B_{12} status, as well as folate and vitamin B_6. If elevated Hcy promotes cognitive dysfunction, then lowering Hcy by means of B vitamin supplementation may protect cognitive function by arresting or slowing the disease process (Troen and Rosenberg 2005). Although elevated plasma Hcy concentrations have been implicated with risk of cognitive impairment and dementia, it is unclear whether low vitamin B_{12} or folate status is responsible for cognitive decline (Clarke 2008).

Various studies have suggested that a generous intake of folate and B vitamins may be beneficial in stroke prevention by reducing the level of plasma Hcy (Robinson et al. 1998). The association between dietary intake of folate and the subsequent risk of stroke and cardiovascular disease (CVD) is well documented.

Participants in the National Health and Nutrition Examination Survey I Epidemiologic Follow-up Study included 9764 US men and women aged 25–74 years who were free of CVD at baseline. The results showed that relative risk of incidence of stroke events was lower among subjects with dietary folate intake in the highest quartile (405.0 μg/day) compared with those in the lowest quartile (99.0 μg/day), after adjustment for established cardiovascular risk factors and dietary factors (Bazzano et al. 2002).

A recent study evaluated whether a combination of folic acid, vitamin B_6 and vitamin B_{12} lowers risk of CVD among high-risk women with and without CVD. A total of 5442 women who were US health professionals aged 42 years or older, with either a history of CVD or three or more coronary risk factors, were enrolled in a randomized, double-blind, placebo-controlled trial to receive a combination pill containing 2.5 mg of folic acid, 50 mg of vitamin B_6 and 1 mg of vitamin B_{12}; they were treated for 7.3 years from April 1998 to July 2005. After 7.3 years of treatment and follow-up, the combination of B vitamins tested did not reduce a combined end point of total cardiovascular events among high-risk women, despite significant Hcy lowering (Albert et al. 2008).

In addition, the results of the Heart Outcomes Prevention Evaluation (HOPE-2) study showed that combined daily administration of 2.5 mg of folic acid, 50 mg of vitamin B_6 and 1 mg of vitamin B_{12} for five years had no beneficial effects on major vascular events in a high-risk vascular disease population (Lonn et al. 2006), although fewer patients assigned to active treatment than to placebo had a stroke (relative risk, 0.75; 95 percent confidence interval, 0.59 to 0.97). Recent findings from this study show that lowering of Hcy with folic acid and vitamins B_6 and B_{12} did reduce the risk of overall stroke, but not stroke severity or disability (Saposnik et al. 2009).

Bønna et al. (2006) evaluated the efficacy of Hcy-lowering treatment with B vitamins for secondary prevention in patients who had had an acute myocardial infarction. The trial included 3749 men and women who had had an acute myocardial infarction within seven days before randomization. Patients were randomly assigned, in a two by-two factorial design, to receive one of the following four daily treatments:

- 0.8 mg of folic acid, 0.4 mg of vitamin B_{12} and 40 mg of vitamin B_6;
- 0.8 mg of folic acid and 0.4 mg of vitamin B_{12};
- 40 mg of vitamin B_6; or
- placebo.

The primary end-point during a median follow-up of 40 months was a composite of recurrent myocardial infarction, stroke and sudden death attributed to coronary artery disease. The authors conclude that treatment with B vitamins did not lower the risk of recurrent cardiovascular disease after acute myocardial infarction.

The folic acid fortification programme in the US has decreased the prevalence of low levels of folate and hyperhomocysteinemia to 1% or lower (Green and Miller 2005). Folate is distributed widely in green leafy vegetables, citrus fruits and animal products. The biologically active form of folic acid is tetrahydrofolate (THF), which plays a key role in the transfer of one-carbon units, such as methyl, methylene and formyl groups, to the essential substrates involved in the synthesis of DNA, RNA and proteins. More specifically, THF is involved with the enzymatic reactions necessary for synthesis of purine, thymidine and amino acid. Not surprisingly, manifestations of folate deficiency thereafter would result in impairment of cell division, accumulation of possibly toxic metabolites such as Hcy, and impairment of methylation reactions involved in the regulation of gene expression. Mechanistically speaking, current theory proposes that folate is essential for synthesis of S-adenosylmethionine, which is involved in numerous methylation reactions. This methylation process is central to the biochemical basis of proper neuropsychiatric functioning.

A large body of evidence suggests that intake of folate and B vitamins may be beneficial in stroke prevention by reducing levels of plasma Hcy; however, limited information is available regarding dietary intake of vitamins or use of vitamin supplements by stroke patients. Beamer *et al.* (1999) collected information regarding the use of vitamins supplements from 231 patients with acute ischemic stroke. These authors also recorded vitamin intake from 94 subjects with similar clinical risk factors for stroke, including hypertension, diabetes and myocardial ischemia, and from 59 healthy community volunteers who denied the presence of stroke risk factors and who were matched in age with the two vascular groups. Fewer subjects who had stroke were taking vitamins, compared with healthy elderly community volunteers. In addition, Hcy levels available from 49 stroke patients, 31 patients with stroke risk factors, and seven control subjects, were significantly lower in subjects taking vitamins. The pattern of lower Hcy values with vitamin use was consistent across groups, including stroke (13.9 ± 6.7 *vs.* $14.6 \pm 4.1\,\mu mol/L$), risk (13.5 ± 4.3 *vs.* $15.0 \pm 6.9\,\mu mol/L$) and healthy elderly subjects (7.1 ± 0.1 *vs.* $10.5 \pm 2.3\,\mu mol/L$). The researchers concluded that Hcy levels are influenced by a complex interaction of gender, dietary levels of protein intake, dietary and/or supplemental vitamin use and cardiovascular risk factors. These results suggest that use of vitamin supplements may be associated with lower levels of Hcy in elderly persons whether or not stroke or stroke-related risk factors are present. Also, these data suggest that less frequent use of vitamin supplementation in the elderly may be associated with an increased risk for stroke (Beamer *et al.* 1999).

Different epidemiological studies suggest that raised plasma concentrations of total Hcy may be a common, causal and treatable risk factor for atherothromboembolic ischemic stroke. Therefore, a study aimed to assess the effect of Vitamins to Prevent Stroke (VITATOPS)—an international, multi-centre, randomized, double-blind, placebo-controlled clinical trial—using a multivitamin therapy (folic acid 2 mg, vitamin B_6 25 mg, vitamin B_{12} 500 µg) has been carried out. A total of 8000 patients were randomized and followed up for a mean period of 2.5 years (range 1–8 years) (VITATOPS Trial Study Group 2002).

Some findings of this clinical trial are that daily administration of folic acid, vitamin B_6 and vitamin B_{12} to patients with recent stroke or transient ischemic attack was safe but did not seem to be more effective than placebo in reducing the incidence of major vascular events. These results do not support the use of B vitamins to prevent recurrent stroke. The results of ongoing trials and an individual patient data meta-analysis will add statistical power and precision to present estimates of the effect of B vitamins (VITATOPS Trial Study Group 2010).

Another important result found was that among 273 people who completed the final assessment after 7.1 ± 2.1 years of follow-up, random assignment to B vitamins was associated with a reduction in the hazard of major depression compared with placebo. If these findings can be validated externally, B vitamin supplementation could offer hope as an effective, safe, and affordable intervention to reduce the burden of post-stroke depression (Almeida *et al.* 2010).

Publication of the results of the landmark Vitamin Intervention for Stroke Prevention (VISP) Trial is the first evidence from a large randomized controlled trial of the effect of lowering total Hcy *via* folic acid-based multiple B vitamin supplementation on the incidence of 'hard' clinical events, such as recurrent stroke, in patients with recent ischemic stroke (Spence *et al.* 2001; Toole *et al.* 2004). Hankey and Eikelboom (2004) believed that Hcy hypothesis of atherothrombotic vascular disease in general, and stroke in particular, remains viable. Two studies using different methods yield consistent results in support of the hypothesis (He *et al.* 2004; Wald *et al.* 2002).

Several studies are in progress to determine whether treatment with folic acid in combination with vitamins B_6 and B_{12} will reduce the risk of stroke in patients with increased serum Hcy. A meta-analysis of 12 randomized trials of vitamin supplements to lower Hcy levels was carried out to determine the optimal dose of folic acid required to lower Hcy levels and to assess whether vitamin B_{12} or vitamin B_6 had additive effects (Clarke and Armitage 2000). This meta-analysis demonstrated that reductions in blood Hcy levels were greater at higher pretreatment blood Hcy levels and at lower pretreatment folate concentrations. After standardization for a pretreatment Hcy concentration of 12 µmol/L and folate concentration of 12 nmol/L (approximate average concentrations for Western populations), dietary folic acid reduced Hcy levels by 25% (95% confidence interval: 23–28%) with similar effects in a daily dosage ranging from 0.5 to 5 mg. Vitamin B_{12} (mean 0.5 mg) produced an additional reduction in blood Hcy of 7%, whereas vitamin B_6 (mean 16.5 mg) did not have any significant effect. Hence, in typical populations, daily supplementation with both 0.5–5 mg folic acid and about 0.5 mg vitamin B_{12} would be expected to reduce Hcy levels by one quarter to one third (from about 12 µmol/L to about 8–9 µmol/l). Large-scale randomized trials of such regimens are now required to determine whether lowering Hcy levels by folic acid and vitamin B_{12}, with or without added vitamin B_6, reduces the risk of vascular disease.

In this sense, it has also to be considered on one hand that high doses of folic acid may mask the megaloblastic anaemia due to vitamin B_{12} deficiency seen in elderly people as a result of atrophic gastritis. On the other hand, we have to be aware that vitamin B_6 status is primarily a determinant of postprandial Hcy levels, but not fasting levels. Thus, in studies of B vitamin supplements and Hcy, it often appears that vitamin B_6 has little effect on Hcy levels because these studies typically only look at fasting Hcy levels.

In addition, it is interesting to consider the issue of folic acid fortification. In the US and Canada, folic acid fortification of enriched grain products was fully implemented by 1998. Yang *et al.* (2006) evaluated trends in stroke-related mortality before and after folic acid fortification in the US and Canada and, as a comparison, during the same period in England and Wales, where fortification is not required. They observed a trend consistent with the hypothesis that folic acid fortification is contributing to a reduction in stroke deaths.

Patients with chronic inflammation as well as those with chronic or acute infection are at elevated stroke risk (Goldstein 2000). Due to the high prevalence of high Hcy among apparently well-nourished populations and the tendency for Hcy concentrations to increase with age, and the effects of Hcy on stroke risk, lowering the levels of Hcy may have profound implications for public health (Perry 1999a,1999b). However, according to Robinson (2000) the benefit of routine measurement of Hcy concentrations remains speculative until the results of the some intervention trials become known. In the same sense, Hankey (2002) concluded that there is not sufficient evidence to recommend routine screening of plasma Hcy and routine treatment of high Hcy concentrations with vitamins to prevent symptomatic cerebrovascular disease.

29.3 Homocysteine

High Hcy concentration is an independent risk factor for coronary artery disease, stroke, peripheral vascular atherosclerosis, and for arterial and venous thromboembolism, although the mechanisms for this effect remain poorly understood. The association between cognitive function and risk of death from stroke suggests that cerebrovascular disease is an important cause of declining cognitive function.

Hcy is believed to cause atherogenesis and thrombogenesis *via* endothelial damage, focal vascular smooth muscle proliferation probably causing irregular vascular contraction, and coagulation abnormalities (Christopher *et al.* 2007). The significance of any association between cardiovascular disease CVD and stroke and circulating Hcy concentrations is attracting considerable attention (Carlsson 2007; Christopher *et al.* 2007; Clarke *et al.* 2007; Cui *et al.* 2008; Loscalzo 2006; McNulty *et al.* 2008; Ntaios *et al.* 2008; Spence 2006).

Morris *et al.* (2000) evaluated the association between serum Hcy concentration and self-report heart attack or stroke on adult male and female participants in the third National Health and Nutrition Examination Survey (NHANES III). The study reported 2.4 times more episodes of heart attack or stroke in men with Hcy concentration of >12 μmol/L than among people with lower values.

According to Kuller and Evans (1998), Hcy and B vitamins levels may contribute to the development of vascular disease through mechanisms independent of the atherosclerosis process. In fact, whereas high Hcy levels are directly related to development of atherosclerosis, a decrease in folate or vitamins B_{12} and B_6 increases the risk of vascular disease independently of atherosclerosis. High Hcy levels could be associated with an enhancement of inflammatory process and increased risk of thrombosis.

Giles *et al.* (1998) found that, in a representative sample of US adults, Hcy concentration was independently associated with an increased likelihood of non-fatal stroke, and this association was present in both black and white adults. In a different study, Perry *et al.* (1995) measured serum Hcy levels in 107 cases and 118 control males, matched for age and without a history of stroke at the time of screening; some did develop stroke or myocardial infarction during follow-up. Levels of Hcy were not very high but lower in controls (11.3–12.6 μmol/L) than among cases (12.7–14.8 μmol/L). Other studies have also indicated that small differences in Hcy levels may significantly contribute to increase the risk of ischemic stroke (Bots *et al.* 1999; Fallon *et al.* 2001; Verhoef *et al.* 1994).

To assess the relationship of Hcy concentrations with vascular disease risk, a meta-analysis of observational studies was carried out, showing that elevated Hcy is at most a modest independent predictor of ischemic heart disease and stroke risk in healthy populations. Studies of the impact on disease risk of genetic variants that affect blood Hcy concentrations will help determine whether Hcy is causally related to vascular disease, as may large randomized trials of the effects on ischemic heart disease and stroke of vitamin supplementation to lower blood Hcy concentrations (Homocysteine Studies Collaboration 2002).

After a long debate, due to conflicting data from clinical studies, Hcy is now largely accepted as a risk factor for CVD including stroke. Zhang and co-workers (Zhang *et al.* 2010) have evaluated the role of elevated homocysteine levels in stroke recurrences, proving that Chinese patients with high Hcy levels had an increased risk of stroke recurrence and of all-cause mortality with respect to patients with lower levels. Remarkably, in their study, high Hcy levels were associated with an increased risk of stroke recurrence for athero-thrombotic stroke and intracerebral hemorrhage, but not lacunar stroke (Sacco and Carolei 2010).

29.4 Concluding Remarks

A large body of evidence suggests that intake of folate and B vitamins may be beneficial in stroke prevention by reducing levels of plasma Hcy. However, limited information is available regarding dietary intake of vitamins or use of vitamin supplements by stroke patients. Several studies are in progress to determine whether treatment with folic acid in combination with vitamins B_6 and B_{12} will reduce the risk of stroke in patients with increased serum. The significance of any association between CV and stroke, and circulating Hcy concentrations is attracting considerable attention.

As a general conclusion we can affirm that epidemiological associations and results of intervention studies suggest that low B vitamin status and elevated blood Hcy are risk factors for stroke. After reviewing the observational and intervention studies, there is an incomplete understanding of mechanisms and some conflicting findings. Therefore the available evidence is insufficient to recommend the routine use of B vitamins for the prevention of stroke. Further well-designed controlled clinical trials are necessary to further progress in this area.

Summary Points

- This chapter focuses on the role of B vitamins—folate, B_6 and B_{12}—on stroke.
- Folic acid, vitamin B_6 and vitamin B_{12} are all cofactors in homocysteine metabolism.

- Poor dietary intake of folate, vitamin B_6 and vitamin B_{12} are associated with increased risk of stroke.
- There is an incomplete understanding of mechanisms and some conflicting findings; therefore the available evidence is insufficient to recommend the routine use of B vitamins for the prevention of stroke.
- Further well-designed controlled clinical trials are necessary to further progress in this area.

Key Facts

Key Features of B Vitamins: Folate, Vitamin B_6 and Vitamin B_{12}

- Folate (folic acid and folacin) is a water-soluble B vitamin that is necessary in forming coenzymes for purine and pyrimidine synthesis, erythropoiesis, and methionine regeneration.
- The current dietary reference intake (DRI) established by the Food and Nutrition Board of the Institute of Medicine for folate is 400 micrograms per day.
- The richest food sources of folate are dark-green leafy vegetables, whole-grain cereals, fortified grain products and animal products.
- Vitamin B_6 refers to a group of nitrogen-containing compounds with three primary forms: pyridoxine, pyridoxal, and pyridoxamine. They are water-soluble and are found in a variety of plant and animal products.
- Vitamin B_6 participates in more than 100 enzymatic reactions and is needed, among other functions, for protein metabolism, conversion of tryptophan to niacin, and neurotransmitter formation.
- The current DRI established by the Food and Nutrition Board of the Institute of Medicine for vitamin B_6 is 1.7 milligrams per day.
- The best dietary sources include poultry, fish, meat, legumes, nuts, potatoes and whole grains.
- Vitamin B_{12} (cyanocobalamin) is a water-soluble B vitamin that acts as a coenzyme for fat and carbohydrate metabolism, protein synthesis, and hematopoiesis.
- The current DRI established by the Food and Nutrition Board of the Institute of Medicine for vitamin B_{12} is 2.4 micrograms per day.
- Vitamin B_{12} is found naturally in animal products such as fish, poultry, meat, eggs or dairy. It is also found in fortified breakfast cereals and enriched soy or rice milk. Most people have plenty of vitamin B_{12} in their diets.

Key Features of Stroke

- According to the World Health Organization (WHO), a stroke is caused by the interruption of the blood supply to the brain, usually because a blood vessel bursts or is blocked by a clot. This cuts off the supply of oxygen and nutrients, causing damage to the brain tissue.
- The most common symptom of a stroke is sudden weakness or numbness of the face, arm or leg, most often on one side of the body.

- Other symptoms include: confusion, difficulty speaking or understanding speech; difficulty seeing with one or both eyes; difficulty walking, dizziness, loss of balance or coordination; severe headache with no known cause; fainting or unconsciousness.
- The effects of a stroke depend on which part of the brain is injured and how severely it is affected. A very severe stroke can cause sudden death.
- Globally, cerebrovascular disease (stroke) is the second leading cause of death.
- It is a disease that predominantly occurs in mid-age and older adults.
- WHO estimated that in 2005, stroke accounted for 5.7 million deaths worldwide, equivalent to 9.9% of all deaths.
- Over 85% of these deaths will have occurred in people living in low and middle income countries and one third will be in people aged less than 70 years.

Definitions of Words and Terms

Atherogenesis. Formation of atheromas, plaques in the intima of arteries that narrow the arterial passage, restricting blood flow and increasing the risk of occlusion.

B vitamins. Among the 13 vitamins, B vitamins are eight water-soluble vitamins (vitamin B_1: thiamine; vitamin B_2: riboflavin; vitamin B_3: niacin; vitamin B_5: pantothenic acid; vitamin B_6: pyridoxine, pyridoxal or pyridoxamine; vitamin B_7: biotin; vitamin B_9: folic acid or folate; and vitamin B_{12}: cyanocobalamin). Folate and vitamins B_6 and B_{12} have joint effects on homocysteine.

Cardiovascular disease. Cardiovascular disease (CVD) is a group of disorders of the heart and blood vessels and includes coronary heart disease, cerebrovascular disease, peripheral arterial disease, rheumatic heart disease, congenital heart disease (malformations of heart structure existing at birth) and deep vein thrombosis and pulmonary embolism.

Cerebrovascular accident. Or stroke.

Cognitive function. Intellectual or mental process by which one becomes aware of, perceives, or comprehends ideas. It involves all aspects of perception, thinking, reasoning and remembering. Aging and some disease implies a cognitive decline or cognitive impairment.

Epidemiological studies. Epidemiology is the study of the distribution and determinants of disease frequency in human populations and the application of this study to control health problems. Therefore, epidemiological studies harvest valid and precise information about the causes, prevention and treatments for disease. Types of epidemiological studies are experimental studies and observational studies (cohort and case–control studies).

Homocysteine. Homocysteine is a sulfur-containing amino acid that occurs naturally in all humans. Elevated plasma homocysteine concentration is linked to an increased risk of ischemic stroke. Lowering homocysteine plasma levels is linked to increasing the intake of folic acid and vitamins B_6 and B_{12}.

Intervention studies. Intervention studies represent the experimental aspect of epidemiology. Intervention studies are prospective studies in which an exposure or exposures are assigned to people and subsequent outcomes observed.

Stroke. A stroke is an interruption of the blood supply to any part of the brain. Stroke can be caused by either a clot obstructing the flow of blood to the brain (ischemic) or by a blood vessel rupturing and preventing blood flow to the brain (hemorrhagic). Ischemic stroke comprises the majority of all strokes (approximately 85%).

Vitamin deficiency. This is the lack of vitamins in the organism in the adequate amount. Vitamins are compounds that cannot be synthesized by humans and therefore must be ingested to prevent metabolic disorders. Inadequate intake or subtle deficiencies in several vitamins are risk factors for chronic diseases such as cardiovascular disease, cancer and osteoporosis.

List of Abbreviations

AD	Alzheimer's disease
CVD	cardiovascular disease
Hcy	homocysteine
MTHF	5-methyltetrahydrofolate
NHANES	National Health and Nutrition Examination Survey
HOPE	Heart Outcomes Prevention Evaluation
DRI	dietary reference intake
THF	tetrahydrofolate
VISP	Vitamin Intervention for Stroke Prevention
VITATOPS	Vitamins to Prevent Stroke
WHO	World Health Organization

References

Albert, C.M., Cook, N.R., Gaziano, J.M., Zaharris, E., MacFadyen, J., Danielson, E., Buring, J.E., and Manson, J.E., 2008. Effect of folic acid and B vitamins on risk of cardiovascular events and total mortality among women at high risk for cardiovascular disease: A randomized trial. *The JAMA-Journal of the American Medical Association.* 299: 2027–2036.

Almeida, O.P., Marsh, K., Alfonso, H., Flicker, L., Davis, T.M., and Hankey, G.J, 2010. B-vitamins reduce the long-term risk of depression after stroke: The VITATOPS-DEP trial. *Annals of Neurology.* 68: 503–510.

Balk, E.M., Raman, G., Tatsioni, A., Chung, M., Lau, J., and Rosenberg, I.H., 2007. Vitamin B6, B12, and folic acid supplementation and cognitive function: a systematic review of randomized trials. *Archives of Internal Medicine.* 167: 21–30.

Bazzano, L.A., He, J., Ogden, L.G., Loria, C., Vupputuri, S., Myers, L., and Whelton, P.K., 2002. Dietary intake of folate and risk of stroke in US men and women: NHANES I Epidemiologic Follow-up Study. National Health and Nutrition Examination Survey. *Stroke*. 33: 1183–1189.

Beamer, N.B., Coull, B.M., Press, R.D., and Anderson, L.R., 1999. Vitamin use in patients with ischemic stroke. *Neurology*. 52: A64.

Bønaa, K.H., Njølstad, I., Ueland, P.M., Schirmer, H., Tverdal, A., Steigen, T., Wang, H., Nordrehaug, J.E., Arnesen, E., and Rasmussen, K., 2006. Homocysteine lowering and cardiovascular events after acute myocardial infarction. *New England Journal of Medicine*. 354: 1578–1588.

Bots, M.L., Launer, L.J., Lindemans, J., Hoes, A.W., Hofman, A., Witteman, J.C.M., Koudstaal, P.J., and Grobbee, D.E., 1999. Homocysteine and short-term risk of myocardial infarction and stroke in the elderly: the Rotterdam Study. *Archives of Internal Medicine*. 159: 38–44.

Carlsson, C.M., 2007. Lowering homocysteine for stroke prevention. *Lancet*. 369: 1841–1842.

Christopher, R., Nagaraja, D., and Shankar, S.K., 2007. Homocysteine and cerebral stroke in developing countries. *Current Medical Chemistry*. 14: 2393–2401.

Clarke, R., and Armitage, J., 2000. Vitamin supplements and cardiovascular risk: review of the randomized trials of homocysteine-lowering vitamin supplements. *Seminars in Thrombosis and Hemostasis*. 26: 341–348.

Clarke, R., Armitage, J., Lewington, S., Collins, R., and Clarke, R., 2007. Homocysteine-lowering trials for prevention of vascular disease: pProtocol for a collaborative meta-analysis. *Clinical Chemistry and Laboratory Medicine*. 45: 1575–1581.

Clarke, R., 2008. B vitamins and prevention of dementia. *Proceedings of the Nutrition Society*. 67: 75–81.

Clarke, R., Armitage, J., Lewington, S., Collins, R., and Clarke, R., 2007. Homocysteine-lowering trials for prevention of vascular disease: Protocol for a collaborative meta-analysis. *Clinical Chemistry and Laboratory Medicine*. 45: 1575–1581.

Cui, R.Z., Moriyama, Y., Koike, K.A., Date, C., Kikuchi, S., Tamakoshi, A., and Iso, H., 2008. Serum total homocysteine concentrations and risk of mortality from stroke and coronary heart disease in Japanese: the JACC study. *Atherosclerosis*. 198: 412–418.

Fallon, U.B., Elwood, P., Ben-Shlomo, Y., Ubbink, J.B., Greenwood, R., and Smith, G.D., 2001. Homocysteine and ischaemic stroke in men: the Caerphilly study. *Journal of Epidemiology and Community Health*. 55: 91–96.

Giles, W.H., Croft, J.B., Greenlund, K.J., Ford, E.S., and Kittner, S.J., 1998. Total homocyst(e)ine concentration and the likelihood of nonfatal stroke: results from the Third National Health and Nutrition Examination Survey, 1988-1994. *Stroke*. 29: 2473–2477.

Goldstein, L.B., 2000. Novel risk factors for stroke: homocysteine, inflammation, and infection. Current Atherosclerosis Reports.4.

Gonzalez-Gross, M., Marcos, A., and Pietrzik, K., 2001. Nutrition and cognitive impairment in the elderly. *British Journal of Nutrition*. 86: 313–321.

Green, R., and Miller, J.W., 2005. Vitamin B12 deficiency is the dominant nutritional cause of hyperhomocysteinemia in a folic acid-fortified population. *Clinical Chemistry and Laboratory Medicine.* 43: 1048–1051.

Hankey, G.J., 2002. Is homocysteine a causal and treatable risk factor for vascular diseases of the brain (cognitive impairment and stroke)? Annals of Neurology. 51: 279–281.

Hankey, G.J., and Eikelboom, J.W., 2004. Folic acid-based multivitamin therapy to prevent stroke: the jury is still out. *Stroke.* 35: 1995–1998.

Hankey, G.J., 2002. Is homocysteine a causal and treatable risk factor for vascular diseases of the brain (cognitive impairment and stroke)? *Annals of Neurology.* 51: 279–281.

He, K., Merchant, A., Rimm, E.B., Rosner, B.A., Stampfer, M.J., Willett, W.C., and Ascherio, A., 2004. Folate, vitamin B6, and B12 intakes in relation to risk of stroke among men. *Stroke.* 35: 169–174.

Henon, H., Lebert, F., Durieu, I., Godefroy, O., Lucas, C., Pasquier, F., and Leys, D., 1999. Confusional state in stroke: relation to preexisting dementia, patient characteristics, and outcome. *Stroke.* 30: 773–779.

Homocysteine Studies Collaboration, 2002. Homocysteine and risk of ischemic heart disease and stroke: a meta-analysis. *The JAMA-Journal of the American Medical Association.* 288: 2015–2022.

Howard, V.J., Sides, E.G., Newman, G.C., Cohen, S.N., Howard, G., Malinow, M.R., and Toole, J.F., 2002. Changes in plasma homocyst(e)ine in the acute phase after stroke. *Stroke.* 33: 473–478.

Iso, H., Moriyama, Y., Sato, S., Kitamura A., Tanigawa, T., Yamagishi, K., Imano, H., Ohira, T., Okamura, T., Naito, Y., and Shimamoto, T., 2004. Serum total homocysteine concentrations and risk of stroke and its subtypes in Japanese. *Circulation.* 109: 2766–2772.

Kase, C.S., Wolf, P.A., Kelly-Hayes, M., Kannel, W.B., Beiser, A., and D'Agostino, R.B., 1998. Intellectual decline after stroke: the Framingham Study. *Stroke.* 29: 805–812.

Kuller, L.H., and Evans, R.W., 1998. Homocysteine, vitamins, and cardiovascular disease. *Circulation.* 98: 196–199.

Lonn, E., Yusuf, S., Arnold, M.J., Sheridan, P., Pogue, J., Micks, M., McQueen, M.J., Probstfield, J., Fodor, G., Held, C., Genest, J., and the Heart Outcomes Prevention Evaluation (HOPE) 2 Investigators, 2006. Homocysteine lowering with folic acid and B vitamins in vascular disease. *The New England Journal of Medicine.* 354: 1567–1577.

Loscalzo, J., 2006. Homocysteine trials—clear outcomes for complex reasons. *The New England Journal of Medicine.* 354: 1629–1632.

McNulty, H., Pentieva, K., Hoey, L., and Ward, M., 2008. Homocysteine, B vitamins and CVD. *Proceedings of the Nutrition Society.* 67: 232–237.

Mesnard, F., Roscher, A., Garlick, A.P., Girard, S., Baguet, E., Arroo, R.R.J., Lebreton, J., Robins, R.J., and Ratcliffe, R.G., 2002. Evidence for the involvement of tetrahydrofolate in the demethylation of nicotine by Nicotiana plumbaginifolia cell-suspension cultures. *Planta.* 214: 911–919.

Molloy, A.M., Mills, J.L., Kirke, P.N., Weir, D.G., and Scott, J.M., 1999. Folate status and neural tube defects. *Biofactors.* 10: 291–294.

Morris, M.S., Jacques, P.F., Rosenberg, I.H., Selhub, J., Bowman, B.A., Gunter, E.W., Wright, J.D., and Johnson, C.L., 2000. Serum total homocysteine concentration is related to self-reported heart attack or stroke history among men and women in the NHANES III. *Journal of Nutrition.* 130: 3073–3076.

Nilsson, K., Gustafson, L., and Hultberg, B. 1999. Plasma homocysteine is a sensitive marker for tissue deficiency of both cobalamines and folates in a psychogeriatric population. *Dementia and Geriatric Cognitive Disorders.* 10: 476–482.

Ntaios, G.C., Savopoulos, C.G., Chatzinikolaou, A.C., Kaiafa, G.D., and Hatzitolios, A., 2008. Vitamins and stroke: the homocysteine hypothesis still in doubt. *Neurologist.* 14: 2–4.

Patel, M.D., Coshall, C., Rudd, A.G., and Wolfe, C.D., 2002. Cognitive impairment after stroke: clinical determinants and its associations with long-term stroke outcomes. *Journal of the American Geriatrics Society.* 50: 700–706.

Perry, I.J., Refsum, H., Morris, R.W., Ebrahim, S.B., Ueland, P.M., and Shaper, A.G., 1995. Prospective study of serum total homocysteine concentration and risk of stroke in middle-aged British men. *Lancet.* 346: 1395–1398.

Perry, I.J., 1999a. Homocysteine, hypertension and stroke. *Journal of Human Hypertension.* 13: 289–293.

Perry, I.J., 1999b. Homocysteine and risk of stroke. *Journal of Cardiovascular Risk.* 6: 235–240.

Perry, I.J., Refsum, H., Morris, R.W., Ebrahim, S.B., Ueland, P.M., and Shaper, A.G., 1995. Prospective study of serum total homocysteine concentration and risk of stroke in middle-aged British men. *Lancet.* 346: 1395–1398.

Pohjasvaara, T., Erkinjuntti, T., Ylikoski, R., Hietanen, M., Vataja, R., and Kaste M., 1998. Clinical determinants of poststroke dementia. *Stroke.* 29: 75–81.

Robinson, K., 2000. Homocysteine, B vitamins, and risk of cardiovascular disease. *Heart.* 83: 127–130.

Robinson, K., Arheart, K., Refsum, H., Brattström, L., Boers, G., Ueland, P., Rubba, P., Palma-Reis, R., Meleady, R., Daly, L., Witteman, J., and Graham, I., 1998. Low circulating folate and vitamin B6 concentrations: risk factors for stroke, peripheral vascular disease, and coronary artery disease. *Circulation.* 97: 437–443.

Robinson, K., 2000. Homocysteine, B vitamins, and risk of cardiovascular disease. *Heart.* 83: 127–130.

Rogers, P.J., 2001. A healthy body, a healthy mind: long-term impact of diet on mood and cognitive function. *Proceedings of the Nutrition Society.* 60: 135–143.

Rosenberg, I.H., 2001. B vitamins, homocysteine, and neurocognitive function. *Nutrition Reviews.* 59: S69–S73.

Sacco, S., and Carolei, A., 2010. Homocysteine and stroke: another brick in the wall. *Clinical Science*. 118: 183–185.

Saposnik, G., Ray, J.G., Sheridan, P., McQueen, M., Lonn, E., and the HOPE 2 Investigators, 2009. Homocysteine-lowering therapy and stroke risk, severity, and disability. Additional findings from the HOPE 2 trial. *Stroke*. 40: 1365–1372.

Sarkar, P.K., and Lambert, L.A., 2001. Aetiology and treatment of hyperhomocysteinaemia causing ischaemic stroke. *International Journal of Clinical Practice*. 55: 262–268.

Selhub, J., Bagley, L.C., Miller, J., and Rosenberg, I.H., 2000. B vitamins, homocysteine, and neurocognitive function in the elderly. *The American Journal of Clinical Nutrition*. 71: 614S–620S.

Singleton, C.K., and Martin, P.R., 2001. Molecular mechanisms of thiamine utilization. *Current Molecular Medicine*. 1: 197–207.

Spence, J.D., 2006. Homocysteine and stroke prevention: Have the trials settled the issue?. *International Journal of Stroke*. 1: 242–244.

Spence, J.D., Howard, V.J., Chambless, L.E., Malinow, M.R., Pettigrew, L.C., Stampfer, M., and Toole, J.F., 2001. Vitamin Intervention for Stroke Prevention (VISP) Trial: rationale and design. *Neuroepidemiology*. 20: 16–25.

Spence, J.D., 2006. Homocysteine and stroke prevention: hHave the trials settled the issue? International Journal of Stroke. 1: 242–244.

Toole, J.F., Malinow, M.R., Chambless, L.E., Spence, J.D., Pettigrew, L.C., Howard, V.J., Sides, E.G., Wang, C.H., and Stampfer, M., 2004. Lowering homocysteine in patients with ischemic stroke to prevent recurrent stroke, myocardial infarction and death. The Vitamin Intervention for Stroke Prevention (VISP) randomized controlled trial. *The JAMA-Journal of the American Medical Association*. 291: 565–575.

Troen, A., and Rosenberg, I., 2005. Homocysteine and cognitive function. *Seminars in Vascular Medicine*. 5: 209–214.

Verhoef, P., Hennekens, C.H., Malinow, M.R., Kok, F.J., Willett, W.C., and Stampfer, M.J., 1994. A prospective study of plasma homocyst(e)ine and risk of ischemic stroke. *Stroke*. 25: 1924–1930.

VITATOPS Trial Study Group, 2002. The VITATOPS (Vitamins to Prevent Stroke) Trial: rationale and design of an international, large, simple, randomised trial of homocysteine-lowering multivitamin therapy in patients with recent transient ischaemic attack or stroke. Cerebrovascular Diseases. 13. 120–126.

VITATOPS Trial Study Group, 2010. B vitamins in patients with recent transient ischaemic attack or stroke in the VITAmins TO Prevent Stroke (VITATOPS) trial: a randomised, double-blind, parallel, placebo-controlled trial. *Lancet Neurology*. 9: 855–865.

Wald, D.S., Law, M., and Morris, J.K., 2002. Homocysteine and cardiovascular disease: evidence on causality from a meta-analysis. *British Medical Journal*. 325: 1202–1208.

Yang, Q., Botto, L.D.J., Erickson, J.D., Berry, R.J., Sambell, C., Johansen, H., and Friedman, J.M., 2006. Improvement in stroke mortality in Canada and the United States, 1990 to 2002. *Circulation*. 113: 1335–1343.

Yoo, J.H., and Lee, S.C., 2001. Elevated levels of plasma homocyst(e)ine and asymmetric dimethylarginine in elderly patients with stroke. *Atherosclerosis*. 158: 425-430.

Zhang, W., Sun, K., Chen, J., Liao, Y., Qin, Q., Ma, A., Wang, D., Zhu, Z., Wang, Y., and Hui, R., 2010. High plasma homocysteine levels contribute to the risk of stroke recurrence and all-cause mortality in a large prospective stroke population. *Clinical Science*. 118: 187–194.

CHAPTER 30
Epilepsy and B Vitamins

TERJE APELAND,*[a] ROALD E. STRANDJORD[b] AND MOHAMMAD AZAM MANSOOR[c]

[a] Department of Medicine, Stavanger University Hospital, Armauer Hansens vei 20, Postbox 8100, 4048 Stavanger, Norway; [b] Department of Neurology, Stavanger University Hospital, Stavanger, Norway; [c] Department of Natural Sciences, University of Agder, Service box 422, 4604 Kristiansand, Norway
*Email: terje.apeland@sus.no

30.1 Introduction

Epilepsy is a term applied to a group of chronic brain disorders characterized by epileptic seizures. Epilepsy may arise from a variety of different neurological conditions and *via* many different pathophysiological mechanisms. Some patients have seizures that are often easy to treat, for instance, as a part of an age dependent syndrome; while in others the seizures may be therapy resistant associated with neurologic disabilities. There are about 50 million individuals with epilepsy in the world and so epilepsy is an important health issue.

30.2 Infrequent B Vitamin Disorders may Induce Epilepsy

Epileptic disorders during the first year of life have a variety of clinical pictures and outcomes. A few of these disorders of infancy are related to the cofactor function of B vitamins. B vitamins play an important role for normal brain function:

(1) Inborn errors of metabolism may occasionally affect B vitamin function in the central nervous system. Data on a few hundred cases have been published and are of special interest.
(2) Vitamin deficiencies due to malnutrition may sometimes cause brain disorders and seizures.

30.2.1 Vitamin B_6

30.2.1.1 Pyridoxine Dependent Seizures (PDS)

Pyridoxine dependent seizures were first described in 1954. They are due to an autosomal recessive inborn error of metabolism, which deranges the lysine metabolic pathway. A mutation in the ALDH7 gene results in dysfunction of an aldehyde dehydrogenase with accumulation of α-aminoadipic semialdehyde (AASA). AASA condenses with pyridoxal 5′-phosphate (PLP) by a Knoevenagel condensation reaction, which depletes the brain of free and active PLP (Mills *et al.* 2010). PLP, the active cofactor form of vitamin B_6, is essential for the homeostasis of some neurotransmitters, most importantly, the homeostasis of the inhibitory transmitter, γ-aminobutyric acid (GABA). If PLP-dependent enzymes become deficient of their cofactor, activity slows down. In the brain, GABA levels may decrease, resulting in seizures. Thus, GABA is the prime suspect in pyridoxine dependent seizures, although some contradictory findings have been reported. Interestingly, the antiepileptic drug (AED) vigabatrin (VGB) inhibits the PLP-dependent enzyme GABA-aminotransferase, which degrades GABA, and thus GABA levels increase.

Typically, seizures debut during the first days of life, more rarely first within the age of two years. Seizures are unresponsive to standard treatment with AEDs. The young patients may die in status epilepticus unless pyridoxine is administered in high doses. Pyridoxine has an immediate effect on the seizures, which disappear. The epilepsy will relapse if vitamin therapy is stopped and the patients are dependent on lifelong pyridoxine in pharmacological doses (Baxter 2003). A recent study of six patients suggests that high daily pyridoxine doses are associated with an improved outcome of IQ scores.

30.2.1.2 Pyridoxal 5′-phosphate (PLP) Responsive Seizures

There have been a few reports of newborn seizures not responding to pyridoxine, but having good response to PLP. In these patients, an impairment of pyridoxine 5′-phosphate oxidase has been postulated. Pyridoxine 5′-phosphate oxidase converts pyridoxine to PLP. Inactivating mutations in the PNPO gene, which encodes this enzyme, have been reported in these patients (Mills *et al.* 2005).

30.2.1.3 Infantile Vitamin B_6 Deficiency

A heat-treated commercial milk formula became popular for breastfeeding infants in the US in the 1950s. However, it was discovered that the formula was deficient in vitamin B_6 and also had some anti-pyridoxine activity. Some of the babies fed on this formula became ill with epileptic symptoms, similar to PDS. Seizures and electroencephalogram (EEG) abnormalities vanished within a few minutes after pyridoxine injection.

30.2.2 Folate

Folate, a water-soluble B vitamin, is important in one-carbon metabolism. It plays a key role in amino acid metabolism and is important for the synthesis of DNA, some neurotransmittors and myelin, which is essential for the function of neurons. *In utero*, there is active transport of folate from mother to baby and, therefore, deficiency is seldom observed in newborns.

30.2.2.1 Folinic Acid Responsive Seizures

Folinic acid responsive seizures have been reported in a few babies. They all presented with seizures within the first few days of life, with no beneficial effect of AEDs. However, the injection of folinic acid led to marked improvement in seizure control. The mechanism behind this neonatal vitamin-responsive epilepsy is poorly understood. Unidentified substances in the cerebrospinal fluid (CSF) have been reported, but their significance is not known (Torres *et al.* 1999).

30.2.2.2 Cerebral Folate Deficiency

Normally, there is an active transport of folate across the blood-brain barrier. The concentration of 5-methyl tetrahydrofolate (5-MTHF) in the cerebrospinal fluid is three times higher than in the serum. The epithelial cells of the choroid plexus, which produce cerebrospinal fluid, have folate receptors, which are part of an active transport system (Figure 30.1).

Recently, some investigators have reported young patients that develop psychomotor retardation and epilepsy at the age of four to six months. The patients have low concentration of folate in cerebrospinal fluid, despite normal serum concentrations. The condition has been named cerebral folate deficiency. In these patients, antibodies against folate receptors have been detected; these are suggested to inhibit the active folate transport across the blood–brain barrier. Some investigators have proposed that soluble folate receptors, present in milk, may induce the formation of blocking autoantibodies. Folinic acid (0.5 mg/kg body weight) is the treatment of choice and may induce considerable clinical improvement with normalisation of 5-MTHF in the cerebrospinal fluid (Gordon 2009).

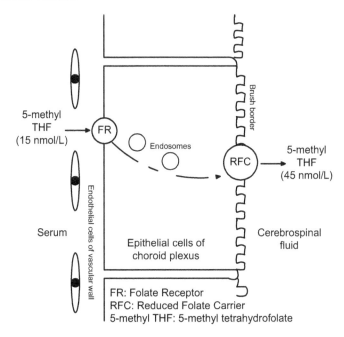

Figure 30.1 Folate transport across the blood–brain barrier. The active transport of folate into the brain takes place in the epithelial cells of the choroid plexus. 5-Methyl tetrahydrafolate (5-MTHF), the main form of folate in serum, is bound to the high affinity folate receptors on the epithelial cells and transported into the cells. Endosomes and transport proteins are involved in the intracellular transfer. Finally, the reduced folate carriers on the brush border deliver 5-MTHF into the CSF. This folate transport system is energy dependent and saturable. It stabilizes cerebrospinal folate concentration whether the serum folate is low or high. Thus, the transport system has an important homeostatic function for brain folate. Likewise, vitamins B_1, B_2, B_6, B_7 and B_{12} have specialized transport systems into the CSF.

30.2.3 Thiamine

Thiamine is a cofactor for some important enzymes that participate in the Krebs cycle and the pentose phosphate pathway. Thiamine deficiency may induce damage in regions of the brain with high metabolic demands. Thiamine is also involved in the synthesis of neurotransmittors. In adults, thiamine deficiency may lead to Wernicke's encephalopathy.

30.2.3.1 Infantile Beriberi

In 2003, investigators reported an outbreak of thiamine deficiency in 20 Israeli children. It was reported that the children had been fed a soy-based milk that was deficient in thiamine. Most of the infants had severe symptoms and signs. Seven of the children also had seizures. Initially, epilepsy was diagnosed and

AEDs administered with some effect. Later, when the diagnosis of vitamin deficiency was made, thiamine administration led to some neurologic improvement and fewer seizures. However, most of the children became persistently neurologic disabled and needed chronic therapy with AEDs (Fattal-Valevski et al. 2009).

In 1910, an autopsy study of 86 children with infantile beriberi concluded that 40% of these children had convulsions during their terminal illness. Nowadays, thiamine deficiency is very rare in developed countries. It may occur in breast-fed infants of thiamine deficient mothers. The onset of symptoms may be very abrupt and dramatic with high mortality rate.

30.2.4 Vitamin B_{12}

Vitamin B_{12} is essential for DNA synthesis, in collaboration with folate. Deficiency for B_{12} is not uncommon among adults, especially the elderly. Haematological and neurological symptoms and signs are the most prominent findings.

30.2.4.1 Inborn Errors of Cobalamin Metabolism

Inborn errors of cobalamin metabolism are rare: cobalamin C/D deficiency impairs the synthesis of methyl or adenosyl cobalamin, and can give rise to combined homocysteinuria and methylmalonic aciduria causing severe neurologic symptoms and epilepsy. Therapy with hydroxycobalamine and betaine is rarely successful.

30.2.4.2 Infantile Vitamin B_{12} Deficiency

Babies are born with a limited hepatic reserve of vitamin B_{12}. The content of vitamin B_{12} in breast milk is important to maintain adequate supplies. Infants of vegetarian mothers or mothers with unrecognized pernicious anaemia may therefore develop vitamin B_{12} deficiency. Vitamin B_{12} deficiency may have detrimental neurological effects such as psychomotor retardation and epilepsy.

30.3 Antiepileptic Drugs and B Vitamins

Antiepileptic drugs (AEDs) provide the best treatment option for most patients with epilepsy (Glauser et al. 2006). The first synthetic anticonvulsant was phenobarbital (PB) (1912). The other older AEDs still in clinical use include phenytoin (PHT), primidone (PRD), carbamazepine (CBZ) and valproate (VPA). There has been a significant influx of new AEDs during the past 20 years and more than 30 are available today. In approximately 75% of the patients, the age of onset is below 20 years and long-lasting therapy is often required. Therefore, the chronic adverse effects of AEDs are an important issue.

30.3.1 AEDs and Folate

In 1952, Mannheim *et al.* described the occurrence of megaloblastic anaemia in patients on AEDs. Several reports followed and megaloblastic anaemia was found to occur in less than 1% of patients on PHT, PB and PRD. The anaemia was associated with low levels of folate and could be corrected with the administration of folic acid (Reynolds 1975).

Since then, at least 47 reports on AEDs and B vitamin status have been published; 43 cross-sectional studies and 4 prospective trials, comprising about 6700 patients. Most of the investigators report on findings in less than 100 patients; the exceptions are two with a larger material of 610 and 2730 patients, respectively (Krause *et al.* 1988; Linnebank *et al.* 2011). The levels of B vitamins vary according to age, gender, intake of multivitamins, use of tobacco and intake of alcohol. However, it may be difficult to evaluate some aspects of the cross-sectional studies due to the lack of properly matched controls.

30.3.1.1 Folate Deficiency in Relation to the Different AEDs

Folate has been extensively studied in epilepsy since the first report in 1952. Most studies on patients with unselected AEDs demonstrate low levels of serum folate compared with healthy control subjects. Studies in patients on monotherapy have revealed that some AEDs are associated with a low folate status (Table 30.1). It appears that AEDs with inducer effects on liver enzymes are most likely to cause low folate levels (Apeland *et al.* 2003; Kishi *et al.* 1997). On the other hand, valproate (VPA) has little effect on total folate concentrations in adult patients, but appears to decrease the folate status in children (Verrotti *et al.* 2000). The reason for this difference is unclear.

30.3.1.2 Mechanisms of Folate Deficiency

There is limited knowledge as to how AEDs influence B vitamin status. In animal studies, PHT and PB have been shown to induce hepatocyte proliferation with hepatomegaly. This leads to increased demand for B vitamin cofactors with possible systemic depletion.

Phenobarbital-like CYP450 inducers modulate the expression of different genes in the liver which increase or decrease the corresponding enzyme activities. Possibly, the activity of some enzymes involved in folate metabolism may change. The reduced folate carrier (RFC) is a membrane-bound protein, which is important for the transport of reduced folates (5-MTHF and THF) in several tissues. In animal studies, inducer AEDs appear to downregulate liver RFC activity by modulating its phosphorylation (Halwachs *et al.* 2007). In human studies, both the liver enzyme inducers PHT and CBZ and the inhibitor VPA decrease the intestinal absorption of folate (Hendel *et al.* 1984). Furthermore, there are data suggesting that VPA may disturb folate metabolism without changing total concentrations. Most laboratories estimate the total concentration of serum folate and therefore do not detect changes in distribution of the different forms.

Table 30.1 Antiepileptic drugs (AEDs) and their effects on B vitamin status. The table presents current knowledge of the effects of some AEDs on B vitamins. When B vitamin concentrations are below normal, there may be a risk of significant deficiency. The effect on plasma homocysteine has been included as this illustrates an actual metabolic effect. The C677T polymorphism of MTHFR appears to be an important determinant of hyperhomocysteinemia in patients on AEDs.

Acronym	Drug name	Homocysteine	Folate	Vitamin B_1	Vitamin B_2	Vitamin B_6	Vitamin B_7	Vitamin B_{12}
PB	Phenobarbital	↑	↓	(↓)	↓	↓	↓	(↓)^a
PHT	Phenytoin	↑	↓	(↓)	↓	↓	↓	0
PRD	Primidone	↑	↓	(↓)	↓	↓	↓	(↓)^a
CBZ	Carbamazepine	↑	↓	(↓)	↓	↓	0	0
OXC	Oxcarbazepine		(↓)^a					0
VPA	Valproate	(↑)^b	(↓)^b	0	0			↑
GBP	Gabapentin		(↓)			(↓)		0
LTG	Lamotrigine	0	0					0
PGB	Pregabalin	(↑)	↓					(↓)^a
VGB	Vigabatrin		0					
TPM	Topiramate	↑	(↓)					(↓)^a
LEV	Levetiracetam	0	0					0
CLB	Clobazam	0	0					0
CZP	Clonazepam	0	0					0
Unselected	AEDs	↑	↓	(↓)	↓	↓	↓	0

^{a}Low concentrations detected only in patients on combination therapy (two or more AEDs).
^{b}Effects detected in children - not in adults. 0 = no effect; ↑ = elevated concentrations; ↓ = low concentrations.

Folate deficiency is most likely to occur in patients on long-lasting AED therapy. Furthermore, folate concentrations are inversely related to the number of prescribed AEDs. Patients on combinations of two or more AEDs tend to have lower folate levels than patients on monotherapy. Combinations of AEDs may have synergistic effects on folate absorption and/or metabolism. Moreover, the dietary intake of B vitamins tends to decrease with an increasing number of AEDs. Patients on combination therapy tend to a have more severe epilepsy, often associated with other neurological deficits and may even need to live in institutions.

30.3.1.3 Pathological Effects of Folate Deficiency

Low levels of serum folate are often associated with low levels of erythrocyte folate (intracellular). The mean corpuscular volume (MCV) of erythrocytes may increase slightly with low folate levels, though overt macrocytic anaemia occurs infrequently. Carbamazepine (CBZ) therapy may cause leukopenia and neutropenia. In a randomized trial in patients on CBZ, subjects on folic acid had higher leucocyte counts and less neutropenia compared with subjects without vitamin supplements. Other studies found no connection between folate and AED-induced haematological abnormalities.

A few studies have examined the folate concentrations in the spinal fluid of patients with epilepsy. 5-MTHF is the main form of folate in both blood and cerebrospinal fluid. The CSF concentrations of 5-MTHF are usually about three times higher than serum concentrations. In the choroid plexus, there is an active transport mechanism which tends to keep cerebrospinal folate concentrations stable despite variable serum concentrations (Figure 30.1) (Ramaekers *et al.* 2002).

If serum concentrations decrease, CSF concentrations will remain unchanged for some time and then possibly decrease. The presence of a reversible neuropsychiatric condition due to folate deficiency has been disputed for several years. Epilepsy in itself is a condition with a range of organic and psychosocial aspects. AEDs have many neurologic side effects, and it is possible that some of them are mediated *via* folate deficiency. It obviously is a complex task to isolate the effects of folate deficiency. Several studies have found an association between mental symptoms and low folate levels (Bottiglieri *et al.* 1995). Interestingly, trials in patients with depression have revealed that there is a subgroup with low folate status and poor response to antidepressive treatment unless they are supplemented with folic acid. Folate is a cofactor of the synthesis of the neurotransmittor serotonin, which plays an important role in depressive diseases (Moore 2005). It appears that there are a few patients with severe folate deficiency that may benefit from folic acid supplementation. In general, there is not much evidence that a low folate status cause neuropsychiatric symptoms in patients with epilepsy. This may be due to the active transport of folate into the brain, which protects the brain against variations in serum folate concentration (Figure 30.1).

30.3.1.4 Folate and Homocysteine Metabolism

Many patients with epilepsy have an elevated concentration of plasma homocysteine, which is a sensitive indicator of B vitamin status. In the general population, plasma total homocysteine concentration has a strong inverse relationship to the folate status. Homocysteine is also inversely related to the levels of vitamins B_2, B_6 and B_{12}. The inverse relationship to folate is most pronounced in patients with the C677T polymorphism of methylenetetrahydrofolate reductase (MTHFR) (Figure 30.2). The polymorphism is associated with reduced activity of the MTHFR enzyme and higher plasma homocysteine concentration than normal. Large meta-analyses have demonstrated that individuals with the homozygous C677T MTHFR have elevated risk of ischemic heart disease, stroke and deep vein thrombosis; this is probably mediated through increased concentrations of homocysteine (Wald *et al.* 2002).

Patients on inducer AEDs often have higher plasma homocysteine and lower serum folate concentrations than normal, a finding that occurs primarily in patients with the C677T polymorphism of MTHFR. It appears to be a synergy between inducer AEDs and C677T MTHFR in disturbing the homocysteine

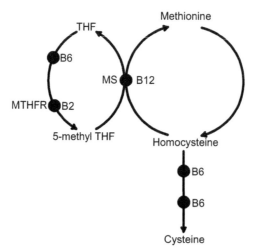

Figure 30.2 Homocysteine metabolism and its B vitamin cofactors. Homocysteine can be reversibly methylated to methionine or irreversibly transsulfurated to cysteine. The remethylation is dependent on vitamin B_{12}, folate and vitamin B_2. The transsulfuration needs vitamin B_6 as cofactor. Thus, plasma homocysteine is inversely related to the levels of the actual B vitamins.

and folate metabolism (Belcastro et al. 2010; Yoo and Hong 1999). Patients with epilepsy have an increased mortality rate compared with the general population caused by a wide range of diseases, including diseases of the circulatory system (Nilsson et al. 1997). There are some indications that long-term AED therapy *per se* may cause or accelerate vascular disease (Tan et al. 2009). Low serum folate and high plasma homocysteine concentrations have been related to elevated levels of markers of vascular disease in patients on AEDs (Apeland et al. 2002). In other words, AEDs appear to increase the effect of C677T MTHFR on homocysteine metabolism and vascular disease.

30.3.1.5 Folate and Foetal Malformations

Folate plays an important role in the development and function of the central nervous system. The high folate levels in cerebrospinal fluid give an indication of its importance. An insufficient folate status during the early weeks of gestation leads to problems with the development of the neural tube. Periconceptual folic acid reduces the rate of neural tube defects in the general population, and is the reason for the mandatory folic acid fortification of cereals in the US and Canada.

The risk of foetal malformation is increased in women with epilepsy compared with the general population. Minor dysmorphic features are most frequent, but more severe forms such as facial clefts and neural tube defects are not uncommon. The two major forms of neural tube defects are (i) anencephaly, a lethal malformation, and (ii) spina bifida—a closure defect in the spinal column that may lead to paralysis of the lower limbs. The rates of malformations are about 3% with CBZ and lamotrigine (LTG), 7% with VPA, and 15% with combinations of two or more AEDs. It is probable that AEDs have several different teratogenic mechanisms. Low folate levels appear to be associated with increased risks of foetal malformations in women on AEDs. Furthermore, it has been suggested that maternal C677T MTHFR polymorphism or some abnormality related to methionine synthetase increase the risk of foetal malformation in patients on AEDs (Mills et al. 1995).

In a few studies, the protective effect of folic acid among pregnant women on AEDs has been assessed. A possible but small reduction in congenital malformations with folic acid supplements has been reported (Hill et al. 2010). It appears that most of the teratogenic effects of AEDs are *via* other mechanism than low folate. Despite the modest protective effect of folate supplements, it is still recommended in women with epilepsy.

30.3.2 AEDs and Vitamin B_6

Pyridoxal 5′-phosphate (PLP) is one of the active forms of vitamin B_6. More than 140 different PLP-dependent enzymes are known, and they are involved in many different cellular processes.

30.3.2.1 Vitamin B_6 in Relation to AEDs

Low vitamin B6 levels have been found in 20–50% of patients on unselected AEDs (Reynolds 1975). Low plasma concentrations of PLP have been found in patients on PRD, PHT, CBZ and possibly VPA monotherapy (Apeland et al. 2003; Krause et al. 1988; Verrotti et al. 2000) (Table 30.1).

30.3.2.2 Mechanisms of Vitamin B_6 Reduction

It is unknown how AEDs may reduce the vitamin B_6 concentrations. Liver enzyme induction may be of importance. There is no evidence that a low vitamin B_6 status may be caused by epilepsy itself (Krause et al. 1988).

30.3.2.3 Pathological Effects of a Low Vitamin B_6 Status

In patients on AEDs, the clinical implications of low vitamin B_6 are unclear. Low PLP concentrations probably contribute to abnormal homocysteine metabolism (Apeland et al. 2003). Elevated homocysteine and low folate as well as a low vitamin B_6 are markers of vascular disease in the general population. In animal studies, vitamin B_6 supplementation decrease the rate of foetal malformations after VPA and CBZ exposure (Elmazar et al. 1992). These observations may be relevant for patients, although, there is no evidence at present.

Clinical observations suggest that pyridoxine supplements have no effects on seizure frequency. Like folate, vitamin B_6 has a specialized transport system into the cerebrospinal fluid.

30.3.3 AEDs and Vitamin B_{12}

Vitamin B_{12} is important for normal brain function.

30.3.3.1 Vitamin B_{12} in Relation to AEDs

Vitamin B_{12} levels are usually within the normal range in patients on unselected AEDs. In patients on monotherapy with AEDs, serum cobalamin is not different from controls, except VPA, which is associated with elevated serum concentrations of cobalamin (Apeland et al. 2003; Belcastro et al. 2010; Krause et al. 1988; Reynolds 1975). Low vitamin B_{12} levels have been detected among patients on two or more AEDs (Linnebank et al. 2011).

30.3.3.2 Mechanisms of Vitamin B_{12} Disturbances

Possibly, the elevated serum cobalamin in patients on VPA may be due to toxic effects on the liver. A higher level of circulating cobalamin does not necessarily indicate higher vitamin B_{12} stores in the liver. The reasons for B_{12} deficiency in patients on several AEDs may be due to the combined metabolic actions of drugs or an inadequate diet. In a study of 12 patients, PHT, PB and PRD had no effect on the absorption of vitamin B_{12} from the gut.

30.3.3.3 Pathological Effects of Vitamin B_{12} Deficiency

Vitamin B_{12} deficiency is associated with neurological and haematological symptoms and signs. It is important to detect and treat patients with vitamin B_{12} deficiency, especially before therapy with folic acid, as this may mask the haematological effects of vitamin B_{12} deficiency.

30.3.4 AEDs and Vitamin B_2

Riboflavin is important in human nutrition. Riboflavin is converted to the active flavin nucleotides: flavin adenine dinucleotide (FAD) and flavin mononucleotide (FMN). Riboflavin deficiency has been related to foetal malformations and anaemia.

30.3.4.1 Vitamin B_2 Deficiency in Relation to AEDs

In a small study of children on PHT, their mean urinary excretion of riboflavin was low, *i.e.* 14% of dietary intake. This may indicate riboflavin deficiency (Lewis *et al.* 1998). Patients on inducer AEDs (PHT, PB, PRD and CBZ) have low plasma riboflavin (Apeland *et al.* 2003; Krause *et al.* 1988). Low plasma riboflavin may indicate increased risk of vitamin B_2 deficiency. Furthermore, patients with low plasma riboflavin also have elevated plasma flavin nucleotides (Apeland *et al.* 2003).

30.3.4.2 Mechanisms of Vitamin B_2 Deficiency

In animals, phenobarbital treatment increases the activity of the enzymes that synthesize flavin nucleotides from riboflavin (Hamajima *et al.* 1979). Therefore, it is possible that the inducer AEDs are associated with non-dietary riboflavin deficiency. Krause *et al.* (1988) estimated that 30–40% of patients were at risk for clinical deficiency.

30.3.4.3 Pathological Effects of Vitamin B_2 Deficiency

The clinical implications of a low riboflavin status in patients on AEDs are largely unknown. There may be an inhibitory effect on C677T MTHFR function and homocysteine metabolism. Importantly, riboflavin deficiency may be related to the elevated risk of foetal malformations in women on AEDs.

30.3.5 AEDs and Vitamin B_7 (Biotin)

30.3.5.1 Vitamin B_7 Deficiency in Relation to AEDs

There is sparse information on the biotin status in patients on AEDs. Krause *et al.* (1988) reported that over 80% of the patients were at risk of biotin deficiency. The biotin levels were influenced by the inducer AEDs and not VPA.

30.3.5.2 Mechanisms of B_7 Deficiency

There are no studies on how AEDs may reduce the levels of biotin, but the association with inducer AEDs suggests that liver enzyme induction may be of importance.

30.3.5.3 Pathological Effects of B_7 Deficiency

Low biotin levels appear to be related to gingival hyperplasia and perhaps cerebellar symptoms. Otherwise, little is known about the clinical significance of low biotin levels.

30.3.6 AEDs and Vitamin B_1

Thiamine (vitamin B_1) is important for some neurotransmittors and the cerebellar function.

30.3.6.1 Vitamin B_1 Deficiency in Relation to AEDs

There are few reports on the thiamine levels in patients on AEDs. Krause *et al.* (1988) reported that inducer AEDs might have negative effects on the thiamine levels. Twenty-five per cent of the patients had low thiamine levels; however, there were no significant difference between patients and controls. Botez *et al.* (1993) found that 31% of patients on PHT and PB had low thiamine levels, and reported improvement in neuropsychological parameters after treatment with thiamine.

In conclusion, the prevalence of B_1 deficiency appears to be the same as in the general population.

30.3.6.2 Mechanisms of B_1 Deficiency

It is uncertain that AEDs influence vitamin B_1 levels.

30.3.6.3 Pathological Effects of Vitamin B_1 Deficiency

Thiamine deficiency may have abrupt and dramatic effects on brain function and should always be treated promptly when there is clinical suspicion.

30.4 AEDs and Vitamin Supplementation

Clearly, many patients on AED therapy have low levels of several B vitamins and are at risk of clinical deficiency. It appears that the AEDs with inducer effects on liver enzymes have stronger effects on B vitamin status than other AEDs. The available information is incomplete: while some information is available for the AEDs that have been in use for decades, there are less data on the latest on the market (Table 30.1). Most scientific reports concentrate on folate and vitamin B_{12}, and less on the other B vitamins. Thus, further studies are needed to improve the knowledge and insight in this area.

Meanwhile, what should we do? Primarily, it seems reasonable to monitor B vitamin status in patients on therapy with AEDs. Plasma homocysteine is a good indicator of deficiency of folate, B_{12}, B_6 and to some extent vitamin B_2. Thus, plasma homocysteine may be employed for routine screening in clinical practice.

30.4.1 Folic Acid Supplements

It is not uncommon to prescribe folic acid to patients on AEDs. Doses from 0.2 to 15 mg folic acid per day have been recommended. However, pharmacological doses of folic acid may be associated with adverse effects, such as increased incidence of tumours. Doses of folic acid above 0.4 mg may cause the absorption of unmetabolized folic acid that can be detected in the circulation for several hours after intake. Furthermore, high doses of folic may increase the breakdown rate of some AEDs such as PHT. The serum concentrations of PHT may drop significantly when folic acid is introduced and seizures may be aggravated. It has been suggested that high doses of folic acid may have direct effects on the nervous system and cause seizures. However, there is no indication that this is a problem in patients with epilepsy and folic acid is not contraindicated for this reason (Moore 2005).

Folate deficiency usually takes 3–5 years to evolve in patients with epilepsy. Therefore, prophylactic B vitamin supplements may be recommended for patients at risk. As long as physiological doses are used, vitamin therapy should be safe.

In general, women of childbearing potential are encouraged to take 0.4 mg of folic acid per day, with the dose increased to 4 mg per day for high-risk women. The recommendation of 0.4 mg folic acid per day applies to most fertile women on AEDs. Four mg per day is recommended for women on inducer AEDs and VPA. However, there is sparse evidence that favour high doses of folic acid. Some reports suggest that supplementation with low-dose folic acid (*i.e.* <0.4 mg) may be sufficient for normalizing homocysteine metabolism in epilepsy (Apeland *et al.* 2002).

30.4.2 Supplements with Vitamins B_1, B_2, B_6, B_7 and B_{12}

It appears reasonable to recommend supplementation with several B vitamins in physiological doses to many patients with epilepsy. Patients on inducer AEDs and VPA, and patients with an insufficient diet, are at risk for deficiency of several B vitamins. Moreover, the evidence that folic acid can prevent adverse effects in patients on AEDs is not much higher than the evidence for a protective effect of vitamins B_2, B_6, B_7 and B_{12} (Ranganathan and Ramaratnam 2005). In animal studies, deficiency of vitamins B_2 and B_6 has been related to the occurrence of foetal malformations. Therefore, women of childbearing potential should supplement their diet with vitamin B_2 and B_6, as well as folic acid.

Summary Points

- This chapter describes how some rare B vitamin disorders may cause brain dysfunctions with seizures and how the treatment of epilepsy with antiepileptic drugs may cause B vitamin disturbances.
- 'Pyridoxine dependent seizures' occur in newborns. They are unresponsive to antiepileptic drugs but may be treated with pyridoxine.
- 'Cerebral folate deficiency' is caused by antibodies that inhibit active folate transport to the brain. It can be treated with folinic acid.
- Antiepileptic drugs are the best treatment option for most patients with epilepsy and long-lasting therapy is often required.
- Antiepileptic drugs, primarily those with inducer effects on liver enzyme activity, may cause a low folate status with risk of deficiency.
- Antiepileptic drugs may be associated with low levels of vitamins B_6 and B_2, but vitamin B_{12} is usually within normal range.
- There some reports of low levels of thiamine and biotin in patients on antiepileptic drugs; but the significance is uncertain.
- Plasma homocysteine is a useful indicator of B vitamin status in patients on antiepileptic drugs.
- Antiepileptic drugs have many adverse effects and some of them are due to disturbances of B vitamin status.
- A low folate status in patients on antiepileptic drugs has been related to the occurrence of megaloblastic anaemia, neuropsychiatric symptoms, foetal malformations and vascular disease.
- Folic acid is commonly prescribed to patients on antiepileptic drugs, primarily to women of childbearing potential.
- Many patients with epilepsy should take prophylactic B vitamin supplements in physiologic doses (multivitamins).
- There is sparse evidence that B vitamin supplements can prevent adverse effects of antiepileptic drugs. More studies are needed.

Key Facts

Key Features about Epilepsy

- Epilepsy comprises a group of brain disorders characterized by seizures, which clinically vary with complex behaviour, spells of impaired consciousness and convulsions.
- Epileptic fits are due to paroxysmal brain dysfunction with excessive neuronal discharges.
- Epileptic disorders may arise from a variety of different neurological conditions.
- Epilepsy is classified according to the proposal made by the International League Against Epilepsy in 1989.
- In general, seizures are a relatively common symptom and may occur during a variety of acute conditions in which brain function becomes temporarily

Epilepsy and B Vitamins 519

deranged—such as hypoglycemia (insulinoma), meningitis, head trauma, fever or drugs. Of special interest: psychogenic seizures are often difficult to separate from genuine epileptic seizures, and are therefore not recognized.
- Epilepsy is more likely to arise early in life or in elderly people above 65 years.
- There are approximately 50 million people with epilepsy in the world.
- In areas of the world with poor health conditions, one in 25 individuals have epilepsy; in Western Europe the prevalence is one in 125.
- Epilepsy, known for more than 4000 years, has been surrounded by ignorance and superstition.

Key Facts about Epilepsy Treatment

- The first synthetic anticonvulsant was phenobarbital (PB), discovered by chance and introduced in 1912.
- Phenytoin (PHT), available from 1938, was the result of scientific research.
- Today, more than 30 different antiepileptic drugs (AEDs) are available.
- Antiepileptic drugs are the best treatment option for most patients with epilepsy.
- The clinical usefulness, safety and efficiency of the different AEDs vary as well as their use in clinical practice.
- Low dose monotherapy should be the first therapy option. Thereafter two or more drugs when needed.

Key Facts about the C677T Polymorphism of MTHFR

- Methylenetetrahydrofolate reductase (MTHFR) is part of the folate/homocysteine metabolism (Figure 30.2).
- Vitamin B_2 is an important cofactor for MTHFR.
- The thermolabile C677T polymorphism of MTHFR is a common genetic variant and occurs in its homozygote form in about 10% of the general population.
- The C677T polymorphism results in a slightly altered configuration of the MTHFR enzyme; and this causes problems for the attachment of the vitamin B_2 cofactor.
- In the general population, the thermolabile C677T polymorphism is associated with reduced activity of the MTHFR enzyme and about 20% higher plasma homocysteine than normal.
- In the general population, elevated plasma homocysteine is a marker of increased risk for vascular disease and foetal malformations.
- Large meta-analyses have demonstrated that the homozygous C677T MTHFR polymorphism is associated with a higher risk of ischemic heart disease, stroke and deep vein thrombosis, which probably is mediated through raised homocysteine.

- The reduced activity of C677T MTHFR may be restored to normal by ensuring a good B vitamin status. Accordingly, the effect on plasma homocysteine is less pronounced in countries where the food is fortified with folic acid.

Definitions of Words and Terms

Blood–brain barrier. A layer of cells that surround the brain. It separates blood from brain tissue and blocks the passage of many substances.
Cerebral. Pertaining to the cerebrum (brain).
Epilepsy. From the Greek, *epilepsia* (attack, seizure). A term applied to a group of brain disorders characterized by attacks of epileptic seizures (fits).
Hyperhomocysteinemia. Designates a state with elevated concentration of plasma homocysteine, above the normal range.
Hypoglycemia. Designates a state with low concentration of blood glucose, less than the normal range.
In utero. Within the uterus (womb), not yet born.
Inborn errors of metabolism. A group of disorders due to defective function of various specific enzymes, genetic in origin.
Inducer AEDs. Antiepileptic drugs that have phenobarbital-like inducer effects on the liver enzymes.
Megaloblastic anaemia. Anaemia with large red blood cells in the bone marrow, often caused by deficiency of vitamin B_{12} or folate.
Neurological. From the Greek, *neuron* (nerve) and *logia* (study of). Pertaining to the nervous system.
Neurotransmittors. Substances that transmit signals between neurons, across the synapses.
Pathophysiological. Related to derangements of body functions, caused by disease.
Pharmacological dose of a vitamin. Large amounts of a vitamin, exceeding normal needs of the body.
Physiological dose of a vitamin. Normal amounts of a vitamin, in accordance with body requirements under normal conditions.
Polymorphism. Genetic variation in the base-pair sequence of DNA occurring so often that it cannot be due only to recurrent mutation.
Status epilepticus. Prolonged or repeated seizures (fits), without pause.
Teratogenic. The ability to cause malformations in a foetus.

List of Abbreviations

AED	antiepileptic drug
AASA	α-aminoadipic semialdehyde
CBZ	carbamazepine
CSF	cerebrospinal fluid
CYP450	cytochrome P450 group of enzymes

DNA	deoxyribonucleic acid
EEG	electroencephalogram
FAD	flavin adenine dinucleotide
FMN	flavin mononucleotide
GABA	γ-aminobutyric acid
LTG	lamotrigine
MCV	mean corpuscular volume (of erythrocytes)
5-MTHF	5-methyl tetrahydrofolate
MTHFR	methylenetetrahydrofolate reductase
PB	phenobarbital
PDS	pyridoxine dependent seizures
PHT	phenytoin
PLP	pyridoxal 5′-phosphate
PRD	primidone
PHT	phenytoin
RFC	reduced folate carrier
THF	tetrahydrofolate
VPA	valproate
VGB	vigabatrin

References

Apeland, T., Mansoor, M.A., Pentieva, K., McNulty, H., Seljeflot, I., and Strandjord, R.E., 2002. The effect of B-vitamins on hyperhomocysteinemia in patients on antiepileptic drugs. *Epilepsy Research*. 51: 237–247.

Apeland, T., Mansoor, M.A., Pentieva, K., McNulty, H., and Strandjord, R.E., 2003. Fasting and post-methionine loading concentrations of homocysteine, vitamin B(2), and vitamin B(6) in patients on antiepileptic drugs. *Clinical Chemistry*. 49: 1005–1008.

Baxter, P., 2003. Pyridoxine-dependent seizures: a clinical and biochemical conundrum. *Biochimica et Biophysica Acta*. 1647: 36–41.

Belcastro, V., Striano, P., Gorgone, G., Costa, C., Ciampa, C., Caccamo, D., Pisani, L.R., Oteri, G., Marciani, M.G., Aguglia, U., Striano, S., Ientile, R., Calabresi, P., and Pisani, F., 2010. Hyperhomocysteinemia in epileptic patients on new antiepileptic drugs. *Epilepsia*. 51: 274–279.

Botez, M.I., Botez, T., Ross-Chouinard, A., and Lalonde, R., 1993. Thiamine and folate treatment of chronic epileptic patients: a controlled study with the Wechsler IQ scale. *Epilepsy Research*. 16: 157–163.

Bottiglieri, T., Crellin, F.C., and Reynolds, E.H., 1995. Folate and Neuropsychiatry. In: Bailey, L.B.E. (ed.) *Folate in Health and Disease*. Marcel Dekker, New York, USA. pp. 435–456.

Elmazar, M.M., Thiel, R., and Nau, H., 1992. Effect of supplementation with folinic acid, vitamin B6, and vitamin B12 on valproic acid-induced teratogenesis in mice. *Fundamentals of Applied Toxicology*. 18: 389–394.

Fattal-Valevski, A., Bloch-Mimouni, A., Kivity, S., Heyman, E., Brezner, A., Strausberg, R., Inbar, D., Kramer, U., and Goldberg-Stern, H., 2009. Epilepsy in children with infantile thiamine deficiency. *Neurology*. 73: 828–833.

Glauser, T., Ben-Menachem, E., Bourgeois, B., Cnaan, A., Chadwick, D., Guerreiro, C., Kalviainen, R., Mattson, R., Perucca, E., and Tomson, T., 2006. ILAE treatment guidelines: evidence-based analysis of antiepileptic drug efficacy and effectiveness as initial monotherapy for epileptic seizures and syndromes. *Epilepsia*. 47: 1094–1120.

Gordon, N., 2009. Cerebral folate deficiency. *Developmental Medicine & Child Neurology*. 51: 180–182.

Halwachs, S., Kneuer, C., and Honscha, W., 2007. Downregulation of the reduced folate carrier transport activity by phenobarbital-type cytochrome P450 inducers and protein kinase C activators. *Biochimica et Biophysica Acta*. 1768: 1671–1679.

Hamajima, S., Ono, S., Hirano, H., and Obara, K., 1979. Induction of the FAD synthetase system in rat liver by phenobarbital administration. *International Journal for Vitamin and Nutrition Research*. 49: 59–63.

Hendel, J., Dam, M., Gram, L., Winkel, P., and Jorgensen, I., 1984. The effects of carbamazepine and valproate on folate metabolism in man. *Acta Neurologica Scandinavica*. 69: 226–231.

Hill, D.S., Wlodarczyk, B.J., Palacios, A.M., and Finnell, R.H., 2010. Teratogenic effects of antiepileptic drugs. *Expert Review of Neurotherapeutics*. 10: 943–959.

Kishi, T., Fujita, N., Eguchi, T., and Ueda, K., 1997. Mechanism for reduction of serum folate by antiepileptic drugs during prolonged therapy. *Journal of Neurological Science*. 145: 109–112.

Krause, K.H., Bonjour, J.P., Berlit, P., Kynast, G., Schmidt-Gayk, H., and Schellenberg, B., 1988. Effect of long-term treatment with antiepileptic drugs on the vitamin status. *Drug Nutrient Interactions*. 5: 317–343.

Lewis, D.P., Van Dyke, D.C., Stumbo, P.J., and Berg, M.J., 1998. Drug and environmental factors associated with adverse pregnancy outcomes. Part I: Antiepileptic drugs, contraceptives, smoking, and folate. *The Annals of Pharmacotheraphy*. 32: 802–817.

Linnebank, M., Moskau, S., Semmler, A., Widman, G., Stoffel-Wagner, B., Weller, M., and Elger, C.E., 2011. Antiepileptic drugs interact with folate and vitamin B12 serum levels. *Annals of Neurology*. 69: 352–359.

Mannheimer, E., Pakesch, F., Reimer, E.E. and Vetter, H., 1952. Hemopoietic complications of hydantoin therapy of epilepsy. Med Klin (Munich) 47(42): 1397–1401.

Mills, J.L., McPartlin, J.M., Kirke, P.N., Lee, Y.J., Conley, M.R., Weir, D.G., and Scott, J.M., 1995. Homocysteine metabolism in pregnancies complicated by neural-tube defects. *Lancet*. 345(8943): 149–151.

Mills, P.B., Surtees, R.A., Champion, M.P., Beesley, C.E., Dalton, N., Scambler, P.J., Heales, S.J., Briddon, A., Scheimberg, I., Hoffmann, G.F., Zschocke, J., and Clayton, P.T., 2005. Neonatal epileptic encephalopathy caused by mutations in the PNPO gene encoding pyridox(am)ine 5′-phosphate oxidase. *Human Molecular Genetics*. 14: 1077–1086.

Mills, P.B., Footitt, E.J., Mills, K.A., Tuschl, K., Aylett, S., Varadkar, S., Hemingway, C., Marlow, N., Rennie, J., Baxter, P., Dulac, O., Nabbout, R., Craigen, W.J., Schmitt, B., Feillet, F., Christensen, E., De Lonlay, P., Pike, M.G., Hughes, M.I., Struys, E.A., Jakobs, C., Zuberi, S.M., and Clayton, P.T., 2010. Genotypic and phenotypic spectrum of pyridoxine-dependent epilepsy (ALDH7A1 deficiency). *Brain*. 133: 2148–2159.

Moore, J.L., 2005. The significance of folic acid for epilepsy patients. *Epilepsy Behaviour*. 7: 172–181.

Nilsson, L., Tomson, T., Farahmand, B.Y., Diwan, V., and Persson, P.G., 1997. Cause-specific mortality in epilepsy: a cohort study of more than 9,000 patients once hospitalized for epilepsy. *Epilepsia*. 38: 1062–1068.

Ramaekers, V.T., Hausler, M., Opladen, T., Heimann, G., and Blau, N., 2002. Psychomotor retardation, spastic paraplegia, cerebellar ataxia and dyskinesia associated with low 5-methyltetrahydrofolate in cerebrospinal fluid: a novel neurometabolic condition responding to folinic acid substitution. *Neuropediatrics*. 33: 301–308.

Ranganathan, L.N., and Ramaratnam, S., 2005. Vitamins for epilepsy. *Cochrane Database of Systematic Reviews*. Issue 2: CD004304.

Reynolds, E.H., 1975. Chronic antiepileptic toxicity: a review. *Epilepsia*. 16: 319–352.

Tan, T.Y., Lu, C.H., Chuang, H.Y., Lin, T.K., Liou, C.W., Chang, W.N., and Chuang, Y.C., 2009. Long-term antiepileptic drug therapy contributes to the acceleration of atherosclerosis. *Epilepsia*. 50: 1579–1586.

Torres, O.A., Miller, V.S., Buist, N.M., and Hyland, K., 1999. Folinic acid-responsive neonatal seizures. *Journal of Child Neurology*. 14: 529–532.

Verrotti, A., Pascarella, R., Trotta, D., Giuva, T., Morgese, G., and Chiarelli, F., 2000. Hyperhomocysteinemia in children treated with sodium valproate and carbamazepine. *Epilepsy Research*. 41: 253–257.

Wald, D.S., Law, M., and Morris, J.K., 2002. Homocysteine and cardiovascular disease: evidence on causality from a meta-analysis. *British Medical Journal*. 325(7374): 1202.

Yoo, J.H., and Hong, S.B., 1999. A common mutation in the methylenetetrahydrofolate reductase gene is a determinant of hyperhomocysteinemia in epileptic patients receiving anticonvulsants. *Metabolism*. 48: 1047–1051.

CHAPTER 31

B Vitamins and Folate in Multiple Micronutrient Intervention: Function and Effects

FARUK AHMED

School of Public Health, Griffith University, Gold Coast Campus, QLD 4222, Australia
Email: f.ahmed@griffith.edu.au

31.1 Introduction

Micronutrient deficiencies are one of the most prevalent public health problems affecting more than two billion people worldwide (UNICEF and MI 2004). The magnitude of the problem is much greater in developing countries, where multiple micronutrient (MMN) deficiencies often occur concurrently as a result of poor quality diet. Although vitamin A, iron and iodine are the major micronutrient deficiencies in populations, deficiencies of zinc, vitamin C, folic acid and other B vitamins (vitamin B_2, B_6, B_{12} and niacin) are often present simultaneously (Huffman *et al.* 1999). MMN deficiencies are common throughout the lifespan, but specially in pregnant women and children. Micronutrient deficiencies, if left untreated, can have considerable consequences on health and economic development (UNICEF and MI 2004).

Although the use of prenatal MMN supplements is nearly universal in industrialized countries, the supplements that are usually provided during pregnancy in the developing countries still focus on iron and folic acid. As the prevalence of micronutrient deficiencies in low-income countries is widespread (Huffman et al. 1999), and given the high requirement for these micronutrients during pregnancy and their important roles in metabolic pathways, it is unlikely that repletion of only one or two of these micronutrients in the multiple deficiencies can be effective in reducing adverse birth outcomes.

Over the last decade, there has been an increasing interest in providing MMN supplements, instead of iron and folic acid supplements, not only for pregnant and lactating women but also for children and adolescent girls in developing countries to prevent anaemia and micronutrient deficiencies (Huffman et al. 1999). To date a good number of studies have investigated the effect of MMN interventions, provided either as a supplement or through fortified food, on various aspects of health and nutrition status in infants, children and pregnant women in developing countries. Furthermore, several studies have been conducted in developed countries to explore the possible benefit of multivitamins, especially B_2, B_6 and folic acid, on cognition and cardiovascular disease among adult and elderly populations, though there is considerable variability in clinical trials.

This chapter provides a brief description of the functions of selected B vitamins followed by an overview of their effect in MMN intervention on various nutrition and health outcomes in different population groups.

31.2 Functions of B Vitamins

The primary function of B vitamins is to help body cells produce energy through metabolism of carbohydrate, fat and protein. These vitamins are required for brain energy metabolism, neurotransmitter synthesis and functioning, and myelination of the spinal cord and brain cells. More specifically, folic acid, vitamin B_6 and B_{12} deficiencies or congenital defects in enzymes that are involved in the metabolism of these B vitamins, are known to be associated with impaired brain function (Rosenberg 2001). Folic acid, vitamin B_6 and B_2 are also involved in homocysteine metabolism known to be important for cardiovascular function.

Folic acid is needed to make DNA and RNA, and thus in producing all new cells. Folate deficiency can result in altered methylation of DNA and possibly increase the risk of cancer. Folate deficiency is also known to have a negative effect on pregnancy outcomes, notably the risk of babies born with neural tube defects, low birth weight (LBW) and preterm delivery. Vitamin B_{12} insufficiency may also adversely influence neural tube formation in early stage of embryogenesis, independent of its role in maintaining folate status. Vitamin B_2 is needed for normal reproduction, cellular growth and repair, and to help accelerate healing of injuries. Vitamin B_2 deficiency can lead to functional deficiency of vitamin B_6. Vitamin B_6 is involved in protein metabolism and cellular growth, and its deficiency can suppress immune response. Both vitamins B_2 and B_6 are needed for the synthesis niacin from tryptophan.

B vitamins also play important roles in haematopoiesis (Fishman *et al.* 2000). For example, folate deficiency by impairing DNA synthesis leads to ineffective erythropoiesis, and as a result a macrocytic anaemia develops. Deficiency of vitamin B_{12}, by impairing functions of folic acid, leads to ineffective erythropoeisis and in turn to macrocytic anaemia. Vitamin B_6 is also required in synthesis of haem, a component of haemoglobin and perhaps for mobilization of iron from stores. Vitamin B_2 is known to mobilize stored iron from the liver and its deficiency is associated with anaemia.

31.3 Effect of Supplementation

31.3.1 Haemoglobin Concentration

In populations with MMN deficiencies, it is likely that iron deficiency anaemia is compounded by insufficiencies of other micronutrients. It has been suggested that the co-existence of these other deficiencies with iron deficiency increases the risk of overt anaemia and limits the haematological response to iron supplementation (Allen *et al.* 2000).

Several studies have been conducted in different population groups investigating whether MMN supplements can increase haemoglobin beyond that of iron with or without folic acid supplements. A recent meta-analysis of 13 trials in pregnant women showed that MMN supplements had the same effect on haemoglobin and iron status as iron with or without folic acid supplements (Allen *et al.* 2009a), although in five studies, MMN supplements contained only half the amount of iron included in the iron and folic acid supplements. The authors have suggested that the addition of other micronutrients did not impair the efficacy of the iron supplements.

On the other hand, a randomized trial among anaemic Indonesian infants showed that daily MMN supplements were more efficacious in improving haemoglobin levels than daily iron and folic acid supplementation (Untoro *et al.* 2005). A systematic review of randomized trials in children from developing countries found that the addition of MMN to iron supplementation resulted in marginal improvement in haemoglobin (0.14 g/dL; 95% CI: 0.00, 0.28, $P = 0.04$) response compared with iron supplementation alone (Gera *et al.* 2009). The type of MMN intervention in this analysis included supplements in the form of tablets or fortified food, and the interventions were given either daily or intermittently for more than two months except two studies where supplementation was carried out for two months or less.

A similar increase (2.4 g/L) in haemoglobin was also reported for intermittent MMN supplementation for 26 weeks in a study among Bangladeshi adolescent girls with nutritional anaemia (Ahmed *et al.* 2010). However, a study among well-nourished Australian non-anaemic children found no effect of MMN fortification on haemoglobin status (Osendarp *et al.* 2007).

In all but a few of the studies described above, the MMN supplements contained at least 11–13 micronutrients including B vitamins added to iron and folic acid. It is apparent from these studies that the addition of MMN to iron supplementation, given either daily or intermittently, can be marginally beneficial in improving haemoglobin concentration except for pregnant women, but it is not known which constituent(s) and what level of each of the micronutrients confers this beneficial effect.

31.3.2 Micronutrient Status

Only a limited number of studies have investigated the effect of MMN supplements, containing folic acid and other B vitamins, on B vitamins status in different population groups. A randomized double-blind study among apparently healthy pregnant women showed that regular intake of MMN supplements, at recommended dietary allowance (RDA) equivalent doses, from $14+2$ weeks of gestation up to delivery improved red blood cell (RBC) folic acid and serum vitamin B_2 and B_6 levels significantly, but not plasma vitamin B_{12} concentration (Hininger *et al.* 2004). Another study among pregnant women from developing countries reported that MMN supplements contained 14 different micronutrients, at approximately the recommended daily intake for each nutrient and taken daily until 32.6 weeks of pregnancy, increased serum vitamin B_{12}, B_2 and vitamin B_6 concentrations significantly compared with iron and folic acid supplements (Christian *et al.* 2006). Despite this overall improvement in these three micronutrients, a high proportion of the women remained deficient in late pregnancy, raising the question of adequacy of MMN supplements containing only RDA equivalent doses to correct micronutrient deficiencies in this population.

A randomized controlled trial among anaemic Bangladeshi adolescent girls showed that twice-weekly MMN supplements, containing iron and 14 other micronutrients at approximately double the RDA equivalent dose for each nutrient taken for 52 weeks, improved vitamin B_2 and folate status significantly compared with iron and folic acid supplementation (Ahmed *et al.* 2010). However, a high proportion of these girls remained vitamin B_2 deficient at the end of the trial indicating that an even higher dosage or more frequent supplementation is needed to reduce riboflavin deficiency in this population. Furthermore, although the MMN and iron–folic acid supplements contained the same level of folic acid (400 µg), the mean increase in RBC folic acid was significantly higher in the MMN group (Ahmed *et al.* 2010), suggesting the presence of other micronutrients in the MMN supplements might have influenced the uptake of folic acid.

In conclusion, it is apparent from the above studies that MMN supplements have the potential for improving the status of various B vitamins in populations where concurrent micronutrient deficiencies are common. However, more-in-depth studies are needed to determine the appropriate dosage of each specific micronutrient and the duration of supplements that is sufficient to restore their micronutrient status.

31.3.3 Growth in Children

A meta-analysis of randomized controlled trials examining the effect of MMN interventions on the growth of children showed improved length or height, and weight compared with a control group, with an average effect size of 0.13 (95% CI: 0.055–0.21; $p=0.0015$) for length or height and 0.14 (95% CI: 0.029–0.25; $p=0.015$) for weight (Allen et al. 2009b). Although the majority of trials have demonstrated positive effects on growth, some effects were limited to specific subgroups. The type of MMN intervention in this analysis included tablets, powders or beverages or fortified food, and all interventions contained more than three micronutrients (majority contained several B vitamins) provided for at least four weeks.

31.3.4 Morbidity in Children

Two systematic reviews of the effects of MMN intervention on childhood morbidity showed equivocal results. One review including studies of either MMN supplements or fortification found only two of seven studies effective in reducing the incidence of morbidity (Allen et al. 2009b). Another review of MMN fortified foods found four of seven studies had some beneficial effects (Best et al. 2011). It is worth mentioning that the most common benefits were reduced incidence or duration of diarrhoea and/or respiratory-related morbidity, which are the most common illnesses in children, for which a difference may be easier to detect (Best et al. 2011). Detecting differences in less common diseases or other health outcomes may require longer durations or larger cohorts. Table 31.1 presents the overall effects of MMN supplementation on various nutrition and health outcomes in children.

Table 31.1 Effects of multiple micronutrient intervention (fortification/supplementation) on various nutrition and health status in children based on the findings of available studies.

Nutrition/health status	Effect[c]
Length/height	(+)
Weight	(+)
Haemoglobin levels	(+)
B-vitamins status	(+)
Morbidity	(+/−)
Motor development	(+)
Fluid intelligence[a]	(+)
Crystallized intelligence[b]	(−)
Working memory	(+)

[a]Reasoning abilities.
[b]Acquired skills and knowledge.
[c](+) = positive; (+/−) = equivocal; (−) = no effect.

31.3.5 Birth Outcomes

A prospective, observational cohort study among teenage mothers in the US demonstrated that multiple vitamin and mineral supplements could reduce the risk of preterm delivery and LBW (Scholl *et al.* 1997). A randomized double-blind study among apparently healthy pregnant women in France also showed that regular intake of MMN supplements, at RDA equivalent doses, from 14 + 2 weeks of gestation up to delivery improved birth weights by 10% and reduced LBW significantly compared with placebo (Hininger *et al.* 2004).

In 1998, the United Nations Children's Fund (UNICEF), the World Health Organization (WHO) and the United Nations University (UNU) jointly developed a MMN supplement preparation for pregnant women containing 15 micronutrients including a range of B vitamins at dosages that approximated the RDA (UNICEF *et al.* 1999). This MMN preparation, known as the United Nations International Multiple Micronutrient Preparation (UNIMMAP), was then tested in ten countries (total of 12 randomized controlled trials) spanning three continents—Asia, Latin America and sub-Saharan Africa. The objective of the trials was to determine whether UNIMMAP could improve maternal micronutrient status, including anaemia, as well as various birth outcomes, such as LBW, stillbirths and neonatal deaths.

A meta-analysis of the effect of these UNIMMAP trials commissioned by UNICEF, WHO and UNU on birth size and duration of gestation (Fall *et al.* 2009) showed a significant, although small, increase (22.4 g) in birth weight across the whole distribution of weights, a reduction in the prevalence of LBW and small-for-gestational age (SGA) birth of about 10%, and a 13% increase in prevalence of large-for-gestational age (LGA) babies compared to iron alone (one trial), iron-folic acid (nine trials), iron-folic acid and vitamin A (one trial) or a placebo (one trial). However, they found no significant effect of MMN supplementation on birth length or head circumference or the prevalence of preterm birth. The effect size of MMN supplements for individual studies included in this meta-analysis are shown in Figure 31.1 for LBW and in Figure 31.2 for preterm birth.

Another meta-analysis of the same 12 trials (Ronsmans *et al.* 2009) showed that UNIMMAP supplementation did not reduce rates of stillbirth, or early or late neonatal deaths. Table 31.2 shows the overall effects of UNIMMAP supplementation trials on various birth outcomes.

A more recent systematic review consist of 17 trials conducted in developing countries, including the UNIMMAP trials, concluded that compared to iron-folic acid supplements, MMN supplementation during pregnancy can reduce significantly the risk of SGA births but has no effect on perinatal mortality (Kawai *et al.* 2011). While a positive effect of MMN supplementation during pregnancy on birth weight is observed, there is insufficient evidence for significant functional benefits. It has therefore been suggested that further research and monitoring is required before MMN supplementation can be recommended to replace the current standard practice of iron–folic acid supplementation in pregnant women (Fall *et al.* 2009; Ronsmans *et al.* 2009).

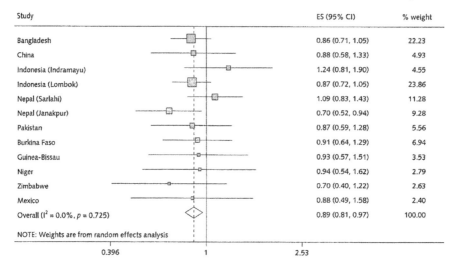

Figure 31.1 Effect of multiple micronutrient supplementation during pregnancy on low birth weight. Random-effects model forest plot shows the effect size (ES) of multiple micronutrient supplementation during pregnancy compared with controls on low birth weight in 12 randomized controlled trials conducted in ten countries using United Nations International Multiple Micronutrient Preparation.
Data are from Fall et al. 2009, with permission from the *Food and Nutrition Bulletin*.

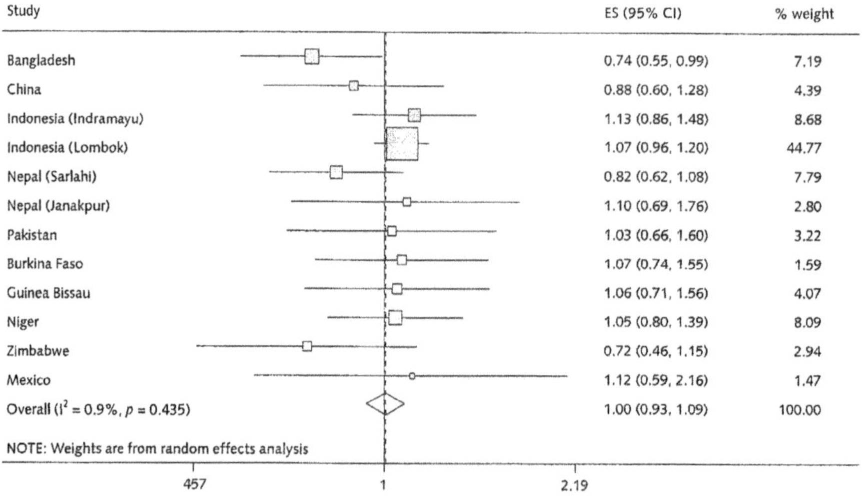

Figure 31.2 Effect of multiple micronutrient supplementation during pregnancy on preterm birth. Random-effects model forest plot shows the effect size (ES) of multiple micronutrient supplementation during pregnancy compared with controls on preterm birth in 12 randomized controlled trials conducted in ten countries using United Nations International Multiple Micronutrient Preparation.
Data are from Fall et al. (2009), with permission from the *Food and Nutrition Bulletin*.

Table 31.2 Effect of multiple micronutrient supplementation during pregnancy on various birth-outcomes based on the findings of available studies.

Birth Outcome	Effect[a]
Birth weight	↑
Prevalence of LBW	↓
Prevalence of SGA birth	↓
Prevalence of LGA birth	↑
Prevalence of preterm delivery	✗
Prevalence of still birth	✗
Prevalence of early neonatal mortality	✗
Prevalence of late neonatal mortality	✗
Prevalence of perinatal mortality	✗

[a] ↑ = significant increase; ↓ = significant decrease; ✗ = no significant effect.

Furthermore, given the growing evidence that pre-pregnancy undernutrition also increases the risk of preterm delivery and reduces foetal growth, even if nutrition improves during pregnancy, both Fall *et al.* (2009) and Ronsmans *et al.* (2009) suggest that future research should include trials of MMN supplementation during as well as before pregnancy.

31.3.6 Cognitive Function

Micronutrient malnutrition in children has been found to impair their cognitive performance and developmental potential. Because micronutrient deficiencies often coexist, supplementing children with MMN could have advantages over single micronutrient supplementation.

Over the past decade, a good number of studies have investigated the effect of MMN supplements including B vitamins on cognitive performance in children. Allen *et al.* (2009b) in their review of four randomized controlled trials showed a positive effect of MMN intervention on the young child's motor development. A recent systematic review of 17 randomized trials, with 5–25 micronutrients and lasting 4–60 weeks, on cognitive performance among healthy schoolchildren aged 5–18 years from both developed and developing countries has shown a marginally positive effect of micronutrients on reasoning abilities (fluid intelligence) and academic performance but no effects on acquired knowledge and other cognitive domains (Eilander *et al.* 2010). In this meta-analysis, children received MMN supplements in 11 trials and a MMN-fortified beverage, biscuits or seasoning powder in six trials. However, the authors concluded that the present evidence is not strong enough to recommend routine MMN supplementation for improving cognitive performance in children (Eilander *et al.* 2010). Another review of investigations of the impact of MMN provided through fortified foods for school age children showed significant beneficial effects on various working memory tests in four of six studies (Best *et al.* 2011).

Despite evidence for an association between low B vitamins status and cognitive decline in the elderly, a role for B vitamins supplementation is not yet convincing. Recent systemic reviews of randomized trials of vitamin B_{12}, B_6 or folic acid supplementation, alone or in combination, concluded that at present there is insufficient evidence for a beneficial effect in elderly with either normal or impaired cognitive function (Balk *et al.* 2007; Malouf and Grimley 2008).

31.3.7 Cardiovascular Disease

Elevated blood homocysteine level is an independent risk factor for cardiovascular disease (CVD). Several randomized controlled trials have shown that supplements combining folic acid and vitamin B_{12} with or without vitamin B_6 can decrease homocysteine levels in people with vascular disease or diabetes and in young adult women (Albert *et al.* 2008; Bronstrup *et al.* 1998; Lonn *et al.* 2006. A study among older men and women showed that multivitamin/ mineral supplements for eight weeks resulted in significantly lower homocysteine levels (McKay *et al.* 2000). However, several large prospective studies have failed to show any reduction in the risk of cardiovascular disease (Albert *et al.* 2008; Lonn *et al.* 2006). In addition, some studies reported a trend toward an increased risk of CVD with combined B vitamin treatment (Bonaa *et al.* 2006). In conclusion, currently available data provides inadequate evidence to support a role for B vitamins in reducing the risk of cardiovascular disease.

31.4 Concluding Remarks

Based on the information available it can be concluded that providing multiple micronutrients, including B vitamins, as a supplement or fortified foods can have a positive effect on growth and some aspects of cognition in children, and on young child's motor development. The effects of MMN intervention on morbidity outcomes in children are somewhat equivocal. Compared with iron and folic acid supplementation, MMN supplements can only marginally improve haemoglobin response in children and adolescent girls, and did not show any significant benefit on maternal anaemia during pregnancy. MMN supplements have the potential to improve status of B vitamins. However, more studies are needed to determine the appropriate dosages and duration for some of the micronutrients delivered in the supplements, both in pregnancy and other population groups.

Supplementation with MMN during pregnancy increases birth weight and significantly reduces the prevalence of LBW and SGA birth compared with iron–folic acid alone, but has no effect on reducing stillbirth, or early or late neonatal deaths. The effect of supplementation with B vitamins on improving cognition among elderly people and reducing the risk of cardiovascular disease is not yet convincing. Overall there is considerable evidence for some beneficial effect of MMN supplementation on child growth and development, improving haemoglobin and micronutrient status in children, and on pregnancy outcomes, but it is not known which constituent(s) at which level among the multiple micronutrients confer(s) this beneficial effect. More carefully designed studies are needed to determine the effect of B vitamins in MMN supplementation.

Summary Points

- This chapter focuses on the effects of B vitamins and folate in multiple micronutrient intervention.
- B vitamins play important roles in energy metabolism and brain function.
- Some B vitamins are involved in haematopoiesis, deoxyribonucleic acid and ribonucleic acid synthesis, and amino acid metabolism.
- Providing multiple micronutrients, including B vitamins, as a supplement or fortified food can have positive effects on child growth, motor development and cognitive performance.
- Multiple micronutrient supplementation can significantly improve micronutrient status including B vitamins.
- Multiple micronutrient supplements can only marginally improve haemoglobin response in children and adolescent girls compared with iron and folic acid alone.
- Supplementation of multiple micronutrients during pregnancy increases birth weight and significantly reduces the prevalence of low birth weight and small-for-gestational age birth compared with iron and folic acid alone.
- Multiple micronutrient supplements have no effect on rates of stillbirth, or early or late neonatal deaths.
- The effect of supplementation with B vitamins on improving cognition among elderly people and reducing the risk of cardiovascular disease is not yet convincing.
- The roles of specific B vitamins in these responses are yet to be determined.

Key Facts

Key Facts about Multiple Micronutrient Intervention

- Multiple micronutrient deficiencies often occur concurrently in populations at risk who usually have poor quality diet.
- Multiple micronutrient intervention is the means of providing various micronutrients in addition to diet to meet requirements.
- Food fortification is one of the ways of intervention that provides micronutrient fortified foods or beverages as well as fortificants that can be added to foods or drinks on-site or at home.
- Supplementation is another way of intervention that provides micronutrients in the form of tablets or syrups.
- There is a possibility of negative interactions with co-administration of several micronutrients, but is still poorly understood.

Key Facts about Anaemia (Low Haemoglobin in Blood)

- The most common cause of nutritional anaemia is iron deficiency.
- Other micronutrients that cause anaemia include deficiencies of folate and vitamins B_2, B_6, B_{12}, A and C.
- These micronutrients are known to affect the synthesis of haemoglobin either directly or indirectly by affecting the absorption and/or mobilization of iron.
- Non-nutritional causes of anaemia such as malaria, worm infestation, infection, chronic diseases and genetic disorders may also be important in some populations.

Key Facts about Homocysteine

- Homocysteine is a sulfur-containing amino acid derived from methionine metabolism.
- Folic acid, vitamin B_{12} and B_6 are involved in homocysteine metabolism.
- High plasma homocysteine can be largely attributed to folic acid and vitamin B_{12} deficiency, and to a lesser extent, vitamin B_6 deficiency.
- Elevated homocysteine levels are associated with vascular diseases, stroke and thrombosis.
- Elevated homocysteine levels are associated with cognitive function.

Definition of Words and Terms

Embriogenesis: The process of formation and development of an embryo.
Fortification: The process of addition of one or more essential nutrients to a food.
Haematopoiesis: The process of formation and development of blood cells.
Large-for-gestational age: When birth weight of a new born baby is above the 90th percentile at gestational age.
Low birth weight: According to WHO, when birth weight of a new-born baby is below 2500 gm.
Macrocytic anaemia: A type of anaemia in which the average size of red blood cell is larger than normal.
Micronutrient: The nutrients that are essential for normal growth and development but needed in small amounts.
Neurotransmitter: A chemical that is released in the brain to help transmit an impulse from a nerve cell to another nerve, muscle, organ, or other tissue.
Small-for-gestational age: When birth weight of a new-born baby is below the 10th percentile at gestational age.
Stillbirth: The birth of an infant who has died in the uterus after having survived through at least the first 28 weeks of pregnancy.

List of Abbreviations

CVD	cardiovascular disease
DNA	deoxyribonucleic acid
LBW	low birth weight
LGA	large-for-gestational age.
MI	Micronutrient Initiative
MMN	multiple micronutrient
RBC	red blood cell
RDA	recommended dietary allowance
RNA	ribonucleic acid
SGA	small-for-gestational age
UNU	United Nations University
UNICEF	United Nations Children's Fund
UNIMMAP	United Nations International Multiple Micronutrient Preparation
WHO	World Health Organization

References

Ahmed, F., Khan, M.R., Akhtaruzzaman, M., Karim. R., Williams, G., Torlesse, H., Darnton-Hill, I., Dalmiya, N., Banu, C.P., and Nahar, B., 2010. Long-term intermittent multiple micronutrient supplementation enhances hemoglobin and micronutrient status more than iron-folic acid supplementation in Bangladeshi rural adolescent girls with nutritional anemia. *Journal of Nutrition.* 140: 1879–1886.

Albert, C.M., Cook, N.R., Gaziano, J.M., Zaharris, E., MacFadyen, J., Danielson, E., Buring, J.E., and Manson, J.E., 2008. Effect of folic acid and B vitamins on risk of cardiovascular events and total mortality among women at high risk for cardiovascular disease: a randomized trial. *The Journal of the American Medical Association.* 299: 2027–2036.

Allen, L.H., Rosado, J.L., Casterline, J.E., Lopez, P., Munoz, E., Garcia, O.P., and Martinez, H., 2000. Lack of hemoglobin response to iron supplementation in anemic Mexican preschoolers with multiple micronutrient deficiencies. *The American Journal of Clinical Nutrition.* 71: 1485–1494.

Allen, L.H., Peerson, J.M., and the Maternal Micronutrient Supplementation Study Group. 2009a. Impact of multiple micronutrient versus iron-folic acid supplements on maternal anaemia and iron status in pregnancy. *Food and Nutrition Bulletin.* 30: S527–S532.

Allen, L.H., Peerson, J.M., and Olney, D.K., 2009b. Provision of multiple rather than two or fewer micronutrients more effectively improves growth and other outcomes in micronutrient-deficient children and adults. *Journal of Nutrition.* 139: 1022–1030.

Balk, E.M., Raman, G., Tatsioni, A., Chung, M., Lau, J., and Rosenberg, I.H., 2007. Vitamin B6, B12, and folic acid supplementation and cognitive function: a systematic review of randomized trials. *Archives of Internal Medicine.* 167: 21–30.

Best, C., Neufingerl, N., Rosso, J.M.D., Transler, C., van den Briel, T., and Osendarp, S., 2011. Can multi-micronutrient food fortification improve the micronutrient status, growth, health, and cognition of schoolchildren? *Nutrition Reviews.* 69: 186–204.

Bonaa, K.H., Njolstad, I., Ueland, P.M., Schirmer, H., Tverdal, A., Steigen, T., Wang, H., Nordrehaug, J.E., Arneseen, E., Rasmussen, K., and NORVIT Trial Investigators, 2006. Homocysteine lowering and cardiovascular events after acute myocardial infraction. *The New England Journal of Medicine.* 354: 1578–1588.

Bronstrup, A., Hages, M., Prinz-Langenohl, R., and Pietrzik, K., 1998. Effects of folic acid and combinations of folic acid and vitamin B12 on plasma homocysteine concentrations in healthy young women. *The American Journal of Clinical Nutrition.* 68: 1104–1110.

Christian, P., Jiang, T., Khatry, S.K., LeClerq, S.C., Shrestha, S.R., and West, K.P. , 2006. Antenatal supplementation with micronutrients and biochemical indicators of status and subclinical infection in rural Nepal. *The American Journal of Clinical Nutrition.* 83: 788–794.

Eilander, A., Gera, T., Sachdev, H.S., Transler, C., van der Knaap, H.C.M., Kok, F.J., and Osendarp, S.J.M., 2010. Multiple micronutrient supplementation for improving cognitive performance in children: systematic review of randomized controlled trials. *The American Journal of Clinical Nutrition.* 91: 115–130.

Fall, C.H.D., Fisher, D.J., Osmond, C., Margetts, B.M., and the Maternal Micronutrient Supplementation Study Group (MMSSG), 2009. Multiple micronutrient supplementation during pregnancy in low-income countries: A meta-analysis of effects on birth size and length of gestation. *Food and Nutrition Bulletin.* 30: S533–S546.

Fishman, S.M., Christian, P., and West, K.P. Jr., 2000. The role of vitamins in the prevention and control of anaemia. *Public Health Nutrition.* 3: 125–150.

Gera, T., Sachdev, H.P.S., and Nestel, P., 2009. Effect of combining multiple micronutrients with iron supplementation on Hb response in children: systematic review of randomized controlled trial. *Public Health Nutrition.* 12: 756–773.

Hininger, I., Favier, M., Arnaud, J., Faure, H., Thoulon, J.M., Hariveau, E., Favier, A., and Roussel, A.M., 2004. Effects of a combined micronutrient supplementation on maternal biological status and newborn anthropometrics measurements: a randomized double-blind, placebo-controlled trial in apparently healthy pregnant women. *European Journal of Clinical Nutrition.* 58: 52–59.

Huffman, S.L., Baker, J., Schumann, J., and Zehner, E.R., 1999. The case for promoting multiple vitamin and mineral supplements for women of reproductive age in developing countries. *Food and Nutrition Bulletin.* 20: 379–394.

Kawai, K., Spiegelman, D., Shankar, A. H., and Fawzi, W.W., 2011. Maternal multiple micronutrient supplementation and pregnancy outcomes in developing countries: meta-analysis and meta-regression. *Bulletin of the World Health Organization.* 89: 402–411B.

Lonn, E., Yusuf, S., Arnold, M.J., Sheridan, P., Pogue, J., Micks, M., McQueen, M.J., Probstfield, J., Fodor, G., Held, C., Genest, J. Jr., and Heart Outcomes Prevention Evaluation (HOPE) 2 Investigators, 2006. Homocysteine lowering with folic acid and B vitamins in vascular disease. *The New England Journal of Medicine*. 354: 1567–1577.

Malouf, R., and Grimley E. J., 2008. Folic acid with or without vitamin B12 for the prevention and treatment of health elderly and demented people. Cochrane Database Systematic Reviews. *Issue 8*: CD004514.

McKay, D.L., Perrone, G., Rasmussen, H., Dallal, G., and Blumberg, J.B., 2000. Multivitamin/mineral supplementation improves plasma B-vitamin status and homocysteine concentration in healthy older adults consuming a folate-fortified diet. *Journal of Nutrition*. 130: 3090–3096.

Osendarp, S.J., Baghurst, K.I., Bryan, J., Calvaresi, E., Hughes, D., Hussaini, M., Karyadi, S.J., van Klinken, B.J., van der Knaap, H.C., Lukito, W., Mikarsa, W., Transler, C., Wilson, C., and NEMO Study Group, 2007. Effect of a 12-mo micronutrient intervention on learning and memory in well-nourished and marginally nourished school-age-children: 2 parallel, randomized, placebo controlled studies in Australia and Indonesia. *The American Journal of Clinical Nutrition*. 86: 1082–1093.

Ronsmans, C., Fisher, D.J., Osmond, C., Margetts, B.M., Fall, C.H.D., and the Maternal Micronutrient Supplementation Study Group (MMSSG), 2009. Multiple micronutrient supplementation during pregnancy in low-income countries: A meta-analysis of effects on stillbirths and on early and late neonatal mortality. *Food and Nutrition Bulletin*. 30: S547–S555.

Rosenberg, I.H., 2001. B vitamins, homocysteine, and neurocognitive function. *Nutrition Reviews*. 59: S69–S74.

Scholl, T.O., Hediger, M.L., Bendich, A., Joan, I., Schall, J.I., Smith, W.K., and Krueger, P.M., 1997. Use of multivitamin/mineral prenatal supplements: influence on the outcome of pregnancy. *American Journal of Epidemiology*. 146: 134–141.

Untoro, J., Karyadi, E., Wibowo, L., Erhardt, M.W., and Gross, R., 2005. Multiple micronutrient supplements improve micronutrient status and anaemia but not growth and morbidity of Indonesian infants: A randomized, double-blind, placebo-controlled trial. *Journal of Nutrition*. 135: S639–S645.

United Nations Children's Fund (UNICEF), World Health Organization (WHO) and United Nations University (UNU), 1999. Composition of a multi-micronutrient supplement to be used in pilot programmes among pregnant women in developing countries. Report of a UNICEF/WHO/UNU workshop, July 9, 1999, 22 pp.

United Nations Children's Fund (UNICEF) and Micronutrient Initiative (MI), 2004. Vitamin and mineral deficiency. A global damage assessment report. The Micronutrient Initiative, Ottawa, Canada, 7 pp.

CHAPTER 32

Wernicke's Encephalopathy caused by Thiamine (Vitamin B_1) Deficiency

ALAN S. HAZELL[a,b]

[a] Department of Medicine, University of Montreal, Montreal, Quebec, Canada; [b] Departamento de Neurologia, Universidade Estadual de Campinas (UNICAMP), Campinas, São Paulo, Brazil
Email: alan.stewart.hazell@umontreal.ca

32.1 Introduction

Thiamine deficiency is the established cause of Wernicke's encephalopathy (WE), an acute neurological disorder constituting one of two components of Wernicke–Korsakoff syndrome (WKS), a neuropsychiatric disorder characterized by ophthalmoplegia, gait ataxia and confusion/memory loss. Up to 80–90% of these patients with WE go on to develop the more debilitating chronic amnesic state, referred to as Korsakoff's psychosis.

Analysis of available data suggests that WE in chronic alcoholics occurs at a frequency of approximately 35%; in the population as a whole, the figure is about 1.5%. Current evidence also suggests that thiamine deficiency occurs in chronic alcoholics at a frequency of at least 25–31% and up to 80%. WE is therefore a common complication of chronic alcoholism, and other disorders associated with grossly impaired nutritional status that include gastrointestinal disorders, hyperemesis gravidarum, HIV AIDS, and malignant disease. WE remains surprisingly difficult to diagnose during life, with the classical clinical features often being absent in both alcoholic and non-alcoholic cases, leading to only a 20% success rate in adults and 60% in paediatric cases. These findings suggest a higher frequency than many other disorders such as epilepsy and Parkinson's disease, making WE an important health care issue.

32.2 Neuroanatomical Damage in Wernicke's Encephalopathy and Thiamine Deficiency

In thiamine deficiency, damage to the brain is focal in nature. During experimental thiamine deficiency, mammillary bodies as well as the medial, midline and intralaminar nuclei of the thalamus, inferior colliculus, periaqueductal area and floor of the fourth ventricle typically show severe damage (Troncoso et al. 1981). Although the thiamine content of the brain is almost uniform at about 13 μg/g dry weight, development of cerebral vulnerability nevertheless occurs. The exact basis for this differential vulnerability, including the precise relationship between thiamine depletion and focal damage, remains a mystery.

WE patients characteristically display gross neuropathological changes that include brain atrophy, hemorrhages, edematous necrosis, white matter damage, gliosis and significant neuronal loss (Victor et al. 1989). A pattern of neuropathological damage as in thiamine deficiency is typical of the disorder, consisting of neuronal cell loss in the mammillary bodies, multiple thalamic nuclei including the medial dosal structures, inferior colliculus, and periaqueductal area and floor of the fourth ventricle, along with damage to the cerebellum, cranial nerve nucleus, pretectal regions and locus coeruleus. The cerebral cortex is also affected in WE, in which atrophy and white matter damage are significant features, and which may be due to alcohol exposure.

32.3 Thiamine-dependent Enzymes

In brain, four major enzyme systems utilize thiamine in the form of thiamine diphosphate (TDP) as a major cofactor, i.e. α-ketoglutarate dehydrogenase complex (KGDHC), pyruvate dehydrogenase complex, branched-chain α-keto acid dehydrogenase complex (BCKDHC) and transketolase.

During experimental thiamine deficiency, onset of neurological symptoms occurs when TDP concentrations fall to less than 15% of normal values. These reductions in TDP levels are accompanied by reductions in KGDHC activity, a key rate-limiting enzyme of the tricarboxylic acid cycle, in both affected animals (Héroux and Butterworth 1995) and in autopsied brain tissue from alcoholic patients with neuropathologically confirmed WE (Butterworth et al. 1993). Thus, KGDHC dysfunction is a probable contributor to the pathophysiology of this disorder, although details of its precise role in the resulting histological damage are unclear. BCKDHC activity is also reduced in brain in thiamine deficiency (Navarro et al. 2008), suggesting that dysfunction of this enzyme complex is a probable contributor to the pathophysiology of this disorder, although details of its role in the resulting histological damage are unknown at the present time.

32.4 Pathophysiology of Thiamine Deficiency

The underlying pathophysiological mechanisms due to thiamine deficiency are complex. Compromised brain energy metabolism due to mitochondrial dysfunction is a major consequence of the disorder (Figure 32.1). Impaired function of thiamine-dependent enzymes leads to a reduction in decarboxylation of pyruvate and α-ketoglutarate, resulting in decreased ATP production, pyruvate accumulation and lactate production (Aikawa *et al.* 1984; Navarro *et al.* 2005). Unless thiamine is rapidly supplemented, refractory acidosis and death can occur within 24 hours. Areas of increased lactic acidosis are localized to brain regions that subsequently develop histological lesions and are likely play a significant role in the pathophysiology in thiamine-deficient animals. Additionally, thiamine and its derivatives are crucial in stabilizing the resting membrane potential, thereby maintaining ionic balance and conduction of the action potential. Chronic deprivation of thiamine impairs this function and the electrophysiological characteristics of the cell, which can lead to severe pathological consequences.

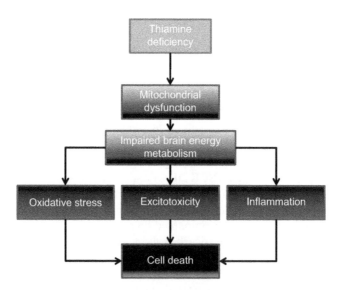

Figure 32.1 Consequences of impaired oxidative metabolism in thiamine deficiency. Thiamine deficiency results in mitochondrial dysfunction in which compromised brain energy metabolism produces oxidative stress, excitotoxicity and inflammatory responses leading to neuronal cell death.

Decreased activity of the rate-limiting enzyme KGDH in thiamine deficiency represents a major probable cause of decreased energy status in vulnerable areas of the brain, with impaired glucose metabolism being a crucial factor contributing to loss of neuronal function, including glutamatergic neurotransmission. This is consistent with the findings of Robertson and colleagues in which levels of incorporation of [^{14}C]-3-O-methyl-D-glucose were depressed in the vulnerable areas of the brain in thiamine deficiency, indicating decreased glucose uptake (Robertson et al. 1975). These findings were reinforced by studies that showed dramatically reduced local cerebral glucose utilization (Hakim and Pappius 1981) and decreased levels of phosphocreatine (Aikawa et al. 1984). However, the exact relationship between KGDHC enzyme activities and focal vulnerability remains uncertain since decreased activities in brain occur in animals in the absence of any change in concentration of protein, with neither activity nor protein levels being predictive of vulnerability (Sheu et al. 1998).

32.5 Glucose Utilization in Thiamine Deficiency

As discussed above, decreased activity of KGDHC represents a major probable cause of reduced energy status in vulnerable areas of the brain in neurodegenerative disease and in thiamine deficiency, with impaired glucose metabolism being a crucial factor contributing to loss of neuronal function, including glutamatergic neurotransmission. This is consistent with the findings of Robertson and colleagues in which levels of incorporation of [^{14}C]-3-O-methyl-D-glucose were depressed in the vulnerable areas of the brain in thiamine deficiency, indicating decreased glucose uptake. These results are reinforced by the dramatically reduced local cerebral glucose utilization and decreased levels of ATP and phosphocreatine as a consequence of thiamine deficiency.

During the development of thiamine deficiency, alterations in brain energy requirements become apparent in the form of an initial generalized reduction in glucose utilization. This initial reduction is followed by a localized increase, then a decrease, which is restricted to histologically vulnerable areas (Hakim and Pappius 1983). These changes in local cerebral glucose utilization occur prior to the onset of neurological symptoms. The lowered ATP and phosphocreatine levels reported by Aikawa and colleagues (1984) are likely to be associated with the ultimate decline in glucose utilization. Overall, this impaired energy metabolism reflects changes in mitochondrial function that can be detected using a gene array in which downregulation of a large number of genes associated with mitochondrial energy metabolism is observed (Figure 32.2). Thiamine replenishment often results in an incomplete recovery of glucose utilization to normal levels and available evidence suggests that more complete improvement results if the replenishment is carried out in advance of the rise in glucose utilization. Such studies suggest that structures vulnerable to thiamine deficiency are likely to develop functional impairments prior to the symptomatic stage of the illness.

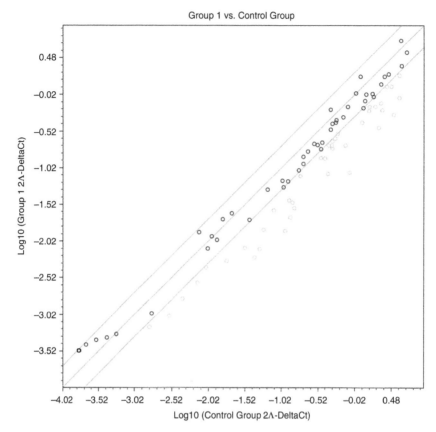

Figure 32.2 Effect of thiamine deficiency on mitochondrial energy metabolism illustrated with a scatter plot of relative expression from mitochondrial energy metabolism gene array. Array analysis shows Group 1 (thiamine deficiency) *versus* control (pair-fed control) expression in which numerous genes are downregulated (gray circles). In general, thiamine deficiency decreased the expression of many genes of the mitochondrial metabolic pathway.

32.6 Excitotoxicity and Thiamine Deficiency

Over the years, a number of studies have provided considerable evidence for the existence of an excitotoxic event in thiamine deficiency. Decreased incorporation of [^{14}C]-glucose into glutamate (Gaitonde *et al.* 1975) is consistent with reduced activities of KGDHC, resulting in reduced energy status in thiamine-deficient animals. In addition, treatment with the noncompetitive *N*-methyl-D-aspartate (NMDA) glutamate receptor antagonist MK-801 has been shown to lead to a reduction in the extent of neuronal damage in thiamine-deficient rats. Focal increases in extracellular glutamate concentration are restricted to vulnerable brain regions in thiamine deficiency (Hazell *et al.* 1993) and have provided the first direct evidence for glutamate-mediated excitotoxicity, along with the presence of excitotoxic-like lesions in damaged areas of the brain (Armstrong-James et al. 1988).

The report of downregulation of the astrocytic glutamate transporters, excitatory amino acid transporters 1 and 2 (EAAT1 and EAAT2; also known as GLAST and GLT-1, respectively) localized to vulnerable brain regions in thiamine deficiency (Hazell et al. 2001) has provided an underlying basis for the elevation of glutamate in the extracellular space. This, along with the demonstration that cultured astrocytes exposed to thiamine deficiency also results in a loss of astrocytic glutamate transporters (Hazell et al. 2003), reinforces the excitotoxicity story and suggests a targeting of astrocytes in this disorder (Hazell 2009). The demonstration that human cases of WE also exhibit loss of the EAAT1 and EAAT2 glutamate transporters and that antioxidants are effective in blocking this effect in experimental animals (Hazell et al. 2010) provides strong evidence that excitotoxicity is indeed a key process underlying the pathophysiology of both thiamine deficiency and WE, and that oxidative stress is a major contributing factor to the loss of glutamate transport function (Figure 32.1).

A key feature of excitotoxic cell death involves an imbalance in Ca^{2+} homeostasis and demonstration of the activation of L-type voltage-sensitive calcium channels (VSCCs) allowing entry of Ca^{2+} that is localized to vulnerable brain regions (Hazell et al. 1998) (Figure 32.3). Along with the rise in

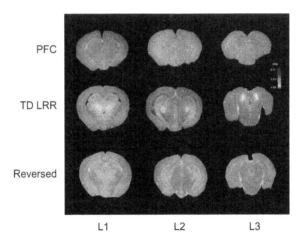

Figure 32.3 Depolarization of vulnerable brain regions in thiamine deficiency. Thiamine deficiency leads to activation of L-type voltage-sensitive calcium channels (VSCCs) in the brain. Representative coronal sections from a pair-fed control (PFC), thiamine-deficient rat at loss of righting reflex (TD LRR), and thiamine-replenished (reversed) animal showing increased specific binding of [^{3}H]-nimodipine, an L-type VSCC antagonist, at the level of the vulnerable thalamus (L1) including the medial geniculate (L2) and the inferior colliculus (L3). Specific binding of nimodipine increases due to activation of L-type VSCCs. The activated channel state occurs as a consequence of membrane depolarization.

interstitial glutamate levels and loss of astrocytic glutamate transporters, this lends further credence to the idea of excitotoxic-mediated damage in affected regions of the brain in thiamine deficiency.

32.7 Oxidative Stress and Thiamine Deficiency

Increased production of reactive oxygen species (ROS) has been reported in brain in thiamine deficiency (Langlais et al. 1997). Other changes consistent with oxidative stress in thiamine deficiency (Figure 32.1) include findings of increased expression of heme oxygenase-1 (HO-1) and intracellular adhesion molecule-1 (ICAM-1) as well as microglial activation. Moreover, selective induction of the endothelial isoform of nitric oxide synthase (eNOS) has been reported in the brains of thiamine-deficient animals, with increased eNOS expression being evident in the medial thalamus in thiamine deficiency. These findings suggest that an early insult in thiamine deficiency is increased production of nitric oxide by cerebrovascular endothelial cells. Using an immunohistochemical approach, increased eNOS immunostaining has been observed in thalamic microvessels of thiamine deficiency mice (Calingasan and Gibson 2000) and increased immunostaining of inducible-NOS (iNOS) in microglia has also been reported in the brains of these animals.

These findings of increased eNOS expression are likely to be of pathophysiological significance since nitric oxide can react with the superoxide radical to form peroxynitrite, which is toxic to neurons. The toxicity results from its ability to nitrate tyrosine residues and to cross-link thiol groups in proteins. Indeed, increased nitrotyrosine immunolabelling has been described in neuronal processes in the thalamus of thiamine-deficient animals and may be indicative of its involvement in the pathogenesis of selective neuronal cell death due to thiamine deficiency. Studies have also demonstrated that targeted disruption of the eNOS gene results in a reduction in extent of the necrotic lesions in the thalamus of thiamine-deficient animals, further evidence in support of an involvement of the gene in the pathophysiologic consequences of this disorder.

Persistent net ROS formation in thiamine deficiency can also initiate a cascade of cell death pathways *via* intracellular messengers, *e.g.* intracellular caspase-3-mediated apoptosis. Development of oxidative stress also leads to other disturbances in brain function, including an inhibition of glutamate uptake due to transporter protein nitrosylation following peroxynitrite formation (Hazell et al. 2003; Trotti et al. 1996; Volterra et al. 1994). Under conditions of oxidative stress, levels of HO-1, eNOS, iNOS, ICAM-1 and microglial activation are increased. Thus, oxidative stress can lead to profound neuropathological consequences in thiamine deficiency and WE.

32.8 Inflammation in Thiamine Deficiency and Wernicke's Encephalopathy

During the 1960s, alterations in glial cell morphology in thiamine deficiency were first reported including evidence of swelling and the appearance of phagocytic vacuoles. These findings are consistent with pathological changes that can be attributable to the presence of an inflammatory process (Figure 32.1). Further evidence has been described in support of this mechanism in thiamine deficiency; increased microglial reactivity, an indication of inflammation, is an early cellular response, while production of pro-inflammatory cytokines in both vulnerable and non-vulnerable regions of brain have been reported (Karuppagounder *et al.* 2007; Vemuganti *et al.* 2006). Levels of these various inflammatory-related gene products in different brain regions may be an important factor(s) in the determination of selective vulnerability in thiamine deficiency.

32.9 Blood–Brain Barrier in Thiamine Deficiency

Considerable evidence for blood–brain barrier (BBB) damage in thiamine deficiency and WE has been described, with disturbances localized to brain regions vulnerable to thiamine deficiency, including the presence of hemorrhagic lesions (Torvik 1985; Vortmeyer and Colmant 1988). Such a process may also contribute to previous reports of brain edema identified in both thiamine deficiency and in cases of WE.

Impairment in oxidative metabolism plays a significant role in BBB breakdown in thiamine deficiency in which oxidative stress mediated by eNOS is involved (Beauchesne *et al.* 2009), and is likely also the case for several neurodegenerative disease states. In addition, inflammatory processes occurring in thiamine deficiency can disrupt the BBB (Guenther and Neu 1984). Furthermore, a number of studies have evaluated BBB integrity, both spatially and temporally, in experimental thiamine deficiency and acute WE. For example, findings have revealed disruption of BBB integrity adjacent to the third ventricle, cerebral aqueduct and fourth ventricle in acute WE, consistent with the location of histological lesions. Increased BBB permeability was also reported in whole brain and in vulnerable brain regions in thiamine deficiency, with a likely factor contributing to this susceptibility being the cerebral energy deficit. Recent evidence also indicates that BBB tight junction proteins such as caveolin-1, occludin and associated scaffolding proteins are decreased, concomitant with increased matrix metalloproteinase-9 levels in thiamine deficiency (Beauchesne *et al.* 2010). The underlying basis for these changes, however, remains currently unresolved.

32.10 Neurodegenerative Disease and Thiamine Deficiency

Previous studies have shown that onset of neurodegenerative diseases such as Alzheimer's disease (AD), amyotrophic lateral sclerosis, Parkinson's disease and Huntington's disease are related to mitochondrial dysfunction brought about by reduction in KGDHC activity (Butterworth and Besnard 1990;

Gibson et al. 2000; Lin and Beal 2006), resulting in decreased NADH production and impaired cerebral energy metabolism, a characteristic feature of thiamine deficiency. This makes experimental thiamine deficiency an important tool for studying the relationship between loss in KGDHC activity and neurodegeneration. Reports have also indicated that levels of thiamine diphosphate are decreased in AD (Héroux et al. 1996; Mastrogiacoma et al. 1996), suggestive of a direct link between this disease and thiamine deficiency, although the exact significance of this finding remains unclear. In addition, oxidative stress, inflammation and excitotoxicity represent pathologic features associated with the majority of the above disorders. Although it is unusual for thiamine deficiency to occur in association with neurodegenerative diseases, most display the presence of one or more of these pathomechanisms, making thiamine deficiency a viable model in which to study these processes.

32.11 Relative Contributions of Thiamine Deficiency and Alcohol to Wernicke's Encephalopathy

Wernicke's encephalopathy is often considered to be an alcohol-related manifestation due to impaired metabolism and associated with major histological lesions. In Korsakoff's psychosis, the neuroanatomical substrate for this neuropsychiatric component of WKS, as in WE, is diencephalic, particularly affecting the anterior thalamic nuclei (Harding et al. 2000). Research on malnutrition, a consequence of poor dietary habits in several alcoholics, indicates that thiamine deficiency can contribute to subcortical damage in these individuals, leading to severe cognitive deficits. Moreover, thiamine deficiency contributes significantly to different forms of alcohol-induced brain injury, including varying degrees of cognitive impairment, and the most severe, alcohol-induced persisting dementia (also known as alcoholic dementia), along with cerebellar atrophy and brain shrinkage.

Over the years, the relationship between thiamine-deficient dependent brain damage and alcoholic brain damage has been well studied. It is now evident that neuropathological damage in thiamine deficiency and alcohol toxicity overlap in frequency and distribution of lesions. In general, alcohol exposure has an adverse effect on the cerebral cortex, diencephalon and neurochemical systems of the brain but the effects are more severe when combined with thiamine deficiency. Conversely, thiamine-deficient dependent brain damage is greater and more severe in the presence of alcohol. In addition, alcoholics with WE show differing results from non-alcoholic cases in terms of brain imaging studies (Zuccoli et al. 2009). As a result of this damaging effect of alcohol and its strong relationship to malnourishment and the associated neurological impairment, the National Institute on Alcohol Abuse and Alcoholism recommends the normal limit for alcohol consumption per day to be two drinks for men and one drink for women—where one drink is equivalent to one 12-ounce (350 mL) bottle of beer, a 5-ounce (150 mL) glass of wine, or a 1 1/2-ounce (4.5 mL) shot of spirits.

Previous findings from different alcoholic subgroups have revealed that alcoholics with WKS had more marked brain shrinkage (87% of control value) than either the uncomplicated alcoholics (94%) or those with cirrhosis (94%). This also suggests that thiamine deficiency plays an active role in brain shrinkage in alcoholics. Moreover, cerebellar degeneration has been reported in many previous neuropathological investigations but there is no direct evidence for its susceptibility in chronic alcoholism. Instead, cerebellar atrophy is related strongly to the thiamine deficiency insult rather than alcohol toxicity (Harper 2009; Mulholland 2006).

Although the ability of alcohol to enhance ROS formation and/or oxidative stress has not been widely studied, a few reports have described conditions under which this may occur. For example, the formation of lipofuscin pigments following chronic ethanol administration in rats in various brains regions has been described, suggesting the presence of ROS damage in alcohol neurotoxicity (Freund 1979). In addition, chronic alcohol exposure results in an inhibition of the activity of superoxide dismutase (SOD), the enzyme that normally scavenges superoxide anion radicals. Thus, oxidative stress, which is likely an effect of acute thiamine deficiency, has been reported as a significant causative factor for alcoholic brain damage. In thiamine deficiency studies in the rat, early increases in SOD immunoreactivity and increased expression of eNOS have been reported. Taken together, these findings suggest that oxidative stress plays a major role in cerebral damage associated with both thiamine deficiency and alcohol toxicity.

32.12 Concluding Remarks

Thiamine deficiency results in an impairment of oxidative metabolism. The consequences of this include a series of events that set the stage for cerebral vulnerability. While thiamine deficiency and Wernicke's encephalopathy involve a variety of functional impairments at several different levels, understanding the mechanisms that underlie the characteristic focal vulnerability associated with this disorder remains a major challenge and is expected to continue to yield important new insight into how metabolic brain disease and neurodegeneration are interrelated.

Summary Points

- Wernickès encephalopathy (WE) due to thiamine (vitamin B1) deficiency can lead to significant brain damage.
- Biochemical pathways involved in energy production and nucleic acid synthesis are affected by thiamine deficiency.
- Altered activity of α-ketoglutarate dehydrogenase complex (KGDHC), a key enzyme of the tricarboxylic acid cycle, is a major contributor to damage resulting from thiamine deficiency.
- Excitotoxic damage involving loss of astrocytic glutamate transporters is a feature of thiamine deficiency and WE.

- Oxidative stress is an important part of the pathophysiology of thiamine deficiency in which key proteins involved in processes such as nitric oxide production, apoptosis, and glutamate transport are affected.
- Microglia-associated induction of cerebral inflammation is a typical feature of thiamine deficiency.
- The blood–brain barrier (BBB) is disrupted in thiamine deficiency and can lead to serious damage in areas of the brain where it occurs.
- Chronic alcoholics constitute the majority of patients with WE.

Key Facts about Wernicke's Encephalopathy

- Thiamine deficiency is the cause of Wernicke's encephalopathy (WE).
- Antemortem diagnosis of WE occurs in approximately 20% of total cases.
- WE is characterized by damage to periventricular areas of the brain including the thalamus, inferior colliculus and vestibular nuclei.
- Thiamine-dependent enzymes consist of pyruvate dehydrogenase complex, α-ketoglutarate dehydrogenase complex (KGDHC), branched-chain α-ketoacid dehydrogenase complex and transketolase.
- Thiamine-dependent enzymes are involved in energy production, nucleic acid synthesis and branched-chain amino acid metabolism.
- Thiamine deficiency provides an important platform for modelling the pathophysiology of neurodegeneration disease.
- Altered glucose metabolism is a key feature of thiamine deficiency and WE.
- Thiamine deficiency and WE are associated with glutamate-mediated excitotoxicity.
- Oxidative stress and inflammation are two major consequences of thiamine deficiency.
- Disruption of the blood–brain barrier is a major consequence of thiamine deficiency and WE.
- The majority of cases of WE involve a combination of the effects of both alcohol and thiamine deficiency in the resulting neuropathology.

Definitions of Words and Terms

Astrocyte. A type of glial cell that is critical to normal brain function. Among its important roles include the ability to take up glutamate effectively and thus protect against excitotoxicity.

Blood–brain barrier. A physical barrier that separates blood from the extracellular fluid of the brain.

Diencephalon. Part of the brain that includes the thalamus, a vulnerable area in Wernicke's encephalopathy and located near the midline.

Excitotoxicity. A pathological mechanism in the brain in which excess extracellular glutamate overstimulates receptors leading to cell death.

Inflammation. A reactive process in which primarily microglia in brain produce chemicals such as cytokines and chemokines as part of a complex protective response to a foreign agent or pathogen.

Oxidative metabolism. Breakdown of glucose in cells that results in ATP production under oxygen rich (aerobic) conditions.

Oxidative stress. An inability to efficiently remove reactive oxygen species which exert damaging effects on proteins, DNA and lipids in the cell.

Reactive oxygen species. Molecules that contain oxygen and are chemically reactive.

Thiamine. A water-soluble B vitamin important as a cofactor for enzymes involved in energy production and nucleic acid synthesis in the cell.

Voltage-sensitive calcium channel. A protein in the cell membrane that acts as a pore, allowing the entry of calcium predominantly, and which opens as a result of depolarization, in which the inside becomes positive relative to the outside. Sustained opening (or activation) of the channel can produce excessive entry of calcium into the cell, leading to cell death.

List of Abbreviations

AD	Alzheimer's disease
AIDS	acquired immunodeficiency virus
ATP	adenosine triphosphate
BBB	blood–brain barrier
BCKDHC	branched-chain α-keto acid dehydrogenase complex
EAAT	excitatory amino acid transporter
eNOS	endothelial nitric oxide synthase
HIV	human immunodeficiency virus
HO-1	heme oxygenase-1
ICAM-1	intracellular adhesion molecule-1
IEG	immediate-early gene
iNOS	inducible nitric oxide synthase
KGDHC	α-ketoglutarate dehydrogenase complex
NMDA	N-methyl-D-aspartate
ROS	reactive oxygen species
SOD	superoxide dismutase
TD	thiamine deficiency
TDP	thiamine disphosphate
VSCC	voltage-sensitive calcium channel
WE	Wernicke's encephalopathy
WKS	Wernicke–Korsakoff syndrome

Acknowledgements

The author is currently a visiting professor in the Department of Neurology at Universidade Estadual de Campinas (UNICAMP), São Paulo, Brazil. Studies performed in the author's laboratory were supported by grants from the Canadian Institutes of Health Research.

References

Aikawa, H., Watanabe, I.S., Furuse, T., Iwasaki, Y., Satoyoshi, E., Sumi, T., and Moroji, T., 1984. Low energy levels in thiamine-deficient encephalopathy. *Journal of Neuropathology and Experimental Neurology*. 43: 276–287.

Armstrong-James, M., Ross, D.T., Chen, F., and Ebner, F.F., 1988. The effect of thiamine deficiency on the structure and physiology of the rat forebrain. *Metabolic Brain Disease*. 3: 91–124.

Beauchesne, E., Desjardins, P., Hazell, A.S., and Butterworth, R.F., 2009. Endothelial NOS mediates blood-brain barrier alterations in the thiamine-deficient mouse brain. *Journal of Neurochemistry*. 111: 452–459.

Beauchesne, E., Desjardins, P., Butterworth, R.F., and Hazell, A.S., 2010. Up-regulation of caveolin-1 and blood-brain barrier breakdown are attenuated by N-acetylcysteine in thiamine deficiency. *Neurochemistry International*. 57: 830–837.

Butterworth, R.F., and Besnard, A.M., 1990. Thiamine-dependent enzyme changes in temporal cortex of patients with Alzheimer's disease. *Metabolic Brain Disease*. 5: 179–184.

Butterworth, R.F., Kril, J.J., and Harper, C.G., 1993. Thiamine-dependent enzyme changes in the brains of alcoholics: relationship to the Wernicke-Korsakoff syndrome. *Alcoholism: Clinical and Experimental Research*. 17: 1084–1088.

Calingasan, N.Y., and Gibson, G.E., 2000. Vascular endothelium is a site of free radical production and inflammation in areas of neuronal loss in thiamine-deficient brain. *Annuals of the New York Academy of Sciences*. 903: 353–356.

Freund, G., 1979. The effects of chronic alcohol and vitamin E consumption on aging pigments and learning performance in mice. *Life Sciences*. 24: 145–151.

Gaitonde, M.D., Fayein, N.A., and Johnson, A.L., 1975. Decreased metabolism *in vivo* of glucose into amino acids of the brain of thiamine-deficient rats after treatment with pyrithiamine. *Journal of Neurochemistry*. 24: 1215–1223.

Gibson, G.E., Park, L.C., Sheu, K.F., Blass, J.P., and Calingasan, N.Y., 2000. The alpha-ketoglutarate dehydrogenase complex in neurodegeneration. *Neurochemistry International*. 36: 97–112.

Guenther, W., and Neu, I.S., 1984. Investigation of the blood-brain barrier for IgG in inflammatory syndromes of the central nervous system. *European Neurology*. 23: 132–6.

Hakim, A.M., and Pappius, H.M., 1981. The effect of thiamine deficiency on local cerebral glucose utilization. *Annuals of Neurology*. 9: 334–339.

Hakim, A.M. and Pappius, H.M., 1983. Sequence of metabolic, clinical, and histological events in experimental thiamine deficiency. *Annals of Neurology*. 13: 365–375.

Harding, A., Halliday, G., Caine, D., and Kril, J., 2000. Degeneration of anterior thalamic nuclei differentiates alcoholics with amnesia. *Brain*. 123: 141–154.

Harper, C., 2009. The neuropathology of alcohol-related brain damage. *Alcohol and Alcoholism.* 44: 136–140.

Hazell, A.S., Butterworth, R.F., and Hakim, A.M., 1993. Cerebral vulnerability is associated with selective increase in extracellular glutamate concentration in experimental thiamine deficiency. *Journal of Neurochemistry.* 61: 1155–1158.

Hazell, A.S., Hakim, A.M., Senterman, M.K., and Hogan, M.J., 1998. Regional activation of L-type voltage-sensitive calcium channels in experimental thiamine deficiency. *Journal of. Neuroscience Research.* 52: 742–749.

Hazell, A.S., Rao, K.V., Danbolt, N.C., Pow, D.V., and Butterworth, R.F., 2001. Selective down-regulation of the astrocyte glutamate transporters GLT-1 and GLAST within the medial thalamus in experimental Wernicke's encephalopathy. *Journal of Neurochemistry.* 78: 560–568.

Hazell, A.S., Pannunzio, P., Rama Rao, K.V., Pow, D.V., and Rambaldi, A., 2003. Thiamine deficiency results in downregulation of the GLAST glutamate transporter in cultured astrocytes. *Glia.* 43: 175–184.

Hazell, A.S., 2009. Astrocytes are a major target in thiamine deficiency and Wernicke's encephalopathy. *Neurochemistry International.* 55: 129–135.

Hazell, A.S., Sheedy, D., Oanea, R., Aghourian, M., Sun, S., Jung, J.Y., Wang, D., and Wang, C., 2010. Loss of astrocytic glutamate transporters in Wernicke encephalopathy. *Glia.* 58: 148–156.

Héroux, M., and Butterworth, R.F., 1995. Regional alterations of thiamine phosphate esters and of thiamine diphosphate-dependent enzymes in relation to function in experimental Wernicke's encephalopathy. *Neurochemical Research.* 20: 87–93.

Héroux, M., Raghavendra Rao, V.L., Lavoie, J., Richardson, J.S., and Butterworth, R.F., 1996. Alterations of thiamine phosphorylation and of thiamine-dependent enzymes in Alzheimer's disease. *Metabolic Brain Disease.* 11: 81–88.

Karuppagounder, S.S., Shi, Q., Xu, H., and Gibson, G.E., 2007. Changes in inflammatory processes associated with selective vulnerability following mild impairment of oxidative metabolism. *Neurobiology of Disease.* 26: 353–362.

Langlais, P.J., Anderson, G., Guo, S.X., and Bondy, S.C., 1997. Increased cerebral free radical production during thiamine deficiency. *Metabolic Brain Disease.* 12: 137–143.

Lin, M.T., and Beal, M.F., 2006. Mitochondrial dysfunction and oxidative stress in neurodegenerative diseases. *Nature.* 443: 787–795.

Mastrogiacoma, F., Lindsay, J.G., Bettendorff, L., Rice, J., and Kish, S.J., 1996. Brain protein and alpha-ketoglutarate dehydrogenase complex activity in Alzheimer's disease. *Annals of Neurology.* 39: 592–598.

Mulholland, P.J., 2006. Susceptibility of the cerebellum to thiamine deficiency. *Cerebellum.* 5: 55–63.

Navarro, D., Zwingmann, C., Hazell, A.S., and Butterworth, R.F., 2005. Brain lactate synthesis in thiamine deficiency: a re-evaluation using 1H-13C nuclear magnetic resonance spectroscopy. *Journal of Neuroscience Research.* 79: 33–41.

Navarro, D., Zwingmann, C., and Butterworth, R.F., 2008. Impaired oxidation of branched-chain amino acids in the medial thalamus of thiamine-deficient rats. *Metabolic Brain Disease*. 23: 445–455.

Robertson, D.M., Manz, H.J., Haas, R.A., and Meyers, N., 1975. Glucose uptake in the brainstem of thiamine-deficient rats. *American Journal of Pathology*. 79: 107–118.

Sheu, K.F., Calingasan, N.Y., Lindsay, J.G., and Gibson, G.E., 1998. Immunochemical characterization of the deficiency of the alpha-ketoglutarate dehydrogenase complex in thiamine-deficient rat brain. *Journal of Neurochemistry*. 70: 1143–1150.

Torvik, A., 1985. Two types of brain lesions in Wernicke's encephalopathy. *Neuropathology and Applied Neurobiology*. 11: 179–190.

Troncoso, J.C., Johnston, M.V., Hess, K.M., Griffin, J.W., and Price, D.L., 1981. Model of Wernicke's encephalopathy. *Archives of Neurology*. 38: 350–354.

Trotti, D., Rossi, D., Gjesdal, O., Levy, L.M., Racagni, G., Danbolt, N.C., and Volterra, A., 1996. Peroxynitrite inhibits glutamate transporter subtypes. *Journal of Biological Chemistry*. 271: 5976–5979.

Vemuganti, R., Kalluri, H., Yi, J.-H., Bowen, K.K., and Hazell, A.S., 2006. Gene expression changes in thalamus and inferior colliculus associated with inflammation, cellular stress, metabolism, and structural damage in thiamine deficiency. *European Journal of Neuroscience*. 23: 1172–1188.

Victor, M., Adams, R.D., and Collins, G.H., 1989. *The Wernicke-Korsakoff Syndrome and Related Neurologic Disorders due to Alcoholism and Malnutrition*. F.A. Davies, Philadelphia, USA. p. 231.

Volterra, A., Trotti, D., Tromba, C., Floridi, S., and Racagni, G., 1994. Glutamate uptake inhibition by oxygen free radicals in rat cortical astrocytes. *Journal of Neuroscience*. 14: 2924–2932.

Vortmeyer, A.O., and Colmant, H.J., 1988. Differentiation between brain lesions in experimental thiamine deficiency. *Virchows Archive A: Pathological Anatomy and Histology*. 414: 61–67.

Zuccoli, G., Santa Cruz, D., Bertolini, M., Rovira, A., Gallucci, M., Carollo, C., and Pipitone, N., 2009. MR imaging findings in 56 patients with Wernicke encephalopathy: nonalcoholics may differ from alcoholics. *American Journal of Neuroradiology*. 30: 171–176.

CHAPTER 33

Disturbances in Acetyl-CoA Metabolism: A Key Factor in Preclinical and Overt Thiamine Deficiency Encephalopathy

ANDRZEJ SZUTOWICZ,* AGNIESZKA JANKOWSKA-KULAWY AND HANNA BIELARCZYK

Department of Laboratory Medicine, Medical University of Gdańsk, Dębinki 7 Str., 80-211 Gdańsk, Poland
*Email: aszut@gumed.edu.pl

33.1 Introduction

Encephalopathy is one of key, although not ubiquitous, symptoms of thiamine deficiency (TD) in humans and animals. It may be caused by lack or decreased provision of thiamine in the diet due to single or combined conditions such as alcoholism, starvation, long-term consumption of low thiamine or thiamine-depleting diets containing high levels of thiaminase (milled rice, row fresh water fish, shellfish) or by thiamine antagonist-rich diets (tea, coffee, betel nuts, sulfite-processed foods). Thiamine deficiency, with or without apparent clinical manifestation, may be also facilitated by physiopathological conditions increasing thiamine utilization such as

Table 33.1 Population groups with increased prevalence of subclinical thiamine diphosphate deficits and at risk of overt disease.

Group	Cause/mechanism
Third world countries/starvation	Populational under-nutrition, decreased intake
Alcoholics/addicts/low socio-economic status/elderly/disabled people, *anorexia nervosa*	Low intake, under-nutrition, inhibition of intestinal and erythrocyte thiamine transport, inhibition of thiamine phosphorylation, reduction in liver stores, hypomagnesemia
Loop diuretics (furosemide, ethacrynic acid, *etc.*)	Used in heart failure, they increase loss of TDP and several other trace elements and minerals with urine
Different drugs	Phenytoin, oral contraceptives, antibiotics
Selective diets (vegetarians)	Low intake, (meats contain highest amounts of TDP/thiamine), sulfites as food preservatives, thiaminase (row fish, shellfish) hydroxyphenols (tannic, caffeic, chlorigenic acids)
Foods containing thiamine antagonists	
Pregnancy/breastfeeding	Increased utilization by foetus, or excretion with milk combined with low thiamine diet
Newborns/infants	Shortages of TDP in foetal life and during breastfeeding by thiamine-deficient mothers
Severe bacterial or parasite infections	TDP consumption and cleavage by microorganisms or parasites
Excessive physical exercises	Increased breakdown, overuse of isotonic 'sport drinks'
Dialysis (peritoneal, extracorporeal)	Removal with dialysis fluids
Diarrhoea/enteropathies	Increased loss with faeces, decreased intestinal absorption

pregnancy, lactation, severe infections and excessive physical exercise as well as hyperthyroidism or folate deficits. Also alcoholism, diarrhoea, loop diuretics, peritoneal and extracorporeal dialysis may deplete several vitamin stores, including thiamine diphosphate (TDP), in the organism. Usually, the combination of multiple causes contributes to development of thiamine deficiency in the individual patient (Butterworth 1989) (Table 33.1).

Symptoms of overt beriberi are usually abrupt with presentation of left ventricular cardiac and peripheral vascular failure yielding water retention due to stimulation of the renin–angiotensin–aldosterone axis by hypovolemia, as well as striated and smooth muscles weakness in the 'wet' form of the disease. At this stage of the disease, the clinical diagnosis is usually apparent but outcome of the supplementary treatment with thiamine may be poor, due to irreversible alterations in muscles and the central nervous system (CNS).

However, most cases of border line TDP deficit remain undiagnosed. Early symptoms of thiamine deficiency are frequently confused with other metabolic, cardiovascular, gastrointestinal or neurological conditions. Such underdiagnosed cases of thiamine deficiency are relatively frequent in developed countries, where an undernourished population is not thought to be a common problem.

33.2 Sources of Neuronal Susceptibility to Thiamine Diphosphate Deficiency

The significance of correct TDP status for the well-being of all living organisms is because TDP is the cofactor for key enzymes regulating the rate of oxidative energy metabolism in practically all cells and tissues (Figure 33.1). These enzymes include:

- Pyruvate dehydrogenase complex (PDHC) is a multienzyme complex which, through a multistep reaction, provides acetyl-CoA for the

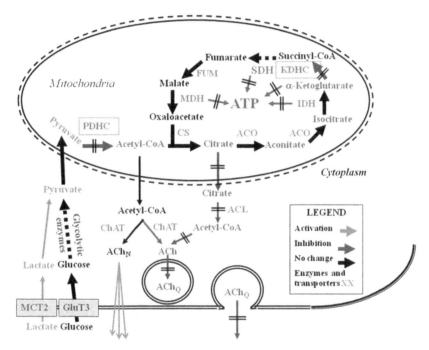

Figure 33.1 Alterations in pathways of energy and acetyl-CoA metabolism in TDP-deficient brain cells: specific features of cholinergic neurons. Inhibition of PDHC due to TD decreases acetyl-CoA synthesis in the mitochondria and inhibits citrate synthesis and metabolic flux through the first steps of the TCA cycle. Lack of citrate suppresses acetyl-CoA transport to cytoplasm and its re-conversion to acetyl-CoA through the ACL pathway. The reduction in acetyl-CoA level in the cytoplasm of cholinergic neurons inhibits ACh synthesis and its quantal release from their depolarized terminals, yielding symptoms of cognitive and motor deficits. These may be reversed by provision of thiamine. Simultaneous inhibition of KDHC aggravates energy deficits that may cause irreversible structural losses of cholinergic and other groups of neurons. Energy deficits result in sustained depolarization of TD neurons which stimulates PTP-dependent direct transport of acetyl-CoA to cytoplasm, but affects energy-dependent ACh loading into synaptic vesicles. Therefore, in thiamine deficiency, cholinergic neurons' release of nonquantal ACh is activated while the quantal one is inhibited.

tricarboxylic acid cycle (TCA) and energy production in the respiratory chain;
- Ketoglutarate dehydrogenase multienzyme complex (KDHC), which is a rate-limiting step of the TCA cycle;
- Transketolase, providing NADPH moieties for glutathione synthesis, a main protectant against free radical excess;
- Branched-chain α-ketoacid dehydrogenase complex (BCKDHC).

The brain utilizes almost exclusively glucose as a principal energy source. It consumes 20% of body glucose and oxygen under postprandial, resting conditions. Pyruvate (the final product of glycolysis) enters the TCA cycle in the mitochondria after its conversion to acetyl-CoA in the oxidative decarboxylation reaction catalysed by PDHC. Activity of the E1 subunit of PDHC (EC 1.2.4.1) is totally dependent on saturation of its catalytic centre with TDP. PDHC activity in the brain is 5–10 times higher than in other tissues in order to meet the high energy demand for maintenance of neuronal action potentials. In addition, the E1 subunit of KDHC (EC 1.2.4.2) requires bound TDP for its catalytic activity. KDHC is a rate-limiting step in the TCA cycle (Figure 33.1). Therefore, TDP deficits and other pathological conditions, which inhibit one or both enzymes, strongly reduce acetyl-CoA supply and its metabolic flux through TCA compromising viability and transmitter functions in brain neurons (Bubber *et al.* 2004; Butterworth 1989; Jankowska-Kulawy *et al.* 2010; Szutowicz *et al.* 2006) (Figure 33.2).

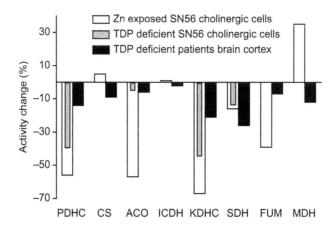

Figure 33.2 TD-evoked changes in PDHC and KDHC display similar patterns in isolated septal cholinergic neuronal cells and in human *post mortem* TD brains. Bars indicate difference between control and experimental groups' cholinergic cells, and non TD and TD human brains. Values recalculated from Bizon-Zygmańska *et al.* (2011), Bubber *et al.* (2004) and Ronowska *et al.* (2007). Sustained depolarization by Zn could aggravate primary TD-inhibitory effects on pyruvate dehydrogenase and several enzymes of the TCA cycle including, aconitase, ketoglutarate and succinate dehydrogenases. ACO, aconitase; CS, citrate synthase; FUM, fumarase; ICDH, isocitrate dehydrogenase; MDH, malate dehydrogenase; SDH, succinate dehydrogenase.

About 80% of the energy produced in the brain is consumed by neurons, which constitute only 10% of all brain cells. Such large amounts of energy are indispensable for the maintenance and restoration of neuronal plasma membrane potentials (of several Hz frequency) during events linked to neurotransmission.

33.3 Brain Thiamine Diphosphate Levels and Encephalopathy

Relatively weak positive correlations have been found between memory performance and IQ test scores and whole blood TDP concentrations (Pitel *et al.* 2011). Other investigators claim a diagnostic and predictive value of this laboratory parameter in combination with multi-questionnaire anamnesis and clinical assessment (Table 33.2).

Erythrocyte transketolase activity was found to be lowered in different groups of TD patients (Herve *et al.* 1995; Khounnorath *et al.* 2011) (Table 33.2). However, *post mortem* brains of alcoholics who died without symptoms of Wernicke's encephalopathy (WE) revealed 20–35% reductions in the activities of all TDP-dependent enzymes in various brain regions (Lavoie and Butterworth 1995). This means that some degree of TD-induced reductions in oxidative metabolism may be tolerated. In fact, individual sensitivity to TDP deficits may be modified by coexisting clinical conditions such as alcoholism, renal insufficiency, dialysis programmes, aging, diabetes, cardiovascular complications, and voluntary or socio-economically dependent dietary habits (Tables 33.1 and 33.2).

33.4 Transmitter Systems in Thiamine Deficiency Encephalopathy

The TD-induced inhibition of metabolic flux through PDHC takes place in all cells of the affected organism (Jankowska-Kulawy *et al.* 2010). It causes primary inhibition of acetyl-CoA synthesis in the mitochondria, yielding a proportional reduction of its metabolic flux through first steps of the TCA cycle and further aggravation of this suppression due to inhibition of KDHC (Jankowska-Kulawy *et al.* 2010; Shi *et al.* 2007) (Figure 33.1). As a result, oxidative phosphorylation process slows down, yielding ATP deficits (Jhala and Hazell 2011).

The sensitivity of particular cell groups to thiamine deficiency is apparently dependent on the ratios of their rates of oxidative metabolism to actual energy expenditures, as well as the possibility of using alternative sources of energy. Therefore, excitable cells, like neurons, with firing rates from few to several tens of Hz are particularly vulnerable to TD-generated energy shortages. Cholinergic neurons, responsible for cognitive functions, as well as cranial and spinal cholinergic motor neurons, are apparently more susceptible to thiamine deficiency than other brain neurons due to the obligatory utilization of an

Table 33.2 Laboratory markers of thiamine diphosphate deficiency in different groups at risk.

Laboratory/clinical marker[a]	Group characteristics/values[b]	References
ETK-A Serum pyruvate	Different psychiatric patients ($n = 326$) normal thiamine status 62%; low thiamine no symptoms 38% (ETK-A > 20%; serum pyruvate > 79 µmol/L) and gross clinical thiamine deficiency ($n = 6$)	Carney and Barry 1985
Erythrocyte TDP/ ETK-A	Healthy ($n = 52$) 176 nM/11%; alcoholics > 50g alcohol per day: I group ($n = 53$) 182 nM/10%, II group ($n = 32$) 91 nM/42%	Herve et al. 1995
ETK	Healthy 25.9 µmol/min/L; renal failure – dialysis: before 19.4, immediately after 25.9, 48 h after 20.9 µmol/min/L	Pietrzak and Baczyk 2001
Thiamine Deficiency Questionnaire (TDQ) Erythrocyte TDP Serum γ-glutamyl transferase	Alcoholics ($n = 58$) TDQ score 68.8 had predictive values for thiamine deficiencydevelopment 74%. Erythrocyte TDP < 165 nmol/L in 31 (53%) subjects	Sgouros et al. 2004
ETK	Heart failure NYHA III/IV with furosemide ($n = 22$); with spironolactone ($n = 11$) ETK 277 nmol/min/L; no spironolactone ($n = 11$) 155 nmol/min/L	Rocha et al. 2008
Neuropsychological deficits Erythrocyte TDP	Alcoholics ($n = 28$): no signs WE ($n = 8$); at risk of WE ($n = 16$); signs of WE ($n = 4$), Erythrocyte TDP levels correlated only with memory performance but not with other neuropsychological scores	Pitel et al. 2011
ETK and ETK-A	Sick infants, no signs of beriberi (Laos, $n = 788$) Low ETK < 0.59 µmol/min/g Hb and high ETK-A > 31% ($n = 40$), low ETK only ($n = 51$); low ETK correlated with higher mortality rate.	Khounnorath et al. 2011

[a]ETK, erythrocyte transketolase activity; TDP, thiamine diphosphate; ETK-A-TDP, activation index for transketolase activity.
[b]Diagnostic cut off values for ETK-A: healthy subjects < 15%; mild deficiency 15–25%; severe deficiency > 25%.

additional pool of acetyl-CoA for acetylcholine (ACh) synthesis (Ronowska et al. 2010; Szutowicz et al. 2006).

33.4.1 Glutamate

There are significant regional and cell type dependent differences in the susceptibility to thiamine deficiency. Most significant, structural and functional losses of neurons are observed in thalamic nuclei, mammillary bodies, hippocampus and some brain stem nuclei. These losses are compatible with severalfold increases in levels of glutamate in extracellular compartments, triggering excitotoxic effects in these regions (Todd and Butterworth 2001). There is also known that TD-inhibited Ca/depolarization evoked quantal release of glutamate from hippocampal slices (Le et al. 1991). This indicates that excitotoxic conditions in thiamine-deficient brain could be brought about by a combination of sustained depolarization of glutamatergic terminals, increased nonquantal release of glutamate, and impairment of its uptake by astrocytes and postsynaptic neurons (Jhala and Hazell 2011).

33.4.2 Acetylcholine

Cognitive functions in the brain require reciprocal interactions between functionally competent glutaminergic and septal cholinergic neurons. Cholinergic neurons differ from other transmitter groups with the presence of specific proteins involved in the synthesis, vesicular accumulation and quantal release of ACh (Tucek 1993). Choline acetyltransferase (ChAT, EC 2.3.1.6), the key enzyme of cholinergic neurons, synthesizes ACh from choline and acetyl-CoA (Figure 33.1). The equilibrium constant for the ChAT reaction is equal to 14; this precludes its shift toward ACh formation and the existence of a direct correlation between choline and acetyl-CoA concentrations and the size of the ACh pool in cholinergic terminals (Szutowicz et al. 1996; Tucek 1993). Therefore, our recent data demonstrating the suppression of acetyl-CoA level in cytoplasmic compartment of thiamine-deficient brain nerve terminals and septal cholinergic cells SN56 provide a simple explanation for the reduction in ACh content and release observed in these conditions (Table 33.3) (Bizon-Zygmańska et al. 2011; Jankowska-Kulawy et al. 2010).

Similar, suppressive effects of thiamine deficiency on ACh levels and turnover varying from 17–60% were repeatedly reported for brain preparations obtained from animals in which thiamine deficiency was previously evoked by different TDP-depriving agents (Gibson et al. 1982). More recent *in vivo* microdialysis studies revealed that thiamine deficiency resulted in significant 40–100% decreases in increases in extracellular ACh concentrations in hippocampus, frontal and retrosplenial cortex induced by memory related tasks (Savage et al. 2007). Indeed, our study on nerve terminals isolated from whole cerebrum demonstrated that thiamine deficiency caused an increase in spontaneous/non-quantal ACh release and 50% inhibition of quantal depolarization-induced Ca-dependent ACh release (Table 33.3). The latter strongly correlated with decreases in the cytoplasmic concentration of acetyl-CoA, which were caused by inhibition of ATP-citrate lyase pathway in thiamine-deficient brain nerve terminals (Jankowska-Kulawy et al. 2010) (Table 33.3, Figure 33.3D).

Suppression of ACh metabolism remains in contrast with no alterations in ChAT expression in thiamine-deficient neurons (Figure 33.3C) (Jankowska-Kulawy et al. 2010, Savage et al. 2007). This apparent discrepancy indicates

Table 33.3 Thiamine diphosphate deficits reduce acetyl-CoA levels in subcellular compartments of the brain neurons, suppressing their viability and cholinergic neurotransmission. Data from Bielarczyk et al. (2005), Bizon-Zygmańska et al. (2011) and Jankowska-Kulawy et al. (2010); expressed as means ± SEM in pmol/mg protein ($n = 5$–20 experiments). Italics refer to fractions of acetyl-CoA and acetylcholine provided through the ATP–citrate lyase pathway (see Figure 33.1). Bold numbers refer to nonviable cell fractions assayed by trypan blue exclusion assay and expressed as a percentage of the whole cell population.

Experimental model	Mitochondrial acetyl-CoA Control TD	Cytoplasmic acetyl-CoA Control TD	Acetylcholine content/ release Control TD
Rat brain synaptosomes Pyrithiamine TDP deficiency	36.2 ± 2.5 17.5 ± 0.5^a	37.9 ± 1.6 22.3 ± 1.8^a *17.6 ± 1.0 1.7 ± 4.5^a*	4.9 ± 0.6 2.3 ± 0.4^a *2.8 ± 0.5 0.3 ± 0.4^a*
Non-differentiated SN56 cholinergic neuroblastoma cells. Amprolium 2 mM evoked TD (−31%)	15.9 ± 1.9 9.0 ± 2.2^a **2.9 ± 0.7 5.3 ± 0.6^a**	29.3 ± 3.3 20.9 ± 0.1^a	21.5 ± 2.8 17.7 ± 2.1
Differentiated SN56 cholinergic neuroblastoma cells. Amprolium 2mM evoked TD (−31%)	4.2 ± 0.5† 6.8 ± 0.5 **3.8 ± 0.9 13.9 ± 2.0** ^{a,b}	39.4 ± 2.6† 22.5 ± 4.3^a	36.8 ± 2.0† 22.0 ± 2.0^a

^{a}Significantly different from respective: control $p < 0.01$.
^{b}Significantly different from respective non-differentiated cells: $p < 0.05$.

that functional acetyl-CoA dependent disturbances in cholinergic transmission precede irreversible structural damage of cholinergic neurons in thiamine-deficient brains.

33.5 Acetyl-CoA Metabolism a Primary Target for TDP Deficiency

33.5.1 Sources of Acetyl-CoA in the Brain

The glucose reaches brain extracellular compartments through high capacity/low affinity ($K_m = 8$ mmol/L) hypoglycemia-inducible glucose transporters (GLUT) on blood–brain barrier (BBB) and is taken up by neurons expressing GLUT1 (low affinity) and GLUT3 (high affinity) transporters ($K_m = 2.8$ mmol/L) (Simpson et al. 2007). Glycolytic metabolism provides pyruvate, which is an almost exclusive source of acetyl-CoA in the brain through PDHC reaction in the mitochondrial matrix compartment.

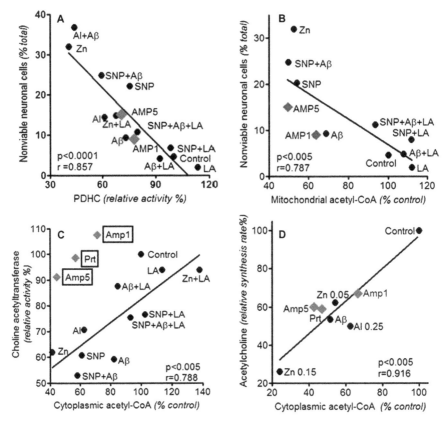

Figure 33.3 Thiamine deficit is one of several pathogenical signals that may reduce pyruvate dehydrogenase complex activity and result in encephalopathy with dominating cholinergic deficits. (A) Inverse correlation between pyruvate dehydrogenase complex activity and nonviable fraction of septal cholinergic neuroblastoma cells in the presence of amprolium-TD (Amp 1 and 5 mmol/L) and other neurotoxic conditions including sodium nitroprusside (SNP) exposure. (B) Inverse correlation between mitochondrial acetyl-CoA and nonviable fraction of septal cholinergic neuroblastoma cells in the presence of amprolium-TD (Amp 1 and 5 mmol/L) and other neurotoxic conditions. (C) Correlation between cytoplasmic acetyl-CoA and ChAT activity in septal cholinergic neuroblastoma cells in different neurotoxic conditions. Amprolium and pyrithiamine (Prt)-evoked acetyl-CoA deficits cause no changes in ChAT activities and are not considered in the correlation equation. (D) Correlation between cytoplasmic acetyl-CoA and ACh synthesis in amprolium and pyrithiamine-evoked TD and other neurotoxic conditions. Data are taken, and converted to relative values, from: Bizon-Zygmanska et al. 2011; Bielarczyk et al. 2005, 2006; Jankowska-Kulawy et al. 2010; Ronowska 2007, 2010; Szutowicz et al. 2000, 2006.

Such a pattern of energy metabolism exists in all types of brain neuronal and glial cells. In addition, neurons are capable of direct uptake and utilization of lactate, either produced intracerebrally by astroglial cells or transported from the circulation through the monocarboxylate transporter (MCT) on the BBB. From the intracerebral extracellular compartment, lactate is preferentially taken up by neurons containing both high affinity MCT2 ($K_m = 0.7$ mmol/L) and medium affinity MCT1 ($K_m = 4$ mmol/L) transporters. Astrocytes containing MCT1 and low affinity MCT4 ($K_m = 8$ mmol/L) take up lactate less avidly, but with higher capacity than neurons. (Simpson et al. 2007) (Figure 33.1).

From the cytoplasm, pyruvate is transported into the mitochondrial matrix through the inner mitochondrial membrane mMCT and different antiporter systems, and decarboxylated to acetyl-CoA in a TDP-dependent PDHC reaction. Therefore, in thiamine deficiency, PDHC inhibition becomes a key element of the pathomechanisms of brain injury through the reduction of acetyl-CoA fuelled energy production.

33.5.2 Acetyl-CoA Compartmentalization in Brain Cells

One may assume that neurons consuming 80% of brain oxygen/glucose also synthesize the equivalent fraction of whole brain acetyl-CoA. Over 95% of this intermediate is utilized in TCA/respiratory chain yielding energy supporting their neurotransmitter activities. This implies that thiamine deficiency and any pathological conditions that reduce PDHC activity and acetyl-CoA provision in the brain mitochondria are more harmful for neurons than for glial cells.

Cholinergic neurons, unlike those producing other neurotransmitters, and glial cells as well, utilize additional amounts of acetyl-CoA for ACh synthesis in their cytoplasmic/synaptoplasmic compartment. During depolarization-induced ACh release, cholinergic neurons that may consume up to 1–3% of whole cell acetyl-CoA for the re-synthesis of neurotransmitter pool (Figure 33.1) (Szutowicz et al. 1996; Tucek 1993). However, it is known that non-excitable cells in the mitochondrial inner membrane are relatively impermeable to acetyl-CoA. Therefore, acetyl-CoA has to be transported out of mitochondria by indirect pathways involving citrate, acetate or acetyl-carnitine, which in the cytoplasm are converted back to acetyl-CoA by pathways driven by ATP-citrate lyase (ACL; EC 2.3.3.8), acetyl-CoA synthase (EC 6.2.1.3) and carnitine acetyltransferase (EC 2.3.1.7), respectively (Szutowicz et al. 1996).

In brain neurons, about 30–50% of acetyl units are transported to the cytoplasm indirectly in the form of citrate through the ACL pathway (Figure 33.1) (Jankowska-Kulawy et al. 2010). The remaining 50% fraction of acetyl-CoA may be transported from their mitochondria as a whole particle directly to the synaptoplasmic compartment through Ca-sensitive high permeability anionic channels, equivalent to permeability transition pores (PTPs) (Bielarczyk and Szutowicz 1989). Therefore, only observations of acetyl-CoA levels in particular sub-compartments of brain cells may provide sufficient detail to explain the acetyl-CoA dependent mechanisms of diverse malfunctions in thiamine-deficient brains.

33.5.3 Redistribution of Acetyl-CoA in Brain Subcellular Compartments in Pathological Conditions

Early findings in whole brains of pyrithiamine–TD rats reported either no change or slight decrease of acetyl-CoA levels. However, they did not take into account cell type-dependent or intracellular compartmentalization of this metabolite (Heinrich et al. 1973, Reynolds, Blass, 1975). So far, only our studies have provided quantitative data on mitochondrial and cytoplasmic pools of acetyl-CoA in brain nerve terminals and cholinergic neuronal cell lines (Table 33.3, Figure 33.3) (Bielarczyk et al. 2003; Bizon-Zygmańska et al. 2011; Jankowska-Kulawy et al. 2010; Ronowska et al. 2007; Szutowicz et al. 2006).

Physiological and pathological action potentials increase $[Ca^{2+}]$ in neuronal cytoplasm. They include depolarization-evoked release of glutamate/Zn from glutamatergic terminals, which constitute about 50% of all brain synapses (Takeda 2011). However, several pathological conditions such as hypoxia/anoxia, hypoglycemia or accumulation of diverse neurotoxic species resulted in direct or indirect sustained neuronal depolarization, causing excessive rise of their cytoplasmic $[Ca^{2+}]$ and secondary depletion of high-energy compounds (Figure 33.3A,B).

TD-induced energy deficits cause sustained depolarization of glutamatergic terminals and the increase of glutamate-Zn release followed by prolonged depolarization and Zn uptake by postsynaptic cholinergic neurons (Ronowska et al. 2007; Takeda 2011). The glutamate content in the synaptic cleft is also increased due to TD-induced downregulation of glutamate aspartate (GLAST) and glutamate 1 (GLT-1) transporters in astrocyte plasma membranes (Jhala and Hazell 2011). In effect, Zn and glutamate co-released from 'gluzinergic' terminals reach high concentrations in the synaptic cleft (up to 0.3 and 1.0 mmol/L, respectively) (Takeda 2011). Subsequently Zn overloads depolarized postsynaptic glial and neuronal cells (including cholinergic ones), yielding inhibition of energy/ATP production and cell loss (Bizon-Zygmańska, 2011; Ronowska et al. 2007, 2010).

The key mechanism for Zn toxicity may be the inhibition of several enzymes of energy metabolism by Zn including PDHC, aconitase (EC 4.2.1.3), KDHC and succinate dehydrogenase (EC1.3.5.1) (Figure 33.2) (Bubber et al. 2004; Jankowska-Kulawy et al. 2010; Ronowska et al. 2010).

A reduction in the intramitochondrial level of acetyl-CoA was found to correlate with loss of cell viability, whereas inhibition of quantal ACh release correlated with decreasing concentration of acetyl-CoA in the cytoplasmic compartment of thiamine-deficient neurons (Figure 33.3). As with any other pathological condition, the susceptibility of cholinergic neurons with high expression of the cholinergic phenotype to thiamine deficiency was significantly greater than those with residual cholinergic activity, being compatible with the lower level of mitochondrial acetyl-CoA in the former (Table 33.3). Residual ACh metabolism in non-differentiated cholinergic neurons was not affected by thiamine deficiency as it required less cytoplasmic acetyl-CoA (Bizon-Zygmańska et al. 2011).

One of serious complications of thiamine deficiency caused by the impairment of pyruvate utilization by PDHC is metabolic lactic acidosis. This is recognized as the sole condition responsible for lethal outcome of some misdiagnosed or lately treated thiamine-deficient patients (Klein et al. 2004) (Figure 33.1).

33.6 Alcoholism and Subclinical Thiamine Deficiency

Alcoholics are the largest human group worldwide at risk of TDP deficiency due to their addictive life style eliminating sufficient supply vitamins and other micronutrients from the diet. Up to 80% of alcoholics were found to develop different degrees of thiamine deficiency, many of them without any, or minimal, clinical manifestations. These appear usually in the advanced stages of alcoholic disease and are linked with the social, economic and physical deterioration of affected persons (Butterworth 1989).

There is a claim that low TDP in blood may be a laboratory marker for alcoholism and a predictor for overt thiamine deficiency (Ceccanti et al. 2005). However, this marker is apparently secondary to earlier, direct and diverse effects of ethanol on all systems of the human body which pave the road to alcoholic–TD encephalopathies of the Wernicke–Korsakoff type (Tables 33.1 and 33.2) (Pitel et al. 2011). Therefore, preventive measures are undertaken in several countries, including mandatory supplementation of bread flour with thiamine. However, there is no evidence as to whether such treatment delays onset of dementia in the population of alcoholics.

The level of TDP in the blood is seen as a weak predictor of Wernicke–Korsakoff syndrome (Ceccanti et al. 2005). Therefore, combination of Thiamine Deficiency Questionnaire and serum γ-glutamyl transferase activity is proposed as a reliable (>80% sensitivity) approach for identification of TDP-deficient alcoholics (Table 33.2) (Sgouros et al. 2004). Such a conclusion is justified by the fact that, apart from interfering with TDP synthesis, alcohol exerts several independent toxic effects. These include direct hyperosmotic and cytotoxic effects of ethanol and derived acetaldehyde (Mira et al. 1995). Ethanol inhibits gluconeogenesis, and in combination with coexisting alcoholic liver cirrhosis, may induce deep hypoglycemia, which itself may be harmful for brain neurons. Cholinergic neurons are particularly prone to such conditions due to their extra demand for acetyl units for ACh synthesis (Figure 33.1).

Acetate is a final product of ethanol oxidation. Its utilization by muscles and brain cells requires its conversion to acetyl-CoA by acetyl-CoA synthase. This consumes ATP, yielding AMP and energy deficits, which in turn stimulate purine nucleotide catabolism and hyperuricemia (Lavoie and Butterworth 1995). Through such mechanisms, alcohol may exert direct toxic effects uncovering pre-existing subclinical TD-dependent energy shortages.

Acetyl-CoA may also contribute to the different susceptibility of particular individuals to the addictive effects of ethanol. Namely, nerve terminals from brains of short-sleeping, ethanol-preferring rats contained 30% less acetyl-CoA and displayed 50% lower overall ACh release than those from long-sleeping, addiction-resistant animals (Jankowska et al. 1997). It remains to be determined if and how these metabolic differences mediate differential susceptibility of the brain to thiamine deficiency in humans.

It is known that chronic treatment with alcohol induces compensatory response in the form of stimulation of thiamine uptake and its conversion to TDP. That is 'why the appearance of clinical symptoms of thiamine deficiency may be markedly delayed despite of systemic shortages of thiamine (Parkohomenko et al. 2011). Increased levels of TDP in the blood of alcoholic people with adequate dietary supply of thiamine seem to support such a hypothesis.

33.7 Screening for Minimal or Asymptomatic Thiamine Deficiency

In light of the difficulties in diagnosing subclinical forms of thiamine deficiency, all patients deemed at risk of encephalopathy should be screened for such possibility. However, there is no clear indicator for laboratory tests or criteria for preclinical thiamine deficiency diagnosis. Decreased TDP levels in blood of alcoholics correlated modestly with worsening of their memory performance (Table 33.2). This may result from large variations in individual susceptibility to low TDP levels. In addition, only 0.5% of the body's whole store of TDP is present in the blood and blood TDP may thus not reflect well its content in the brain and other tissues (Pitel et al. 2011).

In elderly patients with cardiac failure, the reduction in activity of erythrocyte transketolase correlated strongly with cumulative dosage of furosemide (Table 33.2) (Rocha et al. 2008). Correct diagnosis of putative thiamine deficiency is crucial in avoiding complications during the treatment of renal insufficiency by dialysis or chronic heart failure with loop diuretics. The prevalence of this complication was reported to range from as little as 3% to as high as 97%, depending on the local pharmacotherapeutic procedures (Sica 2007). Therefore, patients with acute vomiting and hypovolemia of unclear cause should not be treated with glucose infusion without simultaneous supplementation with TDP. Otherwise, such treatment may worsen the patient's conditions due to increased accumulation of lactic acid (Klein et al. 2004).

However, there are no clear cut-off values for basic laboratory parameters, which could be useful for preselecting patients at risk of potential detrimental conditions. Some parameters that are easy to detect, such as lactic/pyruvic metabolic acidosis, are not specific for thiamine deficiency (Table 33.2). On the other hand, blood TDP and transketolase assays have low sensitivity due to significant overlapping results between healthy and thiamine-deficient individuals. This makes laboratory assessment of borderline thiamine deficiency conditions difficult. Thus, anamnesis and awareness of the socio-demographic conditions of the patient are as important as laboratory tests in the early diagnosis of thiamine deficiency.

Summary Points

- This chapter focuses on acetyl-CoA dependent mechanisms of thiamine deficiency.
- The prevalence of subclinical forms of thiamine deficiency is relatively frequent in several risk groups including alcoholics, elderly, cardiac and renal failure patients, and infants.
- Thiamine diphosphate (TDP) is a cofactor of E1 subunits of pyruvate and ketoglutarate dehydrogenase complexes. Therefore, its deficiency results in shortage of acetyl-CoA and energy deficits in all tissues including brain.
- Thiamine deficiency-evoked energy deficits cause glutamate–zinc induced excitotoxic activation of postsynaptic cells including cholinergic neurons.
- The combination of primary thiamine deficiency-induced acetyl-CoA deficits and excitotoxicity-induced inhibition of pyruvate dehydrogenase causes early functional impairment of cholinergic transmission, yielding symptoms of dementia and progressing to irreversible damage of the brain cholinergic system.
- There are laboratory tests that might serve as markers of early stages of thiamine deficits. These include erythrocyte transketolase activity, blood TDP or serum γ-glutamyl transferase, which in combination with questionnaire anamnesis, may facilitate early diagnosis to impose easy and efficient treatment with thiamine.

Key Facts

- **Energy production in the brain.** The brain of a resting person consumes 20–25% of whole body oxygen and glucose pools. Glucose is first converted to pyruvate in the glycolytic pathway. Pyruvate oxidation in mitochondria provides large amounts of energy in form of ATP. Oxidation of 1 mol glucose generates 36 mols of ATP. Most of this energy (80%) is utilized for maintenance of neuronal membrane potentials, important for their neurotransmitter functions.
- **Mitochondria.** These are intracellular structures, producing the bulk of cell energy. They oxidize glucose-derived pyruvate, keto-acids and fatty acids in the tricarboxylic and beta oxidation cycles, respectively. These pathways are coupled with the oxidative phosphorylation chain in which ATP is synthesized. Neuronal mitochondria do not utilize fatty acids. They almost exclusively utilize glucose or lactate and sometimes keto-acids.
- **Acetyl-CoA.** There is a key metabolite of energy metabolism. It is produced in mitochondria by decarboxylation of pyruvate, beta oxidation of fatty acids, or hydrolysis of acetoacetate. In condensation reaction with oxaloacetate, acetyl-CoA yields citrate, which is the first intermediate in the tricarboxylic acid chain. Acetyl-CoA is also used for the synthesis of acetylcholine and the acetylation of several low molecular weight compounds and proteins. In liver and adipose tissue, acetyl-CoA is used for the synthesis of the fatty acids chain.

- **Alcoholism.** This addiction frequently links with secondary depletion of vitamin B_1, mainly due to insufficient intake. Alcoholic disease is a main cause of vitamin B_1 deficiency worldwide. It damages the liver, pancreas, kidney, heart and brain. The most advanced form of alcoholic brain damage is Wernicke–Korsakoff's dementia, of which the main symptoms are dementia and loss of motor functions of the striatal muscles (nystagmus).
- **Gamma-glutamyl transferase.** This enzyme is located in cell plasma membranes of several tissues, with the highest levels present in liver. Alcohol and other toxic compounds that impair liver cells result in a major increase in the activity of this enzyme in serum. Therefore, it is a good marker of injury to the liver and a diagnostic marker for monitoring the abstinence of alcoholic patients during withdrawal therapy.

Definitions of Words and Terms

Acetyl-CoA. A key energy metabolite in the brain and all tissues serving as a source of dicarboxylic units feeding the tricarboxylic acid cycle.

Beriberi. Clinical manifestations of thiamine deficits in the form of vascular and neurological disturbances including motor impairment, sensory and reflex functions, mental confusion, oedema and cardiomyopathy. Beriberi can be fatal if thiamine is not supplied.

Choline acetyltransferase. Enzyme synthesizing acetylcholine from choline and acetyl-CoA.

Cholinergic neurons. Neurons in which choline acetyltransferase synthesizes acetylcholine, accumulates it in synaptic vesicles and releases it, upon depolarization, in the quantal mode. Those neurons located in the septal/ subcortical and cortical regions of the brain are responsible for cognitive and memory functions. Cholinergic motor neurons of anterior horns of medulla oblongata form neuro-muscular synapses responsible for voluntary movements of striated muscles.

Excitotoxicity. Toxic effects caused by sustained depolarization of glutamatergic neurons due to TD-evoked energy deficits, or other pathological conditions, yielding excessive activation of postsynaptic glutamatergic receptors. These cause increases in intracellular Ca^{2+} and Zn^{2+} concentrations, excessive free radical and nitric oxide (NO) synthesis, leading to functional and structural impairment of adjacent neuronal and glial cells.

Glutamatergic neurons. Neurons using glutamate and/or glutamate–Zn complexes as a neurotransmitter. Glutamatergic neurons constitute about 50% of entire brain neurons.

Mitochondria. Cellular organelle oxidizing energy substrates in the tricarboxylic acid cycle, coupled with the respiratory chain producing the bulk of the high-energy compound ATP, indispensable for maintenance of all vital functions of each cell and the whole organism.

Pyruvate dehydrogenase complex. Catalyses the oxidative decarboxylation of pyruvate. The complex has three different catalytic subunits and various regulatory proteins. It is located in the mitochondrial matrix and provides acetyl-CoA for condensation reaction with oxaloacetate, catalysed by citrate synthase—the first step of tricarboxylic acid cycle. Thus it links the glycolytic and tricarboxylic acid cycles.

Thiamine, thiamine diphosphate. Humans are not able to synthesize the thiazol ring, the key element of thiamine. Therefore thiamine must be supplied from the diet (daily allowance is 1.0–1.5 mg). It is converted to the active compound thiamine diphosphate (vitamin B_1) by thiamine kinases present in all tissues.

Wernicke's encephalopathy. This is most frequent clinical manifestation of thiamine deficiency in developed countries. It is frequently associated with alcoholism and other conditions impairing nutrition. This neuropsychiatric disorder is characterized by eye muscle paralysis, abnormal posture and gait, and impaired cognitive functions. Progressive deterioration of WE patients ends with Korsakoff's psychosis with manifestation of amnesia, stupor and loss of conceptual functions.

List of Abbreviations

ACh	acetylcholine
ACL	ATP–citrate lyase
ACO	Aconitase
BBB	blood–brain barrier
BCKDHC	branched-chain α-ketoacid dehydrogenase complex
ChAT	choline acetyltransferase
CNS	central nervous system
CS	citrate synthase
ETK	erythrocyte transketolase
FUM	fumarase
GLAST	glutamate aspartate transporter
GLT	glutamate transporter
GLUT	glucose transporter
ICDH	NADP isocitrate dehydrogenase
KDHC	2-oxoglutarate dehydrogenase complex
MCT1	2,4-monocarboxylate transporters
MDH	malate dehydrogenase
NMDA	N-methyl-D-aspartate
NO	nitric oxide
PDHC	pyruvate dehydrogenase complex
PTP	permeability transition pore
SDH	succinate dehydrogenase
SNP	sodium nitroprusside
TCA	tricarboxylic acid cycle
TD	thiamine deficiency
TDP	thiamine diphosphate
WE	Wernicke encephalopathy

Acknowledgements

This study was supported by the Ministry of Research and Higher Education projects NN401 1029937 and IP 2010 035370.

References

Bielarczyk H., and Szutowicz A., 1989. Evidence for regulatory function of synaptoplasmic acetyl-CoA in acetylcholine synthesis in nerve endings. *Biochemical Journal.* 262: 377–380.

Bielarczyk, H., Tomaszewicz, M., Madziar, B., Ćwikowska, J., Pawełczyk, T., and Szutowicz, A., 2003. Relationships between cholinergic phenotype and acetyl-CoA level in hybrid murine neuroblastoma cells of septal origin. *Journal of Neuroscience Research.* 73: 717–721.

Bielarczyk, H., Jankowska-Kulawy, A., Gul, S., Pawełczyk, T., and Szutowicz, A., 2005. Phenotype dependent differential effects of interleukin-1β and amyloid-β on viability and cholinergic phenotype of T17 neuroblastoma cells. *Neurochemistry International.* 47: 466–473.

Bielarczyk, H., Gul, S., Ronowska, A., Bizon-Zygmańska, D., Pawełczyk, T., and Szutowicz, A., 2006. RS-α-lipoic acid protects cholinergic cells against sodium nitroprusside and amyloid-β neurotoxicity through restoration of acetyl-CoA level. *Journal of Neurochemistry.* 98: 1242–1251.

Bizon-Zygmańska, D., Jankowska-Kulawy, A., Bielarczyk, H., Pawełczyk, T., Ronowska, A., Marszałł, M., and Szutowicz, A., 2011. Acetyl-CoA metabolism in amprolium-evoked thiamine pyrophosphate deficits in cholinergic SN56 neuroblastoma cells. *Neurochemistry International.* 59: 208–216.

Bubber, P., Ke, Z.J., and Gibson G.E., 2004. Tricarboxylic acid cycle enzymes following thiamine deficiency. *Neurochemistry International.* 45: 1021–1028.

Butterworth, R.F., 1989. Effects of thiamine deficiency on brain metabolism: Implications for pathogenesis of the Wernicke-Korsakoff syndrome. *Alcohol and Alcoholism.* 24: 271–279.

Carney, M.W.P., and Barry S., 1985. Clinical and subclinical thiamine deficiency in clinical practice. *Clinical Neuropharmacology.* 8: 286–293.

Ceccanti, M., Mancinelli, R., Sasso, G.F., Allen, J.P., Binetti, R., Mellini, A., Attilia, F., Toppo, L., and Attilia, M.L., 2005. Erythrocyte thiamine (Th) esters: a major factor of the alcohol withdrawal syndrome or a candidate marker for alcoholism itself? *Alcohol and Alcoholism.* 40: 283–290.

Gibson, G., Barclay, L., and Blass, J., 1982. The role of the cholinergic system in thiamin deficiency. *Annals of the New York Academy of Sciences.* 387: 382–403.

Heinrich, C.P., Stadler, H., and Weiser, H., 1973. The effect of thiamine deficiency on the acetylcoenzymeA and acetylcholine levels in the rat brain. *Journal of Neurochemistry.* 21: 1273–1281.

Herve, C., Beyne, P., Letteron, Ph., and Delacoux E., 1995. Comparison of erythrocyte transketolase activity with thiamine phosphate ester levels in chronic alcoholic patients. *Clinica Chimica Acta.* 234: 91–100.

Jankowska, A., Kisielevski, Y., Oganesjan, N., Tomaszewicz, M., and Szutowicz, A., 1997. Cholinergic mechanisms in inherited and acquired tolerance to ethanol. In: Teelken, A., and Korf, J. (ed.) *Neurochemistry Cellular, Molecular, and Clinical Aspects*. Plenum Press, New York, USA, pp. 847–851.

Jankowska-Kulawy, A., Bielarczyk, H., Pawełczyk, T., Wróblewska, M., and Szutowicz A., 2010. Acetyl-CoA and acetylcholine metabolism in nerve terminal compartment of thiamine deficient rat brain. *Journal of Neurochemistry*. 115: 333–342.

Jhala, S.S., and Hazell, A.S., 2011. Modeling neurodegenerative disease pathophysiology in thiamine deficiency: Consequences of impaired oxidative metabolism. *Neurochemistry International*. 58: 248–260.

Khounnorath, S., Chamberlain, K., Taylor, A.M., Soukaloun, D., Mayxay, M., Lee, S.J., Phengdy, B., Luangxay, K., Sisouk, K., Soumphonphakdy, B., Latsavong, K., Akkhavong, K., Nicholas. J., White, N., and Newton P.N., 2011. Clinically unapparent infantile thiamin deficiency in Vientiane, Laos. *PLOS Neglected Tropical Diseases*. 5: e969.

Klein, M., Weksler, N., and Gurman, M.G., 2004. Fatal metabolic acidosis caused by thiamine deficiency. *The Journal of Emergency Medicine*. 26: 301–303.

Lavoie, J., and Butterworth, R.F., 1995. Reduced activities of thiamine-dependent enzymes in brains of alcoholics in the absence of Wernicke's encephalopathy. *Alcoholism: Clinical and Experimental Research*. 19: 1073–1077.

Le, O., Heroux, M., and Butterworth, R.F., 1991. Pyrithiamine-induced thiamine deficiency results in decreased Ca^{2+}-dependent release of glutamate from rat hippocampal slices. *Metabolic Brain Disease*. 6: 125–132.

Mira, L., Maia, L., Berreira, L., and Manso, C.F., 1995. Evidence for free radical generation due to NADH oxidation by aldehyde oxidase during ethanol metabolism. *Archives of Biochemistry and Biophysics*. 318: 53–58.

Parkhomenko, Y.M., Kudryavtsev, P.A., Pylypchuk, Yu, S., Chekhivska, L.I., Stepanenko, S.P., Sergiichuk, A.A., and Bunik, V.I., 2011. Chronic alcoholism in rats induces a compensatory response, preserving brain thiamine diphosphate, but the 2-oxo- acid dehydrogenases are inactivated despite unchanged coenzyme. Journal of Neurochemistry. 117: 1055–1065.

Pietrzak, I., and Baczyk, K., 2001. Erythrocyte transketolase activity and guanidine compounds in hemodialysis patients. *Kidney International*. 59(Suppl. 78): S97–S101.

Pitel, A.L., Zahr, N.M., Jakson, K., Sassoon, S.A., Rosenbloom, M.J., Pfefferbaum, D., and Sullivan, E.V., 2011. Signs of preclinical Wernicke's encephalopathy and thiamine levels as predictors of neuropsychological deficits in alcoholism without Korsakoff's syndrome. *Neuropsychopharmacology*. 36: 580–588.

Reynolds, S.F., and Blass, J.P., 1975. Normal levels of acetyl coenzyme A and of acetylcholine in the brains of thiamin-deficient rats. *Journal of neurochemistry*. 24: 183–186.

Rocha, R.M., Silva, G.V., de Albuquerque, D.C., Tura, B.R., and Filho, F.M.A., 2008. Influence of spironolactone therapy on thiamine blood

levels in patients with heart failure. *Brazilian Archives of Cardiology*. 90: 324–328.
Ronowska, A., Gul-Hinc, S., Bielarczyk, H., Pawełczyk, T., and Szutowicz, A., 2007. Effects of zinc on SN56 cholinergic neuroblastoma cells. *Journal of Neurochemistry*. 103: 972–983.
Ronowska, A., Dyś, A., Jankowska-Kulawy, A., Klimaszewska-Łata, J., Bielarczyk, H., Romianowski, P., Pawełczyk, T., and Szutowicz, A., 2010. Short-term effects of zinc on acetylcholine metabolism and viability of SN56 cholinergic neuroblastoma cells. *Neurochemistry International*. 56: 143–151.
Savage, M.L., Roland, J.and Klintsova, A., 2007. Selective septohippocampal—but not forebrain amygdalar—cholinergic dysfunction in diencephalic amnesia. *Brain Research*. 1139: 210–219.
Sgouros, X., Baines, M., Bloor, R.N., McAuley, R., Ogundipe, L., and Willmott, S., 2004. Evaluation of a clinical screening instrument to identify states of thiamine deficiency in inpatients with severe alcohol dependence syndrome. *Alcohol and Alcoholism*. 39: 227–232.
Shi, Q., Karuppaguunder, S.S., Xu, H., Pechman, D., Chen, H., and Gibson, G.E., 2007. Responses of mitochondrial α-ketoglutarate dehydrogenase complex to thiamine deficiency may contribute to regional selective vulnerability. *Neurochemistry International*. 50: 921–931.
Sica, D.A., 2007. Loop diuretics therapy, thiamine balance, and heart failure. *Congestive Heart Failure*. 13: 244–247.
Simpson, A.I., Carruthers, A., and Vannucci, J.S., 2007. Supply and demand in cerebral energy metabolism: the role of nutrient transporters. *Journal of Cerebral Blood Flow and Metabolism*. 27: 1766–1791.
Szutowicz, A., Tomaszewicz, M., and Bielarczyk, H., 1996. Disturbances of acetyl-CoA, energy and acetylcholine metabolism in some encephalopathies. *Acta Neurobiologiae. Experimentalis*. 56: 325–34.
Szutowicz, A., Tomaszewicz, M., Jankowska, A., Madziar, B., and Bielarczyk, H., 2000. Acetyl-CoA in cholinergic neurons and their susceptibility to neurotoxic inputs. *Metabolic Brain Disease* 15: 29–44.
Szutowicz, A., Bielarczyk, H., Gul, S., Ronowska, A., Pawełczyk, T., and Jankowska-Kulawy, A., 2006. Phenotype-dependent susceptibility of cholinergic neuroblastoma cells to neurotoxic inputs. *Metabolic Brain Disease*. 21: 149–161.
Takeda, A., 2011. Insight into glutamate excitotoxicity from synaptic zinc homeostasis. *International Journal of Alzheimer's Disease*. Doi:10.4063/2011/491597.
Todd, G.K., and Butterworth, F.R., 2001. In vivo microdialysis in an animal model of neurological disease: thiamine deficiency (Wernicke) encephalopathy. Methods. 23: 56–61.
Tucek, S., 1993. Short-term control of the synthesis of acetylcholine. *Progress in Biophysics and Molecular Biology*. 60: 59–69.

CHAPTER 34
Thiamine Deficiency and Neuronal Calcium Homeostasis

ZUNJI KE[a] AND JIA LUO*[b]

[a] Key Laboratory of Nutrition and Metabolism, Institute for Nutritional Sciences, Shanghai Institutes for Biological Sciences, Chinese Academy of Sciences Institute for Nutritional Sciences, 294 Taiyuan Road, Shanghai 200031, China; [b] Department of Internal Medicine, College of Medicine, University of Kentucky, HSRB 130, 1095 Veterans Drive, Lexington, Kentucky 40536, USA
*Email: jialuo888@uky.edu

34.1 Introduction

Vitamin B_1, also called thiamine, is required for all tissues and high concentrations are found in skeletal muscle, heart, liver, kidneys and brain. Thiamine diphosphate (TDP) is the active form and it serves as a cofactor for several enzymes involved in carbohydrate catabolism. These enzymes are also important in the biosynthesis of many cellular constituents, including neurotransmitters, and for the production of reducing equivalents used in oxidant stress defenses (Ba 2008). Thiamine is considered an 'anti-stress' vitamin because it strengthens the immune system and improves the body's ability to withstand stress conditions (Haas 1988).

Thiamine is essential for brain function. In humans, thiamine deficiency (TD) is a critical factor in the etiology of Wernicke–Korsakoff syndrome (WKS), which is characterized by severe memory loss, cholinergic deficits and selective cell death in specific brain regions (Ke *et al.* 2003; Todd and Butterworth,

2001). In animals, thiamine deficiency induces chronic mild impairment of oxidative metabolism and neuroinflammation, leading to regionally selective neuronal death in the brain (Ke *et al.* 2003). These features of TD-induced brain damage are similarly observed in a number of age-related neurodegenerative diseases, including Alzheimer's disease (AD) (Gibson and Huang 2004). Thiamine-dependent processes play an important role in neurodegeration in Alzheimer's disease, in which thiamine selectively alters the level of oxidative stress (Gibson and Blass 2007; Huang *et al.* 2010).

Thiamine deficiency in animals and cell cultures has been used to model aging associated neurodegenerative processes. These models offer well-controlled systems with which to study cellular/molecular mechanisms of neurodegenerative diseases. The mechanisms underlying TD-induced neuronal damage remain unclear and a number of possibilities have been proposed. These include mitochondrial dysfunction (Singleton and Martin 2001), glutamate excitotoxicity (Hazell and Butterworth 2009), endoplasmic reticulum (ER) stress (Wang *et al.* 2007), impairment of oxidative metabolism and acidosis (Hakim and Pappius 1983; Pannunzio *et al.* 2000).

Calcium ion (Ca^{2+}) is an important intracellular messenger that is essential for neuronal survival and function. It has been demonstrated that thiamine and oxidants interact to modify cellular calcium stores (Huang *et al.* 2010). Ca^{2+} overload can trigger several downstream lethal reactions, including nitrosative stress, oxidative stress, ER stress and mitochondrial dysfunction (Huang *et al.*

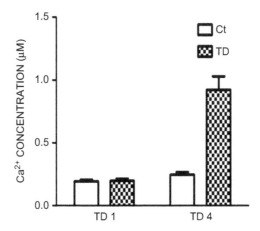

Figure 34.1 Thiamine deficiency (TD) increased intracellular calcium concentration in in cultured cortical neurons. Cortical neurons were isolated from the cerebral cortex of mice and maintained *in vitro* for seven days (DIV7). TD was induced in these neurons by the treatment of amprolium (1 mmol/L) for one or four days (TD1 or TD4). The intracellular calcium concentration was measured. Short-term TD (TD1) had little effect on calcium homeostasis; longer duration of TD (TD4) caused a significant increase in intracellular calcium concentration. The results are the mean ± SEM of three repeated experiments.

Copyright of the figure remains with the authors.

2010; Peng et al. 2006). The perturbation of Ca^{2+} homeostasis has been implicated in the pathogenesis of various neurodegenerative diseases including AD (Supnet and Bezprozvanny 2010). We propose that the disruption of Ca^{2+} homeostasis may be a potential mechanism for TD-induced neuronal damage. Here we discuss the evidence of TD-induced alterations in Ca^{2+} dynamics and its association with selective neurodegeneration.

34.2 Thiamine Deficiency Disrupts Homeostasis of Intracellular Ca^{2+}

We used an *in vitro* neuronal model to investigate the effect of thiamine deficiency on Ca^{2+} homeostasis (Lee et al. 2010). In this system, cultured cortical neurons were made thiamine deficient by removing thiamine from the culture medium and including amprolium, which is a competitive inhibitor of thiamine transport. Our results indicated that four days of thiamine deficiency (TD4), but not TD1 caused a significant increase in resting intracellular Ca^{2+} $[Ca^{2+}]i$ (Figure 34.1, Lee et al. 2010). Since α-amino-3-hydroxyl-5-methyl-4-isoxazole-propionic acid receptors (AMPARs) are important mediators of Ca^{2+} influx, we further examined the effect of thiamine deficiency on AMPA-elicited Ca^{2+} influx. TD4, but not TD1, caused a two-fold increase of AMPA-stimulated Ca^{2+} influx (Lee et al. 2010), suggesting that thiamine deficiency can modulate Ca^{2+} influx by modulating AMPARs.

Calcium ions can enter cells in multiple ways: through ion channels, Ca^{2+} transporters, direct damage of the membrane and from internal stores (ER and mitochondria). One important mechanism is through the activation of glutamate receptors (Hardingham and Bading 2010). There are three types of glutamate receptors: *N*-methyl-D-aspartate receptors (NMDARs), AMPARs and kainate receptors (KRs), each having several subtypes. In the mammalian central nervous system (CNS), AMPARs are widely expressed in neurons and in glia, and mediate the vast majority of fast excitatory synaptic transmission (Krugers et al. 2010). AMPARs are tetramers made up of combinations of four subunits: GluR1, GluR2, GluR3 and GluR4 (also called 'GluRA-D').

The Ca^{2+} permeability of AMPAR channels is determined by the GluR2 subunit (Hume et al. 1991). GluR2 is a critical subunit in determining many of the major biophysical properties of AMPARs including, but not limited to, Ca^{2+} permeability, receptor kinetics, single-channel conductance and blockage by endogenous polyamines (Isaac et al. 2007). The great majority of AMPARs in the CNS exist as heteromers containing GluR2. AMPARs lacking GluR2 are permeable to Ca^{2+} and Zn^{2+}. The subunit composition and Ca^{2+} permeability of AMPARs are dynamically remodelled in a cell- and synapse-specific manner during development and in response to neuronal activity. The subunit composition and permeability of AMPARs are also modulated by neuronal insults, such as seizures, ischemia, excitotoxicity, spinal cord injury and neurological diseases such as AD and amyotrophic lateral sclerosis (ALS) (Liu and Zukin 2007).

The property of GluR2 is altered by pre-mRNA editing, a post-transcriptional modification that involves enzymatic deamination of a specific adenosine in the pre-mRNA prior to splicing (Barbon et al. 2003). Enzymes responsible for RNA editing are termed 'adenosine deaminases acting on RNA' (ADARs), and three structurally related ADARs (ADAR1–ADAR3) have been identified in mammals. RNA editing of GluR2 is predominantly catalysed by ADAR2 (Kawahara et al. 2004). ADAR2 pre-mRNA and mRNA themselves are susceptible to self-editing (Burns et al. 1997).

We have demonstrated that thiamine deficiency inhibited editing of GluR2 and increased the ratio of unedited GluR2 (Lee et al. 2010). The inhibitory effect of thiamine deficiency on GluR2 editing is mediated by ADAR2. Thiamine deficiency selectively decreased ADAR2 expression and its self-editing ability without affecting ADAR1. Overexpression of ADAR2 reduced the AMPA-mediated rise of $[Ca^{2+}]i$ and protected cortical neurons against TD-induced cytotoxicity, whereas downregulation of ADAR2 increased AMPA-elicited Ca^{2+} influx and exacerbated TD-induced death of cortical neurons. The evidence indicates that thiamine deficiency increases $[Ca^{2+}]i$ by inhibiting ADAR2-dependent editing of GluR2.

34.3 Consequences of TD-induced Disruption of Ca^{2+} Homeostasis

Alterations in the cytoplasmic Ca^{2+} concentrations can trigger a range of downstream neurotoxic cascades, including the uncoupling mitochondrial electron transfer from ATP synthesis, and the activation and overstimulation of enzymes such as calpains and other proteases, protein kinases, nitric oxide synthase (NOS), calcineurin and endonucleases (Szydlowska and Tymianski 2010). It has been demonstrated that calcium stores and intracellular calcium-binding proteins are changed in neurons and other non-neuronal cells from patients with Alzheimer's disease (Gibson et al. 1996; Gibson and Huang 2004; Supnet and Bezprozvanny 2010).

In the late stage of thiamine deficiency, an increase in extracellular glutamate is observed in some brain regions (Todd and Butterworth 2001). It has been proposed that the selective vulnerability to thiamine deficiency is mediated by a glutamate-induced excitotoxic process in affected structures (Hazell and Butterworth 2009). Our results indicate that thiamine deficiencystimulates AMPA-elicited Ca^{2+} influx by modulating GluR2 properties. A recent study shows that Ca^{2+}-permeable AMPARs are induced by *in vitro* traumatic mechanical injury in cortical neurons which results in increased $[Ca^{2+}]_i$ (Spaethling et al. 2008). Ca^{2+} permeation through AMPARs is crucial in several forms of synaptic plasticity and cell death associated with neurological diseases and disorders. The changes in GluR2 properties may serve as a 'molecular switch' leading to the formation of Ca^{2+}-permeable AMPARs and enhanced toxicity following neurological insults (Ben-Ari and Khrestchatisky 1998).

Abnormal editing of GluR2 pre-mRNA has been associated with certain neurological insults such as ALS and brain ischemia (Buckingham et al. 2008; Peng et al. 2006). A recent study indicates that inhibiting RNA editing of GluR2 enhances the death of motor neurons through excitotoxicity, whereas enhanced RNA editing reduces calcium permeability and protects motor neurons (Buckingham et al. 2008). Furthermore, the expression of AMPARs with unedited GluR2 is highly toxic in cultured hippocampal neurons (Mahajan and Ziff 2007).

We have previously demonstrated that thiamine deficiency induces ER stress or unfolded protein response (UPR) in animals and in cultured neurons (Wang et al. 2007). ER stress refers to the accumulation of unfolded or misfolded proteins in the ER lumen, resulting in an overall decrease in translation, enhanced protein degradation and increased levels of ER chaperones, which consequently increases the protein folding capacity of the ER. Sustained ER stress ultimately leads to the apoptotic death of the cell (Xu et al. 2005). A major cause of ER stress is the perturbation of calcium homeostasis (Matus et al. 2011). Therefore, it is likely that TD-induced alterations in $[Ca^{2+}]_i$ causes ER stress.

In summary, thiamine deficiency drastically increases neuronal $[Ca^{2+}]_i$ through altering GluR2 properties. TD-mediated disruption of $[Ca^{2+}]_i$ may underlie glutamate excitotoxicity or ER stress-induced neurodegeneration in selective brain structures.

Summary Points

- Thiamine deficiency (TD) causes neurodegeneration.
- TD induces calcium influx to neurons.
- A TD-induced increase in intracellular calcium concentration is toxic to neurons.
- TD-induced calcium influx is mediated by AMPA receptor.
- Prevention of calcium influx may protect TD-induced neurodegeneration.

Key Facts about Glutamate Receptors

- Glutamate receptors are responsible for the glutamate-mediated post-synaptic excitation of neural cells and are important for neural communication, memory formation, learning and regulation.
- There are three types of glutamate receptors: N-methyl-D-aspartate receptors, α-amino-3-hydroxyl-5-methyl-4-isoxazole-propionic acid receptors and kainate receptors. Each has several subtypes.
- Glutamate receptors play an important role in maintaining intracellular calcium homeostasis.
- Over-stimulation of glutamate receptors results in neurotoxicity and neurodegeneration.
- Thiamine deficiency may cause excessive production of neurotransmitters which stimulate glutamate receptors and induce neurotoxicity.

Definitions of Words and Terms

Calcium homeostasis: calcium homeostasis is the mechanism by which the body maintains adequate calcium levels.

Endoplasmic reticulum stress or unfolded protein response: A response to an accumulation of unfolded or misfolded proteins in the lumen of the endoplasmic reticulum. The unfolded protein response (UPR) has two primary aims: initially to restore normal function of the cell by halting protein translation and to activate the signaling pathways that lead to increasing the production of molecular chaperones involved in protein folding. If these objectives are not achieved within a certain time lapse or the disruption is prolonged, the UPR aims to apoptosis.

Glutamate excitotoxicity: The pathological process by which neurons are damaged and killed by excessive stimulation by neurotransmitters such as glutamate and similar substances.

Glutamate receptors: A group of synaptic receptors located primarily on the membranes of neuronal cells. Glutamate is a neurotransmitter and is particularly abundant in the nervous system. Glutamate receptors are responsible for the glutamate-mediated post-synaptic excitation of neural cells and are important for neural communication, memory formation, learning and regulation. Furthermore, glutamate receptors are implicated in the pathologies of a number of neurodegenerative diseases due to their central role in excitotoxicity and their prevalence throughout the central nervous system.

Neurodegeneration: A progressive loss of structure or function of neurons, including death of neurons.

RNA editing: Molecular processes in which the information content in an RNA molecule is altered through a chemical change in the base makeup.

Wernicke's encephalopathy: An acute neurological illness caused by severe deficiency of the vitamin thiamine or vitamin B_1.

List of Abbreviations

AD	Alzheimer's disease
ADARs	adenosine deaminases acting on RNA
ALS	amyotrophic lateral sclerosis
AMPARs	α-amino-3-hydroxyl-5-methyl-4-isoxazole-propionic acid receptors
$[Ca^{2+}]i$	intracellular Ca^{2+} concentration
CNS	central nervous system
ER	endoplasmic reticulum
GLUR	glutamate receptor
KRs	kainate receptors
NMDAR	*N*-methyl-D-aspartate receptor
NOS	nitric oxide synthase
TD	thiamine deficiency
TDP	thiamine diphosphate
UPR	unfolded protein response
WKS	Wernicke–Korsakoff syndrome

References

Ba, A., 2008. Metabolic and structural role of thiamine in nervous tissues. *Cellular and Molecular Neurobiology.* 28: 923–931.

Barbon, A., Vallini, I., La Via, L., Marchina, E., and Barlati, S., 2003. Glutamate receptor RNA editing: a molecular analysis of GluR2, GluR5 and GluR6 in human brain tissues and in NT2 cells following in vitro neural differentiation. Brain Research. *Molecular Brain Research.* 117: 168–178.

Ben-Ari, Y., and Khrestchatisky, M., 1998. The GluR2 (GluRB) hypothesis in ischemia: missing links. *Trends in Neurosciences.* 21: 241–242.

Buckingham, S.D., Kwak, S., Jones, A.K., Blackshaw, S.E., and Sattelle, D.B., 2008. Edited GluR2, a gatekeeper for motor neurone survival? *BioEssays.* 30: 1185–1192.

Burns, C.M., Chu, H., Rueter, S.M., Hutchinson, L.K., Canton, H., Sanders-Bush, E., and Emeson, R.B., 1997. Regulation of serotonin-2C receptor G-protein coupling by RNA editing. *Nature.* 387: 303–308.

Gibson, G.E., Zhang, H., Toral-Barza, L., Szolosi, S., and Tofel-Grehl, B., 1996. Calcium stores in cultured fibroblasts and their changes with Alzheimer's disease. *Biochimica et Biophysica Acta.* 1316: 71–77.

Gibson, G.E., and Huang, H.M., 2004. Mitochondrial enzymes and endoplasmic reticulum calcium stores as targets of oxidative stress in neurodegenerative diseases. *Journal of Bioenergetics and Biomembranes.* 36: 335–340.

Gibson, G.E., and Blass, J.P., 2007. Thiamine-dependent processes and treatment strategies in neurodegeneration. *Antioxidants & Redox Signaling.* 9: 1605–1619.

Haas, R.H., 1988. Thiamin and the brain. *Annual Review of Nutrition.* 8: 483–515.

Hakim, A.M., and Pappius, H.M., 1983. Sequence of metabolic, clinical, and histological events in experimental thiamine deficiency. *Annals of Neurology.* 13: 365–375.

Hardingham, G.E., and Bading, H., 2010. Synaptic *versus* extrasynaptic NMDA receptor signalling: implications for neurodegenerative disorders. Nature Reviews. *Neuroscience.* 11: 682–696.

Hazell, A.S., and Butterworth, R.F., 2009. Update of cell damage mechanisms in thiamine deficiency: focus on oxidative stress, excitotoxicity and inflammation. *Alcohol and Alcoholism.* 44: 141–147.

Huang, H.M., Chen, H.L., and Gibson, G.E., 2010. Thiamine and oxidants interact to modify cellular calcium stores. *Neurochemical Research.* 35: 2107–2116.

Hume, R.I., Dingledine, R., and Heinemann, S.F., 1991. Identification of a site in glutamate receptor subunits that controls calcium permeability. *Science.* 253: 1028–1031.

Isaac, J.T., Ashby, M.C., and McBain, C.J., 2007. The role of the GluR2 subunit in AMPA receptor function and synaptic plasticity. *Neuron.* 54: 859–871.

Kawahara, Y., Ito, K., Sun, H., Aizawa, H., Kanazawa, I., and Kwak, S., 2004. Glutamate receptors: RNA editing and death of motor neurons. *Nature.* 427: 801.

Ke, Z.J., DeGiorgio, L.A., Volpe, B.T., and Gibson, G.E., 2003. Reversal of thiamine deficiency-induced neurodegeneration. *Journal of Neuropathology and Experimental Neurology*. 62: 195–207.

Krugers, H.J., Hoogenraad, C.C., and Groc, L., 2010. Stress hormones and AMPA receptor trafficking in synaptic plasticity and memory. Nature Reviews. *Neuroscience*. 11: 675–681.

Lee, S. Yang, G., Yong, Y., Liu, Y., Zhao, L., Xu., J., Zhang, X., Wan, Y., Feng, C., Fan, Z., Liu, Y., Luo, J., and Ke, Z.J., 2010. ADAR2-dependent RNA editing of GluR2 is involved in thiamine deficiency-induced alteration of calcium dynamics. *Mol. Neurodegener*. 5: 54.

Liu, S.J., and Zukin, R.S., 2007. Ca^{2+}-permeable AMPA receptors in synaptic plasticity and neuronal death. *Trends in Neurosciences*. 30: 126–134.

Mahajan, S.S., and Ziff, E.B., 2007. Novel toxicity of the unedited GluR2 AMPA receptor subunit dependent on surface trafficking and increased Ca2+-permeability. *Molecular and Cellular Neurosciences*. 35: 470–481.

Matus, S., Glimcher, L.H., and Hetz, C., 2011. Protein folding stress in neurodegenerative diseases: a glimpse into the ER. *Current Opinion in Cell Biology*. 23: 239–252.

Pannunzio, P., Hazell, A.S., Pannunzio, M., Rao, K.V., and Butterworth, R.F., 2000. Thiamine deficiency results in metabolic acidosis and energy failure in cerebellar granule cells: an in vitro model for the study of cell death mechanisms in Wernicke's encephalopathy. *Journal of Neuroscience Research*. 62: 286–292.

Peng, P.L., Zhong, X., Tu, W., Soundarapandian, M.M., Molner, P., Zhu, D., Lau, L., Liu, S., Liu, F., Lu, Y., 2006. ADAR2-dependent RNA editing of AMPA receptor subunit GluR2 determines vulnerability of neurons in forebrain ischemia. *Neuron*. 49: 719–733.

Singleton, C.K., and Martin, P.R., 2001. Molecular mechanisms of thiamine utilization. *Current Molecular Medicine*. 1: 197–207.

Spaethling, J.M., Klein, D.M., Singh, P., and Meaney, D.F., 2008. Calcium-permeable AMPA receptors appear in cortical neurons after traumatic mechanical injury and contribute to neuronal fate. *Journal of Neurotrauma*. 25: 1207–1216.

Supnet, C., and Bezprozvanny, I., 2010. The dysregulation of intracellular calcium in Alzheimer disease. *Cell Calcium*. 47: 183–189.

Szydlowska, K., and Tymianski, M., 2010. Calcium, ischemia and excitotoxicity. *Cell Calcium*. 47: 122–129.

Todd, K.G., and Butterworth, R.F., 2001. In vivo microdialysis in an animal model of neurological disease: thiamine deficiency (Wernicke) encephalopathy. *Methods*. 23: 55–61.

Wang, X., Wang, B., Fan, Z., Shi, X., Ke Z.J., and Luo, J., 2007. Thiamine deficiency induces endoplasmic reticulum stress in neurons. *Neuroscience*. 144: 1045–1056.

Xu, C., Bailly-Maitre, B., and Reed, J.C., 2005. Endoplasmic reticulum stress: cell life and death decisions. *The Journal of Clinical Investigation*. 115: 2656–2664.

CHAPTER 35

Role of Thiamine in Obesity-related Diabetes: Modification of the Gene Expression

YUKA KOHDA,* TAKAO TANAKA AND HITOSHI MATSUMURA

Laboratory of Pharmacotherapy, Osaka University of Pharmaceutical Sciences, 4-20-1 Nasahara, Takatsuki, Osaka 569-1094, Japan
*Email: ykohda@gly.oups.ac.jp

35.1 Introduction

This study hypothesized that obesity and diabetes occur because glucose absorbed by cells is not completely metabolized; specifically, metabolic syndrome should not develop if cells completely metabolize glucose. Thiamine (vitamin B_1) acts as a lubricant for carbohydrate metabolism and the amount of catalytic thiamine absorbed must necessarily increase if glucose is to be metabolized in large quantities. We studied thiamine, an important element in combating metabolic syndrome, and the genes concerned with metabolic and functional disorders, as well as the factors related to obesity and diabetes. The aim of this study was to bridge the gap between basic research and medical practice.

Obesity is of major clinical relevance with regard to metabolic disorders; nevertheless, an effective pharmaceutical treatment remains unavailable. Obesity frequently leads to type 2 diabetes mellitus (DM). Substantial evidence suggests that thiamine has beneficial effects with regard to diabetic complications, having possible associations with type 2 DM and metabolic disorders. Thiamine is an indispensable coenzyme that is required for intracellular glucose metabolism. A high dose of thiamine has been reported to be effective in arresting the development of diabetic complications (Babaei-Jadidi *et al.* 2004; Beltramo *et al.* 2008; Berrone *et al.* 2006; Hammes *et al.* 2003; Schmid *et al.* 2008; Thornalley 2005). We previously reported that thiamine attenuates diabetic cardiomyopathy in streptozotocin (STZ) treated rats (Kohda *et al.* 2008). Thiamine has been reported to activate transketolase and improve the abnormally activated minor metabolic pathway (Berrone *et al.* 2006; Hammes *et al.* 2003; Schmid *et al.* 2008). Similar to transketolase, pyruvate dehydrogenase (PDH) requires thiamine as a coenzyme. Diabetic patients have reduced PDH activity, which damages glucose metabolism (Bajotto *et al.* 2004; Schummer *et al.* 2008).

Excess carbohydrate intake could be accompanied by a thiamine-deficient condition. Therefore, thiamine intervention may impact metabolic abnormalities in Otsuka Long–Evans Tokushima Fatty (OLETF) rats, which were extensively studied as a model of human obesity and metabolic disorders. Based on the assumption that excess glucose influx causes relative thiamine deficiency, we provided thiamine-containing water to the OLETF rats. Thiamine attenuated type 2 DM and metabolic abnormalities in the rat model of human obesity and metabolic disorders. We also investigated the influence of thiamine on PDH activity in the OLETF rats.

35.2 Does Thiamine Intervention Decrease the Extent of Weight Gain?

Body weight is one of the parameters of obesity. We previously found that thiamine decreases body and organ weight gains and prevents obesity-related metabolic disorders (Tanaka *et al.* 2010). No difference was observed in body weight between the two groups at the start of the study. Despite equivalent food consumption by the two groups, body weight was lower in the thiamine-treated group than in the untreated control group (Figure 35.1A,B). Body weight gain was more than 20% lower in the thiamine-treated group than in the untreated control group throughout the remaining experimental period.

Despite the equivalent amount of food consumed, thiamine decreased the extent of weight gain. This interesting finding could have great significance for reduction in obesity if thiamine administration shows a similar effect on human metabolism. We therefore examined other obesity parameters in detail, *i.e.* visceral fat mass and adipocyte size.

35.3 Thiamine Intervention Decreases not only Body Weight but also Visceral Fat Mass and Adipocyte Size

Lipid accumulation in visceral white adipose tissue is considered to be a cause of obesity and obesity-related metabolic disorders. We previously found that

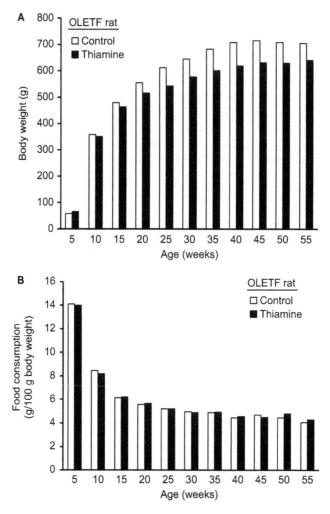

Figure 35.1 Effects of thiamine on body weight and food consumption in OLETF rats. Thiamine decreased body weight (A) despite the equivalent amount of food consumed (B) by the Otsuka Long–Evans Tokushima Fatty (OLETF) rats. Four-week-old male OLETF rats were maintained with water (untreated control group) or water containing 0.2% thiamine (thiamine-treated group). Food consumption relative to body weight, calculated as food consumption (g) divided by 100 g body weight, was measured weekly, and all values are expressed as mean ± SE to 21 weeks ($n=8$) and 51 weeks ($n=8$) in each group. Modified from Tanaka, Kohda et al. (2010).

thiamine treatment profoundly decreased visceral fat mass in OLETF rats (Tanaka et al. 2010). Table 35.1 shows the amount of visceral fat mass in the untreated control and thiamine-treated groups. Tibial length was same in both groups, suggesting no substantial growth retardation in the thiamine-treated

Table 35.1 Effects of thiamine on visceral fat mass in OLETF rats. Thiamine decreased visceral fat mass in Otsuka Long–Evans Tokushima Fatty (OLETF) rats. Retroperitoneal fat, epididymal fat and tibial length were determined at both 25 weeks (25 W) and 55 weeks (55 W) of age. Values are mean ± SE (n = 8). *$P<0.05$, **$P<0.01$ and ***$P<0.001$ by unpaired, two-tailed Student's t test. Modified from Tanaka, Kohda et al. (2010).

Group	Tibia (mm)	Retroperitoneal fat (g)	Epididymal fat (g)
Control (25-W OLETF)	43.7 ± 0.24	26.8 ± 1.95	17.6 ± 0.93
Thiamine (25-W OLETF)	43.7 ± 0.27	21.4 ± 2.51*	13.7 ± 1.53*
Control (55-W OLETF)	46.3 ± 0.29	54.9 ± 4.37	17.1 ± 0.92
Thiamine (55-W OLETF)	45.7 ± 0.28	32.8 ± 2.43***	13.0 ± 0.68**

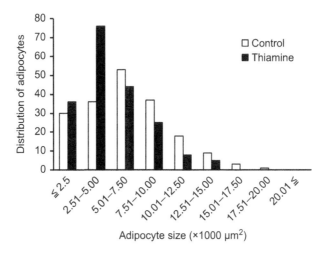

Figure 35.2 Effects of thiamine on adipocyte size in OLETF rats. Thiamine diminished adipocyte size, as shown by the distribution of this size in epididymal white adipose tissue obtained from the 55-week-old Otsuka Long–Evans Tokushima Fatty (OLETF) rats. Open columns, untreated control rats; closed columns, thiamine-treated rats. Modified from Tanaka, Kohda et al. (2010).

group. The thiamine-treated group showed lower epididymal and retroperitoneal fat pad weight than the untreated control group.

Figure 35.2 shows the histograms of adipose cell size obtained from the untreated control and thiamine-treated 55-week-old rats. Histological evaluation revealed that smaller adipocytes in epididymal white adipose tissues were predominant in the thiamine-treated group compared with the untreated control group.

35.4 Thiamine Decreases Hepatic Triglyceride Accumulation and Modulates PDH Activity

We previously found that thiamine ameliorates diabetes-induced PDH inhibition (Kohda et al. 2008, 2010). We then studied the effects of thiamine on the basis of the hypothesis that thiamine deficiency may be a cause of the onset and progression of type 2 DM for the main phenotype of obesity and accompanying metabolic disorders in the OLETF rats.

Thiamine intervention ameliorates body weight gain by suppressing visceral fat in the OLETF rats. We also found an improvement in biochemical findings, degree of fatty liver and obesity-related cardiac and renal pathology and dysfunction (Tanaka et al. 2010). Hepatic triglyceride accumulation was significantly reduced by thiamine treatment in the 25- and 55-week-old rats (Table 35.2). PDH activity reduced in diabetic patients, affecting glucose metabolism. PDH activity was significantly increased by thiamine administration in the OLETF rats (Table 35.3). Thiamine is involved in protection against obesity-related metabolic disorders by activating PDH in the OLETF rats.

Table 35.2 Effects of thiamine on hepatic steatosis in OLETF rats. Thiamine decreased hepatic steatosis in 25-week-old (25 W) and 55-week-old (55 W) Otsuka Long–Evans Tokushima Fatty (OLETF) rats, as shown by quantitative evaluation of hepatic triglyceride content. Values are mean ± SE (n = 8). *$P<0.05$ and **$P<0.001$ by unpaired, two-tailed Student's t test. Modified from Tanaka, Kohda et al. (2010).

Group	Triglyceride (nmol/g liver)
Control (25-W OLETF)	15.7 ± 0.85
Thiamine (25-W OLETF)	4.25 ± 1.13**
Control (55-W OLETF)	46.9 ± 4.11
Thiamine (55-W OLETF)	25.0 ± 2.75*

Table 35.3 Effects of thiamine on PDH activity in OLETF rats. Pyruvate dehydrogenase (PDH) activity in hepatic mitochondria was increased by thiamine intervention in 55-week-old (55 W) Otsuka Long–Evans Tokushima Fatty (OLETF) rats. Values are mean ± SE (n = 4–7). *$P<0.05$ by unpaired, two-tailed Student's t test. Modified from Tanaka, Kohda et al. (2010).

Group	PDH activity (nmol/mg protein/min)
Control (55-W OLETF)	0.426 ± 0.007
Thiamine (55-W OLETF)	0.467 ± 0.013*

35.5 Thiamine Differentially Modifies Transcript Expression Levels of Genes Involved in Carbohydrate Metabolism, Lipid Metabolism, Vascular Physiology and Carcinogenesis in the Liver

To gain more insight into the effects of thiamine on the molecular physiology of obesity and DM, we evaluated the profiles of transcript expression levels of genes in the liver of the untreated control and thiamine-treated rats using gene microarrays. Gene microarray analyses revealed that of 33 849 genes on the array, 76 were expressed at least two-fold higher or lower ($P < 0.05$) in the liver of thiamine-treated rats than that of the untreated control rats. Among these 76 genes, several were involved in carbohydrate metabolism, lipid metabolism, vascular physiology and carcinogenesis (Figure 35.3).

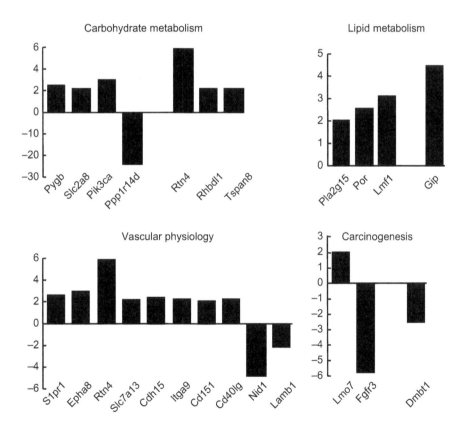

Figure 35.3 Genes modified by thiamine intervention in OLETF rats. Thiamine modified transcript expression levels of genes implicated in carbohydrate metabolism, lipid metabolism, vascular physiology and carcinogenesis in the liver of the Otsuka Long–Evans Tokushima Fatty (OLETF) rats. Tanaka and Kohda unpublished data.

Dysregulation of carbohydrate homeostasis is a characteristic of obesity and DM. Microarray analysis revealed that transcript expression levels of genes such as *Hk1*, *Pygb* and *Slc2a8*, which contributed to carbohydrate metabolism, were higher in the liver of the thiamine-treated rats than that of the untreated control rats. *Slc2a8* encodes the solute carrier family 2 facilitated glucose transporter (GLUT), member 8 GLUT8, which is a novel glucose transporter protein that is widely distributed in tissues, including liver tissue. GLUT8 is predominantly located in an intracellular compartment, primarily in perivenous hepatocytes, which is suggested to be involved in regulation of nutrient flow and glycolytic fluxes (Gorovits *et al.* 2003). Higher transcript expression levels of *Hk1* and *Pygb* in the liver of the thiamine-treated rats, together with higher expression of *Slc2a8*, may indicate that thiamine helps the liver to function as a glucose-supplying organ, rather than a glucose-depositing organ, even under polyphagic conditions. In addition, transcripts of genes such as *Rtn4*, *Rhbdl1* and *Tspan8* whose functions are not yet completely elucidated, although their involvement in carbohydrate metabolism has been suggested, were expressed at higher levels in the liver of the thiamine-treated rats than that of the untreated control rats.

Gip encodes gastric inhibitory polypeptide—also known as glucose-dependent insulinotropic polypeptide (GIP). GIP is a gut-derived incretin secreted in response to nutrient ingestion. Increased GIP signalling has been suggested to promote fat accumulation, rather than fat oxidation, within adipocytes. Hence, GIP is described as the obesity hormone and may contribute to the pathogenesis of type 2 DM (Marks 2006). However, GIP potentiates glucose-dependent insulin secretion and enhances β-cell mass through regulation of β-cell proliferation, neogenesis and apoptosis. Incretin receptor signalling exerts physiologically relevant actions critical for glucose homeostasis and represents a pharmacologically attractive target for the development of agents for treatment of type 2 DM. GIP is released from duodenal endocrine K cells. Although thiamine prevented obesity in the OLETF rats, the transcript expression level of *Gip* was higher in the liver of the thiamine-treated rats than that of the untreated control rats. This result, contradictory to the suggested role of GIP in obesity, requires further study.

Obesity and DM are major risk factors for the development of cardio-, reno- and retinovascular complications, which are major contributors to morbidity and mortality. Failure to maintain a homeostatic balance in endothelial cells or basement membrane components is considered critical to the development and progression of vascular complications of diabetes and obesity. Transcript expression levels of genes such as *S1pr1*, *Epha8*, *Rtn4*, *Slc7a13*, *Cdh15*, *Itga9*, *Cd151* and *Cd40lg* that participate in endothelial homeostasis were higher and those of genes encoding basement membrane components, such as *Nid1* and *Lamb1*, were lower in the liver of the thiamine-treated rats than that of the untreated control rats.

Favourable effects of thiamine have been indicated for renal and retinal angiopathies in STZ-induced diabetic rat models (Babaei-Jadidi *et al.* 2003; Hammes *et al.* 2003). Taking these factors into consideration, thiamine probably maintains the homeostatic balance of vascular integrity in polyphagia-induced obesity followed by DM.

Obesity is also associated with certain types of cancer. Recent international cancer prevention guidelines recommend weight loss, where appropriate, for cancer risk reduction. Transcript expression levels of genes related to carcinogenesis, including *Lmo7*, *Fgfr3* and *Dmbt1*, were modulated by thiamine intervention. Transcript expression levels of *Fgfr3* and *Dmbt1* were lower and those of *Lmo7* was higher in the liver of the thiamine-treated rats than that of the untreated control rats.

In a previous study, although numbers were small, kidney and spleen cancers were detected in two of the eight untreated control OLETF rats, but in none of the eight thiamine-treated rats. Thiamine deficiency may predispose individuals to neoplastic conditions. Thus, excess carbohydrate intake may be accompanied by a relative thiamine deficiency and may become a risk factor for cancer.

35.6 Concluding Remarks

Obesity is a global health problem, and the incidence of obesity is expected to increase significantly in the future on account of overconsumption of foods containing carbohydrates and fats and lack of exercise. Modern diets largely comprise processed foods high in carbohydrates and fats, which are cheap and easily available in supermarkets and fast food outlets. However, some of the techniques used to produce highly refined grains can lead to loss of micronutrients, including thiamine. Epidemiological data have shown that few patients develop diabetes or hypertension even when they eat thiamine-containing brown rice or wheat germ, rather than white rice. Absorption of thiamine from food might be important in preventing metabolic syndrome. However, if thiamine intake is insufficient, this can be increased by consuming dietary supplements and medicines. Thiamine is a soluble vitamin and seems to have no potential side effects, even if consumed in excess.

It is critical to develop a new preventive strategy against obesity. Obesity frequently leads to type 2 DM, which can progress to diabetic complications such as nephropathy and cardiomyopathy. Current pharmaceutical interventions are moderately effective in treating clinical symptoms of metabolic syndrome; nevertheless, a basic pharmaceutical method for preventing obesity and its related metabolic disorders remains to be established. Elucidation of the mechanisms concerning obesity prevention and diabetic complications by improving glucose metabolism may contribute to the prevention of metabolic syndrome and the development of new drugs for treatment of diabetes and obesity.

Summary Points

- This chapter focuses on thiamine (vitamin B_1) and obesity.
- Obesity frequently leads to type 2 diabetes mellitus.

- Diabetic patients have reduced pyruvate dehydrogenase (PDH) activity, which impairs glucose metabolism.
- In this study, PDH activity is significantly increased by thiamine intervention in Otsuka Long–Evans Tokushima Fatty (OLETF) rats.
- Thiamine protects against obesity-related metabolic disorders by activating PDH in OLETF rats.
- In addition, thiamine profoundly decreases visceral fat mass and diminishes adipocyte size in OLETF rats.
- Furthermore, thiamine intervention significantly modifies the gene expression involved in obesity and obesity-related metabolic disorders.
- In this study, we report that thiamine intervention has the potential to prevent obesity and multiple features of obesity-related metabolic disorders in OLETF rats.

Key Facts

Key Features of Otsuka Long–Evans Tokushima Fatty Rats

- OLETF rats were developed by Kawano *et al.* (1991, 1992).
- In this study, four-week-old OLETF rats were provided by the Tokushima Research Institute, Otsuka Pharmaceutical (Tokushima, Japan).
- Polyphagia-induced OLETF rats lack functional cholecystokinin-A receptors.
- Cholecystokinin is associated with satiety control mechanisms.
- OLETF rats exhibit progressive obesity and metabolic disorders similar to human metabolic syndrome.

Key Features of Thiamine Intervention in OLETF Rats

- We chose OLETF rats as models for studying human obesity.
- Four-week-old OLETF rats were randomly divided into two groups: an untreated control group and a thiamine-treated group.
- OLETF rats were given water (untreated control group) or water containing 0.2% thiamine (thiamine-treated group) for 21 weeks ($n=8$) and 51 weeks ($n=8$).
- All rats were housed in cages and provided with standard rodent chow (NMF; Oriental Yeast Co., Japan).
- Body weight and food intake were measured weekly throughout the experimental period.

Key Features of Obesity Worldwide

- Obesity and its related metabolic disorders have reached pandemic proportions and show no signs of improvement.

- The incidence of obesity is expected to significantly increase because of changing food habits (increased overall caloric intake of lipids and simple sugars) and decreasing physical activity.
- In 2005, the World Health Organization (WHO) estimated that more than 400 million adults worldwide were obese.
- Because of the increasing number of dire health problems in many countries, it is of utmost importance to develop novel preventive strategies for obesity.
- Current strategies include diet and exercise; nevertheless, an effective pharmaceutical strategy remains to be established.
- Many factors, including genetics, environment and lifestyle changes, undoubtedly provide impetus to an increase in the morbidities associated with obesity.
- Among these factors, we believe that changes in dietary habits should be given considerable importance when dealing with these pandemic morbidities.

Key Features of Microarray Analysis

- Microarray analysis was performed on 55-week-old OLETF rats.
- Two rats were picked randomly, each from the untreated control and thiamine-treated groups.
- Differential gene expression was analysed using CodeLink Whole Rat Genome Bioarrays (Applied Microarrays).
- GenePix Pro software (MDS Inc., Toronto, Canada) was used to acquire and align the microarray images.
- Genes were analysed by enrichment analysis using the GenMAPP pathway data (www.genmap.org) with Microarray Data Analysis Tool version 3.0 (Filgen Inc., Nagoya, Japan).

Definitions of Words and Terms

Adipocyte size. Adipocyte size may contribute to the prediction of obesity and obesity-related metabolic disorders.

Diabetic complications. Diabetes causes various complications such as retinopathy, nephropathy, neuropathy and microangiopathy.

Diabetes mellitus (DM). Diabetes is characterized by an increased metabolism of fatty acids due to reduced glucose utilization. Mutation of thiamine transporter gene *SLC19A2* is linked to type 2 DM.

Fatty liver. Hepatic steatosis is a common pathological condition of obesity and DM.

Glucose metabolism. Intracellular glucose is used for ATP production through the glycolytic pathway, pro-glucose oxidation (TCA) cycle and mitochondrial electron transport chain; these are referred to as the major metabolic pathways. Furthermore, there are minor pathways involved in glucose metabolism, such as the polyol pathway, hexosamine biosynthetic pathway and diacylglycerol-protein kinase C pathway, which metabolize a part of the cellular glucose.

Metabolic syndrome. Metabolic syndrome can lead to obesity, hypertension and hyperglycemia. A worldwide increase in the incidence of metabolic syndrome can seriously impair the quality of life of patients. Various life-style-related diseases (*e.g.* hyperlipidemia, diabetes and arteriosclerosis) affect the metabolic syndrome.

Obesity. Obesity is a condition with major clinical relevance to metabolic disorders such as insulin resistance, type 2 DM, steatosis, hypertension and dyslipidemia.

Polyphagia. The term *Polyphagia* is of Greek origin, where *poly* means too much and *phagia* means food.

Pyruvate dehydrogenase (PDH). PDH utilizes thiamine as a coenzyme. A follow-up study of subjects with reduced PDH activity reported that reduced PDH activity may lead to a future onset of diabetes.

Thiamine. Thiamine plays crucial roles in cellular metabolism, and impaired thiamine uptake results in a variety of disorders.

List of Abbreviations

DM	diabetes mellitus
GIP	glucose-dependent insulinotropic polypeptide
GLUT	glucose transporter
OLETF	Otsuka Long–Evans Tokushima Fatty
PDH	pyruvate dehydrogenase
STZ	streptozotocin
WHO	World Health Organization

References

Babaei-Jadidi, R., Karachalias, N., Ahmed, N., Battah, S., and Thornalley, P.J., 2003. Prevention of incipient diabetic nephropathy by high-dose thiamine and benfotiamine. *Diabetes.* 52: 2110–2120.

Babaei-Jadidi, R., Karachalias, N., Kupich, C., Ahmed, N., and Thornalley, P.J., 2004. High-dose thiamine therapy counters dyslipidaemia in streptozotocin-induced diabetic rats. *Diabetologia.* 47: 2235–2246.

Bajotto, G., Murakami, T., Nagasaki, M., Tamura, T., Tamura, N., Harris, R.A., Shimomura, Y., and Sato, Y., 2004. Downregulation of the skeletal muscle pyruvate dehydrogenase complex in the Otsuka Long-Evans Tokushima Fatty rat both before and after the onset of diabetes mellitus. *Life Sciences.* 75: 2117–2130.

Beltramo, E., Berrone, E., Tarallo, S., and Porta, M., 2008. Effects of thiamine and benfotiamine on intracellular glucose metabolism and relevance in the prevention of diabetic complications. *Acta Diabetologica.* 45: 131–141.

Berrone, E., Beltramo, E., Solimine, C., Ape, A.U., and Porta, M., 2006. Regulation of intracellular glucose and polyol pathway by thiamine and benfotiamine in vascular cells cultured in high glucose. *The Journal of Biological Chemistry*. 281: 9307–9313.

Gorovits, N., Cui, L., Busik, J.V., Ranalletta, M., Hauguel, de-M.S., and Charron, M.J., 2003. Regulation of hepatic GLUT8 expression in normal and diabetic models. *Endocrinology*. 144: 1703–1711.

Hammes, H.P., Du, X., Edelstein, D., Taguchi, T., Matsumura, T., Ju, Q., Lin, J., Bierhaus, A., Nawroth, P., Hannak, D., Neumaier, M., Bergfeld, R., Giardino, I., and Brownlee, M., 2003. Benfotiamine blocks three major pathways of hyperglycemic damage and prevents experimental diabetic retinopathy. *Nature Medicine*. 9: 294–299.

Kawano, K., Hirashima, T., Mori, S., Saitoh, Y., Kurosumi, M., and Natori, T., 1991. New inbred strain of Long-Evans Tokushima lean rats with IDDM without lymphopenia. *Diabetes*. 40: 1375–1381.

Kawano, K., Hirashima, T., Mori, S., Saitoh, Y., Kurosumi, M., and Natori, T., 1992. Spontaneous long-term hyperglycemic rat with diabetic complications. Otsuka Long-Evans Tokushima Fatty (OLETF) strain. *Diabetes*. 41: 1422–1428.

Kohda, Y., Shirakawa, H., Yamane, K., Otsuka, K., Kono, T., Terasaki, F., and Tanaka, T., 2008. Prevention of incipient diabetic cardiomyopathy by high-dose thiamine. *The Journal of Toxicological Sciences*. 33: 459–472.

Kohda, Y., Umeki, M., Kono, T., Terasaki, F., Matsumura, H., and Tanaka, T., 2010. Thiamine ameliorates diabetes-induced inhibition of pyruvate dehydrogenase (PDH) in rat heart mitochondria: investigating the discrepancy between PDH activity and PDH E1α phosphorylation in cardiac fibroblasts exposed to high glucose. *Journal of Pharmacological Sciences*. 113: 343–352.

Marks, V., 2006. Human obesity: its hormonal basis and the role of gastric inhibitory polypeptide. *Medical Principles and Practice*. 15: 325–337.

Schmid, U., Stopper, H., Heidland, A., and Schupp, N., 2008. Benfotiamine exhibits direct antioxidative capacity and prevents induction of DNA damage *in vitro*. *Diabetes Metabolism Research and Reviews*. 24: 371–377.

Schummer, C.M., Werner, U., Tennagels, N., Schmoll, D., Haschke, G., Juretschke, H.P., Patel, MS., Gerl, M., Kramer, W., and Herling, A.W., 2008. Dysregulated pyruvate dehydrogenase complex in Zucker diabetic fatty rats. *American Journal of Physiology. Endocrinology and Metabolism*. 294: E88–E96.

Tanaka, T., Kono T., Terasaki F., Yasui K., Soyama A., Otsuka K., Fujita, S., Yamane, K., Manabe, M., Usui, K., and Kohda, Y., 2010. Thiamine prevents obesity and obesity-associated metabolic disorders in OLETF rats. *Journal of Nutritional Science and Vitaminology*. 56: 335–346.

Thornalley, P.J., 2005. The potential role of thiamine (vitamin B_1) in diabetic complications. *Current Diabetes Reviews*. 1: 287–298.

CHAPTER 36
Riboflavin Uptake

MAGDALENA ZIELIŃSKA-DAWIDZIAK

Department of Biochemistry and Food Analysis, Faculty of Food Science and Nutrition, Poznań University of Life Sciences, Poland
Email: mzd@up.poznan.pl

36.1 Introduction

Riboflavin (RF), one of the essential vitamins, is obtained by the human organism from the diet through small intestinal absorption or, alternatively, from indigenous bacteria which colonize large intestine (Foraker et al. 2003). The mechanism of uptake is still not completely understood, although many studies have focused on this problem.

Riboflavin is delivered in form of free vitamin, or as its coenzymes, i.e. flavin mononucleotide (FMN) and adenine dinucleotide (FAD), which occurs mainly as a prosthetic group of flavoproteins. Release of coenzymes from flavoproteins by acidification in stomach and proteolysis, both gastric and intestinal, must precede the absorption. This hydrolysis also releases several percentages of covalently bound FAD from 8α-(peptidyl)riboflavins (Chia et al. 1978). Free riboflavin is physiologically preferred form of absorbed vitamin B_2 (Daniel et al. 1983). The upper small intestine enzymes which catalyse reversible reactions of conversion nucleotides into riboflavin are located in the brush-border membrane of enterocytes (Figure 36.1).

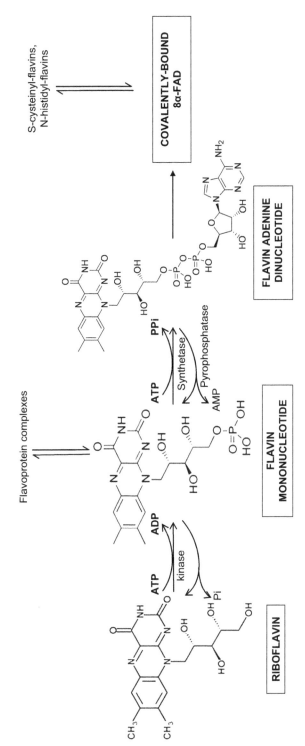

Figure 36.1 Mutual transformations of riboflavin analogues naturally occurring in the body. The figure presents mutual conversions of riboflavin, flavin adenine dinucleotide (FAD) and flavin mononucleotide (FMN) which occur during the absorption of the vitamin in the mammalian intestinal tract.

36.2 Recognized Mechanisms of Riboflavin Transport

Water-soluble vitamins are forced to cross a formidable barrier during absorption—the hydrophobic cell membrane. Three possible mechanisms, which enable riboflavin translocation, are suggested. Two of these mechanisms, a saturable, carrier-mediated process and receptor-mediated endocytosis, occur at the physiological concentration of riboflavin; the third one, passive diffusion, is exploited at higher concentrations of the vitamin (Ball 2004a).

36.2.1 Passive Diffusion

The movement of substances crossing the cell membranes by the passive diffusion (simple diffusion) mechanism is based only on differences in their concentrations. Migration of substances leads to their equal distribution. Small, water-soluble, uncharged molecules pass through narrow aqueous membrane channels. Charged molecules are generally too large to move *via* these channels, although it is possible for some specific inorganic ions (Ball 2004a).

36.2.2 Carrier-mediated Transport

This mechanism of transport has been confirmed for some of water-soluble vitamins (Ball 2004a). Translocation of substances is mediated by transmembrane protein, a type of membrane-bound protein (carrier), which may use energy derived from ATP hydrolysis or co-transport of ions moving down their electrochemical gradient. There are two forms of carrier-mediated transport, *i.e.* facilitated diffusion and active transport. Facilitated diffusion is not dependent on energy, and the driving force is, as in simple diffusion, the concentration gradient. Active transport is energy dependent.

Characteristic for these processes are competition of transported substance with structural analogues, stereospecificity and saturation kinetics. The process of binding the substance transported by the carrier equates to the creation of the transition state and the transport kinetics are analogous with the Michaelis–Menten kinetics (Ball 2004a). The transport process is described, among others by a half-maximal transport constant (K_m) (comparable to the Michaelis–Menten constant).

36.2.3 Receptor-mediated Endocytosis

The mechanism of receptor-mediated endocytosis is based on interactions between a transported substance and a special protein (receptor) bound to a cell membrane. Molecules are directly recognized by the receptor substance, or may be at first attached to a special protein, which in turn forms a complex with the receptor. The complex is locked in coated pits or vesicles, and next transported within the cytosol. The vesicles are uncoated by an ATP-dependent enzyme; the complex from their core is located inside endosomes and, after its dissociation from receptors, it is further conveyed to lysosomes.

36.3 Absorption and Transport of Riboflavin

36.3.1 Small Intestine

Intestinal bioavailability of riboflavin from the diet depends on digestibility of the food, which affects duration of the food in the gut, and is decreased by the consumption of alcohol, which inhibits the activity of FAD and FMN phosphatases, indispensable to their conversion into riboflavin before absorption (Figure 36.1). Simultaneously, the intestinal bioavailability is increased by secretion of bile salts; these extend the transition of food in the intestinal tract and enhance riboflavin solubility and the permeability of the brush-border membrane (Ball 2004b).

Some researchers have shown that absorption of riboflavin is equally efficient in the jejunum and the ileum of the small intestine (Said *et al.* 1993). Nevertheless, others have found that specialized carriers are more intensively expressed in the upper gastrointestinal tract (Jusko and Levy 1975; Sanchez-Pico *et al.* 1989).

In vitro transport of riboflavin across small intestine has been studied in an animal (rabbit) and a human model (Caco-2 cells) (Huang and Swaan 2000; Said *et al.* 1993; Said and Ma 1994; Zielińska-Dawidziak *et al.* 2008). Different values of K_m determined by Huang and Swaan (2000) for uptake in opposite directions (Table 36.1) are an evident indicator of an active transport of riboflavin across small intestine cells, as well as the dependence on temperature (Said and Ma 1994).

The uptake is considered as pH- and Na^+-independent (Said and Ma 1994), which indicates that the system is uniporter—rarely observed for mammalian apical carrier proteins (Huang and Swaan 2000). The inhibitory effect of sulfhydryl group modifying agents, such as *p*-chloromercuribenzene sulfonate (*p*-CMBS) or eosin maleimide (Said and Ma 1994) confirmed the results of earlier *in vivo* and *ex vivo* studies (Jusko and Levy 1975; Spector 1982). It suggests a possible involvement of reducing agent in the process (such as NADH and $FADH_2$).

Inhibition of riboflavin transport across the Caco-2 monolayer by 2,4-dinitrophenol (DNP), reducing the cellular ATP agent, points toward the necessity of delivering energy. However, the initial phase of whole absorption process (*i.e.* transport across the epithelium membrane) does not require a supply of metabolic energy and is a consequence of riboflavin's association with carrier protein which is presented on the brush-border membrane. In the next step, some of the riboflavin absorbed inside the enterocytes' cytosol is phosphorylated by flavokinase and converted by FAD synthetase—enzymes which require ATP molecules for their action (Gastaldi *et al.* 1999).

The inhibitory effect of uptake was observed during the analysis of the absorption of riboflavin analogues: lumiflavin (riboside side chain is exchanged for a methyl group), lumichrome (devoid of a riboside side chain), 8-[NH_2]-riboflavin and iso-riboflavin. Simultaneously, an insignificant effect was noted for D-ribose, panthotenic acid (substances with a pterin ring), suggesting that the riboside side chain has no effect on riboflavin absorption (Said and Ma 1994).

Transport across the membrane and inside the cells is inhibited by phosphorylated analogues of riboflavin (Gastaldi et al. 1999), which proves the importance of phosphorylation processes within the cells. Transporters identified so far in rat and human small intestine cells include transporter RFT1 (GPR172B) and RFT2 (C20orf54). RFT1, also expressed in the placenta, conveys riboflavin in a manner independent of Na^+, pH and difference of potentials, with $K_m = 63.7$ nM. RFT2 is expressed mainly in the testis, but also in the jejunum, ileum and prostate. The transport mediated by this protein was saturable ($K_m = 0.21$ μM), Na^+-independent and poorly pH-sensitive (uptake increased when the pH has changed from 8.0 to 5.4). Both transporters were inhibited by lumiflavine, FMN and FAD, but not by D-ribose, organic ions and other vitamins (Yamamoto et al. 2009; Yao et al. 2010; Yonezawa et al. 2008).

Other studies also indicate a receptor-mediated endocytosis way of riboflavin transport across intestinal cells. This thesis came about from a result observed in a particular condition; a slight absorption dependence on pH, which may suggest pH-dependent uncoupling of ligand–receptor interactions in receptor-mediated endocytosis (Foraker et al. 2003; Yamamoto et al. 2009).

Huang and Swaan (2000) revealed that the receptor-mediated mechanism of riboflavin takes place in the nanomolar range of riboflavin concentrations, whereas carrier-mediated process dominates in the micromolar range. They observed the inhibition of basolateral-apical transport of riboflavin by the same substances as transferrin, which is the typical protein transported through receptor-mediated endocytosis, i.e. brefeldin A and nocodazole. Brefeldin A strongly increased apical to basolateral riboflavin transport, while nocodazole increased apical to basolateral and inhibited basolateral to apical flux of riboflavin.

Under the special growth conditions of intestinal epithelial Caco-2 cells, the transport of riboflavin via passive diffusion processes was reported (Middleton 1990; Zielińska-Dawidziak et al. 2008). When riboflavin was over-supplemented (the incubation was at a concentration higher than in human plasma, i.e. >12 nM), riboflavin uptake was significantly decreased. Conversely, growth in medium deficient in vitamin B_2 caused a significant increase in vitamin uptake.

Researchers' results indicate the existence of an adaptive, down- and upregulatory mechanism of riboflavin absorption. Both the regulatory mechanisms were also observed for colon (Said et al. 2000) and renal cells (Kumar et al. 1998), as well as in in vivo research with rats (Said and Mohammadkhani 1993). During this regulation process, the affinity of riboflavin carriers remains stable (unchanged K_m), while their number and/or activity is changed (increased or decreased value of V_{max}). During the over-supplementation of riboflavin, protein–carrier transcription is restricted.

The small intestine is, therefore, apparently well-suited for the uptake of varied concentrations of riboflavin. The scheme of the uptake process is presented in Figure 36.2. Depending on the concentrations, it uses

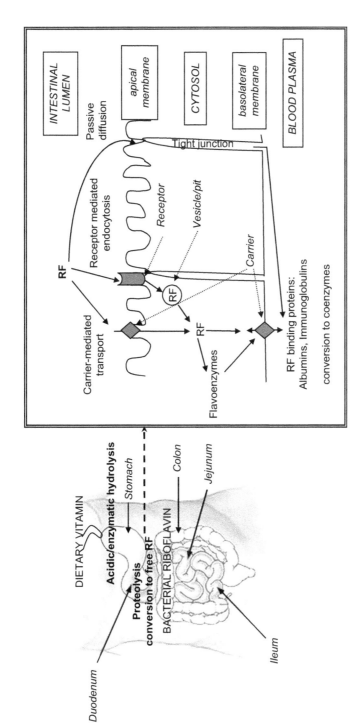

Figure 36.2 Riboflavin (RF) absorption and transport *via* the epithelium of the small intestine. The figure presents the fragments of the digestive tract involved in riboflavin absorption and a scheme of recognized mechanisms of this process.

different mechanisms. Riboflavin's crossing through membranes, both apical and basolateral, is allowed only for the dephosphorylated form of the vitamin. The energy needed for riboflavin absorption is not used for riboflavin uptake, but for riboflavin phosphorylation and its metabolism within enterocytes. Absorption is thus regulated by protein kinase (A and G).

Some of the most important parameters and modulators of riboflavin uptake in small intestine are presented in Tables 36.1–36.4.

36.3.2 Large Intestine

The large intestine is a part of the gastrointestinal tract with diverse microflora which synthesizes significant amounts of vitamin B_2, mainly as free riboflavin. The variability of microflora in the human gut is dependent on dietary habits. It has been proven that vegetable-based diet increases the synthesis of riboflavin in the colon (Iinuma 1955). *In vivo* colonic absorption was demonstrated in rats

Table 36.1 Half-maximal transport constants of carrier mediated riboflavin (RF) uptake. This table presents experimentally determined $K_m \pm$ SEM constants obtained in different cell models during *in vitro* experiments and varying concentrations of riboflavin.

Tissue model	Cell line	Rf concentration [nM]	Km	Reference
Human intestinal epithelium	Caco-2	0.5–160	$9.72 \pm 0.85\,nM^a$	Huang and Swann (2000)
		0.5–160	$4.06 \pm 0.03\,nM^b$	
		3.4–3000	$0.30 \pm 0.03\,\mu M$	Said and Ma (1994)
Human colonic epithelium	NCM460	5.5–1000	$0.14 \pm 0.004\,\mu M$	Said *et al.* (2000)
Peripheral blood cells	PBMC	5–1000	$955 \pm 344\,nM$	Zempleni and Mock (2000)
Human hepatocytes	HEPG2	33–3000	$0.41 \pm 0.08\,\mu M$	Said *et al.* (1998)
Rat brain endothelial	RBE4	nsp[c]	$44\,\mu M$	Patel *et al.* (2010)
Hu Human placental trophoblast	BeWo	1–1000	$1.32 \pm 0.68\,nM$	Huang and Swaan (2001)
Human renal proximal tubule epithelial	HK-2	4.2–3000	$0.67 \pm 0.21\,\mu M$	Kumar *et al.* (1998)
Retinal pigment epithelium cells	ARP-19	8 nM and 100 μM	$80 \pm 14\,nm$	Said *et al.* (2005)
Retinoblastoma cell line	Y-79	0.5 μCi/ml	$19.21 \pm 0.37\,nM$	Kansara *et al.* (2005)

[a]Apical → basolateral direction.
[b]Basolatoral → apical direction of RF transport.
[c]Not specified.

Table 36.2 Main substances inhibiting RF transport.[a,b] This table shows the effect of chemical substances which influenced the velocity of riboflavin uptake, *i.e.* membrane transport, sulfhydryl group and metabolic inhibitors as well as the vitamin analogues.

	Caco-2[1]	NCM460[2]	PBMC[3]	HEP G2[4]	RBE4[5]	BeWo[6]	HK-2[7]	ARPE-19[8]	Y79[9]
Substrate analogues									
Lumiflavin	+	+	NT	+	+	+	+	+	+
Lumichrome	+	+	+	+	+	+	+	+	+
Membrane transport inhibitors									
DIDS	−	−	NT	NT	NT	+	+	+	NT
SITS	−	NT	NT	NT	NT	NT	+	+	NT
Amiloride	+	+	NT	NT	NT	−	NT	+	NT
Furosemide	NT	−	NT	NT	NT	NT	NT	NT	NT
Probenecid	−	−	NT	NT	NT	−	+	NT	NT
Nocodasol	+	NT	NT	NT	NT	NT	NT	NT	NT
Brefeldin	+	NT	NT	NT	NT	NT	NT	NT	NT
Sulfhydryl group inhibitors									
p-CMPS	NT	+	NT	+	+	NT	+	+	NT
p-CMBS	+	NT	NT	NT	NT	NT	NT	NT	NT
Metabolic inhibitors									
Sodium azide	+	+	NT	+	NT	+	+	NT	NT
DNP	+	NT	+	NT	+	NT	+	NT	NT
Iodoacetate	NT	NT	−	+	+	NT	+	NT	NT

[a]Inhibiting effect is: '+' significant; '−' not significant; 'NT' not examined.
[b]References: 1 Huang and Swann (2000), Said and Ma (1994); 2 Said *et al.* (2000); 3 Zempleni and Mock (2000); 4 Said *et al.* (1998); 5 Patel *et al.* (2010); 6 Huang and Swann (2001); 7 Kumar *et al.* (1998); 8 Said *et al.* (2005); 9 Kansara *et al.* (2005).

Table 36.3 Protein kinases involved in the regulation of riboflavin uptake.[a,b] The table lists kinases involved in processes of phosphorylation of riboflavin after its absorption in different examined cell lines.

	Caco-2[1]	NCM460[2]	PBMC[3]	HEP G2[4]	RBE4[5]	BeWo[6]	HK-2[7]	ARPE-19[8]	Y79[9]
Protein kinase C	−	−	NT	−	−	−	−	−	NT
Protein kinase A	+	NT	NT	−	+	NT	−	−	+
Ca^{2+}/calmodulin kinase	−	+	NT	+	+	+	+	+	+
Cyclin-dependent kinases	NT	NT	NT	NT	NT	NT	NT	NT	NT
Tyrosine kinase	NT	NT	NT	−	−	NT	−	NT	NT
Protein kinase G	+	NT	NT	NT	NT	NT	NT	NT	NT

[a]Regulation is: '+' observed; '−' not observed; 'NT' not examined.
[b]References: 1 Huang and Swaan (2000), Said and Ma (1994); 2 Said et al. (2000); 3 Zempleni and Mock (2000); 4 Said et al. (1998); 5 Patel et al. (2010); 6 Huang and Swaan (2001); 7 Kumar et al. (1998); 8 Said et al. (2005); 9 Kansara et al. (2005).

Table 36.4 Extracellular and intracellular factors regulating riboflavin (RF) uptake.[a,b] The table presents modulators naturally occurring in the cells or in culturing medium, which influenced the absorption of riboflavin in various in vitro models (temperature, pH, osmolarity, concentration of RF, Na^+, Cl^- and ATP).

	Cell model system								
Modulator	Caco-2[1]	NCM460[2]	PBMC[3]	HEP G2[4]	RBE4[5]	BeWo[6]	HK-2[7]	ARPE-19[8]	Y79[9]
Temperature	+	+	+	+	+	+	+	+	+
pH	−/+	NT	NT	NT	−	−	+	+	−
Osmolarity	NT	NT	+	NT	NT	NT	NT	NT	NT
Concentration of Na^+ ions	−	−	−	−	+	−	−	−	−
Concentration of Cl^- ions	NT	NT	NT	NT	+	+	NT	NT	NT
Concentration of ATP	+	+	NT	+	+	+	+	+	+
RF deficiency	+	+	NT	+	+	+	+	+	+
RF over-supplementation	+	+	NT	NT	+	NT	+	NT	NT

[a] '+' dependence is observed; '−' not observed dependence; '−/+' dependence observed only for small RF concentrations.
[b] References: 1 Huang and Swaan (2000), Said and Ma (1994); 2 Said et al. (2000); 3 Zempleni and Mock (2000); 4 Said et al. (1998); 5 Patel et al. (2010); 6 Huang and Swaan (2001); 7 Kumar et al. (1998); 8 Said et al. (2005); 9 Kansara et al. (2005).

(Kasper 1970) and in humans by analysis of riboflavin plasma concentration after direct injection into the lumen of the mid-transverse colon (Sorrell et al. 1971).

In vitro absorption was examined with the human colonic epithelial cell line NCM460 (Said et al. 2000). The presented results are similar to those observed for Caco-2 cells; thus active transport ($K_m = 0.14$ μM) is inhibited by structural analogues (lumiflavin and lumichrome), metabolic inhibitors (*i.e.* temperature, sodium azide, DNP) and sulfhydryl group modifying *p*-chloromercuriphenyl sulfonate (*p*-CMPS). Studies have demonstrated Na^+-independency, energy-dependency and regulation by the activity of Ca^{2+}/calmodulin kinase (Tables 36.1–36.4). An adaptive regulation under conditions of deficiency and over-supplementation has also been noted.

36.3.3 Blood

Riboflavin absorbed from the intestine is concentrated in the plasma and erythrocytes of circulating blood. The concentration of this vitamin in plasma determined by Hustad et al. (2002) was 74 nM/L for FAD, 10.5 nM/L for riboflavin and 6.6 nM/L for FMN. These results were compatible with the concentrations noted in previous studies (Schrijver et al. 1991). A much higher concentration of vitamin B_2 forms (4–5 times) is present in the interior of erythrocytes. The dominant fraction of vitamin B_2 in erythrocytes is also FAD (~ 470 nM/L) and FMN (~ 45 nM/L), and only a trace amount of free riboflavin is detected (Hustad et al. 2002). After supplementation, the strongest increase in free riboflavin (83%) and FMN (87%) is observed in plasma and erythrocytes, respectively. The level of flavins in erythrocytes is less dependent on short-term dietary intake of riboflavin compared with the plasma level.

For peripheral blood mononuclear cells, which are components of the immune system (B cells, T cells and granulocytes), saturable kinetics of riboflavin uptake were proved ($K_m = 955$ nM). Absorption was reduced by the addition of, among others, lumichrome and dinitrophenol. The transport was temperature dependent and sodium independent. During proliferation of these cells, uptake of the vitamin increased four times, although the transporter affinity and their number per cells remained unchanged. This suggested that accumulation of riboflavin in the cells is associated with variable cellular volume (osmolarity of environment) (Zempleni and Mock 2000) (Tables 36.1–36.4).

The flavins in plasma may be bound to the plasma proteins: albumins and immunoglobulins. During the examination of mechanisms of riboflavin uptake it was also found that serum albumin acts as a nonspecific transporter for riboflavin (Wangensteen et al. 1996).

36.3.4 Liver cells

Storage of riboflavin in the liver comprises $\sim 30\%$ of total flavin stored in the mammalian body, mainly in FAD form (70–90%). Riboflavin uptake from the blood is preceded by phosphorylation of FMN, which is catalysed in an irreversible reaction by flavokinase involving ATP. Resultant in the reaction, FMN acts as a coenzyme or, with the participation of FAD synthetase, is reversibly converted to FAD (Figure 36.1).

In vitro uptake of riboflavin by liver cells was examined in rats (Aw *et al.* 1983) and human hepatocytes (HEP G2) (Said *et al.* 1998). Carrier-mediated mechanism of riboflavin uptake by HEP G2 cells occurred ($K_m = 0.41$ μM), inhibited by structural analogues of riboflavin as well as metabolic inhibitors and regulated by Ca^{2+}/calmodulin kinase. Details are presented in Tables 36.1–36.4.

36.3.5 Brain

For the uptake of riboflavin by brain cells, it is most important to maintain homeostasis. The concentration of riboflavin in the brain remains constant, both at times of riboflavin deficiency and after massive intravenous applications. In rat brains, more than 90% of total riboflavin is found as FAD and FMN (Ball 2004b).

Cerebrospinal fluid is constantly produced and exchanged, and therefore riboflavin is continuously supplied and crosses the blood–brain barrier. The mechanism of this process was examined *in vitro* in rabbit brain slices (Spector 1980a), rat brain endothelial cells (RBE4 cells) (Patel *et al.* 2010), and isolated rabbit choroid plexus (Spector and Boose 1979).

RBE4 cell line transports riboflavin in a saturable manner ($K_m = 44$ μM). The process is pH independent; however, in contrast to results presented above, it is highly dependent on Na^+ and Cl^- ions. The uptake was inhibited by structural analogues (lumichrome and lumiflavin) and was regulated by kinases A and Ca^{2+}/calmodulin activity (Tables 36.1–36.4).

Later absorption of riboflavin in the choroid plexus progresses against the concentration gradient if its concentration is ≥ 0.7 μM. When the concentration of riboflavin in the medium was 0.08 μM, 28% of riboflavin was intracellularly phosphorylated to FMN or FAD. The process of accumulation of riboflavin in choroid plexus is also recognized as saturable ($K_m = 78$ μM). The inhibition is caused by iodoacetate, low temperature, probenecid and other flavins but not by other vitamins and ribose.

The same phenomenon was observed after injections of riboflavin into the ventricular cerebrospinal fluid of anaesthetized rabbits: only some of the riboflavin was accumulated in the choroid plexus and most was removed (Spector 1980b). Thus, support of brain homeostasis results from transport of riboflavin both from blood into the choroid plexus and in the opposite direction.

The saturable system of crossing the cells was also observed for brain slices (cut in two different directions). The process was ATP dependent (inhibited by DNP and decreased temperature) because absorbed intracellular riboflavin is rapidly converted into FMN and in turn into FAD. Additionally, neither the absorption of FAD nor FMN is possible (Spector 1980a, 1980b).

Relevant in making the maintenance of brain homeostasis possible is the third identified human riboflavin transporter hRFT3 (GPR172A), which is expressed in the brain and salivary gland. It has the same substrate specificity as hRFT1 and hRFT2, and functional characteristics similar to those of hRFT1 (Yao et al. 2010).

36.3.6 Placenta

Placenta is the next organ where uptake of vitamins is against a concentration gradient. In addition, all forms of riboflavin are transported from maternal plasma to foetal circulation where only the concentration of free riboflavin may be higher (even four times) than in maternal plasma. The recognized riboflavin carrier protein (RCP) is synthesized in maternal liver, depending on the concentration of oestrogen. Thus, the synthesis of foetal coenzymes is dependent on maternal RCP synthesis and oestrogen level. The foetal liver also synthesizes RCP; this synthesis increases during the term of pregnancy, but at the beginning of foetal development, resorption of the foetal RCP may take place due to a lack of maternal RCP.

Riboflavin binding protein is not a typical membrane protein because it is soluble in plasma and transports riboflavin to the placental cells by a receptor and/or carrier mediated mechanism (Foraker et al. 2007).

In vitro uptake of riboflavin was studied using the BeWo placental trophoblast cell. The calculated half-maximal transport constant was 1.32 nM (Table 36.1) and inhibition was observed for the same analogues (as discussed above) (Table 36.2). Transport was also temperature and energy dependent, with a possible role of reducing equivalents. Huang and Swaan (2001) also reported the effect of chloride ions (Table 36.4). The presence of riboflavin transporter on the membrane BeWo cells is modulated by calmodulin and cellular cyclic nucleotide level (Table 36.3). This human cell line conveys riboflavin depending on a special RCP protein and internalizes this vitamin *via* receptor-mediated endocytosis (Foraker et al. 2007).

36.3.7 Kidneys

The kidneys are the organs for the elimination of riboflavin from mammalian organisms, maintaining the vitamin homeostasis. Riboflavin passes through the glomeruli and is transported by the tubules, where secretion or reabsorption takes place, depending on riboflavin plasma concentration. Finally, excretion is in the urine.

The first pharmacokinetic and pharmacodynamic experiments on mammalian uptake and mechanism were conducted on renal tubules (Jusko and Levy 1975). *In vitro* experiments on the uptake process were studied using isolated renal brush-border membrane vesicles (Yanagawa et al. 1997) and HK-2 cell line (Kumar et al. 1998).

Studies with isolated renal brush-border membrane indicate that uptake is a carrier-mediated process, resulting in both transport into intravesicular space and binding to membrane surfaces. The process was saturable with $K_m = 25.7 \pm 7.6$ μM, and dependent on physiological pH. The inhibition by probenecid and organic anion transport inhibitor—4,4-diiso-thiocyanatostilbene-2,2-disulfonic acid (DIDS), 4-acetomido-4-isothiocyanatostilbebe-2,2-disulfonic acid (SITS)—was corroborated.

Uptake of riboflavin by HK-2 cells is saturable (with $K_m = 0.67$ μM) (Table 36.1), temperature dependent and energy and Na^+ independent; it is susceptible to inhibition by structural analogues and anion transport inhibitors (Tables 36.2–36.3). Ca^{2+}/calmodulin kinase plays a major role in regulating the absorption (Table 36.3). As the cells adapted to different concentration of riboflavin with time, changes in K_m and V_{max} were observed (Kumar et al. 1998).

36.3.8 Ocular System

The involvement of riboflavin in eyesight is well recognized. It is essential for the maintenance of functions of epithelial ocular tissues and for proper development of surface structures. Riboflavin is delivered from retinal or choroidal blood. Said et al. (2005) examined the process of uptake and its regulation by retinal pigment epithelium cells, using the ARPE-19 cell line. The recognized mechanism has a similar characteristic to that presented above, i.e. the uptake is energy, temperature and pH dependent, but not Na^+ dependent. A saturable kinetic is observed (with $K_m = 80$ nm) (Table 36.1), as well as inhibition by structural analogues, anion transport inhibitors and sulfhydryl group inhibitor. The adaptation to deficiency of the vitamin involves changes in the affinity and number of carriers (Tables 36.2–36.4).

Similar results were obtained with the neural retina cell line Y-79; additionally dependence on chloride ions was observed (Kansara et al. 2005). Detailed data are presented in Tables 36.2–36.4.

36.4 Concluding Remarks

The presented mechanism of active and passive uptake and transport of riboflavin has been suggested by various research groups (Ball et al. 2004b; Foraker et al. 2003; Huang and Swaan 2000). However, the problem is still under discussion. Saturable active transport dominates the riboflavin uptake processes and is indisputably proven in every examined in vivo and in vitro systems. Passive diffusion at high concentrations also appears to be inevitable. However, it may be anticipated that some studies may also confirm involvement of a receptor-mediated mechanism and the role of riboflavin-binding protein in homeostasis maintenance in other mammalian tissues and organs.

Summary Points

- This chapter presents current knowledge on the absorption of riboflavin (mainly by humans).
- The uptake of riboflavin has been extensively studied in animal and human models, in experiments performed *in vivo*, *in vitro* and *ex vivo*.
- A carrier-mediated process was recognized as the mechanism of riboflavin transport in every proposed model.
- Additionally, passive diffusion was proven during riboflavin over-supplementation in small and large intestine, brain and renal cells.
- Receptor-mediated transport is suggested as an additional mechanism of riboflavin uptake and is recognised in both the small and large intestine and in placental cells.
- Uptake of riboflavin depends strongly on its phosphorylation and is regulated by the activity of different kinases.
- Riboflavin uptake is recognized as highly specific, temperature and energy dependent and strongly regulated by the status of the vitamin in different tissues.

Key Facts about Enterocytes

- They form a uni-layer epithelium with a specialized brush-border of microvilli, found in the small intestine and colon.
- They are columnar—their nucleus is situated in the basal part of the cells, under the Golgi apparatus.
- They are polarized—the plasma protein and lipids are distributed into three domains (apical, lateral and basolateral).
- Microvilli formation significantly extends the surface of absorption.
- The major function of these cells includes uptake of nutrients (sugars, lipids, peptides, vitamins, micro- and macroelements) and water.
- They are also responsible for maintaining enterohepatic circulation (*i.e.* the circulation of bile salts between the liver and intestine), as well as for secretion of immunoglobulins A, which protect the mucous membrane of the intestine.
- They possess an energy by glutamine metabolism.
- Some of digestive enzymes are present in the glycoprotein coat which surfaces cells.
- Caco-2 cells are the most commonly used in *in vitro* experiments as a model for enterocytes.

Definition of Words and Terms

Apical membrane: cell membrane domain characteristic for cells with polarized membrane (epithelial, endothelial cells and neurons). A typical epithelial cell has the domain faced to the lumen and often has specialized features such cilia or a brush-border of microvilli.

Basement membrane: named also basal membrane, forms a scaffold for regeneration and growth of epithelial cells.
Basolateral membrane: characteristic especially for epithelial cells, whose basal and lateral membranes remain identical both in activity and in composition.
Carrier: integrated cell membrane protein transporting small molecules, macromolecules or ions *via* the membrane.
Cell junctions: protein complexes characteristic for the epithelium which enable contact between adjoining cells or the cell and extracellular matrix.
Endothelium: cells, which form the inner layer of blood and lymphatic vessels, and face directly to circulating fluids.
Epithelium: an animal tissue covering body surfaces, internal as well external. It is responsible for the protection, secretion, selective absorption and transport of substances and the detection of sensations.
Homeostasis: an ability of organisms to maintain a stable composition of their intra- or extracellular environment.
Receptor: cell membrane or cytoplasmic protein which may attach a specific, compatible molecule (called ligand) such as peptide, hormones, toxins, *etc*. Attached ligand may increase or decrease the activity of the receptor and, in consequence, cause or trigger a specific cellular response.
Uniporter: a special type of carrier, channel or membrane protein. Transports a single substance with a solute gradient without the involvement of additional substances.

List of Abbreviations

ATP	adenosine triphosphate
DIDS	4,4′-diisothiocyanatostilbene-2,2′-disulphonic acid
DNP	2,4-dinitrophenol
FAD	adenine dinucleotide
FMN	flavin mononucleotide
K_m	half-maximal transport constant
NADH	nicotinamide adenine dinucleotide
p-CMBS	*p*-chloromercuribenzene sulfonate
p-CMPS	*p*-chloromercuriphenyl sulfonate
RCP	riboflavin carrier protein
RF	riboflavin
SITS	4-acetamido-4′-isothiocyanatostilbene-2,2′-disulphonic acid

References

Aw, T.Y., Jones, D.P., and McCornick, D.B., 1983. Uptake of riboflavin by isolated rat liver cells. *Journal of Nutrition*. 113: 1249–1254.

Ball, G.F.M., 2004a. *Background physiology and functional anatomy*. Blackwell Science, Ltd, Oxford UK, pp. 12–65.

Ball, G.F.M., 2004b. *Flavins: riboflavin, FMN and FAD (vitamin B2)*. Blackwell Science, Oxford, UK, pp. 289–300.

Chia, C.P., Addison, R., and McCormick, D.B., 1978. Absorption, metabolism, and excretion of 8-α(amino acid)riboflavins in the rat. *Journal of Nutrition*. 108: 373–381.

Daniel, H., Binninger, E., and Rehner, G., 1983. Hydrolysis of FMN and FAD by alkaline phosphatase if the intestinal brush-border membrane. *International Journal of Vitamin and Nutrition Research*. 53: 109–114.

Foraker, A.B., Khantwall, Ch. M., and Swaan, P.W., 2003. Current perspectives on the cellular uptake and trafficking of riboflavin. *Advanced Drug Delivery Reviews*. 55: 1467–1483.

Foraker, A.B., Ray, A., Claro Da Silva, T., Bareford, L.M., Hillgren, K.M., Schmittgen, T.D., and Swaan, P.W., 2007. Dynamin-2 regulates riboflavin endocytosis in human placental trophoblasts. *Molecular Pharmacology*. 72: 553–562.

Gastaldi, G., Laforenza, U., Casirola, D., Ferrari, G., Tosco, M., and Rindi, G., 1999. Energy depletion differently affects membrane transport and intracellular metabolism of riboflavin taken up by isolated rat enterocytes. *Journal of Nutrition*. 129: 406–409.

Huang, S.N., and Swaan, P.W., 2000. Involvement of receptor-mediated component in cellular translocation of riboflavin. *Journal of Pharmacology and Experimental Therapeutics*. 294: 117–125.

Huang, S.N., and Swaan, P.W., 2001. Riboflavin uptake in human trophoblast-derived BeWo cell monolayers: cellular translocation and regulatory mechanisms. *Journal of Pharmacology and Experimental Therapeutics*. 298: 264–271.

Hustad, S., McKinley, M.C., McNulty, H., Schneede, J., Strain, J.J., Scott, J.M., and Ueland, P.M., 2002. Riboflavin, flavin mononucleotide, and flavin adenine dinucleotide in human plasma and erythrocytes at baseline and after low-dose riboflavin supplementation. *Clinical Chemistry*. 48: 1571–1577.

Iinuma, S., 1955. Synthesis of riboflavin by intestinal bacteria. *Journal of Vitaminology*. 1: 6–13.

Jusko, W.J., and Levy, G., 1975. Absorption, protein binding and elimination of riboflavin. In: Tivilin, R. (ed.) *Riboflavin*. Plenum, New York, USA, pp. 99–152.

Kansara, V., Pal, D., Jain, R., and Mitra, A.K., 2005. Identification and functional characterization of riboflavin transport in human-derived retinoblastoma cell line(Y-79): mechanism of cellular uptake and translocation. *Journal of Ocular Pharmacology and Therapeutics*. 21: 275–287.

Kasper, H., 1970. Vitamin absorption in the colon. *American Journal of Proctology*. 21: 341–345.

Kumar, C.K., Yanagawa, N., Ortiz, A., and Said, H.M., 1998. Mechanism and regulation of riboflavin uptake by human renal proximal tubule epithelial cell line HK-2. *American Journal of Physiology*. 274: F104–F110.

Middleton, H.M. 3rd, 1990. Uptake of riboflavin by rat intestinal mucosa *in vitro*. *Journal of Nutrition*. 120: 588–593.

Patel, M.R., Mandava, N., Pal, D., and Mitra, A.K., 2010. Identification and functional characterization of a carrier mediated transport system for riboflavin on rat brain endothelial cells. In: *Pharmaceutical Sciences Word Congress Abstracts*, AAPS Journal M1209.

Said, H.M., and Mohammadkhani, R., 1993. Uptake of riboflavin across brush-border membrane of rat intestine: regulation by dietary levels. *Gastroenterology*. 105: 1294–1298.

Said, H.M., Mohammadkhani, R., and McCloud, E., 1993. Mechanism of transport riboflavin in rabbit intestinal brush border membrane vesicles. *Proceedings of the Society for Experimental Biology and Medicine*. 202: 428–434.

Said, H.M., and Ma, T.Y., 1994. Mechanism of riboflavin uptake by Caco-2 human intestinal epithelial cells. *American Journal of Physiology*. 266: G15–21.

Said, H.M., Ma, T.Y., and Grant, K., 1994. Regulation of riboflavin intestinal uptake by protein kinase A: studies with Caco-2 cells. *American Journal of Physiology*. 267: G955–959.

Said, H.M., Otriz, A., Ma, T.Y., and McCloud, E., 1998. Riboflavin uptake by human-derived liver cells Hep G2: mechanism and regulation. *Journal of Cell Physiology*. 176: 588–594.

Said, H.M., Ortiz, M.P., Moyer, N., and Yanagawa, N., 2000. Riboflavin uptake by human – derived colonic epithelial NCM460 cells. American Journal of Physiology. *Cell Physiology*. 278: C270–C276.

Said, H.M., Wang, S., and Ma, T.Y., 2005. Mechanism of riboflavin uptake by cultured human retinal pigment epithelial ARPE-19 cells: possible regulation by an intracellular Ca^{2+}-calmodulin-mediated pathway. *Journal of Physiology*. 566: 369–77.

Sanchez-Pico, A., Peris-Ribera, J.-E., Toledano, C., Torres-Molina, F., Casabo, V.-G., Martin-Villodre, A., and Pla-Delphina, J.M., 1989. Non-linear absorption kinetics of cefadroxil in the rat. *Journal of Pharmacy and Pharmacology*. 41: 179–185.

Schrijver, J., Speek, A.J., and van den Berg, H., 1991. Flavin adenine dinucleotide in whole blood by HPLC. In: Fidanza, F. (ed.) *Nutritional Status Assessment: A Manual for Population Studies*. Chapman & Hall, London, UK, pp. 251–256.

Sorrell, M.F., Frank, O., Thomson, A.D., Aquino, H., and Baker, H., 1971. Absorption of vitamins from the large intestine *in vivo*. *Nutrition Report International*. 3: 143–148.

Spector, R., and Boose, B., 1979. Active transport of riboflavin by the isolate choroid plexus *in vitro*. *The Journal of Biological Chemistry*. 254: 10286–10289.

Spector, R., 1980a. Riboflavin accumulation by rabbit brain slices *in vitro*. *Journal of Neurochemistry*. 34: 1768–1771.

Spector, R., 1980b. Riboflavin homeostasis in the central nervous system. *Journal of Neurochemistry*. 35: 202–209.

Spector, R., 1982. Riboflavin transport by rabbit kidney slices: characterization and relation to cyclic organic acid transport. *Journal of Pharmacology and Experimental Therapeutics*. 221: 394–398.

Wangensteen, O.D., Bartlett, M.M., James, J.K., Yang, Z.F., and Low, P.S., 1996. Low riboflavin–enhanced transport of serum albumin across the distal pulmonary epithelium. *Pharmaceutical Research.* 13: 1861–1864.

Yanagawa, N., Jo, O.D., and Said, H.M., 1997. Riboflavin transport by rabbit renal brush border membrane vesicles. *Biochimica et Biophysica Acta.* 1330: 172–178.

Yamamoto, S., Inoue, K., Ohta, K.-Y., Fukatsu, R., Maeda, J.-Y., Yoshida, Y., and Yuasa, H., 2009. Identification and functional characterization of rat riboflavin transporter 2. *The Journal of Biochemistry.* 145: 437–443.

Yao, Y., Yonezawa, A., Yoshimatsu, H., Masuda, S., Katsura, T., and Inui, K., 2010. Identification and comparative functional characterization of a new human riboflavin transporter hRFT3 expressed in the brain. *Journal of Nutrition.* 140: 1220–1226.

Yonezawa, A., Masuda, S., Katsura, T., and Inui, K.-I., 2008. Identification and functional characterization of a novel human and rat riboflavin transporter, RFT1. American Journal of Physiology. *Cell Physiology.* 295: C632–C641.

Zempleni, J., and Mock, D.M., 2000. Proliferation of peripheral blood mononuclear cells increases riboflavin influx. *Proceedings of the Society for Experimental Biology and Medicine.* 225: 72–79.

Zielińska-Dawidziak, M., Grajek, K., Olejnik, A., Czaczyk, K., and Grajek, W., 2008. Transport of high concentration of thiamin, riboflavine and pyridoxine across intestinal epithelial cells Caco-2. *Journal of Nutritional Science and Vitaminology.* 54: 423–429.

CHAPTER 37
Riboflavin and β-oxidation Flavoenzymes

BÁRBARA J. HENRIQUES,[†] JOÃO V. RODRIGUES[†] AND CLÁUDIO M. GOMES*

Instituto Tecnologia Química e Biológica, Universidade Nova de Lisboa, Av República EAN, 2785-572 Oeiras, Portugal
*Email: gomes@itqb.unl.pt, URL: http://www.itqb.unl.pt/pbfs

37.1 Riboflavin Metabolism and Chemistry

Riboflavin, commonly known as vitamin B_2, is metabolized inside cells to flavin mononucleotide (FMN) and flavin adenine dinucleotide (FAD), two very important enzyme cofactors. These molecules possess rather unique and versatile chemical properties, which confer on them the ability to be among the most important redox cofactors found in a broad range of enzymes. In this chapter we provide a brief description of riboflavin metabolism and chemistry, overview the different flavoenzymes engaged in fatty acid β-oxidation and their respective roles. We also highlight recent studies shedding light on the cellular processes and biological effects of riboflavin supplementation in the context of metabolic disease.

[†] Equally contributing authors.

Food and Nutritional Components in Focus No. 4
B Vitamins and Folate: Chemistry, Analysis, Function and Effects
Edited by Victor R. Preedy
© The Royal Society of Chemistry 2013
Published by the Royal Society of Chemistry, www.rsc.org

37.1.1 Riboflavin Metabolism

Riboflavin needs to be present in the human typical diet, as animals, unlike many plants, fungi and bacteria, are unable to synthesize this molecule. Dietary intake of this vitamin includes free riboflavin and also its protein bound form, as FAD and FMN in flavoproteins (Figure 37.1A). In the latter case, flavins need to be first released from carrier proteins during digestion and then hydrolysed to riboflavin by alkaline phosphatases and FMN/FAD pyrophosphatase in order to be absorbed at the small intestine.

Apart from dietary intake, riboflavin is also obtained from endogenous synthesis by microflora in the large intestine and is subsequently absorbed. Inside the cell, FMN is formed from vitamin B2 *via* adenosine triphosphate (ATP) phosphorylation and a flavokinase. FMN can be subsequently converted to FAD through a FAD synthetase also in the presence of ATP (Figure 37.1B).

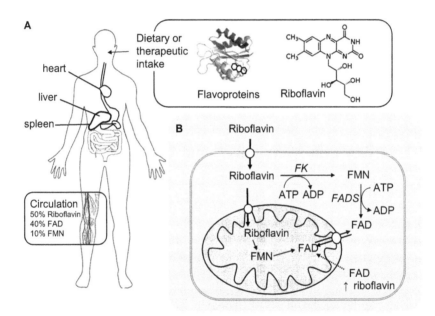

Figure 37.1 Riboflavin metabolism and cellular processing pathways. (A) Riboflavin and flavin intake is made *via* the diet, either in riboflavin-rich aliments or flavoproteins. In the latter, digestion in the stomach releases FAD and FMN cofactors. Riboflavin and flavins achieve a high concentration in the liver, spleen and cardiac muscle; a concentration of about 30 nM riboflavin is also reached in the plasma circulation. (B) Riboflavin is imported into the cell and into the mitochondria *via* specific transporters (white circles in membranes). In the cytoplasm, flavin kinase (FK) and FAD synthetase (FADS) consecutively convert riboflavin into FMN and FAD, at the expense of ATP. An identical mechanism is also thought to be present inside the mitochondria, although a mitochondrial FK remains to be identified. FAD can also be imported into the mitochondria, or diffuse passively when the riboflavin concentrations are high. Figures reprinted from Henriques *et al.* (2010), with permission.

At this stage the two cofactors are available to bind to apo-proteins on the cytosol, or to be transported inside mitochondria where they will be incorporated in the organelle flavoenzymes. It has also been demonstrated that, in rat liver cells, FAD synthesis can occur inside mitochondria, where a riboflavin kinase and FAD synthetase enzymes can be found (Barile *et al.* 1993, 2000). This mitochondrial function is important as a large number of flavoenzymes are located inside this organelle and FAD binding only occurs after protein import.

37.1.2 Flavin Chemistry and Flavoproteins

Flavins are redox active protein cofactors that participate in a broad range of reactions, including oxidation, reduction and dehydrogenation, since they are able to carry out both one- or two-electron and proton transfer reactions. The versatile reactivity includes two-electron dehydrogenation of several substrates, aromatic hydroxylations, activation of molecular oxygen, emission of biologically produced light, signal transduction in programmed cell death, and regulation of biological circadian clocks (Edmondson and Ghisla 1999). It is known that 1–3% of the genes in bacteria and eukaryotic genomes encode for flavoproteins; in most of these proteins, flavins are bound non-covalently (De Colibus and Mattevi 2006).

Flavin cofactors are composed by a catalytic moiety, the isoalloxazine ring system, and a ribityl side chain, a $5'$-terminal phosphate ester in FMN, or a pyrophosphate linkage of FMN with an adenosine monophosphate (AMP) moiety in FAD. The amphipathic structure of the isoalloxazine ring system allows its interaction with a protein either by establishing hydrophobic contacts *via* its xylene moiety, or through the formation of hydrogen bonds engaging the pyrimidine ring (Ghisla and Massey 1986).

The functional diversity of flavoproteins results from the broad range of redox potentials that are accessible to the flavin cofactors, as well as their ability to switch between one or two electron redox chemistry. In solution, flavins are found in equilibrium between the oxidized, reduced and the semi-quinone radical forms, and have a redox potential of about $-210\,\mathrm{mV}$ (*versus* the normal hydrogen electrode) at neutral pH. However, in the protein-bound form, the redox equilibrium can be shifted and the redox potential may span up to 600 mV (Massey 2000). This arises from the fact that flavin–protein interactions may engage a number of non-covalent interactions such as π-stacking, hydrophobic effects, hydrogen bonding and electrostatic interactions, which will ultimately determine the flavin redox potential.

37.2 Mitochondrial β-oxidation Flavoenzymes

Fatty acids are carboxylic acids with straight or branched hydrocarbon chains ranging from four to 28 carbons. These biomolecules are major constituents of biological membranes, as phospholipid esters, and constitute an important source of energy for living organisms, since their oxidation is a highly exergonic process. In addition, fatty acids also play other vital cellular roles as enzyme cofactors, hormones and intracellular messengers. The cell can metabolize fatty acids by three different pathways in the mitochondria or in the peroxisomes. In mammals, mitochondrial β-oxidation of fatty acid provides a major source of ATP for the heart and skeletal muscle. In the liver, kidney, small intestine and also white adipose tissue, β-oxidation provides the formation of ketone bodies used as an energy source by other tissues including the brain. Fatty acid oxidation is particularly important during fasting, sustained exercise, stress, and the neonatal-suckling period, when glucose supplies become limited.

In this section we provide a short overview of the mitochondrial β-oxidation process and focus more specifically on the several flavoenzymes that participate in the pathway, as their adequate function and folding strongly relies on riboflavin metabolism to assure flavin biosynthesis.

37.2.1 Overview of Mitochondrial β-oxidation

To be recruited for β-oxidation, fatty acids are first activated in the cytosol by an ATP-dependent acylation forming an acyl-CoA, and are then transported across the inner mitochondrial membrane as carnitine-derivatives in a process which is mediated by three proteins: carnitine palmitoyl transferase I (CPT I), acyl-carnitine translocase (CAT) and carnitine palmitoyl transferase II (CPT II). Once in the mitochondrial matrix, the acyl-CoA fatty acids undergo dehydrogenation at carbon β (C3) to form the corresponding trans-2-enoyl-CoA. This first reaction is the rate-determining step of β-oxidation and is catalysed by a group of enzymes named acyl-CoA dehydrogenases (ACAD). ACAD constitute quite a large family of flavin-containing enzymes, also comprising other dehydrogenases that participate in amino acid metabolism. These proteins share similar structural properties and operate by comparable mechanisms; nevertheless, each member has different substrate specificities.

For β-oxidation dehydrogenases, the chain-length of the substrate is the major factor governing specificity and individual enzymes have thus been classified accordingly as very long-, long-, medium-, and short-chain acyl-CoA dehydrogenases. Most of these enzymes are located in the matrix, with a few exceptions that are associated with the inner mitochondrial membrane (see below). At this stage, electrons derived from the dehydrogenation reaction are fuelled into the respiratory chain through two key enzymes which act as a hub: electron-transfer flavoprotein (ETF) and electron-transfer flavoprotein: ubiquinone oxidoreductase (ETF:QO). They function as sequential electron carriers: ETF accepts electrons from all ACAD and ETF:QO transfer electrons to ubiquinone, using ETF as substrate. Reduced ubiquinone (ubiquinol) resumes respiration at the level of complex III (ubiquinone:cytochrome c oxidoreductase) (Figure 37.2).

The subsequent steps of β-oxidation are catalysed by enoyl-CoA hydratase, 3-L-hydroxyacyl-CoA dehydrogenase and β-ketoacyl-CoA thiolase, and lead to

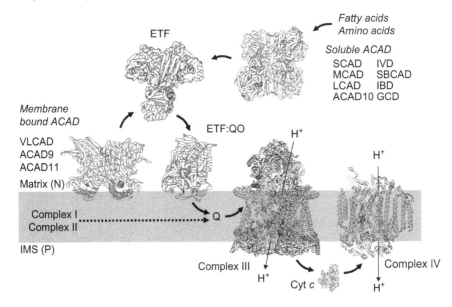

Figure 37.2 Cartoon depicting enzymes participating in mitochondrial β-oxidation and part of the respiratory chain. Acyl-CoA substrates derived from fatty acid and amino acid metabolism are oxidized by several flavin-containing acyl-CoA dehydrogenases (ACAD). Electrons obtained from this reaction are shuttled to the respiratory chain *via* the ETF/ETF:QO hub (electron-transfer flavoprotein and electron-transfer flavoprotein: ubiquinone oxidoreductase). ETF:QO is able to transfer electrons to ubiquinone (Q) (such as respiratory complexes I and II) whose subsequent transfer down to complex IV will result in energy conservation and ATP production. See list of abbreviations for definitions.

the formation of acetyl-CoA and two-carbon shortened fatty acyl-CoA. Unsaturated fatty acids undergo the same reaction until the *cis*-configuration double-bond prevents the formation of a substrate for the acyl-CoA dehydrogenase and enoyl-CoA hydratase. At this point additional enzymes are required such as 3,2 trans-enoyl-CoA isomerase and 2,4-dienoyl CoA reductase 1. Fatty acids with an odd number of carbons are degraded in the same way, but the final product has three carbons, propionyl-CoA, which is converted to succinyl-CoA to enter the citric acid cycle. Overall, each round of β-oxidation results in the formation of acetyl-CoA, acyl-CoA which is two-carbons shorter, NADH and two reducing equivalents, carried by ETF/ETF:QO at the level of reduced flavin ($FADH_2$).

37.2.2 The Flavoprotein Enzymatic Machinery

As outlined above, the mitochondrial fatty acid β-oxidation machinery relies on a variety of enzymes, most of which are strictly dependent on the incorporation of FAD as cofactor for proper functioning. Dietary riboflavin deficiency, or impaired metabolic pathways for the biosynthesis of FAD, is thus expected to have, *a priori*, a major impact on the metabolism of lipids due to deficient degradation of fatty acids *via* β-oxidation. The following sections overview the properties of the different specific flavoenzymes that participate in this mitochondrial β-oxidation machinery.

37.2.2.1 Very long-chain Acyl-CoA Dehydrogenase (VLCAD)

An enzyme specific for the oxidation of long-chain fatty acids was identified in rat liver in the 1990s, and one year later, human VLCAD deficiency (VLCADD) (MIM 201475) was reported. The *ACADVL* gene was mapped to chromosome 17p11.2-11.13. VLCAD catalyses the initial rate-limiting step of the β-oxidation of long-chain saturated fatty acids with a chain length of 14–18 carbons, being responsible for most of palmitoyl-CoA (C16) dehydrogenation in liver, heart, skeletal muscle and skin fibroblasts.

VLCAD is a homodimer associated with the matrix side of the inner mitochondrial membrane (Figure 37.2), as revealed by the crystal structure of this enzyme in complex with substrate myristoyl-CoA (C14-CoA) (McAndrew *et al.* 2008). The catalytic domain, which consists of the first ∼400 residues, has an overall fold similar to other soluble ACAD and harbours a FAD cofactor. The enzymatic mechanisms are also identical—illustrated in Figure 37.3 for medium chain acyl-CoA dehydrogenase (MCAD) (see Section 37.2.2.3). In the active site, a glutamate (Glu-422 in VLCAD, Glu-376 in MCAD) acts as a catalytic base, ideally orientated to abstract a proton from the substrate at C2 position. In addition, the isoalloxazine ring of the flavin moiety is also directly positioned to accept a hydride ion from the substrate (at C3 position). Part of the chain-length specificity of VLCAD may be explained by a significantly longer cavity that accommodates substrate binding (∼24 Å) when comparing with MCAD (∼12 Å) and SCAD (∼8 Å) (McAndrew *et al.* 2008).

In addition to the catalytic domain, VLCAD has an extension of 180 residues at the *C*-terminal that is required for proper membrane binding, especially through interactions *via* a putative amphipathic helix (residues 441–476). Association of VLCAD to the membrane is drastically impaired by mutations in that specific region, as shown by studies on two clinical mutants (A450P and L462P), which nevertheless did not affect the overall protein fold or enzymatic activity (Goetzman *et al.* 2007).

37.2.2.2 Long-chain Acyl-CoA Dehydrogenase (LCAD)

Long-chain acyl-CoA dehydrogenase is a soluble mitochondrial protein that was thought to be essential in humans for the mitochondrial β-oxidation of long chain fatty acids. In mouse, LCAD is highly expressed in most tissues and has a broad activity with acyl-CoA substrates of 6–20 carbons, showing higher activity with lauroyl-CoA (C12-CoA). Mouse mutation models of LCAD deficiency mimic phenotypes of human VLCAD and MCAD deficiency. In humans, however, there are currently no reported cases of human LCAD deficiency. The presence of this enzyme is undetectable in skeletal muscle, where β-oxidation

Riboflavin and β-oxidation Flavoenzymes 617

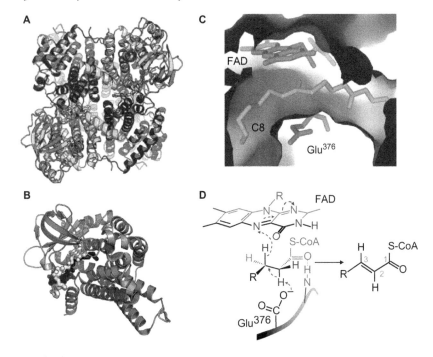

Figure 37.3 Structure of medium-chain acyl-CoA dehydrogenase (MCAD), representative of acyl-CoA dehydrogenases, and general mechanism of α,β-dehydrogenation. (A) Tetrameric structure of medium-chain acyl-CoA dehydrogenase (MCAD). (B) Structure of MCAD monomer showing the FAD cofactor (white) and octanoyl-CoA (black). (C) Surface contour of MCAD active site highlighting substrate binding cavity accommodating an octanoyl-CoA (C8) molecule, and the positioning of the flavin cofactor (top) and the catalytic base -Glu^{376} (bottom). (D) Dehydrogenation mechanism: the hydrogen at carbon 2 (alpha) in the substrate is abstracted as a H^+ by the carboxylate group of an active site glutamate residue, which functions as catalytic base. In a concerted process, the hydrogen at carbon 3 (β) is transferred as an hydride to the N(5) position of the flavin. This results in the formation of a trans-Δ2-enoyl-CoA product, and a reduced flavin cofactor (FADH$_2$). MCAD structure was obtained from Protein Data Bank (PDB: 1udy).

plays an important role. Thus, it is unlikely that function of LCAD could be linked with energy generation through the oxidation of long fatty acids. On the other hand, LCAD was the only one of the long-chain ACAD immunodetected in the lung, possibly having some function in surfactant metabolism (He *et al.* 2007). Its function in humans thus remains unclear.

37.2.2.3 *Medium-chain Acyl-CoA Dehydrogenase (MCAD)*

MCAD is responsible for catalysing the first step of β-oxidation of medium chain fatty acids. Although the maximum catalytic activity is attained with

C6-C8 substrates, MCAD has broad chain-length specificity, exhibiting somewhat lower but still significant activity with fatty acids with chain-lengths ranging from four to 14 carbons. The *ACADM* gene was allocated to chromosome 1p31. Like most of the mitochondrial proteins, MCAD is nuclear encoded and synthesized in the cytosol as a precursor protein of 47 kDa; after import into mitochondria the N-terminal leader peptide is cleaved producing a mature protein of 44 kDa.

Native MCAD is formed by the assembly of four monomers and each monomer is folded into three domains (Kim *et al.* 1993) (Figure 37.3). The N- and C-terminal domains are mainly α-helices packed together and the middle domain is formed by two β-sheets. The flavin cofactor is positioned in the cleft between two adjacent monomers in a way that is deeply buried in the tetramer configuration. Interestingly, the presence of FAD is required for chaperone-assisted MCAD folding into the final tetramer assembly, after mitochondrial import, possibly exerting a nucleating effect (Saijo and Tanaka 1995). Several of the disease-causing missense mutations are believed to influence the folding of the protein inside the mitochondria. Deficiency on MCAD is by far the most frequently detected inheritable defect in β-oxidation.

37.2.2.4 Short-chain Acyl-CoA Dehydrogenase (SCAD)

The *SCAD* gene, located on chromosome 12q22, encodes a cytosolic precursor SCAD that is translocated to mitochondria where the *N*-terminal mitochondrial targeting peptide is proteolytic cleaved. SCAD shows activity mainly with butyryl- and hexanoyl-CoA substrates, and has been named in the past butyryl-CoA dehydrogenase in respect to its preferred substrate. The crystal structure of human SCAD reveals an overall fold very similar with the other ACAD. A special feature of this acyl-CoA dehydrogenase is the presence of a glutamine in position 254 and threonine in position 364 that seem to shorten the substrate binding pocket, contributing to its substrate specificity (Kim *et al.* 1993).

37.2.2.5 Emerging Acyl-coA Dehydrogenases: ACAD9, ACAD10 and ACAD11

Acyl-CoA dehydrogenases 9, 10 and 11 have been classified based on their high sequence similarity with other ACAD, but only a few recent studies about their function have been reported. ACAD9 shares high similarity with VLCAD and is ubiquitously expressed, with particularly high expression in heart, skeletal muscle, kidney, liver and brain (especially in the granular layer) (He *et al.* 2011). Like VLCAD, ACAD9 is associated with the inner mitochondrial membrane, facing the matrix (Ensenauer *et al.* 2005). ACAD9 has been proposed to play a role in the β-oxidation of unsaturated long-chain fatty acids (C16:1-, C18:1-, C18:2-, C22:6-CoA substrates), although it also shows lower activity towards saturated long-chain substrates (Ensenauer *et al.* 2005). More recently, ACAD9 was shown to be required for the biogenesis of complex I (Nouws *et al.* 2010) and mutations in this gene were identified as a cause of complex I deficiency (Haack *et al.* 2010). Clearly, the function of ACAD9 and its role in β-oxidation is far from being completely understood.

The functions of ACAD10 and 11 also remain elusive. Multiple transcripts are made from the genes encoding for ACAD10 and 11, and some forms of alternative splicing are even absent of exons coding for catalytic domains (He *et al.* 2011). ACAD10 has only limited activity with long branched-chain acyl-CoA substrates, whereas ACAD11 utilizes unsaturated substrates with 20–26 carbons. ACAD10 and 11 are mostly expressed in the foetal and adult brain, respectively. It has been suggested that ACAD10 and 11 may serve novel physiological functions in the central nervous system (He *et al.* 2011).

37.2.2.6 Dehydrogenases from Amino Acid Metabolism

Although not being part of fatty acid β-oxidation, it is worth mentioning here the ACAD that participate in amino acid catabolism, *i.e.* isovaleryl-CoA dehydrogenase (IVD) for leucine, short/branched-chain acyl-CoA dehydrogenase (SBCAD) for isoleucine, isobutyryl-CoA dehydrogenase (IBD) for valine and glutaryl-CoA dehydrogenase (GCD) for lysine and tryptophan. These proteins are structurally and mechanistically similar to the dehydrogenases of β-oxidation, and all enzymes have the same electron acceptor and electron-transfer flavoprotein (ETF, see Section 37.2.2.7), so both metabolic pathways are linked at this point.

37.2.2.7 The ETF/ETF:QO Hub

As mentioned above, electrons derived from the dehydrogenation of acyl-CoA substrates are shuttled to the respiratory chain *via* the ETF/ETF:QO hub. Electron-transfer flavoprotein (ETF) accepts electron from all members of ACAD family, constituting a converging point of two distinct metabolic pathways, fatty acid β-oxidation and amino acid degradation. The protein is a heterodimer and the genes encoding the human α and β ETF subunits were mapped to chromosomes 15q23-q25 and 19q13.3, respectively. Both ETF subunits are nuclear encoded; however, the alpha-subunit is synthesized as a precursor protein of 35 kDa, while the β-subunit is synthesized in the cytosol in a form that is indistinguishable from the mitochondrial form. The alpha-subunit precursor sequence is cleaved after import into the mitochondria yielding a mature form with 32 kDa. The completely assembled functional dimer harbours one FAD plus one AMP cofactor (Figure 37.4A). Although the latter does not influence the activity of ETF, it is important for the assembly of the protein.

The crystal structure of human ETF was solved to 2.1 Å resolution, revealing that ETF consists of three distinct domains: domain I is composed of the *N*-terminal portion of alpha-subunit; domain II consists of the *C*-terminal portion of α-subunit and a small *C*-terminal portion of the β-subunit, and domain III is

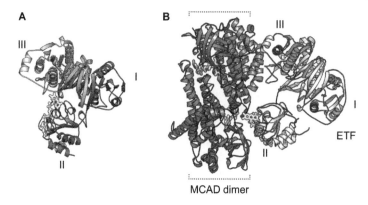

Figure 37.4 Three-dimensional structure of electron-transfer flavoprotein (ETF) alone and in complex with medium-chain acyl-CoA dehydrogenase (MCAD). (A) Human ETF is a dimer of two distinct polypeptide chains, and harbours FAD and AMP cofactors (white sticks). Structure is divided in three sub-domains that are shown in roman numerals. (B) Crystallographic structure of ETF:MCAD complex. ETF domain III is responsible for establishing protein–protein specific interactions. ETF domain II undergoes a dramatic conformational change upon complex formation (compare flavin position in panel A) in order to allow effective electron transfer to the flavin of MCAD. Structures of ETF and ETF:MCAD complexes were obtained from Protein Data Bank (PDB: 1efv and 2A1T, respectively).

made up from the majority of the β-subunit (Roberts et al. 1996) (Figure 37.4A). The AMP cofactor is buried deeply within domain III, making mostly backbone interactions. The FAD cofactor is bound to domain II, positioned in the cleft between the two subunits, and is highly exposed to the solvent.

From the structural/functional point of view, ETF is an interesting enzyme, as it has to interact with several ACAD, plus with ETF:QO (Figures 37.2 and 37.4B). This requires tight protein–protein recognition interactions to ensure specificity. On the other hand, it must also be able to establish more versatile contacts to accommodate structural variations among different partner enzymes. The molecular basis for this behaviour were partially explained upon solving the crystal structure of ETF in complex with MCAD (Toogood et al. 2004) (Figure 37.4B). An anchor region in domain III of ETF, the so called 'recognition loop', which establishes specific interactions with a hydrophobic patch of MCAD, has been identified. Also, domain II which harbours FAD, was found to be highly flexible and capable of sampling different structural conformations until inter-protein electron transfer from the ACAD is allowed.

ETF:QO is the redox partner of ETF in this enzymatic hub. This enzyme will oxidize reduced ETF, mediating electron transfer to the membrane-bound ubiquinone. Thus ETF:QO establishes the link between several mitochondrial oxidative processes taking place in the matrix and the membrane-bound respiratory chain (Figure 37.2). ETF:QO is a monomeric protein of 66 kDa containing a [4Fe–4S] cluster and a FAD cofactor, and is associated with the

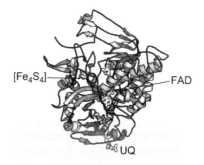

Figure 37.5 Crystallographic structure of pig electron-transfer flavoprotein: ubiquinone oxidoreductase (ETF:QO). The flavin and iron–sulfur cluster cofactors, as well as the ubiquinone substrate are shown in white sticks. An amphipatic region of ETF:QO establishes interactions with membrane and accommodates the ubiquinone substrate.

matrix side of the inner mitochondrial membrane. The gene coding for ETF:QO has been mapped to chromosome 4q32-q35. The crystal structure has revealed that the iron–sulfur cluster is closer to the protein surface while FAD molecule is closer to the ubiquinone; therefore it was postulated that the redox cluster is responsible for accepting electrons from ETF and the flavin cofactor for reduction ubiquinone (Zhang et al. 2006) (Figure 37.5). Two highly hydrophobic peptide segments F114–L131 (β-hairpin) and G427–W451 (α-helix) compose a hydrophobic plateau that is believed to establish interactions with the membrane and, in addition, form the entrance of the ubiquinone-binding pocket.

37.3 Riboflavin Effects in Defective β-oxidation Flavoenzymes

Genetic lesions in the genes encoding for the proteins involved in the mitochondrial β-oxidation machinery result in various human diseases. These inborn errors of fatty acid oxidation, most of which result from missense mutations, arise from functional deficiency as a consequence of decreased biological activity of the affected proteins, either because of mutations affecting the active sites or because of defective protein folding. Dominant negative effects resulting from the sequestering of erroneous conformations by molecular chaperones may also play a role. The clinical phenotypes of fatty acid oxidation disorders are variable, ranging from mild to severe forms, and arise from disease of the affected tissues. The pathology becomes particularly significant under metabolic decompensation resulting from fasting or under febrile illness, especially in infants. Defects in the flavoenzymes addressed in the section above culminate in the accumulation of intermediate metabolites, namely acyl-carnitines, whose detection in blood spot samples from newborns constitutes one of the biochemical hallmarks of these pathological conditions (for a review see Bennett et al. 2000).

One particular case of such pathological conditions is multiple acyl-CoA dehydrogenation deficiency (MADD) or glutaric aciduria type-II, a rare disease arising from defects in either ETF or ETF:QO. In this condition, vitamin B_2 supplementation has in some cases been successfully employed for the treatment of this fatty acid oxidation (FAO) disorder. However, the molecular rational for the beneficial effect of riboflavin supplementation is not fully clearly defined and the reason why only some MADD patients are responsive to riboflavin whereas others are not remains as one of the key challenging issues in the field (Gregersen et al. 1990; Olsen et al. 2007).

In this section we provide an overview of recent approaches that have contributed to clarify riboflavin effects in MADD, one focusing on global proteomic responses upon riboflavin supplementation and the other detailing the molecular rationale for such effects in respect to consequences on the structure, function and folding of ETF.

37.3.1 Proteomics Responses to Riboflavin Supplementation

Gianazza and co-workers have carried a series of elegant proteomic studies aimed at establishing correlations between flavin metabolism and mitochondrial flavoenzyme dysfunction (Gianazza et al. 2006). A detailed investigation was carried out on muscle mitochondria from a patient with profound muscle weakness associated with MADD. The patient received riboflavin supplementation treatment (200 mg/day) in combination with carnitine treatment (2 g/day) which resulted in a substantial improvement, as assessed by biochemical parameters. Prior the therapeutic riboflavin supplementation, the activity of different fatty acid β-oxidation enzymes, respiratory complexes, the ratio between acyl/free carnitine and the levels of intracellular lipids were altered in respect to controls. These data led the authors to evaluate the FAD and FMN concentrations in whole muscle, and the results evidenced a lower amount of available FAD; upon riboflavin therapy the flavin levels were restored to, at least, control levels.

In order to gain a better understanding of the modifications taking place on the protein levels upon supplementation, a series of experiments combining two-dimensional polyacrylamide gel electrophoresis (2D-PAGE) and mass spectrometry methods were designed so as to study the mitochondrial subproteome of muscular tissue before and after riboflavin therapy. The results obtained showed that, under untreated disease conditions, several proteins were downregulated such as the 75 kDa Fe–S subunit of NADH:quinone oxidoreductase, ETF:QO, MCAD, the β subunit of the trifunctional enzyme, 3-hydroxy-isobutyryl-CoA hydrolase, the E2 component of the branched-chain α-ketoacid dehydrogenase complex and the E2 component of pyruvate dehydrogenase complex, among others (Gianazza et al. 2006). Interestingly, several of the affected enzymes were flavoproteins suggesting a dysfunction of flavin metabolism in MADD, in agreement with the reduced amount of FAD in muscle. After riboflavin treatment both protein levels and biochemical parameters reverted to normal levels. Compliance with the riboflavin therapy maintains the improved clinical condition, although this decays after some time upon suspension of the treatment. On a fundamental level, this study showed that MADD results in a functional depletion of several flavoenzymes, which are not restricted to proteins involved in β-oxidation. Also, it puts forward the hypothesis that riboflavin and its derived cofactors could play an important role as transcriptional or translational regulatory factors.

37.3.2 Molecular Basis for Effects of ETF Flavinylation

Riboflavin supplementation affords increased cellular levels of flavins (~ 2.5 fold) but the molecular mechanisms through which the increased availability of the cofactors impact on β-oxidation flavoenzymes only now begins to be more accurately understood. In the early 1990s Nagao and Tanaka (1992) used an *in vitro* system for translation/import into isolated mitochondria of several acyl-CoA dehydrogenases and ETF, and studied the stability of apo and holoenzymes before and after mitochondrial import. The results showed that while mRNA levels of some ACAD were increased during riboflavin depletion, the stability of the precursor proteins was not affected although low mitochondrial levels of riboflavin/FAD resulted in decreased proteolytic stability of the mature acyl-CoA dehydrogenase. These findings led the authors to propose that FAD would bind to the proteins inside mitochondria and that binding of the cofactor decreased the conformational flexibility resulting in higher stability (Nagao and Tanaka 1992).

Subsequent experiments by the Tanaka laboratory using MCAD as model showed that FAD also plays a very important role during the folding process, before the assembly of the subunits into the functional tetramer (Saijo and Tanaka 1995). MCAD depends on the assistance of GroEL/GroES chaperonin system for its folding (Bross *et al.* 1995). Beside these studies focusing in FAD insertion in the acyl-CoA dehydrogenases, Sato and co-workers have studied *in vitro* the assembly of ETF focusing on the roles of FAD and AMP on the folding and dimerization process (Sato *et al.* 1996).

More recently, Henriques and co-workers used an ETF variant associated with MADD, comprising a point mutation at the β-subunit (ETFβ-Asp128Asn), as a model to address the effects of flavinylation resulting from therapeutic vitamin B_2 supplementation, also under heat stress conditions mimicking the febrile conditions that are known to result in metabolic decompensation in patients. In this work it was shown that ETF variant deflavinylates three-fold faster than the wild-type protein during mild heat stress (39 °C) with concurrent loss of activity (Henriques *et al.* 2009). This is in agreement with the fact that a patient with this mutation developed more prominent disease symptoms in connection with a viral infection and fever. Experiments carried out in the presence of a 2.5-fold excess of FAD in respect to ETF, corresponding to the relative increase observed in muscle mitochondria in patients undergoing riboflavin therapy (Gianazza et al. 2006), have shown that flavinylation improves the conformational and proteolytic stability of the protein, also retaining its biological activity (Henriques et al. 2009) (Figure 37.6).

Moreover, flavinylation prevents activity decline and loss of tertiary contacts during heat stress. A noteworthy observation is the fact that ETFβ-Asp128Asn

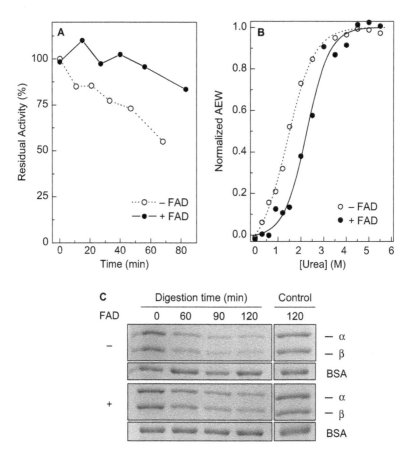

Figure 37.6 Effect of flavin cofactor binding on the stability of the human electron-transfer flavoprotein (ETF) mutant variant Asp128Asn. (A) Activity of the protein is affected by incubation at 39 °C (open circles); however, in the presence of 2.5-fold excess FAD the activity is preserved (black circles). (B) The stability of ETF Asp128Asn to urea-induced chemical denaturation is higher when the flavin is bound to the protein (black circles) than in flavin-depleted ETF (open circles). (C) The presence of flavin cofactor affects the proteolytic susceptibility of ETF Asp128Asn. Upon incubation with trypsin protease ETF Asp128Asn is rapidly degraded (top panel), whereas in the presence of excess flavin, the protein is more resistance to proteolysis.
Figures reprinted from Henriques et al. (2009), with permission.

is not directly located in the FAD binding domain. Therefore, the observations made could be generalized to other mutations in different flavoproteins involved in fatty acid β-oxidation defects. Moreover, the use of this mild mutation, which was modulated by environmental factors, provides a concrete molecular rationale for the efficiency of riboflavin supplementation. However, even though flavinylation can improve the harmful effects of mild destabilized mutants, it is not sufficient to completely rescue protein activity to the level that is required to restore normal β-oxidation: in ETFβ-Asp128Asn fibroblasts cultured with riboflavin-supplemented media, the flux using myristate or palmitate as substrates was only 14% and 28% of controls, respectively (Lundemose *et al.* 1997). This has been also showed for other riboflavin

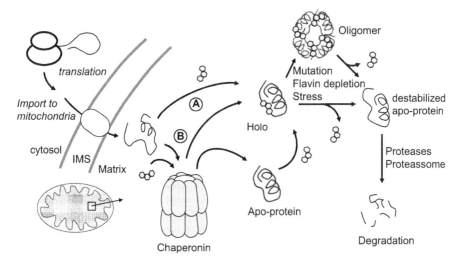

Figure 37.7 Cartoon representing different scenarios for pathways through which FAD may be inserted into proteins conferring structural and functional rescue. After translation and import into the mitochondria the apoprotein form may become flavinylated *via* a chaperonin-independent (A) or chaperonin-dependent (B) pathway. In both cases, steps involving FAD insertion may eventually be mediated by FAD-chaperone proteins. The chaperonin-dependent pathway may involve folding of the apo monomer which then becomes flavinylated upon release or immediately after release. Oligomerization into the functional forms (tetramers or dimers) is made starting from the holo-protein form. Upon an adverse cellular or patho-physiological condition such as a genetic mutation, stress (thermal, oxidative or other) or riboflavin and flavin depletion, cofactor lability may be enhanced thus resulting in an equilibrium of populations in which there is a significant amount of the enzyme in the apo-form. The latter is known to be more conformationally destabilized and susceptible to degradation or misfolding, resulting in loss of function. In some cases, restoring the intra-mitochondrial flavin levels as a result of riboflavin supplementation, results in an increase of the activity of the affected proteins. See text for details and key references.
Figure reprinted from Henriques *et al.* (2010), with permission.

responsive patients with mild forms of MADD, where biochemical and clinical abnormalities were only partially restored upon riboflavin therapy (Amendt and Rhead 1986; Olsen *et al.* 2004). Nevertheless, partial restoration may be sufficient to overcome the disease threshold.

37.4 Concluding Remarks

Riboflavin dietary intake is essential for the biosynthesis of flavin cofactors which are essential for many enzymes in the cell, including the several different flavoenzymes involved in mitochondrial β-oxidation processes. Therapeutic intake of vitamin B_2 at increased levels (typically 200 mg/day) results in higher cellular levels of the cofactors, and studies in conditions of dysfunction as in FAO disease have contributed to establish cellular and molecular mechanisms through which riboflavin and flavins impact on β-oxidation flavoenzymes. Along these lines, we have recently put forward an integrated scenario for possible mechanisms through which this interplay takes place, namely in respect to the structural and functional rescue of faulty enzymes due to flavinylation (Henriques *et al.* 2010) (Figure 37.7). According to this scheme, mitochondria apo-flavoproteins will became flavinylated subsequently to import by two possible mechanisms: *via* a chaperonin- independent (A) or chaperonin-dependent (B) pathway. The insertion of flavin could eventually be mediated by yet unidentified FAD-chaperone proteins. The chaperonin-dependent pathway may involve folding of the apo monomer which then gets flavinylated upon release or immediately after release, as proposed for MCAD. Holo-protein is then available to make oligomerized species like dimers or tetramers. Adverse cellular or pathophysiological conditions such as genetic mutations, stress (thermal, oxidative or other) or riboflavin and flavin depletion, will enhance cofactor lability, resulting in a build-up of conformations in which there is a significant amount of the enzyme in the apo-form. As a result, the enzymes would become conformationally destabilized and susceptible to degradation or misfolding, resulting in loss of function. In agreement, riboflavin supplementation is known to restore intra-mitochondrial flavin levels corresponding to an increase of the activity of the affected proteins (Gianazza *et al.* 2006; Lucas *et al.* 2011).

Overall, dietary riboflavin supplementation impacts directly on higher FAD cofactor bio-availability and this has a direct effect on the functional and structural properties of defective mitochondrial flavoenzymes involved in β-oxidation through the cellular and molecular mechanisms outlined above.

Summary Points

- This chapter is about riboflavin and β-oxidation flavoenzymes.
- Riboflavin, or vitamin B_2, is the biological precursor of the essential redox cofactors FAD and FMN, which are synthesized upon dietary intake of riboflavin.

- Flavins are versatile protein cofactors when inserted into proteins, as these afford a broad range of redox reactions and catalytic properties due to their unique physical and chemical properties.
- The mitochondrial β-oxidation machinery is mainly composed by flavoproteins whose properties are described in this chapter.
- Altered cellular levels of riboflavin and derived flavins impacts on flavoprotein function. This is particularly relevant in the context of inborn errors of metabolism affecting β-oxidation and amino acid catabolism enzymes which are frequently a result of missense mutations and result in protein misfolding or catalytic impairment.
- Therapeutic intake of vitamin B_2 increases cellular FAD concentrations that modulate the expression levels of several flavoproteins, and directly promote the folding, stability and activity of affected flavoproteins
- An integrated scenario for possible mechanisms through which riboflavin and flavins interplay on β-oxidation flavoenzymes is discussed, especially in what concerns flavins as pharmacological chaperones.

Key Facts

Key Facts about Flavoproteins

- Flavoproteins are proteins that contain as prosthetic groups either flavin mononucleotide (FMN) and flavin adenine dinucleotide (FAD), the two biologically active forms of riboflavin.
- 1–3% of the genes in bacteria and eukaryotic genomes encode for flavoproteins.
- The large majority of flavoproteins are oxidoreductases, *i.e.* they are enzymes that catalyse oxidation/reduction reactions.
- Flavins are extremely versatile cofactors because their chemical properties are dramatically influenced by the surrounding environment provided by the protein.
- In most cases flavins are non-covalently attached to the protein. However, in some proteins the flavin cofactor is bound covalently to histidine, cysteine or tyrosine residues, probably to increase saturation of the active site, improve electron transfer or increase protein stability.
- Flavins show unique spectroscopic fingerprints in different redox states (oxidized, semiquinone, reduced) and protein environments. This allows the application of a variety of biophysical methods (*e.g.* visible absorption, visible circular dichroism, resonance Raman and fluorescence emission), to analyse enzymatic reactions and to study flavin chemistry within the flavin-protein complex.
- Most enzymes that participate in β-oxidation of fatty acids are flavoproteins. In all cases the flavin cofactor is non-covalently bound.
- It is believed that riboflavin supplementation corrects some metabolic defects caused by mutations in flavoproteins because it increases the

content of flavins inside cells which then, by binding to the defective protein, exert a structural stabilization effect.

Key Facts about Inborn Errors of Fatty Acid Oxidation

- Mitochondrial fatty acid β-oxidation is a critical metabolic pathway for energy metabolism.
- Over 20 known genetic defects result in impaired fatty acid oxidation, most of which are missense mutations resulting in the change of a single amino acid.
- Defective gene products are either not expressed or exhibit impaired protein folding, structural defects and functional deficiency, among other multi-factorial effects.
- Broad range of clinical phenotypes, ranging from mild to fatal forms, and genotype-phenotype relationships are in many instances unclear.
- The first described inborn error of fatty acid oxidation was muscle CPT II deficiency, which was reported in 1973. MCAD deficiency is the most frequent FAO disorder, with a prevalence of 1 in 12 000 to 20 000 in Caucasians from northern Europe.
- Newborn screening programmes evaluate accumulation of FAO metabolites and guide molecular genetic analysis diagnosis in positive cases.
- Dietary supplementation with high doses of riboflavin and carnitine results in some cases in successful treatments, such as in the case of some MADD patients.

Definitions of Words and Terms

Acyl-CoA dehydrogenases. A family of mitochondrial enzymes containing FAD as redox cofactor that catalyse the dehydrogenation of acyl-CoA thioesters in the initial steps of fatty acid β-oxidation.

Amphipathic helix. Designation of a protein alpha helix in which polar and apolar amino acid side chains align over opposing planes. These secondary structure elements play an important role in proteins that do not interact with membranes *via* transmembrane segments.

Fatty acid β-oxidation. Designation of biochemical pathways through which fatty acids are metabolized, resulting in energy production as their oxidation is a highly exergonic process. In mammals, mitochondrial β-oxidation of fatty acid provides a major source of ATP for the heart and skeletal muscle.

Flavoprotein. Protein that contains a flavin molecule (FAD, FMN) as cofactor, which can have either a covalent or non-covalent attachment to the protein. Flavoproteins are frequently involved in catalytic and electron transfer reactions but biological activity is not limited to these processes.

Isoalloxazine. Designation of the tricyclic heteronuclear organic ring found in flavins.

Ketone bodies. Designations for by-products of fatty acid metabolism such as acetone, acetoacetic acid and β-hydroxybutyric acid, which are used as energy sources in high-energy demand tissues and organs.

MCAD deficiency. The most frequent inborn error of fatty acid oxidation (1 : 12 000 to 20 000 in Caucasians from northern Europe). It affects the medium chain acyl-CoA dehydrogenase and in around 80% of the affected alleles a single common missense mutation has been identified. Symptoms typically appear during infancy or early childhood and can include vomiting, lethargy and hypoglycaemia.

Protein misfolding. A general designation for a defect in the protein folding process that results in a structural conformation which is not identical to the native state of the protein (*e.g.* a destabilized variant or an oligomer such as an amyloid).

Redox potential. A measure of the tendency of a certain molecule to receive or donate electrons, that is to become reduced (decrease in oxidation state) or oxidized (increase in the oxidation state). Synonym of oxidation-reduction potential. SI unit is Volt (V).

Riboflavin-responsive MADD. Designation for patients suffering from multiple acyl-CoA dehydrogenation deficiency, a rare mitochondrial β-oxidation disease, whose clinical and biochemical symptoms recover upon dietary intake of high-doses of the vitamin riboflavin.

List of Abbreviations

ACAD	acyl-CoA dehydrogenases
AMP	adenosine monophosphate
ATP	adenosine triphosphate
CAT	acyl-carnitine translocase
CPT I	carnitine palmitoyl transferase I
CPT II	carnitine palmitoyl transferase II
2D-PAGE	two-dimensional polyacrylamide gel electrophoresis
ETF	electron-transfer flavoprotein
ETF:QO	electron-transfer flavoprotein:ubiquinone oxidoreductase
FAD	flavin adenine dinucleotide
$FADH_2$	flavin adenine dinucleotide, reduced form
FAO	fatty acid oxidation
FMN	flavin mononucleotide
GCD	glutaryl-CoA dehydrogenase
IBD	isobutyryl-CoA dehydrogenase
IVD	isovaleryl-CoA dehydrogenase
LCAD	long chain acyl-CoA dehydrogenase
MADD	multiple acyl-CoA dehydrogenation deficiency
MCAD	medium chain acyl-CoA dehydrogenase
NADH	nicotinamide adenine dinucleotide, reduced form
SBCAD	short/branched-chain acyl-CoA dehydrogenase

SCAD short chain acyl-CoA dehydrogenase
VLCAD very long-chain acyl-CoA dehydrogenase

Acknowledgements

Work in the Gomes laboratory has been supported by CLIMB – Children living with Metabolic Disease (CLIMB, UK), Fundação para a Ciência e Tecnologia (FCT/MCTES PTDC/SAU-GMG/70033/2006, Portugal) through research grants (to C.M.G) and fellowships SFRH/BPD/74475/2010 (to B.J.H.) and SFRH/BPD/34763/2007 (to J.V.R) and by the strategic grant PEst-OE/EQB/LA0004/2011 (to the ITQB Laboratório Associado).

References

Amendt, B.A., and Rhead, W.J., 1986. The multiple acyl-coenzyme A dehydrogenation disorders, glutaric aciduria type II and ethylmalonic-adipic aciduria. Mitochondrial fatty acid oxidation, acyl-coenzyme A dehydrogenase, and electron transfer flavoprotein activities in fibroblasts. *Journal of Clinical Investigation*. 78: 205–213.

Barile, M., Passarella, S., Bertoldi, A., and Quagliariello, E., 1993. Flavin adenine dinucleotide synthesis in isolated rat liver mitochondria caused by imported flavin mononucleotide. *Archives of Biochemistry and Biophysics*. 305: 442–447.

Barile, M., Brizio, C., Valenti, D., De Virgilio, C., and Passarella, S., 2000. The riboflavin/FAD cycle in rat liver mitochondria. *European Journal of Biochemistry*. 267: 4888–4900.

Bennett, M.J., Rinaldo, P., and Strauss, A.W., 2000. Inborn errors of mitochondrial fatty acid oxidation. *Critical Reviews in Clinical Laboratory Sciences*. 37: 1–44.

Bross, P., Jespersen, C., Jensen, T.G., Andresen, B.S., Kristensen, M.J., Winter, V., Nandy, A., Krautle, F., Ghisla, S., Bolundi, L., Kim, J.J., and Gregersen, N., 1995. Effects of two mutations detected in medium chain acyl-CoA dehydrogenase (MCAD)-deficient patients on folding, oligomer assembly, and stability of MCAD enzyme. *The Journal of Biological Chemistry*. 270: 10284–10290.

De Colibus, L., and Mattevi, A., 2006. New frontiers in structural flavoenzymology. *Current Opinion in Structural Biology*. 16: 722–728.

Edmondson, D., and Ghisla, S., 1999. Flavoenzyme structure and function. Approaches using flavin analogues. *Methods in Molecular Biology*. 131: 157–179.

Ensenauer, R., He, M., Willard, J.M., Goetzman, E.S., Corydon, T.J., Vandahl, B.B., Mohsen, A.W., Isaya, G., and Vockley, J., 2005. Human acyl-CoA dehydrogenase-9 plays a novel role in the mitochondrial β-oxidation of unsaturated fatty acids. *The Journal of Biological Chemistry*. 280: 32309–32316.

Ghisla, S., and Massey, V., 1986. New flavins for old: artificial flavins as active site probes of flavoproteins. *Biochemistry Journal*. 239: 1–12.

Gianazza, E., Vergani, L., Wait, R., Brizio, C., Brambilla, D., Begum, S., Giancaspero, T.A., Conserva, F., Eberini, I., Bufano, D., Angelini, C., Pegoraro, E., Tramontano, A., and Barile, M., 2006. Coordinated and reversible reduction of enzymes involved in terminal oxidative metabolism in skeletal muscle mitochondria from a riboflavin-responsive, multiple acyl-CoA dehydrogenase deficiency patient. *Electrophoresis*. 27: 1182–1198.

Goetzman, E.S., Wang, Y., He, M., Mohsen, A.W., Ninness, B.K., and Vockley, J., 2007. Expression and characterization of mutations in human very long-chain acyl-CoA dehydrogenase using a prokaryotic system. *Molecular Genetics and Metabolism*. 91: 138–147.

Gregersen, N., Rhead, W., and Christensen, E., 1990. Riboflavin responsive glutaric aciduria type II. *Progress in Clinical and Biological Research*. 321: 477–494.

Haack, T.B., Danhauser, K., Haberberger, B., Hoser, J., Strecker, V., Boehm, D., Uziel, G., Lamantea, E., Invernizzi, F., Poulton, J., Rolinski, B., Iuso, A., Biskup, S., Figmidt, T., Mewes, H.W., Wittig, I., Meitinger, T., Zeviani, M., and Prokifig, H., 2010. Exome sequencing identifies ACAD9 mutations as a cause of complex I deficiency. *Nature Genetics*. 42: 1131–1134.

He, M., Rutledge, S.L., Kelly, D.R., Palmer, C.A., Murdoch, G., Majumder, N., Nicholls, R. D., Pei, Z., Watkins, P.A., and Vockley, J., 2007. A new genetic disorder in mitochondrial fatty acid β-oxidation: ACAD9 deficiency. *The American Journal of Human Genetics*. 81: 87–103.

He, M., Pei, Z., Mohsen, A.W., Watkins, P., Murdoch, G., Van Veldhoven, P.P., Ensenauer, R., and Vockley, J., 2011. Identification and characterization of new long chain acyl-CoA dehydrogenases. *Molecular Genetics and Metabolism*. 102: 418–429.

Henriques, B.J., Rodrigues, J.V., Olsen, R.K., Bross, P., and Gomes, C.M., 2009. Role of flavinylation in a mild variant of multiple acyl-CoA dehydrogenation deficiency: a molecular rationale for the effects of riboflavin supplementation. *The Journal of Biological Chemistry*. 284: 4222–4229.

Henriques, B.J., Olsen, R.K., Bross, P., and Gomes, C.M., 2010. Emerging roles for riboflavin in functional rescue of mitochondrial β-oxidation flavoenzymes. *Current Medicinal Chemistry*. 17: 3842–3854.

Kim, J.J., Wang, M., and Pafigke, R., 1993. Crystal structures of medium-chain acyl-CoA dehydrogenase from pig liver mitochondria with and without substrate. *Proceedings of the National Academy of Sciences of the United States of America*. 90: 7523–7527.

Lucas, T.G., Henriques, B.J., Rodrigues, J.V., Bross, P., Gregersen, N., Gomes, C.M., 2011. Cofactors and metabolites as potential stabilizers of mitochondrial acyl-CoA dehydrogenases. *Biochim. Biophys. Acta. - Molecular Basis of Disease*. 1812(12): 1658–1663.

Lundemose, J.B., Kolvraa, S., Gregersen, N., Christensen, E., and Gregersen, M., 1997. Fatty acid oxidation disorders as primary cause of sudden and unexpected death in infants and young children: an investigation performed on cultured fibroblasts from 79 children who died aged between 0–4 years. *Molecular Pathology*. 50: 212–217.

Massey, V., 2000. The chemical and biological versatility of riboflavin. *Biochemical Society Transactions*. 28: 283–296.

McAndrew, R.P., Wang, Y., Mohsen, A.W., He, M., Vockley, J., and Kim, J.J., 2008. Structural basis for substrate fatty acyl chain specificity: crystal structure of human very-long-chain acyl-CoA dehydrogenase. *The Journal of Biological Chemistry*. 283: 9435–9443.

Nagao, M., and Tanaka, K., 1992. FAD-dependent regulation of transcription, translation, post-translational processing, and post-processing stability of various mitochondrial acyl-CoA dehydrogenases and of electron transfer flavoprotein and the site of holoenzyme formation. *The Journal of Biological Chemistry*. 267: 17925–17932.

Nouws, J., Nijtmans, L., Houten, S.M., van den Brand, M., Huynen, M., Venselaar, H., Hoefs, S., Gloerich, J., Kronick, J., Hutchin, T., Willems, P., Rodenburg, R., Wanders, R., van den Heuvel, L., Smeitink, J., and Vogel, R.O., 2010. Acyl-CoA dehydrogenase 9 is required for the biogenesis of oxidative phosphorylation complex I. *Cell Metabolism*. 12: 283–294.

Olsen, R.K., Pourfarzam, M., Morris, A.A., Dias, R.C., Knudsen, I., Andresen, B.S., Gregersen, N., and Olpin, S.E., 2004. Lipid-storage myopathy and respiratory insufficiency due to ETF:QO mutations in a patient with late-onset multiple acyl-CoA dehydrogenation deficiency. *Journal of Inherited Metabolic Disease*. 27: 671–678.

Olsen, R.K., Olpin, S.E., Andresen, B.S., Miedzybrodzka, Z.H., Pourfarzam, M., Merinero, B., Frerman, F.E., Beresford, M.W., Dean, J.C., Cornelius, N., Andersen, O., Oldfors, A., Holme, E., Gregersen, N., Turnbull, D.M., and Morris, A.A., 2007. ETFDH mutations as a major cause of riboflavin-responsive multiple acyl-CoA dehydrogenation deficiency. *Brain*. 130: 2045–2054.

Roberts, D.L., Frerman, F.E., and Kim, J.J., 1996. Three-dimensional structure of human electron transfer flavoprotein to 2.1-Å resolution. *Proceedings of the National Academy of Sciences of the United States of America*. 93: 14355–14360.

Saijo, T., and Tanaka, K., 1995. Isoalloxazine ring of FAD is required for the formation of the core in the Hsp60-assisted folding of medium chain acyl-CoA dehydrogenase subunit into the assembly competent conformation in mitochondria. *The Journal of Biological Chemistry*. 270: 1899–1907.

Sato, K., Nishina, Y., and Shiga, K., 1996. In vitro refolding and unfolding of subunits of electron-transferring flavoprotein: characterization of the folding intermediates and the effects of FAD and AMP on the folding reaction. *The Journal of Biochemistry*. 120: 276–285.

Toogood, H.S., van Thiel, A., Basran, J., Sutcliffe, M.J., Scrutton, N.S., and Leys, D., 2004. Extensive domain motion and electron transfer in the human electron transferring flavoprotein medium chain Acyl-CoA dehydrogenase complex. *The Journal of Biological Chemistry*. 279: 32904–32912.

Zhang, J., Frerman, F.E., and Kim, J.J., 2006. Structure of electron transfer flavoprotein-ubiquinone oxidoreductase and electron transfer to the mitochondrial ubiquinone pool. *Proceedings of the National Academy of Sciences of the United States of America*. 103: 16212–16217.

CHAPTER 38

Function and Effects of Niacin (Niacinamide, Vitamin B_3)

AHMED A. MEGAN,[a] SAID O. MUHIDIN,[b] MAHIR A. HAMAD[c] AND MOHAMED H AHMED*[c]

[a] Department of Cardiothoracic Surgery, Cambridge University Hospitals, Papworth Hospital, Everard, Cambridge, UK; [b] Department of Surgery, Barking, Havering and Redbridge University Hospitals, Queen's Hospital, Romford, UK; [c] Division of Acute Medicine, The James Cook University Hospital, Middlesbrough, UK
*Email: elziber@yahoo.com

38.1 Food Sources of Niacin

There are numerous sources of niacin that are essential and these include poultry, fish (tuna, salmon), meat (beef), yeast, legumes, milk and fortified cereals. In addition, niacin is naturally occurring in tiny amounts and the human body can make nicotinic acid from the metabolism of dietary tryptophan (Vosper 2009). The body requires tryptophan for two main reasons: (i) for the synthesis of niacin and (ii) to raise serotonin levels, which is essential for the regulation of sleep, appetite and mood. The vast majority of proteins contain about 1% of tryptophan and it is suggested that approximately 100 g of protein intake a day will be sufficient to ensure optimum levels of niacin in the body. The recommended dose of niacin is higher when there is an increase in physiological states such as pregnancy and lactation. Importantly, the Committee of Medical Aspects of Food Policy (COMA) in the UK stated that the Reference Nutrient Intake (RNI) for niacin was 17 mg/day and 13 mg/day, respectively for men and women (Department of Health 1991). In addition, it is recommended that the daily intake of niacin in breastfeeding women should be increased by 2.3 mg.

38.2 Absorption, Excretion and Clinical Features of Niacin Deficiency

The absorption of niacin in the human body mainly occurs in the stomach and small intestine (Bechgaard and Jespersen 1977). Niacin is taken up by the body swiftly and quickly, and reaches peak plasma levels within 30–60 minutes of being absorbed (Bodor and Offermanns 2008). Furthermore it has a plasma half-life of 60 minutes (Carlson et al. 1968; Svedmyr and Harthon 1970). The enzyme nicotinamide adenine dinucleotide (NAD) glycohydrolase, which is found in the intestine and liver, facilitates the synthesis of nicotinamide from NAD (Henderson and Gross 1979). This is an important step that ensures the availability of nicotinamide for the conversion to NAD. Niacin is metabolized in most tissues in the body and its metabolites are excreted in urine (Jacob et al. 1989; Shibata and Matsuo 1989). Importantly, niacin deficiency occurs mainly as a result of poor diet, but also other conditions such as carcinoid syndrome, Hartnup's disease and drug intake (isoniazid) (Hegyi et al. 2004).

Niacin deficiency leads to unwanted multisystemic problems that are often associated with dermatological changes. Pellagra is a condition in which niacin deficiency causes a symmetrical pigmented rash, thickened skin and superficial scaling which are found in sun-exposed body areas. The classic triad of niacin deficiency are the three D's—dermatitis, diarrhoea and dementia (Hegyi et al. 2004). Some of the effects on the central nervous system include depression, anxiety, restlessness and poor concentration. In addition, alcoholic individuals with poor nutrition can develop pellagroid encephalopathy (Cook et al. 1998). The role of niacin deficiency in cardiovascular disease and lipid metabolism remains to be fully explored.

38.3 Niacin Overdose

Niacin is a safe drug to use and its toxicity is dose related. Niacin toxicity has been shown to manifest itself as altered mental status, nausea and vomiting (Mittal et al. 2007). Evidence shows that excess niacin can cause hepatic dysfunction with associated coagulopathy, jaundice and fulminant liver failure (Etchason et al. 1991; Fischer et al. 1991). In addition niacin overdose can cause impaired glucose control (Mittal et al. 2007).

38.4 Cellular Function and Effects of Niacin

Niacin is an important vitamin for the basic metabolic reactions that occur in the human body and its derivative nicotinamide plays a central role as the co-enzymes nicotinamide adenine dinucleotide (NAD) and nicotinamide adenine dinucleotide phosphate (NADP). These co-enzymes are vital for reduction and oxidative (redox) reactions in which an individual is able to utilize the energy released by these reactions. In addition NAD is a regulator of the redox state in which NAD accepts an electron in its oxidized form (NAD^+) and as an electron donor in its reduced form (NADH) (Benavente and Jacobson 2008).

The term 'redox' is used to describe a chemical reaction where electron transfer occurs that results in reduction, *i.e.* loss of an electron(s), or oxidation, *i.e.* gain of an electron(s), of the substrate. In aerobic respiration NADH transports energy-rich electrons to the electron transport chain that in turn produces ATP (Pollak *et al.* 2007). However, during exercise when the body needs to generate energy in an anaerobic environment, NADH is the driving force behind the conversion of pyruvate to lactic acid, in the process losing an electron (reduction process) to become NAD^+. The recycled NAD^+ ensures the process of glycolysis is maintained and energy production continues; in the absence of NAD, glycolysis would cease to occur. Therefore both NAD and NADP play central roles in basic metabolic reactions that allow the body to function.

NADP is a key component in biosynthesis and is present in high levels in the cytoplasm and mitochondria (Ramirez *et al.* 1995). In addition, it is suggested that NADPH contributes to a number of essential defensive functions and allows detoxification of harmful substances. It is a co-factor for P450 enzymes which help the liver to get rid of biohazardous substances from the body. Furthermore NADPH is the main substrate for the enzyme, NADPH oxidase, which results in the production of free peroxides that are used by the immune system for killing organisms with an oxidative burst (Pollak *et al.* 2007).

Importantly, NADPH also contributes to the metabolism of alcohol and may suggest an important potential role of niacin in treating the associated lipid disorders with excess alcohol intake. Ethanol is primarily oxidized in the liver by three enzymatic pathways: the cytosolic alcohol dehydrogenase (ADH) pathway, the microsomal ethanol oxidizing system (cytochrome P450-3E1) and the peroxisomal catalase pathway (Caballeria *et al.* 1991). Most of the ethanol is oxidized by the cytochrome P450-2EI and ADH pathways. Ethanol is mainly metabolized to acetaldehyde which is extremely reactive and toxic, and therefore the body promptly converts it to less toxic acetate (Paton 2005). The acetate is converted to acetyl-CoA which can enter the Krebs cycle. Ethanol metabolism yields energy; approximately 1 g of ethanol provides 7 kcal of energy and thus 240 g of ethanol per day is sufficient energy to meet basic metabolic needs (Meisenberg and Simmons 1998).

Normally the oxidation of ethanol converts NAD into its reduced form, NADH. In chronic ethanol excess, the NAD that is required normally for the oxidation of fatty acids is depleted and this results in the build-up of fat within the hepatocytes. Beta-oxidation takes place in the mitochondria and it is believed that ethanol affects the normal structure and function of the mitochondria, leading to adaptive changes of lipid composition that may promote increased synthesis of lipids (Meisenberg and Simmons 1998). In addition, ethanol-induced peripheral lipolysis may lead to increased fat deposition to the liver. Subsequently the hepatocytes can use the acetate that comes from ethanol metabolism to further increase lipid synthesis and so the cycle continues with increased risk of hepatomegaly and fatty liver (Meisenberg and Simmons 1998). This illustrates the possible complications that can occur in association of NAD deficiency coupled with excess alcohol consumption. Interestingly, in animal models, Ieraci and Herrera (2006) showed that alcohol consumption during pregnancy led to neural damage in the offspring but that this was prevented by the administration of a single nicotinamide treatment. The authors suggested that niacin mediates this effect in part through decreasing oxidative stress, lipid peroxidation and apoptosis. The subsequent discussion will also elaborate in further important cellular effects of niacin.

38.4.1 Niacin and Cellular Lipid Metabolism

Niacin lowers plasma low density lipoprotein (LDL) and the levels of triglyceride while increasing high density lipoprotein (HDL). It is believed that the lipid-lowering effects of niacin are mediated by a G-protein-coupled receptor present in adipose cells. Niacin binds to the receptor where it produces inhibitory G-proteins and subsequently results in reduction in cyclic adenosine monophosphate (cAMP) and reduced adipose cell sensitivity to lipase action. Ultimately this will decrease peripheral triglyceride catabolism, reduce the supply of free fatty acids to the liver and very low density lipoproteins (VLDL). Interestingly, niacin leads to prolonged half-life of apolipoproteins A1 (components of HDL) and therefore causes raised levels of high density lipoproteins.

In addition to the role of niacin in the degradation of proteins, niacin has a very important role in the catabolism of carbohydrates. We have described the evidence that niacin plays a crucial role in the conversion of glucose to the energy substrate ATP (Bodor and Offermanns 2008; Vosper 2009).

38.4.2 Non-oxidative and Reduction Reactions involving Niacin (NAD Substrate)

Research has shown that NAD is the substrate for three important classes of enzymes: mono-ADP-ribosyltransferases (Hassa *et al.* 2006), poly-ADP-ribose polymerases (PARP) (Virag and Szabo 2002) and sirtuin enzymes (Sauve *et al.* 2006). These enzymes are responsible for protein modifications, DNA repair, endocrine signalling and apoptosis, and act to transfer ADP-ribose to nucleophiles.

38.4.3 DNA Repair

The process of DNA replication is a complex one that is closely monitored in order to allow normal genes to be maintained. There are numerous biochemical pathways that tightly monitor the presence of DNA damage and evidence has suggested PARP is able to pick up DNA strand breaks (Uchida 1993). Of the PARP group of enzymes, it is the PARP1 that plays a very important role in detecting DNA strand breaks. This nuclear enzyme locates and attaches to DNA strand breaks, and sequentially utilizes NAD to generate nicotinamide and ADP ribose (de Murcia and de Menissier 1994). The ADP ribose is then used to make poly-ADP-ribose (nucleic acid polymer) that binds to recipient nuclear proteins (de Murcia et al. 1994; Schreiber et al. 1995; Smith and de Lange 2000).

Over recent decades, a large body of evidence has postulated that it is this process of ribosylation of the recipient proteins that acts in favour of DNA repair through molecular changes of proteins proximal to the damaged DNA strands. In addition it makes it easier for the tightly compacted chromatin structure to loosen up, something that is mandatory for the deployment of DNA repair complexes (Shall 1995). Furthermore, PRAP-1 leads to poly-ADP-ribosylation of many transcription, signalling and DNA replication factors. These factors include DNA-dependent protein kinase (Ariumi et al. 1999), NF-κB (Oliver et al. 1999), p53 (Wesierska-Gadek et al. 1996) and B-MYB (Cervellera and Sala 2000). Evidence has demonstrated that the use of PRAP inhibitors such as 3-aminobenzamide prevent the reconstitution of broken DNA strands and unusually high DNA production (Park et al. 1983). In addition, the use of such inhibitors makes cells more susceptible to cytotoxic injury (Szabo et al. 1997).

38.4.4 Apoptosis and Necrosis

There is a plethora of evidence that supports PARP1 as an important mediator of both programmed cell death (apoptosis) and necrosis (David et al. 2009; Hassa 2009). PARP1 is split by caspases into two components (p89 and p24); this occurs in the process of apoptosis (Germain et al. 1999). Caspases are divided into two groups based on the role either as instigators or executioners of apoptosis; the executioner caspases are caspase-3, caspase-6 and caspase-7 (Slee et al. 2001).

It is widely believed that one of the hallmarks for diagnosing apoptosis in the various tissues is the finding of PARP1 cleaved by caspases (Lazebnik et al. 1994). However, in the presence of significant tissue injury, neuronal trauma leads to hyper-responsiveness of PARP1 and the loss of control of poly-ADP-ribose synthesis, which culminates in extensive non-viable neurol cells (Barr and Conley 2007; Eliasson et al. 1997). Conversely, apoptosis may arise from the diminished cellular energy at a time of tremendous insult to the tissue. The production of NAD for PARP use is an energy-dependent process and increased activity of PARP1 may obstruct cellular energy-dependent processes leading to cellular necrosis (Berger 1985).

38.5 Metabolic Effects of Niacin

Niacin is a water-soluble vitamin that has been available as a lipid-lowering medication and in prevention of atherosclerosis for half a century (Ganji et al. 2003; Olsson 2010). Numerous studies have documented its beneficial effects on reducing cardiovascular disease (CVD) (Olsson 2010). Clinically, the most well-known effect of niacin deficiency is pellagra which, as noted above, is manifested by dermatitis, diarrhoea and dementia (Bodor and Offermanns 2008; Ganji, et al. 2003). In this chapter we look at the effects of niacin on human physiology and its consequent effects on disease states.

38.5.1 Niacin and Lipid Metabolism and Atherosclerosis

In pharmacological doses, niacin has been shown to have several beneficial effects on the lipid profile. It lowers total cholesterol, apolipoprotein B, triglycerides, VLDL, LDL and lipoprotein a [Lp(a)], while increasing the 'good' cholesterol (HDL). Hence it is known as a pan-spectrum lipid agent (Bodor and Offermanns 2008; Ganji et al. 2003; Olsson 2010). Several studies have shown that treatment with niacin reduces fatal coronary events, mainly by stopping the progress of atherosclerosis in the arteries that supply the heart (the coronaries) (Tavintharan and Kashyap 2001).

38.5.1.1 Niacin and Triglyceride and Apo B Metabolism

The effect of niacin on triglyceride and apo B metabolism is mediated through the following two mechanisms (Ganji et al. 2003):

- **Modulating triglyceride lipolysis in adipose tissue.** Adipose tissue is specialized for synthesizing and storing triglycerides, a highly concentrated store of metabolic energy, and also in mobilizing them to the liver in the form of fatty acids and glycerol (i.e. as a fuel). Therefore, through inhibition of lipolysis in adipose tissue, it is possible to modulate plasma lipids (Ganji et al. 2003). Early studies indicated that niacin inhibited the lipolysis of triglycerides, thereby reducing mobilization of fatty acids from adipose tissue (Carlson et al. 1968). The mechanism by which lipolysis takes place involves cAMP-mediated activation of hormone sensitive lipase (Henderson 1983). This links well with previous studies which showed that niacin inhibited adenylate cyclase activity in adipose tissue, resulting in reduced cAMP concentrations (Aktories et al. 1980). It is unclear, however, whether niacin reduces the concentration of cAMP by means of its inhibition of adenylate cyclase, or via its modulation on G-protein receptor dependent/independent mechanisms which can also

regulate production of cAMP (Ganji *et al.* 2003). The overall effect is, of course, inhibition of triglyceride lipolysis.
- **Modulating triglyceride synthesis resulting in overall apo B degradation intracellularly.** Production and secretion of apo B, its associated lipids and the resultant VLDL and Lp(a) particles occurs in the liver, though not exclusively. *Via* peripheral lipoprotein lipase-mediated triglyceride hydrolysis, the secreted VLDLs undergo conversion to intermediate density lipoprotein (IDL) and then LDL (Ganji, *et al.* 2003).

The processing of apo B in the liver plays a central role in controlling apo B lipoprotein secretion (Ganji *et al.* 2003). This is regulated by (Davis 1999; Ginsberg 1995):

1. Localization of newly synthesized apo B as it moves across the endoplasmic reticulum (ER) membrane
2. Degradation of apo B after translocation
3. Mechanisms that govern the synthesis and addition of core lipids to the new VLDL particles for secretion.

Current evidence is that the newly synthesized apo B is not secreted, but rather degraded in hepatocytes. After its production in the ER, it has been suggested that the longer apo B stays in the ER, the more likely it is to get degraded, whereas rapid translocation facilitates its secretion as VLDL particles (Davis 1999). The speed by which it traverses the ER, and therefore whether it undergoes secretion or degradation, is determined by the rate of lipid synthesis and the availability to lipidate apo B. By increasing triglyceride (TG) synthesis and secretion, oleic acid has been shown to stimulate apo B secretion from hepatocytes by facilitating its movement across the ER and away from the proteases, in effect protecting the newly synthesized apo B from degradation (Dixon *et al.* 1991). In general, it has been shown that inhibition of fatty acid and triglyceride synthesis participated in apo B degradation by inhibiting its secretion from hepatocytes (Wu *et al.* 1994). Intracellular apo B processing and degradation are due to the regulation by a protease enzyme, while lipid synthesis, availability and transfer is by microsomal triglyceride transfer protein (MTP).

38.5.1.2 Niacin and Biliary Cholesterol

In a study of patients with hyperlipidaemia, niacin was shown to reduce triglyceride and VLDL significantly, while also being shown to decrease the transport rate (synthesis) of VLDL-TG (Ganji *et al.* 2003). It also increased hepatic excretion of biliary cholesterol, which may in part explain the reason why cholesterol is lost from the body (Grundy *et al.* 1981), which may be due to increased reverse cholesterol transport (RTC). Also, as cholesterol reduction is the result of niacin's effect on LDL, reduction of LDL from VLDL will play an active role in niacin-mediated plasma cholesterol reduction (Grundy *et al.* 1981).

38.5.1.3 Hepatocellular Effects of Niacin on VLDL/LDL Metabolism

In a study assessing its effect on hepatocytes, niacin increased apo B intracellular degradation and decreased its subsequent secretion, but did not seem to alter the 'steady state expression of apo B' or uptake of LDL into hepatocytes. The study also showed that niacin had no effect on MTP activity and therefore no effect on triglyceride transfer in hepatocytes (Ganji et al. 2003).

Other studies showed that niacin had no effect on the protease-mediated degradation of apo B, suggesting that the niacin mechanism is independent of protease. Oleic acid has been shown to increase triglyceride synthesis and decrease intracellular apo B degradation. Niacin was also shown to decrease the inhibition on apo B degradation by oleate. This may suggest that apo B degradation depends on pathways that are responsible for the synthesis and association of triglycerides, before the apo B is processed (Bodor and Offermanns 2008; Ganji, et al. 2003).

Another study showed that niacin inhibited triglyceride production, but had no effect on cholesterol synthesis. It has also been shown to inhibit microsomal diacylglycerol acyltransferase (DGAT), a key rate-limiting enzyme in triglyceride synthesis and therefore availability to hepatocytes, resulting in increased apo B degradation and decreased VLDL/LDL secretion (Ganji et al. 2004; Jin et al. 1999).

38.5.1.4 Role of Niacin in HDL Metabolism

Niacin is the most powerful pharmacological agent in increasing HDL levels. The liver and the small intestine are the major organs for the synthesis and secretion of HDL and its components. Plasma HDL and its components are regulated by many different synthetic and catabolic processes (Bodor and Offermanns 2008; Ganji et al. 2003). There are many forms of HDL particles in plasma, which differ in their chemical and physical composition, with the major two types being HDL2 and HDL3. Apo AI and Apo AII are the major proteins of HDL, and it is the differing amounts of these in HDL that forms the different subfractions. The general subdivisions are those containing apo AI alone and those containing both apo AI and AII. It has been suggested that these have different functions with regard to regulation of metabolism, physiological function and also in their association with atherosclerosis, with studies showing that apo AI HDL decreased the extent of arteriopathy in coronary heart disease (CHD) (Amouyel et al. 1993; Puchois et al. 1987), as it is more effective at ejecting cellular cholesterol than HDL with both apo AI and AII, and is a more efficient donor of cholesterol esters (Rinninger et al. 1998).

Niacin decreases the fractional catabolic rate of apo AI metabolism without altering its synthetic rate (Blum et al. 1977; Shepherd et al. 1979). Furthermore, niacin was selective in its uptake of HDL apo AI but not HDL cholesterol esters, suggesting that niacin may facilitate apo AI-induced cholesterol efflux and reverse cholesterol transport pathway.

Reverse cholesterol transport (RTC) is the process whereby HDL particles augment the efflux of excess cholesterol from foam cells in the artery, with this cholesterol being returned to the liver for excretion in bile. The removal of cholesterol from the arterial macrophages is promoted by ABCA1 and SR-B1 (scavenger receptor class B type I), which promote cholesterol efflux into HDL and transport to liver (Morgan et al. 2007). Studies suggest that niacin has beneficial effect on SR-BI efflux, which is related to changes in HDL (Morgan et al. 2007).

Another hypothesis suggests that niacin inhibits removal of HDL apo AI at an unconfirmed receptor, which works to reduce its catabolism, and is independent of the SR-B1 receptor which is selective for HDL-cholesterol esters. This positive effect of niacin on SR-B1 mediated efflux was observed by demonstration of an increased number of HDL-apo AI particles. The degree of efflux is not mainly influenced by the steady state level of HDL, but is accelerated by HDL with apo AI and higher HDL2–HDL3 ratio (Morgan et al. 2007).

In addition to its effect on SR-B1, niacin can also stimulate ABCA-1 mediated cholesterol efflux from macrophages, which will contribute to niacin-mediated increases in HDL-C (Morgan et al. 2007; Rubic et al. 2004). Niacin also increases the ability of HDL to take up additional cholesterol from peripheral cells and tissues, as well as its ability to mediate cholesterol transport *via* SB-B1, *via* inhibition of hepatic uptake of small HDL by endocytosis—in effect recycling HDL (Morgan et al. 2007). Thus niacin has been shown to promote cellular cholesterol release *via* more than one mechanism.

38.5.2 Effect of Niacin on Atherosclerosis, Inflammation and Vascular Reactivity

It is well established that inflammatory processes play an important role in the causation of cardiovascular disease (CVD). Inflammatory mediators play an important role in the initiation, progression and rupture of atherosclerotic plaque. Therefore, lipid-lowering medication with anti-inflammatory effects may provide better reduction in the incidence of CVD than those with only lipid-lowering effect.

There is increasing evidence that niacin has potential benefit in decreasing inflammation, an important part of atherosclerosis. This may be mediated in part by ability of niacin to increase HDL or direct anti-inflammatory effect independent of the HDL (Vaccari et al. 2007a,b). For instance, administration of extended-release niacin (1000 mg/day) in 52 individuals with the metabolic syndrome was shown to be associated with a marked change of carotid intima-media thickness (IMT), improvement of endothelial function and decrease in high sensitivity C-reactive protein (Thoenes et al. 2007).

Cell adhesion molecules such as intercellular adhesion molecule-1 (ICAM-1), vascular cell adhesion molecule-1 (VCAM-1), platelet endothelial cell adhesion molecule (PECAM) and E-selectin have been associated with atherogenesis and progression of disease. These are molecules which are expressed in endothelial cells, platelets and white blood cells. Their increased expression has been associated with the adhesion, recruitment and migration of white blood cells in vascular surfaces, essential processes in atherosclerosis (Vaccari et al. 2007a,b).

Studies have shown that niacin significantly reduces ICAM-1, VCAM-1, PECAM and E-selectin levels. It also reduced the tumour necrosis factor-*alpha* (*TNF*-α) induced rise in ICAM (it does this by reducing mRNA-induced expression of ICAM). It also decreased the production of ICAM through its reduction in interferon-y (IFN-y) and interleukin-1 (IL-1). Through these mechanisms, niacin induces a reduction of monocyte adhesion in endothelial cells; This may possibly lead to decreased leucocyte adhesion to endothelium, which is an important early event in atherosclerosis (Tavintharan et al. 2009).

Furthermore, Ganji et al. (2009) showed that in cultured human aortic endothelial cells, niacin increased nicotinamide adenine dinucleotide phosphate [NAD(P)H] levels by 54% and reduced glutathione (GSH) by 98%. Niacin inhibited:

- angiotensin II (ANG II) induced reactive oxygen species (ROS) production by 24–86%;
- low density lipoprotein (LDL) oxidation by 60%;
- TNF-α induced NF-κB activation by 46%, VCAM-1 by 77–93%, monocyte chemotactic protein-1 (MCP-1) secretion by 34–124%;
- TNF-α induced monocyte adhesion to HAEC in a functional assay by 41–54%.

The authors concluded that niacin inhibits vascular inflammation by decreasing endothelial ROS production and subsequent LDL oxidation and inflammatory cytokine production (Ganji et al. 2009).

Interestingly, Wu et al. (2010) showed that administration of niacin in New Zealand white rabbits was associated with:

- marked reduction in inflammatory markers VCAM-1, ICAM-1 and MCP-1;
- intima-media neutrophil recruitment and myeloperoxidase accumulation;
- enhanced endothelial-dependent vasorelaxation and cyclic guanosine monophosphate (cGMP) production;
- increased vascular reduced glutathione content;
- protection against hypochlorous acid-induced endothelial dysfunction and TNF-α induced vascular inflammation.

This study showed that niacin inhibits vascular inflammation and protects against endothelial dysfunction independent of these changes in plasma lipid levels.

Digby et al. (2010) showed that nicotinic acid is not only able to suppress pro-atherogenic chemokines but is also able to upregulate the atheroprotective adiponectin through a G-protein-coupled pathway, The authors concluded that nicotinic acid may have pleiotropic effect through positive effect in adiponectin. Similar positive effects of niacin on adiponectin are also seen in humans. For instance, treatment with controlled-release niacin for 52 weeks resulted in sustained improvements in adiponectin levels compared to placebo in patients with metabolic syndrome (Vaccari *et al.* 2007a,b). The SLIM Study showed that Slo-Niacin(R) 1.5 g/day with atorvastatin 10 mg/day improved lipoprotein lipids, apoproteins and inflammation [high-sensitivity C-reactive protein (hsCRP) + TNF-α] markers without hepatotoxicity (Knopp *et al.* 2009).

From the above discussion it is possible to suggest that niacin has shown potential benefits in reducing inflammation in both animal models and human studies. However, It is not yet clear whether the effect of niacin in inflammation may have benefit in patients with chronic kidney disease (CKD).

38.5.3 Niacin and Dyslipidaemia and Hyperphosphatemia with Chronic Kidney Disease

Chronic renal failure is known to be associated with an accelerated process of atherosclerosis. Importantly, cardiovascular disease is the main cause of morbidity and mortality with CKD and kidney transplant recipients. High serum phosphate levels have been shown to be associated with higher mortality for all causes, cardiovascular mortality and vascular calcification (Block *et al.* 2004; Young *et al.* 2004). Phosphorus retention develops in CKD when the glomerular filtration rate (GFR) falls below 25 ml/min. It was estimated that for each 1 mg/dl of serum phosphate increase in CKD, the risk of having myocardial infarction increases by 35% (Kestenbaum *et al.* 2005).

Importantly, several studies suggested the potential benefit of lipid-lowering medication in preventing cardiovascular events in CKD and the transplant population (Muntner *et al.* 2005; Yao *et al.* 2004). In particular, statins were shown to be effective in reducing LDL-cholesterol. It is worth mentioning that evidence of the potential benefit of lipid-lowering medication has started to emerge from clinical trials. In a large *post hoc* analysis of three large trials, pravastatin treatment was shown to reduce the decline in renal function in patients with moderate CKD (GFR 30–60 ml/min). Importantly pravastatin reduced CVD in diabetic patients irrespective of the presence or absence of CKD (Tonelli *et al.* 2005). Interestingly, the Assessment of Lescol in Renal Transplantation (ALERT) study demonstrated that fluvastatin significantly reduces cardiac deaths and myocardial infarction in renal transplant recipients (Jardine *et al.* 2005). These studies demonstrated that patients with CKD would derive benefit from lipid-lowering medication. However, refractory dyslipidaemia and difficulty in lowering LDL to target were reported with CKD and renal transplant. Importantly, niacin has shown potential benefits in treating both high plasma phosphate and dyslipidaemia associated with CKD. A list of those studies that showed potential benefit of niacin in treating dyslipidaemia and high plasma phosphate in CKD is included in Table 38.1.

Table 38.1 Studies showing potential benefit of niacin as treatment for high plasma phosphate and dyslipidaemia associated with CKD.

Study	Treatment of high phosphate	Study	Treatment of dyslipidaemia
Cheng et al. 2008	Prospective, randomized, double-blind, placebo-controlled crossover trial for assessment of the safety and efficacy of niacinamide treatment for eight weeks in 35 hemodialysis patients with phosphorus levels ≥ 5.0 mg/dl. Serum phosphorus fell significantly from 6.26 to 5.47 mg/dl with niacinamide.	Cho et al. 2010	Niacin therapy in animal model attenuated hypertension, proteinuria and tubulo-interstitial injury, reduced renal tissue lipids, CD36, carbohydrate-responsive element binding protein (ChREBP), liver X receptor (LXR), ATP-binding cassette transporter 1 (ABCA-1), ATP-binding cassette sub-family G member 1 (ABCG-1) and SR-B1 abundance, and raised PPAR-alpha and liver fatty acid binding protein (L-FABP). Concluded that niacin administration improves renal tissue lipid metabolism, and renal function and structure.
Muller et al. 2007	Niaspan treatment in 17 patients on dialysis for 12 weeks decreased serum phosphate values from 7.2 ± 0.5 to 5.9 ± -0.6 mg/dl ($p < 0.015$).	Cho et al. 2009	Niacin administration in a rat reduced MCP-1, plasminogen activator inhibitor-1 (PAI-1), transforming growth factor beta (TGF-β), p47(phox), p22(phox), cyclooxygenase-1 (COX-1), and NF-κB activation, ameliorated hypertension, proteinuria, glomerulosclerosis and tubulointerstitial injury.
Sampathkumar et al. 2006	Single treatment of 34 extended release nicotinic acid (375 mg) tablet in patients on haemodialysis for a mean period of eight months. This was associated with a significant decrease in both plasma phosphate and calcium phosphate product.	Restrepo et al. 2008	In nine patients with CKD and on dialysis treated with 1000 mg nictonic acid for eight months decreased LDL-c, triglyceride and increased HDL-c. The authors recommended that nicotinic acid is efficient, very well tolerated and economical in comparison with other drugs, which makes it ideal for the treatment of patients with hyperlipidaemia and refractory hyperphosphatemia to classical treatments.

38.5.4 Effect of Niacin on Insulin Sensitivity and Glucose Metabolism

Several studies have shown different effects of niacin on insulin sensitivity. Administration of long-term niacin in Sprague Dawley rats was associated with marked improvement in insulin sensitivity and decrease in HbA1c (Perricone et al. 2010). Administration of niacin for six months in 30 patients with impaired glucose tolerance was associated with a decrease in circulating Lp(a) by 38% ($p < 0.001$) and fasting triglycerides by 12% ($p < 0.05$). Whole-body insulin sensitivity increased in the niacin treatment group, although this trend was not statistically significant ($p = 0.085$). Six months' niacin led to a significant reduction in mean adipocyte size associated with increased insulin sensitivity in isolated adipocytes and gene expression changes including increased adiponectin, C/EBPalpha, C/EBPdelta, peroxisome proliferator-activated receptor gamma (PPAR-γ) and decreased carnitine palmitoyl transferase 2, hormone sensitive lipase, nicotinic acid receptor (GPR109B) and fatty-acid synthase mRNA expression (Linke et al. 2009).

Furthermore, administration of 2 g/day of extended-release nicotinic acid in 17 men with metabolic syndrome for four months' nicotinic acid therapy gave levels of non-esterified fatty acid (NEFA), glucose and insulin during the oral glucose tolerance test were not significantly different from those before institution of nicotinic acid therapy, suggesting minimal changes in insulin sensitivity (Vega et al. 2005).

Interestingly, in a randomized, placebo-controlled, double-blind study of 30 men with the metabolic syndrome treated for six weeks with 1500 mg extended-release niacin ($n = 20$) or a placebo ($n = 10$), adiponectin increased by 56% ($p < 0.001$) and leptin by 26.8% ($p < 0.012$). Resistin, TNF-α, IL-6 and hsCRP remained unchanged. In spite of the increase in adiponectin, there was no improvement in endothelial function and the Homeostasis Model Assessment (HOMA) index actually deteriorated by 42% ($p < 0.014$). This suggests that short-term treatment with extended-release niacin causes a pronounced increase in adiponectin, but fails to improve atheroprotective functions attributed to it such as insulin sensitivity, anti-inflammation and endothelial function (Westphal et al. 2007). In contrast, the findings of a subgroup analysis of the HDL-Atherosclerosis Treatment (HATS) study showed that aggressive treatment of atherogenic dyslipidemia with niacin and simvastatin therapy, despite modest worsening of insulin sensitivity beyond its generally low baseline level in this population with established coronary disease, reduces the progression of coronary stenosis in subjects classified as being at even greater risk of developing recurrent manifestations of cardiovascular disease by virtue of having the metabolic syndrome, being insulin resistant or having dysglycaemia (Vittone et al. 2007).

In contrast, a number of studies showed worsening of insulin sensitivity with niacin treatment. For instance, in randomised double-blind placebo-controlled treatment study in which 20 men with the metabolic syndrome received 1500 mg niacin for six weeks, low- and medium-molecular weight adiponectin increased by 35% and 33%, respectively, but high molecular weight (HMW) adiponectin by 88% (all $p < 0.05$). The increase in HMW adiponectin was almost twice as large in patients with lower body mass index (BMI) and better insulin sensitivity. However, treatment with niacin induced a deterioration of insulin sensitivity, as assessed by the HOMA insulin resistance (HOMA-IR), independently of the increase in HMW adiponectin. HMW adiponectin is the fraction most affected by treatment with niacin. The niacin-associated deterioration of insulin sensitivity, however, occurs even in subgroups with the greatest increase of HMW adiponectin (Westphal and Luley 2008).

Interestingly, Poynten et al. (2003) have shown that niacin-induced insulin resistance is largely mediated through increased availability of circulating fatty acids to muscle rather than with increased muscle lipid content, while Greenbaum et al. (1996) showed that niacin increased insulin secretion per se; whether this may subsequently deplete the reserve of β pancreatic cells is not yet clear.

Importantly, Libby et al. (2010) found an association between increasing BMI and increasing fasting plasma glucose and diagnosis of new-onset diabetes after initiation of extended-release niacin. Their conclusion was that extended-release niacin may increase fasting plasma glucose into the diabetic range, especially for obese patients (Libby et al. 2010).

It is not yet clear whether the combination of statin and niacin will increase the incidence of diabetes. This is of significance as recent meta-analysis has shown that the association of incidence of diabetes with statin therapy remains uncertain. In particular the authors of the meta-analysis recommended that future statin trials should be designed to investigate whether statin treatment is associated with increased risk of induced diabetes (Rajpathak et al. 2009). Evaluation of the longer term impact of niacin's effect on insulin resistance and glucose metabolism awaits the results of longer-term clinical trials. Perhaps this may elicit or refute the diabetogenic effect of niacin.

38.5.5 Niacin and Cardiovascular Disease

Different clinical trials have shown that nicotinic acid has potential benefit in treating CVD and dyslipidaemia. Importantly, recent meta-analysis showed that niacin was associated with a significant reduction in cardiovascular events and possibly a small but non-significant decrease in coronary and cardiovascular mortality (Duggal et al. 2010).

Administration of nicotinic acid for five years at dose of 3 g/day as monotherapy in the Coronary Drug Project study led to marked secondary prevention of myocardial infarction (Coronary Drug Project 1975). Importantly, niacin is also being shown to have long-term benefit in decreasing mortality after discontinuation of treatment; a follow-up for 15 years of Coronary Drug Project revealed that niacin decreased mortality in patients treated with niacin (Canner et al. 1986).

The administration of niacin in combination with statin or fibrate has shown significant reduction in cardiovascular mortality. For instance, the Stockholm Ischaemic Heart Disease secondary prevention study showed that combined treatment of clofibrate and niacin for five years was associated with marked decrease in ischaemic heart disease (Carlson and Rosenhamer 1988).

In contrast, Brown et al. (2001) showed that the combination of simvastatin and niacin for three years was associated with a marked improvement in coronary stenosis. Interestingly, a study by Taylor et al. (2004) showed that the combination of statin and extended-release niacin slowed the progression of atherosclerosis in individuals known to have coronary heart disease. A similar observation were obtained by the ARBITER 6–HALTS trial (Arterial Biology for the Investigation of the Treatment Effects of Reducing Cholesterol 6–HDL and LDL Treatment Strategies), which concluded that the use of extended-release niacin causes a significant regression of carotid IMT when combined with a statin and that niacin is superior to ezetimibe (Taylor et al. 2009). Furthermore, in a double-blind, randomized, placebo-controlled study of 2 g daily modified-release niacin added to statin therapy in 71 patients (type 2 diabetes with CVD or carotid/peripheral atherosclerosis) with low HDL-cholesterol (HDL-c), niacin increased HDL-c by 23%, decreased LDL-c by 19% and significantly reduced carotid wall area compared with placebo (Lee et al. 2009).

One unique effect of niacin, beside the protective effect on HDL-c, LDL-c and triglyceride is its ability to decrease plasma concentration of Lp(a), which has been suggested to play a role as an independent risk factor for coronary heart disease (Vosper 2009). Interestingly, members of the Oxford research group identified two Lp(a) variants that were strongly associated with both an increased level of Lp(a) and an increased risk of coronary disease in 3145 case subjects with coronary disease (Clarke et al. 2009). Furthermore, Kamstrup et al. (2009) assessed the association of Lp(a) recorded for 31 years for all participants with myocardial infarction (MI) in large studies. Three studies of white individuals from Copenhagen, Denmark, were used:

- Copenhagen City Heart Study (CCHS), a prospective general population study with 16 years of follow-up (1991–2007, $n = 8637$, 599 MI events);
- Copenhagen General Population Study (CGPS), a cross-sectional general population study (2003–2006, n = 29 388, 994 MI events);
- Copenhagen Ischemic Heart Disease Study (CIHDS), a case-control study (1991–2004, $n = 2461$, 1231 MI events).

The authors concluded that a causal association between elevated Lp(a) levels and increased risk of MI exists (Kamstrup et al. 2009).

The unpleasant side of niacin (flushing) is one factor that has limited its widespread use. This effect is likely to be mediated by prostaglandin D_2 (PGD_2), which stimulates PGD_2 receptor 1(DP1) in the skin (Paolini et al. 2008). Activation of this pathway by niacin is independent of the lipid-modifying pathway. Therefore, blocking the flushing pathway will not alter the lipid-lowering effect of niacin. Laropiprant is a potent selective antagonist of prostaglandin 2-receptor subtype-1 and may reduce niacin-induced flushing in large proportions of individuals. The combination of niacin with laropiprant may therefore enable use of niacin at maximum therapeutic doses and may have favourable outcome not only in lipid profile but also in incidence of cardiovascular disease (Parhofer 2009). The outcome of the AIM-HIGH

(Atherothrombosis Intervention in Metabolic Syndrome with Low HDL/High Triglycerides and Impact on Global Health Outcomes) and HPS2-THRIVE trial (Treatment of HDL to Reduce the Incidence of Vascular Events) will increase our understanding of the effect of niacin in the modulation of lipid metabolism and incidence of CVD (AIM-HIGH Investigators 2011).

Summary Points

- Niacin is an important vitamin for the integrity and function of the cell.
- In large doses, niacin can act as a lipid-lowering medication.
- Despite its ability as a wide spectrum lipid lowering medication (decreased LDL-c, triglyceride and increased HDL-c) and ability to decrease Lp(a), its widespread uses is limited by the unpleasant flushing.
- In the majority of clinical trials, niacin has been shown to be effective and safe in decreasing the incidence of cardiovascular event. This may be attributed to the ability of niacin to act as anti-inflammatory, anti-oxidant and modulator of insulin sensitivity (Figure 38.1). This important effect in CVD is achieved with and without the use of statins.

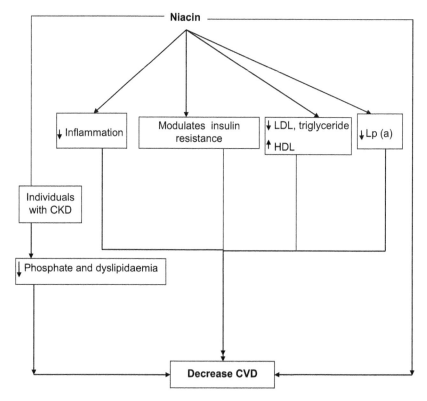

Figure 38.1 Complex effects of niacin on inflammation, insulin sensitivity and dyslipideamia, and potential benefit in treating dyslipidaemia and high plasma phosphate level with CKD. These effects of niacin may ultimately contribute to reducing the incidence of CVD.

- The effect of niacin in insulin resistance remains a controversial issue. The AIM-HIGH (**A**therothrombosis **I**ntervention in **M**etabolic Syndrome with Low **H**DL/**H**igh Triglycerides and Impact on **G**lobal **H**ealth Outcomes) and HPS2-THRIVE trial (Treatment of HDL to Reduce the Incidence of Vascular Events) will increase our knowledge about the clinical impact of niacin.
- The impact of niacin in insulin sensitivity is important as the majority of those who needs to be treated with niacin will have a degree of insulin resistance (individuals with metabolic syndrome and type 2 diabetes). Furthermore, recent meta-analyses suggest that statin administration is associated with the degree of insulin resistance.
- It is possible to speculate that the concern about the wider use of niacin due to associated flushing may extend beyond this point to the impact of niacin in insulin resistance. Interestingly, future studies may reveal whether niacin *per se* or in combination will be safe in type 2 diabetes and in individuals with metabolic syndrome.
- Further analysis of future clinical trials may also reveal an exciting role of niacin as a potential treatment of inflammation, high plasma phosphate and dyslipidaemia in association with CKD.

Key Facts

- Niacin is found in different types of foods (fish, milk, poultry) and is important for regulation of sleep, appetite and mood.
- Niacin is quickly absorbed through the stomach and intestine and severe deficiency is associated with diarrhoea, dementia and dermatitis.
- Niacin at cellular level has protective effects in preventing apoptosis, DNA damage and lipid metabolism.
- Niacin has an important role in the pathway of energy generation within the cell.
- Niacin has wide spectrum anti-lipid lowering effect as it has shown to decrease LDL-c, triglyceride and Lp(a), and increase HDL.
- Niacin is the only lipid-lowering medication that can decrease the level of Lp(a). The increase in the level of Lp(a) is associated with an increase in the incidence of CVD.
- Niacin has shown potential benefit in treating both hyperphosphataemia and hyperlipidaemia associated with chronic kidney disease.
- The protective effect of niacin against cardiovascular disease is subject of ongoing clinical trials.
- The protective effect of niacin against cardiovascular disease is increased with concomitant administration of a statin.

List of Abbreviations

ABCA-1	ATP-binding cassette transporter 1
ABCG-1	ATP-binding cassette sub-family G member 1
ADH	alcohol dehydrogenase

ALERT	Assessment of Lescol in Renal Transplantation
ANG II	angiotensin II (ANG II)
BMI	body mass index
cAMP	cyclic adenosine monophosphate
cGMP	cyclic guanosine monophosphate
CHD	coronary heart disease
ChREBP	carbohydrate-responsive element binding protein
CKD	chronic kidney disease
COMA	Committee of Medical Aspects of Food Policy
COX-1	cyclooxygenase-1
CVD	cardiovascular disease
DGAT	diacylglycerol acyltransferase
ER	endoplasmic reticulum
GFR	glomerular filtration rate
GSH	glutathione
HATS	HDL-Atherosclerosis Treatment
HDL	high density lipoprotein
HDL-c	HDL-cholesterol (HDL-c)
HMW	high molecular weight
HOMA	Homeostasis Model Assessment
HOMA-IR	HOMA insulin resistance
hsCFP	high-sensitivity C-reactive protein
ICAM-1	intercellular adhesion molecule-1
IDL	intermediate density lipoprotein
IFN-y	interferon-y
IL-1	interleukin-1
IMT	intima-media thickness
LDL	low density lipoprotein
L-FABP	liver fatty acid binding protein
Lp(a)	lipoprotein a
LXR	liver X receptor
MCP-1	monocyte chemotactic protein-1
MI	myocardial infarction
MTP	microsomal triglyceride transfer protein
NAD	nicotinamide adenine dinucleotide
NADP	nicotinamide adenine dinucleotide phosphate
NAD(P)H	nicotinamide adenine dinucleotide phosphate, reduced form
NEFA	non-esterified fatty acid
PAI-1	plasminogen activator inhibitor-1
PARP	poly-ADP-ribose polymerases
PECAM	platelet endothelial cell adhesion molecule
PGD_2	prostaglandin D_2
PPAR-γ	peroxisome proliferator-activated receptor gamma
ROS	reactive oxygen species (ROS)
TG	triglyceride
TGF-β	transforming growth factor beta

TNF-α   tumour necrosis factor-alpha
VCAM-1   vascular cell adhesion molecule-1
VLDL   very low density lipoprotein

References

Aktories, K., Schultz, G., and Jakobs, K.H., 1980. Regulation of adenylate cyclase activity in hamster adipocytes. Inhibition by prostaglandins, alpha-adrenergic agonists and nicotinic acid. *Naunyn-Schmiedeberg's Archives of Pharmacology.* 312: 167–173.

Amouyel, P., Isorez, D., Bard, J.M., Goldman, M., Lebel, P., Zylberberg, G., and Fruchart, J. C., 1993. Parental history of early myocardial infarction is associated with decreased levels of lipoparticle AI in adolescents. *Arteriosclerosis, Thrombosis, and Vascular Biology.* 13: 1640–1644.

AIM-HIGH Investigators, 2011. The role of niacin in raising high-density lipoprotein cholesterol to reduce cardiovascular events in patients with atherosclerotic cardiovascular disease and optimally treated low-density lipoprotein cholesterol: baseline characteristics of study participants. The Atherothrombosis Intervention in Metabolic syndrome with low HDL/high triglycerides: impact on Global Health outcomes (AIM-HIGH) trial. *American Heart Journal.* 161: 538–543.

Ariumi, Y., Masutani, M., Copeland, T.D., Mimori, T., Sugimura, T., Shimotohno, K., Ueda, K., Hatanaka, M., and Noda, M., 1999. Suppression of the poly(ADP-ribose) polymerase activity by DNA-dependent protein kinase in vitro. *Oncogene.* 18: 4616–4625.

Barr, T.L., and Conley, Y.P., 2007. Poly(ADP-ribose) polymerase-1 and its clinical applications in brain injury. *Journal of Neuroscience Nursing.* 39: 278–284.

Bechgaard, H., and Jespersen, S., 1977. GI absorption of niacin in humans. *Journal of Pharmaceutical Sciences.* 66: 871–872.

Benavente, C.A., and Jacobson, E.L., 2008. Niacin restriction upregulates NADPH oxidase and reactive oxygen species (ROS) in human keratinocytes. *Free Radical Biology & Medicine.* 44: 527–537.

Berger, N.A., 1985. Poly(ADP-ribose) in the cellular response to DNA damage. *Radiation Research.* 101: 4–15.

Block, G.A., Klassen, P.S., Lazarus, J.M., Ofsthun, N., Lowrie, E.G., and Chertow, G.M., 2004. Mineral metabolism, mortality, and morbidity in maintenance hemodialysis. *Journal of the American Society of Nephrology.* 15: 2208–2218.

Blum, C.B., Levy, R.I., Eisenberg, S., Hall, M., Goebel, R.H., and Berman, M., 1977. High density lipoprotein metabolism in man. *Journal of Clinical Investigation.* 60: 795–807.

Bodor, E.T., and Offermanns, S., 2008. Nicotinic acid: an old drug with a promising future. *British Journal of Pharmacology.* 153(Suppl 1): S68–S75.

Brown, B.G., Zhao, X.Q., Chait, A., Fisher, L.D., Cheung, M.C., Morse, J.S., Dowdy, A.A., Marino, E.K., Bolson, E.L., Alaupovic, P., Frohlich, J., and Albers, J.J., 2001, Simvastatin and niacin, antioxidant vitamins, or the combination for the prevention of coronary disease. *The New England Journal of Medicine*. 345: 1583–1592.

Caballeria, J., Baraona, E., Deulofeu, R., Hernandez-Munoz, R., Rodes, J., and Lieber, C.S., 1991. Effects of H2-receptor antagonists on gastric alcohol dehydrogenase activity. *Digestive Diseases and Sciences*. 36: 1673–1679.

Canner, P.L., Berge, K.G., Wenger, N.K., Stamler, J., Friedman, L., Prineas, R.J., and Friedewald, W., 1986. Fifteen year mortality in Coronary Drug Project patients: long-term benefit with niacin. *Journal of the American College of Cardiology*. 8: 1245–1255.

Carlson, L.A., Oro, L., and Ostman, J. 1968. Effect of a single dose of nicotinic acid on plasma lipids in patients with hyperlipoproteinemia. *Acta Medica Scandinavica*. 183: 457–465.

Carlson, L.A., and Rosenhamer, G., 1988. Reduction of mortality in the Stockholm Ischaemic Heart Disease Secondary Prevention Study by combined treatment with clofibrate and nicotinic acid. *Acta Medica Scandinavica*. 223: 405–418.

Cervellera, M.N., and Sala, A., 2000. Poly(ADP-ribose) polymerase is a B-MYB coactivator. *The Journal of Biological Chemistry*. 275: 10692–10696.

Cheng, S.C., Young, D. O., Huang, Y., Delmez, J.A., and Coyne, D.W., 2008. A randomized, double-blind, placebo-controlled trial of niacinamide for reduction of phosphorus in hemodialysis patients. *Clinical Journal of the American Society of Nephrology*. 3: 1131–1138.

Cho, K.H., Kim, H.J., Rodriguez-Iturbe, B., and Vaziri, N.D., 2009. Niacin ameliorates oxidative stress, inflammation, proteinuria, and hypertension in rats with chronic renal failure. *American Journal of Physiology – Renal Physiology*. 297: F106–F113.

Cho, K.H., Kim, H.J., Kamanna, V.S., and Vaziri, N.D., 2010. Niacin improves renal lipid metabolism and slows progression in chronic kidney disease. *Biochimica et Biophysica Acta*. 1800: 6–15.

Clarke, R., Peden, J.F., Hopewell, J.C., Kyriakou, T., Goel, A., Heath, S.C., Parish, S., Barlera, S., Franzosi, M.G., Rust, S., Bennett, D., Silveira, A., Malarstig, A., Green, F.R., Lathrop, M., Gigante, B., Leander, K., de, F.U., Seedorf, U., Hamsten, A., Collins, R., Watkins, H., and Farrall, M., 2009. Genetic variants associated with Lp(a) lipoprotein level and coronary disease. *The New England Journal of Medicine*. 361: 2518–2528.

Cook, C.C., Hallwood, P.M., and Thomson, A.D., 1998. B Vitamin deficiency and neuropsychiatric syndromes in alcohol misuse. *Alcohol and Alcoholism*. 33: 317–336.

Coronary Drug Project, 1975. Clofibrate and niacin in coronary heart disease. *The Journal of the American Medical Association*. 231: 360–381.

David, K.K., Andrabi, S.A., Dawson, T.M., and Dawson, V.L., 2009. Parthanatos, a messenger of death. *Frontiers in Bioscience*. 14: 1116–1128.

Davis, R.A., 1999. Cell and molecular biology of the assembly and secretion of apolipoprotein B-containing lipoproteins by the liver. *Biochimica et Biophysica Acta*. 1440: 1–31.

Department of Health, Committee on Medical Aspects of Food Policy, 1991. Dietary Reference Values for Food Energy and Nutrients for the UK: Report of the Panel on Dietary Reference Values of the Committee on Medical Aspects of Food Policy. Report on Health and Social Subjects 41. HMSO, London, UK. pp. 210.

de Murcia, G., and de Menissier, M.J., 1994. Poly(ADP-ribose) polymerase: a molecular nick-sensor. *Trends in Biochemical Sciences*. 19: 172–176.

de Murcia, G., Schreiber, V., Molinete, M., Saulier, B., Poch, O., Masson, M., Niedergang, C., and de Menissier, M.J., 1994. Structure and function of poly(ADP-ribose) polymerase. *Molecular and Cellular Biochemistry*. 138: 15–24.

Digby, J.E., McNeill, E., Dyar, O.J., Lam, V., Greaves, D R., and Choudhury, R.P. 2010. Anti-inflammatory effects of nicotinic acid in adipocytes demonstrated by suppression of fractalkine, RANTES, and MCP-1 and upregulation of adiponectin. *Atherosclerosis*. 209: 89–95.

Dixon, J.L., Furukawa, S., and Ginsberg, H.N., 1991. Oleate stimulates secretion of apolipoprotein B-containing lipoproteins from Hep G2 cells by inhibiting early intracellular degradation of apolipoprotein B. *The Journal of Biological Chemistry*. 266: 5080–5086.

Duggal, J.K., Singh, M., Attri, N., Singh, P.P., Ahmed, N., Pahwa, S., Molnar, J., Singh, S., Khosla, S., and Arora, R., 2010. Effect of niacin therapy on cardiovascular outcomes in patients with coronary artery disease. *Journal of Cardiovascular Pharmacology and Therapeutics*. 15: 158–166.

Eliasson, M.J., Sampei, K., Mandir, A.S., Hurn, P.D., Traystman, R.J., Bao, J., Pieper, A., Wang, Z.Q., Dawson, T.M., Snyder, S.H., and Dawson, V.L., 1997. Poly(ADP-ribose) polymerase gene disruption renders mice resistant to cerebral ischemia. *Nature Medicine*. 3: 1089–1095.

Etchason, J.A., Miller, T.D., Squires, R.W., Allison, T.G., Gau, G.T., Marttila, J.K., and Kottke, B.A., 1991. Niacin-induced hepatitis: a potential side effect with low-dose time-release niacin. *Mayo Clinic Proceedings*. 66: 23–28.

Fischer, D.J., Knight, L.L., and Vestal, R.E., 1991. Fulminant hepatic failure following low-dose sustained-release niacin therapy in hospital. *The Western Journal of Medicine*. 155: 410–412.

Ganji, S.H., Kamanna, V.S., and Kashyap, M.L., 2003. Niacin and cholesterol: role in cardiovascular disease (review). *The Journal of Nutritional Biochemistry*. 14: 298–305.

Ganji, S.H., Tavintharan, S., Zhu, D., Xing, Y., Kamanna, V.S., and Kashyap, M.L., 2004. Niacin noncompetitively inhibits DGAT2 but not DGAT1 activity in HepG2 cells. *Journal of Lipid Research*. 45: 1835–1845.

Ganji, S.H., Qin, S., Zhang, L., Kamanna, V.S., and Kashyap, M.L., 2009. Niacin inhibits vascular oxidative stress, redox-sensitive genes, and monocyte adhesion to human aortic endothelial cells. *Atherosclerosis*. 202: 68–75.

Germain, M., Affar, E.B., D'Amours, D., Dixit, V.M., Salvesen, G.S., and Poirier, G.G., 1999. Cleavage of automodified poly(ADP-ribose) polymerase during apoptosis. Evidence for involvement of caspase-7. *The Journal of Biological Chemistry.* 274: 28379–28384.

Ginsberg, H.N., 1995. Synthesis and secretion of apolipoprotein B from cultured liver cells. *Current Opinion in Lipidology.* 6: 275–280.

Greenbaum, C.J., Kahn, S.E., and Palmer, J.P., 1996. Nicotinamide's effects on glucose metabolism in subjects at risk for IDDM. *Diabetes.* 45: 1631–1634.

Grundy, S.M., Mok, H.Y., Zech, L., and Berman, M., 1981b. Influence of nicotinic acid on metabolism of cholesterol and triglycerides in man. *Journal of Lipid Research.* 22: 24–36.

Hassa, P.O., Haenni, S.S., Elser, M., and Hottiger, M.O., 2006. Nuclear ADP-ribosylation reactions in mammalian cells: where are we today and where are we going?. *Microbiology and Molecular Biology Reviews.* 70: 789–829.

Hassa, P.O., 2009. The molecular 'Jekyll and Hyde' duality of PARP1 in cell death and cell survival. *Frontiers of Bioscience.* 14: 72–111.

Hegyi, J., Schwartz, R.A., and Hegyi, V., 2004a. Pellagra: dermatitis, dementia, and diarrhea. *International Journal of Dermatology.* 43: 1–5.

Henderson, L.M., and Gross, C.J., 1979. Metabolism of niacin and niacinamide in perfused rat intestine. *Journal of Nutrition.* 109: 654–662.

Henderson, L.M., 1983. Niacin. *Annual Review of Nutrition.* 3: 289–307.

Ieraci, A., and Herrera, D.G., 2006. Nicotinamide protects against ethanol-induced apoptotic neurodegeneration in the developing mouse brain. *PLoS Medicine.* 3: e101.

Jacob, R.A., Swendseid, M.E., McKee, R.W., Fu, C.S., and Clemens, R.A., 1989. Biochemical markers for assessment of niacin status in young men: urinary and blood levels of niacin metabolites. *Journal of Nutrition.* 119: 591–598.

Jardine, A.G., Fellstrom, B., Logan, J.O., Cole, E., Nyberg, G., Gronhagen-Riska, C., Madsen, S., Neumayer, H.H., Maes, B., Ambuhl, P., Olsson, A.G., Pedersen, T., and Holdaas, H., 2005. Cardiovascular risk and renal transplantation: post hoc analyses of the Assessment of Lescol in Renal Transplantation (ALERT) Study. *American Journal of Kidney Disease.* 46: 529–536.

Jin, F.Y., Kamanna, V.S., and Kashyap, M.L., 1999. Niacin accelerates intracellular ApoB degradation by inhibiting triacylglycerol synthesis in human hepatoblastoma (HepG2) cells. *Arteriosclerosis, Thrombosis, and Vascular Biology.* 19: 1051–1059.

Kamstrup, P.R., Tybjaerg-Hansen, A., Steffensen, R., and Nordestgaard, B.G., 2009. Genetically elevated lipoprotein(a) and increased risk of myocardial infarction. *The Journal of the American Medical Association.* 301: 2331–2339.

Kestenbaum, B., Sampson, J.N., Rudser, K.D., Patterson, D.J., Seliger, S.L., Young, B., Sherrard, D.J., and Andress, D.L., 2005. Serum phosphate levels and mortality risk among people with chronic kidney disease. *Journal of the American Society of Nephrology.* 16: 520–528.

Knopp, R.H., Retzlaff, B.M., Fish, B., Dowdy, A., Twaddell, B., Nguyen, T., and Paramsothy, P., 2009. Slo-Niacin® and atorvastatin treatment of lipoproteins and inflammatory markers in combined hyperlipidemia. *Journal of Clinical Lipidology*. 3: 167–178.

Lazebnik, Y.A., Kaufmann, S.H., Desnoyers, S., Poirier, G.G., and Earnshaw, W.C., 1994. Cleavage of poly(ADP-ribose) polymerase by a proteinase with properties like ICE. *Nature*. 371(6495): 346–347.

Lee, J.M., Robson, M.D., Yu, L. M., Shirodaria, C.C., Cunnington, C., Kylintireas, I., Digby, J.E., Bannister, T., Handa, A., Wiesmann, F., Durrington, P.N., Channon, K.M., Neubauer, S., and Choudhury, R.P., 2009. Effects of high-dose modified-release nicotinic acid on atherosclerosis and vascular function: a randomized, placebo-controlled, magnetic resonance imaging study. *Journal of the American College of Cardiology*. 54: 1787–1794.

Libby, A., Meier, J., Lopez, J., Swislocki, A.L., and Siegel, D., 2010. The effect of body mass index on fasting blood glucose and development of diabetes mellitus after initiation of extended-release niacin. *Metabolic Syndrome and Related Disorders*. 8: 79–84.

Linke, A., Sonnabend, M., Fasshauer, M., Hollriegel, R., Schuler, G., Niebauer, J., Stumvoll, M., and Bluher, M., 2009. Effects of extended-release niacin on lipid profile and adipocyte biology in patients with impaired glucose tolerance. *Atherosclerosis*. 205: 207–213.

Meisenberg, G., and Simmons, W.H., 1998. Principles of Medical Biochemistry. Mosby, St Louis, MO, USA. pp. 742.

Mittal, M.K., Florin, T., Perrone, J., Delgado, J.H., and Osterhoudt, K.C., 2007a. Toxicity from the use of niacin to beat urine drug screening. *Annals of Emergency Medicine*. 50: 587–590.

Morgan, J.M., de la. Llera-Moya and Capuzzi, D.M., 2007. Effects of niacin and Niaspan on HDL lipoprotein cellular SR-BI-mediated cholesterol efflux. *Journal of Clinical Lipidology*. 1: 614–619.

Muller, D., Mehling, H., Otto, B., Bergmann-Lips, R., Luft, F., Jordan, J., and Kettritz, R., 2007. Niacin lowers serum phosphate and increases HDL cholesterol in dialysis patients. *Clinical Journal of the American Society of Nephrology*. 2: 1249–1254.

Muntner, P., He, J., Astor, B.C., Folsom, A.R., and Coresh, J., 2005. Traditional and nontraditional risk factors predict coronary heart disease in chronic kidney disease: results from the atherosclerosis risk in communities study. *Journal of the American Society of Nephrology*. 16: 529–538.

Oliver, F.J., Menissier-de, M.J., Nacci, C., Decker, P., Andriantsitohaina, R., Muller, S., de la, R.G., Stoclet, J.C., and de, M.G., 1999. Resistance to endotoxic shock as a consequence of defective NF-kappaB activation in poly (ADP-ribose) polymerase-1 deficient mice. *EMBO Journal*. 18: 4446–4454.

Olsson, A.G., 2010b. HDL and LDL as therapeutic targets for cardiovascular disease prevention: the possible role of niacin. *Nutrition, Metabolism & Cardiovascular Disease*. 20: 553–557.

Paolini, J.F., Mitchel, Y.B., Reyes, R., Kher, U., Lai, E., Watson, D.J., Norquist, J.M., Meehan, A.G., Bays, H.E., Davidson, M., and Ballantyne, C.M.,

2008. Effects of laropiprant on nicotinic acid-induced flushing in patients with dyslipidemia. *The American Journal of Cardiology*. 101: 625–630.

Parhofer, K.G., 2009. Review of extended-release niacin/laropiprant fixed combination in the treatment of mixed dyslipidemia and primary hypercholesterolemia. *Journal of Vascular Health and Risk Management*. 5: 901–908.

Park, S.D., Kim, C.G., and Kim, M.G., 1983. Inhibitors of poly(ADP-ribose) polymerase enhance DNA strand breaks, excision repair, and sister chromatid exchanges induced by alkylating agents. *Environmental Mutagenesis*. 5: 515–525.

Paton, A., 2005. Alcohol in the body. *British Medical Journal*. 330(7482): 85–87.

Perricone, N.V., Bagchi, D., Echard, B., and Preuss, H.G., 2010. Long-term metabolic effects of different doses of niacin-bound chromium on Sprague–Dawley rats. *Molecular and Cellular Biochemistry*. 338: 91–103.

Pollak, N., Dolle, C., and Ziegler, M., 2007a. The power to reduce: pyridine nucleotides--small molecules with a multitude of functions. *Biochemistry Journal*. 402: 205–218.

Poynten, A.M., Gan, S.K., Kriketos, A.D., O'Sullivan, A., Kelly, J.J., Ellis, B.A., Chisholm, D.J., and Campbell, L.V., 2003. Nicotinic acid-induced insulin resistance is related to increased circulating fatty acids and fat oxidation but not muscle lipid content. *Metabolism*. 52: 699–704.

Puchois, P., Kandoussi, A., Fievet, P., Fourrier, J. L., Bertrand, M., Koren, E., and Fruchart, J.C., 1987. Apolipoprotein A-I containing lipoproteins in coronary artery disease. *Atherosclerosis*. 68: 35–40.

Rajpathak, S.N., Kumbhani, D.J., Crandall, J., Barzilai, N., Alderman, M., and Ridker, P.M., 2009. Statin therapy and risk of developing type 2 diabetes: a meta-analysis. *Diabetes Care*. 32: 1924–1929.

Ramirez, R., Sener, A., and Malaisse, W.J., 1995. Hexose metabolism in pancreatic islets: effect of D-glucose on the mitochondrial redox state. *Molecular and Cellular Biochemistry*. 142: 43–48.

Restrepo Valencia, C.A., and Cruz, J., 2008. Safety and effectiveness of nicotinic acid in the management of patients with chronic renal disease and hyperlipidemia associated to hyperphosphatemia [in Spanish]. *Nefrologia*. 28: 61–66.

Rinninger, F., Kaiser, T., Windler, E., Greten, H., Fruchart, J.C., and Castro, G., 1998. Selective uptake of cholesteryl esters from high-density lipoprotein-derived LpA-I and LpA-I:A-II particles by hepatic cells in culture. *Biochimica et Biophysica Acta*. 1393: 277–291.

Rubic, T., Trottmann, M., and Lorenz, R.L., 2004. Stimulation of CD36 and the key effector of reverse cholesterol transport ATP-binding cassette A1 in monocytoid cells by niacin. *Biochemical Pharmacology*. 67: 411–419.

Sampathkumar, K., Selvam, M., Sooraj, Y. S., Gowthaman, S., and Ajeshkumar, R.N., 2006. Extended release nicotinic acid—a novel oral agent for phosphate control. *International Urology and Nephrology*. 38: 171–174.

Sauve, A.A., Wolberger, C., Schramm, V.L., and Boeke, J. D., 2006. The biochemistry of sirtuins. *Annual Review of Biochemistry*. 75: 435–465.

Schreiber, V., Hunting, D., Trucco, C., Gowans, B., Grunwald, D., de Murcia, G., and de Murcia, J.M., 1995. A dominant-negative mutant of human poly(ADP-ribose) polymerase affects cell recovery, apoptosis, and sister chromatid exchange following DNA damage. *Proceedings of the National Academy of Sciences of the United States of America.* 92: 4753–4757.

Shall, S., 1995. ADP-ribosylation reactions. *Biochimie.* 77: 313–318.

Shepherd, J., Packard, C.J., Patsch, J.R., Gotto, A.M., and Taunton, O.D., 1979. Effects of nicotinic acid therapy on plasma high density lipoprotein subfraction distribution and composition and on apolipoprotein A metabolism. *Journal of Clinical Investigation.* 63: 858–867.

Shibata, K., and Matsuo, H., 1989. Correlation between niacin equivalent intake and urinary excretion of its metabolites, N′-methylnicotinamide, N′-methyl-2-pyridone-5-carboxamide, and N′-methyl-4-pyridone-3-carboxamide, in humans consuming a self-selected food. *The American Journal of Clinical Nutrition.* 50: 114–119.

Slee, E.A., Adrain, C., and Martin, S.J., 2001. Executioner caspase-3, -6, and -7 perform distinct, non-redundant roles during the demolition phase of apoptosis. *The Journal of Biological Chemistry.* 276: 7320–7326.

Smith, S., and de Lange, T., 2000. Tankyrase promotes telomere elongation in human cells. *Current Biology.* 10: 1299–1302.

Svedmyr, N., and Harthon, L., 1970. Comparison between the absorption of nicotinic acid and pentaerythritol tetranicotinate (Perycit) from ordinary and enterocoated tablets. *Acta Pharmacologica et Toxicologica.* 28: 66–74.

Szabo, C., Cuzzocrea, S., Zingarelli, B., O'Connor, M., and Salzman, A.L., 1997. Endothelial dysfunction in a rat model of endotoxic shock. Importance of the activation of poly (ADP-ribose) synthetase by peroxynitrite. *Journal of Clinical Investigation.* 100: 723–735.

Tavintharan, S., and Kashyap, M.L., 2001. The benefits of niacin in atherosclerosis. *Current Atherosclerosis Reports.* 3: 74–82.

Tavintharan, S., Lim, S.C., and Sum, C.F., 2009. Effects of niacin on cell adhesion and early atherogenesis: biochemical and functional findings in endothelial cells. *Basic & Clinical Pharmacology & Toxicology.* 104: 206–210.

Taylor, A.J., Sullenberger, L.E., Lee, H.J., Lee, J.K., and Grace, K.A., 2004. Arterial Biology for the Investigation of the Treatment Effects of Reducing Cholesterol (ARBITER) 2: a double-blind, placebo-controlled study of extended-release niacin on atherosclerosis progression in secondary prevention patients treated with statins. *Circulation.* 110: 3512–3517.

Taylor, A.J., Villines, T.C., Stanek, E.J., Devine, P.J., Griffen, L., Miller, M., Weissman, N.J., and Turco, M., 2009. Extended-release niacin or ezetimibe and carotid intima-media thickness. *The New England Journal of Medicine.* 361: 2113–2122.

Thoenes, M., Oguchi, A., Nagamia, S., Vaccari, C.S., Hammoud, R., Umpierrez, G.E., and Khan, B.V., 2007. The effects of extended-release niacin on carotid intimal media thickness, endothelial function and inflammatory markers in patients with the metabolic syndrome. *International Journal of Clinical Practice.* 61: 1942–1948.

Tonelli, M., Keech, A., Shepherd, J., Sacks, F., Tonkin, A., Packard, C., Pfeffer, M., Simes, J., Isles, C., Furberg, C., West, M., Craven, T., and Curhan, G., 2005. Effect of pravastatin in people with diabetes and chronic kidney disease. *Journal of the American Society of Nephrology*. 16: 3748–3754.

Uchida, K., 1993. Structure of ADP-ribosylating enzyme and DNA repair [in Japanese]. *Nippon Rinsho*. 51: 3051–3061.

Vaccari, C.S., Hammoud, R.A., Nagamia, S.H., Ramasamy, K., Dollar, A.L., and Khan, B.V., 2007a. Revisiting niacin: reviewing the evidence. *Journal of Clinical Lipidology*. 1: 248–255.

Vaccari, C.S., Nagamia, S., Thoenes, M., Oguchi, A., Hammoud, R., and Khan, B.V., 2007b. Efficacy of controlled-release niacin in treatment of metabolic syndrome: Correlation to surrogate markers of atherosclerosis, vascular reactivity, and inflammation. *Journal of Clinical Lipidology*. 1: 605–613.

Vega, G.L., Cater, N.B., Meguro, S., and Grundy, S.M., 2005. Influence of extended-release nicotinic acid on nonesterified fatty acid flux in the metabolic syndrome with atherogenic dyslipidemia. *American Journal of Cardiology*. 95: 1309–1313.

Virag, L., and Szabo, C., 2002. The therapeutic potential of poly(ADP-ribose) polymerase inhibitors. *Pharmacological Reviews*. 54: 375–429.

Vittone, F., Chait, A., Morse, J.S., Fish, B., Brown, B.G., and Zhao, X.Q., 2007. Niacin plus simvastatin reduces coronary stenosis progression among patients with metabolic syndrome despite a modest increase in insulin resistance: a subgroup analysis of the HDL-Atherosclerosis Treatment Study (HATS). *Journal of Clinical Lipidology*. 1: 203–210.

Vosper, H., 2009. Niacin: a re-emerging pharmaceutical for the treatment of dyslipidaemia. *British Journal of Pharmacology*. 158: 429–441.

Wesierska-Gadek, J., Schmid, G., and Cerni, C., 1996. ADP-ribosylation of wild-type p53 in vitro: binding of p53 protein to specific p53 consensus sequence prevents its modification. *Biochemical and Biophysical Research Communications*. 224: 96–102.

Westphal, S., Borucki, K., Taneva, E., Makarova, R., and Luley, C., 2007. Extended-release niacin raises adiponectin and leptin. *Atherosclerosis*. 193: 361–365.

Westphal, S., and Luley, C., 2008. Preferential increase in high-molecular weight adiponectin after niacin. *Atherosclerosis*. 198: 179–183.

Wu, B.J., Yan, L., Charlton, F., Witting, P., Barter, P.J., and Rye, K.A., 2010. Evidence that niacin inhibits acute vascular inflammation and improves endothelial dysfunction independent of changes in plasma lipids. *Arteriosclerosis, Thrombosis, and Vascular Biology*. 30: 968–975.

Wu, X., Sakata, N., Lui, E., and Ginsberg, H.N., 1994. Evidence for a lack of regulation of the assembly and secretion of apolipoprotein B-containing lipoprotein from HepG2 cells by cholesteryl ester. *The Journal of Biological Chemistry*. 269: 12375–12382.

Yao, Q., Pecoits-Filho, R., Lindholm, B., and Stenvinkel, P., 2004. Traditional and non-traditional risk factors as contributors to atherosclerotic cardiovascular disease in end-stage renal disease. *Scandinavian Journal of Urology and Nephrology*. 38: 405–416.

Young, E.W., Akiba, T., Albert, J.M., McCarthy, J.T., Kerr, P.G., Mendelssohn, D.C., and Jadoul, M., 2004. Magnitude and impact of abnormal mineral metabolism in hemodialysis patients in the Dialysis Outcomes and Practice Patterns Study (DOPPS). *American Journal of Kidney Disease*. 44(5 Suppl 2): 34–38.

CHAPTER 39
Pharmacological Use of Niacin for Lipoprotein Disorders

JOHN R. GUYTON,*[a] WANDA C. LAKEY,[a] KRISTEN B. CAMPBELL[b] AND NICOLE G. GREYSHOCK[a]

[a] Department of Medicine, Duke University Medical Center, Durham, NC 27710, USA; [b] Department of Pharmacy, Duke University Medical Center, Durham, NC 27710, USA.
*Email: john.guyton@duke.edu

39.1 Introduction

Nicotinic acid and nicotinamide, collectively known as vitamin B_3, are similar nutritionally, serving a vitamin function in milligram doses. In 1955 Altschul and colleagues discovered the ability of nicotinic acid (niacin) in gram doses to decrease serum cholesterol. Nicotinamide does not reduce cholesterol (Altschul *et al.* 1955). The role of niacin as the first pharmacological agent shown to favourably affect plasma lipoproteins was underscored by expanding recognition of elevated plasma cholesterol levels as a risk factor for atherosclerotic disease. Subsequent clinical trials with niacin demonstrated reductions in cardiovascular disease and death, although the most recent large trial showed no effect (Coronary Drug Project 1975; Carlson and Rosenhamer 1988; AIM-HIGH Investigators 2011).

In this chapter we discuss the multiple targeted and non-targeted pharmacological effects of niacin. We summarize clinical trial results and briefly address potential reasons for the recent null result.

We regard as unfortunate the strict traditional use of 'nicotinic acid' for the lipid-modifying pharmacological agent and 'niacin' for the vitamin including both nicotinic acid and nicotinamide. Medical patients generally want to avoid 'nicotine' and 'acid,' whereas 'niacin' conveys a favourable impression. Since 'niacinamide' is a recognizable term designating nicotinamide, we use 'niacin' throughout this chapter to refer specifically to 'nicotinic acid'.

39.2 Mechanisms of Action

Niacin affects metabolism in adipose tissue, liver, and other tissues *via* multiple pathways (Figure 39.1).

39.2.1 Role of the Niacin Receptor (GPR109A) in Inhibiting Non-esterified Fatty Acid Release from Adipocytes

The first hypotheses for cholesterol reduction with niacin pointed toward a steep niacin-induced decrease in non-esterified fatty acid (NEFA) mobilization from adipose tissue *via* inhibition of lipolysis (reviewed by Carlson 2005). Hormone sensitive lipase mediates lipolysis in response to increased cyclic adenosine monophosphate in adipocytes. A G-protein-coupled cell surface receptor (GPR) inhibiting adenylyl cyclase was proposed, and in 2003, three independent groups identified the human niacin receptors as the low-affinity

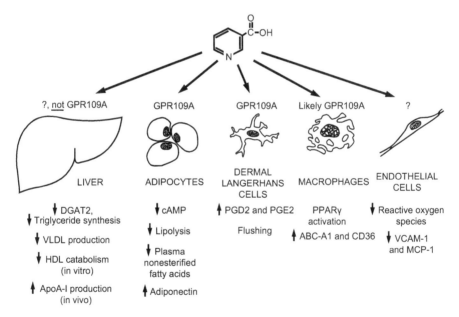

Figure 39.1 Pharmacological effects of niacin relevant to atherosclerosis. Effects on adipocytes and dermal Langerhans cells—and probably effects on macrophages—are dependent on a G-protein coupled receptor, GPR109A. The liver, which mediates most of the changes in lipoprotein metabolism, does not express this receptor. Effects demonstrated in endothelial cells are not known to be receptor-dependent.

HM74 and the high affinity HM74A. The mouse homolog for HM74A is PUMA-G (protein upregulated in macrophages by interferon-γ) (Soga et al. 2003; Tunaru et al. 2003; Wise et al. 2003). The pharmacologically relevant human receptor is HM74A, renamed GPR109A. Tissues from brown and white adipose, spleen, adrenal and lung demonstrated mRNA expression of HM74A/PUMA-G, with absent expression within the liver and intestines (Tunaru et al. 2003; Wise et al. 2003).

β-Hydroxybutyrate was identified as a candidate ligand for GPR109A in transfected CHO cells. Niacin has much greater affinity for the receptor than β-hydroxybutyrate, but concentrations of niacin from dietary sources are still too low to activate the receptor. Levels of β-hydroxybutyrate attained physiologically after an overnight fast and especially after 2–3 days of starvation are high enough to activate GPR109A and participate in a homeostatic feedback loop, whereby prevention of ketoacidosis and promotion of efficient use of energy stores occurs as β-hydroxybutyrate negatively regulates its own production (Taggart et al. 2005).

Although lipolysis inhibition via GPR109A may have some role in niacin's overall pharmacologic effects, it does not explain the changes in lipoprotein metabolism. Discrepancies between the timing and extent of plasma NEFA reductions and suppression of VLDL-TG production have long been noted (Carlson 2005). Most convincingly, MK-0354, a compound specifically developed to activate GPR109A, sharply decreased NEFA, but had no significant effect on plasma lipoprotein levels in humans (Lai et al. 2008).

39.2.2 Effects on Apolipoprotein B-containing Lipoproteins

Reduction in hepatic TG synthesis is the proposed initial step for the effect of niacin in lowering very low density lipoprotein (VLDL), which transports most of the triglyceride (TG) in fasting plasma. In microsomal preparations from human hepatoblastoma (HepG2) cells, niacin demonstrated non-competitive inhibition of diacylglycerol acyltransferase-2 (DGAT2), which catalyses the final step in triglyceride synthesis. Niacin-induced decreases in triglyceride production diminish the ability to lipidate nascent apoB peptide in the endoplasmic reticulum, resulting in intracellular degradation of apoB peptide and ultimately reduction in VLDL secretion (Kamanna and Kashyap 2008). This theory suggests that niacin may bind to DGAT2 or a regulatory protein, but thus far the exact mechanism of inhibition has not been described.

39.2.3 Niacin-Related Mechanisms to Increase HDL and Reverse Cholesterol Transport

Niacin effectively raises high density lipoprotein (HDL) cholesterol levels. In the absence of HDL, peripheral tissues accumulate cholesterol, presumably due to lack of reverse cholesterol transport from peripheral tissues to the liver. Niacin inhibits the degradation of HDL protein by HepG2 cells, potentially by downregulating cell surface expression of the ATP synthase beta chain.

However, niacin does not inhibit the selective uptake of HDL cholesteryl esters into the cells (Kamanna and Kashyap 2008). This combination of effects could raise HDL concentration, enhancing the ability of HDL to shuttle cholesterol from peripheral tissues to the liver. Apolipoprotein A-I (apoA-I) is the major protein associated with HDL. In HepG2 cells, niacin did not increase transcription or synthesis of apoA-I (Kamanna and Kashyap 2008). However, a stable isotope kinetic study in humans showed increased production of apoA-I following the administration of niacin (Lamon-Fava *et al.* 2008).

In peripheral tissues, macrophage uptake of oxidized lipoproteins is mediated largely by CD36, and cholesterol efflux from macrophages to nascent HDL occurs *via* ATP-binding cassette protein A1 (ABCA1). In a monocytoid cell line, ABCA1 expression was increased *via* inhibition of cyclic adenosine monophosphate (cAMP)/protein kinase A, while CD36 was increased *via* activation of the peroxisome proliferator-activated receptor gamma (PPARγ) by a presumed metabolite of prostaglandin D2 (Rubic *et al.* 2004). The proximate effector of these responses was likely GPR109A, but this remains to be shown.

39.2.4 Niacin-Induced Cutaneous Flushing

GPR109A receptors on dermal Langerhans dendritic cells are responsible for niacin-induced skin flushing by upregulating production of prostaglandin D2 (PGD2), which subsequently stimulates DP1 receptors on vascular and possibly other cells. G-protein-coupled receptors can activate heterotrimeric G-proteins and β-arrestins for cellular signalling in a differential manner. Recent results suggest that G-protein signalling independent of β-arrestin is responsible for the niacin-induced decrease in NEFA, while β-arrestin-1 mediates skin flushing (Walters *et al.* 2009).

39.2.5 Niacin-Induced Reduction in Inflammation and Oxidative Stress

Inhibition by niacin of inflammatory and oxidative pathways may exert anti-atherosclerotic effects. In cultured human aortic endothelial cells, niacin increased cellular levels of nicotinamide adenine dinucleotide phosphate (reduced) (NADPH) and glutathione, regulators of redox reactions, and reduced production of reactive oxygen species. Niacin inhibited monocyte adhesion through the inhibition of tumour necrosis factor-alpha (TNF-α) induced expression of vascular cell adhesion molecule-1 (VCAM-1) and monocyte chemoattractant protein-1 (MCP-1) (Ganji *et al.* 2009). Persistent increases in plasma adiponectin, an adipokine with anti-inflammatory properties, were found after 6 months of niacin treatment in humans (Linke *et al.* 2009).

39.3 Lipoprotein Effects of Niacin

Niacin demonstrates a beneficial effect on every lipoprotein class. It reduces the levels of lipoproteins associated with atherosclerosis—chylomicrons,

Table 39.1 Approximate changes in lipids and lipoproteins associated with 2000 mg total daily dose of different types of niacin (Brown et al. 1997; Guyton et al. 2000; Knopp et al. 2009).

Niacin type	Dosing	Total cholesterol	Triglyceride	HDL-C	LDL-C
IR	BID or TID	−12%	−30%	+25%	−15%
ER	HS	−12%	−30%	+25%	−15%
SR	BID (am/pm)	−14%	−30%	+15%	−18%

VLDL, β-VLDL, intermediate density lipoprotein (IDL), low density lipoprotein (LDL), small dense LDL and lipoprotein(a) (Carlson 2005). Niacin is the most effective approved therapeutic agent for raising the atheroprotective lipoproteins HDL and HDL_2 (Guyton et al. 2000). However, slow-release (SR) niacin given twice daily may provide less raising of HDL cholesterol than regular immediate-release (IR) niacin given twice daily or extended-release (ER) niacin given as a single daily dose at bedtime. The effects of different niacin formulations at 2000 mg total daily dose are compared in Table 39.1.

39.4 Adverse Effects and Drug Administration

39.4.1 Flushing

Flushing is experienced as prickly heat experienced mostly on the head and neck, but occasionally over the entire body, often accompanied by visible erythema. Flushing occurs at some time in nearly all subjects who take IR niacin and may cause discontinuation of the drug. ER niacin is associated with decreased frequency of flushing. Cyclooxygenase inhibitors (e.g. aspirin 325 mg or ibuprofen 200 mg) given 30–60 minutes prior to niacin ingestion can reduce flushing. Tachyphylaxis to flushing develops within days with regular administration, and after several months, flushing is no more than a minor annoyance for most patients. Inositol hexanicotinate, sold as 'no-flush' or 'flush-free' niacin, does not appear to be bioavailable and has not been shown to improve lipid levels (Guyton and Bays 2007). A DP1 receptor antagonist, laropiprant, currently marketed in combination with ER niacin in Europe, gives marked reduction but not elimination of skin flushing (Cheng et al. 2006).

Patients often tolerate repeated, minor skin discomfort if they expect it and understand that it is not an allergic reaction that can cause serious harm. Experienced providers can help patients achieve long-term tolerance of either IR or ER niacin in the great majority of cases.

39.4.2 Hepatotoxicity

Serious hepatic toxicity can occur with niacin, but it is almost entirely associated with use of SR formulations. In a randomized clinical trial, 12 out of 23 patients (52%) taking SR niacin developed hepatotoxicity while none of 23 patients taking IR niacin did (McKenney et al. 1994). Therefore, IR or ER niacin is preferred. Nevertheless, SR niacin in doses up to 1000 mg twice daily (750 mg twice daily if concomitant with a statin) can be used with hepatic enzyme monitoring (Guyton and Bays 2007).

ER niacin has shown little hepatotoxicity in clinical trials (Guyton and Bays 2007). However, ER niacin has only been studied as a once-daily dose at bedtime and therefore the safety of twice daily ER niacin is unknown.

39.4.3 Myopathy

Decades of clinical experience have shown that IR niacin or ER niacin monotherapy does not cause myopathy. However, even mild liver toxicity from niacin with co-administration of a statin could cause myopathy due to decreased hepatic clearance of the statin (Guyton and Bays 2007).

39.4.4 Insulin Resistance and Hyperglycemia

When administered at daily doses of 2 g or less, niacin generally is associated with minor (4–5%) increases in fasting glucose and modest increases in hemoglobin A1c (Grundy *et al.* 2002). During the initial 24 weeks of treatment with niacin, glucose elevations may be somewhat higher, followed by a subsequent return toward normal even without specific treatment. Most patients with diabetes mellitus require no or only minor adjustment of their diabetic regimen after starting niacin (Fazio *et al.* 2010).

39.5 Randomized Trials with Cardiovascular and Clinical Endpoints

Key randomized clinical trials employing niacin with cardiovascular and clinical endpoints are summarized in Table 39.2. Most of the trials used niacin in combination with other agents—bile acid sequestrants, fibrates or statins.

39.5.1 Coronary Drug Project, the Only Large Monotherapy Trial

The Coronary Drug Project (CDP) was a placebo-controlled secondary prevention study involving 1110 men randomized to niacin and 2789 randomized to placebo. The average daily niacin dose was 2000 mg and subjects were followed for six years. The niacin group experienced a 26% reduction in non-fatal myocardial infarctions and 24% reduction in cerebrovascular events ($P < 0.05$ in each case) (Coronary Drug Project 1975). A 15-year follow-up study demonstrated an 11% total mortality reduction in subjects originally treated with niacin ($P = 0.0004$) (Canner *et al.* 1986).

Table 39.2 Selected randomized trials of niacin with cardiovascular and clinical endpoints.

Study	Subjects	Treatments (follow-up)	Principal outcomes with niacin
Coronary Drug Project (Coronary Drug Project 1975; Canner et al. 1986)	Men with previous myocardial infarction (MI) and high cholesterol	IR niacin vs. placebo (six years)	MI ↓ 26% ($P<0.05$) and cerebrovascular events ↓ 26% ($P<0.05$) over six years. Total mortality ↓ 11% over 15 years ($P=0.0004$)
Stockholm Ischemic Heart Disease Study (Carlson and Rosenhamer 1988)	Recent MI	Niacin + clofibrate vs. no treatment (five years)	Total mortality ↓ 26% ($P<0.05$) and CHD mortality ↓ 36% ($P<0.01$)
Cholesterol Lowering Atherosclerosis Study (CLAS) (Blankenhorn et al. 1987)	Previous coronary artery bypass graft and hypercholesterolemia	IR niacin + colestipol vs. placebo (2–4 years)	Fewer progressing lesions ($P<0.03$), less new atheroma formation ($P<0.03$), more atherosclerosis regression ($P=0.002$)
Familial Atherosclerosis Treatment Study (FATS) (Brown et al. 1990)	Elevated apolipoprotein B and family history of CHD	Niacin (SR and IR) + colestipol vs. lovastatin + colestipol vs. placebo (2.5 years)	Mean regression of coronary lesions in intensively treated groups vs. progression in control group ($P<0.003$); clinical events reduced by 73% ($P<0.05$)
HDL Atherosclerosis Treatment Study (HATS) (Brown et al. 2000)	CHD and low HDL	Niacin (SR and IR) + simvastatin vs. placebo (three years)	Angiographic regression of coronary lesions with niacin-statin combination therapy ($P<0.001$ vs. placebo); clinical events reduced by 70% ($P=0.03$)
Armed Forces Regression Study (AFREGS) (Whitney et al. 2005)	CHD and low HDL	Niacin, gemfibrozil, and cholestyramine vs. limited use cholestyramine alone (2.5 years)	Mean regression of coronary lesions with intensive treatment vs. progression in controls ($P<0.05$); clinical events reduced by 50% ($P<0.05$)
ARBITER-6/HALTS (Taylor et al. 2009)[a]	CHD and LDL <100 mg/dL and low HDL	ER niacin and statin vs. ezetimibe and statin (14 months)	Reduction in mean and maximal CIMT ($P≤0.001$)
Carotid MRI (Lee et al. 2009)	Low HDL, either type 2 diabetes with CHD or carotid/peripheral atherosclerosis	ER niacin added to statin therapy (12 months)	Reduction in carotid wall area by MRI ($P=0.03$)
AIM-HIGH (AIM-HIGH Investigators 2011)	Established atherosclerotic disease and low HDL	ER niacin vs. placebo added to baseline simvastatin + ezetimibe (three years)	No effect on primary endpoint of combined cardiovascular events.

[a]ARBITER, Arterial Biology for the Investigation of the Treatment Effects of Reducing cholesterol trial. HALTS, HDL and LDL Treatment Strategies; MI, myocardial infarction.

39.5.2 Smaller Randomized Trials with Anatomic Endpoints

In almost all niacin trials since the CDP, the primary endpoint has been anatomic, assessed by a variety of imaging modalities. The Cholesterol-Lowering Atherosclerosis Study (CLAS) was a two-year randomized coronary angiographic trial in 162 men with prior coronary bypass surgery, comparing colestipol and niacin combination therapy *versus* double placebo (Blankenhorn *et al.* 1987). Angiography demonstrated a significant reduction in the average number of progressed lesions per subject ($P<0.03$). In addition, the percentage of new or changed atheroma formation in either native arteries or bypass grafts was reduced ($P<0.03$).

The Familial Atherosclerosis Treatment Study (FATS) randomized 146 men with elevated apolipoprotein B and family history of coronary artery disease to one of three strategies for 2.5 years: niacin and colestipol, lovastatin and colestipol, or conventional therapy. Mean stenosis increased by 2.1% in the control group. In contrast, it decreased by 0.9% with niacin and colestipol and 0.7% with lovastatin and colestipol ($P<0.003$). Although clinical outcomes were not the primary endpoint, there was a 73% reduction in the combined intensive groups *versus* the conventionally treated group (Brown *et al.* 1990).

The HDL-Atherosclerosis Treatment Study (HATS) was a three-year, double-blind angiographic trial of 160 patients with coronary disease, low HDL and normal LDL. Subjects were randomized to simvastatin plus niacin or to placebo. The placebo group demonstrated an increase of 3.9% mean stenosis while the simvastatin–niacin group demonstrated 0.4% regression ($P<0.001$). Major cardiovascular events, including need for revascularization, were significantly decreased in the simvastatin-niacin group ($P=0.04$) (Brown *et al.* 2001).

The ARBITER-6/HALTS study enrolled 208 statin-treated subjects with coronary artery disease (CAD) and randomized them to either ezetimibe 10 mg daily or ER niacin 2000 mg daily. Niacin led to significant reductions in both the mean and maximal carotid intimia-media thickness (CIMT) measurements ($P \leq 0.001$ for all comparisons). The incidence of major cardiovascular events was lower in the niacin group (1% *vs.* 5%, $P=0.04$ by Chi square), although the trial was not intended to study events (Taylor *et al.* 2009).

In 2009 Lee *et al.* reported the effects of ER niacin 2000 mg *versus* placebo added to statin therapy in 71 patients with low HDL and either type 2 diabetes with CAD or carotid/peripheral atherosclerosis. After one year, niacin had significantly reduced the carotid wall area, quantified by magnetic resonance imaging, compared with placebo (Lee *et al.* 2009).

39.5.3 The AIM-HIGH Study

The Atherothrombosis Intervention in Metabolic syndrome with low HDL/ high triglycerides: Impact on Global Health outcomes (AIM-HIGH) study was designed to determine whether the non-LDL (mainly HDL raising) effects of

ER niacin would reduce cardiovascular events in 3414 patients with clinically established atherosclerotic disease and low HDL-cholesterol (HDL-C). LDL-C levels were targeted below 80 mg/dl in both groups with simvastatin supplemented by ezetimibe as needed. After three years mean follow-up, no difference was found in the primary endpoint of time to first major cardiovascular event. Ischemic strokes trended higher in the group assigned to ER niacin ($n=29$) *versus* the placebo group ($n=15$), but the result was not statistically significant (AIM-HIGH Investigators 2011).

The fact that AIM-HIGH showed no trend toward benefit with niacin is surprising in light of the consistency of clinical and/or anatomic benefit in previous niacin studies. The Coronary Drug Project did not measure HDL levels, but it was a large and well-managed randomized clinical trial showing benefit with niacin monotherapy. Among the smaller trials utilizing niacin in combination with bile acid sequestrants, fibrates or statins, nominally significant event reductions were seen in five trials and trials with arterial imaging repeatedly found improvement in anatomic endpoints.

Initial responses to AIM-HIGH include the suggestions that (1) niacin is not effective in reducing cardiovascular events, (2) niacin benefit is mediated only through LDL lowering, or (3) niacin gives no additional benefit in statin-treated patients with LDL-C below 80 mg/dl. Unlike most previous trials that dosed niacin at mealtimes, AIM-HIGH used ER niacin recommended for dosing without food or with only a light snack at bedtime. This opens up a fourth possibility, that an adverse effect of niacin administered in the fasting, or nearly fasting, state might explain the lack of cardiovascular event reduction in AIM-HIGH. Carlson's early work on anti-lipolysis after niacin administration to fasting humans, mentioned near the beginning of this chapter, suggests several avenues of investigation that need attention.

Summary Points

- Niacin has favourable effects on all lipoprotein classes and is the most effective currently available agent for raising HDL cholesterol.
- Marked inhibition by niacin of triglyceride lipolysis in adipocytes is mediated by the G-protein-coupled receptor, GPR109A.
- The same receptor, GPR109A, in skin Langerhans (dendritic) cells initiates production of prostaglandins, mainly PGD2, which mediate niacin-induced flushing by binding DP1 receptors on target vascular and possibly other cells.
- The DP1 receptor antagonist, laropiprant, used clinically in some European countries, gives marked reduction, but not elimination, of skin flushing.
- Pretreatment by oral administration of 325 mg aspirin or another anti-inflammatory agent prior to niacin administration reduces prostaglandin production and inhibits flushing.
- Clinically niacin-induced flushing exhibits tachyphylaxis over days to months, while lipoprotein-modifying effects are persistent.

- Lipoprotein alterations with niacin are thought to be due to changes in hepatic lipid metabolism and are not dependent on GPR109A.
- Multiple randomized clinical trials suggest that niacin, alone or in combination with other lipid drugs, has favourable effects on atherosclerotic lesions.
- Cardiovascular event reductions occurred in randomized trials using twice-daily mealtime niacin, but not in a recent trial of niacin administered at bedtime with minimal or no food intake. This raises the possibility that niacin's effect of blocking the normal fuel supply of fatty acids released from adipocytes might have adverse consequences that restrict the ability of bedtime niacin to reduce cardiovascular events.

Key Facts

Key Facts about Mechanisms in Niacin Pharmacology

- Niacin action at pharmacological doses differs from effects at much lower vitamin doses.
- Two effects of niacin—interruption of fat cell lipolysis resulting in greatly reduced levels of non-esterified fatty acids in blood, and skin flushing—are mediated by a G protein-coupled receptor, GPR109A, discovered in 2003.
- The important effects of niacin on lipoproteins appear not to depend on GPR109A, but instead on incompletely understood hepatic effects of niacin.

Key Facts about Niacin's Effects on Lipoproteins

- Niacin is the most effective clinically available agent for raising levels of HDL.
- Niacin lowers the levels of multiple lipoproteins that promote the development of atherosclerotic plaques: LDL, small dense LDL, VLDL (triglyceride), and lipoprotein(a).

Key Facts about Niacin and Atherosclerosis Prevention

- Randomized clinical trials with arterial imaging endpoints have consistently suggested that niacin or combination therapy including niacin improved atherosclerotic plaques.
- One large and five small randomized clinical trials showed decreased cardiovascular events (heart attacks, serious episodes of coronary chest pain, and cerebrovascular events) with niacin or combination therapy including niacin.
- A recent large randomized trial testing non-LDL (mostly HDL raising) effects of niacin added to baseline statin therapy surprisingly showed no

reduction of cardiovascular events. Multiple explanations for this null result have been proposed.

Definitions of Words and Terms

Apolipoprotein A-I (apoA-I). A small lipid-binding peptide associated with almost all HDL.

Apolipoprotein B (apoB). A very large peptide associated with LDL and VLDL.

Atherosclerosis. A process of cholesterol deposition and tissue reaction that occurs in the inner part of the artery wall. A fibroproliferative tissue reaction may slowly narrow the artery lumen (bore) causing blood flow restriction. A tissue reaction of cell injury weakens the inner lining of the artery, which can erode or rupture inducing a blood clot that suddenly blocks blood flow.

G protein-coupled receptors. An abundant class of cell membrane-spanning receptor molecules whose extracellular domain binds regulatory metabolites or peptides and whose cytoplasmic domain interacts with heterotrimeric GTP-binding proteins.

High density lipoproteins (HDL). Smaller lipoproteins with buoyant density 1.063–1.21 g/mL that carry cholesterol as well as a wide variety of peptides involved in oxidant defence and regulation of inflammation. High HDL levels generally confer protection from atherosclerosis.

β-Hydroxybutyrate. A metabolite made by the liver from fatty acids, especially under conditions of fasting or starvation. β-Hydroxybutyrate can serve as fuel for brain metabolism when glucose availability is compromised.

Lipoproteins. Spherical particles circulating in the blood that contain anywhere from a hundred to more than a million molecules of lipid and one or more protein molecules.

Low density lipoproteins (LDL). Lipoproteins with buoyant density 1.006–1.063 g/mL that carry most of the cholesterol in plasma. LDL infiltrates the inner part of the artery wall and contribute cholesterol to enlarging atherosclerotic plaques.

Niacin-induced flushing. An uncomfortable prickly hot feeling accompanied by redness of the skin.

Non-esterified fatty acids (NEFA). Also known as free fatty acids, these are present in plasma mostly as fatty acid anions bound to albumin, the most abundant protein in plasma.

Reverse cholesterol transport. The movement of cholesterol from peripheral tissues to the liver, dependent on HDL.

Tachyphylaxis. The lessening of a biologic response that may occur after stimulus and response are repeated multiple times.

Very low density lipoproteins (VLDL). Larger lipoproteins with buoyant density less than 1.006 g/ml that remain in solution unless subjected to high g-forces in an ultracentrifuge. VLDL carry triglyceride from the liver to peripheral tissues such as muscle and fat.

List of Abbreviations

ABCA1	ATP-binding cassette protein-A1
apoA-I	apolipoprotein A-I
CAD	coronary artery disease
cAMP	cyclic adenosine monophosphate
CDP	Coronary Drug Project
CIMT	carotid intima-media thickness
CLAS	Cholesterol-Lowering Atherosclerosis Study
DGAT2	diacylglycerol acyltransferase-2
ER	extended-release
FATS	Familial Atherosclerosis Treatment Study
GPR	G-protein-coupled cell surface receptor
HDL	high density lipoprotein
HDL-C	HDL-cholesterol
IDL	intermediate density lipoprotein
IR	immediate-release
LDL	low-density lipoprotein
MCP-1	monocyte chemoattractant peptide-1
MI	myocardial infarction
NADPH	nicotinamide adenine dinucleotide phosphate, reduced form
NEFA	non-esterified fatty acids
PGD2	prostaglandin D2
PGE2	prostaglandin E2
PPARγ	peroxisome proliferator-activated receptor-gamma
SR	slow-release
TG	triglyceride
TNF-α	tumour necrosis factor-alpha
VCAM-1	vascular cell adhesion molecule-1
VLDL	very low density lipoproteins

References

AIM-HIGH Investigators, 2011. Niacin in patients with low HDL cholesterol levels receiving intensive statin therapy. *The New England Journal of Medicine*. 365: 2255–2267.

Altschul, R., Hoffer, A., and Stephen, J.D., 1955. Influence of nicotinic acid on serum cholesterol in man. *Archives of Biochemistry*. 54: 558–559.

Blankenhorn, D.H., Nessim, S.A., Johnson, R.L., Sanmarco, M.E., Azen, S.P., and Hemphill, L.C., 1987. Beneficial effects of combined colestipol-niacin therapy on coronary atherosclerosis and coronary venous bypass grafts. *The Journal of the American Medical Association*. 257: 3233–3240.

Brown, B.G., Albers, J.J., Fisher, L.D., Schaefer, S.M., Lin, J.T., Kaplan, C., Zhao, X.Q., Bisson, B.D., Fitzpatrick, V.F., and Dodge, H.T., 1990. Regression of coronary artery disease as a result of intensive lipid-lowering therapy in men with high levels of apolipoprotein B. *The New England Journal of Medicine.* 323: 1289–1298.

Brown, B.G., Bardsley, J., Poulin, D., Hillger, L.A., Dowdy, A., Maher, V.M., Zhao, X.Q., Albers, J.J., and Knopp, R.H., 1997. Moderate dose, three-drug therapy with niacin, lovastatin, and colestipol to reduce low-density lipoprotein cholesterol < 100 mg/dl in patients with hyperlipidemia and coronary artery disease. *American Journal of Cardiology.* 80: 111–115.

Brown, B.G., Zhao, X.Q., Chait, A., Fisher, L.D., Dowdy, A., Cheung, M.C., Morse, J.S., Marino, E.K., Bolson, E.L., Alaupovic, P., Frohlich, J., and Albers, J.J., 2001. Simvastatin and niacin, antioxidant vitamins, or the combination for the prevention of coronary disease. *New England Journal of Medicine.* 345: 1583–1592.

Canner, P.L., Berge, K.G., Wenger, N.K., Stamler, J., Friedman, L., Prineas, R.J., and Friedewald, W., 1986. Fifteen year mortality in Coronary Drug Project patients: Long-term benefit with niacin. *Journal of the American College of Cardiology.* 8: 1245–1255.

Carlson, L.A., and Rosenhamer, G., 1988. Reduction of mortality in the Stockholm Ischaemic Heart Disease Secondary Prevention Study by combined treatment with clofibrate and nicotinic acid. *Acta Medica Scandinavica.* 223: 405–418.

Carlson, L.A., 2005. Nicotinic acid: The broad-spectrum lipid drug. A 50th anniversary review. *Journal of Internal Medicine.* 258: 94–114.

Cheng, K., Wu, T.J., Wu, K.K., Sturino, C., Metters, K., Gottesdiener, K., Wright, S.D., Wang, Z., O'Neill, G., Lai, E., and Waters, M.G. 2006. Antagonism of the prostaglandin D2 receptor 1 suppresses nicotinic acid-induced vasodilation in mice and humans. *Proceedings of the National Academy of Sciences of the United States of America.* 103: 6682–6687.

Coronary Drug Project Research Group, 1975. Clofibrate and niacin in coronary heart disease. *The Journal of the American Medical Association.* 231: 360–381.

Ganji, S.H., Qin, S., Zhang, L., Kamanna, V.S., and Kashyap, M.L., 2009. Niacin inhibits vascular oxidative stress, redox-sensitive genes, and monocyte adhesion to human aortic endothelial cells. *Atherosclerosis.* 202: 68–75.

Grundy, S.M., Vega, G.L., McGovern, M.E., Tulloch, B.R., Kendall, D.M., Fitz-Patrick, D., Ganda, O.P., Rosenson, R.S., Buse, J.M., Robertson, D.D., and Sheehan, J.P., 2002. Efficacy, safety, and tolerability of once-daily niacin for the treatment of dyslipidemia associated with type 2 diabetes. *Archives of Internal Medicine.* 162: 1568–1576.

Guyton, J.R., Blazing, M.A., Hagar, J., Kashyap, M.L., Knopp, R.H., McKenney, J.M., Nash, D.T., and Nash, S.D., 2000. Extended-release niacin *versus* gemfibrozil for treatment of low levels of high density lipoprotein cholesterol. *Archives of Internal Medicine.* 160: 1177–1184.

Guyton, J.R., and Bays, H.E.,2007. Safety considerations with niacin therapy. *American Journal of Cardiology.* 99: S22–S31.

Fazio, S., Guyton, J.R., Polis, A., Adewale, A.J., Tomassini, J.E.,Ryan, N.W., and Tershakovec, A.M., 2010. Long-term safety and efficacy of triple

combination ezetimibe/simvastatin + extended-release niacin in hyperlipidemic patients. *American Journal of Cardiology.* 105: 487–494.
Kamanna, V.S., and Kashyap, M.L., 2008. Mechanism of action of niacin. *American Journal of Cardiology.* 101: 20B–26B.
Knopp, R.H., Retzlaff, B.M., Fish, B., Dowdy, A., Twaddell, B., Nguyen, T., and Paramsothy, P., 2009. The SLIM study: Slo-Niacin(r) and atorvastatin treatment of lipoproteins and inflammatory markers in combined hyperlipidemia. *Journal of Clinical Lipidology.* 3: 167–178.
Lai, E., Waters, M.G., Tata, J.R., Radziszewski, W., Perevozskaya, I., Zheng, W., Wenning, L., Connolly, D.T., Semple, G., Johnson-Levonas, A.O., Wagner, J.A., Mitchel, Y., and Paolini, J.F., 2008. Effects of a niacin receptor partial agonist, MK-0354, on plasma free fatty acids, lipids, and cutaneous flushing in humans. *Journal of Clinical Lipidology.* 2: 375–383.
Lamon-Fava, S., Diffenderfer, M.R., Barrett, P.H., Buchsbaum, A., Nyaku, M., Horvath, K.V., Asztalos, B.F., Otokozawa, S., Ai, M., Matthan, N.R., Lichtenstein, A.H., Dolnikowski, G.G., and Schaefer, E.J., 2008. Extended-release niacin alters the metabolism of plasma apolipoprotein (apo) A-I and apoB-containing lipoproteins. *Arteriosclerosis, Thrombosis, and Vascular Biology.* 28: 1672–1678.
Lee, J.M., Robson, M.D., Yu, L.M., Shirodaria, C.C., Cunnington, C., Kylintireas, I., Digby, J.E., Bannister, T., Handa, A., Wiesmann, F., Durrington, P.N., Channon, K.M., Neubauer, S., and Choudhury, R.P., 2009. Effects of high-dose modified-release nicotinic acid on atherosclerosis and vascular function: A randomized, placebo-controlled, magnetic resonance imaging study. *Journal of the American College of Cardiology.* 54: 1787–1794.
Linke, A., Sonnabend, M., Fasshauer, M., Hollriegel, R., Schuler, G., Niebauer, J., Stumvoll, M., and Bluher, M., 2009. Effects of extended-release niacin on lipid profile and adipocyte biology in patients with impaired glucose tolerance. *Atherosclerosis.* 205: 207–213.
McKenney, J.M., Proctor, J.D., Harris, S., and Chinchili, V.M., 1994. A comparison of the efficacy and toxic effects of sustained- vs immediate-release niacin in hypercholesterolemic patients. *Journal of the American Medical Association.* 271: 672–677.
Rubic, T., Trottmann, M., and Lorenz, R.L., 2004. Stimulation of CD36 and the key effector of reverse cholesterol transport ATP-binding cassette A1 in monocytoid cells by niacin. *Biochemical Pharmacology.* 67: 411–419.
Soga, T., Kamohara, M., Takasaki, J., Matsumoto, S., Saito, T., Ohishi, T., Hiyama, H., Matsuo, A., Matsushime, H., and Furuichi, K., 2003. Molecular identification of nicotinic acid receptor. *Biochemical and Biophysical Research Communications.* 303: 364–369.
Taggart, A.K., Kero, J., Gan, X., Cai, T.Q., Cheng, K., Ippolito, M., Ren, N., Kaplan, R., Wu, K., Wu, T.J., Jin, L., Liaw, C., Chen, R., Richman, J., Connolly, D., Offermanns, S., Wright, S.D., and Waters, M.G., 2005. (D)-beta-hydroxybutyrate inhibits adipocyte lipolysis *via* the nicotinic acid receptor PUMA-G. *The Journal of Biological Chemistry.* 280: 26649–26652.

Taylor, A.J., Villines, T.C., Stanek, E.J., Devine, P.J., Griffen, L., Miller, M., Weissman, N.J., and Turco, M., 2009. Extended-release niacin or ezetimibe and carotid intima-media thickness. *The New England Journal of Medicine.* 361: 2113–2122.

Tunaru, S., Kero, J., Schaub, A., Wufka, C., Blaukat, A., Pfeffer, K., and Offermanns, S., 2003. PUMA-G and HM74 are receptors for nicotinic acid and mediate its anti-lipolytic effect. *Nature Medicine.* 9: 352–355.

Walters, R.W., Shukla, A.K., Kovacs, J.J., Violin, J.D., DeWire, S.M., Lam, C.M., Chen, J.R., Muehlbauer, M.J., Whalen, E.J., and Lefkowitz, R.J., 2009. Beta-arrestin1 mediates nicotinic acid-induced flushing, but not its antilpolytic effect, in mice. *The Journal of Clinical Investigation.* 119: 1312–1321.

Whitney, E.J., Krasuski, R.A., Personius, B.E., Michalek, J.E., Maranian, A.M., Kolasa, M.W., Monick, E., Brown, B.G., and Gotto, A.M., Jr., 2005. A randomized trial of a strategy for increasing high-density lipoprotein cholesterol levels: Effects on progression of coronary heart disease and clinical events. *Annals of Internal Medicine.* 142: 95–104.

Wise, A., Foord, S.M., Fraser, N.J., Barnes, A.A., Elshourbagy, N., Eilert, M., Ignar, D.M., Murdock, P.R., Steplewski, K., Green, A., Brown, A.J., Dowell, S.J., Szekeres, P.G., Hassall, D.G., Marshall, F.H., Wilson, S., and Pike, N.B., 2003. Molecular identification of high and low affinity receptors for nicotinic acid. *The Journal of Biological Chemistry.* 278: 9869–9874.

CHAPTER 40
Pellagra: Psychiatric Manifestations

RAVI PRAKASH,*[a] PRIYANKA RASTOGI[b] AND SUPRAKASH CHOUDHARY[a]

[a] Psychiatry Department, Ranchi Institute of Neuropsychiatry and Allied Sciences, Kanke, Ranchi, India; [b] Department of Clinical Psychology, Central Institute of Psychiatry, Kanke, Ranchi, India
*Email: drravi2121@gmail.com

40.1 Introduction

Pellagra is a deficiency disease associated with low levels of vitamin B_3 (niacin) and/or tryptophan and often involving other B vitamins. Since the time Gasper Casal first described the disease in 1972, it was observed that patients with pellagra were all poor, subsisted mainly on maize and rarely ate fresh meat. Subsequent occurrences have been recorded in the form of epidemic outbreaks, consequent to either introduction to maize as a major food or increased consumption of other niacin-deficient diets like jowar (*Sorgum vulgare*) (World Health Organization 2000a).

The disease occurs mostly among poorer groups whose diet consists mainly of the cheapest available food, maize, supplemented with salt pork, lard and molasses (World Health Organization 2000a). Sporadic cases continue to be seen globally and are associated with monotonous diets of untreated maize, food faddism, tuberculosis treatment, malabsorption states and alcoholism (Aspinall 1964; Cook *et al.* 1998; Cunha *et al.* 2000; Leung and Ungwari 2004; Rudin 1981). The virtual disappearance of pellagra as an endemic health problem in recent years can be attributed mainly to a general rise in the standard of living of small farmers, accompanied by greater diversification of the diet (World Health Organization 2000a, 2000b; Hegyi *et al.* 2004).

The clinical picture is a combination of multisystem alterations typically involving gastrointestinal, skin and central nervous system abnormalities. The cardinal manifestations have been popularly known as the three D's which are dementia, dermatitis and diarrhoea (World Health Organization 2000a; Hegyi *et al.* 2004). The classical descriptions of pellagra have always incorporated psychiatric features as an integral component of this disorder. However, the psychiatric manifestations are so varied yet overlapping that it is difficult to generate an accurate classification for these features. For this reason, we are keeping this classification as simple as possible in this chapter. For the purpose of this chapter, we organize the psychiatric features into three categories:

(a) Common early psychiatric features
(b) Cognitive deficits
(c) Psychotic features.

40.2 Common Early Psychiatric Features

Psychiatric manifestations are fairly common but are easily overlooked due to their non-specific nature. These are commonly seen as irritability, poor concentration, anxiety, fatigue, restlessness, apathy and depression. The occurrence of psychosis in pellagra is an uncommon finding, especially as an early feature and we have described it under a separate heading. Additionally, the psychotic spectrum is usually seen in advanced stages of pellagroid encephalopathy, commonly found in chronic alcoholics (Cook *et al.* 1998).

In one of the early case series of eight patients, early mental symptoms observed were insomnia (8/8) and anxiety (6/8), followed by confusion (8/8), restless behaviour (5/8), monologue (7/8), inappropriate smile (4/8) and refusal of nourishment (5/8) (Ishi & Nishihara 1985). Psychiatric symptoms are often associated with other features of encephalopathy, including confusion, coma and death.

As mentioned above, pellagra has become so uncommon in developed as well as developing countries that the current incidence in these parts of the world is unknown. Due to such a low prevalence, the psychiatric states in this disorder are inadequately studied.

40.3 Cognitive Deficits in Pellagra

Cognitive impairments are common in this disorder, so much so that they have been included in the cardinal features of this disease ('dementia' in the three D's). These generally include impaired concentration, memory impairments and, in severe cases (pellagraoid encephalopathy), confusion, delirium and coma (Hegyi *et al.* 2004). The cognitive deficits are often been termed as 'reversible dementia' which can be treated by administration of niacin (Mahler et al. 1987). Regarding the progression of dementia in Pellagra, no uniformity has been observed. However, more often, the onset and progression has been reported to be insidious rather than acute (Geschwind et al. 2007).

Although knowledge of pellagra is often deficient in psychiatric settings in view of its low incidence and uncommon psychiatric presentation, it may be important in providing clues for biological underpinnings of psychotic states in general and of delusional parasitosis in particular. The psychotic symptoms due to alcoholism and drugs are myriad and are difficult to correlate with niacin deficiency alone, keeping in view other deficiency states found comorbidly in such conditions.

40.4 Psychotic Spectrum Features in Pellagra

Earlier scientific literature reveals that niacin deficiency has been associated with three main types of psychiatric manifestations, which are (a) schizophreniform, (b) manic depressive types and (c) anxiety and depressive disorders (Rudin 1981). Schizophreniform manifestations include auditory hallucinations and persecutory delusions.

Occurrence of delusory parasitosis is an extremely rare phenomenon in this deficiency state. This usually develops secondary to the patients' complaints of pruritus and paraesthesias due to pellagra skin lesions. This is the reason why patients with such complaints are referred to a dermatologist rather than to a psychiatrist (Leung and Ungwari 2004). It seems that such patients are prone to develop delusory parasitic ideas independent of the skin lesions. However, delusional parasitosis is known for its resistance to antipsychotic treatment and such a rapid resolution of any delusion.

The literature is deficient regarding any effect of niacin on improvement of psychotic conditions. Although the exact relation between nicotinic acid deficiency and pathogenesis of delusions or hallucinations is not clear, it is likely to involve subtle neuronal insult. *Post mortem* examination has revealed chromatolysis in such patients (Ishii and Nishihara 1981). Indirect evidence suggests that niacin antagonism is associated with evident neuroglial (especially astrocytic) degeneration and subsequent disturbances in signal transmission across neurons (Penkowa et al. 2002). Other studies also point towards an association of niacin abnormalities with psychosis.

A recent study showed abnormal niacin sensitivity in schizophrenia patients as evidenced by attenuation of the flush response to niacin in schizophrenia. However, there is still an ongoing debate whether this response is due to altered pharmacological sensitivity to niacin or an inadequate cutaneous vasodilatory response to the stimulus (Messamore et al. 2003). Early studies even attempted to use niacin as an augmenting agent for the treatment of schizophrenia with mixed results (Ananth et al. 1973; Petrie et al. 1981). In a placebo-controlled comparative study by Ramsay et al. (1970), it was found that, while no significant differences were seen in total Brief Psychiatric Rating Scale (BPRS) scores prior to commencement of the clinical trial, statistically significant improvement in 'emotional withdrawal' was seen only with nicotinamide and not with placebo.

40.5 Case Vignette: A Unique case of Pellagra Delusional Parasitosis

Mr B/S is a 45-year-old male, of low socioeconomic status, hailing from a rural background, who presented to the hospital out patients department with complaints of suspiciousness towards family members, muttering and smiling to self, and complaining of insects in his abdomen. A detailed mental status examination on the day of presentation revealed an apathetic patient, retarded psychomotor activity, slow speech of soft quality, dysphoric affect, delusion of parasitosis and poor insight. He did not reveal any perceptual abnormality. A detailed elaboration of his delusion of parasitosis revealed that he was very sure of having caterpillars of 5–6 inches in length inside his abdomen, which entered his body by his foot and travelled all the way to his stomach. He expressed repeatedly that he could feel the crawling of these caterpillars. The main reason of his surety for this infestation was that he was becoming weaker day by day and this was all due to the worms sucking his blood from his intestine. He also had ideas of hopelessness and helplessness secondary to the parasitosis phenomenon in his body. His cognitive functions showed impairments in the Mini Mental Status Examination (MMSE), with a score of 23. He had impairments in recent recall and following of written commands as well as serial subtraction.

Physical examination revealed scaly lesions on his skin in the sun-exposed parts, most prominent being on the dorsum of his both hands, both feet, forehead and below the neck (Figures 40.1–40.3). The neck lesions were very similar to the description of Casal necklace in the literature. He also had mild

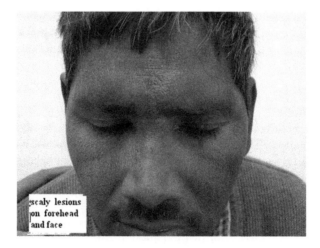

Figure 40.1 Scaly lesions on forehead and face.

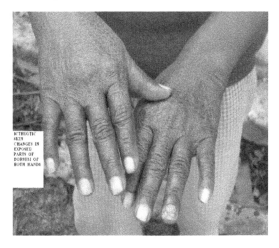

Figure 40.2 Icthyotic skin changes in exposed parts of dorsum of both hands.

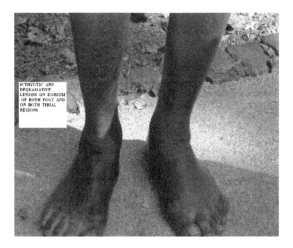

Figure 40.3 Icthyotic and desquamative lesions on dorsum of both feet and on both tibial regions.

anaemia. A physical history revealed that he was suffered frequent diarrhoea in the past couple of months. The patient was subjected to a detailed investigation, including a hemogram, liver function tests, kidney function tests, lipid profile, random and post-prandial blood sugar, Na and K electrolytes, thyroid function tests, electroencephalogram, CT scan of brain, Venereal Disease Research Laboratory (VDRL) test for syphilis, enzyme-linked immunosorbent assay (ELISA) for HIV and tuberculosis, chest X-ray in PA view and stool for occult blood.

The test results pointed towards his poor nutritional state. His hemoglobin was 10.2 mg/dl. His mean corpuscular volume was 80 fl and mean corpuscular hemoglobin was 25 pg. His albumin concentration was also slightly decreased (3.4 g/dl). His electroencephalogram record showed a nonspecific slowing in the alpha range of frequency 7.5–8 Hz. The rest of the investigations including CT scan did not reveal any abnormalities.

On the basis of the mental status examination and history, he was given a working diagnosis of unspecified schizophrenia, which was diagnosed using the International Classification of Diseases (ICD) ICD-10 criteria. He was also given only a provisional diagnosis of pellagra due to lack of confirmatory evidences such as serum niacin estimation or skin biopsy because of lack of laboratory ability or dermatologic consultation.

He was subsequently started on injections of haloperidol 10 mg intramuscular (IM) twice a day (BID) and phenergan 50 mg IM BID, and B complex capsule containing niacin 50 mg/day in four doses. The injections were stopped after two days and he was switched to oral haloperidol 15–20 mg/day and trihexyphenidyl 2 mg/day. Within four days, he had stopped complaining of worms in his intestine and started showing his dysphoria and ideas of hopelessness and helplessness. When confronted about his ideas about parasite infestation, he would say that he did not know what had happened to him to think of such thing. His skin lesions started showing improvement after one week and progressively resolved by the end of two weeks, when he was discharged. He never complained of any gastrointestinal symptoms over his stay in the hospital. He did not show up for follow-up appointments.

40.6 Neurophysiological Mechanisms of Pellagra–Psychiatric Features

There are at least two mechanisms by which niacin deficiency can alter brain functioning. One is by a reduction in the synthesis and release of serotonin (5HT), and the other is by a reduction in the synthesis and release of kynurenic acid (KA). Tryptophan, as the precursor of 5HT, has well-understood importance for normal brain functioning. Abnormalities of serotonergic activity have wide-ranging effects on mood, cognition and other aspects of brain functioning. Diversion of tryptophan to nicotinamide formation can produce deficits in serotonin synthesis. Some of the findings in pellagra, such as depressed mood, may be easily explained by reduced availability of serotonin. Other aspects of altered mentation may be explained by other abnormalities in pellagra. For example, the tryptophan metabolite KA, which is produced and released by astrocytes, is the only known endogenous inhibitor of glutamate activity at the N-methyl-D-aspartate (NMDA) receptor. KA is also an antagonist at the nicotinic cholinergic receptor. Reduced KA synthesis and release could potentiate glutamate activity at the NMDA receptor and cholinergic activity at the nicotinic receptor, causing altered mentation. Furthermore, in rat models, reduced KA synthesis and release enhances the neurotoxicity of a variety of injuries.

Chromatolysis of neurons, and, in particular, of Betz cells and pontine nuclei, may occur in advanced cases of pellagra. The relationships of tryptophan and nicotinic acid metabolism to the psychiatric condition of pellagrins has stimulated interest in their possible involvement in other psychiatric disorders. Tryptophan has been used as an antidepressant though some studies of its use have been inconclusive (Dickerson & Wiryanti 1978; Mendels *et al.* 1975). Young and Sourkes (1974) suggested that the weak or uncertain effect of tryptophan could be due to its metabolism by hepatic L-tryptophan: oxygen oxidoreductase. Since the activity of L-tryptophan:oxygen oxidoreductase is reduced by nicotinamide, it seemed likely that the effectiveness of tryptophan in the treatment of depression would be increased by combined therapy. Studies in 11 newly admitted depressed patients who were given increasing amounts of tryptophan and nicotinamide over four weeks (Chouinard *et al.* 1977) have shown an encouragingly significant reduction of depression and the combination evidently warrants further investigation.

Summary Points

- Pellagra is a deficiency disorder associated with low levels of vitamin B_3 (niacin) and/or tryptophan and often involving other B vitamins.
- The clinical picture is a combination of multisystem alterations with central involvements of gastrointestinal, dermatological and central nervous system abnormalities.
- The cardinal manifestations have been popularly known as the three D's which are dementia, dermatitis and diarrhoea.
- Non-specific psychiatric manifestations are common early presentations of the disorder but are easily overlooked.
- The psychotic spectrum is usually seen in advanced stages of pellagroid encephalopathy, commonly found in chronic alcoholics.
- Psychiatric symptoms are often associated with other features of encephalopathy, including confusion, coma and death.
- Cognitive impairments are common in this disorder. These are non-specific deteriorations in attention and memory domains.
- Cognitive impairments are reversible with administration of niacin.
- Proposed neurophysiological mechanisms of these psychiatric abnormalities are reduction in serotonin and kynurenic acid synthesis.

Key Facts

Key Facts about Pellagra

- Pellagra is a constellation of signs and symptoms which manifest due to deficiency of vitamin B_3 or niacin.
- Psychiatric manifestations are integral components of this disorder.
- Psychosis is a psychiatric condition where the central sign is a loss of insight into the condition.

- Dementia is the deterioration of the intellectual functioning like memory, attention and problem solving abilities.
- Non-specific psychiatric manifestations are irritability, depression, anxiety, *etc.*
- Delusional parasitosis is a psychotic symptom where the patient develops unshakable false belief that they are infested by a living organism, mostly insects or small animals.

Key Facts about Neuropathophysiological Understanding of Pellagra

- There are at least two mechanisms by which niacin deficiency can alter brain function.
- One is by a reduction in the synthesis and release of serotonin (5HT). Serotonin is a neurotransmitter which is concerned with a variety of psychological events.
- The other is by a reduction in the synthesis and release of kynurenic acid (KA). Kynurenic acid is a metabolism product of the amino acid tryptophan. It acts as a cerebral depressant.
- Almost all of the psychiatric manifestations can be explained by alteration in these two neurotransmitter systems.

Definitions of Words and Terms

Cognition: An umbrella term used to denote various psychological processes such as thoughts and appraisal. More specifically, it points to the various processes related to information acquisition, storage and application to solve various problems.

Coma: A critical and often fatal pathological state of diminished consciousness characterized by a persistently decreased awareness of surroundings and self. It usually follows severe injuries to brain such as a head injury.

Delirium: A state of altered consciousness characterized by occurrence of hallucinations, emotional instability, altered awareness of surroundings and sensory disturbances. Follows any serious chemical or physical injury to brain.

Delusion: A disorder of thought characterized by false belief with high conviction and preoccupation.

Delusional parasitosis: A unique type of delusion characterized by the high conviction of belief that the person is being infested by organisms in their body, most commonly by imaginary insects.

Dementia: A progressive neuropsychiatric disorder characterized by deterioration in various cognitive processes such as memory, executive functioning and attention. Can be primary, which is usually found with progressive age or secondary, which is secondary to some other disorder like nutrition deficiency.

Kynurenic acid: A metabolic product of tryptophan, which has been found to have potent neuroprotective effects.
Mini Mental Status Examination: A neuropsychiatric tool to assess various cognitive deteriorations in dementia of various kinds. Contains 21 questions and assess various kinds of memories, executive functioning and constructional skills.
Niacin: A common term used to describe vitamin B_3.
Pellagra: A syndrome complex characterized by the three D's—dementia, dermatitis and diarrhoea—due to deficiency of vitamin B_3 or niacin.
Psychosis: A psychiatric condition characterized by loss of insight and presence of hallucinations, dementia or catatonia.
Serotonin: A vital neurotransmitter responsible for various psychiatric functions, especially implicated in anxiety and depressive disorders.

List of Abbreviations

BID	twice a day
BPRS	Brief Psychiatric Rating Scale
ELISA	enzyme-linked immunosorbent assay
5HT	serotonin
ICD	International Classification of Diseases
IM	intramuscular
KA	kynurenic acid
MMSE	Mini Mental Status Examination
NMDA	*N*-methyl-D-aspartate
VRDL	Venereal Disease Research Laboratory

References

Ananth, J.V., Ban, T.A., and Lehmann, H.E., 1973. Potentiation of therapeutic effects of nicotinic acid by pyridoxine in chronic schizophrenics. *Canadian Psychiatric Association Journal*. 18: 377–383.

Aspinall, D.L., 1964. Multiple deficiency states associated with isoniazid therapy. *British Medical Journal*. 2(5418): 1177–1178.

Cook, C.C.H., Hallwood, P.M., and Thomson, A.D., 1998. B Vitamin deficiency and neuropsychiatric syndromes in alcohol misuse. *Alcohol and Alcoholism*. 33: 317–336.

Cunha, D.F., Monteiro, J.P., Ortega, L.S., Alves, L.G., and Cunha, S.F., 2000. Serum electrolytes in hospitalized pellagra alcoholics. *European Journal of Clinical Nutrition*. 54: 440–442.

Dickerson, J.W.T., and Wiryant, J., 1978. Pellagra and mental disturbance. *Proc. Nutr. SOC*. 37: 167–171.

Geschwind, M.D., Haman, A., and Miller, B.L., 2007. Rapidly progressive dementia. *Neurologic Clinics*. 25: 783–807.

Hegyi, J., Schwartz, R.A., and Hegyi, V., 2004. Pellagra: dermatitis, dementia, and diarrhea. *International Journal of Dermatology*. 43: 1–5.

Ishii, N., and Nishihara, Y., 1981. Pellagra among chronic alcoholics: clinical and pathological study of 20 necropsy cases. *Journal of Neurology, Neurosurgery & Psychiatry*. 44: 209–215.

Ishii, N., and Nishihara, Y., 1985. Pellagra encephalopathy among tuberculous patients: its relation to isoniazid therapy. *Journal of Neurology, Neurosurgery & Psychiatry*. 48: 628–634.

Leung, T.Y., and Ungwari, G.S., 2004. A Chinese adolescent with delusional infestation. *Hong Kong Journal of Psychiatry*. 1: 23–25.

Mahler, M.E., Cummings, J.L., and Benson, D.F., 1987. Treatable dementias. *The Western Journal of Medicine*. 146: 706–712.

Mendels, J., Stinnett, J.L., Burns, D., and Frazer, A., 1975. Amine precursors and depression. *Arch Gen Psychiatry*. Jan;32(1):22–30.

Messamore, E., Hoffman, W.F., and Janowsky, A., 2003. The niacin skin flush abnormality in schizophrenia: a quantitative dose–response study. *Schizophrenia Research*. 62: 251–258.

Penkowa, M., Giralt, M., Camats, J., and Hidalgo, J., 2002. Metallothionein 1 & 2 protects the CNS during neuroglial degeneration induced by 6-aminonicotinamide. *Journal of Comparative Neurology*. 444: 174–189.

Petrie, W.M., Ban, T.A., and Ananth, J.V., 1981. The use of nicotinic acid and pyridoxine in the treatment of schizophrenia. *International Pharmacopsychiatry*. 16: 245–250.

Ramsay, R.A., Ban, T.A., Lehmann, H.E., Saxena, B.M., and Bennett, J. 1970. Nicotinic acid as adjuvant therapy in newly admitted schizophrenic patients. *Canadian Medical Association Journal*. 102: 939–942.

Rudin, D.O., 1981. The major psychoses and neuroses as omega-3 essential fatty acid deficiency syndrome: substrate pellagra. *Biological Psychiatry*. 16: 837–850.

World Health Organization, 2000a. Pellagra and its prevention and control in major emergencies. WHO/ NHD/00.10. Available at: http://whqlibdoc.who.int/hq/2000/who_nhd_00.10.pdf. Accessed 24 May 2012.

World Health Organization, 2000b. Manual on the management of nutrition in major emergencies. IFRC/UNHCR/WFP/WHO. Available at: http://whqlibdoc.who.int/publications/2000/9241545208.pdf. Accessed 24 May 2012.

Young, S.N., Garelis, E., Lal, S., Martin, J.B., Molina-Negro, P., Ethier, R. and Sourkes, T.L., 1974, Tryptophan and 5-Hydroxyindoleacetic acid in human cerebrospinal fluid. *Journal of Neurochemistry*, 22: 777–779.

CHAPTER 41
Pantetheine and Pantetheinase: From Energy Metabolism to Immunity

TAKEAKI NITTO

Laboratory of Pharmacotherapy, Department of Clinical Pharmacy, Yokohama College of Pharmacy, 601 Matano-cho, Totsuka, Yokohama City, Kanagawa 245-0066, Japan
Email: nitto@hamayaku.ac.jp

41.1 The Synthesis and Metabolism of Coenzyme A

Coenzyme A (CoA) is an essential cofactor in more than a hundred synthetic and degradative reactions. The molecule consists of 3′-phosphoadenosine coupled through the 5′ position of the ribose to pantothenic acid by a pyrophosphate linkage; the carboxyl end of pantothenic acid (vitamin B_5) is linked through a peptidic linkage to β-mercaptoethylamine.

The synthesis and metabolism of CoA is shown in Figure 41.1. Pantothenic acid is an important part for CoA synthesis. In the initial step, pantothenic acid is phosphorylated to 4-phosphopantotenic acid by the action of pantothenate kinase. The formation of 4′-phosphopantetheine is a two-step process in which 4′-phosphopantothenic acid and cysteine are first converted to 4′-phosphopantothenoyl-L-cysteine by the formation of a peptide linkage followed by decarboxylation of cysteine. In the final two steps in the pathway, 4′-phosphopantetheine is adenylated to form dephospho-CoA and dephospho-CoA is

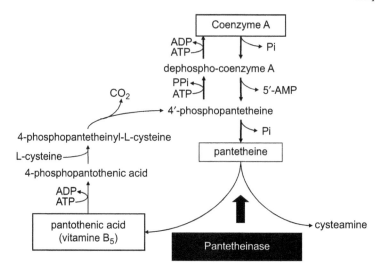

Figure 41.1 Synthetic and metabolic pathway of coenzyme A (CoA). ATP, adenosine triphosphate; ADP, adenosine diphosphate; AMP, adenosine monophosphate; PPi; pyrophosphate; Pi, phosphate.

phosphorylated at the 3' position of ribose to form CoA. The rate of CoA synthesis is determined largely by the regulated pantothenate kinase step.

The degradation of CoA is nearly the reverse of that for the synthetic pathway (Figure 41.1). In the first step, CoA is dephosphorylated at the 3' position of ribose to form dephosphoCoA. DephosphoCoA is then degraded to 4'-phosphopantetheine and 5'-AMP. Dephosphorylation of 4'-phosphopantetheine to pantetheine provides a different intermediate from the CoA-synthesis pathway. In the final step in the degradation pathway, pantetheine is degraded to pantothenic acid (vitamin B_5) and cysteamine (β-mercaptoethylamine). Pantetheinase is the enzyme that catalyses the reaction. Since pantothenic acid generated during the degradation pathway is recycled for another biosynthesis of coenzyme A, the step of hydrolysis of pantetheine by pantetheinase is important for the 'salvage pathway' of the coenzyme A biosynthesis (Figure 41.1).

41.2 Enzymatic Features of Pantetheinase

Panetheinase activity has been determined in several tissues as a pantothenic acid liberating activity in various species including human. Among them, the enzymatic features of pantetheinase have been best characterized using a protein purified from pig kidney. Pig pantetheinase was purified as an ectoenzyme by monitoring the *bona fide* pantetheinase enzymatic activity. The purified pig pantetheinase showed a high specificity for pantothenate moiety but a low specificity for the cysteamine portion using a various pantetheine analogue.

The enzymatic activities of pantetheinase can be irreversibly inhibited by iodoacetamide, iodoacetate, bromopuruvate and *N*-ethylmaleimide, suggesting the presence of an essential sulfhydryl group in the enzyme active site. Disulfides such as panthine, cystamine, 4,4'-dithiopyridine and oxidized mercaptoethanol also have irreversible inhibitory activities to pantetheinase (Pitari et al. 1994). Pantetheinase is a heat-resistant enzyme. Enzymatic activity increases with temperature to an optimum of up to 70 °C. The enzyme retains its enzymatic activity as high as 80 °C (Pitari et al. 1996).

41.3 Pantetheinase Gene Family

Three genes related to pantetheinase have been reported (Aurrand-Lions et al. 1996; Galland et al. 1998; Granjeaud et al. 1999; Maras et al. 1999; Martin et al. 2001; Suzuki et al. 1999) since mouse vanin-1 was firstly cloned. The pantetheinase/vanin family genes are clustered in chromosome 10A2B1 in mice and in chromosome 6q23-24 in human (Galland et al. 1998; Martin et al. 2001).

41.3.1 Pantetheinase/Vanin-1/VNN1

Pantetheinase is also known as the gene name of vanin-1 and VNN1. According to the National Center for Biotechnology Information (NCBI) database, the *vanin-1* genes have so far been identified in mouse, human, rat, chicken and other 16 organisms. The human analogous gene of murine *vanin-1* is called *VNN1* (Galland et al. 1998; Granjeaud et al. 1999). Mouse vanin-1 was first identified in perivascular thymic stromal cells (Aurrand-Lions et al. 1996). A later investigation showed that the expression of vanin-1 mRNA in mouse tissue is found in mainly in the small intestine, kidney and liver (Pitari et al. 2000) in addition to immunological tissues (Table 41.1).

Mouse vanin-1 is induced by the activation of two antioxidant response elements in epithelial cells (Berruyer et al. 2004). Mouse vanin-1 is also expressed male-specifically in Sertoli cells of the developing testis (Bowles et al. 2000). The binding of steroidogenic factor-1 and SOX9 to the specific binding

Table 41.1 Comparison of three pantetheinase family genes in human and mouse.

Gene name	Locus	Length (kbp)	Tissues mainly expressed
HUMAN			
VNN1, pantetheinase	6q23-24	32.4	Spleen, small intestine, peripheral blood leukocyte, liver
VNN2, GPI-80		14.0	Neutrophils, monocytes, colon, spleen, placenta, lung
VNN3		12.0	Spleen, peripheral blood leukocyte, liver
MOUSE			
Vanin-1, pantetheinase	10A2B1	10.6	Kidney, small intestine, liver, testis, heart
Vanin-3		18.4	Spleen, peripheral blood leukocyte, liver, kidney, thymus, heart

element on vanin-1 promoter is involved in the expression in the gonad (Wilson et al. 2005). On the other hand, the pantetheinase protein was purified from porcine kidney by tracing the activity that hydrolyses pantetheine into pantotenic acid and cysteamine.

The amino acid sequences of the purified pig pantetheinase protein (Maras et al. 1999) revealed that the amino acid sequences of mouse vanin-1 and human VNN-1 are quite similar to that of pig pantetheinase. In addition, recombinant proteins made from both mouse vanin-1 and human VNN-1 cDNAs have pantetheinase enzymatic activity (Pitari et al. 2000), confirming that the mouse vanin-1 and the human VNN-1 are homologues of the pig pantetheinase. Mouse pantetheinase/vanin-1 is produced as a glycosylphosphatidyl inositol (GPI)-anchored protein and is located on the plasma membrane of epithelial cells. The role of mouse vanin-1 is well documented using vanin-1-deficient mice (see below).

41.3.2 GPI-80/VNN2

Humans have another pantetheinase-related gene, GPI-80/VNN2. The gene was identified in different ways. Galland et al. (1998) identified two human cDNAs homologous to mouse vanin-1 during screening of cDNA libraries of human kidney, liver and placenta. They named two genes VNN1 and VNN2 after mouse vanin-1. On the other hand, Suzuki et al. (1999) identified one protein recognized by the monoclonal antibody, 3H9, which modulated adhesion and transmigration of activated human neutrophils. The protein was named GPI-80 (a 80 kDa protein with GPI-anchor). Two different genes (cDNAs), VNN2 and GPI-80, turned to be identical according to the public gene database. A mouse homologue of GPI-80/VNN2 has not yet been determined to date. It has been suggested that the gene was divided from the common orthologue of vanin-1/VNN1 and GPI-80/VNN2 during evolution from rodents to primates (Granjeaud et al. 1999). The amino acid sequence of the potential amidohydrolase active centre is conserved between human VNN1 (pantetheinase) and GPI-80/VNN2. GPI-80/VNN2 has pantetheinase enzymatic activity (Martin et al. 2001), although the activity is weaker than VNN1. Both pantetheinase/Vanin-1/VNN1 and GPI-80/VNN2 are biosynthesized as GPI-anchored protein.

GPI-80/VNN2 is expressed in limited cells, primarily found in human neutrophils and occasionally found in human $CD14^+$ monocytes. The human organs expressing GPI-80/VNN2 mRNA other than leukocytes are colon, spleen, placenta and lung. The expression levels of GPI-80/VNN-2 in neutrophil progenitors increase during differentiation and maturation. GPI-80/VNN-2 associates with Mac-1 on the surface of human neutrophils (Yoshitake et al. 2003) and is moving to pseudopodia in front of the cells, while a human neutrophil is migrating toward a chemotactic factor such as formyl-methionyl-leucyl-phenylalanine (fMLP) (Yoshitake et al. 2003) (Figure 41.2).

A lipid raft is important for the physical and functional association between GPI-80/VNN2 and Mac-1 (Huang et al. 2004). While GPI-80/VNN2 is mostly

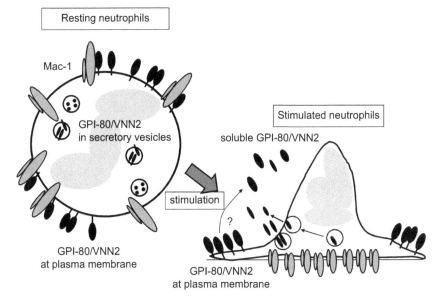

Figure 41.2 Schematics of localization of GPI-80/VNN2 molecule in resting and activated human neutrophils. Resting neutrophils have GPI-80/VNN2 associated with Mac-1 on the cellular surface whereas the soluble form of GPI-80/VNN2 is stored in secretory vesicles. Once activated by stimulators that induce adherence and chemotaxis of neutrophils, neutrophils adhere to cells and supportive tissues *via* Mac-1. Adherence *via* Mac-1 conversely stimulates neutrophils, resulting in the release GPI-80/VNN2. Black ellipses show GPI-80/VNN2 and grey ellipses show dimers of Mac-1 (CD11b/CD18 dimer).

found in plasma membrane (Dahlgren *et al.* 2001; Suzuki *et al.* 1999), the secretory vesicles are a considerable reservoir of intracellular GPI-80/VNN2 (Dahlgren *et al.* 2001). GPI-80/VNN-2 stored in secretory vesicles is considered to be a soluble form without being conjugated with a GPI-anchor to the membrane (Figure 41.2). Stimulation with fMLP (Huang *et al.* 2001) and tumour necrosis factor-alpha (TNF-α) (Nitto *et al.* 2002) induces the release of the soluble form of GPI-80/VNN2, which requires adherence to the ligand of Mac-1 (CD11b/CD18) (Nitto *et al.* 2002) (Figure 41.2). The soluble form of GPI-80/VNN2 are detected in synovial fluids of rheumatoid arthritis patients (Huang *et al.* 2001), and serum derived from coronary sinus of ischemic heart diseases (Inoue *et al.* 2006), suggesting that GPI-80/VNN2 is an indicator of the severity of inflammation especially neutrophil activation.

41.3.3 Vanin-3/VNN3

The third gene named vanin-3 in mouse and VNN3 in humans, respectively, has somewhat different features compared with the two genes described above. Mouse vanin-3 is produced as a soluble protein without a GPI-anchor

(Martin et al. 2001). Mouse vanin-3 expression is widely found in normal mouse tissue such as spleen, peripheral blood leukocyte, liver, kidney, thymus and heart. Mouse vanin-3 expression is induced by oxidative stress, which is involved in antioxidant response elements as well as mouse vanin-1 gene (Berruyer et al. 2004). The cell lysate showed a pantetheine-hydrolysing activity (Martin et al. 2001) as well as mouse vanin-1 when the enzymatic activity was measured using the lysates of the cells transfected with mouse vanin-3 cDNA. On the other hand, VNN3, a putative human gene homologue of mouse vanin-3, seems to encode an incomplete protein (Nitto et al. 2008).

It has been shown that several alternative splice variants are generated from human GPI-80/VNN2 and VNN3 genes (Nitto et al. 2008). Although what this means remains to be elucidated, products derived from these alternative splice variants may have some biological activities other than as pantetheinase enzymes. Human VNN3 mRNA is reported to be expressed in lung, liver and peripheral blood cells in normal individuals (Jansen et al. 2009; Martin et al. 2001). It is demonstrated by Western blotting that human VNN3 protein is detected in skin keratinocytes induced by inflammatory cytokines and that VNN3 expression is increased in psoriatic skin lesions (Jansen et al. 2009). It is also reported that metallothionein-deficient neonatal mice have augmented expression of vanin-3 (Kimura et al. 2000). Given that mouse vanin-3 has comparable pantetheinase activity to vanin-1 to produce cysteamine (Martin et al. 2001), vanin-3 might compensate the role of metallothionein such as an acute phase response to stress in those mice.

41.4 Role of Pantetheinase *in vivo:* Regulation of Inflammation rather than CoA Metabolism?

41.4.1 Vanin-1 Deficient Mice

Among the three kinds of molecules in the pantetheinase gene family, pantetheinase/vanin-1/VNN1 has been best characterized under physiological and pathophysiological conditions. The roles of pantetheinase/vanin-1/VNN1 in certain diseases are the most documented by vanin-1-deficient mice. Vanin-1-deficient mice lack cysteamine (Pitari et al. 2000), one of the metabolites of pantetheine in tissues, suggesting that the pantetheinase/vanin-1 gene product is the major enzyme for the provision of cysteamine in mice. Vanin-1-deficient mice, which fail to produce cysteamine, exhibit resistance to oxidative tissue injury caused by γ-irradiation or by administration of paraquat (Berruyer et al. 2004). Vanin-1-deficient mice show less severe bowel inflammatory reactions in response to non-steroidal anti-inflammatory drugs and to *Schistosoma manosoni* infection (Martin et al. 2004), and to administration of 2,4,6-trinitrobenzene sulfonic acid (Berruyer et al. 2006). Administration of peroxisome proliferator-activated receptor (PPAR) γ antagonist deprived the alleviation of the inflammation (Berruyer et al. 2006), suggesting that pantethienase/vanin-1 antagonizes directly or indirectly PPARγ that is working as an anti-inflammatory factor. This is consistent with the finding that exposure to human mononuclear cells to oxidative stress inducers elicited dramatic upregulation of human VNN1 and down-regulation of PPARγ (Zhang et al. 2011). Lack of vanin-1 prevents occurrence of colon cancer caused by repetitive administration of dextran sodium sulfate in combination with the injection of azoxymethane, a tumour promoter that induces colitis-associated cancer (Pouyet et al. 2010). Deficiency of vanin-1 also affects chodrogenesis in bone marrow stromal cells (Johnson et al. 2008) and granuloma formation against *Coxiella burnetii*, a bacterium that causes Q fever, due to alteration of macrophage function (Meghari et al. 2007).

41.4.2 Cysteamine: A Key Player in Inflammation and Host Defence

Mice lacking pantetheinase exhibit weaker inflammatory responses compared with wild-type mice may be due to less ability to hydrolyse pantetheine followed by the production of cysteamine and pantothenic acid. Extensive investigations using vanin-1-deficient mice show that cysteamine is more responsible for inflammatory response than pantothenic acid. Administration of cysteamine or cystamine (an oxidized dimer of cysteamine) recovers inflammatory response in vanin-1-deficient mice as much as wild-type mice show (Berruyer et al. 2004, 2006; Martin et al. 2004; Meghari et al. 2007; Pouyet et al. 2010; Roisin-Bouffay et al. 2008), suggesting that cysteamine is a key molecule that induces inflammatory response. Since cysteamine has sulfhydryl group in its molecule, it has an ability to cleave the disulfide bonds of cystine (a dimeric amino acid formed covalently by the oxidation of two cysteine residues).

Therefore, cysteamine is currently clinically used for the treatment of nephropathic cystinosis in humans. Cystinosis is an autosomal recessive disorder characterized by an accumulation of cystine in lysosomes throughout the body. The *CTNS* gene, encoding the lysosomal cystine carrier cytonosin, is a responsible gene. Cysteamine treatment dramatically improves the prognosis for children with cystinosis because it deprives cells of cystine. Cysteamine and cystamine are also being investigated to treat Huntington's disease since they inhibit transglutaminase which is suspected of participating in Huntington's disease pathogenesis (Green 1993). In addition, cysteamine and cystamine show a neuroprotective effect by increasing the heat-shock DnaJ-containing protein1b and the brain-derived neurotrophic factor (Borrell-Pages et al. 2006). Therefore, cysteamine is now subject to clinical trial to treat Huntington's disease. Cysteamine shows anti-malaria activity. It has also been shown that the *Char9* locus of mice that encodes pantetheinase/vanin-1 defines susceptibility to malaria (Min-Oo et al. 2007). A/J strain mice, which are not able to produce pantetheinase protein and cysteamine, are susceptible to malaria infection. Administration of cysteamine to the mice partially corrects susceptibility to malaria (Min-Oo et al. 2007), suggesting that cysteamine is expected to be a new medicine for the treatment of malaria infection resistant to currently used medicines.

Vanin-1-deficient mice lack free cysteamine in their tissues and exhibit elevated stores of the reduced form of glutathione (GSH) in multiple tissues instead (Berruyer et al. 2004). Cysteamine directly inhibits γ-glutamylcysteine synthase, the rate-limiting enzyme in the synthesis of GSH. GSH is a redox stress regulator that is the major reduced intracellular thiol. Vanin-1 deficient mice have a γ-glutamylcysteine synthase activity, which results in increased tissue stores of GSH (Berruyer et al. 2004; Martin et al. 2004), suggesting that pantetheinase/vanin-1 regulate GSH-dependent response to oxidative stress. In addition, vanin-1-deficient animals show a decreasing level of selenium-independent GSH peroxidase activity and the GSTA3 protein. These results suggest that pantetheinase has important roles in regulating cellular redox status and that cysteamine is a key molecule for the regulation of redox status in the body.

41.5 Panetetheinase Family Genes and Proteins as Indicators of Human Diseases

Many researchers have investigated whether the expression of pantetheinase family genes and proteins can be used as indicators for certain diseases. Huang et al. (2010) report that monitoring the expression levels of human VNN1 in combination with matrix metalloproteinase-9 may be used as a novel blood biomarker panel for the discrimination of pancreatic cancer-associated diabetes from type 2 diabetes. Fugmann et al. (2011) have shown that the concentrations of VNN1 in human urine distinguished diabetic patients with macroalbumiuria from those with normal albuminuria. Zhang et al. (2011) showed that overexpression of the VNN1 gene in epithelial cells is most strongly associated with progression to chronic immune thrombocytopenia. On the other hand, soluble GPI-80/VNN2 protein is suggested to be an indicator of acute phase neutrophil activation in arthritis (Huang et al. 2001) and ischemic heart disease (Inoue et al. 2006). Human VNN1 and VNN3 expression is increased in psoriatic skin lesion compared with normal individuals (Jansen et al. 2009). Considering that the pantetheinase gene family is suggested to be involved in inflammation, the expression levels of pantetheinase family genes may reflect the severity of the other inflammatory diseases.

41.6 Studies on Pantetheinase in Future

In this decade the pantetheinase family of genes and proteins have become an enhancer for inflammation and an oxidative stress inducer rather than the simple enzymes involved in CoA metabolism. However, the precise roles of the genes and the gene products on inflammatory cell activation and on immune cell functions remain to be further elucidated. Clarification of how to regulate the expression of the pantetheinase family genes should help us understand an unknown regulatory pathway of immune functions. Targeting to the enzymatic activity of pantetheinase, including cysteamine production, may result in the generation of a new type of anti-inflammatory drugs in future.

Summary Points

- This chapter focuses on pantetheinase and its metabolites.
- The pantetheinase gene family consists of three related genes: pantethienase/vanin-1, GPI-80/vanin-2 and vanin-3.
- Pantetheinase, which is also called vanin-1/VNN1, is a GPI-anchored heat-stable enzyme hydrolysing pantetheine into pantothenic acid (vitamin B_5) and cysteamine.
- Pantetheine is produced during the metabolism of CoA.
- Pantetheinase/vanin-1 is expressed in intestine and kidney.
- Mice lacking the pantetheinase/vanin-1 gene are unable to produce cysteamine in tissue and show impaired response to inflammatory stimuli.
- GPI-80/vanin-2 is mainly expressed in human neutrophils and monocytes which may regulate adherence and migration.
- There are two molecular forms of GPI-80/vanin-2: GPI-anchored form and soluble form.
- Soluble GPI-80 is release from activated neutrophils, which depends on adherence *via* Mac-1.
- Vanin-3 is soluble pantetheinase and is induced in inflammatory skin lesions as well as vanin-1.
- Cysteamine is used to treat cystinosis, which is an autosomal recessive disorder characterized by an accumulation of cystine in lysosomes throughout the body.
- Cysteamine enhances inflammatory response.

Key Facts

Key Facts about Neutrophils

- Neutrophils are the leukocytes (white blood cells), which number is about 70% of total white blood cells.
- Neutrophils are named after their stainability to neutral dye (neutro = neutral; phil = favour).
- The major function of neutrophils is to accumulate to tissues being infected by microorganisms or being injured, and to clean up tissue debris and infected microorganisms.
- The clean-up process is called phagocytosis (*i.e.* taking something up into the cells).
- The progenitor of neutrophils is produced in foetal liver and in the bone marrow of individuals.
- The progenitor of neutrophils changes its morphology and function to resemble mature neutrophils that are found in peripheral blood and inflamed tissues under the stimulation of certain bioactive factors (differentiation factors).
- Treatment with antitumor medicines significantly decreases progenitors of neutrophils followed by decrease in mature neutrophils, which induces severe inability to clear microorganisms, *i.e.* susceptibility to infection.

- Gramulocyte-colony stimulating factor (G-CSF) is a well-known neutrophil differentiation factor and is clinically used to increase neutrophil numbers after the treatment with antitumor medicines.
- A congenital disorder that cannot produce neutrophils or that neutrophils are not able to reach an inflamed site shows severe susceptibility to infection leading to death.
- Unusual activation of neutrophils leads to chronic inflammatory diseases such as inflammatory bowel diseases and arthritis.

Key Facts about Cysteamine

- Cysteamine is an aminothiol of the chemical formula H_2N-CH_2-CH_2-SH.
- Hydrolysis of pantetheine by pantetheinase is the best source of cysteamine in the body.
- Cysteamine is an anti-oxidant in the body.
- Oxidation of cysteamine produces hypotaurine.
- Cysteamine is utilized as a medicine to treat cystinosis because it has an ability to cleave the disulfide bond of cystine.
- It is reported that cysteamine is effective for the treatment of Huntington's disease because cysteamine inhibits transglutaminase, one of the enzyme exacerbating the disease.
- Administration of cysteamine enhances inflammatory reactions.
- Cysteamine is used to induce experimental duodenal ulcer model using animals.

Definitions of Words and Terms

Cluster of differentiation (CD). CD is an antigen used for the identification of cell surface molecules present on blood cells, especially immune cells. Each CD antigen is supposed to be recognized by at least one specific monoclonal antibody. There are more than 350 CD antigens for humans.

Coenzyme A (CoA). CoA is a cofactor that fully activates enzymes involved in fatty acid metabolism and synthesis. This molecule has a sulfhydryl group that can bind covalently to the acid portion of a fatty acid.

Cysteamine. Cysteamine is a simple amino thiol that has anti-oxidative activities. Cysteamine has an ability to cleave the disulfide bond between two cysteines and cystine, a dimerized cysteine.

Cystinosis. Cystinosis is a congenital disorder characterized by an accumulation of cystine in lysosomes throughout the body.

Huntington's disease. Huntington's disease is an neurodegenerative disease that affects muscle contraction and mental activity. The disease is a congenital and its responsive gene is called *Hungtintin* and is located on chromosome 4.

Gene-deficient mice. Gene-deficient mice are also called knockout mice. They are genetically engineered mice in which a certain gene has been disrupted artificially. The mice are used to observe the effect of a certain one gene on appearance, behaviour, physical activities, responses to some stimulation, *etc*.

GPI-anchor. GPI-anchor is a glycolipid consisting of phosphatidyl inositol and carbohydrate. GPI-anchor is sometimes conjugated with the carboxy-terminus of certain proteins and retains the protein on the plasma membrane of the cellular surface.

Mac-1. Mac-1 is a dimerized protein consisting of CD11b and CD18, whose function is to enable leukocytes to adhere specifically to other cells expressing its counter ligand and to collagen.

Oxidative stress. Oxidative stress is the condition that arises on exposure to reactive oxygen species (ROS). A reactive oxygen species are sometimes harmful to cells and tissues because it damages biological components such as proteins, nucleic acids and carbohydrates. Cells have several systems to cope with oxidative stress. Generation of GSH is one of these systems.

Pantetheine. Pantetheine is an intermediate during CoA metabolism. Pantetheine consists of a pantothenic acid (vitamin B_5) portion and a cysteamine portion. Pantetheine is a good substrate of pantetheinase, a main gene product of *vanin-1*.

Peroxisome proliferator-activated receptor (PPAR)γ. PPARγ is a transcription factor that induces the genes of several enzymes and biologically active factors related to lipid and sugar uptake at liver and adipose tissues. A compound that is able to bind to this protein is used as an anti-diabetic medicine.

Transglutaminase. Transglutaminase is a calcium-dependent enzyme that catalyses the formation of ε-(γ-glutamyl)lysine isopeptide bonds between a polypeptide-bound glutamine and a lysine of the protein.

List of Abbreviations

CoA	coenzyme A
CD	cluster of differentiation antigens
fMLP	formyl-methionyl-leucyl-phenylalanine
GPI	glycosylphosphatidylinositol
GSH	reduced form of glutathione
NCBI	National Center for Biotechnology Information
PPAR	peroxisome proliferator-activated receptor
TNF-α	tumour necrosis factor-alpha

References

Aurrand-Lions, M., Galland, F., Bazin, H., Zakharyev, V.M., Imhof, B.A., and Naquet, P., 1996. Vanin-1, a novel GPI-linked perivascular molecule involved in thymus homing. *Immunity*. 5: 391–405.

Berruyer, C., Martin, F.M., Castellano, R., Macone, A., Malergue, F., Garrido-Urbani, S., Millet, V., Imbert, J., Dupre, S., Pitari, G., Naquet, P., and Galland, F., 2004. Vanin-1$^{-/-}$ mice exhibit a glutathione-mediated tissue resistance to oxidative stress. *Molecular and Cellular Biology.* 24: 7214–7224.

Berruyer, C., Pouyet, L., Millet, V., Martin, F.M., LeGoffic, A., Canonici, A., Garcia, S., Bagnis, C., Naquet, P., and Galland, F., 2006. Vanin-1 licenses inflammatory mediator production by gut epithelial cells and controls colitis by antagonizing peroxisome proliferator-activated receptor γ activity. *The Journal of Experimental Medicine.* 203: 2817–2827.

Borrell-Pages, M., Canals, J.M., Cordelieres, F.P., Parker, J.A., Pineda, J.R., Grange, G., Bryson, E.A., Guillermier, M., Hirsch, E., Hantraye, P., Cheetham, M. E., Neri, C.,Alberch, J., Brouillet, E., Saudou, F., and Humbert, S., 2006. Cystamine and cysteamine increase brain levels of BDNF in Huntington disease *via* HSJ1b and transglutaminase. *The Journal of Clinical Investigation.* 116: 1410–1424.

Bowles, J., Bullejos, M., and Koopman, P., 2000. A subtractive gene expression screen suggests a role for vanin-1 in testis development in mice. *Genesis.* 27: 124–135.

Dahlgren, C., Karlsson, A., and Sendo, F., 2001. Neutrophil secretory vesicles are the intracellular reservoir for GPI-80, a protein with adhesion-regulating potential. *Journal of Leukocyte Biology.* 69: 57–62.

Fugmann, T., Borgia, B., Revesz, C., Godo, M., Forsblom, C., Hamar, P., Holthofer, H., Neri, D., and Roesli, C., 2011. Proteomic identification of vanin-1 as a marker of kidney damage in a rat model of type 1 diabetic nephropathy. *Kidney International.* 80: 272–281.

Galland, F., Malergue, F., Bazin, H., Mattei, M.G., Aurrand-Lions, M., Theillet, C., and Naquet, P., 1998. Two human genes related to murine vanin-1 are located on the long arm of human chromosome 6. *Genomics.* 53: 203–213.

Granjeaud, S., Naquet, P., and Galland, F., 1999. An ESTs description of the new Vanin gene family conserved from fly to human. *Immunogenetics.* 49: 964–972.

Green, H., 1993. Human genetic diseases due to codon reiteration: relationship to an evolutionary mechanism. *Cell.* 74: 955–956.

Huang, H., Dong, X., Kang, M.X., Xu, B., Chen, Y., Zhang, B., Chen, J., Xie, Q.P., and Wu, Y.L., 2010. Novel blood biomarkers of pancreatic cancer-associated diabetes mellitus identified by peripheral blood-based gene expression profiles. *The American Journal of Gastroenterology.* 105: 1661–1669.

Huang, J., Takeda, Y., Watanabe, T., and Sendo, F., 2001. A sandwich ELISA for detection of soluble GPI-80, a glycosylphosphatidyl-inositol (GPI)-anchored protein on human leukocytes involved in regulation of neutrophil adherence and migration-its release from activated neutrophils and presence in synovial fluid of rheumatoid arthritis patients. *Microbiology and Immunology.* 45: 467–471.

Huang, J.B., Takeda, Y., Araki, Y., Sendo, F., and Petty, H.R., 2004. Molecular proximity of complement receptor type 3 (CR3) and the glycosylphosphatidylinositol-linked protein GPI-80 on neutrophils: effects of cell adherence, exogenous saccharides, and lipid raft disrupting agents. *Molecular Immunology*. 40: 1249–1256.

Inoue, T., Kato, T., Hikichi, Y., Hashimoto, S., Hirase, T., Morooka, T., Imoto, Y., Takeda, Y., Sendo, F., and Node, K., 2006. Stent-induced neutrophil activation is associated with an oxidative burst in the inflammatory process, leading to neointimal thickening. *Thrombosis and Haemostasis*. 95: 43–48.

Jansen, P.A., Kamsteeg, M., Rodijk-Olthuis, D., van Vlijmen-Willems, I.M., de Jongh, G.J., Bergers, M., Tjabringa, G.S., Zeeuwen, P.L., and Schalkwijk, J., 2009. Expression of the vanin gene family in normal and inflamed human skin: induction by proinflammatory cytokines. *The Journal of Investigative Dermatology*. 129: 2167–2174.

Johnson, K.A., Yao, W., Lane, N.E., Naquet, P., and Terkeltaub, R.A., 2008. Vanin-1 pantetheinase drives increased chondrogenic potential of mesenchymal precursors in *ank/ank* mice. *The American Journal of Pathology*. 172: 440–453.

Kimura, T., Oguro, I., Kohroki, J., Takehara, M., Itoh, N., Nakanishi, T., and Tanaka, K., 2000. Metallothionein-null mice express altered genes during development. *Biochemical and Biophysical Research Communications*. 270: 458–461.

Maras, B., Barra, D., Dupre, S., and Pitari, G., 1999. Is pantetheinase the actual identity of mouse and human vanin-1 proteins? *FEBS Letters*. 461: 149–152.

Martin, F., Malergue, F., Pitari, G., Philippe, J.M., Philips, S., Chabret, C., Granjeaud, S., Mattei, M.G., Mungall, A.J., Naquet, P., and Galland, F., 2001. Vanin genes are clustered (human 6q22-24 and mouse 10A2B1) and encode isoforms of pantetheinase ectoenzymes. *Immunogenetics*. 53: 296–306.

Martin, F., Penet, M.F., Malergue, F., Lepidi, H., Dessein, A., Galland, F., de Reggi, M., Naquet, P., and Gharib, B., 2004. Vanin-1$^{-/-}$ mice show decreased NSAID- and *Schistosoma*-induced intestinal inflammation associated with higher glutathione stores. *The Journal of Clinical Investigation*. 113: 591–597.

Meghari, S., Berruyer, C., Lepidi, H., Galland, F., Naquet, P., and Mege, J.L., 2007. Vanin-1 controls granuloma formation and macrophage polarization in *Coxiella burnetii* infection. *European Journal of Immunology* 37: 24–32.

Min-Oo, G., Fortin, A., Pitari, G., Tam, M., Stevenson, M.M., and Gros, P., 2007. Complex genetic control of susceptibility to malaria: positional cloning of the *Char9* locus. *The Journal of Experimental Medicine*. 204: 511–524.

Nitto, T., Araki, Y., Takeda, Y., and Sendo, F., 2002. Pharmacological analysis for mechanisms of GPI-80 release from tumour necrosis factor-α-stimulated human neutrophils. *British Journal of Pharmacology* 137: 353–360.

Nitto, T., Inoue, T., and Node, K., 2008. Alternative spliced variants in the pantetheinase family of genes expressed in human neutrophils. *Gene*. 426: 57–64.

Pitari, G., Antonini, G., Mancini, R., and Dupre, S., 1996. Thermal resistance of pantetheine hydrolase. *Biochimica et Biophysica Acta* 1298: 31–36.

Pitari, G., Malergue, F., Martin, F., Philippe, J.M., Massucci, M.T., Chabret, C., Maras, B., Dupre, S., Naquet, P., and Galland, F., 2000. Pantetheinase activity of membrane-bound Vanin-1: lack of free cysteamine in tissues of Vanin-1 deficient mice. *FEBS Letters.* 483: 149–154.

Pitari, G., Maurizi, G., Ascenzi, P., Ricci, G., and Dupre, S., 1994. A kinetic study on pantetheinase inhibition by disulfides. *European Journal of Biochemistry.* 226: 81–86.

Pouyet, L., Roisin-Bouffay, C., Clement, A., Millet, V., Garcia, S., Chasson, L., Issaly, N., Rostan, A., Hofman, P., Naquet, P., and Galland, F., 2010. Epithelial vanin-1 controls inflammation-driven carcinogenesis in the colitis-associated colon cancer model. *Inflammatory Bowel Diseases.* 16: 96–104.

Roisin-Bouffay, C., Castellano, R., Valero, R., Chasson, L., Galland, F., and Naquet, P., 2008. Mouse vanin-1 is cytoprotective for islet beta cells and regulates the development of type 1 diabetes. *Diabetologia.* 51: 1192–1201.

Suzuki, K., Watanabe, T., Sakurai, S., Ohtake, K., Kinoshita, T., Araki, A., Fujita, T., Takei, H., Takeda, Y., Sato, Y., Yamashita, T., Araki, Y., and Sendo, F., 1999. A novel glycosylphosphatidyl inositol-anchored protein on human leukocytes: a possible role for regulation of neutrophil adherence and migration. *The Journal of Immunology.* 162: 4277–4284.

Wilson, M.J., Jeyasuria, P., Parker, K.L., and Koopman, P., 2005. The transcription factors steroidogenic factor-1 and SOX9 regulate expression of Vanin-1 during mouse testis development. *The Journal of Biological Chemistry.* 280: 5917–5923.

Yoshitake, H., Takeda, Y., Nitto, T., Sendo, F., and Araki, Y., 2003. GPI-80, a β2 integrin associated glycosylphosphatidylinositol-anchored protein, concentrates on pseudopodia without association with β2 integrin during neutrophil migration. *Immunobiology.* 208: 391–399.

Zhang, B., Lo, C., Shen, L., Sood, R., Jones, C., Cusmano-Ozog, K., Park-Snyder, S., Wong, W., Jeng, M., Cowan, T., Engleman, E.G. and Zehnder, J.L., 2011. The role of vanin-1 and oxidative stress-related pathways in distinguishing acute and chronic pediatric ITP. *Blood.* 117: 4569–4579.

CHAPTER 42

Function and Effects of Pyridoxine (Vitamin B_6): An Epidemiological Review of Evidence

JUNKO ISHIHARA[a] AND HIROYASU ISO*[b]

[a] Department of Nutritional Management, Faculty of Nutritional Science, Sagami Women's University, 2-1-1 Bunkyo, Minami-ku, Sagamihara-shi, 252-0383, Japan; [b] Department of Social and Environmental Medicine, Graduate School of Medicine, Osaka University, 2-2 Yamadaoka, Suita-shi, Osaka, 565-0871 Japan
*Email: iso@pbhel.med.osaka-u.ac.jp

42.1 Overall Characteristics and Function

Vitamin B_6 consists of pyridoxine (PN) and five related compounds, namely pyridoxal (PL), pyridoxamine (PM) and their respective 5′-phosphates (PLP, PNP and PMP). PLP is the major form in the human body (Figure 42.1). Vitamin B_6 functions as a coenzyme in more than 100 enzymatic reactions involved in the metabolism of amino acids, glycogen and sphingoid bases.

One-carbon metabolism is a series of biochemical reactions that transfer single methyl groups from one site to another, synthesize nucleotides (purines

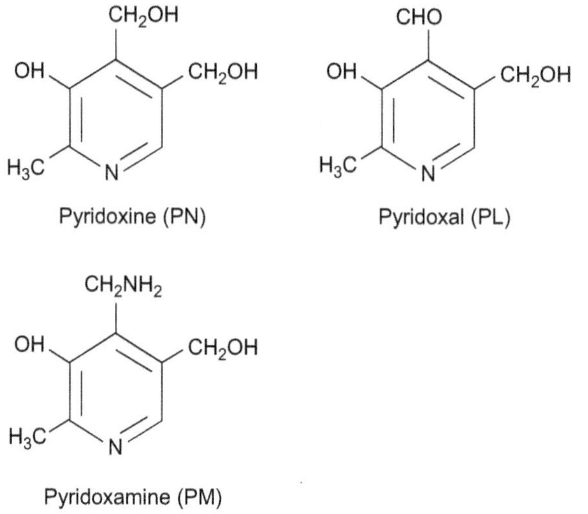

Figure 42.1 Structure of vitamin B_6.

Pyridoxine (PN)

Pyridoxal (PL)

Pyridoxamine (PM)

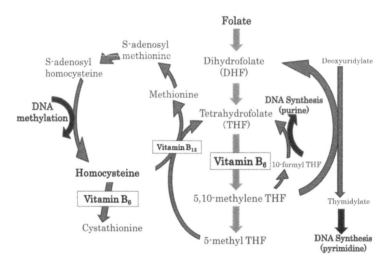

Figure 42.2 Intercellular one-carbon metabolism and the nutrients involved. One-carbon metabolism is a series of biochemical reactions that transfer single methyl groups from one site to another, synthesize nucleotides (purines and thymidilate) for DNA synthesis and repair, and methylate DNA.

and thymidilate) for DNA synthesis and repair, and methylate DNA (Figure 42.2). Folate is primary methyl donor in this pathway and vitamins B_6 and B_{12} are cofactors. Vitamin B_6 acts as a cofactor in the conversion of tetrahydrofolate to 5,10-methylene tetrahydrofolate and also as a cofactor in homocysteine catabolism.

42.2 Current Recommended Intake and Chronic Diseases

The current US recommended intake allowance (RDA) of vitamin B_6 for adults in the US Dietary Reference Intake varies between 1.3 and 2.0 mg per day depending on age, sex and pregnancy/lactation status (Standing Committee on the Scientific Evaluation of Dietary Reference Intakes and its Panel on Folate 1998). Requirements for males and the elderly are higher, as they are for pregnant or lactating females. The RDA values are generally based on depletion-repletion studies which aim to identify the deficiency level which produces a plasma PLP concentration of 20 nmol/L, which is the lower end of normal status (Standing Committee on the Scientific Evaluation of Dietary Reference Intakes 1998).

In addition to the classical clinical symptoms of vitamin B_6 deficiency such as seborrheic dermatitis, microcytic anaemia, epileptiform convulsions, depression and confusion, research in last few decades has focused on the intriguing potential association of vitamin B_6 status and chronic diseases such as vascular diseases and cancer.

Interest in the association between vascular diseases and one-carbon nutrients, including vitamin B_6, arose due to their association with homocysteine. Vitamin B_6 acts as a cofactor for cystathionine-beta-synthase, which synthesizes glutathionine from homocysteine *via* cystathionine and cysteine. Although vitamin B_{12} also acts as a cofactor for methionine synthase, which catalyses the conversion of homocysteine to methionine, a deficiency in one or both nutrients results in the accumulation of homocysteine, and hyperhomocysteinemia is in fact a suggested risk factor for vascular diseases (Selhub *et al.* 1995). Homocysteine promotes atherogenesis by damaging the vascular matrix, increasing the proliferation of endothelial cells, and facilitating oxidative injury to vascular walls (Tsai *et al.* 1994; Welch *et al.* 1997). In addition, vitamin B_6 has been shown to have an antithrombotic effect by inhibiting platelet aggregation and endothelial cell proliferation.

Vitamin B_6 has also attracted interest for its potential protective effect against cancer, as a result of its important role in one-carbon metabolism. Because one-carbon metabolism is the process of DNA methylation and nucleotide synthesis, an insufficiency in the nutrients involved is considered to result in aberrations of DNA due to altered DNA/RNA methylation and disruption of DNA integrity and repair, which then results in carcinogenesis (Choi and Mason 2000).

On the basis of these biological mechanisms, a number of epidemiologic studies have sought evidence for a possible protective role of vitamin B_6 against chronic diseases.

42.3 Epidemiologic Evidence of Vitamin B_6 and Vascular Disease

42.3.1 Findings on Dietary Intake in Prospective Observational Studies

Several large-scale prospective cohort studies have investigated the potential protective effect of vitamin B_6 intake on cardiovascular disease. In the Nurses' Health Study, vitamin B_6 intake was inversely associated with the incidence of coronary heart disease (highest vs. lowest quintile: relative risk (RR) = 0.67, 95% confidence interval (CI): 0.53, 0.85, p trend = 0.002) (Rimm et al. 1998). Inverse associations have also been found for the incidence of myocardial infarction (highest vs. lowest quintile: RR = 0.52, 95% CI: 0.29, 0.91 p trend = 0.04) and total coronary heart disease (highest vs. lowest quintile: RR = 0.60, 95% CI: 0.37, 0.97 p trend = 0.73) among Japanese non-multivitamin supplement users (Ishihara et al. 2008). Additional results for combination of the folate, vitamin B_6 and B_{12} were observed in the study (Figure 42.3). Another cohort study from Japan reported similar results for the association with mortality (Cui et al. 2010).

For stroke, three large-scale prospective cohort studies have investigated the potential protective effect of vitamin B_6 intake. Two reported no association between vitamin B_6 intake and the incidence of stroke (He et al. 2004; Larsson

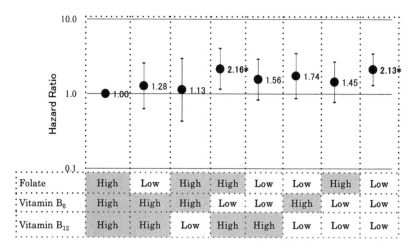

Figure 42.3 Hazard ratio of myocardial infarction among non-multivitamin users by combinations of folate, vitamin B_6 and vitamin B_{12} intake below or above the respective median values. Below-median intakes of all three vitamins conferred an excess risk. Also for persons who had below-median levels of vitamin B_6, but above-median levels of both folate and vitamin B_{12}, there also was a significant excess risk of myocardial infarction. *Statistically significant.
(Ishihara et al. 2008).

et al. 2008), while one reported a significant association between intake and the mortality of stroke (Cui *et al.* 2010).

42.3.2 Findings on Blood Level in Prospective Observational Studies

Findings from a prospective case-cohort study based on a large-scale cohort in four US communities indicated an inverse association between blood PLP and coronary heart disease among both women (highest *vs.* lowest quintile: RR = 0.36, 95% CI: 0.10, 0.98 *p* trend = 0.002) and men (highest *vs.* lowest quintile: RR = 0.48, 95% CI: 0.20, 1.04 *p* trend = 0.02) (Folsom *et al.* 1998). Findings of a nested case-control study also suggested that plasma PLP level is inversely associated with the incidence of coronary and cerebrovascular events (Vanuzzo *et al.* 2007). However, a larger nested case-control study involving the German sub-cohort of the European Prospective Investigation into Cancer and Nutrition (EPIC study) reported that the association of plasma PLP with cardiovascular disease risk is explained by other risk factors such as low-grade inflammation and smoking (Dierkes *et al.* 2007).

42.3.3 Findings from Clinical Trials

A number of clinical trials have attempted to support the observational study findings of a preventive effect of vitamin B_6 against vascular diseases. Most of these clinical trials aimed to determine the long-term effect of homocysteine-lowering treatment and accordingly involved the administration of folic acid and vitamin B_{12} in addition to vitamin B_6. Most findings indicated no significant effect of treatment.

In an early randomized controlled trial (RCT), patients who had undergone percutaneous coronary intervention were assigned to receive homocysteine-lowering therapy (1 mg folic acid, 10 mg vitamin B_6 and 0.4 mg vitamin B_{12}) or placebo for six months. The incidence of major adverse events was lower in those who received treatment (Schnyder *et al.* 2002). However, in a second RCT in patients with advanced chronic kidney disease or end-stage renal disease in which those with a high homocysteine level received high-dose treatment (40 mg folic acid, 100 mg vitamin B_6 and 2 mg of vitamin B_{12}) for approximately three years showed that treatment did not improve mortality or reduce the incidence of vascular disease (Jamison *et al.* 2007). In the Heart Outcomes Prevention Evaluation 2 (HOPE2) study, homocysteine-lowering treatment (2.5 mg folic acid, 50 mg vitamin B_6, 1 mg vitamin B_{12}) for five years significantly decreased plasma homocysteine levels among patients who had vascular disease or diabetes, but did not reduce the risk of death from cardiovascular disease or myocardial infarction (Lonn *et al.* 2006). The same study also reported a 25% decrease in risk of overall stroke, but found no effect on stroke severity or disability (Saposnik *et al.* 2009). An additional trial using the same treatment among subjects with chronic kidney disease reported similar results (Mann *et al.* 2008). Finally, a large-scale RCT in US females with pre-existing cardiovascular disease who received long-term (7.3 years) daily supplementation with a combination of 2.5 mg of folic acid, 50 mg of vitamin B_6 and 1 mg of vitamin B_{12}, or placebo, reported no significant risk reduction for cardiovascular events, despite a significant homocysteine-lowering effect (Albert *et al.* 2008).

Combined analyses of two RCT homocysteine-lowering B vitamin trials showed no significant beneficial effects of vitamin B_6 on cardiovascular events (Ebbing *et al.* 2010). Further, none of the meta-analyses of RCTs has found a protective effect of vitamin B_6 on the progression of atherosclerosis (Bleys *et al.* 2006), cardiovascular events (Marti-Carvajal *et al.* 2009), cerebrovascular diseases or all-cause mortality (Mei *et al.* 2010).

42.3.4 Current Knowledge of the Effect on Vascular Disease

Recent evidence from observational studies indicates that vitamin B_6 status, particularly blood level, is highly likely to be associated with reduced vascular disease. However, clinical trials has not supported those observational findings, possibly indicating that low blood levels of vitamin B_6 may be a consequence of low-grade inflammation or smoking rather than the cause of cardiovascular disease.

42.4 Epidemiologic Evidence of Vitamin B_6 and Cancer

42.4.1 Findings on Dietary Intake in Observational Studies

The association between dietary intake of vitamin B_6 and cancer has been extensively investigated for colorectal cancer in prospective studies. Results from a meta-analysis of nine prospective studies indicated that there is no significant association between vitamin B_6 intake and colorectal cancer risk (Larsson *et al.* 2010) (Figure 42.4). For some groups such as those who drink alcohol, however, possible beneficial effect of vitamin B_6 has been observed (Ishihara *et al.* 2007) (Figure 42.5). Regarding case-control studies, in contrast, a meta-analysis of six studies indicated a significant association between vitamin B_6 intake and colorectal cancer (Theodoratou *et al.* 2008).

With regard to other cancers, data from prospective studies are still limited. Dietary and supplemental intake of vitamin B_6 was not associated with non-Hodgkin lymphoma or multiple myeloma in a cohort study of Finnish male smokers (Lim *et al.* 2006). Dietary vitamin B_6 intake was not significantly associated with pancreatic cancer in a large cohort of Swedish women (Larsson *et al.* 2007). Finally, dietary and supplemental intake of vitamin B_6 was not associated with breast (Cho *et al.* 2007) or ovarian cancer (Kotsopoulos *et al.* 2010) risk in the Nurses' Health Study.

A number of case-control studies have investigated the association between vitamin B_6 intake and specific cancers, including esophageal (Balbuena and Casson 2010; Galeone *et al.* 2006), pharyngeal (Negri *et al.* 2000), gastric (Pelucchi *et al.* 2009), lung (Shi *et al.* 2005), breast (Ma *et al.* 2009), prostate

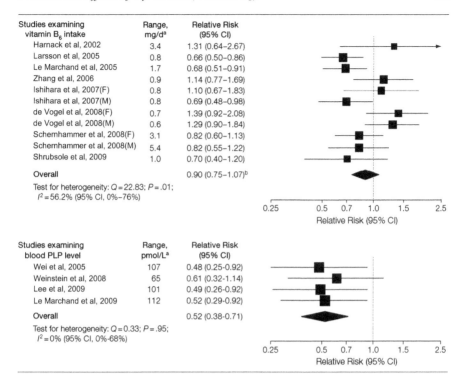

Figure 42.4 Adjusted relative risks of colorectal cancer for the highest *vs.* lowest categories of vitamin B_6 intake or blood PLP level. *Source*: Larsson *et al.* (2010), cited with permission of *The Journal of the American Medical Association*. The size of each square is proportional to the study's weight (inverse of variance). [a]The range is the difference in the midpoint between the highest and lowest categories of exposure. [b]Exclusion of the study by de Vogel *et al.* which appeared to explain the study heterogeneity, yielded a pooled relative risk of 0.80 (95% CI, 0.69–0.92) with no heterogeneity among studies ($P = 0.23$; I2 = 24%; 95% CI, 0–64%).

(Key *et al.* 1997; Pelucchi *et al.* 2005), endometrial (Xu *et al.* 2007) and renal cancer (Bosetti *et al.* 2007), but found no significant association.

A few case-control studies have indicated a risk reduction for cancer with vitamin B_6 intake. A hospital-based case-control study in Spain reported a strong inverse association between intake and bladder cancer (highest *vs.* lowest quintile: odds ratio (OR) = 0.60, 95% CI: 0.40, 0.80, *p* trend = 0.0006) (Garcia-Closas *et al.* 2007), while a US population-based case-control study of non-African-American adults indicated that intake was inversely associated with non-Hodgkin's lymphoma (highest *vs.* lowest quartile: OR = 0.57, 95% CI: 0.34, 0.95, *p* trend = 0.01) (Lim *et al.* 2005). These latter authors further analysed gene-nutrient interactions and found that the protective association seen on comparison of high *vs.* low intake was limited to certain genotypes (Lim *et al.* 2007).

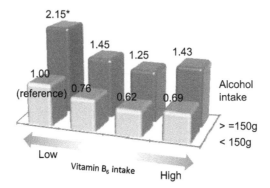

Figure 42.5 Hazard ratio of colorectal cancer according to quartiles of vitamin B_6 intake by alcohol intake among men. Results from a prospective cohort study indicated that higher intake of vitamin B_6 may be beneficial in men with higher alcohol intake. The risk of colorectal cancer associated with alcohol intake was significantly higher in those who had the lowest level of vitamin B_6 intake, but the risk decreased in those with higher vitamin B_6 intake. *Statistically significant.
(Ishihara et al. 2007).

42.4.2 Findings for Blood Level in Observational Studies

The association between blood level of vitamin B_6 and colorectal cancer has been prospectively investigated in nested case-control studies. Findings from a meta-analysis of four nested case-control studies indicated a reduced risk of colorectal cancer with an increased blood PLP level (Larsson et al. 2010), with the dose-response analysis indicating a 49% decrease in risk for every 100 pmol/mL increase in blood PLP level. Results from a subsequent nested case-control study published after the meta-analysis also indicated an inverse association between the sum of vitamin B_6 species (PLP, PL, PA), in addition to PLP and PL alone, with colorectal cancer (Eussen et al. 2010a). This study also found that the inverse association was stronger among males who drank ≥ 30 g of alcohol a day (Eussen et al. 2010a).

Only a few nested case-control studies have investigated the association between blood vitamin B_6 and other cancers. The European Prospective Investigation into Cancer and Nutrition (EPIC study) reported a strong inverse association between blood vitamin B_6 (sum of vitamin B_6) and gastric cancer (RR per increase in quartile of 0.78; p trend <0.01) (Eussen et al. 2010b). A marginally significant inverse association was found between PLP and breast cancer (Zhang et al. 2003), but no association was found for prostate (Johansson et al. 2009) or pancreatic cancer (Schernhammer et al. 2007).

42.4.3 Findings from Clinical Trials

Evidence from randomized controlled trials of the effect vitamin B_6 intake on cancer is still limited. A recent randomized trial of supplementation with a combination of B vitamins (2.5 mg of folic acid, 50 mg of vitamin B_6 and 1 mg of vitamin B_{12}) found no significant effect on the risk of total invasive cancer, colorectal cancer, or any specific cancer among women in the US after 7.3 years of treatment (Zhang et al. 2008).

42.4.4 Current Knowledge of Effect on Cancer

Recent evidence from observational studies indicates a possible association between vitamin B_6 status and cancer, particularly colorectal cancer, albeit that the results from trials are still limited.

In the *Second Expert Report of Food, Nutrition, Physical Activity and the Prevention of Cancer: a Global Perspective* (WCRF/ACRF 2007), which comprehensively reviewed the existing literature on various cancer risk factors of diet and physical activity, vitamin B_6 was not included in the summary of conclusions, which was based on the Panel's interpretation of exposures to potential risk factors for cancer. In their systematic literature reviews, the Panel judged that the evidence for esophageal cancer and vitamin B_6, obtained from a single case-control study only, was still sparse and limited (WCRF/ACRF 2007). The report made no mention of the association between vitamin B_6 and other cancers.

Summary Points

- This chapter focuses on the function and effects of vitamin B_6, particularly its potentially protective effect against chronic diseases.
- Vitamin B_6 functions as a coenzyme in more than 100 enzymatic reactions involved in metabolism.
- In addition to the classical clinical symptoms of vitamin B_6 deficiency, the potential association of vitamin B_6 status with chronic diseases has attracted significant research interest over the last few decades.
- Recent evidence from observational studies indicates that vitamin B_6 status is highly likely to be associated with vascular diseases, albeit that the findings from clinical trials were inconsistent.
- Evidence from observational studies indicates a possible association between vitamin B_6 status and cancer, particularly colorectal cancer, although the results of clinical trials remain limited.

Key Facts: Levels of Evidence Reliability

Evidence ranking level for etiology depends on the design and quality of studies. The highest level is ranked 1a and decreases downward. Table modified from the Oxford Centre for Evidence-based Medicine Levels of Evidence website (www.cebm.net/index.aspx?o=1025), produced since November 1998 by Bob Phillips, Chris Ball, Dave Sackett, Doug Badenoch, Sharon Straus, Brian Haynes, Martin Dawes. Updated by Jeremy Howick in March 2009.

Level	Study design/methods
1a	Systemic review (SR) (with homogeneity) of RCTs
1b	Individual RCT (with narrow confidence interval)
1c	All or none[a]
2a	SR (with homogeneity) of cohort studies
2b	Individual cohort study (including low quality RCT; *e.g.*, <80% follow-up)
2c	'Outcomes' research; ecological studies
3a	SR (with homogeneity) of case-control studies
3b	Individual case-control study
4	Case series (and poor quality cohort and case-control studies)
5	Expert opinion without explicit critical appraisal, or based on physiology, bench research or 'first principles'

[a]Met when all patients died before the Rx became available, but some now survive with it; or when some patients died before the Rx became available, but none now die with it.

Definitions of Words and Terms

Case-control study: A type of epidemiological research design. Case subjects (patients with a disease) are matched with control subjects (people without the disease). Their past exposure (diet, smoking, drinking, *etc.*) is then assessed retrospectively and the differences between the two groups are compared. Although case-control studies are relatively inexpensive and can be carried out over a short period of time, their retrospective study design means that the evidence they provide is weaker than that of cohort studies and randomized controlled clinical trials.

Cohort study: A type of observational epidemiological research design which investigates the causal association between a risk factor and disease. A group of people is followed after the assessment of exposure (diet, smoking, drinking, *etc.*) and the incidence of new disease or death in individuals in the group is assessed over an extended period of time. Thanks to their prospective design, cohort studies provide the strongest evidence of all observational studies.

Confidence interval: A statistical interval estimate which is used to indicate the reliability of an estimate.

Heterogeneity: Mathematical compatibility of the results of individual studies included in the meta-analysis.

Nested case-control study: A type of epidemiological research design which investigates the causal association between a risk factor and disease. Both case subjects (people who develop a disease) and control subjects (those who do not develop the disease) are selected from among a defined cohort in which the bio-sample or data of the individuals have been collected at the beginning of the study. Detailed exposure information, such as *via* the use of biomarkers, is analysed for cases and controls only, not for the entire cohort. The advantages of this design are that it is prospective and is less expensive than the full cohort approach.

Observational study: A type of study design in which the exposure of study participants is observed and compared. Study participants are not subject to any interventions by the investigators. Typical observational designs include case-series, case-control studies and cohort studies.

One-carbon metabolism: A sequential series of biochemical reactions that first transfers single methyl groups from one site to another, synthesizes nucleotides (purines and thymidilate) for DNA synthesis and repair, and then methylates DNA.

Randomized trial: An epidemiological experiment in which subjects in a population are randomly allocated into groups to receive or not receive an experimental preventive or therapeutic procedure, manoeuvre or intervention. The results are assessed by rigorous comparison of rates of disease, death, recovery or other appropriate outcome in the study and control groups (Porta 2008).

Relative risk: The ratio of the risk of an event among the exposed to the risk among the unexposed (Porta 2008).

Systematic review: The application of strategies that limit bias in the assembly, critical appraisal and synthesis of all relevant studies on a specific topic. Systematic reviews focus on peer-reviewed publications about a specific health problem and use rigorous, standardized methods to select and assess articles. A systematic review differs from a meta-analysis in not including a quantitative summary of the results (Porta 2008).

List of Abbreviations

CI	confidence interval
OR	odds ratio
PL	pyridoxal
PLP	pyridoxal 5′-phosphate
PM	pyridoxamine
PMP	pyridoxamine 5′-phosphate
PN	pyridoxine
PNP	pyridoxine 5′-phosphate
RCT	randomized controlled trial
RDA	recommended dietary allowance
RR	relative risk
SR	systematic review

References

Albert, C.M., Cook, N.R., Gaziano, J.M., Zaharris, E., MacFadyen, J., Danielson, E., Buring, J.E., and Manson, J.E., 2008. Effect of folic acid and B vitamins on risk of cardiovascular events and total mortality among women at high risk for cardiovascular disease: a randomized trial. *The Journal of the American Medical Association.* 299: 2027–2036.

Balbuena, L., and Casson, A.G., 2010. Dietary folate and vitamin B6 are not associated with p53 mutations in esophageal adenocarcinoma. *Molecular Carcinogenesis.* 49: 211–214.

Bleys, J., Miller, E.R., 3rd, Pastor-Barriuso, R., Appel, L.J., and Guallar, E., 2006. Vitamin-mineral supplementation and the progression of atherosclerosis: a meta-analysis of randomized controlled trials. *The American Journal of Clinical Nutrition.* 84: 880–887; quiz 954–955.

Bosetti, C., Scotti, L., Maso, L.D., Talamini, R., Montella, M., Negri, E., Ramazzotti, V., Franceschi, S., and La Vecchia, C., 2007. Micronutrients and the risk of renal cell cancer: a case-control study from Italy. *International Journal of Cancer.* 120: 892–896.

Cho, E., Holmes, M., Hankinson, S.E., and Willett, W.C., 2007. Nutrients involved in one-carbon metabolism and risk of breast cancer among premenopausal women. *Cancer Epidemiology, Biomarkers & Prevention.* 16: 2787–2790.

Choi, S.-W., and Mason, J.B., 2000. Folate and carcinogenesis: an integrated scheme1–3. *The Journal of Nutrition.* 130: 129–132.

Cui, R., Iso, H., Date, C., Kikuchi, S., and Tamakoshi, A., 2010. Dietary folate and vitamin b6 and B12 intake in relation to mortality from cardiovascular diseases: Japan collaborative cohort study. *Stroke.* 41: 1285–1289.

de Vogel, S., Dindore, V., van Engeland, M., Goldbohm, R.A., van den Brandt, P.A., and Weijenberg, M.P., 2008. Dietary folate, methionine, riboflavin, and vitamin B-6 and risk of sporadic colorectal cancer. *J Nutr.* 138: 2372–2378.

Dierkes, J., Weikert, C., Klipstein-Grobusch K., Westphal, S., Luley, C., Mohlig, M., Spranger, J., Boeing H., 2007. Plasma pyridoxal-5-phosphate and future risk of myocardial infarction in the European Prospective Investigation into Cancer and Nutrition Potsdam cohort. *The American Journal of Clinical Nutrition.* 86: 214–220.

Ebbing, M., Bonaa, K.H., Arnesen, E., Ueland, P.M., Nordrehaug, J.E., Rasmussen, K., Njolstad, I., Nilsen, D.W., Refsum, H., Tverdal, A., Vollset, S.E., Schirmer, H., Bleie, O., Steigen, T., Midttun, O., Fredriksen, A., Pedersen, E.R., and Nygard, O., 2010. Combined analyses and extended follow-up of two randomized controlled homocysteine-lowering B-vitamin trials. *Journal of Internal Medicine.* 268: 367–382.

Eussen, S.J., Vollset, S.E., Hustad, S., Midttun, O., Meyer, K., Fredriksen, A., Ueland, P.M., Jenab, M., Slimani, N., Boffetta, P., Overvad, K., Thorlacius-Ussing, O., Tjonneland, A., Olsen, A., Clavel-Chapelon, F., Boutron-Ruault, M.C., Morois, S., Weikert, C., Pischon, T., Linseisen, J., Kaaks, R., Trichopoulou, A., Zilis, D., Katsoulis, M., Palli, D., Pala, V., Vineis, P., Tumino, R., Panico, S., Peeters, P.H., Bueno-de-Mesquita, H.B., van Duijnhoven, F.J., Skeie, G., Munoz, X., Martinez, C., Dorronsoro, M., Ardanaz, E., Navarro, C., Rodriguez, L., VanGuelpen, B., Palmqvist, R., Manjer, J., Ericson, U., Bingham, S., Khaw, K.T., Norat, T., and Riboli, E., 2010a. Plasma vitamins

B2, B6, and B12, and related genetic variants as predictors of colorectal cancer risk. *Cancer Epidemiology, Biomarkers & Prevention.* 19: 2549–2561.

Eussen, S.J., Vollset, S.E., Hustad, S., Midttun, O., Meyer, K., Fredriksen, A., Ueland, P.M., Jenab, M., Slimani, N., Ferrari, P., Agudo, A., Sala, N., Capella, G., Del Giudice, G., Palli, D., Boeing, H., Weikert, C., Bueno-de-Mesquita, H.B., Buchner, F.L., Carneiro, F., Berrino, F., Vineis, P., Tumino, R., Panico, S., Berglund, G., Manjer, J., Stenling, R., Hallmans, G., Martinez, C., Arrizola, L., Barricarte, A., Navarro, C., Rodriguez, L., Bingham, S., Linseisen, J., Kaaks, R., Overvad, K., Tjonneland, A., Peeters, P.H., Numans, M.E., Clavel-Chapelon, F., Boutron-Ruault, M.C., Morois, S., Trichopoulou, A., Lund, E., Plebani, M., Riboli, E., and Gonzalez, C.A., 2010b. Vitamins B2 and B6 and genetic polymorphisms related to one-carbon metabolism as risk factors for gastric adenocarcinoma in the European prospective investigation into cancer and nutrition. *Cancer Epidemiology, Biomarkers & Prevention.* 19: 28–38.

Folsom, A., Nieto, F., McGovern, P., Tsai, M., Malinow, M., Eckfeldt, J., Hess, D., and Davis, C., 1998. Prospective study of coronary heart disease incidence in relation to fasting total homocysteine, related genetic polymorphisms, and B vitamins: the Atherosclerosis Risk in Communities (ARIC) study. *Circulation.* 98: 204–210.

Galeone, C., Pelucchi, C., Levi, F., Negri, E., Talamini, R., Franceschi, S., and La Vecchia, C., 2006. Folate intake and squamous-cell carcinoma of the oesophagus in Italian and Swiss men. *Annals of Oncology.* 17: 521–525.

Garcia-Closas, R., Garcia-Closas, M., Kogevinas, M., Malats, N., Silverman, D., Serra, C., Tardon, A., Carrato, A., Castano-Vinyals, G., Dosemeci, M., Moore, L., Rothman, N., and Sinha, R., 2007. Food, nutrient and heterocyclic amine intake and the risk of bladder cancer. *European Journal of Cancer.* 43: 1731–1740.

Harnack L., Jacobs D.R. Jr, Nicodemus K., Lazovich D., Anderson K., and Folsom A.R., 2002. Relationship of folate, vitamin B-6, vitamin B-12, and methionine intake to incidence of colorectal cancers. *Nutr Cancer.* 43: 152–158.

He, K., Merchant, A., Rimm, E.B., Rosner, B.A., Stampfer, M.J., Willett, W.C., and Ascherio, A., 2004. Folate, vitamin B6, and B12 intakes in relation to risk of stroke among men. *Stroke.* 35: 169–174.

Ishihara, J., Otani, T., Inoue, M., Iwasaki, M., Sasazuki, S., and Tsugane, S. for the JPHC Study Group, (2007). Low intake of vitamin B-6 is associated with increased risk of colorectal cancer in Japanese men. *The Journal of Nutrition.* 137: 1808–1814.

Ishihara, J., Iso, H., Inoue, M., Iwasaki, M., Okada, K., Kita, Y., Kokubo, Y., Okayama, A., Tsugane, S., and JPHC Study Group, 2008. Intake of folate, vitamin B6 and vitamin B12 and the risk of CHD: the Japan Public Health Center-Based Prospective Study Cohort I. *Journal of the American College of Nutrition.* 27: 127–136.

Jamison, R.L., Hartigan, P., Kaufman, J.S., Goldfarb, D.S., Warren, S.R., Guarino, P.D., and Gaziano, J.M., 2007. Effect of homocysteine lowering on mortality and vascular disease in advanced chronic kidney disease and end-stage renal disease: a randomized controlled trial. *The Journal of the American Medical Association.* 298: 1163–1170.

Johansson, M., Van Guelpen, B., Vollset, S.E., Hultdin, J., Bergh, A., Key, T., Midttun, O., Hallmans, G., Ueland, P.M., and Stattin, P., 2009. One-carbon metabolism and prostate cancer risk: prospective investigation of seven circulating B vitamins and metabolites. *Cancer Epidemiology, Biomarkers & Prevention.* 18: 1538–1543.

Key, T.J., Silcocks, P.B., Davey, G.K., Appleby, P.N., and Bishop, D.T., 1997. A case-control study of diet and prostate cancer. *British Journal of Cancer.* 76: 678–687.

Kotsopoulos, J., Hecht, J.L., Marotti, J.D., Kelemen, L.E., and Tworoger, S.S., 2010. Relationship between dietary and supplemental intake of folate, methionine, vitamin B6 and folate receptor alpha expression in ovarian tumors. *International Journal of Cancer.* 126: 2191–2198.

Larsson S.C., Giovannucci E., and Wolk A., 2005. Vitamin B6 intake, alcohol consumption, and colorectal cancer: a longitudinal population-based cohort of women. *Gastroenterology.* 128: 1830–1837.

Larsson, S.C., Giovannucci, E., and Wolk, A., 2007. Methionine and vitamin B6 intake and risk of pancreatic cancer: a prospective study of Swedish women and men. *Gastroenterology.* 132: 113–118.

Larsson, S.C., Mannisto, S., Virtanen, M.J., Kontto, J., Albanes, D., and Virtamo, J., 2008. Folate, vitamin B6, vitamin B12, and methionine intakes and risk of stroke subtypes in male smokers. *American Journal of Epidemiology.* 167: 954–961.

Larsson, S.C., Orsini, N., and Wolk, A., 2010. Vitamin B6 and risk of colorectal cancer: a meta-analysis of prospective studies. *The Journal of the American Medical Association.* 303: 1077–1083.

Le Marchand L., Wilkens L.R., Kolonel L.N., and Henderson B.E., 2005. The MTHFR C677T polymorphism and colorectal cancer: the Multiethnic Cohort Study. *Cancer Epidemiol Biomarkers Prev.* 14: 1198–1203.

Le Marchand L., White, K.K., and Nomura, A.M., *et al.*, 2009. Plasma levels of B vitamins and colorectal cancer risk: the multiethnic cohort study. *Cancer Epidemiol Biomarkers Prev.* 18: 2195–2201.

Lee, J.E., Li, H., and Giovannucci, E., *et al.*, 2009. Prospective study of plasma vitamin B6 and risk of colorectal cancer in men. *Cancer Epidemiol Biomarkers Prev.* 18: 1197–1202.

Lim, U., Schenk, M., Kelemen, L.E., Davis, S., Cozen, W., Hartge, P., Ward, M.H., and Stolzenberg-Solomon, R., 2005. Dietary determinants of one-carbon metabolism and the risk of non-Hodgkin's lymphoma: NCI-SEER case-control study, 1998–2000. *American Journal of Epidemiology.* 162: 953–964.

Lim, U., Weinstein, S., Albanes, D., Pietinen, P., Teerenhovi, L., Taylor, P.R., Virtamo, J., and Stolzenberg-Solomon, R., 2006. Dietary factors of one-carbon metabolism in relation to non-Hodgkin lymphoma and multiple myeloma in a cohort of male smokers. *Cancer Epidemiology, Biomarkers & Prevention.* 15: 1109–1114.

Lim, U., Wang, S.S., Hartge, P., Cozen, W., Kelemen, L.E., Chanock, S., Davis, S., Blair, A., Schenk, M., Rothman, N., and Lan, Q., 2007. Gene-nutrient interactions among determinants of folate and one-carbon metabolism on the risk of non-Hodgkin lymphoma: NCI-SEER case-control study. *Blood.* 109: 3050–3059.

Lonn, E., Yusuf, S., Arnold, M.J., Sheridan, P., Pogue, J., Micks, M., McQueen, M.J., Probstfield, J., Fodor, G., Held, C., and Genest, J., Jr., 2006. Homocysteine lowering with folic acid and B vitamins in vascular disease. *The New England Journal of Medicine.* 354: 1567–1577.

Ma, E., Iwasaki, M., Junko, I., Hamada, G.S., Nishimoto, I.N., Carvalho, S.M., Motola, J., Jr., Laginha, F.M., and Tsugane, S., 2009. Dietary intake of folate, vitamin B6, and vitamin B12, genetic polymorphism of related enzymes, and risk of breast cancer: a case-control study in Brazilian women. *BMC Cancer.* 9: 122.

Mann, J.F., Sheridan, P., McQueen, M.J., Held, C, Arnold, J.M., Fodor, G., Yusuf, S., and Lonn, E.M., 2008. Homocysteine lowering with folic acid and B vitamins in people with chronic kidney disease--results of the renal Hope-2 study. *Nephrology Dialysis Transplantation.* 23: 645–653.

Marti-Carvajal, A.J., Sola, I., Lathyris, D., and Salanti, G., 2009. Homocysteine lowering interventions for preventing cardiovascular events. Cochrane Database *System Reviews.* 7: CD006612.

Mei, W., Rong, Y., Jinming, L., Yongjun, L., and Hui, Z., 2010. Effect of homocysteine interventions on the risk of cardiocerebrovascular events: a meta-analysis of randomised controlled trials. *International Journal of Clinical Practice.* 64: 208–215.

Negri, E., Franceschi, S., Bosetti, C., Levi, F., Conti, E., Parpinel, M., and La Vecchia, C., 2000. Selected micronutrients and oral and pharyngeal cancer. *International Journal of Cancer.* 86: 122–127.

Pelucchi, C., Galeone, C., Talamini, R., Negri, E., Parpinel, M., Franceschi, S., Montella, M., and La Vecchia, C., 2005. Dietary folate and risk of prostate cancer in Italy. *Cancer Epidemiology, Biomarkers & Prevention.* 14: 944–948.

Pelucchi, C., Tramacere, I., Bertuccio, P., Tavani, A., Negri, E., and La Vecchia, C., 2009. Dietary intake of selected micronutrients and gastric cancer risk: an Italian case-control study. *Annals of Oncology.* 20: 160–165.

Porta, M. (ed.) for the International Epidemiological Association, 2008. *A Dictionary of Epidemiology*, 5th ed. Oxford University Press, Oxford. 316 pp.

Rimm, E., Willett, W., Hu, F., Sampson, L., Colditz, G., Manson, J., Hennekens, C., and Stampfer, M., 1998. Folate and vitamin B6 from diet and supplements in relation to risk of coronary heart disease among women. *The Journal of the American Medical Association.* 279: 359–364.

Saposnik, G., Ray, J.G., Sheridan, P., McQueen, M., and Lonn, E., 2009. Homocysteine-lowering therapy and stroke risk, severity, and disability: additional findings from the HOPE 2 trial. *Stroke.* 40: 1365–1372.

Schernhammer, E., Wolpin, .B, Rifai, N., Cochrane, B., Manson, J.A., Ma, J., Giovannucci, E., Thomson, C., Stampfer, M.J., and Fuchs, C., 2007. Plasma folate, vitamin B6, vitamin B12, and homocysteine and pancreatic cancer risk in four large cohorts. *Cancer Research.* 67: 5553–5560.

Schernhammer, E.S., Giovannuccci, E., Fuchs, C.S., and Ogino, S., 2008. A prospective study of dietary folate and vitamin B and colon cancer according to microsatellite instability and KRAS mutational status. *Cancer Epidemiol Biomarkers Prev.* 17: 2895–2898.

Schnyder, G., Roffi, M., Flammer, Y., Pin, R., and Hess, O.M., 2002. Effect of homocysteine-lowering therapy with folic acid, vitamin B12, and vitamin B6 on clinical outcome after percutaneous coronary intervention: the Swiss Heart study: a randomized controlled trial. *The Journal of the American Medical Association.* 288: 973–979.

Selhub, J, Jacques, P.F., Bostom, A.G., D'Agostino, R.B., Wilson, P.W., Belanger, A.J., O'Leary, D.H., Wolf, P.A., Schaefer, E.J., and Rosenberg, I.H., 1995. Association between plasma homocysteine concentrations and extracranial carotid-artery stenosis. *The New England Journal of Medicine.* 332: 286–291.

Shi, Q., Zhang, Z., Li, G., Pillow, P.C., Hernandez, L.M., Spitz, M.R., and Wei, Q., 2005. Sex differences in risk of lung cancer associated with methylene-tetrahydrofolate reductase polymorphisms. *Cancer Epidemiology, Biomarkers & Prevention.* 14: 1477–1484.

Shrubsole, M.J., Yang, G., and Gao, Y.T., *et al.*, 2009. Dietary B vitamin and methionine intakes and plasma folate are not associated with colorectal cancer risk in Chinese women. *Cancer Epidemiol Biomarkers Prev.* 18: 1003–1006.

Standing Committee on the Scientific Evaluation of Dietary Reference Intakes and its Panel on Folate, OBV, and Choline and Subcommittee on Upper Reference Levels of Nutrients, Food and Nutrition Board, Institute of Medicine, 1998. *Dietary Reference Intakes for Thiamin, Riboflavin, Niacin, Vitamin B6, Folate, Vitamin B12, Pantothenic Acid, Biotin, and Choline.* Institute of Medicine, National Academy Press, Washington DC, USA. 592 pp.

Theodoratou, E., Farrington, S.M., Tenesa, A., McNeill, G., Cetnarskyj, R., Barnetson, R.A., Porteous, M.E., Dunlop, M.G., and Campbell, H., 2008. Dietary vitamin B6 intake and the risk of colorectal cancer. *Cancer Epidemiology, Biomarkers & Prevention.* 17: 171–182.

Tsai, J., Perrella, M., Yoshizumi, M., Hsieh, C., Haber, E., Schlegel, R., and Lee, M., 1994. Promotion of vascular smooth muscle cell growth by homocysteine: a link to atherosclerosis. *Proceedings of the National Academy of Sciences of the United States of America.* 91: 6369–6373.

Vanuzzo, D., Pilotto, L., Lombardi, R., Lazzerini, G., Carluccio, M., Diviacco, S., Quadrifoglio, F., Danek, G., Gregori, D., Fioretti, P., Cattaneo, M., and De Caterina, R., 2007. Both vitamin B6 and total homocysteine plasma levels predict long-term atherothrombotic events in healthy subjects. *European Heart Journal.* 28: 484–491.

Welch, G., Upchurch, G.J., and Loscalzo, J., 1997. Hyperhomocyst(e)inemia and atherothrombosis. *Annals of the New York Academy of Sciences.* 811: 48–58.

Wei, E.K., Giovannucci, E., Selhub, J., Fuchs, C.S., Hankinson, S.E., and Ma J., 2005. Plasma vitamin B6 and the risk of colorectal cancer and adenoma in women. *J Natl Cancer Inst.* 97: 684–692.

Weinstein, S.J., Albanes, D., and Selhub, J., *et al.*, 2008. One carbon metabolism biomarkers and risk of colon and rectal cancers. *Cancer Epidemiol Biomarkers Prev.* 17: 3233–3240.

World Cancer Research Fund / American Institute for Cancer Research., 2007. Food, Nutrition, Physical Activity, and the Prevention of Cancer: a Global Perspective. Washington DC: AICR.

Xu, W.H., Shrubsole, M.J., Xiang, Y.B., Cai, Q., Zhao, G.M., Ruan, Z.X., Cheng, J.R., Zheng, W., and Shu, X.O., 2007. Dietary folate intake, MTHFR genetic polymorphisms, and the risk of endometrial cancer among Chinese women. *Cancer Epidemiology, Biomarkers & Prevention.* 16: 281–287.

Zhang, S.M., Willett, W.C., Selhub, J., Hunter, D.J., Giovannucci, E.L., Holmes, M.D., Colditz, G.A., and Hankinson, S.E., 2003. Plasma folate, vitamin B6, vitamin B12, homocysteine, and risk of breast cancer. *Journal of the National Cancer Institute.* 95: 373–380.

Zhang S.M., Moore S.C., and Lin J., *et al.*, 2006. Folate, vitamin B6, multivitamin supplements, and colorectal cancer risk in women. *Am J Epidemiol.* 163: 108–115.

Zhang, S.M., Cook, N.R., Albert, C.M., Gaziano, J.M., Buring, J.E., and Manson, J.E., 2008. Effect of combined folic acid, vitamin B6, and vitamin B12 on cancer risk in women: a randomized trial. *The Journal of the American Medical Association.* 300: 2012–2021.

CHAPTER 43
Function and Effects of Biotin

JEAN-JACQUES HOURI,[a] PHILIPPE MOUGENOT,[a] FRANÇOIS GUYON[a,b] AND BERNARD DO*[a,b]

[a] Agence Générale des Equipements et Produits de Santé, Assistance Publique – Hôpitaux de Paris, 7, rue du Fer à Moulin, 75005 – Paris, France; [b] Faculty of Pharmaceutical and Biological Sciences, Paris-Descartes University, 4, avenue de l'Observatoire, 75006 – Paris, France
*Email: bernard.do@eps.aphp.fr

43.1 Biotin and Biochemical Pathways

Biotin is known to be a cofactor for several biotin-containing enzymes, which can be distinguished according to the nature of their activities. They intervene in (1) the anabolic incorporation of CO_2, (2) the catabolic release of CO_2 and (3) the transfer of a carboxyl group (Attwood and Wallace 2002).

The biotin-dependent enzymes are activated by biotinylation in two steps: (1) activation of biotin by substitution of the hydroxyl moiety by adenosine monophosphate, leading to the formation of biotinyl-5′adenylate and (2) connexion to the enzyme through a lysyl residue and the AMP departure (Figure 43.1) (McAllister and Coon 1966).

Biotin is released consecutively to the degradation of the enzyme. The biotin-containing enzymes are degraded into either biocytin (ε-N-biotinyl-L-lysine) or short oligopeptides containing biotin-linked lysyl residues (Mock 2007). This reaction is catalysed by biotinidase (EC 3.5.1.12), an amide hydrolase that releases free biotin from this oligopeptide for reuse.

In mammalian tissues, four biotin-dependent carboxylases are enzymes of intermediate metabolism (Samols *et al.* 1988). Pyruvate carboxylase (PC; EC

Function and Effects of Biotin

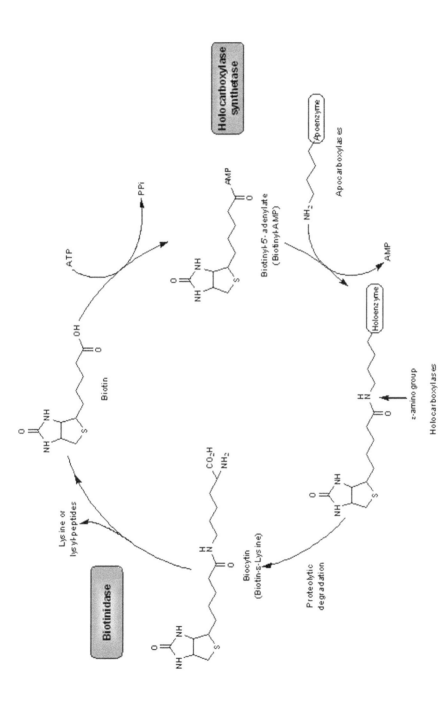

Figure 43.1 Apocarboxylase biotinylation by holocarboxylase synthetase. Holocarboxylases are formed by biotinylation of apocarboxylases and free biotin is subsequently released by catabolism of biotin-containing enzymes.

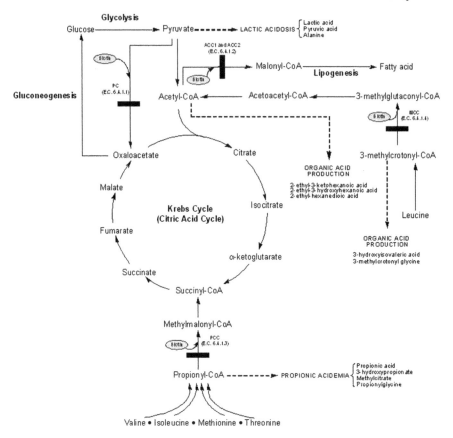

Figure 43.2 Biotin carboxylases and relationship with intermediate metabolism. Alternative pathways responsible of organic acids accumulation (dashed arrows) take place with deficient carboxylases (*black bars*) as a consequence of biotin deficiency. ACC1 and ACC2: acetyl-CoA carboxylase 1 and acetyl-CoA carboxylase 2; PC: pyruvate carboxylase; PCC: propionyl CoA carboxylase; MCC: methylcrotonyl CoA carboxylase.

6.4.1.1), propionyl-CoA carboxylase (PCC, EC 6.4.1.3), and methylcrotonyl-coenzyme A carboxylase (MCC; EC 6.4.1.4) are mitochondrial enzymes. Acetyl-CoA carboxylase (ACC; EC 6.4.1.2) occurs in two isoforms: ACC1 is found in the cytoplasm and ACC2 is located on the outer mitochondrial membrane. A representative scheme of the intermediate metabolism depicting the implications of these carboxylases and the consequence of multiple carboxylase deficiency is presented in Figure 43.2.

43.2 Biotin and Regulation of Gene Expression

In addition to its role as a prosthetic group, recent studies have also suggested that biotin is implicated in various processes such as growth, viability and cellular differentiation. Biotin has been shown to be present in different compartments involved in biotin signalling: cytoplasm, mitochondria and nuclei (Zempleni 2005). Recently, a direct effect of biotin, at the transcriptional level, was shown for a large number of mammalian genes. Transcriptional regulation can be defined as the change in gene expression levels by altering transcription rates.

Thus, the key enzymes implicated in glucose metabolism are regulated by biotin. Glucokinase involved in the glycolysis pathway and phosphoenolpyruvate carboxykinase (PECK; EC 4.1.1.32), a gluconeogenic enzyme, are regulated but with contrary effects (Dakshinamurti 2005). Biotin-dependent gene expression requires first the transformation of biotin into biotinyl-$5'$-AMP by holocarboxylase synthetase and secondly the activation of guanylate cyclase (GC; EC.4.6.1.2) that catalyses the transformation of guanosine triphosphate (GTP) to $3',5'$-cyclic guanosine monophosphate (cGMP) and a cGMP-dependent protein kinase (Leon-Del-Rio 2005). Several studies have shown that holocarboxylase synthetase (HCS) mRNA levels, AAC1, propionyl CoA carboxylase (PCC) and acetyl CoA were significantly reduced during biotin deficiency (Leon-Del-Rio 2005; Pacheco-Alvarez et al. 2002; Rodriguez-Melendez et al. 2001; Rodriguez-Melendez and Zempleni 2003; Solorzano-Vargas et al. 2002). A recent study on mice showed that administering biotin at pharmacological doses decreased serum triglyceride concentrations and lipogenic gene expression in liver and adipose tissues (Larrieta et al. 2010).

The transcriptional factor NF-κB is a protein complex that controls the transcription of DNA. NF-κB is found in almost all animal cell types and is involved in cellular responses to different stimuli, *e.g.* stress, cytokines, free radicals, ultraviolet irradiation, oxidized low density lipoprotein (LDL) and bacterial or viral antigens. NF-κB plays a key role in regulating the immune response to infection and is widely used by eukaryotic cells as a regulator of genes that controls cell proliferation and cell survival. Biotin deficiency is associated with increased nuclear translocation of NF-κB, enhancing the transcriptional activity of antiapoptotic genes in human lymphoid cells (Rodriguez-Melendez et al. 2004).

Histones H1, H2A, H2B, H3 and H4 are DNA-binding proteins that mediate the folding of DNA into chromatin. Various posttranslational modifications of histones (*e.g.* methylation, citrullination, acetylation, phosphorylation, Sumoylation, ubiquitination and ADP-ribosylation) regulate processes such as transcription, replication and repair of DNA. Recently, a novel posttranslational modification has been identified: covalent binding of the vitamin biotin to lysine residues in histones, mediated by biotinidase and holocarboxylase synthetase (Kothapalli et al. 2005). Biotinylation of histones might play a role in cell proliferation, gene silencing and cellular response to DNA damage (Kothapalli et al. 2005; Stanley et al. 2001). Posttranslational modifications of histones with biotinylation by HCS found in greater quantities in the nucleus than in the cytosol or mitochondria (Ballard et al. 2002) and possible debiotinylation by biotinidase (Gravel and Narang 2005) suggest an analogue regulation of histones to that which was observed for the biotinylation of apocarboxylases (Wood and Kumar 1985).

43.3 Sources of Biotin and Human Biotin Requirements

Humans and other mammals do not synthesize the biotin that is brought from dietary source and from microorganisms of the normal micro flora colonizing the large intestine. *In vivo* studies have shown that the human large intestine is capable of absorbing installed biotin, but the contribution of biotin from microorganisms to the totally absorbed biotin has not been fully established and needs to be further clarified (Said 2009).

Biotin is widely present in natural foodstuffs but its absolute content is lower compared with the content of most other water-soluble vitamins. In nature, the biotin molecule possesses several stereoisomers but the D-biotin isomer is the only one to be biologically active. Discrepancies have been noted on published values for biotin content in foods (Höller *et al.* 2006; Staggs *et al.* 2004). The observed gaps might arise from differences in sources and in the performance of the quantification method used. The microbiological assay method involving *Lactobacillus plantarum* is widely used (Bitsch *et al.* 1989a) but was pointed out as suffering from interferences and lack of specificity, giving an over- or underestimated biotin content in foods. Recent analytical development with chromatographic methods successfully overcame the drawbacks of the measurement by bioassay (Höller *et al.* 2006; Staggs *et al.* 2004). Nevertheless, all published sources agree in highlighting that meat, fish, poultry, eggs, dairy products and some vegetables are rich dietary sources of biotin (Combs 1992; Hardinge and Crooks 1961). The natural biotin content of most feed and food is low and typically in the range of a few µg per kg in many vegetables to several hundred µg per kg in pork liver and egg yolk.

The Food and Nutrition Board of the US National Research Council has released adequate intakes for infants, adults and pregnant women (Mock 2004; National Research Council 1989; Yates *et al.* 1998). These recommendations (Table 43.1) refer to studies assuming that, with a daily dose of 60 µg of biotin, adults on parenteral nutrition declared they were symptom-free for six months and with diets supplying 28–42 µg/day, no inadequate biotin status was observed. The dietary biotin intake in Western populations has been estimated to be 35–70 µg per day (143–287 nmoles per day).

Table 43.1 Biotin requirements and current recommendations for dietary intake of biotin.

	Adequate intakes of biotin per day
Infants (0–5 months)	5 µg (20 nmoles)
Adults and pregnant women	30 µg (123 nmoles)
Women during lactation	35 µg (143 nmoles)

43.4 Bioavailability

Contrary to its free form, dietary biotin is not ready for intestinal absorption. For this reason, its bioavailability can vary from 5% to close to 100% depending on foodstuffs ingested (Said et al. 1993).

The intervention of gastrointestinal proteases and peptidases transforms the protein bound forms of dietary biotin into biocytin or biotinyl-L-lysine and biotin-short peptides. In humans, it was assumed that biotinyl proteins and peptides arose from only two sources: protein bound in food and degradation of endogenous biotin-containing carboxylases.

Free biotin is subsequently released by the action of another and specific enzyme, namely biotinidase. The latter step is critical as it conditions efficient absorption and optimal bioavailability of dietary biotin. The carboxyl group of the valeric acid moiety of the biotin molecule must be free for its recognition by the involved transport mechanism through the intestinal enterocytes (Said and Redha 1987). Intestinal biotinidase is found in pancreatic juice, secretion of the intestinal glands, bacterial flora and the brush-border membranes.

For biotin coming from endogenous biotinyl proteins, biotin-containing carboxylases are degraded to biotinyl peptides which leads to the formation of biocytin through sequential hydrolysis. The intervention of biotinidase releases lysine and biotin.

As for the microbial source of biotin, the microflora of the large intestine synthesizes a substantial amount of the vitamin in its free form.

The intestine and the liver then play an important role in maintaining and regulating normal biotin nutrition body homeostasis.

A comparison of bioavailability was undertaken in healthy adults, comparing the urinary recovery of oral doses (2.1, 8.2 or 81.9 µmoles) to that of an intravenous dose (18.4 µmoles) as a standard (Zempleni and Mock 1999b). The choice of urinary determination was based on the fact that biliary excretion of biotin and biotin metabolites is quantitatively minor, above 98% of biotin excretion being urinary (Zempleni et al. 1997a). The urinary excretions of biotin and biotin metabolites were selectively measured by the HPLC/avidin-binding method (Zempleni and Mock 1999c). Overall, the outcomes suggest that free biotin is absorbed nearly completely when pharmacological doses of biotin are administered (Bitsch et al. 1989b). Pioneering studies estimated biotin bioavailability at 24–58% based on urinary recovery, but might have underestimated biotin and biotin metabolite concentrations in urine by using less accurate methods (Zempleni et al. 1996).

43.5 Pharmacological Effects of Biotin

Pharmacological doses of 15 mg/day of biotin administered to diabetic and non-diabetic subjects increased Acetyl-CoA carboxylases by approximately 100% and pyruvate carboxylase by approximately 200% (Báez-Saldana et al. 2004). Biotin excess administration to rats lowered postprandial glucose level and improved tolerance to glucose, ameliorating their diabetic state. Reduced hyperglycaemia was also observed in a group of type I diabetic patients receiving 16 mg/day of biotin for one week (Coggeshall et al. 1985) and increased insulin sensitivity in type 2 diabetics was noted with a treatment of 15 mg/day of biotin during 28 days (Mejia 2005).

A negative correlation between deficient biotin status and blood lipid concentrations was found in rats (Marshall et al. 1976) as well as in humans (Marshall et al. 1980). A decrease in plasma lipids was observed in human healthy volunteers within 30 min of absorption of 100 mg of biotin infusion. It was shown that oral biotin supplementation affected plasma lipid concentrations. The administration of 5 mg/day of biotin decreased hypercholesterolemia in atherosclerosis and hyperlipidemia patients (Dukusova and Krivoruchenko 1972). A 15 mg/day treatment by biotin for 28 days decreased hypertriglyceridemia of subjects whose triacylglycerol concentrations were more than 25% above the normal of 1.8 mmol/L (Báez-Saldana 2004).

In rats, a diet containing biotin reduced acetyl-CoA carboxylase 2, which plays an essential role in the inhibition of fatty acid oxidation (Munday and Hemingway 1999). Therefore, pharmacological doses of biotin may increase fatty acid oxidation by decreasing the acetyl-CoA carboxylase 2 activity.

43.6 Physiopathological Aspects of Biotin Deficiency

There are very few recent publications about the clinical manifestations of a biotin deficiency, probably because the needs in biotin are usually satisfied. Nevertheless, certain categories of people may be particularly subjected to biotin deficiency, for instance pregnant women, patients being co-administered certain drugs, or having deficiencies of the enzymes involved in biotin homeostasis, *etc.*

43.6.1 Causes of Biotin Deficiency

It is now clear that pregnant women have a decreased biotin status (Mock 2009; Stratton et al. 2009; Takechi 2008; Zempleni et al. 2008). Some authors reported that the marginal biotin status was not severe enough to observe typical clinical manifestations of biotin deficiency. Biotin deficiency was established as teratogenic in rodents, particularly in mice, but it is less clearly established for humans. Very low biotin plasma levels were reported in the animal studies compared to what was observed in pregnant women. The transport of biotin from mother to foetus and the catabolism of biotin in the foetus are likely explain the difference in the teratogenic potential of biotin deficiency between animal species. Half of pregnant women have marginal biotin deficiency but there is no evidence that such a deficiency should lead to a change in health policies. The link between biotin deficiency and teratogenic effects is still uncertain for humans and further studies are required to clarify it (Zempleni et al. 2008). Nevertheless, in developed countries, supplementary multivitamins including biotin are administered during pregnancy.

An inadequate nutrition may cause biotin deficiency. Such deficiency would not appear if human alimentation was varied and regular. Biotin deficiency might occur with inadequate feeding formulas or with parenteral nutrition free of vitamin and mineral supplements.

A case of a five-month-old child fed with only an amino acid formula presenting biotin deficiency has been reported (Fujimoto et al. 2005). Urinary biomarkers of biotin status were significantly elevated while decreased serum and urine levels of biotin were observed. The authors underlined the importance of dermatological signs that must warn the physician about a deficiency of biotin. The patient has easily recovered with a daily dose of 1 mg of biotin orally prescribed. A case of two infants presenting biotin deficiency caused by a hypoallergenic formula used in Japan was reported by Watanabe et al. (2010). Feeding formulas should be supplemented with biotin to avoid biotin deficiency, which is reported as frequent in Japan because of a non-systematic supplementation of biotin in the amino acid formulas. A biotin deficiency in dizygotic twins caused by amino acid formula was pointed out by Teramura et al. (2010). As a result, the oral use of amino acid formula must be completed with an adequate dose of biotin supplementation.

Concerning parenteral nutrition, the problem is the same but the existing formulas contain amino acids with adequate vitamin and mineral supplementation. Continuous consumption of raw eggs containing the protein avidin is well-known for causing biotin deficiency as the solid combination avidin/biotin is not absorbed. Nevertheless, such dietary intake is rather rare. Biotin deficiency can be considered as a public health problem in developing countries where severely malnourished children lack of multivitamins (Zempleni et al. 2008).

Anticonvulsants are known for being responsible for biotin deficiency. In 1997, Mock and Dyken concluded that such long-term treatments increase biotin catabolism and cause lower biotin status (Mock and Dyken 1997). The following year, Mock and co-workers confirmed that biotin catabolism increases in patients under long-term treatment with carbamazepine and/or phenytoin or with phenobarbital. Notwithstanding this, more data are necessary to confirm a global decrease of biotin status because the biomarkers of biotin status conflicted in the study (Mock et al. 1998). Valproic acid treatment was shown to decrease biotidinase enzyme activities. The patients were treated with 10 mg of biotin per day to improve skin lesions (Schulpis et al. 2001). Valproic acid induced comparable effects in rats, disturbing biotin homeostasis (Arslan et al. 2009). Nevertheless, further investigations are required to confirm that anticonvulsants must be systematically prescribed with a supplementation of biotin.

Lipoic acid was used in the past as an antidote or to treat diabetic neuropathy. Its use seemed to affect biotin-dependent carboxylase activities, likely due to a competition between lipoic acid and biotin and causing the same disagreements as a real biotin deficiency (Zempleni et al. 1997b; Zempleni and Mock 1999c).

Long-term use of antibiotics might cause a biotin deficiency by affecting the microflora involved in biotin synthesis (Zempleni et al. 2008). Chronic alcohol intake might also cause biotin deficiency. In rats and transgenic mice, chronic alcohol exposure inhibited the biotin intestinal absorption process by disturbing the functions of the SLC5A6 gene. However, a chronic exposure to alcohol has no consequences on the biotin renal reabsorption process (Subramanian et al. 2011a, 2011b; Zempleni et al. 2008). Sealey and co-workers have shown that smoking accelerates biotin catabolism in women, leading to a marginal biotin deficiency.

Recently, a biotin plus zinc deficiency after surgery (pancreatico duodenectomy) was reported in a 16-year old patient. She was diagnosed as biotin-deficient because of dermatological clinical signs (alopecia, skin scales). The authors recommend an early diagnosis to prevent sequel of biotin deficiency with an adequate supplementation of biotin and zinc as treatment (Yazbeck et al. 2010).

Biotin deficiency may be caused by inborn errors on other proteins involved in biotin homeostasis: biotinidase, the sodium-dependent multivitamin transporter and holocarboxylase synthetase (Zempleni et al. 2008). A congenital deficiency of either of these proteins may create impairments in essential metabolisms, causing clinical signs with various intensities.

43.6.2 Biotin Deficiency Assessment

Although frank biotin deficiency is rarely seen, it has been described in patients receiving parenteral nutrition without biotin supplementation and in those with biotinidase deficiency (Zempleni and Mock 1999c). Consuming large amounts of raw egg-white could also provoke biotin deficiency, decreasing biotin absorption tightly bound to avidin.

To allow estimation of human biotin requirements and evaluation of potential deleterious effects of marginal degrees of biotin deficiency, indicators of biotin status need to be determined and validated. Several explored directions include serum concentrations and urinary excretion rates of biotin and biotin metabolites, activities of the biotin-dependent decarboxylases in peripheral blood mononuclear cells, and urinary excretion rates of 3-hydroxyisovaleric acid 3-methylcrotonyl glycine and 2-methylcitric acid.

Excretion rates of biotin and biotin metabolites are early and sensitive indicators of biotin deficiency. A decrease in excretion to the lower limit of normal or even below the limit was observed in normal adults in whom biotin deficiency was experimentally induced.

Unlike the urinary excretion rates, serum concentrations of biotin and biotin metabolites seem not to reflect accurately the biotin deficiency and are therefore inadequate candidates to track it (Mock 1999). Egg-white feeding studies (Mock et al. 1997) and biotin-free diets in patients under total parenteral nutrition (Velazquez et al. 1990) showed that the serum concentrations of biotin, bisnorbiotin and biotin sulfoxide did not decrease significantly during a period of 20 days, whereas other indicators confirmed the presence of biotin deficiency.

Carboxylase activities are sensitive indicators of biotin deficiency. In patients on biotin-free total parenteral nutrition for 24–40 days, propionyl-CoA carboxylase activity in lymphocytes decreased to less than 50% of their level before parenteral nutrition (Velazquez et al. 1990). It is now well-known that available biotin modulates activities of acetyl-CoA carboxylase and pyruvate carboxylase, and also regulates their genetic expression either at the mRNA level (Salorzano-Vargas 2002) or at the posttranscriptional level (Rodriguez-Melendez 2001). But as keys enzymes in central and essential metabolic pathways, acetyl-CoA carboxylase and pyruvate carboxylase are under tight hormonal and allosteric control and their decreased activity in response to marginal biotin deficiency is less likely (Zempleni and Mock 1999a).

Reduced activities of carboxylase enzymes can cause a metabolic block of certain substrates and a use of alternative pathways for catabolism. Therefore, 3-hydroxyisovaleric acid and 3-methylcrotonyl glycine are formed consequently to a shunt of 3-methylcrotonyl carboxylase counterbalancing its activity decrease. Marginal biotin deficiency experimentally induced by 20 days of free biotin diets in human increased 3-hydroxyisovaleric acid excretion in urine above the upper limit of normal. The normal urinary excretion of 3-hydroxyisovaleric acid in healthy adults is 112 ± 38 µmol per 24 hours (Mock et al. 1997). This suggests that 3-hydroxyisovaleric acid urinary excretion is a good indicator of marginal biotin deficiency.

A rise in propionic acid consecutively to the reduced activity of propionyl-CoA decarboxylase can favour the synthesis of odd-chain fatty acids. Their accumulation in hepatic tissue, cardiac tissue and serum phospholipids has been detected in biotin-deficient rats (Suchy et al. 1986) and in biotin-deficient patients on parenteral nutrition (Mock et al. 1988).

Vlasova et al. (2005) suggested a molecular biology technique that consists of the quantification of the gene SLC19A3 expression, considered a relatively sensitive indicator of marginal biotin deficiency. However, the authors considered that this method may not be more sensitive than established indicators of marginal biotin deficiency (Vlasova et al. 2005). Recently, the quantitative measurement of plasma 3-hydroxyisovaleryl-carnitine by liquid chromatography–tandem mass spectrometry (LC-MS/MS) was presented as a novel bio-indicator of biotin status in human. The authors indicated that 3-hydroxyisovaleryl-carnitine concentration in plasma may increase when the patient has a biotin deficiency (Horvath et al. 2010; Stratton et al. 2011).

43.7 Consequences of Biotin Deficiency

Biotin is a co-factor for many major metabolisms in cells and the consequences of a frank biotin deficiency are serious. Dermatological, neurological and biological disorders are classically described as the main clinical consequences of biotin deficiency. Clinical manifestations can be severe and irreversible. Therefore, biotin deficiency should be postulated and explored in patients who present unexplained dermatological or neurological symptoms. With an accurate diagnosis at an early stage, daily supplementation with biotin can be prescribed in time in order to avoid severe consequences of a long-term biotin deficiency.

Dermatological disorders such as scaly dermatitis or hair loss are frequently described as a key for the diagnosis of biotin deficiency. In 1991, Mock published a review concerning the dermatological manifestations of biotin deficiency or biotinidase deficiency. He underlined that alopecia and characteristic scaly erythematous periorificial dermatitis are observed in patients of all ages (infants, children, adults) and suggested that this disorder might be caused by the impairment of fatty acids due to disturbances of the biotin-dependent carboxylases. He reported that Candida infections caused by impaired immune function might increase skin disorders (Mock 1991). More recently, Seymons et al. (2004) reported the case of a four-year-old child who had developed skin problems after unexplained seizures. She was diagnosed as biotin-deficient after histological investigations and the clinical observation that she was not gaining weight although she was eating constantly. The authors concluded that biotin deficiency must be explored in patients with unexplained seizures that are difficult to treat (Seymons et al. 2004). In 2010, Gehring and Dinulos established that skin manifestations such as periorificial or acral dermatitis are signs of the most common nutritional deficiencies: zinc, biotin, protein or essential fatty acids (Gehrig and Dinulos 2010).

Biotin deficiency may lead to neurological disorders. Seizures, hypotonia, ataxia and developmental delay are the main neurological signs due to biotin deficiency. These disorders might appear with a severe biotin deficiency, but they remain exceptional. However, biotin deficiency may be a direct consequence of biotinidase deficiency. Neurological disorders caused by such a deficiency are the same as for biotin deficiency: it is therefore important to treat patients with biotin at an early stage and definitively so as to avoid irreversible neurological damage such as mental retardation (Zempleni et al. 2008).

Biotin deficiency causes disturbances in a variety of carboxylase-mediated metabolic reactions. As a result, such a deficiency may induce ketolactic acidosis and organic aciduria (Zempleni et al. 2008). Organic acids such as 3-methylcrotonylglycine, 3-hydroxyvaleric acid or methylcitric acid are excreted in urine in case of biotin deficiency (Figure 43.2).

There are some more consequences suggested by studies conducted in vitro. In 2004, Griffin and Zempleni observed that biotin deficiency may increase resistance of cancer cells to certain antineoplasic drugs (resistance of human lymphoma cells to vinblastine and doxorubicine) (Griffin and Zempleni 2005). More recently, biotin deficiency appeared to be associated with decreased serum availability of insulin-like growth factor-I in mice. This lower availability has diminished long bone growth and elongation (Báez-Saldana et al. 2009).

Clinical consequences from marginal biotin deficiency have to be distinguished from disorders caused by severe biotin deficiency. A frank biotin deficiency appears to be rare now in developed countries. But marginal biotin deficiency might be more frequent than believed in the past, particularly in certain situations. Such a deficiency has to be considered because of its teratogenic potential, which has been clearly established in rodents. More research has to be performed in pregnant women in order to create guidelines of biotin supplementation if necessary. Moreover, biotin involved in the regulation of the genome through histone biotinylation has to be investigated.

Summary Points

- This chapter focuses on biotin function and effects.
- Biotin originates from plants and microorganisms.
- Biotin acts as a prosthetic group of carboxylases involved in the metabolism of fatty acids, amino acids and glucose.
- Biotin is involved in carboxylase genetic expression.
- Biotin deficiency implies biochemical and clinical disturbances of mammals' vital functions.

Key Facts about Prosthetic Groups

- A prosthetic group is required for the protein's biological activities.
- It is a non-protein component.
- It is either inorganic (metal ions or iron sulfur clusters) or organic (vitamins, sugar, lipids, *etc.*).
- It differs from coenzymes in that it is tightly and permanently bound to proteins and may even be attached through a covalent bond.
- It is continuously recycled as part of the biotin-containing protein's metabolism.

Definitions of Words and Terms

Avidin bound complex: Avidin is a tetrameric biotin-binding protein produced in egg whites. The tetrameric protein contains four identical subunits (homotetramer), each of which can bind to biotin with a high degree of affinity and specificity. The dissociation constant of avidin is measured to be $K_d \approx 10^{-15}$ M, making it one of the strongest known non-covalent bonds.

Biotinidase: (EC number: 3.5.1.12) is a ubiquitous mammal enzyme. Biotinidase extracts biotin from food because the body needs biotin in its free, unattached form. Moreover, biotinidase removes biotin from biocytin and makes it available to be reused by other enzymes.

Histones: Histones are highly alkaline proteins found in eukaryotic cell nuclei that compact and order the DNA into structural units called nucleosomes. They are the chief protein components of chromatin, acting as spools around which DNA winds and they play a role in gene regulation.

***Citrullination*:** Also known as deimination, citrullination is the term used for the posttranslational modification of the amino acid arginine in a protein into the amino acid citrulline. This reaction, shown below, is performed by enzymes called peptidylarginine deiminases (PADs).

***SUMOylation*:** is a post-translational modification involved in various cellular processes, such as nuclear-cytosolic transport, transcriptional regulation, apoptosis, protein stability, response to stress and progression through the cell cycle. Small ubiquitin-like modifier or SUMO proteins are a family of small proteins that are covalently attached to and detached from other proteins in cells to modify their function.

***Ubiquitination*:** or ubiquitinylation is one biochemical post-translational modification requiring several stages to finally lead to the covalent fixing of one or several proteins of ubiquitin (8 kDa) on one or more lysins of the protein substrate. These biochemical modifications have several functions, the best known of which is the degradation of ubiquitinated protein by proteasome.

List of Abbreviations

ACC	acetyl-CoA carboxylase
cGMP	cyclic guanosine monophosphate
GC	guanylate cyclase
GTP	guanosine triphosphate
HCS	holocarboxylase synthetase
LC-MS/MS	liquid chromatography-tandem mass spectrometry
LDL	low density lipoprotein
MCC	methylcrotonyl CoA carboxylase
NF-κB	nuclear factor kappa-light-chain-enhancer of activated B cells
PAD	peptidylarginine deiminase
PC	pyruvate carboxylase
PCC	propionyl CoA carboxylase
PECK	phosphoenolpyruvate carboxylase

References

Arslan, M., Vurucu, S., Balamtekin, N., Unay, B., Akin, R., Kurt, I., and Ozcan, O., 2009. The effects of biotin supplementation on serum and liver tissue biotinidase enzyme activity and alopecia in rats which were administrated to valproic acid. *Brain & Development.* 31: 405–410.

Attwood, P.V., and Wallace, J.C., 2002. Chemical and catalytic mechanisms of carboxyl transfer reactions in biotin-dependent enzymes. *Accounts of Chemical Research.* 35: 113–120.

Báez-Saldana, A., Zendejas-Ruiz, I., Revilla-Monsalve, C., Islas-Andrade, S., Cardenas, A., and Rojas-Ochoa, A., 2004. Effects of biotin on pyruvate carboxylase, acetyl-CoA carboxylase, propionyl CoA carboxylase, and markers for glucose and lipid homeostasis in type 2 diabetic patients and in non diabetic subjects. *The American Journal of Clinical Nutrition.* 79: 238–243.

Báez-Saldaña A., Gutiérrez-Ospina G., Chimal-Monroy J., Fernandez-Mejia, C., and Saavedra R., 2009. Biotin deficiency in mice is associated with decreased serum availability of insulin-like growth factor-I. *European Journal of Nutrition.* 48: 137–144.

Ballard, T.D., Wolff, J., Griffin, J.B., Stanley, J.S., van Calcar, S., and Zempleni, J., 2002. Biotinidase catalyzes debiotinylation of histones. *European Journal of Nutrition.* 41: 78–84.

Bitsch, R., Salz, I., and Hötzel, D., 1989a. Biotin assessment in foods and body fluids by a protein binding assay. *International Journal for Vitamin and Nutrition Research.* 59: 59–64.

Bitsch, R., Salz, I., and Hötzel, D., 1989b. Studies on bioavailability of oral biotin doses for humans. *International Journal for Vitamin and Nutrition Research.* 59: 65–71.

Coggeshall, J.C., Heggers, J.P., Robson, M.C., and Baker, H., 1985. Biotin status and plasma glucose levels in diabetics. *Annals of the New York Academy of Sciences.* 447: 389–392.

Combs, G.F., 1992. Biotin. In: The Vitamins: Fundamental Aspects in Nutrition and Health. Academic Press; San Diego, CA, USA. pp. 329–343.

Dakshinamurti, K., 2005. Biotin-a regulator of gene expression. *The Journal of Nutritional Biochemistry.* 16: 419–423.

Dukusova, O.D., and Krivoruchenko, I.V., 1972. The effect of biotin on the blood cholesterol levels of atherosclerotic patients in idiopathic hyperlipidemia. *Kardiologiia.* 12: 113.

Fujimoto, W., Inaoki, M., Fukui, T., Inoue, Y., and Kuhara, T., 2005. Biotin deficiency in an infant fed with amino acid formula. *The Journal of Dermatology.* 32: 256–261.

Gehrig, K.A., and Dinulos, J.G., 2010. Acrodermatitis due to nutritional deficiency. *Current Opinion in Pediatrics.* 22: 107–112.

Gravel, R.A., and Narang, M.A., 2005. Molecular genetics of biotin metabolism: old vitamin, new science. *The Journal of Nutritional Biochemistry.* 16: 428–431.

Griffin, J.B., and Zempleni, J., 2005. Biotin deficiency stimulates survival pathways in human lymphoma cells exposed to antineoplastic drugs. *The Journal of Nutritional Biochemistry.* 16: 96–103.

Hardinge, M.G., and Crooks, H., 1961. Lesser known vitamins in foods. *Journal of the American Dietetic Association.* 38: 240–245.

Höller, U., Wachter, F., Wehrli, C., and Fizet, C., 2006. Quantification of biotin in feed, food, tablets, and premixes using HPLC-MS/MS. *Journal of Chromatography B: Analytical Technologies in the Biomedical and Life Sciences.* 831: 8–16.

Horvath, T.D., Stratton, S.L., Bogusiewicz, A., Pack, L., Moran, J., and Mock, D.M., 2010. Quantitative measurement of plasma 3-hydroxyisovaleryl carnitine by LC-MS/MS as a novel biomarker of biotin status in humans. *Analytical Chemistry.* 82: 4140–4144.

Kothapalli, N., Camporeale G., Kueh, A., Chew, A Y.C., Oommen, M., Griffin, J.B., and Zempleni, J., 2005. Biological functions of biotinylated histones. *The Journal of Nutritional Biochemistry.* 16: 446–448.

Larrieta, E., Velasco, F., Vital, P., Lopez-Aceves, T., Lazo-de-la-Vega-Monroy M.L., Rojas, A., and Fernandez-Mejia, C., 2010. Pharmacological concentrations of biotin reduce serum triglycerides and the expression of lipogenic genes. *European Journal of Pharmacology.* 644: 263–268.

Leon-Del-Rio, A., 2005. Biotin-dependent regulation of gene expression in human cells. *The Journal of Nutritional Biochemistry.* 16: 432–434.

Marshall, M.W., Haubrich, M., Washington, V.A., Chang, M.W., Young, C.W., Wheeler, M.A., 1976. Biotin status and lipid metabolism in adult obese hypercholesterolemic inbred rats. *Nutritional Metabolism.* 20: 41–61.

Marshall, M.W., Kliman, P.G., and Washington, V.A., 1980. Effects of biotin on lipids and on other constituents of plasma of healthy men and women. *Artery.* 7: 330–351.

McAllister, H.C., and Coon. M.J., 1966. Further studies on the properties of liver propionyl coenzyme A holocarboxylase synthetase and the specificity of holocarboxylase formation. *The Journal of Biological Chemistry.* 241: 2855–2861.

Mejia, C.F., 2005. Pharmacological effects of biotin. *The Journal of Nutritional Biochemistry.* 16: 424–427.

Mock, D.M., Mock, N.I., Johnson, S.B., and Holman, R.T., 1988. Effects of biotin deficiency on plasma and tissue fatty acid composition: evidence for abnormalities in rats. *Pediatric Research.* 24: 396–403.

Mock, D.M., 1991. Skin manifestations of biotin deficiency. *Seminars in Dermatology.* 10: 296–302.

Mock, D.M., and Dyken, M.E., 1997. Biotin catabolism is accelerated in adults receiving long-term therapy with anticonvulsants. *Neurology.* 49: 1444–1447.

Mock, D.M., Malik, M., Stumbo, P., Bishop, W., and Mock, D.M., 1997. Increased urinary excretion of 3-hydroxyisovleric acid and decreased urinary excretion of biotin are sensitive early indicators of decreased status in experimental biotin deficiency. *The American Journal of Clinical Nutrition.* 65: 951–958.

Mock, D.M., Mock, N.I., Nelson, R.P., and Lombard, K.A., 1998. Disturbances in biotin metabolism in children undergoing long-term anticonvulsant therapy. *Journal of Pediatric Gastroenterology and Nutrition.* 26: 245–250.

Mock, D.M., 1999. Biotin status: which are valid indicators and how do we know? *The Journal of Nutrition.* 129: 498S–503S.

Mock, D.M., 2004. Biotin: physiology, dietary sources and requirements. In: Caballero, B., Allen, L., and Prentice, A. (ed.) Encyclopaedia of Human Nutrition; 2nd ed. Academic Press, London, UK: 206–209.

Mock, D.M., 2007. Biotin. In: Zempleni, J. (ed.) Handbook of Vitamins. CRC Press, New York, USA, pp. 361–383.

Mock, D.M., 2009. Marginal biotin deficiency is common in normal human pregnancy and is highly teratogenic in mice. *The Journal of Nutrition*. 139: 154–157.

Munday, M.R., and Hemingway, C.J., 1999. The regulation of acetyl-CoA carboxylase—a potential target for the action of hypolipidemic agents. *Advanced in Enzyme Regulation*. 39: 205–234.

National Research Council, 1989. Recommended dietary allowances. National Academy Press, Washington DC, USA: 165–169.

Pacheco-Alvarez, D., Solorzano-Vargas, R.S., and Del Rio, A.L., 2002. Biotin in metabolism and its relationship to human disease. *Archives of Medical Research*. 33: 439–447.

Rodriguez-Melendez, R., Cano, S., Mendez, S.T., and Velazquez, A., 2001. Biotin regulates the genetic expression of holocarboxylase synthetase and mitochondrial carboxylases in rats. *The Journal of Nutrition*. 131: 1909–1913.

Rodriguez-Melendez, R., and Zempleni, J., 2003. Regulation of gene expression by biotin (review). *The Journal of Nutritional Biochemistry*. 14: 680–690.

Rodriguez-Melendez, R., Schwab, L.D., and Zempleni, J., 2004. Jurkat cells respond to biotin deficiency with increased nuclear translocation of NF-kappaB, mediating cell survival. *International Journal for Vitamin and Nutrition Research*. 74: 209–216.

Said, H.M., and Redha, R., 1987. A carrier-mediated system for transport of biotin in rat intestine *in vitro*. *American Journal of Physiology*. 252: G52–G55.

Said, H.M., Thuy, L.P., Sweetman, L., and Schatzman, B., 1993. Transport of the biotin dietary derivative biocytin (N-biotinyl-L-lysine) in rat small intestine. *Gastroenterology*. 104: 75–80.

Said, H.M., 2009. Cell and molecular aspects of human intestinal biotin absorption. *The Journal of Nutrition*. 138: 158–162.

Samols, D., Thornton, C.G., Murtif, V.L., Kumar, G.K., Haase, F.C., and Wood, H.G., 1988. Evolutionary conservation among biotin enzymes. *The Journal of Biological Chemistry*. 263: 6461–6464.

Schulpis, K.H., Karikas, G.A., Tjamouranis, J., Regoutas, S., and Tsakiris, S., 2001. Low serum biotinidase activity in children with valproic acid monotherapy. *Epilepsia*. 42: 1359–1362.

Sealey, W.M., Teaque, A.M., Stratton, S.L., and Mock, D.M., 2004. Smoking accelerates biotin catabolism in women. *The American Journal of Clinical Nutrition*. 80(4): 932–935.

Seymons, K., De Moor, A., De Raeve, H., and Lambert, J., 2004. Dermatologic signs of biotin deficiency leading to the diagnosis of multiple carboxylase deficiency. *Pediatric Dermatology*. 21: 231–235.

Solorzano-Vargas, S., Pacheco-Alvarez, D., and Leon-Del-Rio, A., 2002. Holocarboxylase synthetase is an obligate participant in biotin-mediated regulation of its own expression and of biotin-dependent carboxylases mRNA levels in human cells. *Proceedings of the National Academy of Sciences of the United States of America*. 99: 5325–5330.

Stanley, J.S., Griffin, J.B., and Zempleni, J., 2001. Biotinylation of histones in human cells. Effects of cell proliferation. *European Journal of Biochemistry.* 268: 5424–5429.

Staggs, C.G., Sealey, W.M., McCabe, B.J., Teague, A.M., and Mock, D.M., 2004. Determination of the biotin content of select foods using accurate and sensitive HPLC/avidin. *Journal of Food Composition and Analysis.* 17: 767–776.

Stratton, S.L., Matthews, N.I., Bogusiewicz, A., Zhu, S., and Mock, D.M., 2009. Evidence that biotin status is impaired in human pregnancy. *The FASEB Journal.* 23: 103.6.

Stratton, S.L., Horvath, T.D., Bogusiewicz, A., Matthews, N.I., Henrich, C.L., Spencer, H.J., Moran, J.H., and Mock, D.M., 2011. Urinary excretion of 3-hydroxyisovaleryl carnitine is an early and sensitive indicator of marginal biotin deficiency in humans. *The Journal of Nutrition.* 141: 353–358.

Subramanian, V.S., Subramanya, S.B., and Said, H.M., 2011a. Chronic alcohol exposure negatively impacts the physiological and molecular parameters of the renal biotin reabsorption process. *American Journal of Physiology – Renal Physiology.* 300: F611–617.

Subramanian, S.B., Subramanian, V.S., Kumar, J.S., Hoiness, R., and Said H.M., 2011b. Inhibition of intestinal biotin absorption by chronic alcohol feeding: cellular and molecular mechanisms. American Journal of Physiology. *Gastrointestinal and Liver Physiology.* 300: G494–501.

Suchy, S.F., Rizzo, W.B., and Wolf, B., 1986. Effect of biotin deficiency and supplementation on lipid metabolism in rats: saturated fatty acids. *The American Journal of Clinical Nutrition.* 44: 475–480.

Takechi, R., Taniguchi, A., Ebara, S., Fukui, T., and Watanabe, T., 2008. Biotin deficiency affects the proliferation of human embryonic palatal mesenchymal cells in culture. *The Journal of Nutrition.* 138: 680–684.

Teramura, K., Fujimoto, N., Tachibana, T., and Tanaka, T., 2010. Biotin deficiency in dizygotic twins due to amino acid formula nutrition. *European Journal of Dermatology.* 20: 856–857.

Vlasova, T.I., Stratton, S.L., Wells, A.M., Mock, N.I., and Mock, D.M., 2005. Biotin deficiency reduces expression of SLC19A3, a potential biotin transporter, in leukocytes from human blood. *The Journal of Nutrition.* 135: 42–47.

Velazquez, A., Zamudio, S., Baez, A., Murguia-Corral, R., Rangel-Peniche, B. and Carrasco, A., 1990. Indicators of biotin status: a study of patients on prolonged total parenteral nutrition. *European Journal of Clinical Nutrition.* 44: 11–16.

Watanabe, Y., Ohya, T., Ohira, T., Okada, J., Fukui, T., Watanabe, T., Inokuchi, T., Yoshino M., and Matsuishi, T., 2010. Secondary biotin deficiency observed in two Japanese infants due to chronic use of hypoallergic infant formula. *Journal of Inherited Metabolic Disease.* 33(Suppl 1): S169: 549-P.

Wood, H.G., and Kumar, G.K., 1985. Transcarboxylase: its quaternary structure and the role of the biotinyl subunit in the assembly of the enzyme and in catalysis. *Annals of the New York Academy of Sciences.* 447: 1–22.

Yates, A.A., Chlicker, S.A., and Suitor, C.W., 1998. Dietary reference intakes: the new basis for recommendations for calcium and related nutrients, B vitamins, and choline. *Journal of the American Dietetic Association.* 98: 699–706.

Yazbeck, N., Muwakkit, S., Abboud, M., and Saab, R., 2010. Zinc and biotin deficiencies after pancreaticoduodenectomy. *Acta Gastro-Enterologica Belgica.* 73: 283–286.

Zempleni, J., McCormick, D.B., Stratton, S.L., and Mock, D.M., 1996. Lipoic acid (thioctic acid) analogs, tryptophan analogs, andurea do not interfere with the assay of biotin and biotin metabolites by high-performance liquid chromatography/avidin-binding assay. *The Journal of Nutritional Biochemistry.* 7: 518–523.

Zempleni, J., Green, G.M., Spannagel, A.U., and Mock, D.M., 1997a. Biliary excretion of biotin and biotin metabolites quantitatively minor in rats and pigs. *The Journal of Nutrition.* 127: 1496–1500.

Zempleni, J., Trusty, T.A., and Mock, D.M., 1997b. Lipoic acid reduces the activities of biotin-dependent carboxylases in rat liver. *The Journal of Nutrition.* 127: 1776–1781.

Zempleni, J., and Mock, D.M., 1999a. Biotin biochemistry and human requirements. *The Journal of Nutritional Biochemistry.* 10: 128–138.

Zempleni, J., and Mock, D.M., 1999b. Bioavailability of biotin given orally to humans in pharmacologic doses. *The American Journal of Clinical Nutrition.* 69: 504–508.

Zempleni, J., and Mock, D.M., 1999c. Advanced analysis of biotin metabolites in body fluids allows a more accurate measurement of biotin availability and metabolism in humans. *The Journal of Nutrition.* 129: 494S–497S.

Zempleni, J., 2005. Uptake, localization, and noncarboxylase roles of biotin. *Annual Review of Nutrition.* 25: 175–196.

Zempleni, J., Hassan, Y.I., and Wijeratne, S.S., 2008. Biotin and biotinidase deficiency. *Expert Review of Endocrinology & Metabolism.* 3: 715–724.

CHAPTER 44
The Importance of Folate in Health

ABALO CHANGO,*[a] DAVID WATKINS[b] AND LATIFA ABDENNEBI-NAJAR[a]

[a] Department of Nutritional Sciences and Health, UPSP 2007.05.137 EGEAL, Institut Polytechnique Lasalle Beauvais, 19, rue Pierre Waguet F-60026 Beauvais Cedex, France; [b] Department of Human Genetics, McGill University, Health Centre, Montreal General Hospital, Room L3-31, 1650 Cedar Avenue, Montreal, QC H3G 1A4, Canada
*Email: abalo.chango@lasalle-beauvais.fr

44.1 Introduction

In 1931 an English hematologist, Lucy Wills, discovered a nutritional factor in yeast that both prevented and cured macrocytic anaemia in pregnant women (Hoffbrand and Weir 2001). Meanwhile, various authors discovered a substance that they referred to as vitamin M, vitamin B_9 or vitamin Bc. These were subsequently shown to be folate, a naturally occurring form of folic acid that is abundant in the leaves of certain plants. Mammals do not synthesize folate in their tissues and so it must be obtained from exogenous sources: folate-rich natural foods, or supplements and fortified foods that contain synthetic folic acid. Foods such as liver and dark green, leafy vegetables are rich sources of natural folate. The recent commercial fortification of cereals, grains and bread with folic acid now represents an important source of folate for humans in certain countries.

The well-known clinical feature of folate deficiency is megaloblastic anaemia. Although folate has been recognized since the 1930s, interest in this vitamin has been growing in the past two decades, partly because of reports that link inadequate folate levels or intake with increased blood homocysteine (which in turn is associated with vascular diseases), as well as an increased risk of neural tube defects (NTDs) in developing foetuses (Eskes 2000; Motulsky 1996). Suboptimal or low maternal folate levels have also been implicated in other adverse maternal and foetal outcomes, including preeclampsia, placental abruption, and early pregnancy loss (Molloy et al. 2008). Although the most vulnerable populations are pregnant women, nursing mothers, young children, elderly people and adults in general are also affected by the clinical consequences of abnormal folate metabolism because of its involvement in neurological and neuropsychiatric disorders, as well as in certain types of cancer (Table 44.1).

Folate plays a major role in carrying one-carbon units within cells. It acts as both a donor and receiver of one-carbon moieties in a variety of reactions. Although folate deficiency is multifactorial (environmental factors, certain disease states and genetic determinants are involved), it is mainly of nutritional origin. Adequate folate is vital for cell division and homeostasis. Dietary

Table 44.1 Folate involvement in human diseases.

Diseases	Folate involvement
Megaloblastic anaemia	• Folate is critical in maintaining normal cell growth and division. The deficiency is related to the premature death of many haemopoietic cells in the bone marrow.
Cancer	• Folate is an essential factor in the DNA synthesis, DNA CG methylation as well as in the DNA repair.
Cardiovascular disease	• Folate deficiency leads to an elevation of blood homocysteine, a risk factor for cardiovascular disease.
Congenital disorders • NTD • Spina bifida • Anencephaly • Encephalocele • Cleft lip • Palate	• A poor maternal folate status in early pregnancy is related to congenital disorders; • The periconceptional supplementation of folic acid status is associated with the reduction of pregnancies with NTDs, etc.
Reproduction • Early pregnancy loss • Preeclampsia • Miscarriage • Placental abruption	• The decline in maternal blood folate level is caused by increasing folate demand of the growing foetus and placenta. • Folate deficiency increases the risk of adverse maternal and foetal outcomes.
Neurological and neuropsychiatric disorders • Alzheimer's disease • Parkinson's disease • Depression	• Folate is recognized for its role in reducing macular degeneration and neurodegenerative disorders. • Folate deficiency is associated with depression, psychogeriatric disease.

Table 44.2 Folate dietary reference intakes according to the age in mg Dietary Folate Equivalents (DFE/day).[a]

Group/age (years)	Adequate Intake (AI)	Recommended dietary allowance (RDA)	
	Men and women	Men	Women
Infants			
0–5	65		
6–11	80		
Children and adolescents			
1–3		150	150
4–8		200	200
9–13		300	300
14–18		400	400
Adults			
≥19		400	400
Pregnant women			
All ages		–	600
Lactating women			
All ages		–	500

[a]DFE reflects the higher bioavailability of synthetic folic acid found in supplements and fortified foods compared to that of naturally occurring food folate (Combs 2000; Food and Nutrition Board 1998).

Reference Intakes for folate have been reported (Food and Nutrition Board 1998) (Table 44.2). The high frequency of folate deficiency has led the Food and Drug Administration in the United States to require folic acid fortification of all enriched cereals and grain products since January 1998. Folate deficiency is a major public health concern both northern and southern countries, and affects both industrialized and non-industrialized nations. In non-industrialized countries, it is particularly accentuated by poverty, limited access to food resources, and infectious diseases (Chango and Abdennebi-Najar, 2011).

In this chapter, we briefly present folate functions. We cover nutritional, genetic and non-genetic condition leading to abnormal folate metabolism. Currently, several scientific approaches are undertaken to deepen our understanding of the relationship between optimal folate intake, human folate status and health. Among these, OMICs approaches and nutritional epigenetics are promising research domains that may give more insight in this area.

44.2 Folate Absorption, Transport and Metabolism

44.2.1 Absorption

There are two major sources of folate for humans: folate that is naturally available in the diet and synthesized folate, which is typically supplied by bacteria in the large intestine. Folic acid is not a natural physiological form and is not commonly available in nature. Dietary folate compounds in the form of polyglutamate are usually associated with dietary protein, from which they are released by digestive proteases (Shane 1989). These compounds must first be hydrolysed to monoglutamate folate in the small intestine before absorption. The enzyme responsible for this deconjugation is folylpoly-γ-glutamate carboxypeptidase (FGCP), which is anchored to the intestinal apical brush border. The monoglutamates are then converted into 5-methyl tetrahydrofolate THF (5-CH$_3$-THF) crossing the intestinal barrier and passing into the blood of the portal vein. Monoglutamate folate is absorbed in the proximal small intestine through a specialized carrier-mediated process (Chango et al. 2000; Zhao et al. 2009). A folate transporter present in the small intestine, proton-coupled folate transporter (PCFT), has been characterized (Zhao and Goldman 2007). This protein was previously identified as an intestinal heme carrier protein (HCP1). PCFT functions in the microenvironment of the duodenal brush border membrane, which is the main absorption site of dietary folate. A number of mutations have been identified in the SLC46A1 gene, which encodes PCFT, in patients suffering from hereditary familial folate malabsorption, including missense, nonsense, splicing, insertions and deletions (Zhao et al. 2007). Expression of the RFC has also been observed in the intestine but its role here is not clear.

44.2.2 Folate Transport

The methylated and reduced 5-CH$_3$-THF is the main form of circulating folate in humans. Folate concentrations vary from 5 to 15 μg/L in plasma and are 20 times higher in red blood cells. Peripheral cells internalize folate through specific transport molecules: folate receptor (FR), reduced folate (RFC), and for specific tissues, PCFT. Each transport molecule has a unique role in mediating folate transport across epithelia and into systemic tissues. Certain tissues (such as enterocytes) lack folate receptor, while in other tissues (e.g. placental trophoblast), functional coordination between those transport proteins has been demonstrated. Transport of folate across the blood–brain barrier at the choroid plexus depends on both folate receptor and PCFT, and is impaired in patients with mutations affecting FOLR1 and SLC46A1 genes (Steinfeld et al. 2009; Zhao and Goldman 2007).

The ubiquitous existence of folate receptor in mammalian cells, tissues and body fluids is now well-established. Folate receptor occurs in isoforms equipped with a hydrophobic glycosylphosphatidyl inositol tail enabling anchorage to plasma membranes. The RFC1, which is an anion exchanger, has the properties of a classical facilitative carrier with high affinity for reduced folates. The 5-CH$_3$-THF is taken up by the folate receptor with high affinity ($K_m = 1$–10 nM) (Table 44.3) and by the RFC with relatively low affinity ($K_m = 1$–10 μM).

The newly identified PCFT functions optimally at pH (6.0–6.2) and shows high specificity for reduced folates and anti-folates (methotrexate) (Yokooji et al. 2009). The characterization of transporters' expression patterns is especially important because, although the transporters have been identified and

Table 44.3 Affinities of folate receptor, reduced folate carrier and proton-coupled folate transporter for folate (oxidized and reduced form) and antifolate.

Folate	Folate receptor	Reduced folate carrier	Proton-coupled folate transporter
Folic acid	High affinity K_m 0.1–1 nM	Low affinity K_m 200–400 μM	High affinity K_m 0.5–2 μM
Reduced folate (5-methylTHF)	High affinity K_m 1–10 nM	High affinity K_m 1–10 μM	High affinity K_m 0.5–2 μM
Antifolate (methotrexate)	Low affinity K_m > 100 nM	High affinity K_m 1–10 μM	High affinity K_m 0.5–2 μM

well-studied, we still need to know the precise biochemical mechanisms underlying folate transport. The need to characterize the expression of these transporters in different tissues, and their regulation under cellular environmental conditions, is a concern for folate nutrigenomics and health.

44.2.3 Distribution, Storage and Excretion

The liver rapidly absorbs from 10 to 20% of dietary folate, with a preference for non-methylated and non-reduced derivatives, while peripheral tissues are enriched in reduced and methylated functional derivatives. Folate is mainly stored in the liver. Hepatic folates are partly excreted into the bile enterohepatic circulation and reabsorbed (Steinberg *et al.* 1979). This is one of the mechanisms involved in the recirculation of folate. Regarding renal elimination, folate is filtered by the glomerulus and reabsorbed into the proximal tubule. The daily urinary excretion of intact folates is between 1 to 12 μg. When the serum plasma folate concentration is very high, it is possible to overwhelm the renal reabsorption capacity; in this case, folate derivatives are excreted in the urine. Due to the possible production by the gut microflora, fecal folate levels are quite high.

44.2.4 Metabolism

Cellular 5-CH$_3$-THF is converted to THF by transfer of the methyl group to homocysteine, forming methionine and THF (Figure 44.1). Methionine can then be converted to *S*-adenosyl-methionine (AdoMet). AdoMet is involved in more than 100 reactions and at least 80 AdoMet-dependent enzymes have been identified (Kagan and Clarke 1994). *S*-Adenosyl-homocysteine (AdoHcy), the by-product of methyl transfer reactions, is hydrolysed, thus regenerating homocysteine, which then becomes available to start a new cycle of methyl-group transfer. Homocysteine has several metabolic fates, depending on the tissue in which its regeneration takes place. For example, in liver tissue homocysteine is remethylated back to methionine through two pathways: the methionine synthase/methionine synthase reductase (MS/MSR) pathway and the betaine-homocysteine methyltransferase (BHMT) pathway. The MS/MSR

The Importance of Folate in Health

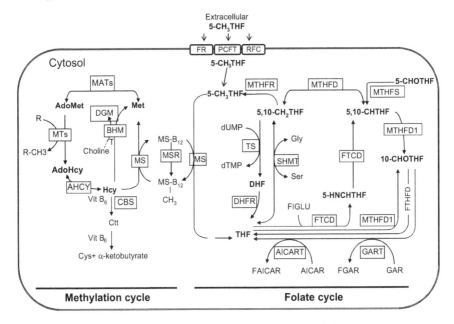

Figure 44.1 Folate-mediated one carbon metabolism network. Enzymes and transport proteins are enclosed in rectangular boxes. AHCY: S-adenosylhomocysteine hydrolase; AICART: 5-aminoimidazole carboxamide ribonucleotide transferase; BHMT: betaine homocysteine methyltransferase; CBS: cystathionine beta-synthase; DHFR: dihydrofolate reductase; FR: folate receptor; FTCD: formimidoyltransferase cyclodeaminase; GART: glycinamide ribonucleotide transformylase; MATs (MATI/MATIII): adenosylmethionine transferase enzyme I/III; MS: methionine synthase; MSR: methionine synthase reductase; MT: methyltransferase; MTHFD: methylenetetrahydrofolate dehydrogenase; MTHFR: 5,10-methylenetetrahydrofolate reductase; MTHFS: 5,10-methylenetetrahydrofolate synthase. RFC: reduced folate; AdoMet: S-adenosylmethionine; AdoHcy: S-adenosylhomocysteine; Hcy: homocysteine; SHMT: serine hydroxymethyltransferase; TS: thymidylate synthase.

pathway is widely distributed, and this remethylation pathway is thought to be ubiquitous. In contrast, the BHMT pathway has limited tissue distribution: it is found primarily in the liver, but may also occur in the kidney (Mudd *et al.* 2007). Homocysteine is also converted into cystathionine and cysteine, and thereafter into glutathione. The first catabolic step is a transsulfuration reaction catalysed by cystathione beta synthase (CBS). In the transsulfuration pathway, homocysteine condenses with serine to form cystathionine in an irreversible reaction catalysed by the pyridoxal phosphate (PLP) containing enzyme, CBS. Cystathionine is hydrolysed by a second PLP-containing enzyme, γ-cystathionase, to form cysteine and α-ketobutyrate.

In the folate cycle, the THF reacts with serine synthetizing $N_{5,10}$-methylene-THF (5,10-CH_2-THF) in a reaction catalysed by serine hydroxymethyltransferase (SHMT). The 5,10-CH_2-THF is reduced to 5-CH_3-THF by the enzyme methylenetetrahydrofolate reductase (MTHFR), or oxidized to 5,10-CH_2-THF by a reversible reaction catalysed by methylenetetrahydrofolate dehydrogenase (MTHFD). 5,10-Methenyl-THF can be converted to 10-formyl-THF by 5,10-methenyl-THF cyclohydrolase. During the synthesis of purine nuclei, C2 and C8 are formed by 10-formyltetrahydrofolate. During pyrimidine ring synthesis, thymidine monophosphate (dTMP) is produced from deoxyuridine monophosphate (dUMP) and 5,10-CH_2-THF by thymidylate synthase (TS). The product of this reaction is dihydrofolate (DHF), which is subsequently reduced to THF by dihydrofolate reductase (DHFR). The unsubstituted THF which is released then reacts with formate (via 10-formylTHF synthetase) to produce 10-formyl-THF, which donates two carbons to the synthesis of the purine ring in reactions mediated first by glycinamide ribonucleotide transformylase (GART) and then by 5-amino-4-imidazole-carboxamide ribonucleotide formyltransferase (AICART).

Folate metabolism is not limited to the cytoplasmic compartment. Most of the folate in tissues is found in the mitochondrion and cytosol (Horne et al. 1997). Individual folate-dependent pathways are compartmentalized within organelles. The cytoplasmic and mitochondrial compartments each possess a parallel array of enzymes catalysing the interconversion of folate coenzymes that carry one-carbon units. The mitochondrial folate metabolism favours incorporation of one-carbon groups from serine and release of formate, while the cytoplasmic metabolism favours incorporation of one-carbon units from formate with purine and thymidine synthesis and homocysteine remethylation.

44.3 Biochemical Function, Consequences of Folate Deficiency and Health Alteration

44.3.1 Biochemical Function

Folate mediates the transfer of single-carbon units to interrelated biochemical reactions. Therefore, the folate derivatives in cells are used as co-substrates that donate single-carbon units to a variety of synthetic reactions which are central to cell viability, including the synthesis of sulfur-containing amino acids from homocysteine, DNA-cytosine methylation; and DNA synthesis, replication and repair.

44.3.1.1 Homocysteine Remethylation

Homocysteine is an intermediate sulfur amino acid. Normal fasting homocysteine levels in plasma are between 5 and 15 µmol/l. Hyperhomocysteinemia are considered to be moderate between 16 and 30 µmol/l, intermediate (31–100 µmol/l) and severe >100 µmol/l. In plasma, 70–80% of homocysteine is bound to plasma proteins and 20–30% is free homocysteine circulating as homocysteine disulfide (homocystine) or mixed disulfides (cysteine-homocysteine). Folate deficiency leads to the decrease of homocysteine remethylation. This results in an increase in homocysteine and/or a decrease in the synthesis of AdoMet and therefore hypomethylation, and in some cases to global DNA hypomethylation and/or misincorporation of uracil into DNA.

A hypothesis has been proposed to explain all known causes of hyperhomocysteinemia by a single, biochemical principle (Selhub and Miller 1992). The hypothesis emphasizes the existence of coordinate regulation by AdoMet of the partitioning of homocysteine between methionine synthesis and catabolism through cystathionine synthesis Elevated homocysteine levels in blood can be caused by a number of factors, including folate and B vitamin (B_{12}, B_6) deficiency and pre-existing diseases such as atherosclerotic disease, diabetes and by various drugs.

44.3.1.2 Transmethylation Reactions and Epigenetic Regulation

Transmethylation is essential for the synthesis or post-translation modifications of many molecules such as DNA, RNA, phospholipids, histones, neurotransmitters and many other molecules (Mudd et al. 2007).

DNA-cytosine methylation is a well-documented epigenetic process in the context of carcinogenesis (Kim 2005). DNA methyltransferases (DNMTs) are the enzymes responsible for methyl group transfer from AdoMet to cytosine in CG sequences. Thus, folate metabolism on such epigenetic processes has an obvious impact and the folate-mediated single-carbon metabolism network is an important biological research topic which is fundamental to public health.

In the epigenetics context, folate is also important for histone methylation. Recently, Luka et al. (2011) showed that nuclear lysine-specific demethylase 1 (LSD1) from HeLa cells is a folate-binding protein. LSD1 is a flavin-containing enzyme that removes the methyl groups from lysines 4 and 9 of histone 3 with the generation of formaldehyde from the methyl group. Using a natural pentaglutamate form of THF, Luka et al. observed binding with the highest affinity ($K_d = 2.8$ μM) to LSD1. The fact that folate participates in the enzymatic demethylation of histones provides a new vision of folate role in the epigenetic control of gene expression at histones level.

44.3.1.3 De Novo Pathway for Production of dTMP, DNA Synthesis and Genomic Stability

When the methylation of dUMP by thymidylate synthetase form dTMP isinadequate, dUMP accumulates and, as a result, DNA polymerases misincorporate uracil into the newly synthetized DNA strand instead of thymine (Chango et al. 2009a) (Figure 44.2). James et al. (1989) showed that alterations in nucleotide metabolism and DNA damage are induced when methyl donor pools are stressed by dietary deficiency. There is now evidence suggesting that this excessive incorporation of uracil in DNA leads to point mutations and results in the generation of single- and double-stranded DNA breaks and chromosome breakage (Figure 44.3). The relative activities of thymidylate synthetase and

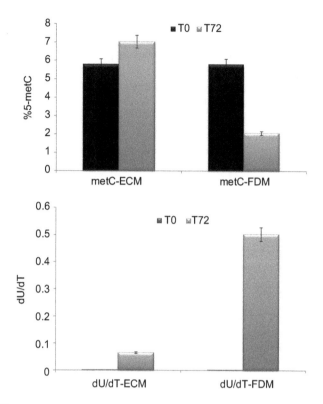

Figure 44.2 Analysis of 5-methyl cytosine (% 5-metC) and dUTP/dTTP (dU/dT) ratio in genomic DNA of HepG2 cell line. Folate depletion results in decrease of 5-metC% and elevation of dU/dT into genomic DNA extracted from HepG2 cell lines grown in complete medium (ECM) or in folate depleted medium (FDM) for 72h (T72) vs. control (T0). (Chango et al. 2009a).

MTHFR may determine the probability of MTHF (methyltetrahydrofolate) donating its single-carbon group for uracil synthesis or homocysteine remethylation pathway. This suggests that genetic defects in one or both enzymes can affect the pathway. It is interesting to note that presence of the common 677C > T polymorphism in MTHFR, which reduces enzyme activity and increases the risk for moderate hyperhomocysteinemia, neural tube defects and Down's syndrome, also results in increased risk for some types of cancer (Mason 2011).

44.3.2 Folate Deficiency and Health Alteration

44.3.2.1 Anaemia, Hyperhomocysteinemia, Cancer, Cognitive Decline in Adults

Megaloblastic anaemia was the first common problem associated with vitamin deficiency (Hoffbrand and Weir 2001). Folate deficiency causes a slowing of cell

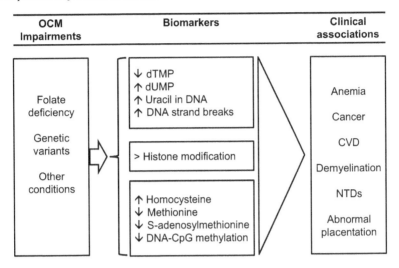

Figure 44.3 Biochemical markers and clinical associations of impaired in folate mediated-one carbon metabolism.

division noted especially in systems with rapid growth of the intestinal mucosa cells. Severe consequences of folate deficiency on intestinal absorption and in the immune system have been observed in children fed exclusively from goat's milk, which is poor in folate (Navarro et al. 1980). Care must take in children with celiac sprue disease and other malabsorption disorders that affect the proximal small intestine, because these conditions can induce folate deficiency.

McCully made the first connection between homocysteine levels and cardiovascular disease in 1969. Since then there have been many studies showing that higher plasma homocysteine levels are associated with a higher risk of cardiovascular, cerebrovascular and peripheral arterial disease. (This topic is developed in the chapter "Cardiovascular effects of B vitamins and folate in the context of fortification").

In regard to the cancer, studies have shown both *in vitro* and *in vivo* that the methylation of specific CG islands can be changed in differentiated cells, with hypermethylation repressing tumour suppressor genes or hypomethylation activating proto-oncogenes (Duthie 2011). Colorectal, uterine, cervical, lung and breast cancer are associated with folate deficiency. DNA hypomethylation and uracil misincorporation/repair are not exclusive mechanisms and both could be important in diagnostics (Chango et al. 2009a).

In regard to cognitive decline and neurodegenerative diseases, folate is important in neurogenesis, in cell growth and proliferation, and in myelination. AdoMet is involved in the synthesis of some neurotransmitters such as catecholamine and serotonin (Fernstrom 2000). Melatonin, norepinephrine, phosphatidylethanolamine and proteins such as myelin basic protein among others are methylated and play an important role in the integrity of the nervous system. Thus, reducing the synthesis of AdoMet and/or the presence of hyperhomocysteinemia secondary to folate deficiency has been associated with neurological disorders such as depression, mental retardation and cognitive impairment (Botez 1976; Herrmann and Obeid, 2011).

Inborn errors of folate metabolism also shed light on the role of folates and products of folate metabolism in neurological function. Disorders that result in decreased levels of 5-CH_3-THF in the central nervous system are frequently associated with development of seizures (*e.g.* MTHFR deficiency, DHFR deficiency, MTHFD1 deficiency, and autoimmune or genetic cases of cerebral folate deficiency due to malfunction of folate receptor alpha). Adult onset cases of MTHFR deficiency have in some cases been associated with psychosis.

44.3.2.2 Folate, Homocysteine and Pregnancy Outcomes

Hyperhomocysteinemia secondary to folate deficiency is a risk factor for pregnancy complications: increased risk of preeclampsia, intrauterine growth restriction, abrupto placenta and stillbirth (Eskes 2000; Chango and Abdennebi-Najar 2011; Tamura and Picciano 2006). Pregnancy is associated with increased folate demand and in some cases can lead to overt folate deficiency. Total folate concentration in foetal plasma is generally 5–6 fold higher than that in the mother, with the placental concentration several fold higher.

The increase in folate requirement during pregnancy is due to the growth of the foetus and uteroplacental organs. Normal placental function is pivotal for optimal foetal growth and development. Although it is a transient organ that persists for only nine months, the placental effects on offspring remain for a lifetime. Indeed, placental dysfunction predisposes to obstetric complications and is commonly associated with abnormalities in foetal programming.

Some outcomes associated with folate deficiency reflect abnormal placental development. Others result when exogenous stressors in the maternal environment combine with maladaptation of the placental response to result in a small placenta, as is typical of intrauterine growth restriction and preeclampsia (Longtine and Nelson 2011). Deficiencies in folate and vitamin B_{12} have been more commonly observed among women who experience recurrent spontaneous abortions, as well as those who experience uterine bleeding during pregnancy. Supplementation with folic acid during the very early stages of pregnancy may therefore be beneficial by promoting trophoblastic invasion (Williams *et al.* 2011). Invasion of extravillous trophoblasts into the maternal decidua and inner third of the myometrium is crucial for the remodelling of the maternal spiral arteries, which allow the establishment of uteroplacental circulation; this is suggested to be the primary defect that occurs in preeclampsia and intrauterine growth restriction (Khong *et al.* 1986). A possible benefit of periconceptual and maintained folic acid supplementation on increased extravillous invasion of trophoblast *via* increased secretion of metalloproteases MMP2, MMP3 and MMP9 (Williams *et al.* 2011), as well as altered placental/trophoblast proliferation and apoptosis, and a doubling of vascular density has been suggested.

It has recently been shown in animal experiments that the effect of methyl deficiency in the mother's diet on offspring may persist into the second generation (transgenerational effect) (Waterland et al. 2008). Epigenetic events occurring in utero can lead to persistent changes in gene expression. The diet provided to female mice of two strains during pregnancy modified the coat colour (agouti versus yellow) and DNA methylation patterns in offspring. These studies demonstrated that feeding a methyl-supplemented diet to the mice increased levels of DNA methylation and the proportion of agouti to yellow mice. Expression of the yellow coat colour is linked with increased risk of obesity, diabetes and cancer. These data point to the likelihood that in utero exposure to methyl group donors such as folate can lead to modified genomic imprinting in the offspring and potentially modify disease risk. Although methyl-donor supplementation can change the epigenetic profile, the timing of dietary supplementation may be critical in determining the ameliorating effects of folate.

Folate deficiency during pregnancy is a well-known risk factor for neural tube defects and is associated with the occurrence of orofacial clefts and other dysmorphologies in offspring. NTDs are a group of serious birth defects that affect the developing nervous system and include anencephaly, spina bifida and encephalocele. The pioneering studies of Smithells et al. (1980) showed the reduction of recurrent NTDs after periconceptional folic acid-containing multivitamin supplementation. Recent studies show that diet supplementation with a folic acid-containing multivitamin or high dose of folic acid alone in the periconceptional period reduced the recurrence of NTDs by 83–91% and 71%, respectively.

In 1992, the US Public Health Service issued a recommendation that all women capable of becoming pregnant consume 400 μg of folic acid daily to reduce the risk of having a child born with a neural tube defect (Centers for Disease Control and Prevention 1992). Food fortification was implemented to reduce the number of pregnancies affected by NTDs (Centers for Disease Control and Prevention 2005). Estimates have shown that the additional intake of folic acid through food fortification has been effective in reducing the prevalence of NTDs at birth.

44.4 Common Genetic Variation in Folate Metabolism

More than 30 enzymes and three transport proteins are involved in the folate-mediated one-carbon metabolism network (Figure 44.1) (Chango et al. 2009b). The enzyme MTHFR is the most studied of these proteins. A common sequence variant in the gene coding MTHFR has been involved in several different complex diseases, including cardiovascular disease (CVD), neural tube defects, pregnancy complications and cancer.

The MTHFR gene maps to chromosome 1 (1p36) carries two common polymorphisms: the single nucleotide polymorphism (SNP) 677C>T, in which the C nucleotide at the position 677 in the cDNA is changed to T, and the SNP 1298A>C (Leclerc and Rozen 2007). Suboptimal intake dietary folate, together with the MTHFR SNP 677C>T in the homozygous state (677TT genotype), results in moderate hyperhomocysteinemia that may be associated to the global genomic DNA hypomethylation. The frequency of this polymorphism in the homozygous state varies on average from five to 25%. The highest frequencies have been reported in southern Mediterranean and North American Hispanic populations, and the lowest in sub-Saharan Africa. Under conditions of inadequate dietary intake of folic acid, homozygous individuals have moderate hyperhomocysteinemia, with possible clinical consequences. The SNP 1298A>C, present in 9–10% of the population, may not be sufficient to disrupt enzymatic function to alter homocysteine remethylation.

Other candidate SNPs in different genes may also influence the folate metabolism network through transport systems, methylation cycle or folate cycle, and impact on health risk. Many SNPs have been identified. Some of these have been studied, although not all reported studies have confirmed association with health alteration.

44.5 Synthetic Folic Acid Use for Health: Supplementation, Fortification and Adverse Effect

The evidence supports both folic acid supplementation and fortification as effective in reducing neonatal mortality from NTDs. A review by Ray *et al.* (2004) of the prevalence of folic acid supplement use both pre- and periconceptionally in community programmes worldwide found that fewer than 50% of women appeared to take folic acid periconceptionally and noted that the rate continues to be much lower than desired. In some programmes, users of supplement tend to take higher than recommended doses.

Fortification with folic acid has produced positive results, reducing the frequency of neural tube defects in the USA, Canada and Chile. In addition to preventing the occurrence of NTDs, folic acid fortification has been associated with reducing the severity of NTDs with potential beneficial effects on health (Hoey *et al.* 2007).

Although these results are impressive, it has recently been suggested that folic acid can be a double-edged sword. High folate status might inhibit colorectal carcinogenesis, but high folate status may promote colorectal growth of already established neoplasms (Kim 2006; Mason 2011). The most likely mechanism by which folic acid supplementation may promote the progression of established preneoplastic and precursor lesions of colorectal cancer is provision of nucleotide precursors to rapidly replicating neoplastic cells for accelerated proliferation and growth (Kim 2006). The other possible mechanism suggested by Kim (2006) is *de novo* methylation of promoter CG islands of tumour suppressor genes with consequent gene inactivation leading to tumour progression.

44.6 Folate Nutrigenetics, Nutrigenomics and Epigenetics For Future Investigations

It is now recognized that genetic factors play an important role in folate metabolism and in determining folate status. However, many unknowns remain concerning the mechanisms and optimal human health.

Nutrigenomics is an emerging and important new field of nutritional science. There are now greater opportunities to identify genes and gene products (transcripts, proteins, metabolites) and to study genetic (hereditary mutation analysis) or epigenetic (DNA methylation, histones modification) changes associated with diseases with more accuracy.

In the context of nutrigenetics, some candidate genes have been identified and studied for their role in folate metabolism. The role of other candidate genes is unclear. Continuing studies on functional impact of SNPs will helpful to identify those which alter the function of genes involved directly or indirectly in folate dependent health maintenance. The relationship between MTHFR SNP 677C>T and dietary folate is one of the best examples of the interaction between genetic and nutritional factors. In most of individuals with the 677TT genotype, an optimal supply of folic acid or an adequate diet (rich in yeast, leaves, grains, fruits, vegetables, liver) is beneficial. Nutrigenetics may help to determine the optimal concentration of folic acid for preventing epigenetic alterations and genome damage dependent on genetic polymorphisms. This will lead to diet recommendations that match to an individual's functional genome to optimize genome health maintenance.

Nutrigenomics approaches are needed to identify new markers of diseases, to highlight the mechanisms involved in the functional benefit of folate for human health. Issues on which this approach will provide answers could include the following: What is the functional impact of different chemical forms of folate: folic acid (oxidized form) and 5,10-MTHF (reduced form)? How physiologically or genetically these forms are optimal for health? A nutrigenomics approach will certainly allow us to learn more about the precise mechanisms of the inductive effect of cancer by folate deficiency and to be precise about the conditions in which high folate could exacerbate carcinogenesis.

Well-known in the experimental context, epigenetics is a continuing domain of investigation in nutritional science. Epigenetic modulation of genes involved in foetal growth by folate metabolism (but also by other dietary methyl source such as choline, methionine and methylcobalamin) are the most plausible candidates for the adaptation of the individual to diet. Nutritional epigenetics may give us a clear picture of the modulation of gene expression by folate and explain how folate or folate deficiency programmes phenotypes established during the early stage of life.

44.7 Concluding Remarks

Since the discovery of folate, there has been scientific progress on its place in health risks. Studies on the role of this vitamin in the pathogenesis of vascular diseases, neural tube defects, neurological diseases and certain types of cancer are ongoing. However, several important aspects remain unexplored. Nutrigenetics, nutrigenomics and nutri-epigenetics are now contributing to deepen our knowledge on the interaction between folic acid and our genes, and to better define the functionality of this important vitamin for health risks.

Summary Points

- This chapter focuses on folate and human health.
- Folate mediates the transfer of one-carbon units into different inter-related biochemical reactions. It is required for the synthesis of sulfur containing amino acids, epigenetic methylation and DNA synthesis.
- Low maternal folate status in early pregnancy increases risks of neural tube defects, placental abruption, and early pregnancy loss.
- Folate deficiency may lead to DNA hypomethylation and/or uracil misincorporation into DNA, both promoting cancer.
- Folate deficiency may promote neurological and neuropsychiatric disorders.
- Diet supplementation or food fortification with folic acid in the periconception period allows reducing the recurrence and severity of NTDs and other adverse outcomes.
- Functional SNPs in key genes of folate metabolism are genetic determinants in the folate metabolism balance.

Key Facts: Folate Discovery and Neural Tube Defects

- Lucy Wills and her colleagues reported in 1931 that yeast contained a substance that could cure macrocytic anemia in pregnant women.
- It wasn't until the early 1940s that folate was finally isolated and identified.
- Richard Worthington Smithells, pediatrician born in 1924, was one of the first to recognize the possible role of folic acid metabolism, for human embryopathy.
- It was at Liverpool that he began his lifelong interest in the prevention of disease and congenital malformations.
- He demonstrated the importance of vitamin supplements in the prevention of spina bifida and other malformations of the spine and brain.

Definitions of Words and Terms

Epigenetics: Epigenetics is the study of heritable changes in gene expression without that they are accompanied by alteration of the DNA sequence. DNA-cytosine methylation is one type of epigenetic mechanism. It consists to the addition of a methyl group to the $5'$-carbon of 2nd cytosine in –CCGG- (CpG) sequences. DNA-methylcytosine residues are often found in short stretches of CpG-rich regions (i.e., CpG islands) found in the 5' region of approximately 60% of genes. Most CpG islands are unmethylated, with the exception of certain imprinted genes. DNA methylation aberrations can occur as either hypo- or hypermethylation. Both forms can lead to transcriptional gene silencing as observed in variety of human malignancies. Histones (a protein) modification such as methylation is another type of epigenetic mechanism that occures at the chromatine level.

Genomic instability: An abnormal cell state associated with an increased rate of heritable genomic alterations including mutations, chromosomal rearrangements, deletions and inversions.

Nutrigenetics: Nutrigenetics interested in the influence of genetic variation between individuals on metabolic or physiological responses induced by nutrient intake, in this case, folic acid. It helps to understand the cause of the variability of responses from one individual to the other side to a given environment such as folate metabolism.

Nutrigenomics: Nutrigenomics is the study of the effects of diet on gene expression, their translation into proteins and metabolic changes that result. It corresponds to understanding how of genes are involved in the metabolism, the variation in the expression of these genes under the influence of these dietary components and their respective functions.

Megaloblastic anemia: Megaloblastic anemia is a blood disorder in which there is decrease in number of red blood cells. Anemia red blood cells are larger-than-normal red blood cells. The two most common causes of megaloblastic anemia are vitamin B_{12} (cobalamin) deficiency and folate deficiency. Although their clinical settings differ considerably, no hematologic finding can distinguish between the two conditions. Folate deficiency causes a slowing of cell division that is especially noted in systems featuring rapid growth of intestinal mucosa cells.

Single Nucleotide Polymorphism (SNP): SNP is a genetic variation that can occur within DNA sequence. SNP variation occurs when a single nucleotide such as an A (adenine), replaces one of the other three nucleotide letters C (cytosine), T (thymine), or G (guanine). SNPs occur every 100 to 300 bases along the human genome and are stable from an evolutionarily standpoint.

List of Abbreviations

AdoHcy	S-adenosylhomocysteine
AdoMet	S-adenosylmethionine
AHCY	S-adenosylhomocysteine hydrolase
AICAR	aminoimidazole carboxamide ribonucleotide
AICART	5-amino-4-imidazolecarboxamide ribonucleotide formyltransferase
BHMT	betaine homocysteine methyltransferase
CBS	cystathionine beta-synthase
CVD	cardiovascular disease

DFE	Dietary Folate Equivalent
DHF	dihydrofolate
DHFR	dihydrofolate reductase
DNMT	DNA methyltransferase
dTMP	thymidine monophosphate
dTTP	2′-deoxythimidine triphosphate
dUMP	deoxyuridine monophosphate
dUTP	2′-deoxyuridine triphosphate
FGCP	folylpoly-γ-glutamate carboxypeptidase
FR	folate receptor
FTCD	formimidoyltransferase cyclodeaminase
GART	glycinamide ribonucleotide transformylase
HCP'	heme carrier protein 1
LSD1	lysine-specific demethylase 1
MAT	adenosylmethionine transferase gene I/III
MS	methionine synthase
MSR	methionine synthase reductase
MT	methyltransferase
MTHFD	methylenetetrahydrofolate dehydrogenase
MTHFR	5,10-methylenetetrahydrofolate reductase.
MTHFS	5,10-methylene-tetrahydrofolate synthase
NTD	neural tube defect
PCFT	proton-coupled folate transporter
PLP	pyridoxal phosphate
RFC	reduced folate
SHMT	serine hydroxymethyltransferase
SNP	single nucleotide polymorphism
THF	tetrahydrofolate
TS	thymidylate synthase

Acknowledgments

A.C. thanks Professors Jean-Pierre Nicolas and David S. Rosenblatt for giving me this interest in folate. Due to the limited space for references, all resources have not been cited. All our regrets to the authors who have inspired this chapter and was not cited here.

References

Botez, M.I., 1976. Folate deficiency and neurological disorders in adults. *Medical Hypotheses*. 2: 135–140.

Centers for Disease Control and Prevention, 1992. Recommendations for the use of folic acid to reduce the number of cases of spina bifida and other neural tube defects. *Morbidity and Mortality Weekly Recommendations and Reports*. 41: 1–7.

Centers for Disease Control and Prevention, 2005. Use of dietary supplements containing folic acid among women of childbearing age—United States. *Morbidity and Mortality Weekly Recommendations and Reports.* 54: 955–958.

Chango, A., Emery-Fillon, N., de Courcy, G.P., Lambert, D., Pfister, M., Rosenblatt, D.S., and Nicolas, J.P., 2000. A polymorphism (80G->A) in the reduced folate carrier gene and its associations with folate status and homocysteinemia. *Molecular Genetics and Metabolism.* 70: 310–315.

Chango, A., Abdel Nour, A.M., Niquet C., and Tessier, F.J., 2009a. Simultaneous determination of genomic DNA methylation and uracil misincorporation. *Medical Principles and Practice.* 18: 81–84.

Chango, A., Nour, A.A., Bousserouel S., Eveillard, D., Anton, P.M., and Gueant, J.L., 2009b. Time course gene expression in the one-carbon metabolism network using HepG2 cell line grown in folate-deficient medium. *The Journal of Nutritional Biochemistry.* 20: 312–320.

Chango, A., and Abdennebi-Najar, L., 2011. Folate metabolism pathway and *Plasmodium falciparum* malaria infection in pregnancy. *Nutrition Reviews.* 69: 34–40.

Combs, G.F., 2000. Vitamins. In: Mahan, L.K., and Escott-Stump, S. (ed.) Krause's Food, Nutrition and Diet Therapy. 10th ed. W.B. Saunders Company, Philadelphia, USA, pp. 67–109.

Duthie, S.J., 2011. Epigenetic modifications and human pathologies: cancer and CVD. *Proceedings of the Nutrition Society.* 70: 47–56.

Eskes, T.K., 2000. Homocysteine and human reproduction. *Clinical & Experimental Obstetrics & Gynaecology.* 27: 157–167.

Fernstrom, J.D., 2000. Can nutrient supplements modify brain function? *The American Journal of Clinical Nutrition.* 71: 1669S–1675S.

Food and Nutrition Board, National Academy of Sciences, Institute of Medicine, 1998. Folate. In: Dietary Reference Intakes for Thiamin, Riboflavin, Niacin, Vitamin B_6, Folate, Vitamin B_{12}, Pantothenic Acid, Biotin, and Choline. National Academy Press, Washington DC, USA. pp. 196–305.

Herrmann, W., and Obeid, R., 2011. Homocysteine: a biomarker in neurodegenerative diseases. *Clinical Chemistry and Laboratory Medicine.* 49: 435–441.

Hoey, L., McNulty, H., Askin, N., Dunne, A., Ward, M., Pentieva, K., Strain, J., Molloy, A.M., Flynn, C.A., and Scott, J.M., 2007. Effect of a voluntary food fortification policy on folate, related B vitamin status, and homocysteine in healthy adults. *The American Journal of Clinical Nutrition.* 86: 1405–1413.

Hoffbrand, A.V., and Weir, D.G., 2001. The history of folic acid. *British Journal of Haematology.* 113: 579–589.

Horne, D.W., Holloway, R.S., and Wagner C., 1997. Transport of S-adenosylmethionine in isolated rat liver mitochondria. *Archives of Biochemistry and Biophysics.* 343: 201–206.

James, S.J., Yin, L., and Swendseid, M.E., 1989. DNA strand break accumulation, thymidylate synthesis and NAD levels in lymphocytes from methyl donor-deficient rats. *The Journal of Nutrition.* 119: 661–664.

Kagan, R.M., and Clarke, S., 1994. Widespread occurrence of three sequence motifs in diverse S-adenosylmethionine-dependent methyltransferases suggests a common structure for these enzymes. *Archives of Biochemistry and Biophysics.* 310: 417–427.

Khong, T.Y., De Wolf, F., Robertson, W.B., and Brosens, I., 1986. Inadequate maternal vascular response to placentation in pregnancies complicated by pre-eclampsia and by small-for-gestational age infants. *British Journal of Obstetrics and Gynaecology.* 93: 1049–1059.

Kim, Y.I., 2005. Nutritional epigenetics: impact of folate deficiency on DNA methylation and colon cancer susceptibility. *The Journal of Nutrition.* 135: 2703–2709.

Kim, Y.I., 2006. Folate: a magic bullet or a double edged sword for colorectal cancer prevention? *Gut.* 55: 1387–1389.

Leclerc, D., Rozen, R., 2007. Molecular genetics of MTHFR: polymorphisms are not all benign [in French]. *Medical Science Paris.* 23: 297–302.

Longtine, M.S., and Nelson, D.M., 2011. Placental dysfunction and fetal programming: the importance of placental size, shape, histopathology, and molecular composition. *Seminars in Reproductive Medicine.* 29: 187–196.

Luka, Z., Moss, F., Loukachevitch, L.V., Bornhop, D.J., and Wagner, C., 2011. Histone demethylase LSD1 is a folate-binding protein. *Biochemistry.* 50: 4750–4756.

Mason, J.B., 2011. Unraveling the complex relationship between folate and cancer risk. *Biofactors.* 37: 253–260.

McCully K.S., 1969. Vascular pathology of homocysteinemia: implications for the pathogenesis of arteriosclerosis. *Am J Pathol.* 56: 111–128.

Molloy, A.M., Kirke, P.N., Brody, L.C., Scott, J.M., and Mills, J.L., 2008. Effects of folate and vitamin B_{12} deficiencies during pregnancy on fetal, infant, and child development. *Food and Nutrition Bulletin.* 29: S101-111; discussion S112-105.

Motulsky, A.G., 1996. Nutritional ecogenetics: homocysteine-related arteriosclerotic vascular disease, neural tube defects, and folic acid. *American Journal of Human Genetics.* 58: 17–20.

Mudd, S.H., Brosnan, J.T., Brosnan, M.E., Jacobs, R.L., Stabler, S.P., Allen, R.H., Vance, D.E., and Wagner, C., 2007. Methyl balance and transmethylation fluxes in humans. *The American Journal of Clinical Nutrition.* 85: 19–25.

Navarro, J., Goutet, J.M., Roy, C., Bonnet-Gajdos, M., and Polonovski, C., 1980. Folic acid deficiency and depression of cellular immunity [in French]. *Archives françaises de pédiatrie.* 37: 279.

Ray, J.G., Singh, G., and Burrows, R.F., 2004. Evidence for suboptimal use of periconceptional folic acid supplements globally. *BJOG: An International Journal of Obstetrics and Gynaecology.* 111: 399–408.

Selhub, J., and Miller, J.W., 1992. The pathogenesis of homocysteinemia: interruption of the coordinate regulation by S-adenosylmethionine of the remethylation and transsulfuration of homocysteine. *The American Journal of Clinical Nutrition.* 55: 131–138.

Shane, B., 1989. Folylpolyglutamate synthesis and role in the regulation of one-carbon metabolism. *Vitamins & Hormones*. 45: 263–335.

Smithells, R.W., Sheppard, S., Schorah, C.J., Seller, M.J., Nevin, N.C., Harris, R., Read, A.P., and Fielding, D.W., 1980. Possible prevention of neural-tube defects by periconceptional vitamin supplementation. *Lancet*. 1(8164): 339–340.

Steinberg, S.E., Campbell, C.L., and Hillman, R.S., 1979. Kinetics of the normal folate enterohepatic cycle. *The Journal of Clinical Investigation*. 64: 83–88.

Steinfeld, R., Grapp, M., Kraetzner, R., Dreha-Kulaczewski, S., Helms, G., Dechert, P., Wevers, R., Grosso, S., and Gärtner J., 2009. FDolate receptor alpha defect causes cerebral folate transport deficiency: a treatable neurodegenerative disorder associated with disturbed myelin metabolism. *American Journal of Human Genetics*. 85: 354–363.

Tamura, T., and Picciano, M.F., 2006. Folate and human reproduction. *The American Journal of Clinical Nutrition*. 83: 993–1016.

Waterland, R.A., Travisano, M., Tahiliani, K.G., Rached, M.T., and Mirza, S., 2008. Methyl donor supplementation prevents transgenerational amplification of obesity. *International Journal of Obesity*. 32: 1373–1379.

Williams, P.J., Bulmer, J.N., Innes, B.A., and Broughton Pipkin, F., 2011. Possible roles for folic acid in the regulation of trophoblast invasion and placental development in normal early human pregnancy. *Biology of Reproduction*. 84: 1148–1153.

Yokooji, T., Mori, N., and Murakami, T., 2009. Site-specific contribution of proton-coupled folate transporter/haem carrier protein 1 in the intestinal absorption of methotrexate in rats. *Journal of Pharmacy and Pharmacology*. 61: 911–918.

Zhao, R., and Goldman, I.D., 2007. The molecular identity and characterization of a Proton-coupled Folate Transporter--PCFT; biological ramifications and impact on the activity of pemetrexed. *Cancer and Metastasis Reviews*. 26: 129–139.

Zhao, R., Matherly, L.H., and Goldman, I.D., 2009. Membrane transporters and folate homeostasis: intestinal absorption and transport into systemic compartments and tissues. *Expert Review in Molecular Medicine*. 11: e4.

CHAPTER 45

Homocysteine and Vascular Disease: A Review of the Published Results of 11 Trials involving 52 260 Individuals

ROBERT CLARKE* AND JANE ARMITAGE

Clinical Trial Service Unit, Richard Doll Building, Old Road Campus, Roosevelt Drive, Oxford OX3 7LF, UK
*Email: robert.clarke@ctsu.ox.ac.uk

45.1 Introduction

Homocysteine is sulfur-containing amino acid that has been linked with higher risks of coronary heart disease (CHD), stroke and other occlusive vascular diseases (Clarke *et al.* 1991). Untreated individuals with homocystinuria, who have extreme elevations of plasma homocysteine (typically > 100 μmol/L), have high rates of vascular disease in their second and third decades (McCully 1969; Mudd *et al.* 1989). Homocystinuria is caused by deficiency of the enzyme cystathionine ß-synthase, required for the conversion of homocysteine to cystathionine. In addition to the associations with vascular risk, persons homozygous for homocystinuria may also have ocular, skeletal and neurologic complications of the disease.

In 1969, McCully proposed the 'homocysteine hypothesis' of vascular disease after demonstrating extensive atherosclerosis and arterial thrombosis at autopsy in two children with homocystinuria (McCully 1969). On the basis of these observations, McCully proposed that moderate elevations of homocysteine (typically 8–15 µmol/L) may cause atherosclerotic vascular disease in the general population.

The 'homocysteine hypothesis' of vascular disease has attracted considerable interest since homocysteine levels are readily lowered by daily dietary supplementation with folic acid, vitamin B_6 and vitamin B_{12} (Homocysteine Lowering Trialists' Collaboration 2005), raising the prospect that dietary supplementation with these B vitamins could prevent vascular disease. Indeed, dietary supplementation with B vitamins to lower homocysteine levels of affected individuals is remarkably effective for the prevention of cardiovascular disease and other complications of homocystinuria (Yap et al. 2001). This review examines the evidence from the observational studies of homocysteine and vascular disease and from the randomized trials of B vitamin supplementation for the prevention of vascular disease.

Classical homocystinuria was first described in 1962 by Carson and colleagues in Northern Ireland (Carson et al. 1963) and simultaneously by Gerritsen and colleagues in Wisconsin, USA (Gerritsen et al. 1962). Within two years, Mudd and colleagues discovered that the metabolic basis of homocystinuriua was due to a deficiency of cystathione ß-synthase (Mudd et al. 1964), a vitamin B_6 dependent enzyme required for transulfuration of homocysteine to cystathionine. A review of the natural history of homocystinuria due to cystathionine ß-synthase reported high risks of suffering a vascular event before age 30 years (Mudd et al. 1985). Subsequently, long-term follow-up of 158 cases of homocystinuria concluded that appropriate treatment to lower homocysteine levels is remarkably effective in reducing life-threatening vascular events in patients with homocystinuria despite imperfect biochemical control of elevated homocysteine concentrations (with homocysteine concentrations after treatment remaining several fold higher than those found in the general population) (Yap et al. 2001).

45.2 Observational Studies of Homocysteine and Cardiovascular Disease

In 1976, Wilcken and Wilcken (1976) first demonstrated an association of circulating homocysteine levels with CHD and, in 1991, Clarke et al. (1991) demonstrated that the associations of homocysteine with CHD, stroke and peripheral vascular disease were independent of age, sex, smoking, elevated cholesterol and systolic blood pressure. These associations were subsequently replicated by many other similar case-control studies. A meta-analysis of the initial case-control studies by Boushey et al. (1995) reported that a 5 µmol/L higher measured homocysteine was associated with an odds ratio for CHD of 1.6 [95% confidence interval (CI0: 1.3–1.9). Indeed, this initial meta-analysis of retrospective case-control studies suggested that the increased risk of CHD associated with a 5 µmol/L higher homocysteine was comparable to that for a 0.5 mmol/L higher total cholesterol (Boushey et al. 1995). Boushey and colleagues concluded that elevated plasma homocysteine was likely to be a causal factor and advocated clinical trials to test the 'homocysteine hypothesis' that folic acid-based vitamin supplements to lower homocysteine levels could prevent vascular disease (Boushey et al. 1995).

The results of prospective or cohort studies of homocysteine and vascular disease (where blood for homocysteine determinations was collected before the onset of disease in cases) in the late 1990s were less extreme than those of retrospective studies (where blood for homocysteine determinations was collected after the onset of vascular disease) (Danesh and Lewington 1998). In 1998, Danesh and Lewington reported that a 5 µmol/L higher measured homocysteine was associated with an odds ratio for CHD of 1.6 (1.4–1.7) in retrospective studies but only 1.3 (95%CI: 1.1–1.5) in prospective studies (Danesh and Lewington 1998). The discrepancy between the results of retrospective and prospective studies was interpreted to indicate 'reverse causality' (*i.e.* the effect of vascular disease on homocysteine concentrations) (Danesh and Lewington 1998).

In order to estimate reliably the associations of homocysteine with CHD and stroke outcomes, individual participant data were collected from all observational studies of homocysteine with CHD and stroke outcomes for the Homocysteine Studies Collaboration (Homocysteine Studies Collaboration 2002). With individual participant data, the Homocysteine Studies Collaboration meta-analysis was able to examine the shape and strength of association of homocysteine with vascular disease after adjustment for bias and confounding due to other risk factors (Homocysteine Studies Collaboration 2002). After excluding individuals with prior disease at enrolment and adjustment for smoking, blood pressure and cholesterol, a 25% lower usual (*i.e.* long-term) homocysteine concentration (about 3 µmol/L, a difference typically achieved by folic acid supplementation in populations without mandatory fortification of grain products with folic acid) was associated with an 11% (95% CI: 4–17%) lower risk of CHD and a 19% (5–31%) lower risk of stroke (Homocysteine Studies Collaboration 2002).

The subsequent evidence from the Homocysteine Studies Collaboration meta-analysis demonstrated the importance of homocysteine for cardiovascular risk was much less extreme than had been previously believed (Table 45.1). The Homocysteine Studies Collaboration meta-analysis did not adjust for the effects of creatinine, as these data were not available, and so was unable to assess the extent to which the association of homocysteine with vascular disease could have been confounded by renal function.

45.3 Homocysteine, Folate and Cancer

While observational studies had reported inverse associations of dietary intake of folate (or blood levels of folate) with overall risks of cancer, colorectal cancer or breast cancer (Giovannucci et al. 2002; Larsson et al. 2007; Lin et al. 2008; Zhang et al. 2003), one small trial involving 1021 individuals with a prior history of colorectal adenoma suggested a possible adverse effect on both the risk of recurrent colorectal adenoma (Cole et al. 2007) and on prostate cancer

Table 45.1 Summary of main findings of the observational studies of homocysteine and risk of coronary heart disease, by year of publication.

Year of publication	Author	No. of CHD cases	Main findings for CHD risk
1976	Wilcken	25	Identified association of homocysteine (tHcy) with CHD
1991	Clarke	60	tHcy association with CHD identified as being independent of other risk factors
1995	Boushey	2458	5 µmol/L lower tHcy →**60%** lower CHD risk in retrospective studies
1999	Danesh and Lewington	3740	5 µmol/L lower tHcy →**30%** lower CHD risk in retrospective studies
2002	Homocysteine Studies Collaboration	5073	25% lower usual tHcy →**10%** lower CHD risk in prospective studies

(Figueiredo *et al.* 2009). In addition, an analysis of secular trends of colorectal cancer incidence in the United States also prompted concerns by showing transient upward fluctuations in colorectal cancer after 1995 that were attributed to the introduction of mandatory folic acid fortification in February 1996 (although the fluctuations in colorectal cancer incidence could equally reflect improved screening introduced at the same time) (Mason *et al.* 2007).

45.4 Trials of B Vitamins for Prevention of Cardiovascular Disease

A number of trials of B vitamin supplementation to lower homocysteine levels for the prevention of cardiovascular disease were designed and initiated in the late 1990s (B-Vitamin Treatment Trialists' Collaboration 2006). This chapter describes a meta-analysis of the published results of 11 homocysteine-lowering trials for prevention of cardiovascular disease involving a total of 52 260 individuals (WAFACS, Albert *et al.* 2008; CHAOS-2, Baker *et al.* 2002; NORVIT, Bonaa *et al.* 2006; FAVORIT, Bostom *et al.* 2011; WENBIT, Ebbing *et al.* 2008; SU.FOL.OM3, Galan *et al.* 2010; HOST, Jamison *et al.* 2007; HOPE-2, Lonn *et al.* 2002; SEARCH, SEARCH Collaborative Group 2010; VISP, Toole *et al.* 2004; VITATOPS, VITATOPS Trial Study Group 2010).

45.4.1 Homocysteine-lowering Trials for Prevention of Cardiovascular Disease

45.4.1.1 Major Trials

Randomized trials were eligible if (i) they involved a randomized comparison of folic acid based B vitamin supplements for prevention of cardiovascular disease *versus* placebo (irrespective of whether any other treatment was administered factorially); (ii) the relevant treatment arms differed only with respect to the homocysteine-lowering intervention (*i.e.* they were unconfounded); and (iii) the trial involved 1000 or more participants for a scheduled treatment duration of at least one year. Summary data were extracted from all trials completed before 2012 for the present meta-analysis.

45.4.1.2 Chief Outcomes of Included Trials

The comparisons were intention-to-treat analyses of first events during the scheduled treatment period in all participants allocated to folic acid-based B vitamins or control (irrespective of any other treatment allocated factorially). The main outcomes were coronary heart disease events, stroke, cancer and all-cause mortality. Coronary heart disease events were defined as the first occurrence of non-fatal myocardial infarction or coronary death; although for several trials the definition of coronary events was restricted to fatal or non-fatal myocardial infarction. Stroke was defined as the first occurrence of either ischemic or hemorrhagic or unspecified strokes. Cancer was defined as any cancer, excluding non-melanoma skin cancers.

45.4.2 Methodological Considerations

For each trial, data were abstracted on the number allocated to each treatment and the number of coronary events, stroke events, cancer events, and deaths from any cause by treatment allocation. The expected number of events assuming treatment had no effect and the observed minus expected (o-e) statistics and their variances (v) were calculated for each trial and summed to produce, respectively, a grand total observed minus expected (G) and its variance (V) (Early Breast Cancer Trialists' Collaborative Group 1990). The one-step estimate of the log of the event rate ratio is G/V. The χ^2 test statistic (χ^2_{n-1}) for heterogeneity between n trials is $S-(G^2/V)$, where S is the sum over all the trials of $(o-e)^2/v$ (Cochran 1954). All analyses were carried out using SAS (Version 9.1).

45.5 Effects of B Vitamins on Cardiovascular Disease, Cancer and Mortality

45.5.1 Characteristics of the Participating Trials

Selected characteristics of the 11 large homocysteine-lowering trials are shown in Table 45.2. Seven trials largely recruited people with prior coronary heart disease or high-risk of coronary disease (Albert *et al.* 2008; Baker *et al.* 2002; Bonaa *et al.* 2006; Ebbing *et al.* 2008; Galan *et al.* 2010; Lonn *et al.* 2002; SEARCH Collaborative Group 2010), two trials recruited people with prior stroke or a transient ischaemic attack (VITATOPS Trial Study Group 2010; Toole *et al.* 2004) and two trials recruited people with end-stage renal disease (Bostom *et al.* 2011; Jamison *et al.* 2007).

Table 45.2 Summary of 11 homocysteine-lowering trials for prevention of coronary heart disease and stroke, by year of publication involving 52 260 individuals.

Year of publication	Trial	No of participants	Daily dose of B vitamins		
			Folic acid (mg/day)	B_{12} (mg/day)	B_6 (mg/day)
2002	CHAOS-2	1882	5.0	–	–
2004	VISP	3680	2.5	0.4	25
2006	NORVIT	3749	0.8	0.4	40
2006	HOPE-2	5522	2.5	1.0	50
2007	HOST	2056	40	0.5	100
2007	WENBIT	3090	0.8	0.4	40
2008	WAFACS	5442	2.5	1.0	50
2008	SEARCH	12 064	2.0	1.0	–
2010	VITATOPS	8164	2.0	0.5	25
2010	SU.FOL.OM3	2501	0.6	0.02	3
2011	FAVORIT	4110	5.0	1.0	50

All the trials compared the effects of folic acid with placebo, except the Vitamin Intervention for Stroke Prevention (VISP) trial which compared the effects of 2.5 mg with 0.02 mg of folic acid (Toole et al. 2004). The daily doses of folic acid used in the trials ranged from 0.8 mg to 40 mg. All trials included vitamin B_{12} (dose range 0.4–1 mg) with folic acid, except the CHAOS-2 trial (Baker et al. 2002) which used a daily dose of 5 mg of folic acid alone.

Seven trials (Albert et al. 2008; Bostom et al. 2011; Galan et al. 2010; Jamison et al. 2007; Lonn et al. 2002; Toole et al. 2004; VITATOPS Trial Study Group 2010) assessed the effects of combinations of folic acid (dose range: 2.5–40 mg), vitamin B_{12} (dose range: 0.4–1 mg) and vitamin B_6 (dose range: 5–100 mg). Two trials (Bonaa et al. 2006; Ebbing et al. 2008) assessed the effects of vitamin B_6 (40 mg) vs. placebo independently of folic acid (0.8 mg) plus vitamin B_{12} (0.4 mg) versus placebo using a factorial design.

The pre-treatment plasma folate concentration was substantially higher in fortified populations compared with that in non-fortified populations, and allocation to B vitamin treatment was associated with a greater proportional reduction in homocysteine concentrations in non-fortified compared with that in fortified populations (Clarke et al. 2010). The meta-analysis assessed the effects of B vitamins on risk of vascular and non-vascular events associated with about a 25% reduction in homocysteine levels for about five years (Clarke et al. 2010).

45.5.2 Effects on CHD and Stroke Outcomes

Among the 50 378 participants in ten trials with CHD events, there were 4516 CHD events; one trial did not report results for CHD events (Baker et al. 2002). Figure 45.1 shows that allocation to B vitamin treatment had no significant effect on risk of CHD events (either beneficial or hazardous) with 2282 events

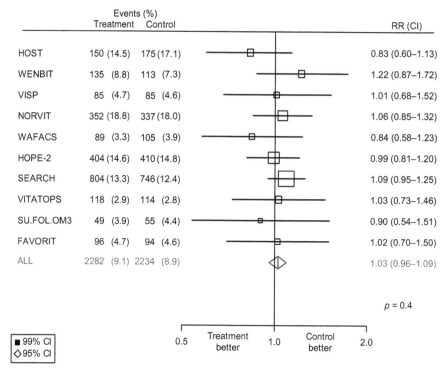

Figure 45.1 Effects of folic acid on CHD in published studies.

(9.1%) in the treatment group and 2234 (8.9%) in the control group, with a hazard ratio (95%CI) of 1.03 (0.96–1.09). There was no heterogeneity in the effects on CHD between the results of individual trials (test for heterogeneity, $p = 0.4$). Despite consistent evidence in observational studies for stronger associations of blood homocysteine levels with risk of stroke than for CHD (Homocysteine Studies Collaboration 2002), Figure 45.2 shows that B vitamins had no significant effect on the overall risk of stroke [1166 *vs.* 1240 stroke events, hazard ratio (HR) 0.94; 95%CI: 0.86–1.02) either. There was no heterogeneity in the effects on stroke risk between the results of individual trials (test for heterogeneity, $p = 0.13$).

45.5.3 Effects on Cancer and All-cause Mortality Outcomes

Data were available on 3010 incident cancers among the 35 603 individuals included in the seven vascular disease trials with published data available on cancer outcomes (Albert *et al.* 2008; Bonaa *et al.* 2006; Ebbing *et al.* 2008; Jamison *et al.* 2007; Lonn *et al.* 2006; SEARCH Trial Collaborative Group 2010; Toole *et al.* 2004). As indicated in Figure 45.3, there were slightly more reported cancers among those allocated active *versus* placebo, but the differences were not statistically significant [1541 (8.7%) *vs.* 1469 (8.2%); HR

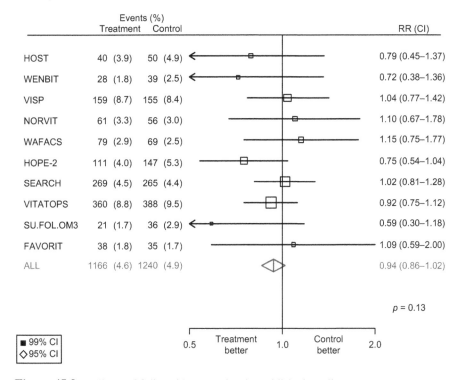

Figure 45.2 Effects of folic acid on strokes in published studies.

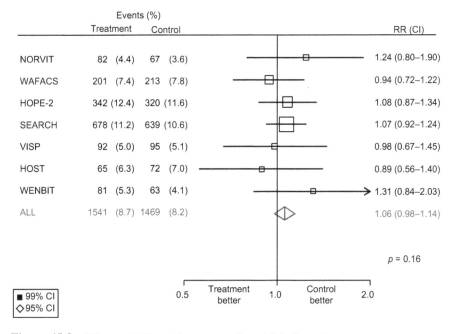

Figure 45.3 Effects of folic acid on cancer in published studies.

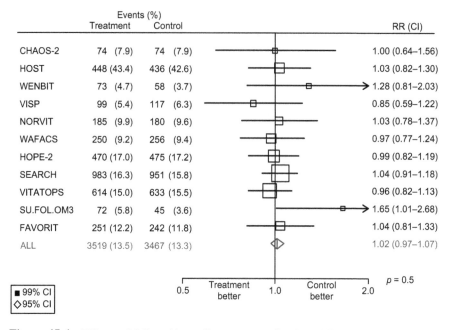

Figure 45.4 Effects of folic acid on all-cause mortality in published studies.

(95%CI) 1.06 (95%CI: 0.98–1.14)]. There was no heterogeneity in the effects on cancer between the results of individual trials (test for heterogeneity, $p = 0.16$).

Data were available on 6982 deaths among the 52 260 participants in ten trials. Figure 45.4 shows that allocation to B vitamins was not associated with any significant differences in overall mortality: 3519 deaths (13.5%) in the treatment group and 3467 deaths (13.3%) in the control group, with a hazard ratio (95%CI) of 1.02 (95%CI: 0.97–1.07). There was no heterogeneity in the effects on mortality between the results of individual trials (test for heterogeneity, $p = 0.5$).

45.6 Analysis of the Role of B Vitamins for Prevention of Cardiovascular Disease

Although the observational studies suggested modest associations of homocysteine with risk of vascular disease that were biologically plausible, such studies could not establish if these associations were causal. The randomized trials assessed the effectiveness of dietary supplementation with B vitamins to lower homocysteine levels on risk of cardiovascular morbidity and mortality. The initial trials were designed in the mid-1990s before the results of the Homocysteine Studies Collaboration meta-analysis (Homocysteine Studies Collaboration 2002) were reported in 2002. Consequently, few of the individual trials had sufficient statistical power to confirm or refute the 10% difference in risk of CHD that was predicted by the meta-analysis of the Homocysteine Studies Collaboration.

The data presented in this chapter involving 4516 CHD and 2406 stroke events in a meta-analysis of 50 378 participants (and 6986 deaths in 52 260 participants) showed that lowering homocysteine levels for about five years had no beneficial effects on either CHD or stroke or overall mortality. These results highlight the importance of meta-analysis of all the available evidence from large-scale trials when assessing moderate differences in risk such as those predicted for homocysteine lowering by the observational studies. Moreover, the findings from this meta-analysis refute the findings of a previous meta-analysis of B vitamin trials that purported to demonstrate that folic acid was effective for the prevention of stroke (Wang *et al.* 2007). The results of the present meta-analysis also differ from an earlier report on secular trends in stroke mortality in the US and the UK which attributed the greater reduction in stroke mortality between 1990 and 2002 in the US compared with that in the UK the introduction of fortification (Yang *et al.* 2006). However, ecological studies such as this cannot be used to infer causal associations. In contrast, the present meta-analysis demonstrated that folic acid supplementation did not influence the risk of stroke.

Concerns have been expressed about the safety of long-term use of folic acid supplements, with possible increased risks of cancer (Kim 2004). Taken together, the present meta-analysis, involving 35 603 individuals with 3010 incident cancer events, demonstrated no statistically significant adverse effects on cancer incidence, overall or in any individual trial. The results for cancer of this meta-analysis are consistent with an earlier meta-analysis from the B Vitamin Treatment Trialists' Collaboration (B Vitamin Treatment Trialists' Collaboration 2010) and differ from those of a previous trial involving 1021 adenoma participants that suggested a significant hazard for cancer (Cole *et al.* 2007), particularly prostate cancer (Figueiredo *et al.* 2009).

Large-scale trials and meta-analysis of such trials are required to assess moderate treatment benefits. However, within trials or between trials within meta-analysis, sub-group analysis may result in spurious results due to the play of chance. Hence, it is important to avoid sub-group analyses to reduce the risk of spurious results except those sub-groups that have been pre-specified on the basis of some prior hypothesis.

Replication of the results of trials is also important before making recommendations for clinical practice or public health. The B Vitamin Treatment Trialists' Collaboration was set up as a prospective meta-analysis involving individual participant data from all large-scale homocysteine-lowering trials (Clarke *et al.* 2010) to assess the effects of B vitamin supplementation on vascular disease, cancer and mortality, overall and in pre-specified sub-groups and to use time to event analysis rather than comparison of summary results by treatment allocation (Clarke *et al.* 2010). The results of the previous meta-analysis, based on individual participant data from eight trials involving 37 000 participants, demonstrated no material differences in the effects of folic acid on vascular outcomes in pre-specified sub-groups, such as folate status, homocysteine levels, dose of folic acid and intensity of homocysteine reduction. Further reports of the B Vitamin Treatment Trialists' Collaboration will provide more reliable estimates of the effects of treatment on site-specific and overall cancer.

If homocysteine levels are not causally related to vascular disease, presumably homocysteine levels are related to disease by being a correlate of underlying disease or risk factors that are related to CHD. Apart from the strong correlation of homocysteine levels with folate and vitamin B_{12}, homocysteine levels are strongly correlated with renal function. In retrospect, most of the observational epidemiological studies did not adjust for the effects of renal function when assessing the associations of homocysteine with CHD and stroke (Homocysteine Studies Collaboration 2002). Thus, the discrepant findings of the observational studies of homocysteine and CHD and stroke, and the homocysteine-lowering trials suggest that elevated homocysteine may be a marker of underlying disease or systemic inflammation rather than being causal. Thus, with the exception of the use of B vitamins in patients with homocystinuria, the available evidence would suggest that routine use of B vitamins cannot be recommended for prevention of vascular disease in the general population.

Summary Points

- Epidemiological studies suggested that moderately elevated homocysteine levels are a potentially modifiable risk factor for cardiovascular disease.
- The risks of vascular disease associated with elevated homocysteine levels are stronger in retrospective than in prospective studies, consistent with elevated homocysteine levels resulting from the disease rather than being causal.
- Elevated homocysteine levels are highly correlated with known risk factors for cardiovascular disease. The strength of the association of homocysteine with cardiovascular disease is attenuated after adjustment for known cardiovascular risk factors.
- The initial homocysteine-lowering trials were designed in the mid-1990s when many people believed homocysteine to be a very important risk factor for cardiovascular disease.
- Some small trials also suggested possible adverse effects of folic acid on risk of cancer, raising concerns about both the efficacy and safety of folic acid.
- A meta-analysis of published results of 11 trials, involving 52 260 individuals treated with folic acid or placebo, on average for five years, assessed the effects on vascular outcomes, cancer and all-cause mortality.
- Allocation to B vitamins had no beneficial effects on any cardiovascular events, with relative risks (95% confidence intervals) of 1.03 (0.96–1.09) for CHD and 0.94 (0.86–1.02) for stroke.
- Allocation to B vitamins had no significant adverse effects on cancer incidence (1.06; 0.98–1.14) and all-cause mortality (1.02; 0.97–1.07).

- The available evidence would suggest that routine use of B vitamins cannot be recommended for prevention of vascular disease in the general population.

Key Facts

- Homocysteine is a sulfur-containing amino acid that plays an essential role in one-carbon metabolism required for protein synthesis in all cells.
- Individuals with rare, albeit extreme elevations of blood homocysteine develop heart disease and stroke at a young age, prompting some experts to question whether members of the general population with moderate elevations of homocysteine might also have an increased risk of heart disease or stroke.
- Observational studies indicate that moderately elevated homocysteine levels were associated with a higher risk of heart disease and stroke.
- Blood levels of homocysteine are easily lowered by folic acid and vitamin B_{12}.
- Large trials of folic acid have demonstrated that lowering blood homocysteine levels for an average of five years had no significant effect on risk of heart disease, stroke, cancer or all-cause mortality.

List of Abbreviations

CHD   coronary heart disease
CI   confidence interval
G   grand total of observed (o) minus expected (e)
tHcy   homocysteine
V   variance
VISP   Vitamin Intervention for Stroke Prevention

References

Albert, C.M., Cook, N.R., Gaziano, J.M., Zaharris, E., MacFadyen, J., Danielson, E. Buring, J.E., and Manson J.E., 2008. Effect of folic acid and B vitamins on risk of cardiovascular events and total mortality among women at high risk for cardiovascular disease: a randomized trial. *The Journal of the American Medical Association.* 299: 2027–2036.

Baker, F., Picton, D., Blackwood, S., Hunt, J., Erskine, M., Dyas, M. Dyas, M., Siva, M., and Brown M.J., 2002. Blinded comparison of folic acid and placebo in patients with ischaemic heart disease: an outcome trial. *Circulation.* 106(Suppl II): 741.

Bonaa, K.H., Njolstad, I., Ueland, P.M., Schirmer, H., Tverdal, A., Steigen, T., Wang, H., Nordrehaug, J.E., Arnesen, E., Rasmussen, K., and the NORVIT Trial Investigators, 2006. Homocysteine lowering and cardiovascular events after acute myocardial infarction. *The New England Journal of Medicine.* 354: 1578–1588.

Bostom, A.G., Carpenter, M.A., Kusek, J.W., Levey, A.S., Hunsicker, L., Pfeffer, M.A., Selhub, J., Jacques, P.F., Cole, E., Gravens-Mueller, L., House, A.A., Kew, C., McKenney, J.L., Pacheco-Silva, A., Pesavento, T., Pirsch, J., Smith, S., Solomon, S., and Weir, M., 2011. Homocysteine-lowering and cardiovascular disease outcomes in kidney transplant recipients: primary results from the Folic Acid for Vascular Outcome Reduction in Transplantation trial. *Circulation.* 123: 1763–1770.

Boushey, C.J., Beresford, S.A., Omenn, G.S., and Motulsky, A.G., 1995. A quantitative assessment of plasma homocysteine as a risk factor for vascular disease. Probable benefits of increasing folic acid intakes. *The Journal of the American Medical Association.* 274: 1049–1057.

B-Vitamin Treatment Trialists' Collaboration, 2006. Homocysteine-lowering trials for prevention of cardiovascular events: a review of the design and power of the large randomized trials. *American Heart Journal.* 151: 282–287.

Carson, N.A., Cusworth, D.C., Dent, C.E., Field, C.M., Neill, D.W., and Westall R.G., 1963. Homocystinuria: a new inborn error of metabolism associated with mental deficiency. *Archives of Disease in Childhood.* 38: 425–436.

Clarke, R., Daly, L., Robinson, K., Naughten, E, Cahalane, S., Fowler, B., and Graham, I., 1991. Hyperhomocysteinemia: an independent risk factor for vascular disease. *The New England Journal of Medicine.* 324: 1149–1155.

Clarke, R., Halsey, J., Lewington, S., Lonn, E., Armitage, J., Manson, J.E. Bønaa, K.H., Spence, J.D., Nygård, O., Jamison, R., Gaziano, J.M., Guarino, P., Bennett, D., Mir, F., Peto, R., and Collins, R.; B-Vitamin Treatment Trialists' Collaboration, 2010. Effects of lowering homocysteine levels with B vitamins on cardiovascular disease, cancer and cause-specific mortality: meta-analysis of 8 randomized trials involving 37 485 individuals. *Archives of Internal Medicine.* 170: 1622–1631.

Cochran, W.G., 1954. Some methods for strengthening the common Chi-squared tests. *Biometrics.* 10: 417–451.

Cole, B.F., Baron, J.A., Sandler, R.S., Haile, R.W., Ahnen, D.J., Bresalier, R.S. McKeown-Eyssen, G., Summers, R.W., Rothstein, R.I., Burke, C.A., Snover, D.C., Church, T.R., Allen, J.I., Robertson, D.J., Beck, G.J., Bond, J.H., Byers, T., Mandel, J.S., Mott, L.A., Pearson, L.H., Barry, E.L., Rees, J.R., Marcon, N., Saibil, F., Ueland, P.M., and Greenberg, E.R., and Polyp Prevention Study Group (2007). Folic acid for the prevention of colorectal adenomas: a randomized clinical trial. *The Journal of the American Medical Association.* 297: 2351–2359.

Danesh, J., and Lewington, S., 1998. Plasma homocysteine and coronary heart disease: systematic review of published epidemiological studies. *Journal of Cardiovascular Risk.* 5: 229–232.

Early Breast Cancer Trialists' (CTT) Collaborative Group, 1990. Treatment of Early Breast cancer: A Systematic Overview of All Available Randomized Trials of Adjuvant Endocrine and Cytotoxic Therapy. Vol 1: Worldwide evidence 1985–1990. Oxford University Press, Oxford, UK, pp. 224.

Ebbing, M., Bleie, Ø., Ueland, P.M., Nordrehaug, J.E., Nilsen, D.W., Vollset, S.E. Refsum, H., Pedersen, E.K., and Nygård, O., 2008. Mortality and cardiovascular events in patients treated with homocysteine-lowering B vitamins after coronary angiography: a randomized controlled trial. *The Journal of the American Medical Association.* 300: 795–804.

Figueiredo, J.C., Grau, M.V., Haile, R.W., Sandler, R.S., Summers, R.W., Bresalier, R.S. et al., 2009. Folic acid and risk of prostate cancer: results from a randomized clinical trial. *Journal of the National Cancer Institute.* 101: 432–435.

Galan, P., Kesse-Guyot, E., Czernichow. S., et al, 2010. Effects of B-vitamins and omega-3 fatty acids on cardiovascular diseases: a randomised placebo-controlled trial. *British Medical Journal.* 341: c6273.

Gerritsen. T., Vaughn. J.G., and Waisman. H.A., 1962. The identification of homocystine in the urine. *Biochemical and Biophysical Research Communications.* 9: 493–496.

Giovannucci, E., 2002. Epidemiologic studies of folate and colorectal neoplasia: a review. *The Journal of Nutrition.* 132(8 Suppl): 2350S–2355S.

Homocysteine Lowering Trialists' Collaboration, 2005. Dose-dependent effects of folic acid on blood concentrations of homocysteine: a meta-analysis of the randomized trials. *The American Journal of Clinical Nutrition.* 82: 806–812.

Homocysteine Studies Collaboration, 2002. Homocysteine and risk of ischemic heart disease and stroke a meta-analysis. *The Journal of the American Medical Association.* 288: 2015–2022.

Jamison, R.L., Hartigan, P., Kaufman, J.S., Goldfarb, D.S., Warren, S.R., Guarino, P.D. et al., 2007. Effect of homocysteine lowering on mortality and vascular disease in advanced chronic kidney disease and end-stage renal disease: a randomized controlled trial. *The Journal of the American Medical Association.* 298: 1163–1170.

Kim, Y.I., 2004. Will mandatory folic acid fortification prevent or promote cancer? *The American Journal of Clinical Nutrition.* 80: 1123–1128.

Larsson, S.C., Giovannucci, E., and Wolk, A., 2007. Folate and risk of breast cancer: a meta-analysis. *Journal of the National Cancer Institute.* 99: 64–76.

Lin, J., Lee, I.M., Cook, N.R., Selhub, J., Manson, J.E., Buring, J.E. et al., 2008. Plasma folate, vitamin B-6, vitamin B-12, and risk of breast cancer in women. *The American Journal of Clinical Nutrition.* 87: 734–743.

Lonn, E., Yusuf, S., Arnold. M.J., Sheridan, P., Pogue, J., Micks, M., Sheridan, P., Pogue, J. et al., 2006. Homocysteine lowering with folic acid and B-vitamins in vascular disease. *The New England Journal of Medicine.* 354: 1567–1577.

Mason. J.B., Dickstein, A., Jacques, P.F., Haggarty, P., Selhub, J., Dallal, G. et al., 2007. A temporal association between folic acid fortification and an increase in colorectal cancer rates may be illuminating important biological principles: a hypothesis. *Cancer Epidemiology, Biomarkers & Prevention.* 16: 1325–1329.

McCully, K.S., 1969. Vascular pathology of homocysteinemia: implications for the pathogenesis of arteriosclerosis. *American Journal of Pathology.* 56: 111–128.

Mudd, S.H., Finkelstein, J.D., Irreverre, F., and Laster, L., 1964. Homocystinuria: an enzymatic defect. *Science*. 143: 1443–1445.

Mudd, S.H., Skovby, F., Levy, H.L., Pettigrew, K.D., Wilcken, B., Pyeritz, R.E., Andria, G., Boers, G.H., Bromberg, I.L., Cerone, R., Fowler, B., Gröbe, H., Schmidt, H., and Schweitzer, L., 1985. The natural history of homocystinuria due to cystathionine ß-synthase deficiency. *American Journal of Human Genetics*. 37: 1–31.

Mudd, S.H., Levy, H.L., and Skovby, F., 1989. Disorders of transsulfuration. In: Schriver, C.F., Baudet, A.L., Sly, W.S., and Valle, D. (ed.) The Metabolic Basis of Inherited Disease. McGraw Hill, New York, USA, pp. 693–734.

Study of the Effectiveness of Additional Reductions in Cholesterol and Homocysteine (SEARCH) Collaborative Group, 2010. Effects of homocysteine-lowering with folic acid plus vitamin B12 *vs.* placebo on mortality and major morbidity in myocardial infarction survivors: a randomized trial. *The Journal of the American Medical Association*. 303: 2486–2494.

Toole, J.F., Malinow, M.R., Chambless, L.E., Spence, J.D., Pettigrew, L.C., Howard, V.J., Sides, E.G., Wang, C.H., and Stampfer, M., 2004. Lowering homocysteine in patients with ischemic stroke to prevent recurrent stroke, myocardial infarction, and death: the Vitamin Intervention for Stroke Prevention (VISP) randomized controlled trial. *The Journal of the American Medical Association*. 291: 565–575.

VITATOPS Trial Study Group, 2010. B-vitamins in patients with recent transient ischaemic attack or stroke in the vitamins to prevent stroke (VITATOPS) trial: a randomised, double-blind, parallel, placebo-controlled trial. *Lancet Neurology*. 9: 855–865.

Wang, X., Qin, X., Demirtas, H., Li, J., Mao, G., Huo, Y., Sun, N., Liu, L., and Xu, X., 2007. Efficacy of folic acid supplementation in stroke prevention: a meta-analysis. *Lancet*. 369: 1876–1882.

Wilcken, D.E.L., and Wilcken, B., 1976. The pathogenesis of coronary artery disease: a possible role for methionine metabolism. *The Journal of Clinical Investigation*. 57: 1079–1082.

Yang, Q., Botto, L.D., Erickson, J.D., Berry, R.J., Sambell, C., Johansen, H. and Friedman, J.M., 2006. Improvement in stroke mortality in Canada and the United States, 1990 to 2002. *Circulation*. 21: 2080–2085.

Yap, S., Boers, G.H., Wilcken, B., Wilcken, D.E., Brenton, D.P., Lee, P.J. Walter, J.H., Howard, P.M., and Naughten, E.R., 2001. Vascular outcome in patients with homocystinuria due to cystathionine beta-synthase deficiency treated chronically: a multicenter observational study. *Arteriosclerosis, Thrombosis, and Vascular Biology*. 21: 2080–2085.

Zhang, S.M., Willett, W.C., Selhub, J., Hunter, D.J., Giovannucci, E.L., Holmes, M.D., Colditz, G.A., and Hankinson, S.E., 2003. Plasma folate, vitamin B6, vitamin B12, homocysteine, and risk of breast cancer. *Journal of the National Cancer Institute*. 95: 373–380.

CHAPTER 46
Vitamin B_{12} and Folate in Dementia

RACHNA AGARWAL

Department of Neurochemistry, Institute of Human Behaviour and Allied Sciences, Dilshad Garden, Delhi, India
Email: rachna1000@gmail.com

46.1 Introduction

Dementia, one of the world's most common neuropathic disorders, is also the most common cause of cognitive decline, affecting more than 50% of the world's population over the age of 85 years (Miller 2000). Alzheimer's Disease International (ADI) estimates that there are currently 30 million people with dementia in the world and this figure will increase to over 100 million by the year 2050. With increasing life expectancy, the prevalence of dementia will increase dramatically in the next few years which may increase the cost of care of patients with dementia more than the care of cancer and vascular patients put together.

Dementia, defined as significant memory impairment and loss of intellectual functions, interfering with the patient's work, usual social activities or relationship with others (Gottfries *et al.* 1998), is a multifactorial disorder in which genetic, environmental and life style factors are involved in the onset and progression of disease. In addition to the increasing age, nutrition-related risk factors (vitamin B_{12}, B_6, folate and antioxidants) and nutrition related disorders such as hypercholesterolemia, hypertriglyceridemia, hypertension and diabetes are associated with dementia (Figure 46.1). Now it is evident that

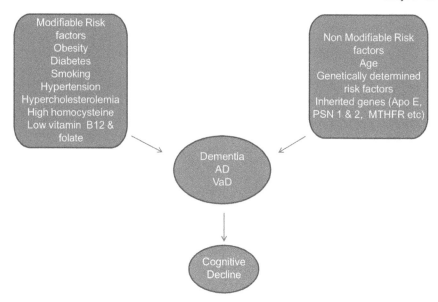

Figure 46.1 Modifiable risk factors of dementia.

cognitive impairment in old age is further exacerbated by low levels of vitamin B_{12} and folate leading to hyperhomocysteinemia. The most probable cause of deficiency could be dietary change including malnutrition occurring at this age. As low serum vitamin B_{12} and folate associated with cognitive decline constitute the modifiable risk factors, it is most practical to assess their role as therapeutic agents that may inhibit or delay the onset of dementia and slow its progression.

In 1884, Leichtenstern and Lichtheim reported, for the first time, neurological association of megaloblastic anaemia. The typical neurological lesion was in the posterior and lateral column of the spinal cord which was termed as 'subacute combined degeneration of the spinal cord' (Russell et al. 1900). Before the discovery of vitamin B_{12} and folate, megaloblastic anaemia was regarded as 'pernicious anaemia', the diagnosis of which was associated with achlorohydria. In 1945, synthesis of folate led to its use in the treatment of pernicious anaemia and showed encouraging results. Further trials showed aggravation or precipitation of neurological complications of pernicious anaemia by folate treatment for the next five years (Reynolds 2006). Three years after folate synthesis, vitamin B_{12} was discovered. It showed beneficial effects on both blood and nervous system. This led to the assumption that neuropsychiatric symptoms of anaemia were caused by the deficiency of vitamin B_{12} alone and not folate (Reynolds 1979). In the late 20th century, use of vitamin B_{12} and folate assays to assess patients with neuropsychiatric disorders, with or without megaloblastic anaemia, and introduction of homocysteine assays (1990s), slowly removed this misconception (Carmel et al. 2001). Today the role of vitamin B_{12}, folate and homocysteine in brain metabolism

Table 46.1 Neurological dysfunction associated with low vitamin B_{12} and folate.

Clinical implication	Mechanism
Depression or psychiatric disorders	• Impaired non-genomic methylation (*e.g.* myelin, protein, phospholipids, polysaccharides, catecholamines)
Subacute combined degeneration of spinal cord	• Impaired DNA synthesis, transcription and methylation • Homocysteine mediated DNA damage • Impaired non-genomic methylation
Cardiovascular disease and stroke	• Homocysteine-related vascular mechanism and oxidative stress • Impaired glutathione metabolism
Cognitive decline and dementia	• All the above mechanisms • Genomic and non-genomic methylation, homocysteine neurotoxicity • Failure of repair mechanism

and function, in relation to nervous system development and repair, mood changes, ageing, cognitive function and dementia (Reynolds 2006) is well proven (Table 46.1).

46.2 Vitamin B_{12}, Folate and Homocysteine

46.2.1 Biochemistry

Vitamin B_{12} and folate are involved in the synthesis of methionine from homocysteine in the presence of methionine synthase for which 5-methyl tetrahydrofolate (THF) and methyl-vitamin B_{12} are cofactors. Methionine in turn along with adenosine triphosphate (ATP) forms *S*-adenosyl methionine (SAM), which is the only methyl group donor in the human central nervous system (CNS). SAM donates methyl groups in number of methylation reactions, involving synthesis of myelin, phospholipids, proteins and neurotransmitters (catecholamine and indoleamines). The methylation cycle is also of importance for the generation of active folate forms. When the methyl group is transferred, SAM is converted to *S*-adenosyl homocysteine (SAH) which is subsequently converted to homocysteine and adenosine. Hence, homocysteine, vitamin B_{12} and folate are linked together in one carbon cycle (Moretti *et al.* 2008).

46.2.2 One-carbon Metabolism and Brain Functions

The one-carbon cycle involves generation of one-carbon units by converting tetrahydrofolate to 5,10-methylene-THF from serine. 5,10-Methylene THF is subsequently used for synthesis of thymidine from uracil and purines (DNA synthesis). It is also utilized in regeneration of methionine from homocysteine by donation of the methyl group from 5-methyl THF which is formed by irreversible reduction of 5,10-methylene THF in the presence of flavin-containing methylene tetrahydrofolate reductase (MTHFR). 5-Methyl THF serves as a substrate to methylate homocysteine in a reaction catalysed by a vitamin B_{12} containing methyl transferase.

Homocysteine metabolism involves three key enzymes: methionine synthase, betaine homocysteine methyl transferase (BHMT) and cystathione β-synthase. Both vitamin B_{12} and folate are required in the methylation of homocysteine to methionine *via* metheonine synthase after donation of a methyl group from SAM during the methylation process. Homocysteine is also methylated by betaine in a reaction catalysed by BHMT and does not involve vitamin B_{12} and folate. The other metabolic fate for homocysteine is the transsulfuration pathway which degrades homocysteine to cysteine and taurine, and is catalysed by cystathione β-synthase with vitamin B_6 as coenzyme.

46.2.3 Interrelationship between vitamin B_{12}, Folate and Homocysteine

Vitamin B_{12} is an essential coenzyme for the methylation cycle by activating THF. An activated folate is not only a coenzyme for methionine synthase but also plays a pivotal role in one-carbon metabolism; it promotes the generation of methionine from homocysteine, which is a cytotoxic sulfur-containing amino acid that can induce DNA strand breakage, oxidative stress and apoptosis. The methylation cycle is very important in the brain and depends on the SAM concentration.

In vitamin B_{12} deficiency, methionine synthase is inhibited causing increased levels of homocysteine and SAH. SAH in turn inhibits SAM mediated methylation, thereby leading to toxic levels of homocysteine causing direct damage to the vascular endothelium and inhibition of *N*-methyl-D-aspartate receptors (NMDA) (Moretti *et al.* 2008). Homocysteine is produced entirely from the methylation cycle, as it is totally absent from any dietary source (Pietrzik and Bronstrup 1997). Hence, an elevated plasma homocysteine concentration is a sensitive marker for vitamin B_{12} and folate deficiency (Parnetti *et al.* 1997). In addition, it can be due to increased frequency of impaired genetic capacity to metabolize homocysteine (Nilsson *et al.* 1996).

46.3 Status of B Vitamins in Dementia

Hyperhomocysteinemia has long been identified as a risk factor for dementia including Alzheimer's disease (AD) and vascular dementia (VaD) (Morris 2003). The relationship of homocysteine metabolism (methylation and transsulfuration pathways) to deficiencies of the vitamin B complex suggests that hypervitaminosis (B_6, B_{12} and folate) could contribute to hyperhomocysteinemia (González-Gross *et al.* 2001).

The large prospective Framingham community study showed that a high plasma homocysteine concentration doubled the risk of developing dementia including AD (Seshadri et al. 2002). Wang et al. (2001) also reported that low concentrations of serum vitamin B_{12} and folate increased the risk of AD in a Swedish community. A prospective study conducted in an Italian population also confirmed that high plasma homocysteine and low serum folate concentrations were independent predictors of dementia and AD, whereas the association with vitamin B_{12} was not significant (Ravalgia et al. 2005). Clarke et al. (1998) observed significantly higher serum homocysteine levels in patients with clinically diagnosed dementia of Alzheimer's type and histologically confirmed AD than in controls. Snowdon et al. (2000) also found a significant correlation between several neuropathological indicators of AD at autopsy and low folate levels that had been measured while these patients were alive. Studies performed by Arkai et al. (1999) on elderly diabetic subjects showed that homocysteine levels correlate negatively with cognitive test scores suggesting that elevated homocysteine levels may lead to cognitive impairment. Similarly, McCaddon et al. (1998) reported significantly higher homocysteine and lower vitamin B_{12} levels in AD patients than control subjects. Low levels of vitamin B_{12} correlated negatively with cognitive scores assessed using Cambridge Mental Disorders of the Elderly Examination (CAMDEX) and its Cambridge Cognition Examination (CAMCOG) scale (Roth et al. 1986).

However, serum vitamin B_{12} concentration may not be the best indicator of metabolic vitamin B_{12} deficiency. Low serum vitamin B_{12} concentration may have normal metabolic status, whereas high serum concentration does not always indicate sufficient vitamin B_{12} availability at tissue level. Though in clinical practice, serum concentration of vitamin B_{12} and folate are routinely used in the assessment of dementia and are included in the clinical guidelines, Mooijaart et al. (2005) and Agarwal et al. (2010) showed that serum concentration vitamin B_{12} and folate is not useful as a tool for identifying the subjects at risk of cognitive decline. Also, subjects with deficiency of the metabolically active form of vitamin B_{12} and presenting with a normal serum vitamin B_{12} concentration are likely to have a higher serum concentration of homocysteine. A number of case-control studies have shown that an elevated homocysteine level is the only common feature among cognitively impaired patients, but also indicate associated deficiency of vitamin B_{12} and folate in patients with cognitive impairment (Selhub et al. 2000). In a large population-based survey of elderly non-demented persons, Morris et al. (2001) reported poor recall in participants aged 60 years and above when homocysteine levels were higher than 13.7 µmol/L. Another large prospective study performed by Dufouil et al. (2003) in people aged over 60 years found an almost three-fold increased risk of cognitive decline if homocysteine levels were 15 µmol/L or more.

46.4 Old Age and Decline in Vitamin B Status

It is well-known that age is a risk factor for dementia and prevalence of dementia increases exponentially after 65 years age (Carr et al. 1997). AD is the most common cause of dementia in the elderly followed by vascular dementia (VaD) (Figure 46.2) (Fassenbender et al. 1999).

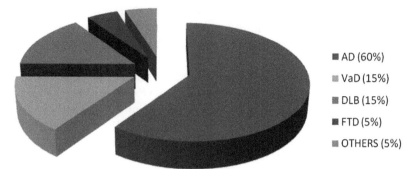

Figure 46.2 Prevalence of subtypes of dementia. AD = Alzheimer's disease; DLB = Lewy body dementia; FTD = frontotemporal dementia; VaD = vascular dementia.

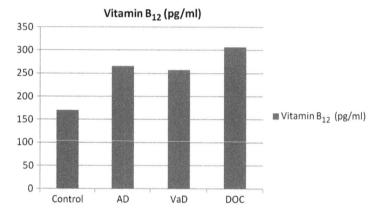

Figure 46.3 Vitamin B_{12} status in different types of dementia. AD = Alzheimer's disease; DOC = dementia due to other causes; VaD = vascular dementia.

Goodwin et al. (1983) undertook an epidemiological survey to study an association between nutritional status and cognitive functioning in a healthy elderly population. This was the first epidemiological evidence that linked low vitamin B status with decline in neurocognitive function in the elderly. They showed that healthy elderly subjects who had low blood concentration of vitamin B_{12}, folate, vitamin C and riboflavin scored poorly on tests of memory and non-verbal abstract thinking. A number of studies have also reported improvement in cognitive performance after supplementation with these vitamins (Martin et al. 1992). Hence, there is a possibility that high homocysteine and reduced vitamin B_{12} and folate levels secondary to poor vitamin status are partially responsible for the cognitive decline in old age (Mooijaart et al. 2005). However it is not possible at this stage to support the assumption whether low vitamin B_{12} and folate status is risk factor for dementia in elderly persons or a consequence of the cognitive decline in subjects with dementia.

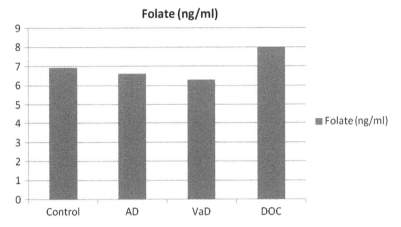

Figure 46.4 Folate status in different types of dementia. AD = Alzheimer's disease; DOC = dementia due to other causes; VaD = vascular dementia.

The age-related decline in B vitamin status has been attributed to multiple factors including poor nutrition due to declining appetite, decline in olfactory functions that may influence eating behaviour and the taste threshold of the elderly, and atrophic gastritis with hypochlorohydria and achlorohydria (Parnetti et al. 1997). In the Framingham heart study, the prevalence of atrophic gastritis among 60–69 years old and those > 80 years was found to be 24% and 37%, respectively (Krasinski et al. 1986). Atrophic gastritis leads to a delay in gastric emptying and decreased secretion of intrinsic factor, thereby reducing vitamin B_{12} absorption. Another consequence of atrophic gastritis is bacterial overgrowth in the stomach and proximal small bowel, which in turn can reduce vitamin B_{12} availability because some types of bacteria take up the vitamin B_{12} for their own use. In addition, increased pH in the stomach significantly limits folic acid absorption; the optimum pH for active folate uptake is 6.3 (Russell et al. 1979).

46.5 Vitamin B and Cognitive Decline Mechanism

Deficiencies of vitamin B_6, vitamin B_{12} and folate may play a role in the pathogenesis of cognitive impairment in the elderly. There are two biological mechanisms by which low levels of vitamin B_{12} or folate might result in dementia. First, vitamin B_{12} and folate are essential factors of methylation cycles involved in the synthesis of methionine from homocysteine. With decline in vitamin B status, homocysteine levels increase; this is associated with increased prevalence of poor cognitive decline and increased risk of development of dementia and AD. The sequential inability to methylate homocysteine leads to SAH accumulation which is a strong inhibitor of all methylation reactions, competing with SAM for the active site on the methyl transferase enzyme (White et al. 1996). Second, a low serum concentration of vitamin B_{12} or folate might result in reduced availability of methyl groups in the brain which can result in an impaired formation of myelin, various neurotransmitters and membrane phospholipids (Hutto 1997). In addition, impaired DNA methylation may increase amyloid β-peptide production and toxicity, as seen in AD (Kruman et al. 2002). It is not clear whether involvement of the methylmalonyl succinic acid pathway contributes to neural damage in vitamin B_{12} deficiency or not (Riedel et al. 1999). A number of mechanisms have been proposed. The first mechanism supports the homocysteine hypothesis of dementia postulated by Clarke et al. (1998). As previously stated, homocysteine levels increase with age due to impaired enzymatic activity, low vitamin B status, drugs and interacting conditions (González-Gross et al. 2001).

Vitamin B status is an important determinant of serum homocysteine levels as it mirrors the functional status of vitamins and interaction between vitamins and genetic factors. Although high plasma homocysteine concentration can be mainly attributed to inadequate vitamin B status, several studies indicate that plasma homocysteine increases with age independent of vitamin B status as hyperhomocysteinemia has been found highly prevalent in the elderly (Selhub et al. 1993). The association between a moderate elevation of plasma homocysteine and cognitive impairment is supported by a number of observations made by Clarke et al. (1998). As per their report, homocysteine levels increase in demented patients and patients with impaired cognitive function, there being a positive association between the degree of cognitive impairment and homocysteine levels. Also, elderly with elevated levels of homocysteine are at an increased risk of developing dementia and demented patients with high homocysteine levels exhibit more rapid disease progression. Hyperhomocysteinemia can affect the CNS directly as well as indirectly. On accumulation homocysteine inhibits methylation reactions by competing with SAM for the active site on methyl transferase enzyme (Moretti et al. 2008). Homocysteine-mediated oxidative damage to the CNS may occur on oxidation of homocysteine to homogentisic acid.

Homocysteine-related impairment of glutathione metabolism and oxidative stress also plays a major role in ageing and dementia. Through oxidative reactions, homocysteine induces direct damage to endothelial cells, increased platelet activity, pro-coagulant effects, increased collagen synthesis and enhanced proliferation of smooth muscles (Bolander-Gouaille 2005). In addition to its oxidant role, Arkai et al. (1999) have suggested that homocysteine injures the small penetrating cerebral arteries and arterioles rather than large brain supplying arteries leading to CNS ischemia and neuronal hypoxia from such vascular dysfunction. This increases the risk of stroke and stroke associated cognitive impairment. Another indirect effect of hyperhomocysteinemia is activation of NMDA receptors by homocysteine, or its oxidation metabolite homocyteic acid, as it is also an agonist at the glutamate site of the NMDA receptor. Being a potential excitotoxin, homocysteine over-stimulates NMDA receptor and contributes to neuronal damage leading to neuropsychiatric disturbances.

46.6 Role of Laboratory Indicators in Detecting Vitamin B Status

The apparent association of low vitamin B status and hyperhomocysteinemia and dementia raises the question: which markers to select that are sensitive and specific enough to reflect the functional status of vitamin B_{12} and folate in the tissues so that they can be used to measure the risk of developing the disease (Table 46.2).

Many studies have found that elderly and cognitively impaired patients may have high homocysteine levels in spite of normal blood concentration of folate and vitamin B_{12}. Hence, serum homocysteine is a more sensitive indicator of low vitamin B_{12} and folate status at tissue level. It is further supported by the fact that homocysteine levels can often be normalized by supplementing the diet with vitamin B_{12} and folate (Diaz-Arrastia 2000).

Low cerebrospinal fluid (CSF) levels of vitamin B_{12} and folate in dementia patients with normal serum levels have been demonstrated in a number of studies. Hence, CSF vitamin B_{12} and folate may be more sensitive indicators of their deficiencies at tissue level. This can be due to formation of inactive vitamin analogues at tissue level which are not detected by current analytical methods and disturbed transport of vitamin across the blood–brain barrier.

Serum methylmalonic acid, an intermediary product of vitamin B_{12} metabolism, is a sensitive indicator of vitamin B_{12} deficiency reflecting the functional status of vitamin B_{12} and folate in the tissues.

Holotranscobalamin, the biologically active form of vitamin B_{12}, is a more reliable marker of vitamin B_{12} status than serum vitamin B_{12} levels.

Red cell folate is a more sensitive indicator of serum folate deficiency. Low red cell folate accompanied by high plasma concentration of homocysteine is a better guide to the degree of deficiency than serum folate (Reynolds 2006).

46.7 Treatment-related Issues

Cognitive impairment has traditionally been considered the inevitable part of ageing for which nothing can be done. Now after substantial attempts

Table 46.2 Diagnostic approach to vitamin B_{12} deficiency.

Suspected vitamin B_{12} deficiency	Suggested tests
Patient presenting with mild to severe hematological or neurological signs or symptoms, or both	Serum vitamin B_{12}
Patient presenting with haematological or neurological signs or symptoms, or both, unlikely due to vitamin B_{12} deficiency	Serum vitamin B_{12} Serum methylmalonic acid Serum homocysteine
Asymptomatic patient accidentally found to have low B_{12} or high homocysteine levels	Serum methylmalonic acid
Asymptomatic patient associated with condition known to cause vitamin B_{12} deficiency	Serum methylmalonic acid (as metabolic changes precede over low vitamin B_{12} levels)

to study this, realization has dawned that most cases of cognitive impairment are the consequence of age-related disease processes. Attempts have been made to identify the factors that cause these diseases, modification of which might be able to prevent the consequential cognitive impairment or delay the progression into dementia. Studies show that vitamin B_{12} and folate are such modifiable factors, which can be associated with cognitive impairment in old age, as their deficiency is very common in elderly persons (Pennypacker et al. 1992). Also, low levels of vitamin B_{12} and folate in conjunction with increased homocysteine have been reported to correlate with decreased performance on cognitive test (Riggs et al. 1996; Wang et al. 2001).

46.7.1 Vitamin B_{12} Supplementation

Two thirds of patients with vitamin-B_{12} deficiency and neurological manifestations occur independently at the time of diagnosis. In addition, the severity of the neurological disorders correlated with the duration of the symptoms and inversely with the hemoglobin concentration (Savage and Lindenbaum 1996). At present, the treatment of neuropsychiatric disorders is based mainly on hematological studies and not on neurological studies (Savage and Lindenbaum 1996). Remission of hematological symptoms (megaloblastosis) can occur within few weeks of parentral vitamin B_{12} supplementation in small doses.

A number of intervention studies (Bryan et al. 2002; Eussen et al. 2006; Hvas et al. 2004; Kral et al. 1970; Kwok et al. 1998; Seal et al. 2002) have been undertaken to assess the effect of vitamin B supplementation on cognitive function. In such studies, vitamin B_{12} interventions ranged from 0.02 mg/day orally to 1 mg/day intramuscularly. The duration of supplementation ranged from four to six months. There is a large heterogeneity among trials in terms of dose, administration route and duration of treatment. Overall, half of the tests found a net improvement in cognitive function and the other half a worsening of cognitive function with vitamin B_{12} supplementation (Balk et al. 2007). Studies performed by Nilsson (1998) showed that vitamin B_{12} supplementation led to no significant improvement, and no less deterioration in their neuropsychological function in patients presenting with dementia, than their matched group, but a significant improvement was demonstrated among the patients presenting with cognitive impairment on the verbal frequency test. At present, a weekly injection of 1000 µg of hydroxycobalamin or cyanocobalamin has been recommended for more than three months to saturate body stores of the vitamin. But adequate studies have not been carried out to claim that the nervous system requires large doses for longer periods than the blood. In addition, the duration and degree of neurological recovery correlates with the duration of symptoms and the severity of disability before treatment. Therefore early diagnosis and treatment is imperative (Savage and Lindenbaum 1996).

46.7.2 Folate Supplementation

Early studies indicated that treatment of patients of vitamin B_{12} deficiency with folic acid led to aggravation of neurological complications or allowed them to progress by masking the anaemia (Reynolds 2006).

A number of trials have been conducted on the effect of folic acid supplementation on cognitive function among participants with dementia or cognitive impairment. The folate doses varied from 0.05 to 20 mg/day and trial duration ranged from five to ten weeks (Bryan *et al.* 2002; Fioravanti *et al.* 1997; Sommer *et al.* 2003). These studies found cognitive improvement after folic acid intervention; it correlated in a linear fashion with the low levels of folate (<3 µg/ml to 6.8 nmol/L) at baseline. However, high folate intake may increase cognitive decline. One possibility is that high intake may be masking unrecognised vitamin B_{12} deficiency. This complication is a potential pitfall of high folate (≥ 1000 µg/day) in food and multivitamins, which is well above the dietary reference intake of 400 µg/day (Morris 2005). As with vitamin B12, the response to folate treatment is detected over a period of 3 months. Hence, small doses over long term may be preferable to larger doses in the short or long term. But which formulation out of folic acid, folinic acid and methyl folate is best for the nervous system is still unknown.

46.8 Concluding Remarks

Based on the available information it can be concluded that prevalence of dementia increases with the advancing age and is growing at an alarming rate, becoming a major public health problem. Hence, preservation of cognitive ability in old age is essential not only to promote an adequate health status but also to delay the onset of dementia and slow its progress, thereby, reducing the societal costs for chronic care and lost productivity.

Low levels of folate, vitamin B_{12} and B_6 in conjunction with increased homocysteine levels have been found in the ageing population and correlate with decreased cognitive performance in old age. Deficiency of the B vitamins (vitamin B_{12} and folate) may play a role in pathogenesis of cognitive impairment in the elderly through hyperhomocysteinemia. On the other hand, elevated plasma homocysteine concentrations are a sensitive marker for vitamin B_{12} and folate deficiency. Homocysteine is produced from the methylation cycle, as it is totally absent from any dietary source. Both folate and vitamin B_{12} are required in the methylation of homocysteine to methionine and in the remethylation and synthesis of *S*-adenosyl methionine, a major methyl donor in the central nervous system. Hence, increased homocysteine levels in association with low levels of vitamin B_{12} and folate have been reported to correlate with decreased performance on cognitive tests. For these reasons, B vitamin supplementation has been shown to prevent or reverse cognitive decline. However, effect of vitamin B_{12} supplementation shows heterogeneous results. The levels of vitamin B along with homocysteine are the modifiable non-genetic risk factors of dementia which may be relevant to the course of AD and VaD, and should be considered for therapeutic intervention. However, it remains an open question whether or not these interventions in terms of B vitamin supplementation (a combination of folate, vitamin B_{12}, and vitamin B_6) will improve cognitive functions or retard the rate of cognitive decline in older adults with or without dementia.

Summary Points

- Dementia, a cognitive disorder, will be a major public health problem by 2050.
- B vitamins (B_{12}, B_6 and folate) play an important role in the pathogenesis of cognitive impairment in the old age.
- Age-related decline in vitamin B status occurs due to poor nutrition and atrophic gastritis with achlorohydria.
- Vitamin B_{12} and folate are involved in methionine synthesis from homocysteine, which in turn is converted to SAM, a major methyl group donor in CNS.
- High homocysteine level is the most sensitive indicator of low vitamin B status.
- Vitamin B_{12} supplementation shows improvement in some cognitive functions (*e.g.* language function), but rarely reverses dementia.
- High doses of folate taken leads to decline in cognitive function.
- Vitamin B_{12} and folate supplementation is preferred in small doses over long periods.

Key Facts

- Leichtenstern and Lichtheim synthesized folate in 1884 and reported for the first time neurological association of megaloblastic anaemia. Treatment of megaloblastic anaemia with folate showed aggravation of neurological complications. After the discovery of vitamin B_{12} three years later, it was shown that the neurological symptoms of megaloblastic anaemia were due to deficiency of vitamin B_{12}.
- Vitamin B_{12} and folate are involved in synthesis of methionine from homocysteine.
- *S*-Adenosyl methionine, synthesized from methionine, is the only methyl group donor in the human central nervous system and is involved in methylation of phospholipids, myelin and catecholamines.
- Low vitamin B_{12} and folate and hyperhomocysteinemia induce cognitive decline by impaired DNA synthesis, impair genomic and non-genomic methylation, homocysteine-induced DNA synthesis, neurotoxicity and oxidative stress.
- Laboratory indicators of low vitamin B status are serum homocysteine, methylmalonic acid and holotranscobalamin and CSF vitamin B_{12} and folate reflecting its functional status at tissue level.
- At present, the treatment of neuropsychiatric disorders by vitamin B_{12} supplementation is based mainly on hematological studies.

- Remission of hematological symptoms (megaloblastosis) can occur within few weeks of parenteral vitamin B_{12} supplementation in small doses.
- Treatment of patients with vitamin B_{12} deficiency with folic acid leads to aggravation of neurological complications.
- Vitamin B and folate supplementation in small doses over the long term is preferable to larger doses in the short term and leads to improvement in cognitive function.

Definition of Words and Terms

Achlorohydria: The absence of hydrochloric acid in the gastric secretions of the stomach.
Dementia: Significant memory impairment and loss of intellectual functions such as memory, concentration and judgment resulting from an organic disease or a disorder of the brain.
Homocysteine: A sulfur containing non-proteinogenic amino acid synthesized from methionine. It is a risk factor for dementia and cardiovascular disorders.
Megaloblastic anaemia: Megaloblastic anaemia is a blood disorder in which there is anaemia with larger-than-normal red blood cells due to an abnormality of red cell development in the bone marrow.
Pernicious anaemia: It is a type of anaemia caused by failure of the stomach to absorb vitamin B_{12} and is characterized by abnormally large red blood cells, gastrointestinal disturbances, and lesion of the spinal cord.

List of Abbreviations

AD	Alzheimer's disease
ADI	Alzheimer's Disease International
ATP	adenosine triphosphate
BHMT	betaine homocysteine methyl transferase
CAMDEX	Cambridge Mental Disorders of the Elderly Examination
CAMCOG	Cambridge Cognition Examination
CNS	central nervous system
CSF	cerebrospinal fluid
DLB	Lewy body dementia
DNA	deoxyribose nucleic acid
DOC	dementia due to other causes
FTD	frontotemporal dementia
MTHFR	methylene tetrahydrofolate reductase
NMDA	*N*-methyl-D-aspartate
SAH	*S*-adenosyl homocysteine
SAM	*S*-adenosyl methionine
THF	tetrahydrofolate
VaD	vascular dementia

References

Agarwal, R., Chhillar, N., Kushwaha, S., Singh, N.K., and Tripathi, C.B., 2010. Role of vitamin B_{12}, folate and TSH in dementia- a hospital based study in North Indian population. *Annals of Indian Academy of Neurology.* 13: 257–262.

Arkai, A., Sako, Y., Itoh, H., and Orimo, H., 1999. Plasma homocysteine, brain MR lesions, and cognitive functions in elderly diabetic patients [abstract]. *Amino Acids.* 17: 44.

Balk, E.M, Raman, G., and Tatsioni, A., 2007. Vitamin B_6, B_{12} and folic acid supplementation and cognitive function: a systematic review of randomised trials. *Archives of Internal Medicine.* 167: 21–30.

Bolander-Gouaille, C., 2005. Homocysteine in the development of dementia: an overview. Business: Briefing: European Pharmacotherapy. 8 pp.

Bryan, J., Calvaresi, E, and Hughes, D., 2002. Short-term folate, vitamin B12, or vitamin B6 supplementation slightly affects memory performance but not mood in women of various ages. *The Journal of Nutrition.* 132: 1345–1356.

Carmel, R., and Jacobsen, D.W. (ed.), 2001. Homocysteine in Health and Disease. Cambridge University Press, Cambridge, UK, 526 pp.

Carr D.B., Goate, A., Phil, D., and Morris, J.C., 1997. Current concepts in the pathogenesis of Alzheimer's disease. *Americal Journal of Medicine.* 103: 3s–10s.

Clarke, R., Smith, A.D, Jobst, K.A, Refsum, H., Sutton, L., and Ueland P.M., 1998. Folic acid, vitamin B12, and serum homocysteine levels in confirmed Alzheimer's disease. *Archives of Neurology.* 55: 1449–1455.

Diaz-Arrastia, R., 2000. Homocysteine and neurologic disease. *Archives of Neurology.* 57: 1422–1427.

Dufouil, C., Alperovich, A., Ducros, V., and Tzourio, C., 2003. Homocysteine, white matter hyperintensities and cognition in elderly healthy people. *Annals of Neurology.* 53: 214–221.

Eussen, S.J., de Groot, L.C., Joosten, L.W., Bloo, R.J, Clarke, R., Ueland, P.M., Schneede, J., Blom, H.J., Hoefnagels, W.H., and van Stavern, W.A., 2006. Effect of oral vitamin B_{12} with without folic acid on cognitive function in older people with mild vitamin B_{12} deficiency: a randomized, placebo controlled trial. *American Journal of Clinical Nutrition.* 84: 361–370.

Fassender, K., Mielke, O., Bertsch, T., Nafe, B., Froschen, S., and Hennerici, M., 1999. Homocysteine in cerebral macroangiography and microangiography. *Lancet.* 353: 1586–1587.

Fioravanti, M., Ferrario, E., Massaia, M., Cappa, G., Rivolta, G., Grossi, E., and Buckley, A.E., 1997. Low folate levels in the cognitive decline of elderly patients and the efficacy of folate as a treatment for improving memory deficits. *Archives of Gerontology and Geriatrics.* 26: 1–13.

Goodwin, J.S, Goodwin, J.M., and Garry, P.J., 1983. Association between nutritional status and cognitive functioning in a healthy elderly population. *The Journal of the American Medical Association.* 249: 2917–2921.

González-Gross, M., Marcos, A., and Pietrzik, K., 2001. Nutrition and cognitive impairment in the elderly. *British Journal of Nutrition*. 86: 313–321.

Gottfries, C.G., Lehmann, W., and Regland, B., 1998. Early diagnosis of cognitive impairment in the elderly with the focus on Alzheimer's disease. *Journal of Neural Transmission*. 105: 773–786.

Hutto, B.R., 1997. Folate and cobalamin in psychiatric illness. *Comprehensive Psychiatry*. 6: 305–314.

Hvas, A.M., Juul, S., Lauritzen, L., Nexo, E., and Ellegaard, J., 2004. No effect of vitamin B-12 treatment on cognitive function and depression: a randomized placebo controlled study. *Journal of Affective Disorder*. 81: 269–273.

Krasinski, S.D., Russell, R.M., Samloff, I.M., Jacob, R.A., Dallal, G.E., McGandy, R.B., and Hartz, S.C., 1986. Fundic atrophic gastritis in an elderly population: effect on hemoglobin and several serum nutritional indicators. *Journal of the American Geriatrics Society*. 34: 800–806.

Kruman, I.I., Kumeravel, T.S., Lohani, A., Pedersen, W.A., Cutler, R.G., Kruman, Y., Haughey, N., Lee, J., Evans, M., and Mattson, M.P., 2002. Folic acid deficiency and homocysteine impair DNA repair in hippocampal neurons and sensitise them to amyloid toxicity in experimental models of Alzheimer's disease. *Journal of Neuroscience*. 22: 1752–1762.

Kral V.A., Solyom L., Enesco, H., and Ledwidge, B., 1970. Relationship of vitamin B_{12} and folic acid to memory function. *Biological Psychiatry*. 2: 19–26.

Kwok, T., Tang, C., Woo, J., Lai, W.K., Law, L.K., and Pang, C.P., 1998. Randomised trial of the effect of supplementation on the cognitive function of older people with subnormal cobalamin levels. *International Journal of Geriatrics Psychiatry*. 13: 611–616.

Martin, D.C., Francis, J., Protetch, J., and Huff, F. J., 1992. Time dependency of cognitive recovery with cobalamin replacement: report of a pilot study. *Journal of the American Geriatrics Society*. 40: 168–172.

McCaddon, A., Davies, G., Hudson, P., Tandy, S., and Cattell, H., 1998. Total serum homocysteine in senile dementia of Alzheimer's type. *International Journal of Geriatric Psychiatry*. 13: 235–239.

Miller, J.W., 2000. Homocysteine, Alzheimer's disease, and cognitive function. *Nutrition*. 16: 675–677.

Mooijaart, S.P., Gusssekloo, J., Frolich, M., Jolles, J., Stott, D.J., Westerndrop, R.G.J., and de Craen, A.J.M., 2005. Homocysteine, vitamin B 12 and folic acid and the risk of cognitive decline in old age: Leiden 85-Plus study. *The American Journal of Clinical Nutrition*. 82: 866–871.

Moretti, R., Torre, P., and Antonello, R.M., 2008. Vitamin B_{12}, folate depletion and homocysteine: What do they mean for cognition? In: Vitamin B: New Research. Nova Science Publishers, Hauppage, NY, USA, pp. 139–152.

Morris, M.C., Evans, D.A., Bienias, J.L., Tagney, C.C., Hebert, L.E., Scherr, P.A., and Schneider, J.A., 2005. Dietary folate and vitamin B_{12} intake and cognitive decline among community-dwelling older persons. *Archives of Neurology*. 8: 657–678.

Morris, M.S., Jacques, P.F., Rosenberg, I.H., and Selhub, J., 2001. Homocystenemia associated with poor recall in the third National Health and Nutrition Examination survey. *The American Journal of Clinical Nutrition.* 73: 927–933.

Morris, M.S., 2003. Homocysteine and Alzheimer's disease. *Lancet Neurology.* 2: 425–428.

Nilsson E.H., 1998. Age- related changes in cobalamin handling: implications for therapy. *Drugs & Aging.* 12: 277–292.

Nilsson, K., Gustafson,L., Faldt, R., Andersson, A., Brattstrom, L., Lindgren, A., Issraelsson, B., and Hulterg, B., 1996. Hyperhomocystenemia—a common finding in a psychogeriatric population. *European Journal of Clinical Investigation.* 26: 853–859.

Parnetti, L., Bottiglieri, T., and Lowenthal, D., 1997. Role of homocysteine in age- related vascular and non-vascular diseases. *Aging.* 9: 241–257.

Pennypacker, L.C., Allen, R.H., Kelly J.P., Mathews, L.M., Grigsby, J., Kaye, K., Lindenbaum, J., and Stabler, S.P., 1992. High prevalence of cobalamin deficiency in elderly outpatients. *Journal of the American Geriatrics Society.* 40: 1197–1204.

Pietrzik, K., and Bronstrup, A, 1997. Folic acid in preventive medicine: a new role in cardiovascular disease, neural tube defects and cancer. *Annals of Nutrition and Metabolism.* 41: 331–343.

Ravalgia, G., Forti, P., Maioli, F., Maioli, F., Martelli, M., Servadei, L., Brunetti, N., Porcellini, E., and Licastro, F., 2005. Homocysteine and folate as risk factors for dementia and Alzheimer's disease. *American Journal of Clinical Nutrition.* 82: 636–643.

Reynolds E.H., 1979. Folic acid, vitamin B and the nervous system: historical aspects. In: Botez, M.I., and Reynolds, E.H. (ed.) *Folic Acid in Neurology, Psychiatry, and Internal Medicine.* Raven Press, New York, USA, pp. 1–5.

Reynolds, E.H., 2006. Vitamin B_{12}, folic acid, and the nervous system. *Lancet Neurology.* 5: 949–960.

Riedel, B., Fiskerstrand, T., Refsum, H., and Ueland, P.M., 1999. Co-ordinate variations in methylmalonyl CoA mutase and in human glioma cells during nitrous oxide exposure and the subsequent recovery phase. *Biochemical Journal.* 341: 133–138.

Riggs, K.M., Spiro, A. 3rdI, Tucker, K., and Rush, D., 1996. Relations of vitamin B12, vitamin B6, folate and homocysteine to cognitive performance in the normative Aging Study. *The American Journal of Clinical Nutrition.* 63; 306–314.

Roth, M., Mountjoy, C.Q., Huppert, F. A., Hendrie, H., Verma, S., and Goddard, R., 1986. CAMEX: A standardized instrument for the diagnosis of medical disorder in the elderly with special reference to the early detection of dementia. *British Journal of Psychiatry.* 149; 698–701.

Russell, J.S.R., Batten, F.E., and Collies, J., 1900. Subacute combined degeneration of the spinal cord. *Brain.* 23: 39–110.

Russell, R.M., Dhar, G.J., Dutta, S.K., and Rosenberg, I.H., 1979. Influence of intramural pH on folate absorption: studies in control subjects and in patients with pancreatic insufficiency. *Journal of Laboratory and Clinical Medicine.* 93: 428–436.

Savage, D.G., and Lindenbaum J., 1996. Neurological complications of acquired cobalamin deficiency: clinical aspects. *Bailliere's Clinical Haematology.* 8: 657–678.

Seal, E.C., Metz, J., Fickler, L., and Melny, J., 2002. A randomized, double-blind, placebo- controlled study of oral vitamin B 12 supplementation in older patients with subnormal or borderline serum vitamin B 12 concentrations. *Journal of the American Geriatrics Society.* 50: 146–151.

Selhub, J., Jacques, P.F., Wilson, P.W.F., Rush, D., and Rosenberg T.H., 1993. Vitamin status and intake as primary determinants of homocysteinemia in an elderly population. *The Journal of the American Medical Association.* 270: 2693–2698.

Selhub, J., Baglay, C.G., Miller, J., and Rosenberg, I.H., 2000. B vitamins, homocysteine and neurocognitive function in the elderly. *The American Journal of Clinical Nutrition.* 71 (Suppl): 614s–620s.

Seshadri, S., Beiser, A., Selhub, J., Jacques, P.F., Rosenberg, I.H., D'Agostino, R.B., Wilson, P.W., and Wolf, P.A., 2002. Plasma homocysteine as a risk factor for dementia and Alzheimer's disease. *New England Journal of Medicine.* 346: 476–483.

Snowdon, D.A., Tully, C.L., Smith, C.D., Perez, R.K., and Markesberry, W.R., 2000. Serum folate and the severity of atrophy of the neocortex in Alzheimer's disease: findings from the Nun study. *The American Journal of Clinical Nutrition.* 71: 993–998.

Sommer, B.R., Hoff, A.L., and Costa, M., 2003. Folic acid supplementation in dementia: a preliminary report. *Journal of Geriatrics, Psychiatry and Neurology.* 16: 156–159.

Wang, H.-X., Wahlin, A., Basun, H., Fastborn, J., Winblad, B., and Frattelioni, L., 2001. Vitamin B12 and folate in relation to the development of Alzheimer's disease. *Neurology.* 56: 1188–1194.

White, L., Petrovich, H., Ross, G.W., Masaki, K.H., Abbott, R.D., Teng, E.L., Rodriguez, B.L., Blanchette, P.L., Havlik, R.J., Wergowske, G., Chiu, D., Foley, D.J., Murdaugh, C., and Curb, J.D., 1996. Prevalence of dementia in older Japanese American men in Hawaii: the Honolulu-Asia Aging study. *JAMA.* 276: 955–960.

CHAPTER 47
Cobalamin and Nutritional Implications in Kidney Disease

KATSUSHI KOYAMA

Department of Nephrology, Kariya-Toyota General Hospital,
5-15 Sumiyoshi-cho, Kariya, Aichi 448-8505, Japan
Email: nephkidedta@do9.enjoy.ne.jp

47.1 Intake and Absorption of Vitamin B_{12}

Vitamin B_{12} is not abundantly contained in foods and is especially scarce in plant foods. To absorb this rare vitamin efficiently, the human body has an ingenious system. Dietary vitamin B_{12} binds to gastric intrinsic factor and is transported *via* a receptor present in the small intestine to the epithelial cells; here it and is taken up in the circulation where it binds to transcobalamin II. The complex of transcobalamin II and B_{12} is then transported to the cells of target organs by endocytosis *via* a receptor expressed in the membrane of the cells.

There have been not been any reports indicating that renal dysfunction affects the intake and absorption of vitamin B_{12}. However, renal failure is associated with gastric epithelial cell dysfunction, which may reduce the release of gastric intrinsic factor. Moreover, decreased renal function may induce anorexia, which results in a reduction in food intake (Kopple *et al.* 1999).

47.2 Intracellular Metabolism of Vitamin B_{12}

47.2.1 Cyanocobalamin Trafficking Chaperone and CKD

Vitamin B_{12} must be converted into its coenzyme forms, adenosylcobalamin and methylcobalamin, in the cell. These coenzymes function as cofactors of methylmalonyl-CoA mutase and methionine synthase, respectively. Chronic kidney disease (CKD) may affect the conversion from vitamin B_{12} to the coenzyme forms. This section describes the intracellular metabolism of cyanocobalamin, which is included in many dietary supplements, in particular, referring to a recently discovered trafficking chaperone called methylmalonic aciduria cdlC type with homocystinuria (MMACHC). Cyanocobalamin is first converted to cob(II)alamin, which has no cyanogen group on the ligand occupying the upper axial position of the cobalamin structure. Cob(II)alamin is further reduced to cob(I)alamin, which can function as a coenzyme in the body.

It has traditionally been considered that removal of a cyanide molecule from cyanocobalamin is mediated by the formation of glutathionylcobalamin. However, Rosenblatt's group recently reported that direct reduction by NADPH and flavoprotein in the presence of a cyanocobalamin trafficking chaperone is more important (Lerner-Ellis *et al.* 2006). This chaperone was named 'MMACHC,' but Rosenblatt and colleagues have proposed calling it 'cyanocobalamin decyanase' (Figure 47.1).

47.2.2 Abnormal Cyanide Metabolism in CKD and Oxidative Stress

We measured the proportion of each vitamin B_{12} analogue in total vitamin B_{12} in patients with end-stage renal disease and demonstrated an increase in the cyanocobalamin fraction and a decrease in the methylcobalamin fraction (Koyama *et al.* 1997) (Figure 47.2). The fraction ratio of cyanocobalamin was as extremely high as $10.5 \pm 2.6\%$, which is equivalent to the ratio ($10.4 \pm 5.2\%$) observed in patients with Leber's disease (Figure 47.3). Patients with Leber's disease have an inborn error in cyanide metabolism and are unable to detoxify cyanide to thiocyanate (SCN) (Wilson 1965a). Leber's disease is a hereditary optic atrophy. The onset and severity of Leber's disease is related to the smoking pattern in many patients (Wilson 1965a, 1965b). We also reported that the increase in blood concentration of thiocyanate, a detoxication product of cyanide, in patients with end-stage renal disease is attributed to reduced urinary clearance of thiocyanate (Koyama *et al.* 1997). Hasuike *et al.* (2004) also demonstrated the accumulation of cyanide in patients with end-stage renal disease, by measuring the cyanide in red blood cells. Tobacco smoke may also cause chronic cyanide poisoning.

Cyanide in the cells can be rapidly detoxified by cyanide sulfurtransferase (known as rhodanese). Some of the cyanide reacts with cytochrome oxidase and damages the function of ATP production. The majority of cyanide *in vivo* is

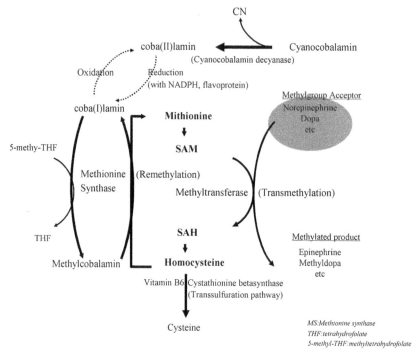

Figure 47.1 Intracellular metabolism of vitamin B_{12}. Cyanocobalamin is first converted into cob(II)alamin, which has no cyanogen group on the ligand occupying the upper axial position of the cobalamin structure. Cob(II)alamin is further reduced to cob(I)alamin, which can function as a coenzyme in the body. Removal of a cyanide molecule from cyanocobalamin is directly reduced by NADPH and flavoprotein in the presence of a cyanocobalamin trafficking chaperone. Cobalamin is reportedly converted into its inactive form, cob(II)alamin, under oxidative stress (Lerner-Ellis et al. 2006). NADPH: nicotinamide adenine dinucleotide phosphate.

enzymatically converted to SCN, and is excreted in the urine (Figure 47.4). The remaining cyanide is mainly metabolized *via* two other routes: (i) production of 2-amino-4-thiazolinecarboxylic acid from cystine and cyanide, and (ii) synthesis of cyanocobalamin *via* the combination of cyanide with some other vitamin B_{12} analogues, such as hydroxycobalamin or methylcobalamin (Boxer and Richards 1952).

Our findings (Koyama et al. 1996, 1997) that the smokers undergoing dialysis have peripheral neuropathy and that the fraction ratio of cyanocobalamin is comparable to that observed in patients with Leber's disease indicate the possibility of cyanide metabolism disorder in patients with CKD. In patients with CKD, thiocyanate (SCN) accumulates due to the decrease in its clearance. It impairs the major metabolic pathway of cyanide and the cyanide pool therefore increases. This increase accelerates cyanide detoxication *via* cyanocobalamin synthesis using vitamin B_{12}, resulting in an increase in the

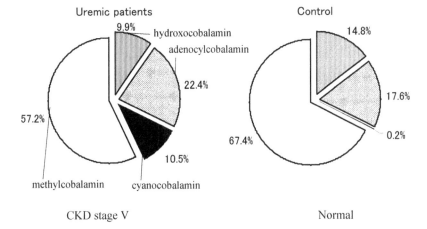

Figure 47.2 Proportion of each vitamin B_{12} analogue in total vitamin B_{12} in patients with CKD stage V. Measurement of the proportion of each vitamin B_{12} analogue in total vitamin B_{12} in patients with end-stage renal disease demonstrated an increase in the cyanocobalamin fraction and a decrease in the methylcobalamin fraction. Reproduced with permission from Koyama et al. (1997).

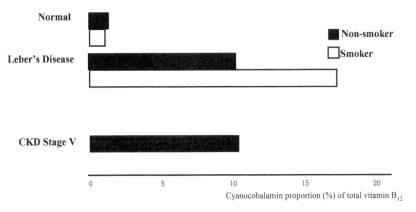

Figure 47.3 Cyanocobalamin fraction in uremia and neuro-ophthalmological diseases. The fraction ratio of cyanocobalamin was as extremely high as $10.5 \pm 2.6\%$, which is equivalent to the ratio ($10.4 \pm 5.2\%$) observed in patients with Leber's disease. Patients with Leber's disease have an inborn error in cyanide metabolism and are unable to detoxify cyanide to thiocyanate (Wilson 1965a). Leber's disease is a hereditary optic atrophy. The onset and severity of Leber's disease is related to the pattern of smoking in many patients (Wilson 1965a, 1965b).

proportion of the cyanocobalamin fraction and a decrease in the proportion of the methylcobalamin fraction (Figure 47.5). This disorder, which may affect the conversion from vitamin B_{12} to the coenzyme forms, is clinically important in CKD patients.

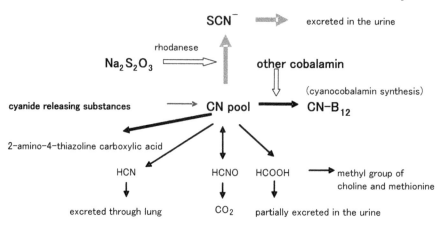

Figure 47.4 Metabolic pathway of cyanide. The cyanide in the cells can be rapidly detoxified by cyanide sulfurtransferase (known as rhodanese). The majority of cyanide *in vivo* is enzymatically converted to SCN and excreted in the urine. The remaining cyanide is metabolized mainly *via* two other routes: (i) production of 2-amino-4-thiazolinecarboxylic acid from cystine and cyanide, and (ii) synthesis of cyanocobalamin *via* the combination of cyanide with some other of vitamin B_{12} analogue, such as hydroxycobala-min or methylcobalamin (Boxer and Richards 1952). CN: cyanide; $Na_2S_2O_3$: sodium thiosulfate; SCN: thiocyanate; $CN-B_{12}$: cyanocobalamin; HCN: hydrocyanic acid; HCNO: cyanic acid; HCOOH: formic acid.

Cyanocobalamin must be decyanated before conversion into the enzyme forms to exert its activity. In CKD patients with cyanide metabolism disorder, the safety of cyanocobalamin is unclear. This mechanism may be associated with the occurrence of adverse events in patients with CKD who have been treated with cyanocobalamin in recent large-scale studies (Armitage *et al.* 2010; House *et al.* 2010). It has also to be noted that cobalamin cannot become its coenzyme form unless it is reduced to cob(I)alamin (Gregory Kelly 1997). Cobalamin is reportedly converted into its inactive form, cob(II)alamin, under oxidative stress (Chen *et al.* 1995). CKD is well known to be associated with increased oxidative stress (Annuk *et al.* 2001) and it is highly likely that CKD interferes with the conversion of cobalamin to its reduced forms (Figure 47.1).

47.3 Does CKD Induce Vitamin B_{12} Deficiency?

Vitamin B_{12} deficiency is generally assessed based on serum vitamin B_{12} concentration, plasma homocysteine concentration and serum methylmalonic acid (MMA) concentration (Savage *et al.* 1994). Serum vitamin B_{12} concentration has been reported by many researchers, including us, to be similar in patients with CKD and healthy individuals (Koyama *et al.* 2002). However, plasma total homocysteine level is elevated in an inverse relationship with the reduction in renal function (Bostom and Lathrop 1997).

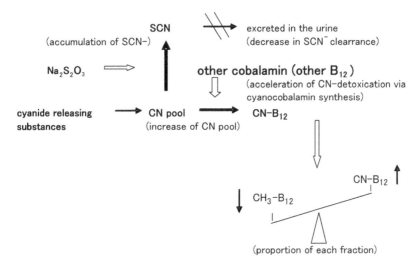

Figure 47.5 Metabolic pathway of cyanide in patients with CKD. In patients with CKD, SCN is accumulated due to the decrease in SCN clearance. This impairs the major metabolic pathway of cyanide and so the cyanide pool increases. This in turn increase accelerates cyanide detoxication *via* cyanocobalamin synthesis using vitamin B_{12}, resulting in an increase in the proportion of cyanocobalamin fraction and a decrease in the proportion of methylcobalamin fraction. Reproduced with permission from Koyama et al. (1997).

Impairment of remethylation is strongly implicated as the cause of hyperhomocysteinemia in patients with CKD, which has been demonstrated in a radioisotope study in patients with CKD (Van Guldener et al. 1999), although reduced clearance of homocysteine has been suggested as the possible cause in some reports. We reported that supplementation with folic acid and methylcobalamin normalized the remethylation pathway (Koyama et al. 2002). Compared with the decreases in homocysteine of $17.3 \pm 8.4\%$ after supplementation with folic acid alone and $18.7 \pm 7.5\%$ after that with methylcobalamin alone, a combination of folic acid and methylcobalamin decreased homocysteine by approximately 60% and normalized the findings of the methionine loading test Table 47.1. This result suggest that both coenzymes, folic acid and methylcobalamin, were insufficient due to reduced availability of these coenzymes in patients with CKD. There is also a report that increased MMA in dialysis patients was reduced by the administration of methylcobalamin (Nakamura et al. 2002).

Given our findings (Koyama et al. 1996, 1997) that patients undergoing dialysis have peripheral neuropathy and that the fraction ratio of cyanocobalamin is comparable to that observed in patients with Leber's disease (manifested by cyanide metabolism disorder), CKD possibly induces vitamin B_{12} deficiency by reducing the availability of this coenzyme.

Table 47.1 Results from a comparative study of homocysteine-lowering therapy with folic acid alone and remethylation pathway-normalizing therapy with folic acid and methylcobalamin. The acceleration of transmethylation due to reduced homocysteine was similar in the folic acid alone group (15 mg/day orally for three weeks) and the folic acid + methylcobalamin (500 μg after each hemodialysis session) group (SAM/SAH, an indicator of transmethylation activity was elevated similarly in both groups). However, the reduction in ADMA was significantly greater in the folic acid + methylcobalamin group than in the folic acid alone group. Reproduced with permission from Koyama et al. (2010).

	Folic acid			Folic acid + methylcobalamin			
	Before	After	P^*	Before	After	P^*	$P^\dagger$
Hcy (μmol/L)	26.0 ± 10.2	18.8 ± 5.8	<0.05	29.9 ± 12.0	10.7 ± 3.4	<0.001	<0.001
SAM (nmol/L)	391.6 ± 118.2	651.0 ± 349.9	<0.05	355.7 ± 121.4	625.5 ± 381.3	<0.01	ns
SAH (nmol/L)	325.5 ± 97.3	456.5 ± 197.7	<0.05	335.6 ± 152.3	448.6 ± 248.5	<0.01	ns
SAM/SAH	1.22 ± 0.23	1.41 ± 0.23	<0.001	1.16 ± 0.33	1.44 ± 0.36	<0.001	ns
ADMA(μmol/L)	0.63 ± 0.07	0.55 ± 0.07	<0.001	0.67 ± 0.09	0.50 ± 0.07	<0.001	<0.01
DMA(μmol/L)	28.8 ± 4.7	26.5 ± 5.9	<0.05	30.5 ± 9.1	29.3 ± 9.8	ns	ns
MA/ADMA	46.1 ± 9.3	48.9 ± 10.1	ns	45.4 ± 12.8	60.1 ± 22.0	<0.001	<0.05

Values are expressed as mean ± SD. P^*: *versus* 'before' by paired Student's *t*-test. $P^\dagger$: *versus* folate-group by ANCOVA.

47.4 Vitamin B_{12}-related Biomarkers and CKD

47.4.1 Homocysteine—the Best Known Biomarker Associated with Vitamin B_{12}

The best known biomarker associated with vitamin B_{12} is homocysteine. In patients with CKD, plasma total homocysteine level is elevated in an inverse relationship with the reduction in renal function (Bostom and Lathrop 1997) and homocysteine is a cardiovascular risk factor (Arnesen et al. 1995). Hyperhomocysteinemia in patients with CKD is thought to be associated with an impaired remethylation pathway (Van Guldener et al. 1999), but the cause of this disorder has not been fully clarified.

We reported that hyperhomocysteinemia in dialysis patients was primarily attributed to a deficiency of folic acid and vitamin B_{12}, since it was corrected by high doses of folic acid and methylcobalamin. However, it has been reported by many researchers, including us, that the blood concentrations of folic acid and vitamin B_{12} in dialysis patients are normal (Koyama et al. 2002; Nakamura et al. 2002). Thus, the mechanism by which the reduced availability of folic acid and vitamin B_{12} occurs in dialysis patients remains to be elucidated. In cells, homocysteine is either remethylated to methionine (via methionine synthase) or is transsulfurated to cysteine (via cystathionine beta synthase). During remethylation, homocysteine receives a methyl group from 5-methyltetrahydrofolate or betaine. Vitamin B_{12} is a necessary cofactor in the folate-dependent remethylation. Since human vascular cells lack cystathionine beta synthase and the transsulfuration pathway therefore does not work (Finkelstein 1998), vascular endothelium may be extremely vulnerable to homocysteine loading induced by impaired remethylation. Considering that dialysis patients have many risk factors that induce vascular endothelial dysfunction such as oxidative stress, hypertensive stress and calcium-phosphate metabolism disorder, homocysteine metabolism disorder should not be ignored in these patients.

47.4.2 Manifestation of Hyperhomocysteinemia

The toxicity of homocysteine is manifested by the following three reactions.

A Reaction directly inducing oxidative stress
B Reaction associated with the accumulation of S-adenosylhomocysteine (SAH) due to persistent slowing of the hydrolysis of SAH
C Reactions due to the inhibition of dimethylarginine dimethylaminohydrolase (DDAH)

The associations between issues B, C and vitamin B_{12} are discussed here. Concerning issue A, oxidative stress by free radicals formed during the reducing process of homocysteine is reported to injure endothelial cells directly (Mansoor et al. 1995).

47.4.2.1 Reaction Associated with the Accumulation of S-adenosylhomocysteine (SAH)

Homocysteine is produced during methionine metabolism *via* the adenosylated compounds *S*-adenosylmethionine (SAM) and *S*-adenosylhomocysteine (SAH). Accumulation of homocysteine causes slowing in the hydrolysis of SAH, resulting in an accumulation of SAM and a decrease in the ratio of SAM to SAH (an indicator of transmethylation activities) (McKeever *et al.* 1995). This transmethylation produces various metabolites. The possible influences of the reduced production of these metabolites are as follows (Perna *et al.* 1996):

(I) Lymphocyte chemokinetic response
(II) Insulin release
(III) Interferon synthesis
(IV) Norepinephrine uptake and release
(V) Catecholamine degradation
(VI) Conversion of phosphatidylethanolamine to phosphatidylcholine
(VII) Brain histamine content
(VIII) Serotonine and dopamine turnover.

Since homocysteine-lowering therapy, which activates the remethylation pathway, accelerates the transmethylation reaction (Koyama *et al.* 2010) (Figure 47.6, Table 47.1, the clinical assessment of homocysteine-lowering therapy may have to include examinations of wide-ranging factors (dementia, prevalence of cancer, *etc.*) as listed above (i–viii).

47.4.2.2 Reactions due to the Inhibition of Dimethylarginine Dimethylaminohydrolase (DDAH)

Asymmetric dimethylarginine (ADMA), a potent endogenous nitric oxide (NO) synthase inhibitor, is thought to be an independent predictor of cardiovascular mortality in end-stage renal disease (Zoccali *et al.* 2004); the circulating concentration of ADMA is greatly increased in hemodialysis patients (Annuk *et al.* 2001; Kielstein *et al.* 2002). A cohort study of patients with end-stage renal disease showed that a concentration-dependent relationship exists between plasma ADMA and the rates of mortality and adverse cardiovascular events (Zoccali *et al.* 2001). ADMA is synthesized by protein arginine methyltransferase (PRTMT) type I through transmethylation (*via* SAM) of the guanidinonitrogens of L-arginine and is degraded by DDAH through hydrolysation to dimethylamine (DMA) and L-citrulline (Leiper *et al.* 1999; Ogawa *et al.* 1987). Since DDAH is inhibited by homocysteine (Stühlinger *et al.* 2001), homocysteine metabolism is involved both in the synthesis and degradation of ADMA (Figure 47.6).

Therefore, discussion of whether or not homocysteine-lowering therapy reduces ADMA is complicated and this may be one of the reasons why a direct

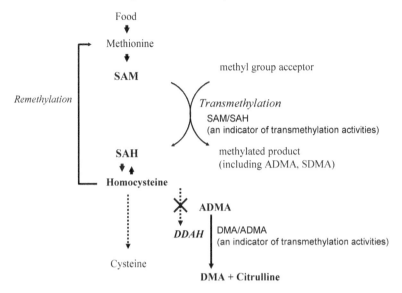

Figure 47.6 Metabolic relationship between homocysteine and asymmetric dimethylarginine (ADMA). Homocysteine is produced during methionine metabolism *via* the adenosylated compounds *S*-adenosylmethionine (SAM) and *S*-adenosylhomocysteine (SAH). An accumulation of homocysteine causes a slowing in the hydrolysis of SAH, resulting in an accumulation of SAM and a decrease in the ratio of SAM to SAH (an indicator of transmethylation activities). ADMA is synthesized by protein arginine methyltransferase (PRTMT) type I through transmethylation (*via* SAM) of the guanidinonitrogens of L-arginine, and it is degraded by DDAH through its hydrolysation to dimethylamine (DMA) and L-citrulline. Since DDAH is inhibited by homocysteine, homocysteine metabolism is involved both in the synthesis and degradation of ADMA.

association between homocysteine-lowering therapy and a reduction in the risk of cardiovascular disease has not been clearly demonstrated. We have obtained interesting findings from a comparative study of homocysteine-lowering therapy with folic acid alone and remethylation pathway-normalizing therapy with folic acid and methylcobalamin Table 47.1. The acceleration of transmethylation due to reduced homocysteine was similar in the folic acid alone group and the folic acid + methylcobalamin group. On the other hand, the reduction in ADMA was significantly greater in the folic acid + methylcobalamin group than in the folic acid alone group (% reduction in ADMA: 13.2 ± 11.2 *vs.* 25.4 ± 10.2, $p < 0.001$). These results suggest that, in hemodialysis patients receiving combined treatment with methylcobalamin and folate, the degradation of ADMA by DDAH is increased (possibly through its homocysteine-lowering effect) more strongly than the production of ADMA by PRTMT type I (through their action on transmethylation activity). The overall effect is thus a reduction in the serum ADMA concentration (Figure 47.6, Table 47.1).

Table 47.2 Efficacy of supplementations on homocysteine-lowering performed in CKD Stage V. In order to assess the efficacy on lowering plasma homocysteine levels, 21 hemodialysis (HD) patients were randomly assigned and provided folic acid supplementation: 15 mg/day orally (group I, $n=7$); methylcobamain 500 μg after each hemodialysis session, in addition to folic acid (group II, $n=7$); or vitamin B_6 60 mg/day orally, in addition to folic acid and methylcobalamin (group III, $n=7$). All patients were treated for three weeks. A methionine-loading test was conducted before and after supplementation. The measurement of amino acid at fasting and 2 hours and 4 hours after methionine load (0.05 g/kg orally). Seven normal controls and 12 HD controls (treated only with methylcobalamin) were also recruited. Mean percentage reduction in homocysteine level (fasting) were $17.3 \pm 8.4\%$ in group I, 57.4 ± 13.3 in group II, $59.9 \pm 5.6\%$ in group III and $18.7 \pm 7.5\%$ in HD patients treated with only methylcobalamin. Group II and III showed normal findings of methionine loading test after treatment whether with vitamin B6 or not. Reproduced with permission from Koyama et al. (2002).

	Vitamin Supplementation					
	Before Time course			After Time course		
	Fasting (0hr)	2hr	4hr	Fasting (0hr)	2hr	4hr
Homocysteine (nmol/ml)						
Control	8.6 ± 1.7	12.3 ± 2.3	14.0 ± 2.5			
Group 1	19.2 ± 2.9	23.8 ± 4.4	15.8 ± 2.3	15.8 ± 2.3	20.0 ± 2.9	23.9 ± 3.6
Group II	20.9 ± 5.7	26.0 ± 6.3	30.4 ± 7.5	$8.3 \pm 1.4^*$	$11.6 \pm 1.7^*$	$14.0 \pm 2.4^*$
Group III	21.3 ± 7.3	26.4 ± 8.4	31.1 ± 9.4	$8.2 \pm 1.9^*$	$11.1 \pm 2.6^*$	$13.6 \pm 3.1^*$
Group HD	25.6 ± 8.4	30.9 ± 9.4	36.2 ± 10.6	21.0 ± 8.1	25.8 ± 9.25	30.7 ± 10.5

*These values were not significantly different from control. HD: hemodialysis.

Based on these findings, we recommend that the efficacy of homocysteine-lowering therapy in the reduction of cardiovascular risk should be evaluated paying attention to the following:

(i) The efficacy should be evaluated in patients with moderate to severe hyperhomocysteinemia.
(ii) The therapy should produce a high homocysteine-lowering rate (preferably 50% or higher).
(iii) ADMA concentration should be measured simultaneously.

47.5 Association with Vitamin B_6

Reduced bone quality in patients with CKD has become a significant issue. It has been reported that hyperhomocysteinemia and vitamin B_6 deficiency reduce

Table 47.3 Methionine loading on patients with CKD Stage V. Group A was treated with folic acid 15 mg/day orally, methylcobamain 500 μg intravenously after each hemodialysis session and vitamin B_6 60 mg/day orally. Group B was treated with folic acid and methylcobalamin (without vitamin B_6). All patients were treated for three weeks. A methionine-loading test was conducted before and after supplementation. Amino acid level was measured at fasting and two hours and four hours after methionine load (0.05 g/kg orally). Both groups showed normal findings of homocysteine profile during the methionine loading test after treatment whether with vitamin B_6 or not. However, profiles of methionine and cysteine were not normalized. Reproduced with permission from Koyama (2011).

	Vitamin Supplementation					
	Before Time course			After Time course		
	Fasting (0hr)	2hr	4hr	Fasting (0hr)	2hr	4hr
Homocysteine (nmol/ml)						
Control	8.8 ± 2.0	12.5 ± 2.7	14.2 ± 3.1			
Group A	21.1 ± 8.9	26.2 ± 10.3	30.8 ± 11.5	8.2 ± 2.3*	11.1 ± 3.1*	13.4 ± 3.7*
Group B	20.5 ± 6.9	25.7 ± 7.6	29.8 ± 9.1	8.2 ± 1.7*	11.7 ± 2.0*	14 ± 2.9*
Methionine (nmol/ml)						
Control	30.9 ± 3.5	308 ± 29.0	208.9 ± 50.1			
Group A	23.6 ± 3.2	361.3 ± 33.3	258.4 ± 31.9	22.9 ± 2.2	351 ± 67.6	263.4 ± 51.8
Group B	22.1 ± 3.1	358.5 ± 41.1	267.5 ± 51.2	22.7 ± 2.0	344.6 ± 67.4	248.9 ± 58.6
Cystine (nmol/ml)						
Control	37.5 ± 2.5	43 ± 6.0	41.5 ± 3.1			
Group A	55.6 ± 4.4	71.5 ± 9.8	80.4 ± 12.8	59.4 ± 7.2	72.7 ± 7.6	75.5 ± 6.5
Group B	66.4 ± 21.5	83.5 ± 18.4	93.5 ± 25.5	57.5 ± 18.5	72.8 ± 21	82.6 ± 24.6

*$p < 0.01$ vs. before supplementation. These values were not significantly different from control.
HD: hemodialysis.

the quality of collagen cross-links that are closely associated with bone strength, and that the formation of such impaired cross-links (advanced glycation end products, pentosidine) is accelerated by vitamin B_6 deficiency and excessive oxidative stress (McLean and Hannan 2007; Saito et al. 2006).

Homocysteine is an intermediate metabolic product at the junction of two metabolic pathways, transsulfuration (requiring vitamin B_6 as a coenzyme) and remethylation (requiring vitamin B_{12} as a coenzyme). When vitamin B_{12} and vitamin B_6 are viewed in terms of homocysteine metabolism, both vitamins may be linked with each other at the degradation of homocysteine. The reduced remethylation in patients with CKD may stimulate sulfur transfer. We consider that the linkage between vitamin B_{12} and vitamin B_6 may be mediated by SAM, since it has been demonstrated that accumulated SAM accelerates the transfer of sulfur (Purohit et al. 2007). Based on the fact as observed in patients with CKD that the transsulfuration pathway did not deteriorate and that the

Table 47.4 Estimation of remethylation and transsulfuration activities. The ratio of methionine to homocysteine (M/H) was significantly decreased in patients with CKD stage V, but elevated after each supplementation regimen supplementation. On the other hand, the ratio of cysteine to homocysteine (C/H) did not deteriorate and increased considerably after treatment whether accompanied by vitamin B_6 or not. M/H represents remethylation activity and C/H represents transsulfuration activity. Reproduced with permission from Koyama (2011).

	Before treatment			After treatment		
	M/H(0 hour)	M/H(2 hour)	M/H(4 hour)	M/H(0 hour)	M/H(2 hour)	M/H(4 hour)
Control	3.6±0.6	25.4±4.8	14.8±2.5	3.6±0.6	25.4±4.8	14.8±2.5
Group A	1.3±0.5*	15.4±5.3*	9.3±3.1*	3±0.9	33.5±9	20.5±4.7*
Group B	1.2±0.3*	15.0±4.6*	9.7±3.1*	2.9±0.6	29.8±5.2	17.9±2.5*
	C/H(0 hour)	C/H(2 hour)	C/H(4 hour)	C/H(0 hour)	C/H(2 hour)	C/H(4 hour)
Control	4.4±1.1	3.6±1.3	3.1±1	4.4±1.1	3.6±1.3	3.1±1
Group A	3±1.2	3±1.1	2.8±1	7.7±2.5*	7.1±2.5*	6.1±2.1*
Group B	3.5±1	3.4±0.9	3.3±0.9	6.9±1.2*	6.2±0.9*	5.9±0.9*

*$p <0.01$ vs. control. M/H: methionine/homocysteine; C/H: cysteine/homocysteine.

transfer of sulfur was accelerated by homocysteine-lowering therapy which increases serum SAM level at the same time (Tables 47.1–47.4), we have proposed the above-mentioned linkage between vitamin B_{12} and vitamin B_6.

47.6 Concluding Remarks

We have primarily described here how the metabolism of vitamin B_{12} is influenced by the progression of CKD (impairment of renal function) and what happens in the body as a result of it. Due to poor accumulation of evidence in this field of medicine, we have had to rely primarily on our own clinical data. We expect further development of this field.

Summary Points

- This chapter primarily describes how the metabolism of vitamin B_{12} is influenced by the progression of chronic kidney disease (CKD), i.e. decreased kidney function, and what happens subsequently in the biological process.
- CKD may affect the conversion from vitamin B_{12} to the coenzyme forms. That may be the reason why CKD induces vitamin B_{12} deficiency by reducing the availability of this coenzyme.
- The best known biomarker associated with vitamin B_{12} is homocysteine.
- Homocysteine metabolism is involved both in the synthesis and degradation of asymmetric dimethylarginine (ADMA), a potent endogenous NO-synthase inhibitor, which is thought to be an independent predictor of cardiovascular mortality in end-stage renal disease.

- Since human vascular cells lack cystathionine beta synthase, vascular endothelium may be extremely vulnerable to homocysteine loading induced by impaired remethylation. Hence, homocysteine management is clinically important in CKD patients.
- When vitamin B_{12} and vitamin B_6 are viewed in terms of homocysteine metabolism, both vitamins may be linked with each other at the degradation of homocysteine.

Key Facts

Key Facts about Chronic Kidney Disease

- A patient with chronic kidney disease (CKD) is defined as having a decreased renal function of 60%, persisting abnormal urine analysis for three months or longer, or a number of (or persisting) renal morphological abnormalities.
- Patients with CKD have been increasing in number worldwide, which has become a major social problem.
- The main cause of CKD is lifestyle-related diseases such as diabetes mellitus and hypertension.
- Advanced CKD may lead to end-stage renal failure that requires renal replacement therapy including hemodialysis, peritoneal dialysis and renal transplantation.
- Cardiovascular complications may occur in proportion to the decrease in renal function. Not only will patients' activity of daily living (ADL) and quality of life (QOL) greatly deteriorate if they have such complications, but there will be a major problem in health and healthcare economics.
- In preventing incidence of cardiovascular complications in patients with CKD, no report has been made on the effectiveness of a single-modality therapy such as hypertensive therapy or monotherapy with statins.
- Comprehensive treatment is required; *i.e.* nutritional therapy (including supplements), or patient education for improvement of lifestyle and advice for medication.

Key Facts about Homocysteine

- Homocyateine is the best known biomarker associated with vitamin B12.
- In patients with CKD, plasma total homocysteine level are elevated in an inverse relationship with the reduction in renal function.
- Homocysteine is a cardiovascular risk factor.
- Homocysteine is formed as an intermediate metabolic product of methionine at the junction of two metabolic pathway: remethylation and transsulfuration.
- Hyperhomocysteinemia in patients with CKD is thought to be associated with an impaired remethylation pathway.

- The toxicity of homocysteine is manifested by the following three reactions are as followed: (A) reaction directly inducing oxidative stress; (B) reaction associated with the accumulation of S-adenosylhomocysteine (SAH) due to persistent slowing of the hydrolysis of SAH; and (C) reactions due to the inhibition of dimethylarginine dimethylaminohydrolase (DDAH).
- An accumulation of homocysteine decreases transmethylation activities.

Key Facts about Asymmetric Dimethylarginine

- Asymmetric dimethylarginine (ADMA), a potent endogenous nitric oxide (NO) synthase inhibitor.
- ADMA level is greatly increased in patients with CKD.
- ADMA is thought to be an independent predictor of cardiovascular mortality in end-stage renal disease.
- The advent of an effective ADMA-reduction therapy might give us a useful tool for the decrease in cardiovascular risk in CKD patients.
- ADMA potently impairs the vascular endothelial function by inhibiting the production of nitric oxide (NO) through the antagonism with arginine, a substrate of NO.
- ADMA is synthesized by protein arginine methyltransferase (PRTMT) type I through transmethylation (*via* SAM) of the guanidinonitrogens of L-arginine, and is degraded by dimethylarginine dimethylaminohydrolase (DDAH) through its hydrolysation to dimethylamine and L-citrulline.
- Since an accumulation of homocysteine decreases transmethylation activities and DDAH is inhibited by homocysteine, homocysteine metabolism is involved both in the synthesis and degradation of ADMA.

Definitions of Words and Terms

Asymmetric dimethylarginine (ADMA). Asymmetric dimethylarginine (ADMA) potently impairs the vascular endothelial function by inhibiting the production of nitric oxide (NO) through the antagonism with arginine, a substrate of NO.

Chronic kidney disease (CKD). A patient with chronic kidney disease (CKD) is defined as having a decreased renal function of 60%, persistent abnormal urine analysis for three months or longer, or a number of (or persisting) renal morphological abnormalities.

Collagen cross-links. The quality of collagen cross-links in the bone structure is closely associated with bone strength. A formation of such impaired cross-links is accelerated by vitamin B_6 deficiency and excessive oxidative stress.

Cyanocobalamin tracking chaperone. Cyanocobalamin trafficking chaperone is a molecule which removes a cyanide molecule from cyanocobalamin mediated by a direct reduction by NADPH and flavoprotein.

Homocysteine. Homocysteine is a cardiovascular risk factor. This substance is formed as an intermediate metabolic product of methionine at the junction of two metabolic pathway: remethylation and transsulfuration.

Leber's disease. Leber's disease is a hereditary optic atrophy. The onset and severity of Leber's disease is related to the pattern of smoking in many patients, who are unable to detoxify cyanide to thiocyanate. The clinical features of neurological disorders manifested by Laber's disease resemble those of uremic nephropathy.

Nitric oxide (NO) synthase. NO is the most potent endogenous vasodilator. Thus, endothelial dysfunction as a result of reduced NO activity is an early step in the progression of atherosclerotic vascular disease. NO synthases with endothelial converts L-arginine to NO and citrulline by means of stereospecific oxidation of the terminal guanidine nitrogen of the amino acid L-arginine. This process can be inhibited selectively by competitive blockade of the NOS active site with such naturally occurring guanidino-substituted analogues of L-arginine as G-monomethyl-L-arginine (L-NMMA) and ADMA.

Remethylation pathway. In cells, homocysteine is either remethylated to methionine (*via* methionine synthase) or is transsulfurated to cysteine (*via* cystathionine beta synthase). In remethylation, homocysteine receives a methyl group from 5-methyltetrahydrofolate or from betaine. Vitamin B_{12} is a necessary cofactor in the folate-dependent remethylation.

Transmethylation. The sulfur-containing amino acid homocysteine is produced during methionine metabolism *via* the adenosylated compounds *S*-adenosylmethionine (SAM) and *S*-adenosylhomocysteine (SAH). SAM is also the main methyl donor in numerous enzymatic transmethylation reactions, which lead to formation of SAH. Important substances such as monoamine neurotransmitters are synthesized in transmethylation reactions.

Transsulfuration pathway. Homocysteine is either remethylated to methionine (*via* methionine synthase) or is transsulfurated to cysteine (*via* cystathionine beta synthase). Transsulfuration requires vitamin B_6 as a cofactor.

List of Abbreviations

ADMA	asymmetric dimethylarginine
ATP	adenosine triphosphate
CKD	chronic kidney disease
CoA	coenzyme A
DDAH	dimethylarginine dimethylaminohydrolase
DMA	dimethylamine
MMA	methylmalonic acid
MMACHC	methylmalonic aciduria cdlC type with homocystinuria
NADPH	nicotinamide adenine dinucleotide phosphate
PRTMT	protein arginine methyltransferase
SAM	*S*-adenosylmethionine
SAH	*S*-adenosylhomocysteine
SCN	thiocyanate

References

Annuk, M., Zilmer, M., Lind, L., Linde, T., and Fellstrom B., 2001. Oxidative stress and endothelial function in chronic renal failure. *Journal of the American Society of Nephrology.* 12: 2747–2752.

Armitage, J.M., Bowman, L., Clarke, R.J., Wallendszus, K., Bulbulia, R, Rahimi, K., Haynes, R., Parish, S., Sleight, P., Peto, R., and Collins R., 2010. Study of the Effectiveness of Additional Reductions in Cholesterol and Homocysteine (SEARCH) Collaborative Group. Effects of homocysteine-lowering with folic acid plus vitamin B12 *vs.* placebo on mortality and major morbidity in myocardial infarction survivors: a randomized trial. *The Journal of the American Medical Association.* 303: 2486–2494.

Arnesen, E., Refsum, H., Bønaa, K.H., Ueland, P.M., Førde, O.H., and Nordrehaug, J.E., 1995. Serum total homocysteine and coronary heart disease. *International Journal of Epidemiology.* 24: 704–709.

Bostom, A.G., and Lathrop, L., 1997. Hyperhomocysteinemia in end-stage renal disease: prevalence, etiology, and potential relationship to arteriosclerotic outcomes. *Kidney International.* 52: 10–20.

Boxer, G.E., and Rickards, J.C., 1952. Studies on the metabolism of the carbon cyanide and thiocyanate. *Archives of Biochemistry and Biophysics.* 39: 7–26.

Chen, Z., Chakraborty, S., and Banerjee, R., 1995. Demonstration that mammalian methionine synthases are predominantly cobalamin-loaded. *The Journal of Biological Chemistry.* 270: 19246–19249.

Finkelstein, J.D., 1998. The metabolism of homocysteine: pathways and regulation. *European Journal of Pediatrics.* 157(Suppl 2): S40–S44.

Gregory Kelly, N.D., 1997. The coenzyme forms of vitamin B_{12}: toward an understanding of their therapeutic potential. *Alternative Medicine Review.* 2: 459–471.

Hasuike, Y., Nakanishi, T., Moriguchi, R., Otaki, Y., Nanami, M., Hama, Y., Naka, M., Miyagawa, K., Izumi, M., and Takamitsu, Y., 2004. Accumulation of cyanide and thiocyanate in haemodialysis patients. *Nephrology Dialysis Transplantation.* 19: 1474–1479.

House, A.A., Eliasziw, M., Cattran, D.D., Churchill, D.N., Oliver, M.J., Fine, A., Dresser, G.K., and Spence, D., 2010. Effect of B vitamin therapy on progression of diabetic nephropathy: a randomized controlled trial. *The Journal of the American Medical Association.* 303: 1603–1609.

Kopple, J.D., 1999. Pathophysiology of protein-energy wasting in chronic renal failure. *The Journal of Nutrition.* 129(1S Suppl): 247S–251S.

Kielstein, J.T., Böger, R.H., Bode-Böger, S.M., Frölich, J.C., Haller, H., Ritz, E., and Fliser, D., 2002. Marked increase of asymmetric dimethylarginine in patients with incipient primary chronic renal disease. *Journal of the American Society of Nephrology.* 13: 170–176.

Koyama, K., Yoshida, A., Takeda, A., Morozumi, K., and Fujinami, T., 1996. Efficacy of methylcobalamin in subclinical uremic neuropathy as detected by measuring vibration perception thresholds. *Nephrology.* 2: 25–28.

Koyama, K., Yoshida, A., Takeda, A., Morozumi, K., Fujinami, T., and Tanaka, N., 1997. Abnormal cyanide metabolism in uraemic patients. *Nephrology Dialysis Transplantation.* 12: 1622–1628.

Koyama, K., Usami, T., Takeuchi, O., Morozumi, K., and Kimura, G., 2002. Efficacy of methylcobalamin on lowering total homocysteine plasma concentrations in haemodialysis patients receiving high-dose folic acid supplementation. *Nephrology Dialysis Transplantation.* 17: 916–922.

Koyama, K., Ito, A., Yamamoto, J., Nishio, T., Kajikuri, J., Dohi, Y., Ohte, N., Sano, A., Nakamura, H., Kumagai, H., and Itoh, T., 2010. Randomized controlled trial of the effect of short-term coadministration of methylcobalamin and folate on serum ADMA concentration in patients receiving long-term hemodialysis. *American Journal of Kidney Diseases.* 55: 1069–1078.

Koyama, K., 2011. Reduced remethylation in patients with chronic kidney disease (CKD) induces vitamin B_6 deficiency by enhanced trans-sulfuration activity. Abstracts of the 422th vitamin B scientific conference. *Vitamins.* 85(2): 92–94.

Leiper, J.M., Santa Maria, J., Chubb, A., MacAllister, R.J., Charles, I.G., Whitley, G.S., and Vallance, P., 1999. Identification of two human dimethylarginine dimethylaminohydrolases with distinct tissue distributions and homology with microbial arginine deiminases. *Biochemical Journal.* 343: 209–214.

Lerner-Ellis, J.P., Tirone, J.C., Pawelek, P.D., Doré, C., Atkinson, J.L., Watkins, D., Morel, C.F., Fujiwara, T.M., Moras, E., Hosack, A.R., Dunbar, G.V., Antonicka, H., Forgetta, V., Dobson, C.M., Leclerc, D., Gravel, R.A., Shoubridge, E.A., Coulton, J.W., Lepage, P., Rommens. J.M., Morgan, K., and Rosenblatt, D.S., 2006. Identification of the gene responsible for methylmalonic aciduria and homocystinuria, cblC type. *Nature Genetics.* 38: 93–100.

Mansoor, M.A., Bergmark, C., Svardal, A.M., Lønning, P.E., and Ueland, P.M., 1995. Redox status and protein binding of plasma homocysteine and other aminothiols in patients with early-onset peripheral vascular disease. Homocysteine and peripheral vascular disease. *Arteriosclerosis, Thrombosis, and Vascular Biology.* 15: 232–240.

McKeever, M., Molloy, A., Weir, D.G., Young, P.B., Kennedy, D.G., Kennedy, S., and Scott, J.M., 1995. An abnormal methylation ratio induces hypomethylation *in vitro* in the brain of pig and man, but not in rat. *Clinical Science.* 88: 73–79.

McLean, R.R., and Hannan, M.T., 2007. B vitamins, homocysteine, and bone disease: epidemiology and pathophysiology. *Current Osteoporosis Reports.* 5: 112–119.

Nakamura, T., Saionji, K., Hiejima, Y., Hirayama, H., Tago, K., Takano, H., Tajiri, M., Hayashi, K., Kawabata, M., Funamizu, M., Makita, Y., and Hata, A., 2002. Methylenetetrahydrofolate reductase genotype, vitamin B12, and folate influence plasma homocysteine in hemodialysis patients. *American Journal of Kidney Diseases.* 39: 1032–1039.

Ogawa, T., Kimoto, M., and Sasaoka, K., 1987. Occurrence of a new enzyme catalyzing the direct conversion of NG,NG-dimethyl-L-arginine to L-citrulline in rats. *Biochemical and Biophysical Research Communications*. 148: 671–677.

Perna, A.F., Ingrosso, D., Galletti, P., Zappia, V., and De Santo, N.G., 1996. Membrane protein damage and methylation reactions in chronic renal failure. *Kidney International*. 50: 358–366.

Purohit, V., Abdelmalek, M.F., Barve, S., Benevenga, N.J., Halsted, C.H., Kaplowitz, N., Kharbanda, K.K., Liu, Q.Y., Lu, S.C., McClain, C.J., Swanson, C., and Zakhari, S., 2007. Role of S-adenosylmethionine, folate, and betaine in the treatment of alcoholic liver disease: summary of a symposium. *The American Journal of Clinical Nutrition*. 86: 14–24.

Saito, M., Fujii, K., Mori, Y., and Marumo, K., 2006. Role of collagen enzymatic and glycation induced cross-links as a determinant of bone quality in spontaneously diabetic WBN/Kob rats. *Osteoporosis International*. 17: 1514–1523.

Savage, D.G., Lindenbaum, J., Stabler, S.P., and Allen, R.H., 1994. Sensitivity of serum methylmalonic acid and total homocysteine determinations for diagnosis of cobalamin and folate deficiencies. *The American Journal of Medicine*. 96: 239–246.

Stühlinger, M.C., Tsao, P.S., Her, J.H., Kimoto, M., Balint, R.F., and Cooke, J.P., 2001. Homocysteine impairs the nitric oxide synthase pathway: role of asymmetric dimethylarginine. *Circulation*. 104: 2569–2575.

Van Guldener, C., Kulik, W., Berger, R., Dijkstra, D.A., Jakobs, C., Reijngoud, D.J., Donker, A.J., Stehouwer, C.D., and De Meer, K., 1999. Homocysteine and methionine metabolism in ESRD: a stable isotope study. *Kidney International*. 56: 1064–1071.

Wilson, J., 1965a. Leber's hereditary optic atrophy. A possible defect of cyanide metabolism. *Clinical Science*. 29: 505–515.

Wilson, J., 1965b. Leber's hereditary optic atrophy. Some clinical and etiological consideration. *Brain*. 86: 347–362.

Zoccali, C., Bode-Böger, S., Mallamaci, F., Benedetto, F., Tripepi, G., Malatino, L., Cataliotti, A., Bellanuova, I., Fermo, I., Frölich, J., Böger, R., 2001. Plasma concentration of asymmetrical dimethylarginine and mortality in patients with end-stage renal disease: a prospective study. *Lancet*. 358(9299): 2113–2117.

Zoccali, C., Mallamaci, F., and Tripepi, G., 2004. Novel cardiovascular risk factors in end-stage renal disease. *Journal of the American Society of Nephrology*. 15(Suppl 1): S77–S80.

Subject Index

Illustrations and figures are in **bold**. Tables are in *italics*.

3σ method 184
5-MHTF *see*
  5-methyltetrahydrofolate
5HT *see* serotonin

α/β-fold 65
AASA (α-aminoadipic
  semialdehyde) 78, 505
ABC transporters 88
aberic acid *see* thiamin
absorption, dietary 103–4
  biocytin 721
  biotin 721
  biotinidase 721
  cobalamins (B_{12}) 459
  flavin adenine dinucleotide
    (FAD) 592, **593**
  flavin mononucleotide
    (FMN) 592, **593**
  flavins (flavoproteins) 612
  folates 736–7
  niacin (nicotinic acid) 634
  riboflavin *601*, **612**
  riboflavin (B_2) 99, 100, 103, 104,
    593, 595, **597**
  thiamin (B_1) 80
ACAD *see* acyl-CoA dehydrogenases
ACC *see* acetyl-CoA carboxylases
accuracy of analysis 433
acetaldehyde 635
acetate 564, 635, 636
acetate kinase 65
acetoin dehydrogenase 63

acetyl-CoA carboxylases
  (ACC) 148–9, 718, 719, 721,
  722, 725
acetyl-CoA synthase 562
acetyl-coenzyme A (acetyl-CoA)
  acetylcholine (ACh) 559–60, **561**
  adipose tissue 566
  biotin 719
  brain sources 560–2
  energy metabolism 318, 560–4,
    566, 567
  Krebs cycle **73**, **635**, **718**
  liver 566
  metabolism **555**
  pantothenic acid 318
  pathological conditions 563–4
  pyruvate 560, 562
  pyruvate dehydrogenase (PDH) 65
  structure **149**
  thiamin diphosphate (ThDP) *560*,
    566
  tricarboxylic acid (TCA) cycle 58
acetylcholine (ACh) 142, **555**,
  558, 566
  acetyl-coenzyme A (acetyl-CoA)
    559–60, **561**
  cognitive function 559–60
  thiamin deficiency 559–60
acetylcholinesterase (AChE) 141, 142
achlorhydria 13, 459, 470, 780, 781
acid hydrolysis
  biotin 382
  cobalamins 423, 447, **448**

acid hydrolysis (*continued*)
  niacin (nicotinic acid) 286–7
  riboflavin (B$_2$) 272, 273, **597**
  thiamin *260–1*, 262
ACL *see* ATP–citrate lyase (ACL) pathway
aconitase 563
acridans 444–5
acridium ester 445
activation energy 221
active transport 594, 595, 605
activity coefficient (ETK AC) 230, 241
acyl-carnitine translocase (CAT) 614
acyl-carnitines 621–2
acyl-CoA 614, 615
acyl-CoA dehydrogenases (ACAD) 614, **615**, 616–19, 620, 628, 629
acyl transferase 132
acyloins 56, 65
ADAM *see* 9-anthryldiazomethane
ADARs *see* adenosine deaminases acting on RNA
Addison, Thomas 13
addition sequences 447–9
adenosine, folates 24
adenosine deaminases acting on RNA (ADARs) 575
adenosine diphosphate (ADP) 114–15
adenosine diphosphate (ADP)-ribose 109, 119–21, 123, 637
adenosine diphosphate (ADP)-ribosyltransferases 114–15, **116**, 119
adenosine monophosphate (AMP) 128, 564, 613, 619–20
  biotin 716
  coenzyme A metabolism 686
  niacin (nicotinic acid) 636
adenosine thiamin diphosphate (AThDP)
  distribution 81
  structure 76–7

adenosine thiamin triphosphate (AThTP) 71
  biosynthesis **82**, 84
  distribution 81
  structure 76–7
adenosine triphosphates (ATP)
  see also electron transport chain
  alcohol 564
  bacteria anaerobic growth 65
  carboxylase biotinylation 148
  cyanide 787
  fatty acid oxidation 614, 628
  hydrolysis 88
  mitochondria 89, 115, 557, 566
  NAD biosynthesis 113, 114
  phosphorolysis 612
  research history 8
  riboflavin metabolism 595, *601*, 603, 612
  synthesis 566, 575, 787
  thiamin deficiency 540, 541
adenosylcobalamin (AdoCbl)
  cofactor behaviour 167–8, 787
  dietary sources 166
  light sensitivity 166, 424
  natural forms 166–7
  research history 14
  structure **13**, 165, **203**
S-adenosylhomocysteine (SAH or AdoHcy) **25**, 801
  cobalamins (B$_{12}$) **788**
  homocysteine metabolism **38**, 738, **739**, 775–6, 794
S-adenosylmethionine (SAM or AdoMet) 167, 169
  B$_6$ 797
  cobalamins (B$_{12}$) **788**, 797
  homocysteine metabolism **38**, **484**, 738, **739**, 741, 771, 794, 801
  metabolism **25**
  as methyl donor 24, 396, 491, 775–6
  nervous system 771
  neurotransmitters 743
adenylate kinase **82**, 83, 84

Subject Index

adenylyl cyclase (adenylate cyclase) 86, 638, 661
adenylyl thiamin diphosphate transferase **82**
adequate intake (AI) 101, *102*, 104, 720
adipocytes
   niacin (nicotinic acid) **661**
   size 581–3, 589
adiponectin 643, 645–6, 663
adipose tissue
   acetyl-coenzyme A (acetyl-CoA) 566
   gastric inhibitory polypeptide 586
   ketone bodies 614
   niacin (nicotinic acid) 636, 638–9, 661
   thiamin 581–4
ADMA *see* asymmetric dimethylarginine
adolescents
   multivitamin supplements 525, 526, 527, 532
   riboflavin intake 102
ADP *see* adenosine diphosphate
ADP-ribosyl cyclases **120**, 121
adsorptive stripping voltammetry (AdSV) 264
AEDs *see* antiepileptic drugs
aequorin (AEQ) 356, 359
affinity biosensors 441–2
affinity chromatography 392, 399, 411
age related differences, vitamin status 461, 774–5
AGEs *see* glycation end-products
AI *see* adequate intake (AI)
AIM-HIGH study 647–8, 649, *666*, 667–8
alanine 113, **127, 308**
Albizia julibrisson 140
albumen 6, 10, 99, 378
albumins **597**, 602, 680
albuminuria 692
alcohol
   fermentation 58, 65, 72
   metabolism 635–6

nicotinamide adenine dinucleotide phosphate (NADP) 635–6
riboflavin digestion 99, 595
thiamin digestion 80
alcohol dehydrogenase 635
alcoholism 88, 243, 567 *see also* liver; Wernicke's encephalopathy
   biotin 723–4
   cobalamins 470
   dementia 546, 567
   encephalopathy 564–5, 634
   niacin 110, 111, 634
   pellagra 676
   riboflavin 103
   thiamin 72, 233, 234, 242, 538, 546–7, 564–5
   thiamin diphosphate (ThDP) *554*, 557
aldoses 65
aldosterone 100
alkaline hydrolysis 286, 287, 423
alkaline phosphatase (ALP) 99, 338, 339, 341, 444, 612
alkaloids 143
1-alkylthio-2-alkylisoindole. 308–9
allicin 79
allithiamin (thiamin allyl disulfide) 79
alloxazine 6
alopecia 726
ALP *see* alkaline phosphatase
Alzheimer's disease (AD) *see also* dementia
   AMPARs 574
   folates 397, *735*
   holotranscobalamin (holoTC) 464
   hyperhomocysteinemia (HHcy) 772–3
   thiamin 79, 239, 240, 573
   thiamin diphosphate (ThDP) 545–6
   vascular dementia 481
   vitamin status 490
American Type Culture Collection (ATCC) 313
2-amino-3-carboxy muconaldehyde, structure **112**

α-amino-3-hydroxy-5-methyl-4-
isoxazolepropionic acid receptors
see AMPARs
5-amino-4-imidazolecarboxamide
ribonucleotide formyltransferase
(AICART) **739**, 740
2-amino-4-thiazolinecarboxylic
acid 788, **790**
D-amino acid oxidase 7
amino acids
coenzyme requirements 14–15
metabolism 138–9, 149, 242, **615**, 619
tetrahydrofolate (THF) 491
α-aminoadipic semialdehyde
(AASA) 505
γ-aminobutyric acid (GABA) 505
δ-aminolevulinic acids synthase 139
aminophthalate 443, **444**, 445, 450
aminopyrimidine 77
aminotransferase 350
Amish lethal microcephaly 79
amnesia, Korsakoff syndrome 242
AMP *see* adenosine monophosphate
AMPARs (α-amino-3-hydroxy-5-
methyl-4-isoxazolepropionic acid
receptors) 574, 575
amphipathic helix 628
ampolium 574
α-amylase 358
amyloid plaques 79, 776
amyotrophic lateral sclerosis
(ALS) 574, 576
anaemia 12–13, 46, 419, 534 *see also*
megaloblastic anaemia
cobalamins 459, 484, 770, *771*
folates 484, 770, *771*
macrocytic 453, 526, 534
pernicious 46, 459, 470, 781
randomized controlled trials
(RCTs) 526–7, 532
anencephaly 513
angina 43
angioplasty, folic acid
supplementation 41–2
Anhagen, Ernst 7
animal assays 264–5

animal studies 22, 588
anion-exchange chromatography 382
anorexia 103, 110, *554*, 786
9-anthryldiazomethane
(ADAM) 383, 384
antibiotics
assay interference 161, 425
biotin 723
thiamin diphosphate (ThDP) *554*
antibodies
biotin analysis 360, 362–4, 385
cobalamins analysis 429, 462
anticoagulants 434
anticonvulsants 153, 723
antiepileptic drugs (AEDs) 505,
508–18, 519, 520
antineoplasic drugs 726
antioxidants definition 143
AOAC International 312–13
Aoyama, Tanemichi 16
APCI *see* atmospheric pressure
chemical ionization
apical membrane 596, **597**, 598,
598, 606
apo-proteins 434, 613
apocarboxylases **717**
apoenzymes 76, 81, 89, 168, 392, 623
apolipoproteins 636, 638, 639–40,
662, 663, 670
apoptosis 123
endoplasmic reticulum stress 576
poly-ADP-ribose polymerases
(PARPs) 119–21, 637–8
thiamin deficiency 544
aquacobalamin 165, 424
arginine 728, 794, 800
β-arrestin 663
arthritis 689, 692
ARTs *see* mono-ADP-
ribosyltransferases
arylformamidase (AFMID) 111, **112**
ascorbic acid (vitamin C)
honey samples *178*, 180, 182–3,
187, 188
oxidation 97
riboflavin digestion 99

Subject Index

aspartame 97
astrocytes 543, 548
asymmetric dimethylarginine
  (ADMA) 36, **38**, 46, *792*, 794–6,
  798, 800
ATCC (American Type Culture
  Collection) 313
atherogenesis 497, 701
atherosclerosis 670 *see also* coronary
  heart disease (CHD)
  B_6 139, 490, 703
  biotin 722
  folates 490
  folic acid supplementation 40,
    41–2
  high-density lipoproteins
    (HDL) 640
  homocysteinaemia 36–7
  homocysteine (Hcy) 41–2, 48, 139,
    482, 490, 493, 494
  homocystinuria 35, 754–5
  inflammation 641
  niacin 109, 638, 640–3, 647, **661**,
    663, 665–8
  niacin (nicotinic acid) 669–70
  nitric oxide 801
  supplementation 46
AThTP *see* adenosine thiamin
  triphosphate
atmospheric pressure chemical
  ionization (APCI) 320, 327–8
ATP *see* adenosine triphosphates
  (ATP)
ATP binding cassette protein A1
  (ABCA1) 663
ATP–citrate lyase (ACL)
  pathway **555**, 559, 562
atrophic gastritis 484, 493, 775, 780
avidin 10, 392, 727
  biotin analysis 355, 356, 381, 383,
    384, 386, 721
  biotin deficiency 723, 724
avidin bound complex 721, 727

B_1 *see* thiamin
B_2 *see* riboflavin

B_3 *see* niacin
B_5 *see* pantothenic acid
B_6 *see also* pyridoxal-5′-phosphate
  (PLP); pyridoxal (PL);
  pyridoxamine-5′-phosphate
  (PMP); pyridoxamine (PM);
  pyridoxine-5′-phosphate (PNP);
  pyridoxine (PN)
  S-adenosylmethionine
    (SAM or AdoMet) 797
  antimicrobial effects 141
  antioxidant function 139
  assay methods 198, 199, 201–2,
    336–49
  bioavailability 140–1
  biological function 699–701
  bone strength 796–7
  cardiovascular disease
    (CVD) 25–6
  cell growth 525
  coenzyme function 139, 142, 699
  coronary heart disease
    prevention 139
  cysteine reaction 136
  deficiency 196, 349, 701
  derivatives 140–1
  dietary reference intake (DRI) 496
  dietary sources 136, 496
  epilepsy 506, 513–14
  excess consumption 350
  extraction 201–2
  gene expression 139
  glycosylated forms 140
  heat treatment 136
  homocysteine analysis 493
  homocysteine (Hcy) 797
  homocysteine metabolism **38**, 39,
    482, **484**, 701
  honey samples *178*, 179
  mortality *29*
  natural forms 335, 699
  niacin biosynthesis 113
  nomenclature 135
  one-carbon metabolism 700
  oxidative stress 139, 142, 143
  oximes 141

B_6 (*continued*)
   pH 136
   pharmaceutical use 779–80
   protein metabolism 525
   recommended dietary allowance (RDA) 701
   research history 10–12, 136–7
   solubility 136
   stability 96, 136
   status measurement 23, 493
   stroke studies *485–9*
   structures **336**
   supplementation 39
   synthesis 142
   tryptophan conversion 110
B_7 *see* biotin
B_9 *see* folates
B_{12} *see* cobalamins
bacteria
   B_6 derivatives 141
   biotin 720, 721
   cobalamins absorption 459
   flavins (flavoproteins) 627
   infections *554*
   intestines 723
   riboflavin (B_2) 592, 612
Barker, Horace 14
barley 4, 16
basement membrane 586, 607
basolateral membrane 596, **597**, 598, *598*, 607
BBB *see* blood-brain barrier
BCKDHC *see* branched-chain α-keto acid dehydrogenase complex
bead injection **278**, 280, 281, 282
beef *411*
beer
   riboflavin *272*
   sunlight-flavour 271
bees 173
behavioural differences 22
benfotiamine (*S*-benzoylthiamin *O*-monophosphate) 79
benzaldehyde, structure **74**
benzoin, structure **74**
benzothiazoles 74
beriberi 15–16, 71–2, 88, 554
   early discovery of cause 3
   future research 15
   infantile 507–8
   research history 3–4, 14
   Russo-Japanese War 15–16
   symptoms 243
   thiamin levels 233
   thiaminases 77
   urine analysis 229
betaine **25**, 508, 801
betaine-homocysteine methyltransferase (BHMT) **38**, **484**, 738–9, 772
Betz cells 681
beverages, riboflavin *272*
BH_4 *see* tetrahydrobiopterin
bicarbonate 148–9, 447, 451
bifid shunt 65
bile acids 117
bile salts 99, 104, 595, 606
binding assays 435
   biotin 355
   cobalamins 426–9, *431*, 442
   folates 357, 399
biocytin 149, 377, 382, 383, 386, 716, **717**
   digestion 721
bioluminescence 356, 359
"bios" 9
biosensors 387–8, 392, 399, 439–42, 454
   classification **441**
biosynthesis
   adenosine thiamin triphosphate (AThTP) **82**, 84
   biotin 147
   chlorophyll 139
   cholesterol 117
   cobalamins (B_{12}) 166, 420
   coenzyme A (CoA) 685–6
   fatty acids 58, 117
   flavin adenine dinucleotide (FAD) 613
   hemoglobin 139

niacin (nicotinic acid) 634
nicotinamide adenine dinucleotide (NAD) 111–14
nicotinamide adenine dinucleotide phosphate (NADP) 114
purine 77, 491
pyridoxal-5′-phosphate (PLP) 138
steroids 58
thiamin (B_1) 77
thiamin diphosphate (ThDP) 77, 83
thiamin monophosphate (ThMP) 77, 81, **82**, 83
thiamin triphosphate (ThTP) 81, **82**, 84
biotin *see also* carboxylases
  assay methods 198, 355–7, 358–64, 378–91, 720, 725
  biochemical function 353–4, 377, 716–20
  biosynthesis 147
  catabolism 147–8, 153
  chromatograms **384**
  coenzyme function 148–9, 153, 353, 716, 725
  consumption levels 391
  deficiency 354, 378, 719, 722–7
  dietary sources 377, 719–20
  digestion 721
  epilepsy 515–16
  excretion 724
  extraction 378, 379–80, 381, 382, 384
  glucose metabolism **718**, 719
  inborn errors 724
  labelled samples 385–7
  natural forms 377–8, 720
  nomenclature 10
  pharmaceutical use 721–2
  recommended dietary allowance (RDA) *378*, 389, 720
  research history 10, 146
  signalling pathways 146
  silyl esters 382
  structure **10**, **147**, **354**, 377–8, **717**
  supplements 721–2
biotin-4-amidobenzoate 386
biotin-ε-lysine *see* biocytin
biotin sulfone
  catabolism 148
  structure **147**
biotin-sulfoxide, catabolism 148
biotin-sulfoxide, structure **147**
biotinidase 149, 152, 153, 716, **717**, 727
  deficiency 726
  digestion 721
  inborn errors 724
  infants 378
  lysine 719
ε-N-biotinyl-L-lysine *see* biocytin
birth weight 525, 529, **530**, *531*, 532, 534
bis-(trimethylsilyl) acetamide (BSA) 356, 382
bisnorbiotin, structure **147**
bisnorbiotin methyl ketone, catabolism 148
bisnorbiotin sulfone 148
bityrosine 97
black-tongue disease 8
bladder cancer 705
blood *see also* haematopoiesis; plasma; serum
  folic acid 354
  pantothenic acid measurement 303
  riboflavin (B_2) 602
  thiamin measurement 230–4
blood-brain barrier (BBB) 520, 548, 562
  folates 506, **507**, 737
  riboflavin (B_2) 603
  thiamin 227
  thiamin deficiency 545, 548
  Wernicke's encephalopathy 545
bone marrow 484, 691
bone strength 796–7, 800
boronic acid 364, **365**
bowel diseases, riboflavin 103
brain *see also* cholinergic neurons; cognitive decline; epilepsy; neurodegenerative diseases; Wernicke's encephalopathy
  B vitamins function 525

brain (*continued*)
  B$_6$ phosphorolysis 138
  cobalamins deficiency 464–5, 469–70
  energy metabolism **555**, 556–7, 561–2, 566, 614
  glucose 556
  glutamate 575
  glutamatergic neurons 541, 567
  ischemia *see* stroke
  one-carbon metabolism 490, 771–2
  riboflavin (B$_2$) 603–4
  thiamin 80, 240
  thiamin deficiency 72, 539–40, 545–6, 558–9, 572–6
  thiamin diphosphate (ThDP) 81–2, 85
  thiamin triphosphate (ThTP) 84
branched-chain α-keto acid dehydrogenase complex 556
branched-chain α-keto acid dehydrogenase complex (BCKDHC) 539–40
Braunstein, Alexander 11
bread
  folate quantification 161, *410*, 411
  pantothenic acid *303*
  riboflavin *272*
breast cancer 704, 706
4-bromomethylmethoxy-coumarin 383
brush-border membrane 593, 595, 604–5, 606, 721
BSA (bis-(trimethylsilyl) acetamide) 356, 382
Bunge, Gustav von 4–5
burning feet syndrome 302

Ca^{2+}/calmodulin kinase *600*, 602, 603, 605
cadmium selenide, quantum dots 217
cadmium telluride, nanorods 217
caffeine, riboflavin digestion 99
calcineurin 575

calcium
  cytoplasm 563
  homeostasis 543, 572–6, 577
  mobilization 119, 121, 122
  stores 575
calcium binding proteins 575
calcium channels **543**, 549, 562
calcium ion 573–6
calcium pantothenate
  solubility 130
  structure **129**
  supplements 318
calcium transporters 574
calibrators, cobalamins 424–5, 427, 433
calpains 575
Calvin–Benson cycle 59, 65
cAMP *see* cyclic adenosine monophosphate
Canada, food fortification 42–3, 491, 513
cancer 27–8
  B$_6$ 701, 704–7
  biotin 726
  bladder 705
  breast 704, 706
  colon 27, *29*, 398, 691
  colorectal 27, *29*, 355, 704, **705**, 706, 707, 746, 756–7
  diabetes mellitus 587
  DNA methylation 701, 741
  esophageal 707
  folates *735*, 743, 747, 756–7
  folic acid 355, 398, 525, 746, 760–2, 763
  gastric 706
  meta-analysis studies 27, *29*, 46, 704, 706, 760–2, 763, 764
  methylenetetrahydrofolate reductase (MTHFR) 742
  multiple myeloma 704
  niacin deficiency 110
  non-Hodgkin lymphona 704, 705
  ovarian 704
  pancreatic 692, 704, 706
  prostate 706, 756–7

Subject Index 813

pyridoxal-5′-phosphate (PLP) 706
randomized controlled trials
  (RCTs) 27, *29*, 706–7
supplementation risks 46, 517
thiamin 585, 587
capillary electrophoresis
  biotin 382
  cobalamins 198
  folates 399
carbamazepine (CBZ) 153, 508, 509,
  511, 513, 514, 515, 723
carbamyl phosphate 147
carbohydrate metabolism 139
  niacin (nicotinic acid) 636
  thiamin 210, 572, 580, 585–6
carbonate enhancement effect 447,
  450–1
1′-N-carboxybiotinyl 148
carboxyfluorescein (CF) 360, 365
carboxylases 146, **718**
  biotin 377, 716–18, 725
  biotinidase 152, 153
  biotinylation 148, 153, **717**
carcinoid syndrome 111, 634
cardiovascular disease (CVD) 497
  see also cholesterol; statins
  B_6 494, 702–4
  C677T MTHFR 512
  cobalamins (B_{12}) 494, *771*
  deficiency effect 197
  epilepsy 512, 513
  folates 397, 490, *771*
  homocysteine (Hcy) 25, 35, 36–7,
    139, 354, 494, 743, 754–65
  homocysteinuria 754–5
  inflammation 641, **648**
  kidney disease 643, 794, 799
  meta-analysis studies 25, *29*, 43,
    489, 495, 757–64
  niacin 21, 645, 646–8
  pyridoxal-5′-phosphate
    (PLP) 514, 703
  smokers 472
  stroke 494
  supplementation 25, 35–6, *44–5*,
    488, 490–1, 532, 757–64

carnitine 622
carnitine acetyltransferase 562
carnitine palmitoyl transferases
  (CPT I & II) 614, 645
carotenoids, oxidation 97
carotid artery intima-media
  thickness (cIMT) 41, 46, 641,
  647, 667
carriers 594, 595, **597**, *598*, 606, 607
CAS numbers 133
Casal, Gaspar 675
casein 389–91
caspases 544, 637–8
Castle, William 13
CAT *see* acyl-carnitine translocase
catalase 97, 119
catalysts 221
cathepsins 141
caveolin-1 545
CBZ (carbamazepine) 153, 508, 509,
  511, 513, 514, 515, 723
CD *see* cluster of differentiation
CDP-*Star* 443–4
celiac sprue disease 103, 104, 743
cell junctions 607
cell membranes 606–7
  fatty acids 613, 614
  glycosylphosphatidyl inositol
    (GPI) 688–9
  riboflavin 594, 596, **597**, 598, *598*
  thiamin triphosphate (ThTP) 85
β-cells 586
cells
  B vitamins function 482
  signalling pathways 146, 152–3
Central Food Technological Research
  Institute (CFTRI) 447
cercosporin 139
cereals
  folates *406*, 410, 736
  niacin 290–2
  pantothenic acid *303*
  processed foods 587
  riboflavin *272*, 277, 279
  thiamin 252
cerebral folate deficiency 506, 518

cerebrospinal fluid (CSF)
  cobalamins (B_{12}) 777
  folates 777
  5-methyltetrahydrofolate
    (5-MHTF or 5MT) 506, **507**,
    511
  riboflavin (B_2) 603
  thiamin 239
cerebrovascular accident *see* stroke
certified reference materials
  (CRM) 186, 204
CF *see* carboxyfluorescein
CFRTI *see* Central Food
  Technological Research Institute
chaperonin **625**, 626
ChAT *see* choline acetyltransferase
cheese, biotin 391
chemical warfare agents 141, 142
chemiluminescence (CL) 442–52,
  454
children
  biotin requirements *378*
  cognitive function 531–2
  folates 743
  growth 528
  morbidity 528
  motor development 531
  multiple micronutrient (MMN)
    deficiency 524–5, 531
  riboflavin requirements 101, *102*
  supplements 525, 528, 531, 532
chloride ions *601*, 603, 604
chlorophyll, biosynthesis 139
cholesterol *see also* cardiovascular
    disease (CVD)
  biosynthesis 117
  metabolism 149
  niacin 26–7, 638, 639, 641, 660,
    661, 662–3
choline acetyltransferase
  (ChAT) 559–60, **561**, 567
cholinergic neurons **555**, 557–8,
  559–60, **561**, 567
  acetyl-coenzyme A
    (acetyl-CoA) 562, 563
  alcoholism 564

choroid plexus 506, **507**, 511, 603,
  737
Christian, W. 8
chromatin 150–1, 154
  histones 719, 727
  holocarboxylase synthetase
    (HLCS) 151
  ribosylation 637
chromatography 392 *see also* high
  performance liquid
  chromatography
  biotin 381–4
  folates 399
chromatolysis 677, 681
chromosomes 123
chronic atrophic gastritis 197
chronic kidney disease
  (CKD) 786–800 *see also* renal
  disease
CID *see* collision-induced
  dissociation
cIMT *see* carotid artery intima-media
  thickness
cirrhosis 111
citrate **73**, **555**, 566
citrate synthase 568
citric cycle 615
citrullination 719, 728
citrulline 794
citrus fruit 3
CL *see* chemiluminescence
cloud point extraction (CPE) 216
cluster of differentiation (CD) 694
CoA *see* coenzyme A
cobalamin binding proteins 426–7,
  435 *see also* haptocorrin; intrinsic
  factor; transcobalamin
cobalamins (B_{12}) 419 *see also*
  adenosylcobalamin;
  aquacobalamin; cyanocobalamin;
  holotranscobalamin;
  hydroxycobalamin;
  methylcobalamin
  S-adenosylmethionine
    (SAM or AdoMet) **788**, 797
  analogues 421, 434

Subject Index

assay methods 198, 419–20, 421–33, 440–2, 445–52, 777
biosynthesis 166, 420
biotinylation 428
cardiovascular disease (CVD) 25–6, *29*
cobalt-carbon bond 164, 166, 167, 168, 446
coenzyme function 419, 420–1, 798
cofactor behaviour 167–8
cognitive effect 24–5, 28, *29*, 770, *771*, 780
corrin ring 165
deficiency 24–5, 196–7, 420, 435, 453, 459–60, 469–75, 508, 749, 772, 790
dietary reference intake (DRI) 496
dietary sources 166, 459, 496
epilepsy 508, 514–15
extraction 202, 422–4, 429, *431*
folding 165–6
folic acid 167
homocysteine metabolism **25**, **38**, 39, 453, **484**, 771, **788**
intrinsic factor 786
labelled samples 427, 428
metabolism 432, **459**, 508, 787–90
methionine 24
methionine synthase 167
mortality *29*
nervous system 420, 471, 482
one-carbon metabolism 700
pharmaceutical use 778, 779–80
remethylation pathway 801
research history 14, 419
serum 420–33, 434
stability 166
status measurement 23
stroke studies *485–9*
structure 164–6, **203**, **422**
supplementation 39, 778
tetrahydrofolate (THF) 473
cobalt 164, 212, 446, **448**
coeliac disease 103, 104, 743

coenzyme A (CoA) 127, 132
 biosynthesis 685–6
 cofactor function 15, 685
 metabolism 686
 nomenclature 128
 pyruvate dehydrogenase 58
 structure **129**, **304**, 685
coenzymes 89, 196, 350
 necessity 14–15
 research history 6–7
cofactors 65, 168, 196, 350, 392 *see also* enzymes
cognition definition 682
cognitive decline *see also* dementia
 caused by deficiency 24–5, 490, 531–2
 cobalamins (B_{12}) 770, *771*, 780
 folates 28, *29*, 743, 770, *771*, 780
 folic acid supplementation 47
 homocysteine levels 28, 494, 775
 randomized controlled trials (RCTs) 28, *29*, 532, 778, 779
 risk factors **770**
 stroke 481–2
 vascular dementia 482
cognitive function 497
 acetylcholine (ACh) 559–60
 cobalamins (B_{12}) 24–5, 28, *29*
 folates 24–5
 multiple micronutrient (MMN) supplements 531–2
 pellagra 676–7
cohort studies 22, 23, 708
 cancer 27, 704
 cardiovascular disease (CVD) 25–6, 702–3, 756
 coronary heart disease (CHD) 702–3, 756
 pregnancy 529
 stroke (brain ischemia) 702–3
colitis 111
collagen 97, 797
 cross-links 800
collision-induced dissociation (CID) 330
colon cancer 27, *29*, 398, 691

colorectal cancer 27, *29*, 355, 704, **705**, 706, 707, 746, 756–7
colorimetric methods
  biotin 381
  disadvantages 285
  niacin 285
column switching 283, 286
columns 259, *260–1*, 266, 319, 324–5, 327
coma 682
Combe, James 13
computational chemistry 202, 206
confidence intervals 708
congenital disorders *735*
consumption levels
  biotin 391
  folates 397–8
contraception, oral 111, *554*
control samples 432
copper, use in analytical methods 212
corn *see* maize
coronary artery by-passes, folate supplementation 40
Coronary Drug Project (CDP) 665, *666*, 668
coronary heart disease (CHD) 37
  see also atherosclerosis
  B_6 702
  flow mediated dilatation (FMD) 39
  glycosylphosphatidyl inositol (GPI) 689
  homocysteine 139
  homocysteine (Hcy) 755–6, *757*, 764
  meta-analysis studies *29*, 37, 755–6, *757*, 759–60, 762–4
  niacin (nicotinic acid) 647
correlation coefficient R^2 184
corrin family 164
cortical necrosis 77
cortisol 111
cost of analysis
  biotin 361, 365, 368, *379*, 380, 387
  cobalamins (B_{12}) 430, 443
  folates 403

  niacin (nicotinic acid) 286, 295
  pantothenic acid *306*, 310, 326
  thiamin 219, 257, 262, 263
coulometry 199, 200, 203–4, 205, 206
CPE *see* cloud point extraction
CPT *see* carnitine palmitoyl transferases
creatinine 756
CRM *see* certified reference materials
cryopreservation 398
CVD *see* cardiovascular disease
CYaB-thiamin triphosphatase (CYTH) 86
cyanide 472, 787–90, **791**, 800, 801
cyanide sulfurtransferase 787, **790**
cyanocobalamin
  analysis 199, 202, 424, 442
  chronic kidney disease (CKD) 788–90
  dietary sources 166
  medical use 166, 778
  metabolism 787–90
  research history 14
  stability 166
  structure 165, **203**, **483**
  supplements 787
cyanocobalamin-β-(5-aminopentylamide) 442
cyanocobalamin decyanase 787, **788**
cyanocobalamin tracking chaperone 800
cyanogen bromide 212
cyclic adenosine monophosphate (cAMP) 636, 638–9, 663
cyclic guanosine monophosphate (cGMP) 642, 719
cyclin-dependent kinases *600*
cystathionine **25**, **38**, **484**, **700**, 739
cystathionine β-synthase (CBS) **38**, 470, **484**, 701, 739, 772, 801 *see also* homocysteinuria
  deficiency 754, 755
cysteamine 686, 690, 691–2, 693, 694
  Huntington's disease 691
  malaria 691

cysteine
  B$_6$ reaction 136, 139
  cleavage 694
  coenzyme A synthesis 685
  flavins (flavoproteins) 627
  homocysteine metabolism **484**, 739, 801
  metabolism **25**, **38**
  riboflavin (B$_2$) 99
  thiamin biosynthesis 77
cystinosis 691, 693, 694
CYTH (CYaB-thiamin triphosphatase) 86
cytochrome oxidase 787
cytochrome P450 inducers 509
cytonisin 691
cytoplasm 166, 168–9
  acetyl-CoA carboxylase 1 148–9
  folates 740
  riboflavin (B$_2$) **612**
cytosine, methylation 24, 748–9
cytosol
  fatty acids 614
  thiamin diphosphate (ThDP) 80, 81, **82**
  thiamin triphosphate (ThTP) 84

DAD *see* diode array detector
dairy products *see also* milk
  biotin 377, 380, 386, 389–91
  pantothenic acid *303*
  riboflavin *272*, *276*
dalton (Da) 297
Day, Paul L. 12
DDAH *see* dimethylarginine dimethylaminohydrolase
de novo synthesis 123, 143
deep vein thrombosis 37, *45*
deficiency
  B$_2$ 138
  B$_6$ 196, 701
  biotin 354, 378, 719, 722–7
  biotinidase 726
  chronic atrophic gastritis 197

cobalamins (B$_{12}$) 24–5, 196–7, 420, 435, 453, 459–60, 469–75, 749, 772, 790–1
cystathionine β-synthase (CBS) 754, 755
epilepsy 505, 506, 508, 509–11
folates 197, 453, 484, 506, 509–11, 735–6, 740–1, 742–4, 749
multiple micronutrient (MMN) 524–5
niacin (nicotinic acid) 110, 634 *see also* pellagra
research history 5
riboflavin 103, 138, *601*
screening 565
thiamin 71–2, 88, 228, 539–47, 558–9
thiamin diphosphate (ThDP) 554–7, *558*, 560–4, 565, 572–6
dehydrongenases 350
deimination 719, 728
delirium 682
delusion 677, 682
delusional parasitosis 677, 678, 680, 682
dementia 28, 682 *see also* Alzheimer's disease
  alcoholism 546
  cobalamins (B$_{12}$) 769–81
  definition 769–70
  folates 769–81
  hyperhomocysteinemia (HHcy) 772–3
  niacin deficiency 676–7
  niacin (nicotinic acid) 634
  pellagra 676–7, 682
  risk factors **770**, 772–3, 777–8
  stroke relationship 481
  subtypes **774**
  thiamin deficiency 566
  vascular 481–2
density functional theory (DFT) 169, 202, 206
5′-deoxy adenosyl-cobalamin 421

deoxyuridine monophosphate
 (dUMP) 740, 741–2
depression
 cobalamins (B_{12}) *771*
 folates 511, *735*, *771*
 niacin 109, 680
 pellagra 680
 supplementation 492, 511
dermatological disorders 723, 724, 726
δ-6-desaturase 139
desorption 221
dethiobiotin 147
developing countries
 biotin 723
 cobalamins 468
 folates 736
 multiple micronutrient (MMN)
  deficiency 524–5, 526
 niacin deficiency 675–6
 thiamin deficiency *554*
DGAT (diacylglycerol
 acyltransferase) 640, 662
diabetes mellitus *see also* glycation
 end-products; insulin sensitivity
 biotin 721–2
 cancer 587
 cardiomyopathy 581
 complications 586–7, 589
 kidneys 643, 799
 lipoic acid 153, 723
 niacin supplementation 27
 obesity 580–9
 pyruvate dehydrogenase
  (PDH) 581, 590
 riboflavin 103
 statins 643, 646
 supplementation trials *44*
 thiamin 79, 580–9
 VNN1 gene 692
diacylglycerol acyltransferase
 (DGAT) 640, 662
diallyl thiosulphinate *see* allicin
dialysis
 cobalamins (B_{12}) 788, 791
 hyperhomocysteinemia
  (HHcy) 793

niacin (nicotinic acid) 111, *644*
thiamin excretion *554*, 565
diarrhoea
 niacin (nicotinic acid) 111, 634
 pellagra 676, 679
 riboflavin absorption 103
 thiamin diphosphate (ThDP) *554*
 tropical sprue 105
diastase 200, 205
Diels–Alder reaction 137, 142, 143
diencephalon 548
dienes 143
2,4-dienoyl CoA reductase 1 615
dietary intake levels
 B_6 702–3, 704
 folates *486*
 lifetime consumption 26
dietary reference intake (DRI) *see also*
 recommended dietary allowance
 B_6 496
 cobalamins (B_{12}) 496
 folates 496
dietary sources 196
 adenosylcobalamin (AdoCbl) 166
 B_6 136, 496
 biotin 377, 719–20
 cobalamins (B_{12}) 166, 459, 496
 cyanocobalamin 166
 flavin adenine dinucleotide
  (FAD) 99
 flavin mononucleotide (FMN) 99
 folates 160, 475, 491, 496, 734
 hydroxycobalamin 166
 methylcobalamin (MeCbl) 166
 niacin (nicotinic acid) 110, 122,
  633–4
 pantothenic acid (B_5) 317
 riboflavin (B_2) 97, *98*, 598
 suphitocobalamin 166
 thiamin (B_1) 227, 252
 thiamin monophosphate
  (ThMP) 252
 thiamin pyrophosphate (TPP) 252
diets
 Mediterranean 47, 472
 modern 587

diffusion 594, 596, 605, 606
digestion 103–4 *see also* intestines
 biocytin 721
 biotin 721
 biotinidase 721
 flavin adenine dinucleotide (FAD) 592, **593**
 flavin mononucleotide (FMN) 592, **593**
 flavins (flavoproteins) 612
 folates 736–7
 niacin (nicotinic acid) 634
 riboflavin *601*, **612**
 riboflavin (B_2) 99, 104, **593**, 595, **597**
 thiamin (B_1) 80
7,8-dihydrofolate *159*
dihydrofolate (DHF) **739**, 740
2,4-dihydroxy-3,3-dimethylbutyric acid *see* pantoic acid
dihydroxyacetone phosphate 138
2,4-dimethyl-5-methoxyoxazole 137
dimethylamine (DMA) 794, **795**
p-dimethylaminocinnamaldehyde 381
dimethylarginine dimethylaminohydrolase (DDAH) 793, 794–5
dimethylglycine **25**
diode array detector (DAD) **278**, *279*, 328
 folates 357
 pantothenic acid 320, 323
 riboflavin *276*
disease
 effect on metabolism 22
 mechanisms of B vitamins 24–5
dislipidemias, niacin 109
diuretics 229, 243
 thiamin excretion *554*, 565
diurnal variability, holotranscobalamin (holoTC) 461
DNA methylation 748–9
 B_6 139, **700**, 701
 biotin 719
 cobalamins (B_{12}) 776
 folates 24, 197, 525, 741, 776

DNA methyltransferases (DNMTs) 741
DNA repair 123
 adenosine diphosphate (ADP)-ribose transfer reactions 119
 cancer 701
 niacin (nicotinic acid) 637
DNA synthesis *see also* one-carbon metabolism
 B_6 139, **700**
 cancer 46
 deoxyuridine monophosphate (dUMP) 741–2
 folic acid **38**, 167, 525
 nicotinamide adenine dinucleotide phosphate (NADP) 117
 tetrahydrofolate (THF) 491
DNA transcription 719
Donath, William 5
dopamine 482
DRI *see* dietary reference intake
Drummon, Jack 5
du Vigneaud, V. 10
Dumas, Jean 4
Dutch East Indies 4, 5
dyslipidemia 21, 643, *644*, 645, **648**

E-selectin 642
ECL *see* electrochemiluminescence
ectoenzymes 123
education level, stroke risk 481
eggs 6, 10 *see also* avidin
 albumen consumption 378, 723, 724
 biotin 720, 723, 724
Eijkman, Christiaan 4, 14
elderly subjects *see also* cognitive decline
 B_6 701
 brain function 490
 cobalamins deficiency 460, 464–5, 466, **468**, 469–70, 471, 472, 777
 dementia 773–5, 776
 folates 777
 homocysteine levels 777
 riboflavin intake 102

elderly subjects (*continued*)
  stroke risk 481
  thiamin diphosphate (ThDP) *554*, 565
  thiamin levels 230, 233, 234, 240, 254
electrochemical detection (ED) 205, 206, 264, 266
  B_6 199, 200
  biotin 355, 360–4
  cobalamins 198, 199, 200, 442
  folates 357–8, 399
  high performance liquid chromatography (HPLC) 198, 199, 205, 262
  simultaneous analysis 262
electrochemical immunosensors 361–2, 364
electrochemiluminescence 443, 445
electron-transfer flavoprotein (ETF) 614, **615**, 619–20, 622, 623–4
electron-transfer flavoprotein:ubiquinone oxidoreductase (ETF:QO) 614, **615**, 620–1, 622
electron transport chain 589, 635
  see also adenosine triphosphates (ATP)
electrospray ionization (ESI)
  niacin (nicotinic acid) 289, *291*, 297
  pantothenic acid 320, 325, 328
  supplement analysis 384–5
ELISA *see* enzyme-linked immunosorbent assay
Ellinger, P. 6
ELS *see* evaporative light-scattering (ELS) detection
eluent 282, 283
Elvehjem, C.A. 8, 9
embriogenesis 534
emission 221
encephalomyelopathy 88

encephalopathy 79, 88, 89, 553, 557–60, **561** *see also* Wernicke's encephalopathy
  alcoholism 564–5, 676
  pellagra 676
  screening 565
endocytosis 594, 596, **597**, 605, 606, 607
endonucleases 575
endoplasmic reticulum (ER) **120**, 121
  apolipoproteins 639, 662
  biotin catabolism 148
  calcium ions 574
  niacin (nicotinic acid) 662
  stress 573, 576
  thiamin metabolism 85
endothelial cells 36, 37
  atherosclerosis 642, **661**
  dementia 776
  diabetes mellitus 586
  function 39–40
  homocysteine (Hcy) 701, 776, 793
  niacin (nicotinic acid) **661**
endothelial nitric oxide synthase (eNOS) **38**, 48
  homocysteinaemia 36
  5-methyltetrahydrofolate 40
  phosphorylation 36
  thiamin deficiency 544, 545, 547
  vascular walls **41**
endothelium 607, 642, 799
energy metabolism 196, 392, 525
  see also carbohydrate metabolism
  brain **555**, 556–7, 561–2, 566, 614
  fatty acids 613–21
  nicotinamide adenine dinucleotide (NAD) 115–17, 635
  thiamin deficiency 540–1, **542**, 546, **555**
  thiamin diphosphate (ThDP) 72, **73**, **555**
enhanced chemiluminescence *see* electrochemiluminescence
eNOS *see* endothelial nitric oxide synthase
enoyl-CoA hydratase 614–15

enteritis, infectious 103, 104
enterocytes 104, 606, 721
enterohepatic circulation 104
enzyme-linked immunosorbent assay (ELISA) 313
  cobalamins 426, 429
  folates *412*
  holotranscobalamin (holoTC) 462–3
  pantothenic acid 307
enzyme protein binding assay (EPBA) 386–7
enzymes 221, 350, 392 *see also* cofactors
  apo- 76, 89, 168, 392, 623
  holo- 89, 623
  hydrolysis **597**
  olo- 168
EPBA *see* enzyme protein binding assay
epidemiology
  definition 497
  homocysteine 492
epigenetics 154, 748–9
  folates 741, 745, 747
  histones 150
epilepsy
  B_6 513–14
  biotin 515–16
  cobalamins (B_{12}) 508, 514–15
  folates 506, 509–13
  pharmaceuticals 508–18, 519
  riboflavin (B_2) 515
  supplements 516–17
  symptoms 504, 518–19
  thiamin (B_1) 507–8, 516
  vitamin B disorders 504–8
epithelial cells
  folic acid 354, 506, **507**
  membranes 606–7
  retina 605
  VNN1 gene 692
equilibration 287, 295, 401
ER *see* endoplasmic reticulum
erythrocyte transketolase (ETK)
  activation assay 230–3, 241, 243, 254

erythrocytes 243
  folates 511
  mean corpuscular volume (MCV) 511
  riboflavin (B_2) 602
  transaminase coefficient 23
erythropoiesis 526
esophageal cancer 707
estroprogestinic therapy 466, **468**, 472, 475
ethanol *see* alcohol
ethnic variations, cobalamins 464, 465, 471
ETK *see* erythrocyte transketolase (ETK) activation assay
ETK AC *see* activity coefficient
eukaryotic cells 719
Euler-Chelpin, Hans von 8
European Prospective Investigation into Cancer and Nutrition (EPIC study) 706
evaporative light-scattering (ELS) detection 320, 328
evidence ranking level 707, *708*
excitation 221
excitatory amino acid transporters 543
excitotoxicity 548, 567, 574, 575, 577
  calcium homeostasis 543
  thiamin deficiency **540**, 542–4, 559
excretion
  biotin 724
  folates 738
  riboflavin (B_2) 100
  thiamin (B_1) 77, 229, *554*, 565
exercise 121, *554*, 587, 635
extraction 221 *see also* microwave-assisted extraction; solid phase extraction
  B_6 201–2
  biotin 378, 379–80, 381, 382, 384
  cobalamins 202, 422–4, 429, *431*
  effect on results 200–1
  flavins (flavoproteins) 279
  fluorimetry 258
  folates 159–60

extraction (*continued*)
  internal standards 401–3
  microbiological assay 256
  niacin 285–6, 296
  pantothenic acid 303–4, 322
  riboflavin 273–80, 281
  thiamin 215–16, 255–6, 258, *260*
  trienzyme method 358
eyes 605

facilitated diffusion 594
FAD *see* flavin adenine dinucleotide
fasting, sirtuins 121
fatty acids *see also* triglycerides
  biosynthesis 58, 117
  biotin 722, 725, 726
  cell membranes 613, 614
  disorders 621–6
  inborn errors 621–2
  isomerization 97
  metabolism 139, 149, 483
  non-esterified (NEFA) 670
  oxidation 149, 613–26, 627, 628, 635–6
  oxidative stress 722, 725
  transport 614
FAVORIT trial *45, 759, 760, 761–2*
FBP *see* folate binding protein
ferredoxins 65
FID *see* flame ionization detection
FILIA *see* flow injection liposome immunoanalysis
fish
  pantothenic acid *303*
  riboflavin *272*
flame ionization detection (FID) 382
flavin adenine dinucleotide (FAD)
  analysis 273, 279
  apo proteins 613
  biosynthesis 613
  blood 602
  cofactor behaviour 615–16
  dietary sources 99
  digestion 592, **593**
  electron-transfer flavoprotein (ETF) 619–20

electron-transfer flavoprotein:ubiquinone oxidoreductase (ETF:QO) 620–1
  epilepsy 515
  flavins (flavoproteins) 627
  function 101
  MCAD deficiency 622–3, 625
  protein insertion **625**
  pyridoxine analysis 341
  pyruvate dehydrogenase 58
  redox behaviour 611
  research history 7
  riboflavin metabolism 100
  structure **7, 94**, 613
flavin adenine dinucleotide (FAD) phosphatase 595, 612
flavin adenine dinucleotide (FAD) pyrophosphatase 99
flavin adenine dinucleotide (FAD) synthetase 100, 595, 603, 612
flavin dependent oxidase 138
flavin mononucleotide (FMN)
  analysis 273, 279
  apo proteins 613
  blood 602
  dietary sources 99
  digestion 592, **593**
  epilepsy 515
  flavins (flavoproteins) 627
  function 101
  MCAD deficiency 622
  phosphorolysis 603
  redox behaviour 611
  research history 6–7
  riboflavin metabolism 100
  structure **7, 94**, 613
flavin mononucleotide (FMN) phosphatase 99, 595, 612
flavins (flavoproteins) 100 *see also* flavin adenine dinucleotide (FAD); flavin mononucleotide (FMN); pyruvate oxidase; riboflavin
  cobalamins metabolism **788**
  cofactor function 613, 623–5, 627
  digestion 612

distribution 94
extraction 279
forms 95
redox behaviour 94–5, 613, 627
riboflavin (B$_2$) 627–8
spectroscopic values 627
structure 613
flavocoenzymes 100 see also flavin adenine dinucleotide (FAD); flavin mononucleotide (FMN)
flavokinase 100, 595, 603, 612
flavoproteins see flavins
flour
  folate fortification 42–3, *410*
  niacin *292*
  pantothenic acid *303*
flow injection analysis (FIA) 217–18, 221, 282, 370
  cobalamins 446
  on-line 275–80
  off-line 274–5
  riboflavin 274–80, 281
  spectrofluorimetry 212
flow injection liposome immunoanalysis (FILIA) 358–9, **360**, 370
flow injection–solvent extraction (FI-SE) 218–19
flow mediated dilatation (FMD)
  folic acid supplementation 39
  homocysteinaemia 37
fluorescein isothiocyanate 383
fluorescence 104, 221, 266, 350, 454
  resonance Rayleigh scattering (RRS) 217
  sensitized 221
  synchronous 221
fluorimetry 220, 328
  biotin 360–1
  blood samples 233–4
  folates 358, 399, 409, 410–11
  honey samples 181–2
  matrix effects 273
  niacin 286
  pantothenic acid 308–10
  riboflavin 103, *274*, *277*

supplement analysis 320
thiamin 211–19, *257*, *258*, 265
flush response 677
flushing 647, 649, 663, 664
FMD see flow mediated dilatation
FMN see flavin mononucleotide
foetal development
  biotin 722
  cobalamins 468
  folates 744–5
  folic acid 21, *29*, 30, 484, 513, 735
  riboflavin (B$_2$) 100, 604
  thiamin diphosphate (ThDP) *554*
folate binding protein (FBP) 357, 399
folate conjugase 158, 160, 358
folate monoglutamate 158, 160
folate polyglutamates 160
folate receptors 506, **507**, 737, *738*, **739**
folate trap 473
folates see also folic acid
  adenosine 24
  analysis 160–2
  assay methods 357–8, 364–7, 398–412, 777
  binding activity **369**
  biochemical function 740–5
  biological function 24, 396, 475
  blood-brain barrier 506, **507**
  cognitive effect 24–5, 770
  deconjugation 405, 413
  deficiency 24, 197, 453, 484, 506, 509–11, 735–6, 740–1, 742–4, 749
  dementia 28
  dietary intake levels *486*
  dietary reference intake (DRI) 496
  dietary sources 160, 475, 491, 496, 734
  digestion 736–7
  epilepsy 506, 509–13, 517
  excretion 738
  extraction 159–60
  glycation end-products 405
  guanosine 24
  homocysteine metabolism 47, 512–13, 735, 740–1, 771

folates (*continued*)
  inborn errors 744
  isotopologues 403, **404**, 405–6
  labelled samples 357
  metabolism 738–40, 745–6, 747
  methionine 24
  monoglutamates 403, 405
  natural forms 475
  nomenclature 158, 734
  one-carbon metabolism 700, 740
  pharmaceutical use 779–80
  plasma levels 161
  polyglutamates 403, 405
  recommended dietary allowance (RDA) *736*
  research history 12, 158, 734, 780
  riboflavin function 101
  serotonin 511
  serum levels 161
  stroke studies *485–9*, 490–1
  structure **397**
  supplementation 779
  thymidine 24
  transport 737–8
folic acid *see also* folates
  assay methods 198
  binding activity **369**
  biochemical function 354
  brain function 469–70
  cancer 27, *29*
  cardiovascular disease (CVD) 25–6, *29*, 354, 490
  cobalamins 167
  coenzyme function 196, 482
  cognitive decline 28, *29*
  dietary sources 160
  foetal development 21
  homocysteine metabolism **38**, 47, **484**
  honey samples *178*, 179–80
  isotopologues 403, 405–6
  light sensitivity 160
  medical use *792*, 795, *796*, *797*
  methionine synthase 167
  molecular weight 160
  mortality *29*
  nomenclature 158, 160, 162
  recommended dietary allowance (RDA) 745
  research history 12, 158, 734
  solubility 160
  stability 46, 96, 160
  structure **13**, 26, 158–9, 160, **354, 367, 397, 482**
  supplementation **38**, 469–70, 517, 744, 745, 746, 779
  synthetic form 475
  vascular walls **41**, 42
folinic acid responsive seizures 506
Folkers, K. 10, 14
folylpoly-γ-glutamate carboxypeptidase (FGCP) 737
food adulteration, honey samples 174
food analysis *see also* matrix effects
  B$_6$ 342–8, 349
  biotin 356, 378–91
  cobalamins 445, *452*
  folates 161, 358, 366, 405, *406*, 410–12
  honey 174–83, 188
  method comparison *379*
  niacin 285–97
  pantothenic acid *303*, 304, 307, 308, 310, 311–12
  riboflavin 271–82
  simultaneous 197–205
  thiamin 254–65
food fortification
  analysis 199, 255, 405, *406*
  B$_6$ 136
  folates 405, *406*, 410, 491, 493, 734, 736, 745
  folic acid 37–42, 42–3, 46–7, 161, 354–5, 397–8, 746
  multiple micronutrient (MMN) 532
  niacin 110, 286, 290–2
  pantothenic acid 318
  safety 46–7, 48
  shelf life 410
  thiamin 252–3, 564
food irradiation 96, 104

food preservatives 75, *554*
formic acid 322
5-formiminotetrahydrofolate *159*, **397**
N-formylkynurenine 111, **112**
5-formyltetrahydrofolate *159*, **397**, 399, 403
10-formyltetrahydrofolate *159*, **397**, 399, 740
free radicals 143
fruit juice
  analysis 199–200, *201*, 262, 264
  riboflavin *272*
fungi
  cercosporin 139
  pyridoxal-5′-phosphate (PLP) synthesis 138
  pyridoxine (PN) 139
Funk, Casimir 5, 195
furosemide 565
fursultiamine (thiamin tetrahydrofurfuryl disulfide) 79

γ-glutamylcysteine synthase 691
G-protein coupled receptors (GPR) 636, 638, 643, **661**, 662, 663, 670
GABA (γ-aminobutyric acid) 505
GABA-aminotransferase 505
gamma-glutamyl transferase 567
garlic 79
gas chromatography-mass spectrometry (GC-MS)
  biotin 356, 382
  folates 357, 405
  pantothenic acid 310
gastrectomy 459
gastric cancer 706
gastric inhibitory polypeptide (GIP) 586
gastric intrinsic factor *see* intrinsic factor
gastrointestinal tract, folate deficiency 484
GC-MS *see* gas chromatography-mass spectrometry

gender differences
  B_6 701
  cobalamins 461
gene-deficient mice 694–5
gene expression *see also* epigenetics
  B_6 139
  biotin 152–3, 718–20, 725
  folate deficiency 24, 509, 745
  homozygous defects 490
  liver 509, 585–6
  nutrition 747, 749
  tetrahydrofolate (THF) 491
  thiamin 585–6
  transgenerational effect 745
genomic instability 749
gingival hyperplasia 516
ginkgotoxin (4′-*O*-methylpyridoxine) 140
Ginko biloba 140
GIP *see* gastric inhibitory polypeptide
glomerular filtration rate (GFR) 465–6, 643
GLS *see* β-glucosidase
glucokinase 719
glucokinase gene 152–3
gluconeogenesis 149
  biotin 354, **718**
glucose
  biotin **718**, 719
  brain 556
  energy metabolism **73**, 541, 548, 549, 581, 589–90
  gastric inhibitory polypeptide 586
  glycolytic cycle 566, **718**
  level affects sirtuin 121
  liver 586
  metabolic syndrome 580
  mitochondria 566
  niacin (nicotinic acid) 636, 645
  oxidation 72
  synthesis 564
  thiamin deficiency 541, 548
  tolerance test 645
glucose-dependent insulinotropic polypeptide *see* gastric inhibitory polypeptide

glucose transporters 560, 586
β-glucosidase (GLS) 338, 339, 341, 347, 350
glutamate 117, 575, 577
    niacin deficiency 680
    thiamin deficiency 542–3, 558–9, 563
glutamate 1 (GLT-1) transporter 563
δ-glutamate-1-semialdehyde aminomutase 139
glutamate aspartate (GLAST) transporter 563
glutamate receptors (GluR) 574–6, 577 see also AMPARs; kainate receptors; N-methyl-D-aspartate receptors
glutamate transporters 543, 547
glutamatergic neurons 541, 567
glutamic acid, structure **159**
glutamine 138, 606
γ-glutamyl hydrolase see folate conjugase
γ-glutamyl transferase 564
glutaric aciduria type-II see MCAD deficiency
glutathione 119, 556, 642, 663, 691, 739, 776
glutathione reductase 119
glutathionine 701
gluten 104
glycation end-products (AGEs) 336, 350, 797
glyceraldehyde 3-phosphate **73**, 77, 138
glycinamide ribonucleotide transformylase (GART) **739**, 740
glycine **25**, 77
glycogen 139
glycogen phosphorylase 139
glycolysis 560, 566, 589, **718**
    bifid shunt 65
    glucokinase 719
    nicotinamide adenine dinucleotide (NAD) 635
    pyruvate dehydrogenase 58, 64, 568

glycosides 143
glycosylphosphatidyl inositol (GPI) 119, 688–9, 693, 737
goats milk 743
gold nanoparticles 361–2, **363**, 364
Goldberger, Joseph 8
Golgi apparatus
    NMNAT 114
    thiamin metabolism 85
gonad 688
GPI see glycosylphosphatidyl inositol
GPI-anchor 695
Grijns, Gerrit 4
guanidinonitrogens 794
guanosine, folates 24
guanosine triphosphate (GTP) 719
guanylate cyclase (GC) 719
Gunsalus, Irwin 11
György, Paul 6, 10

HABA see p-hydroxy-azobenzene-2′-carboxylic acid
haematopoiesis 13–14, 526, 534
haemoglobin 526, 532
haptocorrin (HC) 434
    genetic absence 459–60
    human milk 428–9, 433
    metabolism **459**
    protein binding assays 426, 428–9
    serum 420, 421, 465
    surface plasmon resonance 429
Harris, S.A. 10
Hartnup disease 111, 634
Hayaishi, O. 8
HC see haptocorrin
Hcy see homocysteine
HDL see high-density lipoproteins
headaches, niacin 109
heart, thiamin deficiency 228
heart disease see coronary heart disease
heat treatment
    B_6 136
    pantothenic acid 130
    thiamin 75
heme oxygenase-1 (HO-1) 544

Subject Index

hemodialysis 43
hemoglobin, biosynthesis 139
Henderson, L.M. 8
hepatic steatosis 589
hepatocytes *see* liver
hepatomegaly 509
hepatotoxicity 664–5
heterogeneity 708
heterolysis 169
HHcy *see* hyperhomocysteinemia
high-density lipoproteins
  (HDL) 26–7, 636, 640–1, 647,
  662–3, 664, 670
high-performance affinity
  chromatography 384
high performance liquid
  chromatography (HPLC) 206,
  241–2, 313, 328–9
  advantages *379*
  B_6 337, 339–47
  biotin 356, *379*, 382–3
  blood samples 233–4
  cobalamins 430
  diode array detector (DAD) **278**,
    279, 320
  disadvantages *379*
  electrochemical detection
    (ED) 198, 199, 205, 262
  folates 161, 357–8, 399
  inductively coupled plasma-mass
    spectrometry 198–9
  mass spectrometry (MS) 199, 399
  pantothenic acid 307–10, **309**,
    319–20
  reversed phase *see* reversed-phase
    high performance liquid
    chromatography
  riboflavin *274, 276, 277,* **278**, *279*
  simultaneous analysis 262, 325–6
  thiamin 79–80, 233–4, 259, *260–1*,
    265
histidine 99, 627
histones 727, 749
  biotinylation 150–1, 152, 153,
    719–20
  structure 154

HLCS *see* holocarboxylase synthetase
Hodgkin, Dorothy 14
Hogan, Albert 12
Holiday, E.R. 6
holo-proteins 434, 626
holocarboxylase synthetase (HLCS)
  biotin 148, 153, **717**, 719
  chromatin 151
  domains 152
  fatal deficiency 151
  glucose metabolism 719
  histone biotinylation 150–1, 152
  inborn errors 724
  lysine 719
  regulation 151–2
  single nucleotide polymorphisms
    (SNP) 151
holocarboxylases 148–9, **717**
holoenzymes 89, 623
holotranscobalamin (holoTC) 435,
  475
  anaemia 777
  analysis 460–9
  cobalamin deficiency
    correlation 432, 433, 463–4,
    471–2, 777
  diurnal variability 461
  homocysteine metabolism 466,
    467
  transport function 429, 458–9
homeostasis 603, 607
Homeostasis Model Assessment
  (HOMA) index 645, 646
homocysteinaemia 48, 139, 167
  atherosclerosis 36–7
  factors 472
  supplementation 39, 47
  vascular wall inflammation 39
homocysteine (Hcy) 472, 476,
  497, 534
  asymmetric dimethylarginine
    (ADMA) 794–5
  atherogenesis 701
  atherosclerosis 41–2, 48, 139, 482,
    493, 494
  B_6 493, **700**, 701, 797

homocysteine (Hcy) (*continued*)
  cardiovascular disease (CVD) 25, 35, 36–7, 139, 354, 494, 743, 754–65
  chronic kidney disease (CKD) 793, *798*
  cobalamins 23, 419, 453, 473, 771, 772, **788**
  cognitive effect 24–5
  dementia 28, 776, 779
  factors affecting levels 472, 492, 494
  folates 47, 512–13, 735, 771, 772
  holotranscobalamin (holoTC) 466, *467*
  methylation 24, 396, 779
  5-methyltetrahydrofolate 482
  pregnancy 468, *469*
  remethylation 740–1, 791, *792*, 793, 795, *798*, 801
  renal disease 471
  stroke 139, 482–95, 497, 755, 760, 776
  studies *485–9*, 490, 491
  supplementation 35, 41, 532, 755
  toxicity 793, 800
  transmethylation 794, 801
  transsulfuration pathway 801
homocysteine metabolism 25, **38**, **41**, **484**, **512**
  S-adenosylhomocysteine (SAH or AdoHcy) **38**, 738, **739**
  S-adenosylmethionine (SAM or AdoMet) **38**, 169, **484**, 738, **739**, 741, 771, 801
  cobalamins (B$_{12}$) 771, 772
  folates 47, 512–13, 735, 771, 772
  methionine synthase 167, 772
  pyridoxal-5′-phosphate (PLP) 514
homocysteinuria 49, 765 *see also* cystathionine β-synthase
  atherosclerosis 35, 754–5
  cardiovascular disease (CVD) 754–5
  inborn errors 508
  supplements 764
  vascular thrombosis 37, 47

homolysis 169
homozygous defects 490
honey 173–83, 187–8
HOPE-2 trial 43, *44*, 46, 491, 703, *759*, *760*, *761–2*
Hopkins, Sir Frederick G. 5, 8
hormone replacement therapy 466, **468**, 472, 475
hormones 117
horseradish peroxidase (HRP) 213–15, 355, 383, 443, 444–5
HOST trial *45*, *759*, *760*, *761–2*
HRP *see* horseradish peroxidase
human milk *see also* lactation
  biotin 386, 389
  cobalamins 428–9, 433, 508
  thiamin 508
Huntington's disease 691, 694
hydrophilic interaction liquid chromatography (HILIC) 289
p-hydroxy-azobenzene-2′-carboxylic acid (HABA) 355, 381
3-L-hydroxyacyl-CoA dehydrogenase 614–15
2-hydroxyacyl-CoA ligase 80
3-hydroxyanthranilate dioxygenase (HAAO) **112**, 113
3-hydroxyanthranilic acid 8
3-hydroxyanthranylate **112**, 113
β-hydroxybutyrate 662, 670
hydroxycobalamin
  dietary sources 166
  epilepsy 508
  pharmaceutical use 778
  structure 165, **203**
hydroxyethyl thiamin diphosphate 72
10-hydroxyethylflavin 100, **101**
hydroxyethylthiamin 72
3-hydroxyisovaleric acid **149**, 153, 725
3-hydroxyisovaleric acid-carnitine **149**, 725
3-L-hydroxykynurenine **112**, 113
hydroxymethyl pyrimidine phosphate 77

hydroxymethylriboflavin 100, **101**
hydroxypyrimidine 77
3-hydroxyvaleric acid 726
hyperglycaemia 665, 721–2
hyperhomocysteinemia (HHcy) 139, 465, 472–3, 475, 520, 740
  biological effects 793–6
  chronic kidney disease (CKD) 791, 793–6
  dementia 772–3
  elderly subjects 770, 776
  food fortification 491
  methylenetetrahydrofolate reductase (MTHFR) 746
  pregnancy 744
  vascular diseases 701
hyperlipidemia 197, 722
hyperphosphatemia 643
hypertension 197, 799
hypertriglyceridemia 722
hyperuricemia 564
hypoglycemia 520, 564
hypothyroidism 100, 104
hypovolemia 554, 565

IAC *see* immunoaffinity chromatography
Ichiba, A. 10
ICP-MS (inductively coupled plasma-mass spectrometry) 198
imine 143 *see also* Schiff's bases
immune system
  biotin 719
  nicotinamide adenine dinucleotide phosphate (NADP) 119, 635
  riboflavin (B_2) 602
  thiamin 572
immunoaffinity chromatography (IAC) 359–61, 364–6, **368**, 370
immunoassays
  advantages 385–6
  biotin 359, 362, 385–6
  pantothenic acid 311–12
immunoglobulins 99, **597**, 602, 606

immunosensors 369–70
  biotin 358–64
  folic acid 364–7
Imperial Japanese Army 3–4, 15–16
Imperial Russian Army 15–16
imprecision of analysis 433
inborn errors
  biotin 724
  cyanide metabolism 787, **789**
  fatty acids 621–2
  folates 744
  homocysteinuria 508
incretins 586
indolamine-pyrrole 2-3 dioxygenase 111
indoleamine dioxygenase (INDO) **112**
inducible nitric oxidase synthase (iNOS)
  homocysteinaemia 36
  thiamin deficiency 544
inductively coupled plasma-mass spectrometry (ICP-MS) 198
infant formula
  B_6 deficiency 506
  biotin 389, 391, 723
  epilepsy 506
  folates *406*, *410*
  hypoallergenic 723
  simultaneous analysis 442
  thiamin 262, 507
infants
  beriberi 507–8
  biotin *378*, 391, *720*, *723 see also* lactation
  biotinidase 378
  birth weight 525, 529, **530**, *531*, 532, 534
  cobalamins 469, 508
  epilepsy 504–6, 507–8
  folic acid 30
  thiamin *554*
infection stroke risk 493
infections, thiamin deficiency *554*
infectious enteritis 103, 104

inflammation 548
　atherosclerosis 641–3
　cardiovascular disease (CVD) 641
　cysteamine 691–2, 693
　glycosylphosphatidyl inositol
　　(GPI) 689
　homocysteinaemia 39
　metallothionein 690
　niacin (nicotinic acid) 641–3,
　　648, 663
　pantetheinase 690–1, 692
　stroke risk 493, 494
　thiamin deficiency **540**, 544–5
　VNN1 gene 692
　VNN2 gene 689, 692
　VNN3 gene 690, 692
　Wernicke's encephalopathy 544–5
iNOS see inducible nitric oxidase
　synthase
inositol 9
insecticides 141, 142
insulin sensitivity 645–6, **648**, 649,
　665 see also diabetes mellitus
interferon-y (IFN-y) 642
interleukin-1 (IL-1) 642
internal standards
　isotopologues 400, 401, 403, 413
　labelled samples 286, 310, 358, 400
intervention trials *487–8*, 494,
　498, 778
intestines 104, **597**
　bacteria 592, 598, 612, 720, 721,
　　723, 736–7
　biocytin 377
　biotin 389, 720, 721, 723
　cobalamins 420, 470, 484, 786
　folates 736–7
　ketone bodies 614
　malabsorption 459, 470
　niacin (nicotinic acid) 634
　riboflavin (B_2) 99, 100, 592, **593**,
　　595–602, 612
intracellular adhesion molecule-1
　(ICAM-1) 544, 642
intrinsic factor 14, 426–7, 428, 429,
　459, 786

ion channels 574
ion chromatography (IC) 381–2
ion-pair chromatography
　(IPC) 319–20, 329
ion suppression 320–1, 329
ion traps (ITMS) 321, 326, 329
IPC see ion-pair chromatography
iron
　B_6 526
　riboflavin (B_2) 526
　tryptophan conversion 110
iron tetrasulfonatophthalocyanine
　(FeTSPc), use in analytical
　methods 212–13
irradiation 96, 104
irritable bowel syndrome (IBS) 103, 104
ischemia, brain see stroke
ischemic heart disease 689, 692
　C677T polymorphism 512, 519
　homocysteine (Hcy) 495
　niacin (nicotinic acid) 647, *666*
Ishiguro, Tadanori 16
iso-coenzyme A (CoA) 132
iso-riboflavin 595
isoalloxazine 613, 616, 628
isomers 413–14
isoniazid 111, 634
isotope dilution mass spectrometry
　(IDMS) 330–1
　biotin 385
　niacin 286–97
　pantothenic acid 322, 326
　thiamin 262–3
isotope ratios 287, 289–90
isotopes 297, 399–400
isotopologues 400, 401, 403, **404**,
　405–7, 413
isotopomers 400, 414
ITMS see ion traps
IUPAC 133

Jansen, Barend 5
Japan 3–4, 5, 15–16
Jones reductor 446, 454
jowar 675
Jukes, T.H. 9

Subject Index

KA *see* kynurenic acid
kainate receptors (KR) 574
Karrer, P. 6
KDHC *see* 2-oxoglutarate dehydrogenase
Keresztesy, J.C 10, 14
keto-acids, mitochondria 566
2-keto acids 56, 63, 65
α-keto acids *see* 2-oxo acids
ketoacidosis 662
β-ketoacyl-CoA thiolase 614–15
β-ketobiotin 148
β-ketobisnorbiotin 148
α-ketobutyrate 739
α-ketoglutarate 540
α-ketoglutarate dehydrogenase (KGDHC)
  thiamin deficiency 541, 542, 545–6, 547
  thiamin diphosphate (ThDP) 242, 539
ketone bodies 614, 629
ketoses 65
KGDHC *see* α-ketoglutarate dehydrogenase
kidneys
  B_6 703
  cobalamins 465–6
  disease 28–30, 43, *45*, 643, *644*, **648**, 703, 786–800
  flavins (flavoproteins) 100
  folates 77
  homocysteine (Hcy) 703, 764
  intrinsic factor 786
  ketone bodies 614
  niacin (nicotinic acid) 643
  pantetheinase 688
  riboflavin (B_2) 596, 604–5
  thiamin excretion 77
Knight, Bert C.J.G. 8
knockout mice 694–5
Knoevenagel condensation 505
Koch, Robert 4, 14
Kögl, F. 10
kohlrabi, use in analytical methods 213

Korsakoff syndrome 72, 242, 243, 538, 546
Koschara, W. 6
KR *see* kainate receptors
Krebs cycle **73**, 507, 635, **718**
Kuhn, Richard 6, 10
kynurenic acid (KA) 680, 683
kynureninase 113
kynurenine 111, **112**
kynurenine-3 monooxygenase 111
kynurenine-anthranilate pathway 111–13
kynurenine monooxygenase (KMO) **112**

lab-on-valve **278**, 279–80, 283
lactase 104
lactate
  brain 540, 561–2
  mitochondria 566
lactation *see also* human milk
  B_6 701
  biotin 389, *720*
  niacin (nicotinic acid) 634
  riboflavin 101, *102*
  thiamin diphosphate (ThDP) *554*
  thiamin requirements 254
lactic acid 565
lacto-flavin *see* riboflavin
Lactobacilli
  biotin analysis 720
  cobalamin analysis 425, 446
  folate analysis 398, 411–12
  pantothenic acid analysis 304–5
  thiamin analysis 256
lactose intolerance 103, 104
lamotrigine (LTG) 513
Langerhans cells **661**, 663
last universal common ancestor (LUCA) 86
LCAD *see* long-chain acyl-CoA dehydrogenase
LDL *see* low-density lipoproteins
Leber's disease 787, **789**, 801

legumes
  pantothenic acid *303*
  thiamin 252
Leigh's disease 88
Lepkovsky, Samuel 10
leucine metabolism 149, **718**
leukopenia 511
ligand-binding *379*, 385–8, 392
limit of detection (LOD) 184, *185*, 188, 221, 297
limit of quantification (LOQ) 184, *185*, 188
Lind, James 3
linearity 184–6, *185*, 189, 433, 434, 440
lipid metabolism 585
lipid oxidation 97
lipid rafts 688–9
lipofuscin 547
lipogenesis **718**
lipoic acid 153, 723
lipoproteins (Lp) 670
  niacin (nicotinic acid) 636, 638, 647, 660, 663–4, 669
liposomes 358, 359–67, 370
liquid chromatography (LC) 206, 266 *see also* atmospheric pressure chemical ionization; stable isotope dilution assays
  folates 358, 405, 406
  hydrophilic interaction 289
  mass spectrometry (MS) 286, 288–9, 310–11
  mobile phases 322
  niacin 286, 288–9
  pantothenic acid 310–11, 320–6
  tandem mass spectrometry 263, 323, 358, 405, 406
  ultra-performance 289
liver 104 *see also* alcoholism
  acetyl-coenzyme A (acetyl-CoA) 566
  S-adenosylmethionine 169
  anaemia treatment 12, 13, 419
  apolipoproteins 639–40
  B_6 138, 514

biotin 721
cobalamins (B_{12}) 514
fatty acids 635–6
foetal development 604
folates 509, 738
gamma-glutamyl transferase 567
gene expression 509, 585–6
hepatic steatosis 584, 589
hepatomegaly 509
ketone bodies 614
NAD/NADP biosynthesis 111
niacin (nicotinic acid) 634, 639–40, **661**
P450 enzymes 118–19, 635
pregnancy 604
riboflavin metabolism 100, 602–3
thiamin 584
thiamin diphosphate (ThDP) 80, 85
triglycerides 584, 639
Lohman, K. 7
long-chain acyl-CoA dehydrogenase (LCAD) **615**, 616–17
low-density lipoproteins (LDL) 26–7, 636, 642, 643, 647, 670
LTG (lamotrigine) 513
LUCA (last universal common ancestor) 86
luciferin 74
lumichrome 96
lumiflavin 6, 96, 100, 595
Lumigen PS-3 444–5
luminol 443, **444**, 445–6, 447–50, 451
lungs 617
Lunin, Nicholas 4
lymphoid cells 719
lysine 138
  biotin 719
  deacetylation 121
  pyridoxal-5′-phosphate (PLP) 139
  pyridoxine dependent seizures (PDS) 505
lysozyme 97

Mac-1 695
macrocytic anaemia 453, 526, 534
MADD *see* MCAD deficiency

magnesium (Mg) 83, 85, 86, 100, 113, 139
Maillard reactions 75
maize 110, 122, 123
  pellagra 8, 675
maleria 691
malnutrition, epilepsy 505
malonyl-CoA 149, **149**
Martius, F. 13
mass analysers 321, 329
mass spectrometry (MS) 266, 313, 329, 370 *see also* isotope dilution mass spectrometry
  cobalamins 430
  folates 399
  folic acid 358
  high performance liquid chromatography (HPLC) 199, 399
  liquid chromatography (LC) 286, 288
  pantothenic acid 310–11
  selected ion recording 286, 288, *289*, 299
  tandem 329
  time-of-flight (TOF) 331
mass-to-charge ratio (m/z) 297–8
matrix effects 200, 204, 281, 285–6 *see also* ion suppression
  folates assays 401
  pantothenic acid assays 319
  riboflavin assays 272–3
matrix metalloproteinase-9 545, 692
MCAD *see* medium-chain acyl-CoA dehydrogenase
MCAD deficiency (MADD) 622–6, 629
MCC *see* 3-methylcrotonyl-CoA carboxylase
MCT *see* monocarboxylate transporters
meat
  biotin 720
  cobalamins 166
  folates 411
  pantothenic acid *303*
  riboflavin *272*
  thiamin 253

medication
  assay interference 161
  biotin effects 146, 153
  niacin effects 111
Mediterranean diet 47, 472
medium-chain acyl-CoA dehydrogenase (MCAD) **615**, 616, 617–18, **620**
megaloblastic anaemia 471, 484, 493, 520, 735, 742–3, 749, 781
  antiepileptic drugs (AEDs) 509
  research history 770
memory, cobalamins 471
mercury, use in analytical methods 212
meta-analysis studies
  C677T MTHFR 512
  cancer 27, *29*, 46, 704, 706, 760–2, 763, 764
  cardiovascular disease (CVD) 25, *29*, 43, *489*, 495, 757–64
  cognitive decline 28, *29*
  coronary heart disease (CHD) *29*, 37, 755–6, *757*, 759–60, 762–4
  diabetes mellitus 646
  mortality *29*
  multiple micronutrient (MMN) supplements 526–7, 529, **530**, 531, *531*
  stroke *29*, 37, *489*, 492, 493, 495, 756, 759–60, 763, 764
metabolic lactic acidosis 564, 565
metabolic syndrome 580, 587, 590, 641, 645
metabolism, effect of disease 22
metalloproteases 744
metallothionein 690
5,10-methenyltetrahydrofolate *159*, **397**, 403
methionine 97, 167
  chronic kidney disease (CKD) *797*
  cobalamins 24, 483
  folates 24, 396
  homocysteine metabolism **484**, 738
  metabolism 24, **25**, 476

methionine synthase 419, 801
  cobalamins 167, 701, 787, **788**
  folic acid 167
  homocysteine metabolism 25, **38**, **484**, 738–9, 772
  5-methyltetrahydrofolate 482
  pregnancy 513
5-methyl-5,6-dihydrofolate *159*
N-methyl-D-aspartate receptors (NMDARs) 542, 574, 680, 772, 776
methyl malonic acid (MMA) 453, 472
methyl malonyl-coA-mutase 419
N^1-methyl-nicotinamide 111
methyl-tetrahydrofolate reductase (MTHFR) 37
methylation
  S-adenosylmethionine (SAM or AdoMet) 771
  B_6 139
  brain function 490, 772, 775–6
  cytosine 24
  folates 24, 197, 491
  homocysteine (Hcy) 24, 396, 779
methylcitric acid 726
methylcobalamin (MeCbl) 420–1, 472
  cofactor behaviour 167
  dietary sources 166
  light sensitivity 424
  medical use 791, *792*, 795, *796*, *797*
  metabolism 787, **788**, 789
  natural forms 166–7
  photodissociation 166
  research history 14
  structure **13**, 165, **203**
3-methylcrotonyl-CoA, structure **149**
3-methylcrotonyl-CoA carboxylase (MCC) 725
  biotin 148, **718**
  leucine metabolism 149, **718**
  lipoic acid 153
3-methylcrotonyl glycine **149**, 725, 726
5,10-methylenetetrahydrofolate 24, *159*
  B_6 700–1
  folate cycle 739–40

homocysteine metabolism 25, **484**
isotopologues 405
structure **397**
methylenetetrahydrofolate dehydrogenase (MTHFD) **739**, 740
methylenetetrahydrofolate reductase (MTHFR) **484**
  C677T polymorphism 512, 513, 515, 519–20, 742, 745–6
  deficiency 744
  DNA synthesis 742
  folate cycle **739**, 740
  homocysteine metabolism 25, 512
3-methylglutaconyl-CoA, structure **149**
methylmalonic acid (MMA)
  cobalamins 23, 419, 776, 777, 790, 791
  hyperhomocysteinemia 465
methylmalonic aciduria 508
methylmalonic aciduria cdlC type with homocystinuria (MMACHC) 787, **788**
methylmalonyl-CoA **149**, 167–8
methylmalonyl-CoA mutase 167–8, 787
methylmalonylacidemia 167–8
5-methyltetrahydrofolate (5-MHTF or 5MT) 24, 49, *159*, 167
  antioxidant effect 40
  assay methods 357–8, 405–7
  cerebrospinal fluid (CSF) 506, **507**, 511
  cobalamins metabolism **788**
  DNA synthesis 742
  flow mediated dilatation (FMD) 39
  fluorescence 411
  folate digestion 737
  folate supplementation 40
  folate transport 737, *738*
  homocysteine metabolism 25, **38**, 396, 482, **484**
  isotopologues 403, 405–6, **408**
  metabolism 738

plasma 408–9
remethylation pathway 801
structure **397**
vascular walls **41**, 46
methyltetrahydrofolate-reductase (MTHFR) **38**
Meyerhof, Otto 8
micellar liquid chromatography (MLC) 263–4
Michaelis–Menten kinetics 594
Michi, K. 10
micro beads 216
microarray analysis 585–6, 589
microbiological assays 198, 313–14, 351, 435
　accuracy 257
　advantages *379*, 398
　B_6 336
　biotin 379–80, 720
　cobalamins 420, *431*
　disadvantages 285, 305, *379*, 380, 398
　electrochemical detection (ED) 442
　folates 161, 398, 408–9
　honey samples 181
　niacin 285
　pantothenic acid 302, 304–7, 319
　SIDA comparison 408–9, 410–12
　thiamin 256–7, 265
microperoxidase 443
microRNA (miR) signalling 146
microwave-assisted extraction (MAE) 279
migraine, niacin 109
milk *see also* dairy products
　artificial 4
　biotin 377, 380, 386, 388, 389–91
　cobalamins 166
　goats 743
　human 386, 389, 428–9, 433, 508
　niacin 292–5
　pantothenic acid *303*
　pasteurization 130, 136
　riboflavin 96–7, 271, *272*, 277, 279
　sunlight-flavoured 97

Mini Mental Status Examination (MMSE) 678, 683
Minot, G.R. 13
MIP *see* molecularly imprinted polymer
miscarriage 468, *735*
mitochondria 89, 567 *see also* acyl-CoA dehydrogenases; tricarboxylic acid (TCA) cycle
　acetyl-CoA carboxylase 2 149
　acetyl-coenzyme A (acetyl-CoA) 557, 563
　adenosine triphosphates (ATP) 89, 115, 557, 566
　adenosylcobalamin (AdoCbl) 167
　electron transport chain 589, 635
　energy metabolism 115, **555**, 566
　fatty acids oxidation 566, **612**, 613–26, 635–6
　folates 740
　glucose 566
　keto-acids 566
　Krebs cycle **73**
　lactate 566
　3-methylcrotonyl-CoA carboxylase (MCC) 149
　NMNAT 114
　propionyl-CoA carboxylase (PCC) 149
　pyruvate 562
　pyruvate carboxylase (PC) 149
　pyruvate dehydrogenase **73**, 557
　riboflavin (B_2) **612**, 613
　thiamin deficiency 540, **542**, **555**, 563
　thiamin diphosphate (ThDP) 79, 80, 81, **82**, **555**
　thiamin triphosphate (ThTP) 81, 84
MLC *see* micellar liquid chromatography
MMA *see* methyl malonic acid; methylmalonic acid
MMSE *see* Mini Mental Status Examination
mobile phase 283

molar extinction coefficient 351
molecular weight
  folic acid 160
  niacin (nicotinic acid) *109*
  niacinamide (nicotinamide) *109*
  nicotinamide adenine dinucleotide (NAD) *117*
  nicotinamide adenine dinucleotide phosphate (NADP) *117*
molecularly imprinted polymer (MIP) 279, 280, 281, 283
mono-ADP-ribosyltransferases (ARTs) 119, **120**
monocarboxylate transporters (MCT) 153, 562
monoclonal antibodies 360, 399, 463, 688, 694
monocyte chemoattractant protein-1 (MCP-1) 663
monoglutamate folate 737
monoglutamates 403
monomethyl L-arginine (LNMMA) 39
morbidity, children 528
Mori, Rintaro (Ogai) 16
4-morpholineethanesulfonic acid 405
mortality 29
  B_6 702, 703
  epilepsy 513
  folic acid *762*
motor development 531
MRM *see* multiple reaction monitoring
MTHFR *see* methyltetrahydrofolate-reductase
multiple micronutrient (MMN) deficiency 524–5, 526
multiple myeloma 704
multiple reaction monitoring (MRM) 298, 329–30
  chromatograms *324*
  columns 324–5
  isotopic effect 290
  niacin 286, 288, *289*, 290, 296
  pantothenic acid 321, 325

quadrupole traps 321
sample preparation 288
multivitamin analysis methods 262, 265
Murphy, W.P. 13
muscles
  thiamin diphosphate (ThDP) 80, 83
  thiamin triphosphate (ThTP) 81
myelin 776
  cobalamins 469–70, 483
  folates 743
myeloperoxidase 642
myocardial infarction *485*
  B_6 702, 703
  folic acid supplementation 39
  kidney disease 643
  niacin (nicotinic acid) 646, 647
  randomized controlled trials (RCTs) 43, *44–5*, 491
myopathy 665

NAD *see* nicotinamide adenine dinucleotide
NAD kinase 114
NAD+ synthetase-1 (NADSYN1) **112**, 114
NADP *see* nicotinamide adenine dinucleotide phosphate
NADPH
  function 118–19
  niacin 111, 113, 642, 663
  thiamin 211
  transketolase (TK) 556
NADPH oxidase 36, **41**, 119
nanomaterials 220
  biotin analysis 361–2, **363**, 364
  thiamin analysis 217
National Health and Nutrition Examination Survey (NHANES)
  folates 490
  homocysteine (Hcy) 494
  riboflavin 102
National Institute of Standards and Technology (NIST) 327
necrosis 123, 637–8
NEFA *see* non-esterified fatty acids

Subject Index 837

nerve agents 141
nervous system 88 see also dementia; neural tube defects
   S-adenosylmethionine (SAM or AdoMet) 771, 780
   AMPARs 574
   cobalamins 420, 471, 482–3, 484
   epilepsy 505
   folates 484, 517, 743–4
   niacin (nicotinic acid) 634
   organophosphorus compounds 142
   thiamin 87, 210, 240, 241
Neuberg, Carl 7
neural tube defects
   cobalamins 197, 468
   folates 21, *29*, 30, 47, 197, 355, 396–7, 484, 513, 735, 745, 748
   food fortification 746
neurodegenerative diseases 577
   see also Alzheimer's disease; Huntington's disease
   cobalamins (B_{12}) *771*
   folates 743, *771*
   mechanisms of B vitamins 573
   thiamin 545–6, 573
neurons
   chromatolysis 681
   energy requirements 557
   pellagra 681
neurotransmitters 482, 520, 534
   acetyl-coenzyme A (acetyl-CoA) 562
   cobalamins (B_{12}) 776
   folates 743, 776
   riboflavin 101
   thiamin 572
   transmethylation reactions 801
neutrigenetics 749
neutropenia 511
neutrophils 642, 688, 692, 693–4
NF-κB (nuclear factor kappa-light-chain-enhancer of activated B cells) 719
NHANES see National Health and Nutrition Examination Survey

niacin adenine dinucleotide
   NAD biosynthesis 113–14
   structure **112**
niacin adenine dinucleotide phosphate 121
niacin adenine mononucleotide, structure **112**
niacin-induced flushing 670
niacin (nicotinic acid) see also pellagra
   absorption spectra 109
   amphoterism 109, **110**
   appearance 109
   assay methods 198, 285–97
   biological function 634–8, **661**
   biosynthesis 634
   cardiovascular disease (CVD) 21, 26–7, *29*
   CAS number *109*
   cholesterol reduction 26–7
   chromatograms **293**
   deficiency 110, 634 see also pellagra
   diabetes mellitus 27
   dietary sources 110, 122, 633–4
   digestion 634
   dyslipidemia 21
   extraction 285–6, 296
   fluorimetry 663, 664
   flushing 647, 649
   food fortification 110, 286, 290–2
   honey samples 177–9, *179*
   labelled samples 287, 288, 295
   lipoproteins 636
   medical use 109
   metabolism 111–13, **112**, 634
   molecular weight *109*
   NAD biosynthesis 111–13, 114
   natural forms 286
   nomenclature 108, 122
   pharmaceutical use 638, 641, 643, 660–9
   recommended dietary allowance (RDA) 110, 633–4
   research history 8, 122
   riboflavin 99, 101, 525
   salts 109

niacin (nicotinic acid) (*continued*)
  serotonin 680
  solubility 109
  spectra *291*
  stability 109
  structure 9, **109**, **112**
  toxicity 634, 664–5
  tryptophan 633
  zwitterionic behaviour 109
niacinamide (nicotinamide)
  absorption spectra 109
  appearance 109
  CAS number *109*
  depression 681
  kynurenine-anthranilate pathway 111–13
  metabolism **112**
  molecular weight *109*
  NAD biosynthesis 111–13, 114
  NAD recycling 114–15, **116**
  nomenclature 108
  sirtuins 121
  solubility 109
  stability 109
  structure 9, **109**, **112**
  tryptophan conversion 113
nicotinamide *see* niacinamide
nicotinamide adenine dinucleotide (NAD)
  absorption spectra 115
  adenosine diphosphate (ADP)-ribose transfer reactions 15, 119–21
  appearance 115
  biological function 108–9, 635, 636
  biosynthesis 111–14
  CAS number *117*
  energy metabolism 115–17, 635
  molecular weight *117*
  nomenclature *117*
  pyridoxal-5′-phosphate (PLP) analysis 341
  recycling 114–15, **116**
  reduced form NADH 58, 115, **118**
  research history 8
  solubility 115
  structure 9, **112**, 115, *117*
nicotinamide adenine dinucleotide (NAD) glycohydrolase 634
nicotinamide adenine dinucleotide phosphate (NADP)
  absorption spectra 115
  adenosine diphosphate (ADP)-ribose transfer reactions 119–21
  antioxidant behaviour 118–19
  appearance 115
  biological function 108–9, 635
  biosynthesis 114
  biotin catabolism 148
  CAS number *117*
  cobalamins metabolism **788**
  energy metabolism 115–17
  molecular weight *117*
  nomenclature *117*
  reduced form *see* NADPH
  research history 8
  solubility 115
  structure 9, 115, *117*
nicotinamide mononucleotide
  NAD biosynthesis 114
  NAD recycling 115, **116**
  structure **112**
nicotinamide/nicotinate mononucleotide adenyltransferases (NMNATs) **112**, 114, 115, **116**
nicotinamide phosphoribosyltransferase (NAMP/PBEF1) **112**, 114, 115, **116**
nicotinamide riboside 115, **116**
nicotinamide riboside kinase (NRK) 115, **116**
nicotinic acid *see* niacin
nicotinic acid mononucleotide 9
nicotinic acid phosphoribosyltransferase (NAPRT1) **112**, 114
Nishizuka, Y. 8
NIST (National Institute of Standards and Technology) 327

nitric oxide synthase (NOS) 39, 575, 801
nitrotyrosine immunolabelling 544
nixtamalization 122, 123
NMDARs *see* *N*-methyl-D-aspartate receptors
NMNAT *see* nicotinamide/nicotinate mononucleotide adenyltransferases
nomenclature
   B_6 135
   biotin 10
   coenzyme A (CoA) 128
   folates 158, 734
   folic acid 158, 160, 162
   IUPAC 133
   niacin (nicotinic acid) 108, 122
   niacinamide 108
   nicotinamide adenine dinucleotide (NAD) *117*
   nicotinamide adenine dinucleotide phosphate (NADP) *117*
   pantothenic acid 9, 127–8, 133, 317
   riboflavin (B_2) 6, 93–104
   tetrahydrofolate (THF) 160
   thiamin 5, 71
   thiamin diphosphate (ThDP) 55
non-esterified fatty acids (NEFA) 670
non-Hodgkin lymphona 704, 705
NORVIT trial 43, *44*, *759*, *760*, *761–2*
nuclear factor kappa-light-chain-enhancer of activated B cells (NF-κB) 719
nuclear lysine-specific demethylase 1 (LSD1) 741
nucleoplasm 169
nucleosomes 727
nucleotide diphosphatase 99
Nurses' Health Study 702–3
nutrigenetics 747
nutrigenomics 747, 749
nutritional research, study types 22–3

obesity 197, 476, 590
   cobalamins 466, **468**, 472
   diabetes mellitus 580–9, 646
   gastric inhibitory polypeptide 586
   insulin sensitivity 646
   thiamin 580–90
observational studies 709
   cardiovascular disease (CVD) 702–3, 755–6, *757*
   multiple micronutrient (MMN) supplements 529
   stroke *485–7*, 495
occludin 545
ocular system 605
offline definition 283
OGDH *see* 2-oxoglutarate dehydrogenase
olefins 143
oleic acid 147
oloenzymes 168
one-carbon metabolism 709
   brain function 490, 771–2
   folates 699–701, 735, **739**, 740, 772
online definition 283
optic atrophy 801
optical biosensor inhibition immunoassay 311–12
optical sensing 218
oral contraception 111, *554*
organometallic compounds 169
organophosphorus compounds 142
orthologs 123
oryzanin *see* thiamin
osmolarity *601*
ovarian cancer 704
overconsumption 587
ovo-flavin *see* riboflavin
oxaloacetate **73**, **149**
oxazoles 137
oxidase 351
oxidation 220
   fatty acids 149, 613–26, 627, 628, 635–6
   mitochondria 566
oxidative metabolism 545, 549, 557
oxidative stress 549, 695 *see also* reactive oxygen species (ROS)
   alcohol 547
   B_6 139, 142, 143

oxidative stress (*continued*)
  chronic kidney disease (CKD) 790
  cobalamins (B_{12}) 790
  niacin (nicotinic acid) 663
  pantetheinase 692
  thiamin deficiency **540**, 544, 545, 547
  VNN3 gene 690
oximes 141, 142, 143
2-oxo acids 71, 89 *see also* 2-oxoglutarate; pyruvate
2-oxoglutarate **73**, 89
2-oxoglutarate dehydrogenase complex (KDHC) **555**, 556, 557, 563
2-oxoglutarate dehydrogenase (OGDH)
  Krebs cycle **73**
  thiamin diphosphate (ThDP) 72, 79, 80, 81, **82**
oxygen 143, 446

P450 enzymes, NADPH 118–19, 635
pABA *see para*-aminobenzoic acid
PADs *see* peptidylarginine deiminases
palmitoyl-CoA 616
pancreatic cancer 692, 704, 706
pantetheinase 686–93
pantetheine 131, 303, **304**, 691, 695
pantethine 131, 687
panthenol 130–1
pantoic acid, structure **128**, **308**
pantolactone 130
pantothenic acid 595–6
pantothenic acid (B_5)
  adequate intake (AI) 127, 133
  appearance 129
  assay methods 304–12, 318–27
  CAS number 128
  cleavage 130
  coenzyme A (CoA) 685, 686
  deficiency 133, 302–3, 317–18
  dietary sources 317
  extraction 303–4, 322
  food analysis *303*
  function 318
  heat treatment 130
  honey samples *178*, 179
  inflammation 691
  labelled samples 310
  nomenclature 9, 127–8, 133, 317
  research history 9–10, 127, 318–19
  salts 129–30
  solubility 129
  stability 96, 129, 130, 318
  stereoisomers 128
  structure **128**, **304**, 318
  supplements 318–27
  synthesis 127
  synthetic form 133
papain 200, 205, 382
para-aminobenzoic acid (pABA)
  folate analysis 357, 358
  structure **159**
paraesthesias 677
parenteral nutrition
  biotin 720, 723, 724, 725
  cobalamins (B_{12}) 778
  riboflavin (B_2) 103, 720
  thiamin 720
Paris 4
Parkinson's disease (PD)
  folates *735*
  thiamin 239
PARPs *see* poly-ADP-ribose polymerases
passive diffusion 594, 596, **597**, 605, 606
pasteurization 104
PB (phenobarbital) 508, 509, 514, 515, 516, 519, 723
PBEF *see* pre-B-cell colony-enhancing factor
PC *see* pyruvate carboxylase
PCC *see* propionyl-CoA carboxylase
PDC *see* pyruvate decarboxylase
PDH *see* pyruvate dehydrogenase
PDHC *see* pyruvate dehydrogenase complex
PDS (pyridoxine dependent seizures) 505, 518
Pekelharing, Cornelius 5

pellagra 8, 110, 122, 634, 638, 675–82
pentose phosphate pathway 59, 65, **73**, 507
peptidylarginine deiminases (PADs) 728
permeability transition pores (PTP) **555**, 562
pernicious anaemia 46, 459, 470, 781
peroxidase, use in analytical methods 213–15
peroxisome proliferato-activated receptor (PPAR)γ 645, 663, 690–1, 695
peroxisomes, thiamin diphosphate (ThDP) 80
peroxynitrates
  homocysteinaemia 36
  thiamin deficiency 544
Peters, R. 7
PFOR *see* pyruvate:ferredoxin oxidoreductase
pH
  B_6 136
  chemiluminescence 447
  folic acid absorption 775
  riboflavin 95–6, 596, *601*
  stomach 775
  thiamin 253
phenobarbital (PB) 508, 509, 514, 515, 516, 519, 723
phenytoin (PHT) 508, 509, 514, 515, 516, 517, 519, 723
phosphatases 351
phosphatidyl inositol 695
2-phospho-cyclic ADP-ribose 121
phosphocreatine 541
phosphoenolpyruvate carboxykinase (PECK) 719
phosphohydrolases 84, 85, 86
4-phosphohydroxy-L-threonine 138
phosphoketolase (PK)
  catalysis 64, 65
  pathway 60–1
  structure 62
phospholipid esters 613
phosphopantetheine **127, 304**, 685, 686
phosphopantetheinyladenylate phosphate *see* coenzyme A
4-phosphopantotenic acid 685, 686
phosphorescence 454
5-phosphoribosyl-1-pyrophosphate (PRPP) 8
phosphorolysis 65, 596, 598, 603, 606
  adenosine triphosphates (ATP) 612
  biotin 719
photochemical reactions 221
photodissociation 169
photosensitivity 105, 139
PHT (phenytoin) 508, 509, 514, 515, 516, 517, 519, 723
pimelyl-CoA 147
PK *see* phosphoketolase
pKa 133
PL *see* pyridoxal
placenta 47, 604, 744
plasma 435
  folates 357–8, 406, *407*, 408–9
  holotranscobalamin (holoTC) 461
  niacin (nicotinic acid) 634
  pantothenic acid 304, 307
  riboflavin 99
  riboflavin (B_2) 602, **612**
plasmon surface resonance *see* surface plasmon resonance
platelet endothelial cell adhesion molecule (PECAM) 642
PLDH *see* pyridoxal 4-dehydrogenase
PLP *see* pyridoxal-5′-phosphate
PM *see* pyridoxamine
PMP *see* pyridoxamine-5′-phosphate
PN *see* pyridoxine
PNG *see* pyridoxine-β-glucoside
PNOX *see* pyridoxine oxidase
PNP *see* pyridoxine-5′-phosphate
poisoning therapy 141
poly-ADP-ribose polymerases (PARPs) 119–21, 636, 637
polyglutamate folate 737
polyglutamates 403
polymorphs 520
polyneuritis gallinarum 4

polyphagia 590
pork, thiamin 253
Port Arthur 15
positive electrospray ionization 289, *291*
PPAT *see* pyridoxamine-pyruvate aminotransferase
PRD (primidone) 153, 508, 509, 514, 515
pre-B-cell colony-enhancing factor (PBEF) 114
precision 186, 189, 433
preeclampsia 468, *469*, 471, 474, 476, 735, 744
pregnancy *see also* placenta
  alcohol 636
  anaemia 532
  B_6 701
  biotin 146, *378*, 389, *720*, 722
  C677T polymorphism 513
  cobalamins 461, 468, 471, 474
  epilepsy 513
  folic acid *29*, 30, 47, 484, 513, 735, 744–5, 746, 748
  homocysteine metabolism 468, *469*
  hyperhomocysteinemia (HHcy) 744
  methionine synthase 513
  multiple micronutrient (MMN) deficiency 524–5
  preeclampsia 468, *469*, 471, 474, 476, 735, 744
  riboflavin (B_2) 99–100, 101, *102*, 515, 604
  supplements 524–5, 526, 527, 529, **530**, *531*, 532
  thiamin 233, 254
  thiamin diphosphate (ThDP) *554*
Preiss-Handler pathway **112**
premenstrual syndrome 101
preservatives 75, *554*
primidone (PRD) 153, 508, 509, 514, 515
prions 89
PRMTs *see* protein methyltransferases
processed foods 587
proline 117
propionyl-CoA **149**, 615
propionyl-CoA carboxylase (PCC) 148, 149, **718**, 719, 725
proprionate 167
prostaglandin D2 (PGD2) 647, 663
prostate cancer 706, 756–7
prosthetic group 105, 727
protease 358
proteasome 728
protein, misfolding 618
protein arginine methyltransferase (PRTMT) 794, **795**
protein binding assays *see* binding assays
protein kinases 575, *600*, 603, 663
protein methyltransferases (PRMTs) **38**
proteins
  apo- 434, 613
  coenzyme requirements 14–15
  holo- 434, 626
  misfolding 621, **625**, 626, 629
proteolysis 424
proteolytic degradation, deconjugation 412
proton-coupled folate transporter (PCFT) 737, *738*
PRPP *see* 5-phosphoribosyl-1-pyrophosphate
pruritus 677
psoriatic skin lesions 691, 692
psychosis 677–8, 683
PteGlu *see* folic acid
pteridine ring, structure **159**
pteroic acid, structure **159**
pteroylglutamates 158–9, 162
pteroylglutamic acid *see* folic acid
pteroylheptaglutamate 405, 412
pteroylpolyglutamic acid *see* folic acid
PTP *see* permeability transition pores
pulsed differential voltammetry 264
pulsed electric fields 96
purine, biosynthesis 77, 491, 740
pyridinium aldehyde 141

Subject Index

pyridoxal 4-dehydrogenase (PLDH) 337–8, 341
pyridoxal-5′-phosphate (PLP)
  assay methods *340*, 341, 344, 347
  biochemical function 336
  biosynthesis 138
  cardiovascular disease (CVD) 514, 703
  chromatograms *345*
  coenzyme function 138, 142, 335
  coronary heart disease (CHD) 703
  cystathionine β-synthase (CBS) 739
  diabetes mellitus 336, 350
  enzymes 138–9, 142
  epilepsy 505, 513–14
  homocysteine metabolism **25**
  isomers **138**
  plasma levels 701, 703, 706
  responsive seizures 505–6
  structure **136**, **336**
pyridoxal-5′-phosphate (PLP) synthase 138, 505
pyridoxal kinase, cobalamins deficiency 470
pyridoxal phosphate
  research history 12
  structure **11**
pyridoxal (PL)
  assay methods 339, *340*, 343
  biochemical function 138, 335
  chromatograms *345*
  derivatives 141
  food fortification 136
  phosphorolysis 138
  research history 11
  structure **11**, **136**, **336**, **483**, **700**
pyridoxamine-5′-phosphate (PMP)
  assay methods *340*, 341, 344
  biochemical function 336
  chromatograms *346*
  coenzyme function 138
  structure **136**, **336**
pyridoxamine phosphate, structure **11**

pyridoxamine (PM)
  assay methods *340*, 341, 343
  biochemical function 336
  chromatograms *345*
  diabetes mellitus 350
  phosphorolysis 138
  research history 11
  structure **11**, **136**, **336**, **483**, **700**
pyridoxamine-pyruvate aminotransferase (PPAT) 337, 338–9, 341
4-pyridoxic acid (4-PA) 335, **336**
pyridoxine-5′-phosphate (PNP)
  assay methods *340*, 341, 344
  chromatograms *346*
  structure **136**, **336**
pyridoxine-5′-phosphate (PNP) synthase 138
pyridoxine-β-glucoside (PNG) 140–1, 335–6, **336**, 339, 341, 344–7, 350
pyridoxine dependent seizures (PDS) 505, 518
pyridoxine hydrochloride 199
pyridoxine oxidase (PNOX) 337, 338, 341
pyridoxine phosphate, structure **11**
pyridoxine (PN) *see also* B$_6$
  assay methods 200, 341, 343–4
  cardiovascular disease (CVD) *29*
  cercosporin 139
  chromatograms *345*
  derivatives 140
  honey samples *178*, 179
  phosphorolysis 138
  research history 10–12, 137
  riboflavin function 101
  structure **11**, 135, **136**, **336**, **482**, **700**
  synthesis 137
4-pyridoxolactone (4-PLA) **336**, 337–8, 339, 342–4, 348
pyrimidine 740
pyruvate 89
  acetyl-coenzyme A (acetyl-CoA) 560, 562
  assay methods 341

pyruvate (*continued*)
  decarboxylaton 64, 65, 567–8
  mitochondria 562
  nicotinamide adenine dinucleotide (NAD) 635
  oxidation 566
  pyridoxal-5′-phosphate (PLP) 138
  research history 7
  structure **149**
  thiamin deficiency 540
  tricarboxylic acid (TCA) cycle 556
pyruvate carboxylase (PC)
  biotin 148, 381, **718**, 721, 725
  gluconeogenesis 149
  lipoic acid 153
pyruvate decarboxylase (PDC)
  alcohol fermentation 72
  catalysis 65
  reactions 57–8, 64
  structure 62
  thiamin diphosphate (ThDP) 55
pyruvate dehydrogenase complex (PDHC) 567–8
  acetyl-coenzyme A (acetyl-CoA) 555–6, 557
  metabolic lactic acidosis 564
  thiamin diphosphate (ThDP) 556, **561**, 562
  tricarboxylic acid (TCA) cycle 557
  zinc 563
pyruvate dehydrogenase (PDH)
  classification 63
  diabetes mellitus 581, 590
  energy metabolism 58, 64–5, **73**
  flavin adenine dinucleotide (FAD) 58
  glycolysis 58, 64, 568
  inhibition 566
  Leigh's disease 88
  reactions **57**, 64
  structure 62
  'swinging domain' 58
  thiamin 581, 584
  thiamin diphosphate (ThDP) 72, 79, 80, 81, **82**, 242, 539

pyruvate oxidase (POX) 72
  classification 63
  oxidation 58–9
  reactions **57**, 64
  structure **61**, 62
  types 59
pyruvate:ferredoxin oxidoreductase (PFOR) 58–9, 63, 64

quadrupole traps 321, 325, 330
quality control programmes 433
quantum dots, thiamin analysis 217
quantum mechanics 169
quinolinate
  NAD biosynthesis 113–14
  structure **112**
quinolinate phosphoribosyltransferase (QPRT) **112**, 113
quinolinic acid 8

radioimmunoassays (RIA) 313, 435
  folates 162, 408
  holotranscobalamin (holoTC) 429, 462
  pantothenic acid 307
Ramasarma, G.B. 8
randomized controlled trials (RCTs) 22–3, 709
  anaemia 526–7, 532
  atherosclerosis *666*, 667, 669–70
  cancer 27, *29*, 706–7
  cardiovascular disease (CVD) 25–6, *29*, 46, 48, 532, 646–7, 665–8, 703–4, 757–64
  cognitive decline 28, *29*, 532, 778, 779
  criticisms 43–7
  flow mediated dilatation (FMD) 39
  metabolic syndrome 645–6
  myocardial infarction 491
  post coronary stenting 42
  renal disease 43, 643, *644*
  stroke 42–3, 493
  subject variation 42

Subject Index

RCP *see* riboflavin carrier protein
RDA *see* recommended dietary allowance
reactive oxygen species (ROS) 549, 695 *see also* oxidative stress
  alcohol 547
  cobalamins analysis 451
  homocysteine (Hcy) 36
  niacin (nicotinic acid) 642
  thiamin deficiency 544
receptor-mediated endocytosis 594, 596, **597**, 605, 606, 607
recommended dietary allowance (RDA) 49, 105, 123 *see also* dietary reference intake (DRI); Reference Nutrient Intake
  B_6 701
  biotin *378*, 389
  folates *736*
  folic acid 40, 47, 745
  niacin 110, 633–4
  riboflavin 101–2
  thiamin 253–4
recovery levels in assays *185*, 186–7
recovery time in assays 440
red blood cells *see* erythrocytes
redox 49, 123
  nicotinamide adenine dinucleotide (NAD) 635
  nicotinamide adenine dinucleotide phosphate (NADP) 635, 663
  pantetheinase 691
redox potential 629
redox sensitive transcriptional factors 49
reduced folate carrier (RFC-1) 78, 509
reduced folate (RFC) 737, *738*, **739**
reference intervals 434, 435
reference materials (RM) 186, 204, 298, 322, 327
Reference Nutrient Intake (RNI) 633–4, *736*
refractive index (RI) detection 320, 330
relative risk 709
remethylation pathway 801

renal disease 28–30, *45*
  B_6 703
  cardiovascular disease (CVD) 643, 794, 799
  cobalamins 465–6, 470, 471, 786–800
  homocysteine (Hcy) 43, 703
  niacin (nicotinic acid) 643, *644*, **648**
  statins 643
renin–angiotensin–aldosterone axis 554
repeatability *185*, 186, 189
reproducibility *185*, 186, 189, 434, 440
resonance Rayleigh scattering (RRS) 217
respiratory chain 58, 562, 614, **615**, 619
response time 440
restenosis 41–2, 47
retina 605
reverse cholesterol transport (RTC) 639, 641, 662–3, 670
reverse isotope dilution 298
reversed-phase high performance liquid chromatography (RP-HPLC) 266, 298
  biotin 356, 382–3
  cobalamins 430
  folates 357
  honey samples 181, 183
  riboflavin *276*, *277*, *279*
  simultaneous analysis 262
  thiamin 259
rhodanese 787, **790**
RIA *see* radioimmunoassays
ribitol 93
riboflavin-5′-phosphate *see* flavin mononucleotide
riboflavin (B_2)
  absorption spectra 97
  adequate intake (AI) 101, *102*
  alcohol 99, 595
  analogues 595–6
  appearance 93

riboflavin (B$_2$) (*continued*)
  assay methods 271–82
  bacteria 592
  blood 602
  brain 603–4
  deficiency 103, 138, *601*
  dietary sources 97, *98*, 598
  digestion 99, 104, **593**, 595, **597**, *601*, **612**
  epilepsy 515
  excretion 100
  extraction 273–80, 281
  eyes 605
  flavins 627–8
  function 101, 525
  homeostasis 603, 604
  honey samples 177, *178*
  immune system 602
  intestines 99, 100, 592, **593**, 595–602, 612
  kidneys 604–5
  metabolism 100, 592, **593**, 611–13
  natural forms 271–2, 281, 592
  niacin (nicotinic acid) 99, 101, 525
  nomenclature 6, 93–104
  oxidation of other vitamins 97
  pH 596, *601*
  pharmaceutical use 622–4, 626
  phosphorolysis 596, 598, 606
  photodissociation 96–7
  placenta 604
  recommended dietary allowance (RDA) 101–2
  research history 93
  stability 95–7
  structure **7**, 94
  temperature *601*, 602
  transport 594, 595, 596, *599*, 604, 605
  tryptophan conversion 110
riboflavin binding protein 604
riboflavin carrier protein (RCP) 604
riboflavin-responsive MADD 622–3, 625–6, 629
riboflavin transporters 99, 596
riboflavinyl peptide ester 100

D-ribose 595–6
ribose 5-phosphate **73**, 138
riboswitches 77–8
ribulose 5-phosphate 138
rice 3, 4, 5, 16, 72, 137, 140
RM *see* reference materials
RNA synthesis
  folic acid 525
  tetrahydrofolate (THF) 491
ROS *see* reactive oxygen species
Rossman fold 115
Royal Navy 3
RTC (reverse cholesterol transport) 639, 641, 662–3, 670
Russo-Japanese War 4, 15–16
ryanodine receptors **120**, 121

salts 4
salvage pathway 123
sample preparation *see* extraction
sarcoplasmic reticulum 121
sarin 141
SCAD *see* short-chain acyl-CoA dehydrogenase
scaffolding proteins 545
scaly erythematous periorificial dermatitis 726
scavenger receptor class B type I (SR-BI) 641
Schiff's bases 138, 139, 143, 350
  see also imine
schizophrenia 109, 677, 680
Schuster, P. 7
SCN *see* thiocyanate
screen mode 204
screen printed electrodes 361–2, **363**, 364
scurvy 3, 15
seafood
  analysis 200–5
  riboflavin *272*
SEARCH trial 43, *45*, *759*, *760*, *761–2*
sedoheptulose 7-phosphate **73**
seizures 726, 744 *see also* epilepsy
selected ion monitoring (SIM) 321, 330

Subject Index

selected ion recording (SIR) 286, 288, 289, 299
selectivity of analysis 204, 298, 440
sensitivity of analysis 298–9, 401, 434, 440
sensitized fluorescence 221
serine **25**, 739
serine hydroxymethyltransferase (SHMT) 739
serotonin 482, 683
  folates 511, 743
  niacin deficiency 680
  tryptophan 111, 633
Sertoli cells 687
serum 435
  cobalamins 420–33, 434, *451*, 460–9
  folates 406, *407*, 511
shelf life 410
Shorb, Mary 14
short-chain acyl-CoA dehydrogenase (SCAD) 618
SIDA *see* stable isotope dilution assays
Siege of Paris 4
signalling pathways, biotin 146, 152–3
silyl esters 382
SIM *see* selected ion monitoring
simultaneous analysis 199–205, 254, 262, 265, 325–6, 440
single nucleotide polymorphisms (SNP) 154, 749
  holocarboxylase synthetase (HLCS) 151
  methionine synthase gene 167
  methylenetetrahydrofolate reductase (MTHFR) 746, 747
singlet molecular oxygen 143, 351
  quenching 139, 336, 350
SIR *see* selected ion recording
sirtuins **120**, 121, 122, 636
small ubiquitin-like modifier (SUMO) proteins 728
smokers
  biotin 146, 724
  cobalamins deficiency 466, **468**, 472

cyanide 788
Leber's disease 787, 788, **789**, 801
stroke risk 481
Snell, Esmond 9–10, 11, 12
SOD *see* superoxide dismutase
sodium-dependent multivitamin transporter (SMVT) 153
sodium ions *601*, 602, 603
sodium nitroprusside (SNP) exposure **561**
solid phase extraction (SPE) 281–2
  automation 274–80, 281 *see also* flow injection analysis
  batch methods 273–4
  biotin 384
  cobalamins 427–8
  disadvantages 275
  fluorimetry 215–16
  niacin 288, 296
  riboflavin 273–80, 281
  thiamin 215–16, 255–6, *260*
solubility
  B_6 136
  coenzyme A (CoA) 132
  folic acid 160
  niacin 109
  niacinamide 109
  nicotinamide adenine dinucleotide (NAD) 115
  nicotinamide adenine dinucleotide phosphate (NADP) 115
  pantethine 131
  panthenol 131
  pantothenic acid 129
  thiamin 72–3
soman 141
sorbent 283
sorghum 675
soybean 140
specific optical rotation 133
specificity of analysis 401, 434
spectrofluorimetry *see* fluorimetry
spectrophotometry
  biotin 355
  cobalamins 420
Spies, T. 14

spina bifida 484, 513, *735*, 745, 748
  see also neural tube defects
spinach 12, 158, 162
spongiform encephalopathies 89
sports drinks *554*
SPR see surface plasmon resonance
SRM see standard reference materials
stability 197
  B$_6$ 96, 136
  biosensors 440, 453
  cobalamins 166, 453
  coenzyme A (CoA) 132
  folic acid 46, 96, 160
  niacin (nicotinic acid) 109
  niacinamide (nicotinamide) 109
  pantothenic acid 96, 129, 130
  riboflavin (B$_2$) 95–7
  tetrahydrofolate (THF) 160
  thiamin 96, 253
stable isotope dilution assays (SIDA) 399–412, 413, 414
standard reference materials (SRMs) 327
statins 643, 645, 646, 647, 665
status epilepticus 505, 520
stents 41–2, 47
stereoisomers 133, 392
Stern, K.G. 6
steroids 58, 117
Stevens, J.R. 10
stillbirth 529, 532, 534
stomach 104, 775
Strecker degradation 97
streptavidin
  biotin analysis 355, 383, 386, 387, 392
  cobalamins analysis 428
stroke (brain ischemia) 496–7, 498
  B$_6$ 493, 702–3
  C677T MTHFR 512
  cardiovascular disease (CVD) 494
  cobalamins 493
  dementia 481
  folates 490, 493, 756
  folic acid 760, *761*, 763
  food fortification 42–3, 48, 493
  homocysteine (Hcy) 139, 482–95, 497, 755, 760, 776
  meta-analysis studies *29*, *37*, *489*, 492, 493, 495, 756, 759–60, 763, 764
  niacin (nicotinic acid) 668
  nutritional status 482
  recurrences 495
  studies 490–1, 492, 493
  supplementation 490, 493
  thiamin 574, 576
  trials *44*, *485–9*, 493, 494, 668
structure
  acetyl-CoA **149**
  adenosine diphosphate (ADP) ribose **120**
  adenosine thiamin diphosphate (AThDP) 76–7
  adenosine thiamin triphosphate (AThTP) 76–7
  adenosylcobalamin 13, **203**
  β-alanine **127**, **308**
  2-amino-3-carboxy muconaldehyde **112**
  aminophthalate **444**
  benzaldehyde **74**
  benzoin **74**
  biocytin **717**
  biotin 10, **147**, **354**, 377–8
  biotin-d,l-sulfoxide **147**
  biotin sulfone **147**
  bisnorbiotin **147**
  calcium pantothenate **129**
  cobalamins 164–6, **203**, **422**
  coenzyme A (CoA) **129**, **304**, **685**
  cyanocobalamin 165, **203**, **483**
  flavin adenine dinucleotide (FAD) **7**, **94**, 613
  flavin mononucleotide (FMN) **7**, **94**, 613
  folate isotopologues **404**
  folates **397**
  folic acid 13, 26, 158–9, 160, **354**, **367**, **397**, 482
  food vs. supplements 26
  5-formiminotetrahydrofolate **397**

N-formylkynurenine 112
10-formyltetrahydrofolate 397
5-formyltetrahydrofolate 397
glutamic acid 159
histones 154
3-hydroxyanthranylate 112
hydroxycobalamin 203
10-hydroxyethylflavin 101
3-hydroxyisovaleric acid 149
3-hydroxyisovaleric acid-carnithine 149
3-L-hydroxykynurenine 112
hydroxymethylriboflavin 101
iso-coenzyme A (CoA) 132
kynurenine 112
luminol 444
malonyl-CoA 149
5,10-methenyltetrahydrofolate 397
methylcobalamin 13, 203
3-methylcrotonyl-CoA 149
3-methylcrotonyl glycine 149
5,10-methylenetetrahydrofolate 397
3-methylglutaconyl-CoA 149
methylmalonyl-CoA 149
5-methyltetrahydrofolate (5-MHTF or 5MT) 397
niacin adenine dinucleotide 112
niacin adenine mononucleotide 112
niacin (nicotinic acid) 9, 109, 112
niacinamide mononucleotide 112
niacinamide (nicotinamide) 9, 109, 112
nicotinamide adenine dinucleotide (NAD) 9, 112, 115, *117*
nicotinamide adenine dinucleotide phosphate (NADP) 9, 115, *117*
oxaloacetate 149
pantetheine 131, 304
pantethine 131
panthenol 130
pantoic acid 128, 308
pantolactone 130
pantothenic acid 128, 304, 318
para-aminobenzoic acid 159

phosphoketolase (PK) 62
phosphopantetheine 127, 304
propionyl-CoA 149
pteridine ring 159
pteroic acid 159
pyridoxal 11, 136, 336, 483, 700
pyridoxal-5′-phosphate (PLP) 11, 136, 336
pyridoxamine 11, 136, 336, 483, 700
pyridoxamine-5′-phosphate (PMP) 11, 136, 336
4-pyridoxic acid (4-PA) 336
pyridoxine 11, 135, 136, 336, 482, 700
pyridoxine-5′-phosphate (PNP) 11, 136, 336
pyridoxine-β-glucoside (PNG) 336
4-pyridoxolactone (4-PLA) 336
pyruvate 149
pyruvate decarboxylase (PDC) 62
pyruvate dehydrogenase (PDH) 62
pyruvate oxidase (POX) 61, 62
quinolinate 112
riboflavin 7, 94
8α-sulfonylriboflavin 101
tetrahydrofolate (THF) 13, 397
tetranorbiotin 147
thiamin 6, 72–4, 210, 211, 253
thiamin diphosphate (ThDP) 6, 56, 76
thiamin monophosphate (ThMP) 76
thiamin pyrophosphate 253
thiamin triphosphate (ThTP) 76
thiochrome 74, 258
transketolase 61–2, 61
tryptophan 112
study types 22–3
  variation in results 26, 27, 30
succinate 73
succinate dehydrogenase 563
succinic acid 776
succinyl-CoA 167, 615
sugars
  catabolism 242
  in honey 173–4

sulbutiamine (o-isobutyrylthiamin disulfide) 79
sulfites, thiamin 75
8α-sulfonylriboflavin 100, **101**
SUMOylation 719, 728
'sunlight-flavour' 97, 271
superoxide dismutase (SOD) 547
suphitocobalamin, dietary sources 166
supplements
  analysis 197–200, 213–15, 217, 317–27, 366, *367*
  biotin 381, 721–2
  cardiovascular disease (CVD) 25, 35–6, *44–5*, *488*, 490–1, 532, 757–64
  children 525, 528, 531, 532
  cobalamins analysis 450, *451*
  cobalamins (B_{12}) 778
  cognitive function 490, 531–2
  electrospray ionization (ESI) 384–5
  epilepsy 516–17
  folates 366, *367*, 779
  folic acid 744, 745, 746
  haemoglobin 526–7
  homocysteine levels 35, 41, 492, 532
  intervention trials *487–8*
  multiple micronutrient (MMN) 526–33
  pantothenic acid 318–27
  popularity 197
  side effects 318, 517
  thiamin 253
surface plasmon resonance (SPR) 387, 441–2, 454
  biotin 388
  cobalamins 429, *431*
  folates 399
Suzuki, Umetaro 5
synaptosomes 89
synchronous fluorescence 221
systematic reviews *489*, 490, 709

tabun 141
tachyphylaxis 670
takadiastase 382
Takagi, Kanehiro 3–4
tandem mass spectrometry 329
  liquid chromatography (LC) 263, 323, 358, 405, 406
TC *see* thiochromes; transcobalamin
TDP effect 230–3
telomeres 121, 123
temperature, riboflavin (B_2) *601*, 602, 604
teratogens 513, 520, 722
testis 687
tetrahydrobiopterin (BH4) 36, 40, **41**, 49
tetrahydrofolate (THF) **38**, 159, 160, 167
  B_6 700–1
  biological function 491
  cobalamins 473
  cobalamins metabolism **788**
  homocysteine metabolism **25**, **484**, 738
  isotopologues 403
  metabolism 738–9
  nomenclature 160
  stability 160
  structure **13**, **397**
  substrate behaviour 15
tetranorbiotin
  catabolism 148
  structure **147**
tetranorbiotin methyl ketone, catabolism 148
tetrapyrroles 164
thalamus 548
ThDP *see* thiamin diphosphate
ThDP riboswitch (Thi-box) 77–8
theophylline
  cobalamins deficiency 470
  riboflavin digestion 99
Theorell, Hugo 7
THF *see* tetrahydrofolate
Thi-box (ThDP riboswitch) 77–8
thiamin (B_1)
  active forms 227–8, 572
  alcohol 546–7

Subject Index   851

antagonists 553
assay methods 200, 211–19, 228–41, 254–65
binding proteins 86–7
biological function 73–4, 210–11, 572, 580
biosynthesis 77
brain 240
carbohydrate metabolism 580–1
cerebrospinal fluid (CSF) 239
deficiency 71–2, 88, 228, 539–47, 558–9
degradation 77
derivatives distribution 80–1
dietary sources 227, 252
digestion 80
epilepsy 507–8, 516
excretion 77, 229, *554*, 565
extraction 215–16, 255–6, 258, *260*
food analysis 254–65
heat treatment 75
homeostasis 80, 88
honey samples 177, *179*
lactate degradation 7
Maillard reactions 75
metabolic syndrome 580–1
metabolism 81–7
nomenclature 5, 71
pentose phosphate pathway 230
phosphorylation 75–6
physiopathological conditions 553–4
recommended dietary allowance (RDA) 253–4
research history 5, 7, 72
solubility 72–3
spectra **214**
stability 74–5, 96, 253
structure **6**, 72–4, 210, **211**, 253
sulfite action 75
synthetic forms 79
thermal processing 253
thiolate **75**
tissue samples 234, 239–41
transport 78–9

Wernicke's encephalopathy 233, 538–48
'yellow' form **75**
ylide form 74, 88
thiamin allyl disulfide (allithiamin) 79
thiamin (B_1), dietary sources 227
thiamin diphosphatases (ThDPases) 81, 85
thiamin diphosphate (ThDP) 14–15, 242, 243
  acetyl-coenzyme A (acetyl-CoA) *560*, 566
  assay methods 230–4, 239–41
  biosynthesis 77, 83
  brain 240, 539, 546
  coenzyme function 55–6, 72, 74, 87, 228
  cofactor behaviour 566, 572
  cognitive function 557
  deficiency 554–7, *558*, 560–4, 565, 572–6
  distribution 80
  energy metabolism 72, **73**, 228
  erythrocyte transketolase (ETK) activation assay 230–3
  hydrolysis 85
  metabolism 81–3
  nomenclature 55
  phosphoketolase **60**, 61
  physiopathological conditions 554
  research history 64
  structure **6**, **56**, **76**
  synthesis 75
  tissue samples 234, 239–41
  transketolase **60**
  transport 79
  V-conformation 56
  ylide form 56, **57**, 63–4
thiamin diphosphate (ThDP) dependent enzymes 55–65
  classification 62–3
  research history 64
thiamin diphosphokinase 83
thiamin disulfide (TDS) 212
thiamin kinase 83

thiamin monophosphatase
  (ThMPase) 84
thiamin monophosphate (ThMP)
  biosynthesis 77, 81, **82**, 83
  cerebrospinal fluid (CSF) 239
  dietary sources 252
  hydrolysis 84–5
  structure **76**
  synthesis 75
  transport 78
thiamin phosphate kinase 83
thiamin pyrophosphate (TPP)
  dietary sources 252
  structure **253**
thiamin pyrophosphokinase
  (TPK) 83
thiamin-responsive megaloblastic
  anaemia (TMRA) 78
thiamin transporters (ThTR) 78–9
thiamin triphosphatase
  (ThTPase) 81, 83, 84, 86
thiamin triphosphate (ThTP) 71
  biosynthesis 81, **82**, 84
  cell membranes 85
  distribution 80–1
  hydrolysis 85–6, 234
  membranes, cellular 85
  phosphorylation properties 75–6
  stability 234
  structure **76**
  synthesis 75
thiaminases 77, 553
thiamine hydrochloride, analysis
  199
thiamine pyrophosphate *see* thiamin
  diphosphate
thiazole 74, 76, 87
thin layer chromatography (TLC),
  biotin 356–7, 381
thiochromes (TC) 266
  flow injection analysis (FIA) 218
  fluorescence 80
  HPLC analysis 80, 255, 259
  spectrofluorimetry 211–13, 216,
    217, 219, 258
  structure **74, 258**

thiocyanate (SCN) 787–8, **790, 791,**
  801
ThMP *see* thiamin monophosphate
thrombocytopenia 692
thrombosis 29, 36–7, 43
  B_6 701
  C677T MTHFR 512
  homocysteine (Hcy) 494, 512
  supplementation trials 45
ThTP *see* thiamin triphosphate
thymidine, folates 24, **38**, 491
thymidine monophosphate
  (dTMP) 740, 741–2
thymidylate synthase (TS) 740, 741–2
thymine 14
thyroxine 100, 104
time-of-flight mass spectrometry
  (TOF) 321, 326, 331
tissue samples 234, 239–41
TK *see* transketolase
TLC *see* thin layer chromatography
TMRA *see* thiamin-responsive
  megaloblastic anaemia
tobacco 472, 787 *see also* smokers
TOF *see* time-of-flight mass
  spectrometry
Tokyo Chemical Society 5
Tönnis, B. 10
toxicokinetics 123
TPK *see* thiamin pyrophosphokinase
TPP *see* thiamin pyrophosphate
3,2 trans-enoyl-CoA isomerase 615
transcobalamin (TC) 434, 435
  binding assays 442
  genetic disorders 459–60
  genotypes 461–2
  metabolism **459**
  serum 420–1, 465
  transport 786
transferrin 596
transgenerational effect 745
transglutaminase 691, 695
transketolase (TK) 59, **60**, 72 *see also*
  erythrocyte transketolase (ETK)
  activation assay
  catalysis 64

classification 63
furosemide 565
pentose phosphate pathway 65, 73, 230
structure 61–2, **61**
thiamin 233, 581
thiamin diphosphate (ThDP) 81, **82**, 83, 242, 539, 556, 557
transmembrane proteins 594
transmethylation 801
transminase *see* aminotransferase
transport
  fatty acids 614
  folates 737–8
  riboflavin (B_2) 594, 595, 596, *599*, 604, 605
  thiamin (B_1) 78–9
  thiamin diphosphate (ThDP) 79
  thiamin monophosphate (ThMP) 78
transsulfuration pathway 801
tri-iodothyronine 104
tricarboxylic acid (TCA) cycle 58, 64, 568, 589
  acetyl-coenzyme A (acetyl-CoA) 556, 562, 566
  biotin 353
  2-oxoglutarate dehydrogenase complex (KDHC) 556
  pyruvate 556
  pyruvate dehydrogenase complex (PDHC) 557
  thiamin deficiency 547, **555**
  thiamin diphosphate (ThDP) 539
trienzyme extraction 358
triethanolamine 446
trifluoroacetic acid (TFA) 322
triglycerides (TG) 636, 638–9, 640, 645, 647, 662 *see also* fatty acids
triiodothyronine 100
triple quadrupole mass spectrometry 325
tropical sprue 103, 105
trueness 186–7, 189
trypsin 382

tryptophan
  analysis 286–7
  antidepressive 681
  function 633
  intake 110
  metabolism **112**
  niacin (nicotinic acid) 108, 633
  niacinamide biosynthesis 111–13
  pellagra 8, 675, 680, 681
  research history 5, 8
  riboflavin 97, 99, 525
  structure **112**
tryptophan dioxygenase (TDO2) 111, **112**
L-tryptophan:oxygen oxidoreductase 681
tuberculosis 111
tumour necrosis factor-alpha (TNF-α) 642, 663, 689
turbidimetry 425, 426
tyrosine
  flavins (flavoproteins) 627
  riboflavin (B_2) 97
  thiamin biosynthesis 77
  thiamin deficiency 544
tyrosine kinase *600*

ubiquinone 59, 614, **615**, 621
ubiquitination 719, 728
ultra high-performance liquid chromatography (UHPLC) 289, 319, 328–9, 383
ultraviolet detector 266
unfolded protein response (UPR) 576, 577
uniporters 595, 607
United Nations International Multiple Micronutrient Preparation (UNIMMAP) 529, **530**
upper limit approach 1 (ULA1) 184
uracil 24, 741–2
urine analysis 216
  B_6 342, 348
  biotin 721, 726
  folates 406, *407*

urine analysis (*continued*)
   folic acid 405
   HPLC **309–10**
   pantothenic acid 303, 307, 308–10
   pyridoxal 339
   sample times 229
   thiamin 228–9, 254
USA, food fortification 42–3, 491, 513
UV detection
   biotin 383
   disadvantages 307
   honey samples 181, 182
   juice samples 200
   pantothenic acid 307–8
   simultaneous analysis 262
   thiamin 259

validation procedures 183–7, 188, 189, 433–4
valproate (VPA) 508, 509, 513, 514, 517, 723
vanin-1 *see* pantetheinase
vanin genes 687–90, 692
vascular cell adhesion molecule-1 (VCAM-1) 642, 663
vascular dementia 481–2
vascular redox 49
vascular thrombosis 36–7
vascular walls
   folic acid **41**, 42, 46
   homocysteine (Hcy) 701
vegans
   cobalamins deficiency 420, 460, 465, 466, **468**, 470, 473
   thiamin diphosphate (ThDP) *554*
vegetables
   biotin 720
   folates 411, *412*
   pantothenic acid *303*
vegetarians
   cobalamins 460, 470
   riboflavin intake 101–2
   thiamin diphosphate (ThDP) *554*
venous thrombosis 29, 43

very long-chain acyl-CoA dehydrogenase (VLCAD) **615**, 616
very low-density lipoproteins (VLDL) 26–7, 636, 639, 662, 670
VGB (vigabatrin) 505
vigabatrin (VGB) 505
villi 104
VISP trial 42, 43, *44*, *487*, 493, 759, *760*, *761–2*
VitaFast® analysis 257
vitamers 123, 196, 206
vitamin A, oxidation 97
vitamin C *see* ascorbic acid
vitamin status 23–4, 30, 196
VITATOPS trial 39, 42, *44*, *487–8*, 492, 758–9, *760*, *761–2*
VITRO trial 43, *45*
VLCAD *see* very long-chain acyl-CoA dehydrogenase
VLDL *see* very low-density lipoproteins
VNN1 gene 687, 690–1, 692
VNN2 gene 688–9, 692
VNN3 gene 689–90, 692
Voetglin, Carl 8
voltage-sensitive calcium channel **543**, 549
voltammetric analysis methods 264, 265
von Bunge, Gustav 4–5
von Lubarsch, O. 13
VPA (valproate) 508, 509, 513, 514, 517, 723
VX poison 141

WAFACS trial *44*, *759*, *760*, *761–2*
Wagner-Jauregg, Theodor 6
Warburg, Otto 6–7, 8
water retention 554
weight *see* adipose tissue; obesity
WENBIT trial *45*, *759*, *760*, *761–2*
Wendt, G. 10
Wernicke-Korsakoff syndrome 72, 88, 242–3, 538, 546–7, 567, 572
   thiamin diphosphate (ThDP) 564

Wernicke's encephalopathy 72, 89, 228, 242, 243–4, 577
  alcoholism 538, 546–7
  symptoms 539
  thiamin levels 233, 538–48
whey 391
Whipple, G. 13
Willcock, E.G. 8
Williams, Robert 5
Williams, Roger 9, 10
Wills, Lucy 12, 158, 734, 748
wine, sunlight-flavour 271

X-ray crystallographic analysis 65, 169
xenobiotics 123
xylulose 5-phosphate **73**

yeast
  anaemia treatment 12
  fermentation 8, 65, 72
  flavin mononucleotide (FMN) 6–7
  folates 158
  pyridoxine 137
  pyruvate decarboxylase 55
  research history 6–7, 8, 9
  thiamin 252
  thiamin pyrophosphokinase (TPK) 83

zinc
  B_6 138, 139
  biotin 724
  riboflavin metabolism 100
  thiamin deficiency 563
zwitterions 123, 221